Abiotic Stress in Plants

Edited by Shah Fahad, Shah Saud,
Yajun Chen, Chao Wu and Depeng Wang

Published in London, United Kingdom

IntechOpen

Supporting open minds since 2005

Abiotic Stress in Plants
http://dx.doi.org/10.5772/intechopen.91549
Edited by Shah Fahad, Shah Saud, Yajun Chen, Chao Wu and Depeng Wang

Contributors
Malik Ghulam Asghar, Anam Bashir, Nawab Ali, Mohammad Akmal, Noor Ul Haq, Samina N. Shakeel, Marzena Kurowska, Suphia Rafique, Silvana Aparecida Barbosa de Castro, Vinícius Coelho Kuster, Sinan Meriç, Alp Ayan, Çimen Atak, Ayman EL Sabagh, Akbar Hossain, Mohammad Sohidul Islam, Muhammad Aamir Iqbal, Ali Raza, Çetin Karademir, Emine Karademir, Abdul Rehman, Md. Atikur Rahman, Rajesh Singhal, Analía Llanes, Muhammad Ali Raza, Muhammad Mubeen, Wajid Nasim, Hirofumi Saneoka, Celaleddin Barutçular, Ram Swaroop Meena, Xin He, Cüneyt Uçarlı, Tinashe Zenda, Huijun Duan, Songtao Liu, Arvind Kumar, Nitika Sandhu, Shailesh Yadav, Sajal Roy, Nasrin Chowdhury, Amanpreet Singh, Harmandeep Singh Chahal, Navdeep Kaur, Deeksha Marothia, Pratap Kumar Pati, Yajun Chen, Jian Cui, Shah Saud, Shah Fahad, T.V. Vineeth, Shrvan Kumar, Monika Shukla, Anil Chinchmalatpure, Parbodh Chander Sharma, Mona H. H. Soliman, Haifa Abdulaziz S. Alhaithloul, Abdelghafar M. Abu-Elsaoud, Sivaji Mathivanan, Fuchun Xie, Rahul Datta, Dong Qin, Sujata Harlapur, Suneeta Harlapur, Shanthbasavareddi F. Harlapur, Hanen Enneb, Leila Ben Yahya, Datta Asaram Dhale, Mohamed Ilyas, Mohamed Bagues, Kamel Nagaz

First published in London, United Kingdom, 2021 by IntechOpen
IntechOpen is the global imprint of INTECHOPEN LIMITED, registered in England and Wales, registration number: 11086078, 5 Princes Gate Court, London, SW7 2QJ, United Kingdom
Printed in Croatia

British Library Cataloguing-in-Publication Data
A catalogue record for this book is available from the British Library

Additional hard and PDF copies can be obtained from orders@intechopen.com

Abiotic Stress in Plants
Edited by Shah Fahad, Shah Saud, Yajun Chen, Chao Wu and Depeng Wang
p. cm.
Print ISBN 978-1-83881-055-9
Online ISBN 978-1-83881-062-7
eBook (PDF) ISBN 978-1-83881-063-4

For EU product safety concerns:
IN TECH d.o.o., Prolaz Marije Krucifikse Kozulić 3, 51000 Rijeka, Croatia, info@intechopen.com or visit our website at intechopen.com.

Meet the editors

Dr. Shah Fahad is an assistant professor in the Department of Agronomy, University of Haripur, Khyber Pakhtunkhwa, Pakistan. He obtained his Ph.D. in Agronomy from Huazhong Agriculture University, China, in 2015. After completing his postdoctoral research in Agronomy at Huazhong Agriculture University (2015–2017), he accepted the position of assistant professor at the University of Haripur. He has published more than 290 peer-reviewed papers (impact factor = 910.45) with more than 260 research and 30 review articles on important aspects of climate change, plant physiology and breeding, plant nutrition, plant stress responses, and tolerance mechanisms, and exogenous chemical priming-induced abiotic stress tolerance. He has also contributed fifty book chapters to various volumes published by well-renowned publishing houses. He has also edited fifteen book volumes, including this one. Dr. Fahad received the Young Rice International Scientist award and distinguished scholar award in 2014 and 2015, respectively. He won fifteen projects from international and national donor agencies. Dr. Fahad was named among the top 2 percent of scientists in a global list compiled by Stanford University, California. His areas of interest include climate change, greenhouse emission gasses, abiotic stresses tolerance, roles of phytohormones and their interactions in abiotic stress responses, heavy metals, and regulation of nutrient transport processes.

Dr. Shah Saud received his Ph.D. in Horticulture (turf grasses) from Northeast Agricultural University, Harbin, China. He is currently working as a postdoctoral researcher in the Department of Horticulture, Northeast Agricultural University. Dr. Saud has published more than 125 research publications in peer-reviewed journals. He has edited three books and written twenty-five book chapters on important aspects of plant physiology, plant stress responses, and environmental problems in relation to agricultural plants. According to Scopus, Dr. Saud's publications have received roughly 2500 citations with an h-index of 24. He is an editor and reviewer for more than twenty peer-reviewed international journals. His outstanding work has been recognized with five awards for scientific and technological progress in Heilongjiang province, China. Dr. Saud has also won five international projects.

Dr. Yajun Chen obtained a Ph.D. in Crop Cultivation and Farming System from Northeast Agricultural University, Harbin, China, in 2005. Later, she completed her postdoctoral research in biology at Northeast Forestry University, Harbin, China. Currently, she is a professor at the Horticulture College, Northeast Agricultural University. Over the course of a dozen years, Dr. Chen has studied in Australia (2003–2004), the United States (2010–2011), and Norway (2015–2016) as a visiting scholar funded by the Chinese government. Her research work encompasses ornamental plant adverse physiology and ecology, coping with ecological environment adaptation, and restoration of garden plants. Dr. Chen has published more than 110 research papers in peer-reviewed journals in her field at home and abroad. As a chief editor she has edited more than twenty books on important aspects of local flora, turf and flower culture, and plant stress physiology. She has also trained more than eighty master's students and five Ph.D. candidates in these fields. Her outstanding work has been recognized with eight awards for scientific and technological progress in Heilongjiang province, China. She has also won two international and thirteen national projects.

Dr. Depeng Wang obtained a Ph.D. in Agronomy and Crop Physiology from Huazhong Agriculture University, Wuhan, China, in 2016. He is currently a professor at the College of Life Science, Linyi University, Linyi, China. He is the principal investigator of the Crop Genetic Improvement, Physiology & Ecology Center at Linyi University. His current research focuses on agronomy and crop ecology and physiology, including characteristics associated with high-yield crops, the effect of temperature on crop grain yield and solar radiation utilization, morphological plasticity to agronomic manipulation in leaf dispersion and orientation, and optimal integrated crop management practices for maximizing crop grain yield. Dr. Wang has published more than thirty-six papers in reputed journals. He has edited one book and written four book chapters on important aspects of crop physiology, environmental stress, and crop quality formation. According to Google Scholar Citation, his publications have received about 100 citations. He is a reviewer for five peer-reviewed international journals. Dr. Wang is a provincial crop expert in green, high-quality, and efficient technology, and has participated in six national projects with more than $4 million in research funding.

Dr. Chao Wu engages in the field of crop cultivation and physiology, and plant phenomics. He received a Ph.D. from Huazhong Agricultural University, Wuhan, China, in 2016, and completed his postdoctorate at Nanjing Agricultural University, Nanjing, China, in 2019. Currently, he is an associate research fellow at Guangxi Institute of Botany, Guangxi Zhuang Autonomous Region, and the Chinese Academy of Sciences, Guilin, China. He chairs the Natural Science Foundation of Jiangsu Province and two postdoctoral science research foundations. His main research focus is physiological mechanisms of abiotic-stress tolerance (heat, drought) in crops and medicinal plants.

Contents

Preface XIII

Chapter 1 1
A Review on Kentucky Bluegrass Responses and Tolerance to Drought Stress
by Jian Cui, Saud Shah, Shah Fahad and Yajun Chen

Chapter 2 13
Management of Abiotic Stress in Forage Crops
by Amanpreet Singh and Harmandeep Singh Chahal

Chapter 3 29
Responses of Neotropical Savannah Plant Species to Abiotic Stresses: A Structural and Functional Overview
by Silvana Aparecida Barbosa de Castro and Vinícius Coelho Kuster

Chapter 4 59
Influence of Water Stress on Growth, Chlorophyll Contents and Solute Accumulation in Three Accessions of *Vicia faba* L. from Tunisian Arid Region
by Hanen Enneb, Leila Ben Yahya, Mohamed Ilyas, Datta Asaram Dhale, Mohamed Bagues and Kamel Nagaz

Chapter 5 73
Abiotic Stress Responses in Plants: Current Knowledge and Future Prospects
by Deeksha Marothia, Navdeep Kaur and Pratap Kumar Pati

Chapter 6 91
Molecular Abiotic Stress Tolerans Strategies: From Genetic Engineering to Genome Editing Era
by Sinan Meriç, Alp Ayan and Çimen Atak

Chapter 7 127
Elevated CO_2 Concentration Improves Heat-Tolerant Ability in Crops
by Ayman EL Sabagh, Akbar Hossain, Mohammad Sohidul Islam, Muhammad Aamir Iqbal, Ali Raza, Çetin Karademir, Emine Karademir, Abdul Rehman, Md Atikur Rahman, Rajesh Kumar Singhal, Analía Llanes, Muhammad Ali Raza, Muhammad Mubeen, Wajid Nasim, Celaleddin Barutçular, Ram Swaroop Meena and Hirofumi Saneoka

Chapter 8 145
Ecological and Economic Potential of Major Halophytes and Salt Tolerant Vegetation in India
by T.V. Vineeth, Shrvan Kumar, Monika Shukla, Anil Chinchmalatpure and Parbodh Chander Sharma

Chapter 9 177
Protagonist of Mineral Nutrients in Drought Stress Tolerance of Field Crops
by Malik Ghulam Asghar and Anam Bashir

Chapter 10 199
An Insight into the Responses of Early-Maturing *Brassica napus* to Different Low-Temperature Stresses
by Xin He

Chapter 11 211
Effects of Salinity on Seed Germination and Early Seedling Stage
by Cüneyt Uçarlı

Chapter 12 233
Abiotic Stress Tolerance in Crop Plants: Role of Phytohormones
by Haifa Abdulaziz S. Alhaithloul, Abdelghafar M. Abu-Elsaoud and Mona H. Soliman

Chapter 13 257
Morphophysiological Traits, Biochemical Characteristic and Productivity of Wheat under Water and Nitrogen-Colimitation: Pathways to Improve Water and N Uptake
by Nawab Ali and Mohammad Akmal

Chapter 14 285
Adapting Cereal Grain Crops to Drought Stress: 2020 and Beyond
by Tinashe Zenda, Songtao Liu and Huijun Duan

Chapter 15 315
Abiotic Stress-Induced Molecular and Physiological Changes and Adaptive Mechanisms in Plants
by Sivaji Mathivanan

Chapter 16 337
HSPs under Abiotic Stresses
by Noor ul Haq and Samina N. Shakeel

Chapter 17 353
Plant Growth and Morphophysiological Modifications in Perennial Ryegrass under Environmental Stress
by Fuchun Xie, Rahul Datta and Dong Qin

Chapter 18 371
Advances in Developing Multigene Abiotic and Biotic Stress-Tolerant Rice Varieties
by Nitika Sandhu, Shailesh Yadav and Arvind Kumar

Chapter 19 391
Salt Stress in Plants and Amelioration Strategies: A Critical Review
by Sajal Roy and Nasrin Chowdhury

Chapter 20 423
TIP Aquaporins in Plants: Role in Abiotic Stress Tolerance
by Marzena Małgorzata Kurowska

Chapter 21 445
Drought Responses on Physiological Attributes of *Zea mays* in Relation to Nitrogen and Source-Sink Relationships
by Suphia Rafique

Chapter 22 467
Ecofriendly Marigold Dye as Natural Colourant for Fabric
by Sujata F. Harlapur, Suneeta Harlapur and Shantabasavareddi F. Harlapur

Preface

A biological system is an endogenic and dynamic system that always seeks a constant in-flow of energy to maintain its meta-stable condition called homeostasis. Any environmental disruption of this homeostasis may be explicated as biological stress. Biological stress limits the growth and yield potential of a crop and is broadly grouped into biotic and abiotic stresses. Biotic stress is due to interactions between organisms starting from predation to allelopathy. Abiotic stress appears due to the adverse effects of non-living environmental factors (i.e., water, temperature, light, metal, mineral nutrients, etc.) on plants, which are often sporadic and highly localized. Understanding abiotic stress responses in plants are critical for the development of new varieties of crops that are better adapted to harsh climate conditions. This book provides a comprehensive overview of the agronomical, physiological, and molecular basis of plant responses to external stress like salinity, drought, temperature, and heavy metals to aid in the engineering of stress-resistant crops.

Over twenty-two chapters, *Abiotic Stress in Plants* presents a comprehensive picture of plant responses to external stress. It is designed for researchers, technologists, policy makers, and undergraduate and postgraduate students studying sustainable crop production and crop protection. This book is a worthy addition to university libraries and research establishments where agricultural and agronomical sciences are studied and taught. We are thankful to all the authors who contributed their valuable chapters to this book. We are also extremely grateful to IntechOpen's Publishing Process Manager Ms. Sara Gojevic-Zrnic for helping us to publish this book in an excellent form in the shortest possible time. We owe our sincere thanks and irreparable gratitude to our families whose consistent encouragement and love have been a tremendous impetus for completing this book.

Dr. Shah Fahad
Agriculture Department,
University of Haripur,
Haripur Kyhber Paktunkhwa, Pakistan

Dr. Shah Saud and Dr. Yajun Chen
Horticulture Department,
Northeast Agricultural University,
Harbin, China

Dr. Depeng Wang
College of Life Science,
Linyi University,
Linyi, China

Dr. Chao Wu
Guangxi Institute of Botany,
Guangxi Zhuang Autonomous Region and the Chinese Academy of Sciences,
Guilin, China

Chapter 1

A Review on Kentucky Bluegrass Responses and Tolerance to Drought Stress

Jian Cui, Saud Shah, Shah Fahad and Yajun Chen

Abstract

Kentucky bluegrass (*Poa pratensis* L.) is an excellent cool-season turfgrass and is extensively used in urban green space, parks and sports fields worldwide, but it is sensitive to drought stress. Drought reduces turf quality of Kentucky bluegrass by influences on the shoot density, texture, uniformity, color, growth habit and recuperative capacity. It has been a challenge for breeding water saving cultivars and enhances water use efficiency in Kentucky bluegrass. Many studies have revealed the mechanisms of drought stress tolerance in Kentucky bluegrass via multiple approaches. The morphological and physiological attributes as well as molecular information were discovered for better understanding and improving its drought tolerance. In this chapter, we will draw a systematic literature review about Kentucky bluegrass in response to drought stress and provide future perspectives of Kentucky bluegrass drought resistance research.

Keywords: Kentucky bluegrass, drought stress, tolerance, mechanism

1. Introduction

Drought is one of the major environmental factors that affect plant growth and survival worldwide. Turfgrass as an important part of the green ecosystem in the urban area which provides many kinds of environmental functions such as beautify and green the city, cooling warm weather, and soil stabilization. As urban rapid expansion nowadays, the percentage of land converted into turfgrass has been increasing. Turfgrasses are generally classified into cool- and warm- season groups based on their adaptation to specific ranges in temperature and precipitation, which are mainly governed by latitude and altitude [1, 2]. Kentucky bluegrass (*Poa pratensis* L.) is an excellent cool-season turfgrass and is extensively used in public parks, golf courses and residential lawns in temperate and cold temperate zones and cool plateau areas. This is a highly variable, rhizomatous species. Many improved cultivars of this species have been developed and used in landscaping areas in the world. A wide range of diverse cultivars and accessions of Kentucky bluegrass have been characterized based on pedigree, common turf performance, and morphological characteristics and were grouped into different genotypes such as Common, Compact, Compact-America, Julia, Mid-Atlantic and Midnight types [3]. Although these germplasm possess prominent ornamental value, however, Kentucky bluegrass extremely limiting used by water scarcity in practice, especially in semiarid, arid regions as well as the areas with the increase demand on water for agricultural, residential and industrial use [4–6]. It has been a challenge to select water saving cultivars and reduce water input in turf management [7]. Drought reduces the turf quality of Kentucky bluegrass by influences on the shoot density, texture, uniformity, color, growth habit and recuperative capacity [8, 9], in response to drought stress, different genotypes of Kentucky bluegrass performed various adaptive mechanisms and strategies in respective to their morphology, physiology and molecular bases. Understanding these mechanisms of Kentucky bluegrass tolerance to drought stress is a key step for improving drought resistant cultivars and reduces water input in management. Here, we summarize research progress in drought stress of Kentucky bluegrass for providing an overview of the field to readers and also for providing guidelines for practical management strategies under limited water availability.

2. Responses to drought stress in Kentucky bluegrass

Kentucky bluegrass, like other agronomic, horticultural, and landscape vegetation, requires water for growth and provides esthetics functional benefits for environments. Main causes of water deficiency may result from low rainfall, inadequate irrigation, as well as summer heat, which could greatly limit growth, and turf quality of Kentucky bluegrass. Like other plants, the ability of Kentucky bluegrass to maintain growth and survive under drought stress is broadly considered as drought resistance. Three major strategies of plant drought resistance are considered as escape, avoidance, and tolerance [2, 10] and are illustrated in **Figure 1**. Drought resistance traits vary genetically by exhibiting survival strategies among Kentucky bluegrass cultivars under water limited conditions [11]. However, these strategies are not mutually exclusive, and Kentucky bluegrass may utilize more than one when facing to water shortage.

2.1 Turf performance and morphological responses

Turf quality (TQ) refers to two aspects including visual quality and functional quality [1]. Turfgrass density, texture, uniformity, color, growth habit and smoothness are the most visible factors influence turf appearance quality, while functional quality such as playability in a particular sport turf is determined not only by some of the visual determinants, but also by other characteristics as well, including rigidity, resiliency, verdure, rooting and recuperative capacity. The drought resistance characteristics of specific species in morphology, growth patterns largely determine the turf quality under drought conditions [12, 13].

Morphological traits on Kentucky bluegrass often change with soil moisture [14], and plants often exhibit a severe decline in TQ and may become dormant during extended drought conditions [15]. Although Kentucky bluegrass can escape by going into dormancy under severe drought conditions especially during summer periods [16]. People still desire to sustain a green surface during drought periods for esthetics, sports, and other eco-functions. Therefore, drought escape is only considered a viable alternative for Kentucky bluegrass in those areas where irrigation is not available and survival of the turfgrass following drought is the primary objective [17].

Water use of the turfgrass canopy is influenced by water loss via shoot transpiration and soil evaporation, and by water uptake from the soil through the root system. The differences among Kentucky bluegrass cultivars in shoot and root characteristics such as leaf orientation and canopy configuration, tiller or shoot density, rooting depth, and root density are associated with water-use rate [18].

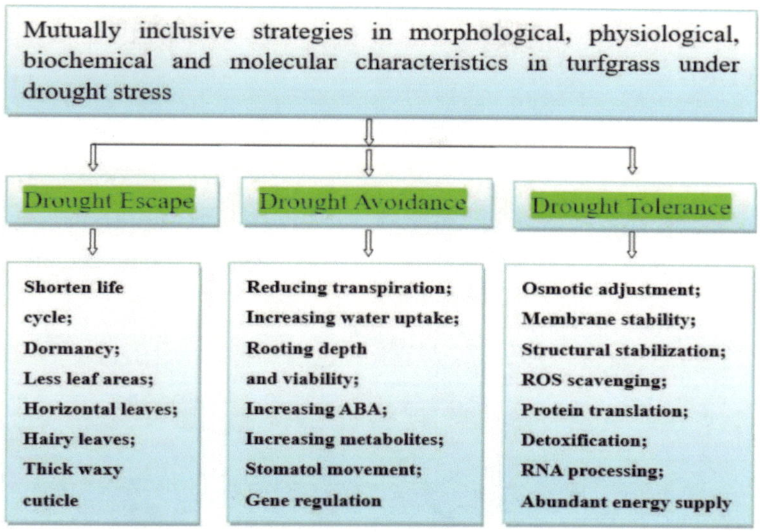

Figure 1.
Drought resistance strategies of turfgrass under drought conditions (illustrated by Jian Cui).

Lower ET rates of grass species were generally characterized with comparatively a high shoot density and relatively horizontal leaf orientation; and also with a low leaf area, including a slow vertical leaf extension rate and a narrow leaf texture [19]. Shoot vertical extension rate was positively correlated with water-use rate for Kentucky bluegrass with upright growth cultivars [20]. Based on a random spaced plant of 61 Kentucky bluegrass cultivars under untrimmed conditions, the morphological properties and comparative water use rate were strongly correlations by use discriminant analysis [21]. Low-water use cultivars had 13% more horizontal leaf orientation, 6% narrower leaf texture, 13% more lateral shoots per plant, 12% slower vertical leaf extension rate, 2% more leaves per shoot, and 7% shorter leaf blades and sheaths than the high-water use cultivar [21]. Low soil water content resulted significantly different in shoot-to-root ratio, vertical growth and survival rate among 11 Kentucky bluegrass cultivars, and more influence on tillering rate than other morphological indicators [11]. The components of leaf epidermis including stomatal apparatus, silica cells and cuticle are very important to the drought resistance of turfgrass canopy. Chen et al. found that leaf epidermis characteristics were associated with drought resistance among Kentucky bluegrass cultivars [22]. The opening or closing status of stomatal apparatus, the silica cell size and density, the thickness of the wax layer on leaf surface were related to the drought resistance and varied among Kentucky bluegrass cultivars (**Figure 2**).

Generally, an extensive deep root system of plants is important for efficient water uptake from the soil. Deep rooting enables plant to avoid water stress by taking up water from deeper in the soil profile when the surface soil is dry. The increased drought resistance of Kentucky bluegrass cultivars was correlation to the increased water uptake activity at the 15 to 30 cm soil depth [23]. Deeper root system in Kentucky bluegrass can avoid drought stress more consistently than total root mass [24]. However, other research found that not all turfgrasses with extensive root system are necessarily high water use. Hence, improvements in drought tolerance in Kentucky bluegrass may not necessarily occur by selecting germplasm with deep rooting or high root to shoot ratios due to no correlation between deep rooting and the ability to withstand long periods of water deficit [18]. Study on Kentucky bluegrass with the relation of Carbon-13 discrimination and water use efficiency (WUE) showed that turfgrass performance under drought may be

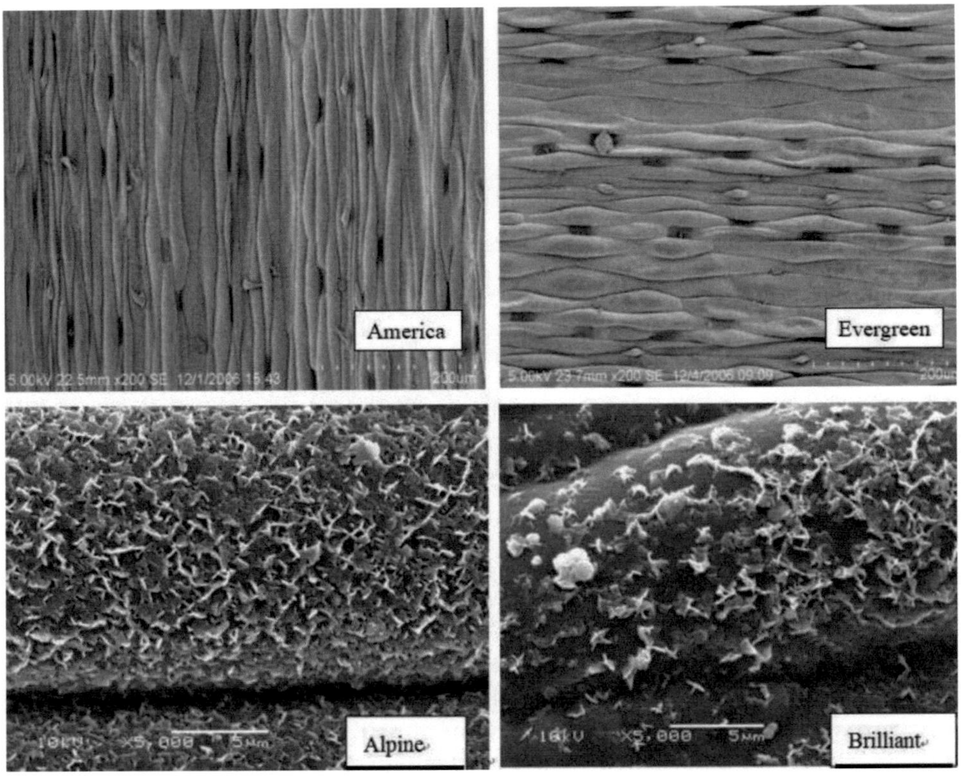

Figure 2.
Leaf epidermis characteristics among Kentucky bluegrass cultivars (photos provided by Yajun Chen).

improved for low Carbon-13 discrimination, which has been shown to be negatively correlated with WUE. Low Carbon-13 discrimination values were associated with less wilt and leaf firing, suggesting that Carbon-13 discrimination may be a useful selection criterion for superior performance under limiting soil moisture [25]. Kentucky bluegrass is a typically rhizomatous perennial, and ramets distributions in the grass clonal system connected by rhizomes largely rely on water status in soil. Drought stress severely influenced all agronomical, anatomical attributes of Kentucky bluegrass especially the seeking water behavior of rhizomes in both homogenous and heterogeneous environments [26]. Understanding these morphological characteristics respond to drought may assist to selecting low-water use Kentucky bluegrass cultivar in future for better adaptation water scarcity environment and saving water in turf management [27].

2.2 Bio-physiological and molecular in responses to drought

Severe drought can cause Kentucky bluegrass to go into dormancy and lose its greenness, which is a survival strategy as we described above, but we do not expect Kentucky bluegrass to adopt escape drought strategy to lose its landscape value during water deficiency. However, three drought resistance strategies including escape, avoidance and tolerance may be used simultaneously in different drought settings for specific turfgrass, and so does Kentucky bluegrass [2]. Drought-avoiding traits on Kentucky bluegrass can exhibit many different morphological characteristics and water absorption ability in response to drought, including previous reports as discussed before on leaf orientation and canopy configuration, tiller or shoot density, rooting system, leaf epidermis, lower ET and the like. Drought avoidance and tolerance are critical strategies for maintenance turf greenness and extending green period, fitness turf visual and functional quality under short or long drought stress. Many desirable drought resistance attributes on Kentucky bluegrass involve in physiological and biochemical metabolic activities as well as molecular regulators can serve as selection criteria for improving drought resistance cultivars using for turfgrass practical managements. For maintenance structural stabilization under drought stress, strategies on physiological responses of Kentucky bluegrass exhibited by increasing cellular osmotic potential and antioxidant enzyme activities, suppressing reactive oxygen species (ROS) and ABA production, etc.

The sucrose, fructose, glucose, starch as well as carbohydrate metabolism enzymes associated with plant tolerance to severe drought stress and post-drought recovery in two Kentucky bluegrass cultivars 'Midnight' and 'Brilliant' [28]. The differential in accumulation of different types of soluble carbohydrates could be related to the genetic variability and biological functions during drought and post-drought recovery in Kentucky bluegrass. These soluble sugars as osmolytes play critical role in maintenance cellular turgor by increasing osmotic potential and energy sources in photosynthetic process, and sucrose also plays protective roles in proteins and membranes from drought damages [29]. Sucrose, proline, as well as inorganic ions are important osmolytes contributing to in cell osmotic adjustment when turfgrass faced to drought stress [30]. In addition, another interesting result was that superior drought resistance for 'Midnight' Kentucky bluegrass could be characterized by the accumulation of sucrose in association with increased activity of sucrose-synthesizing enzymes (sucrose phosphate synthase and sucrose synthase), suggesting that increased sucrose accumulation resulting from the maintenance of active sucrose synthesis could relate to superior turf performance during water loss in Kentucky bluegrass [28]. Two Kentucky bluegrass cultivars contrasting in drought tolerance were evaluated the photosynthetic responses and underlying enzyme activities during drought stress and re-watering [31]. Compared to 'Brilliant', drought-tolerant 'Midnight' maintained significantly higher net photosynthetic rate, higher enzymatic activity and transcript level of ribulose-1,5-bisphosphate carboxylase (Rubisco), higher enzymatic activity of glyceraldehyde phosphate dehydrogenase (GADPH) during 10-d drought stress and in responses to re-watering, as well as higher Rubisco activation state upon re-watering, suggesting that carboxylation controlled by Rubisco and carbon reduction regulated by GAPDH could be the key metabolic processes imparting genetic variation in photosynthetic responses to drought stress while active Rubisco, GAPDH and Rubisco activase could all be involved in the superior post-drought recovery in Kentucky bluegrass [32]. Drought-tolerance 'midnight' and drought-sensitive 'Brilliant' as drought research model cultivars of Kentucky bluegrass have been detected 88 drought-responsive proteins by gel electrophoresis and mass spectrometry analyses. Many proteins involved in amino acid or energy metabolism were down regulated under drought stress, but most of those proteins had higher abundance in 'Midnight' than in 'Brilliant'. These proteins may serve as drought stress responsive proteins imparting Kentucky bluegrass adaptation to drought [33].

One of the major drought avoidance strategies in plant is the antioxidant physiological response by increasing antioxidant enzymes activity or by developing an effective reactive oxygen species (ROS) scavenging system to suppress ROS production which is a major cause of cellular injuries induced by drought [34]. Researches have revealed that availability of antioxidant enzyme activity and induced gene relative expression, the accumulation of abscisic acid and the level of membrane fatty acid saturation were important metabolic factors contributing to enhance drought resistance in Kentucky bluegrass although the responses in cultivars, organs and growth stages were varied [35–38]. Some studies found that the activity of superoxide dismutase (SOD) increased while that of catalase (CAT) and peroxidase (POD) decreased in leaves of Kentucky bluegrass when only surface soil drying was simulated [39]. The increased SOD activity in Kentucky bluegrass leaf cannot inhibit formation and accumulation of free radicals, but only delay accumulation of free radicals to a certain extent for alleviate active oxygen damage to cells [40, 41]. Under drought stress, the increased superoxide (O2•-) and hydrogen peroxide (H2O2) in leaves and roots of Kentucky bluegrass were associated with lipid peroxidation. In addition, the increases in the activities of ascorbate peroxidase (APX), monodehydroascorbate reductase (MHAR), dehydroascor bate reductase (DHAR) in leaves and that of CAT, glutathione reductase (GR), and MHAR in roots, but reduction in the activities of SOD and DHAR in roots [42]. After 6 days of rewatering, 'Midnight' displayed significantly higher activity levels of CAT, POD, and APX compared with 'Brilliant'. The differential responses of the activities of antioxidant enzymes to drought stress and post-drought rewatering between 'Midnight' and 'Brilliant' indicated that antioxidant enzymes including APX, SOD, GR, MR (monodehydroascorbate reductase), and DR (dehydroascorbate reductase) in the AsA-GSH (ascorbate–glutathine) cycle may play important protective roles involved in scavenging oxidant stress induced reactive oxygen species in Kentucky bluegrass for cellular survival of severe water deficit and post-drought recovery [43]. The increased content of unsaturated lipids in Kentucky bluegrass leaves under drought stress was crucial to maintain cell membrane fluidity and reduce ROS production. Leaf dehydration tolerance and postdrought recovery in Kentucky bluegrass was associated with their ability to maintain relative higher proportion and level of unsaturated fatty acids, particularly linolenic acids and linoleic acids [44, 45]. These researches on antioxidant metabolic responses revealed key information for controlling drought tolerance in turfgrass species and would facilitate the development of drought-tolerant germplasm through biotechnology.

So far, the important physiological mechanism of plant adaptation to drought resistance is ABA (Abscisic acid) accumulation, the amount of the primary chemical of ABA can reach up to more than 50-fold in plants under drought conditions [46, 47]. ABA is one of the most drastic changes observed hitherto in the concentration of a plant hormone responding to an environmental stimulus [37]. The functions and the signaling pathways of ABA in plants' responses to drought stresses have been extensively studied, and it is now well accepted that ABA plays important roles in plant including turfgrass adaptation to environmental stresses [5, 48, 49]. The relationship of ABA accumulation and drought resistance in different genotypes of Kentucky bluegrass varied. Leaf ABA content in drought susceptible cultivars increased sharply after 2 days of drought stress to as much as 34-fold the controls, while in drought resistant cultivars, the content ABA in leaves also increased with drought, but to a lesser extent than in drought sensitive cultivars. In addition, the stomatal conductance, photosynthetic rate, leaf water potential and turf quality in drought resistant cultivars performed less severe decline during drought than drought sensitive cultivars, indicating that stomatal conductance was negatively related to ABA accumulation and the ABA concentration during drought could regulate by stomatal behavior of Kentucky bluegrass [31]. The exogenous ABA application on Kentucky bluegrass demonstrated that ABA sprayed on leaves can help maintain higher turf quality and delayed the quality decline during drought stress. ABA treated grass had higher cell membrane stability indicated by less electrolyte leakage of leaves, and higher photochemical efficiency [50]. ABA effects on shoot growth and stomata, it may also facilitate osmotic adjustment and expression of specific proteins [6]. Further study suggesting that ABA accumulation in response to drought stress could be used as a metabolic factor to select for drought tolerance in Kentucky bluegrass [35]. Some studies on Kentucky bluegrass discovered that the interaction of endogenous hormones could contribute to increase drought resistance. Under drought conditions, the concentrations of ABA, JA (Jasmonate) and BR (Brassinolide) increased significantly compared to well water plants. Drought stressed Kentucky bluegrass had higher leaf ABA, lower leaf trans-zeatin riboside (ZR), isopentenyl adenosine (iPA), and indole-3-acetic acid (IAA), but similar level of leaf gibberellin A4 (GA4) when contrasted to the control, suggested that drought stress-induced injury to Kentucky bluegrass may be associated with hormonal alteration and may have better photosynthetic function and performance [5, 51].

Recently, a great progress on biological omics such as metabolomics, proteomic and transcriptomic have revealed many drought response metabolites, proteins and molecular factors in major crops [2]. However, information on Kentucky bluegrass involved in molecular mechanisms and genes underlying drought resistance is lacking. Limited researchers have identified some drought-responsive genes and gene signaling transduction pathways as described below. Under drought stress and water recovery treatments, the gene expression patterns of antioxidant enzymes of Kentucky bluegrass were differentially or cooperatively involved in the defense mechanisms in the leaves and roots. For the leaves, the expressions of iron SOD (*FeSOD*), cytosolic copper/zinc SOD (*Cu/ZnSOD*), chloroplastic *Cu/ZnSOD*, and *DHAR* were down-regulated by drought stress but recovered to control level after rewatering, while the expressions of *GR* and *MDHAR* were up-regulated and remained that levels after recovery. For the roots, the expressions of cytosolic *Cu/ZnSOD*, manganese SOD (*MnSOD*), cytosolic *APX*, *GR*, and *DHAR* were down-regulated under drought stress but recovered except for *GR* and *DHAR*, while *MDHAR* expression was up-regulated. No differences in *CAT* transcript abundance were noted among the treatments [42]. Transcriptome sequencing on cuticular wax deposition of Kentucky bluegrass have discovered that a number of genes involved in very longchain fatty acids and cuticular wax biosynthesis, transportation and regulation pathways, and these genes presented differentially expressed patterns between the leaf non-elongation zone and the emerged blade zone [52]. There were 9 candidate reference genes expression stability in the leaves and roots of Kentucky bluegrass under different stresses (drought, salt, heat, and cold), and were evaluated using the GeNorm, NormFinder, BestKeeper, and RefFinder bio-tools. Among 9 genes, *ACT* and *SAM* maintained stable expression in drought-treated leaves, and *GAPDH* combined with *ACT* was stable in drought-treated roots. The expression stability of reference genes in Kentucky bluegrass will be particularly useful in the selection of stress-tolerance genes and the identification of the molecular mechanisms conferring stress tolerance in this species [53]. A comparative transcriptomic study also found that many differentially expressed genes were enriched in 'Plant hormone signal Transduction' and 'MAPK signaling pathway-Plant'. Some key up-regulated genes, including *PYL*, *JAZ*, and *BSK*, were involved in hormone signaling transduction of abscisic acid, jasmonic acid, and brassinosteroid and possibly these genes play important roles in coping with drought stress in Kentucky bluegrass under drought stress [5]. These studies suggest importance of molecular functions and related genes for protection and improvement in Kentucky bluegrass tolerance to drought, however, molecular mechanisms underlying drought tolerance in Kentucky bluegrass are largely unknown.

3. Summary and future perspectives

Global climate change and the continuously growing population in the world, result increasingly limited and more costly in water availability. Water conservation in turf management becomes extremely important. Kentucky bluegrass is one the most popular and widely used cool-season turfgrass for amenity, sports and environmental conservation. Turf industry has released many improved Kentucky bluegrass cultivars in the world since last century. Among these germplasm, a wide range in water consumption variation represented by the daily evapotranspiration rate [20]. Hence, to identify ideal water conserving properties on Kentucky bluegrass in reducing irrigation requirements by enhance drought tolerance has been pursued by scientists. Many researchers have revealed a great deal of results in the understanding of morphological and physiological traits and mechanisms responding to drought resistance with Kentucky bluegrass. In response to drought stress, various adaptation strategies of Kentucky bluegrass were taken based on morphology, physiology and genetics. Narrow leaf texture, leaf orientation and canopy configuration, shoot and root systems are associated with Kentucky bluegrass drought resistance. Moreover, the physiological characteristics with the increased ABA and metabolites to enhance cell water potential and regulating stomatal movement, the protein translation and ROS scavenging to protect membrane stability are vital strategies for Kentucky bluegrass survive and maintain greenness under drought conditions. However, although some drought-responsive genes and gene signaling transduction pathways as well as proteins for Kentucky bluegrass tolerance to drought stress have been identified, our knowledge of the response is still limited information on how whole plants perceive and conductive these signals under long period drought. On Kentucky bluegrass, systemic study on the molecular mechanisms of drought resistance is still lagging. Identification of novel candidate genes, proteins, metabolites, and molecular markers, and integrating transcriptomics, proteomics and metabolomics to explore intercellular communication and drought resistance manipulating signals are crucial for molecular breeding and marker-assisted selection of Kentucky bluegrass with superior drought tolerance in future.

Author details

Jian Cui[1], Saud Shah[2], Shah Fahad[3] and Yajun Chen[2]*

1 School of Architecture and Built Environment, The University of Adelaide, Adelaide, Australia

2 College of Horticulture and Landscape Architecture, Northeast Agricultural University, Harbin, China

3 Department of Agronomy, The University of Haripur, Haripur, Pakistan

*Address all correspondence to: chenyajun622@163.com

References

[1] Turgeon AL, editor. Turfgrass Management. 8th ed. New Jersey: Pearson education; 2008. p. 436

[2] Huang B, DaCosta M, Jiang Y: Research advances in mechanisms of turfgrass tolerance to abiotic stresses: from physiology to molecular biology. Crit. Rev. Plant Sci. 2014; 33: 141-189. DOI: 10.1080/07352689.2014.870411

[3] Honig JA, Averello V, Bonos, SA, Meyer WA: Classification of Kentucky bluegrass (*Poa pratensis* L.) cultivars and accessions based on microsatellite (Simple Sequence Repeat) markers. HortSci 2012; 47: 1356-1366.

[4] Fan J, Zhang W, Amombo E, Hu L, Kjorven JO, Chen L: Mechanisms of environmental stress tolerance in turfgrass. Agronomy. 2020; 10: 522. DOI: 10.3390/agronomy10040522 www.mdpi.com/journal/agronomy

[5] Chen Y, Chen Y, Shi Z, Jin Y, Sun H, Xie F, Zhang L: Biosynthesis and signal transduction of ABA, JA, and BRs in response to drought stress of Kentucky bluegrass. Int. J. Mol. Sci. 2019; 20: 1289. DOI: 10.3390/ijms20061289 www.mdpi.com/journal/ij

[6] Fry J, Huang B, editors. Advanced Turfgrass Science and Physiology. 1st ed. New Jersey: John Wileys and Sons; 2004. 310p.

[7] Chen Y, Pettersen T, Kvalbein A, Aamlid TS: Playing quality, growth rate, thatch accumulation and tolerance to moss and annual bluegrass invasion as influenced by irrigation strategies on red fescue putting greens. J. Agro. Crop Sci. 2018; 204: 185-195.

[8] Shah S, Li X, Chen Y, Zhang L, Shah F, Saddam H, Arooj S, Chen Y: Silicon application increases drought tolerance of Kentucky bluegrass by improving plant water relations and morphophysiological functions. The Sci. World J. Volume 2014, Article ID. 2014; 368694. DOI: 10.1155/2014/368694

[9] Shah S, Chen Y, Shah F, Hussain S, Li N, Li X, Alhussien SAAFE: Silicate application increases the photosynthesis and its associated metabolic activities in Kentucky bluegrass under drought stress and post-drought recovery. Environ. Sci. Pollut. Res. 2016; 23: 17647-17655.

[10] Levitt J, editor. Responses of Plants to Environmental Stress. 2nd ed. New York: Academic Press; 1980.

[11] Chai Q, Guo Z, Ren J, Nan Z: Assessment of drought resistance of Kentucky bluegrass (*Poa pratensis*) varieties at seedling stage. New Zeal. J. Crop Hort. 2006; 34: 319-328.

[12] Su K, Bremer DJ, Keeley SJ, Fry JD: Effects of high temperature and drought on a hybrid bluegrass compared with Kentucky bluegrass and tall fescue. Crop Sci. 2007; 47: 2152-2161.

[13] Song Y, Huang B: Differential effectiveness of doubling ambient atmospheric CO2 concentration mitigating adverse effects of drought, heat, and combined stress in Kentucky bluegrass. J. Am. Soc. HortSci. 2014; 139: 364-373.

[14] Michael DR, Douglas EK, Kenneth H, Debra R: Drought tolerance and rooting capacity of Kentucky bluegrass cultivars. Crop Sci. 2008; 48: 2429-2436.

[15] Leksungnoen N, Johnson PG, Kjelgren RK: Physiological responses of turfgrass species to drought stress under high desert conditions. HortSci. 2012; 47: 105-111.

[16] Domenghini JC, Bremer DJ, Fry JD, Davis GL: Prolonged drought

and recovery responses of Kentucky bluegrass and ornamental groundcovers. HortSci. 2013; 48: 1209-1215.

[17] Merewitz E, Meyer W, Bonos S, Huang B: Drought stress responses and recovery of Texas × Kentucky hybrids and Kentucky bluegrass genotypes in temperate climate conditions. Agron. J. 2010; 102: 258-268.

[18] Beard JB, Kenna MP, editors. Water Quality and Quantity Issues for Turfgrasses in Urban Landscapes. 1st ed. Belmond: PSI Printing Services, Inc. 2006. 298p.

[19] Kim KS, Beard JB: Comparative turfgrass evapotranspiration rates and associated plant morphological characteristics. Crop Sci. 1988; 28: 328-331.

[20] Shearman RC: Kentucky bluegrass cultivar evapotranspiration rates. HortSci. 1986; 21: 455-457.

[21] Ebdon JS. Petrovic AM: Morphological and growth characteristics of low- and high-water use Kentucky bluegrass cultivars. Crop Sci., 1998; 38: 143-152.

[22] Chen Y, Zu Y, Liu H, Chen Y, Liu D: Leaf epidermis ultra-microstructure associated with drought resistance in *Poa pratensis* cultivars. Acta Agrestia Sinica. 2005; 13: 339-343.

[23] Bonos SA, Murphy JA: Growth responses and performance of Kentucky bluegrass under summer stress. Crop Sci. 1999; 39: 770-774.

[24] Keeley SJ, Koski AJ: Root distribution and ET as related to drought avoidance in *Poa pratensis*. In: Science and Golf IV: Proceedings of the World Scientifific Congress of Golf. Pp. 2002; 555-563. Thain, E., Ed. Routledge, New York.

[25] Ebdon JS, Kopp KL: Relationships between water use efficiency, carbon isotope discrimination, and turf performance in genotypes of Kentucky bluegrass during drought. Crop Sci. 2004; 44: 1754-1762.

[26] Shah S, Shah F, Chen Y, Ihsan MZ, Hammad HM, Nasim W, Amanullah J, Muhammad A, Alharby H: Effects of nitrogen supply on water stress and recovery mechanisms in Kentucky bluegrass plants. Front. Plant Sci. 2017; 8: 983.

[27] Richardson MD, Karcher DE, Hignight K, and Rush D: Drought tolerance and rooting capacity of Kentucky bluegrass cultivars. Crop Sci. 2008; 48: 2429-2436.

[28] Yang Z, Xu L, Yu J, DaCosta M, Huang B: Changes in carbohydrate metabolism in two Kentucky bluegrass cultivars during drought stress and recovery. J. Amer. Soc. Hortic. Sci. 2013; 138: 24-30.

[29] DaCosta M, Huang B: Osmotic adjustment associated with variation in bentgrass tolerance to drought stress. J. Amer. Sco. Hort. Sci. 2006; 131: 338-344.

[30] Jiang Y, Huang B: Osmotic adjustment and root growth associated with drought preconditioning-enhanced heat tolerance in Kentucky bluegrass. Crop Sci. 2001; 41: 1168-1173.

[31] Xu L, Yu J, Han L, Huang B: Photosynthetic enzyme activities and gene expression associated with drought tolerance and post-drought recovery in Kentucky bluegrass. Environ. Exp. Bot. 2013; 89: 28-35.

[32] Wang Z, Huang B: Genotypic variation in abscisic acid accumulation, water relations, and gas exchange for Kentucky bluegrass exposed to drought stress. J. Amer. Soc. Hort. Sci. 2003; 128: 349-355.

[33] Xu C, Huang B: Comparative analysis of drought responsive proteins

in Kentucky bluegrass cultivars contrasting in drought tolerance. Crop Sci. 2010; 50: 2543-2552.

[34] Mittler R: Oxidative stress, antioxidants and stress tolerance. Trends Plant Sci. 2002; 7: 405-410.

[35] Wang Z, Huang B, Bonos S, Meyer W: Abscisic acid accumulation in relation to drought tolerance in Kentucky bluegrass. HortSci. 2004; 39: 1133-1137.

[36] Wang Z, Huang B: Physiological recovery of Kentucky bluegrass from simultaneous drought and heat stress. Crop Sci. 2004; 44: 1729-1736.

[37] DaCosta M, Huang B: Drought survival and recuperative ability of bentgrass associated with abscisic acid and cytokinin production. J. Amer. Hortic Sci. 2007; 132: 60-66.

[38] Yu J, Liu M, Yang Z, Huang B: Growth and physiological factors involved in interspecific variations in drought tolerance and post drought recovery in warm- and cool-season turfgrass species. J. Am. Soc. Hortic. 2015; 140: 459-465.

[39] Fu J, Huang B: Involvement of antioxidants and lipid peroxidation in the adaptation of two cool-season grasses to localized drought stress. Environ. Exp. Bot. 2001; 45: 105-114.

[40] Zhang XZ, Schmidt RE: Antioxidant response to hormone containing product in Kentucky bluegrass subjected to drought. Crop Sci. 1999; 39: 545-551.

[41] Liu J, Xie X, Du J, Sun J, Bai X: Effects of simultaneous drought and heat stress on Kentucky bluegrass. Sci. Hort. 2008; 115: 190-195.

[42] Bian S, Jiang Y: Reactive oxygen species, antioxidant enzyme activities and gene expression patterns in leaves and roots of Kentucky bluegrass in response to drought stress and recovery. Sci. Hort. 2009; 120: 264-270.

[43] Xu L, Han L, Huang B: Antioxidant enzyme activities and gene expression patterns in leaves of Kentucky bluegrass in response to drought and post-drought recovery. J. Amer. Soc. Hort. Sci. 2011a; 136: 247-255.

[44] Xu L, Han L, Huang B: Membrane fatty acid composition and saturation levels associated with leaf dehydration tolerance and post-drought rehydration in Kntucky bluegrass. Crop Sci. 2011b; 51: 273-281.

[45] Peng Y, Xu C, Huang B: Improved heat tolerance through drought preconditioning associated with changes in lipid composition, antioxidant enzymes and protein expression and in Kentucky bluegrass. Crop Sci. 2012; 52: 807-817.

[46] Zeevaart JAD: Changes in the levels of abscisic acid and its metabolites in excised leaf blades of *Xanthium strumarium* during and after water stress. Plant Physiol. 1980; 66: 672-678.

[47] Auler PA, Amaral MN, Rossatto T, Crizel RL, Milech C, Chaves FC, Souza GM, Braga EJB: Metabolism of abscisic acid in two contrasting rice genotypes submitted to recurrent water deficit. Physiol. Plant. 2020; DOI: https://doi.org/10.1111/ppl.13126

[48] Zhu JK: Abiotic stress signaling and responses in plants. 2016: Cell. 167: 313-324.

[49] Gong Z, Xiong L, Shi H, Yang S, Herrera Estrella LR, Xu G, Chao DY, Li J, Wang PY, Qin F, Li J, Ding Y, Shi Y, Wang Y, Yang Y, Guo Y, Zhu JK: Plant abiotic stress response and nutrient use efficiency. SCI. CHINA Life Sci. 2020; 63: 635-674. DOI: https://doi.org/10.1007/s11427-020-1683-x

[50] Wang Z, Huang B, and Xu Q: Effects of abscisic acid on drought responses of Kentucky bluegrass. J. Amer. Soc. Hort. Sci. 2003; 128: 36-41.

[51] Zhang X, Goatley M, Wu W, Ervin E, Shang C: Drought-induced injury is associated with hormonal alteration in Kentucky bluegrass. Plant Sig. and Beh. 2019; e1651607-3.DOI: https://doi.org/10.1080/15592324.2019.1651607.

[52] Ni Y, Guo N, Zhao Q, Guo Y: Identification of candidate genes involved in wax deposition in *Poa pratensis* by RNA-seq. BMC Genomics. 2016; 17: 314. DOI: 10.1186/s12864-016-2641-2

[53] Niu K, Shi Yi, Ma H: Selection of candidate reference genes for gene expression analysis in Kentucky bluegrass (*Poa pratensis* L.) under abiotic stress. Frontiers in Plant Sci. 2017; 8: 193. DOI: 10.3389/fpls.2017.00193.

Chapter 2

Management of Abiotic Stress in Forage Crops

Amanpreet Singh and Harmandeep Singh Chahal

Abstract

Forage plays a key role in rearing ruminants and protecting the environment. Apart from serving as the primary source of food for domestic and wild animals, forages also contribute to human civilization in different ways like protecting soil through crop over and fertility by addition of organic matter. It also provides habitat for wild animals. A survival strategy plays a more important role than a growth strategy to improve the sustainability of forage production, especially in extreme environmental conditions . Climate change is likely to affect the forage production and nutritional food security for domestic animals. Long-term rainfall data in India indicate that rainfed areas experience 3 to 4 years of drought in every 10 years. Of these, one or two of it occur in severe form. Forage crop production is largely affected by abiotic factors related stress such as drought, salinity, etc. There is need to adopt various conventional and genetic approaches to improve stress tolerance of forage crops.

Keywords: forage crops, abiotic stress, management, breeding and micronutrients

1. Introduction

In the agricultural context, stress has been defined as the conditions in which plants are prevented from fully expressing their genetic potential for growth, development, reproduction, and, ultimately, crop productivity [1]. Abiotic stress negatively affects the livelihoods of farmers and their families, the sustainability of livestock, as well as national economies and food security. Forages are generally described as plants and its parts consumed by domestic livestock. Forage plays a key role in rearing ruminants and protecting the environment. Apart from serving as the primary source of food for domestic and wild animals, forages also contribute to human civilization in different ways like protecting soil through crop over and fertility by addition of organic matter. It also provides habitat for wild animals. In the biological soil–plant–animal system, forage is highly demanded by livestock. Escalation in the human population in the coming decades will put the higher burden on land for food crops and fiber production. As a result, we may face forced forage cultivation in those areas having poorer soils regarding fertility and management [2]. The water use for irrigation is incredibly high and this trend could increase considerably in the future leading to shortage of water availability [3]. For perennial forage and natural vegetation, the ability to survive during adverse environmental periods is a life saving feature. A survival strategy plays a more important role than a growth strategy to improve the sustainability of forage production, especially in extreme environmental conditions [4]. Forage crop production is largely affected by abiotic factors related stress such as drought, salinity, etc. There is need to adopt various conventional and genetic approaches to improve stress tolerance of forage crops.

2. Forage status

Currently, India faces a critical imbalance in its natural resource base: around 18 percent of humans and 15 percent of the world's animal population are only served by 2.4 percent of the geographical area, 1.5 percent of forests and pastures, and 4.2 percent of water resources [5]. The three main sources of forage supply in India are crop residues, cultivated forage, and forage from common property resources such as forests, permanent pastures, and pastures. Due to the multiplicity of forage crops produced in different seasons and regions, the surplus and deficit in different regions, the non-commercial nature of crops and forage production with minimal inputs from degraded and marginal land, there has been a large gap in the availability and need for forage. Currently, the country faces a net deficit of 35.6 percent of green forage, 10.95 percent of residues from dry crops, and 44 percent of concentrated ingredients for animal feed [6]. Supply and demand for the forage scenario are presented in **Figure 1**. Furthermore, in the case of forage, regional and seasonal deficiencies are more important than national deficiencies, since it is not economical to transport forage over long distances. Furthermore, the available forages are of low quality and deficient in available energy, protein, and minerals. Farmers maintain large herds of animals to compensate for low productivity, adding pressure on forage and other natural resources [7]. Almost two-thirds of the total cost of animal production is due to food and fodder. Consequently, any attempt to improve the availability of food and fodder and save the cost of food would result in better remuneration for farmers. The area under cultivated forage is only 8.4 million hectares and has been static for the past two decades. The potential for further increases seems very small due to demographic pressure for food crops. Recent crop diversification, where cash crops replace traditional cereal crops, especially coarse grains, is likely to have an impact on the availability of crop residues for animal production [8]. Likewise, the productivity of certain important cultivated forages is highly variable. Among Kharif forages, sorghum, corn, cowpea, Napier-bajra hybrid, and guinea they have a wide range. However, during rabi, the choice is limited to oats, alfalfa, and berseem. Emphasis should be placed on new area-specific crops that can break down yield barriers and meet the challenges of the food deficit.

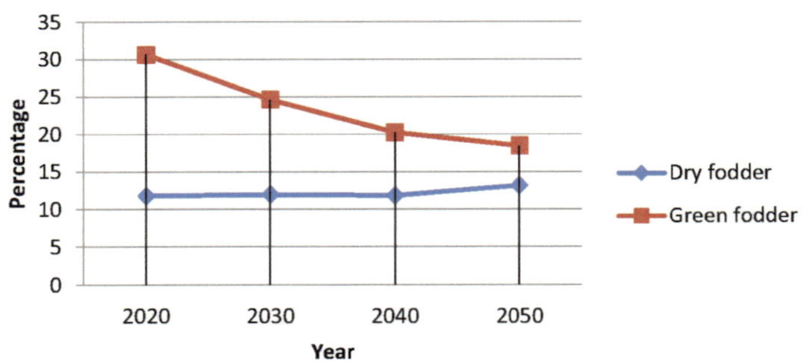

Figure 1.
*Deficient trend of fodder crop concerning future demand. *IGFRI vision, 2050.*

3. Different types of abiotic stress faced by crops

3.1 Temperature restriction

The tropical climate is cursed by higher temperatures and radiation that limit the growth and development of plants. High temperatures cause burns, sunburn, and discoloration of the leaves, reducing plant growth [9]. Limiting growth, metabolism, and performance potential due to exposure to a temperature below or above the thermal threshold for optimal biochemical, physiological, and morphological development is called thermal stress [10]. Plants are classified into psychophilic, mesophilic, and thermophilic according to their tolerance to low, medium, and high temperatures [1]. The adversity of heat stress varies with the duration, stage, and intensity of stress [11]. Increased heat stress adversely affects the spikelets number, the number of florets per plant in rice crop, and the seeds in forage crop like sorghum [12]. It also reduces quality due to reduced production of oil, starch, and protein [13]. Stress at low temperatures causes wilting, bleaching, darkening, necrosis, and death of plants [1]. Approximately 15percent of arable land is said to be affected by frost stress [14].

3.2 Moisture stress

About 28 percent of the world's land is too dry for agricultural support [15]. The estimated annual yield loss due to extraction in the tropics is almost 17 percent [16]. Increasing the draft with the changing climate scenario leads to a decrease in plant physiology, growth, and reproduction [17]. The moisture deficit causes greater transpiration and reduces the availability of water from the roots of the plants [18], which tends to balance the water on the negative side that affects growth, the relationship between nutrients and water, photosynthesis and assimilation of sharing and, ultimately, performance [19]. The stress response plan in plants varies according to the species according to its stages and other growth factors [20]. High-temperature stress affects enzyme activity, cell division in plants [21] and also changes the growth period and distribution of crops [22].

3.3 Heavy metal stress

Heavy metals are those metals that have a specific weight greater than 5 g cm^{-3} or an atomic mass greater than 20 and are generally toxic even at low concentrations [23]; some of heavy elements or metals are cadmium (Cd), lead (Pb), arsenic (As), silver (Ag), etc. Heavy metal contamination in the soil is mainly due to human activities such as mining, smelting, intensive agricultural practices, fuel production, electroplating, etc. [24] and may also be due to natural processes such as soil erosion, excessive weathering of rocks and minerals, and volcanic eruption. Among heavy metals, some have known physiological functions in the plant system called non-essential heavy metals, namely arsenic (As), lead (Pb), cadmium (Cd), mercury (Hg), and selenium. (Sc) and some are involved in different physiological functions of plants as a cofactor of enzymatic reactions [25] or role in redox reactions [26] called essential heavy metals, namely cobalt (Co), copper (Cu), manganese (Mn), zinc (Zn), iron (Fe), molybdenum (Mo) and nickel (Ni).

3.4 Salt stress

Crops are said to be subject to salt stress when they cannot express their full genetic potential in terms of growth, development, and reproduction, since the salinity of the soil exceeds the critical level [27] and dissolved salts in the soil and irrigation water vary from place to place [28]. The detrimental effect of soils affected by salt may be due to a high concentration of salt in the soil solution, i.e. osmotic effects or a high concentration of specific ions such as sodium or chloride that can damage sensitive crops, i.e. a specific ionic effect. The harmful effect of saline soil is due to the concentration of soluble salt, while the harmful effects of sodium soil are due to deterioration of the physical state of the soil [29]. The harmful effect of salt stress may be due to a specific ionic effect, that is, Na^+ and Cl^- [30] or to interact with other dynamics of mineral nutrients [31].

3.5 Nutritional stress

Several mineral elements contribute to the growth and development of a plant, 17 of which are called essential nutrients according to the essentiality criteria defined by Arnon and Stout. Since mineral nutrition is discipline independent of plant physiology [32, 33] divide essential minerals into four groups according to their biological structures and metabolic functions. There is some nutritional stress (deficiency or excess) reported by various scientists in different plants. Nitrate plays pivotal role cytokinin biosynthesis and transport, and a higher level of nitrate (NO^{3-}) inhibits root growth and the root: shoot ratio [34]. Phosphorus deficiency limits the lengthening of the primary roots and improves the formation of lateral roots, decreases the proportion of the dry weight of the roots of the shoots [35], reduces the leaves [36] and affects the reproductive organs formation [37], plants with potassium deficiency (K^+) are sensitive to lodging and airflow [38]. A sulfur deficiency decreased the net photosynthesis and the hydraulic conductivity of the roots [39], the reduction in the dry weight ratio of the roots of the shoots [40], an alteration in the metabolism of carbohydrates followed by an induced accumulation of starch [41].

4. Impact of abiotic stress on physiology of forage

4.1 Photosynthesis

Moderate stress in water deficit plants reduces photosynthesis which is accompanied by closing of stoma [42]. Measurement of the photosynthetic response and the activity of the ribosose bisphosphate carboxylase vifro (RUBISCO) in alfalfa (*Medicago sativa* L.) exposed to an increasing water deficit and found evidence of adverse osmotic effects [43].

4.2 Forage quality

The digestibility of legumes and their fiber have been largely affected by water availability [44, 45]. Drought affects the forage composition and quality by altering plant maturity and ratio of leaf mass to stem mass [46].

4.3 Establishment of seedlings in forages

Water availability highly affects the forage seedlings growth and maturity [47]. Seminal roots support seedlings for a short time. Seminal root system absorbs by the hydraulic conductivity of the suboptoptic internode. Redmann and Qi (1992) found that the diameter of the xylem vessels in warm-season grass seedlings that emerged from different planting depths and length of suboptoptic internode plays an important role in transport of water from the root to the shoot and reducing hydraulic conductivity.

5. Impact of climatic anomalies on forages in terms of stress

Climate change has become a serious threat to life on earth. There is also a global trend of increased storms on most lands. Glaciers are continuously melting, while daily high temperature with heat waves became more common [48]. Coping with climate variability is becoming a major challenge for human civilization. Higher seasonal variability regarding the distribution of precipitation, extreme events of temperature, and precipitation cause damage to crops and raise serious concerns about agricultural production. Among adverse weather events, drought is the major factor to directly affect the population. A warmer climate with increasing climatic variability will increase the risk of climatic extremes. Meteorological data analyzed over 5 decades from CRIDA's Gunegal research farm, a typical rain region, showed low precipitation. Climate change is likely to affect the forage production and nutritional food security for domestic animals. Long-term rainfall data in India indicate that rainfed areas experience 3 to 4 years of drought in every 10 years. Of these, one or two of it occur in severe form [49].

6. Abiotic stress management in major fodder crops

6.1 Sorghum

6.1.1 Water stress and its management

Sorghum with its persistent green character, well developed root system, higher water-use efficiency and epicuticular wax represents a good system for studying physiological features related to drought tolerance. Depending on stress development at any growth stage, sorghum shows a stress response before flowering and after flowering, respectively. All these different responses are affected by various genetic processes [50]. Pre-flowering stress affects plant biomass, panicle size, kernel quantity, and grain yield [51], whereas posttesthetic dryness leads to premature senescence of leaves and stems, lodging and the reduction of seed size [52]. Post-synthesis drought also increases plant sensitivity to biotic stress, such as charcoal rot (*Macrophomina phaseolina*) and fusarium stem rot (*Fusarium moniliforme*) [52]. For drought tolerance before flowering, six distinct genomic regions were Recombinant inbred sorghum lines (RIL) derived from the cross between the genotypes Tx7078 (tolerant before flowering, sensitive to post-flowering) and B35 (sensitive to pre-flowering, tolerant after flowering) [53]. The response to dryness after flowering is associated with the persistent green character of sorghum. Staying green is essentially the retention of the surface of mature green leaves (GLAM). Maintaining the remaining green character during the grain filling phase under stress conditions of soil water deficit constitutes an important element of drought tolerance [54].

6.1.2 Epicuticular wax

Epicuticular wax (EW) forms a glaucous upper coating that is visible on many cultivated plants called waxy bloom. Species, organ, stage of development, and environmental conditions are all those things that affect buildup of wax.

Composition and structure of epicuticular wax is very diverse which is considered a potential useful trait and has been related to resistance against different adverse environmental conditions [55]. Sorghum differs from other field crops in its ability to produce sufficient amount of EW that is placed on the leaf blade as well as leaf sheath generally during pre-flowering and stages of maturity. Sorghum leaf sheath bloom is composed of large amount of free fatty acids with a 16 to 33 carbon chain length [56].

6.1.3 Osmotic adjustment

Two traits named osmotic adjustment and antioxidant capacity have been related with drought tolerance mechanisms. Osmotic adjustment has been associated with sustained performance under water limiting conditions in many crops and is an inherited characteristic. Two major independent genes namely OA1 and OA2 in sorghum have been reported to control Osmotic adjustment inheritance.

6.1.4 Cold tolerance

Sorghum from the tropical and subtropical regions of Africa [57] is well adapted to warm growing conditions. Cool temperatures at the beginning of the growing season are therefore an important limitation for the growth of temperate sorghum areas [58]. Cross developed from local Chinese races, ShanQui Red (SQR, cold-tolerant), and SRN39 (cold-sensitive) was used for QTL analysis of early-season cold yields on sorghum [59].

6.2 Bajra

Bajra [*Pennisetum glaucum*] is a C4 plant with very high photosynthetic efficiency. Bajra also have high dry matter production capacity. It is generally cultivated under the most adverse agroclimatic conditions, where other crops such as sorghum and corn do not stand well.

6.2.1 Selecting genotypes is a good approach to managing abiotic stress

Pearl millet germplasm screening helped in the development of highly advanced breeding techniques, an improvement in the population, including OPVs, genetic pools and compounds, possible parental hybrids, and accessions of the high-throughput genetic material of cereals and forages, presumably with a high degree of salt tolerance (**Table 1**).

6.2.2 Low soil fertility

Soils in the areas where pearl millet is grown are often poor infertility because they contain a small amount of organic matter (0.05–0.40percent) due to low ground cover, coarse soil texture, and prevailing high temperatures [63]. Soils also contain low to moderate levels of available phosphorus (10–25 kg ha^{-1}). This problem was mainly solved through nutrient management. The possibilities of genetic improvement for the efficient use of nutrients are increasingly explored in some cultures [64]. Only recently has strategic research been launched at ICRISAT in the West and Central Africa region to identify QTL to increase the efficiency of phosphorus and examine the stability of its expression across genetic environments.

Abiotic stress	Genotypes	References
Drought	CZP 9802; 863B and PRLT 2/89-33ICMP 83,720	[60]
Heat	H77/833–2, H77/29–2 and CVJ 2–5–3-1-3, 77/371XBSECT CP1	[61]
Salinity	33, 10,876 and 10,878 (Sudan), 18,406 and 18,570 (Namibia), and ICMV93753 and ICMV 94474 (India); 863-B, CZI 98–11, CZI 9621, HTP 94/54	[62]

Table 1.
Available genotypes for abiotic stress tolerance in pearl millet.

6.3 Forage corn

Corn forage (*Zea mays* L.) has become an important component of ruminant rations in recent years. It is the only crop among non-leguminous forages that combines better nutritional quality. With a large amount of biomass [65]. Although the crop has great adaptability [66], it is the least tolerant of abiotic stress among cereals. Drought, salinity, and high temperatures are among the major abiotic stresses that negatively impact corn production in most regions of global corn production [67]. Soils with saline stress are present on all continents and in almost all climatic conditions. However, its distribution is relatively more extensive in arid and semi-arid regions than in humid regions [68]. Mohammed and Mohammed 2019 stated that the appropriate genotype based on stress selection is the inexpensive and manageable stress method based on salt, water, and heat or combined form and also concluded that the reduction in stress performance would be reduced to 20–40 parents.

6.4 Cowpea

The cowpea (*Vigna unguiculata*) is one of the most important legumes cultivated by subsistence farmers for human and animal consumption, mainly in the semi-arid regions of Africa and Brazil. In Africa, it is used for the livelihood of millions of people in the semi-arid regions of the West and Center [69] and is considered the most important grain legume crop in the sub-Saharan region.

6.4.1 Reproductive improvements

Cowpea is relatively drought tolerant. Despite this feature, however, drought can cause a considerable loss of performance. Efforts have been made to select the cowpea genetic material to identify lines with better drought tolerance than currently available varieties. According to Watanabe et al. [70], certain lines of genetic material, in particular, TVu 11,979 and TVu 14,914, were consistently very drought tolerant under real field conditions. Drought can occur at the beginning of the season, mid-season, or the crop development stage. Studies have shown that cowpea plants can show drought tolerance in the vegetative stage [71] and the reproductive stage [32]. Some cowpea lines exhibit a green persistence feature, also called delayed leaf senescence (DLS), which can help plants tolerate terminal and mid-season drought [32].

6.4.2 Gene selection

In cowpea plants, overexpression of the CPRD 8, CPRD12, CPRD14, CPRD22 and CPRD46 genes that confer tolerance to water stress [72], as well as the production of VusAPX genes connected to VucAPX, VupAPX and VutAPX of antioxidant enzymes [73], it is reported, in addition to the expression of the high level of the PvP5CS gene associated with the production of proline, an amino acid that fulfills the function of osmotic adjustment between species during drought.

6.5 Abiotic stress tolerance mechanism

Climate and soil determine many plant adaptations and the ecogeographic distribution of species and ecotypes show differences in physiology and development patterns that provide good evidence of adaptation mechanisms. Plants respond to environmental change as individuals through phenotypic plasticity and in populations through the selection and associated evolutionary processes. Determining the genetics underlying adaptation processes is not always easy because environmental factors can be complex or poorly defined. However, extreme environmental pressures, such as heavy metal contamination from the soil or harsh winter conditions [74] can produce detectable genetic changes. Multiple genes may be responsible for a response to a certain factor, or the same gene may be involved in different adaptive responses specific genetic interactions can be in a state of change or become fixed, limiting the possibilities for future evolution. Phenotypic plasticity acts as a buffer to prevent excessive gene flow in response to short-term changes.

7. Improving forages for abiotic stress response based on breeding techniques

7.1 Greater tolerance to stress through genetic transformation

Genetic improvement of forages through the selection of conventional plants is slow because most forage species are self-incompatible, limiting inbreeding to concentrate the desired genes to be used in the rapid development of new cultivars. Genetic transformation allows the direct introduction of desirable genes, thus offering new opportunities for forage molecular selection. Like many other crops, drought tolerance is an important goal in improving alfalfa. Since cuticle waxes play a central role in limiting the breathable loss of water from the plant surface, the genetic engineering of plant waxes is expected to eventually increase tolerance to environmental stress in crops such as agronomic importance [75].

7.2 Improvement of stress tolerance through intergeneric hybridization

Extensive hybridization with relative species followed by introgression of chromosomes and/or chromosome fragments has been considered an effective means of transferring salt and other stress tolerance genes to target species to extend the gene pool. Intergeneric hybrids between species of Lolium (Ryegrass) and Festuca (Fescue) have attracted much attention from forage breeders. Rye grasses are considered ideal grasses due to their fast establishment, their ability to resist intense grazing, their good palatability, and their high nutritional value [76].

Alfalfa (*Medicago sativa* L.) is widely cultivated in temperate and tropical regions for green forage, hay, silage, and grass. As a perennial forage plant, alfalfa is a fairly hardy species and has a relatively high level of drought tolerance compared too many other legume forage plants [77]. Alfalfa's increased drought tolerance is due in part to deeper roots and the ability to extract more available water from the root zone [78]. Detection of salt-sensitive proteins in two contrasting alfalfa cultivars using a comparative proteome approach revealed two new proteins, NAD synthetase, and biotin carboxylase-3, as being salt sensitive. These results provide new information on alfalfa salt stress tolerance [79]. The effects of rhizobia strains on the amino acid composition of alfalfa under salt stress indicate that proline, glutamine, arginine, GABA, and histidine accumulate significantly in salt-stressed nodules, suggesting increased production of amino acids associated with osmoregulation, nitrogen storage, or energy metabolism to counteract salt stress [80] is a widely allogeneous forage legume species distributed worldwide due to its wide range of climate adaptation [81]. But it is less drought tolerant than other temperate perennial forage legumes due to its shallow root system and its inability to effectively control transpiration [82]. Biochemical studies have indicated that when white clover was stressed by a water deficit, De novo synthesis of amino acids, including proline, has increased in both leaves and roots [83]. This phenomenon may serve as an adaptive response during the first days of drought since the transient increase in amino acid concentration has been followed by a decrease in protein synthesis that slows plant growth.

Cowpea (*Vigna unguiculata* L.) growing in a variety of environments from tropical to arid/semi-arid regions, increased tolerance to drought and heat would be desirable. The cowpeas (*Vigna marina*) that grow on sandy beaches in the tropical and subtropical regions closest to the sea have the potential to be a source of genes for breeding salt-tolerant cultivars. Chankaew et al., [84] first reported QTL mapping for salt tolerance in the *Vigna marina*, and multiple internal mapping consistently identified an important QTL that can explain 50 percent of the phenotypic variation. The flanking marker can facilitate the transfer of salt tolerance of this subspecies in related Vigna cultures.

8. Micronutrient stress management

8.1 Sorghum

Sorghum (*Sorghum bicolor*) is one of the important forage crops for high agricultural production and good nutritional value for animals. Nutrient requirements for growing sorghum are high; they are grown for forage, in part from organic sources, and are supplemented primarily with inorganic fertilizers. The growth, development, and biological yield of crops affected by balanced fertilization have shown positive effects. Micronutrients increase crop productivity and also maintain soil health. A very small amount is required. Soil application of micronutrients is preferable for what is desired. Choudhary et al. [85] observed that the combined application of micronutrients, that is, a considerably higher yield of cereals, stems, and organic, is obtained through a soil + leaf application. The results showed a significant increase in grain yield (14.15 and 12.13 percent), biological yields (11.37 and 9.31 percent), and in stem yield (10.75 and 8.60 percent) and through the combined spraying of soil and foliar on the soil and foliar application, respectively.

8.2 Pearl millet

Pearl millet (*Pennisetum glaucum* L.) is one of the main millet crops in arid and semi-arid areas. Weather. Due to the drought-tolerant nature, it grows well in poor sandy soils. Sustainable production of pearl millet can be achieved through the balanced use of nutrients in crops with the fusion of organic and inorganic sources. Intensive farming is followed in the current system, most farmers use high-yielding whole crop varieties, ultimately a significant removal of nutrients from the soil in recent years, and the consumption of fertilizers has remained well less than the elimination one. So that the qualitative and quantitative improvement of the crop yield goes through mineral fertilization and that its quality can be improved through adequate practices of nutrient management and soil cultivation [86].

8.3 Maize

The third most important cereal crop is maize (*Zea mays* L.) worldwide and India. It is cultivated in temperate and tropical regions of the world. It is the most important cereal for animal feed. In India, 45 percent of maize production is used in various forms of staple foods [87]. Corn, rice, and wheat are estimated to provide at least 30percent of food calories to more than 4.5 billion people in 94 developing countries. The demands for animal feed and biofuels can be met by increasing maize production [88]. The application of micronutrients can be carried out in several ways, such as seed treatment, soil and foliar application [89], which depends on the characteristics of the soil and the climate of the region. Corn productivity can be improved by applying Zn and B to the soil.

8.4 Cowpea

Cowpea (*Vigna unguiculata*) is a legume and is used as a forage crop that is grown during the Kharif season, requiring only an initial dose of nitrogen (15–25 kg N ha^{-1}). Most nitrogen requirements are met by symbiotic nitrogen fixation. The strong application of NPK fertilizers has led to micronutrient deficiencies in many parts of the country. To achieve high yields and maintain them over the years, it becomes highly relevant to predict emerging nutrient deficiencies and to develop appropriate breeding technologies. Balanced fertilization is inevitable to increase the productivity of the crop. Among the micronutrients, Zn, Fe, B, Mn, and Mo significantly improved yield, and micronutrient foliar spray is economical on legumes.

Author details

Amanpreet Singh* and Harmandeep Singh Chahal
Department of Agriculture, Khalsa College Amritsar, Punjab, India

*Address all correspondence to: nandgharia@gmail.com

References

[1] Levitt J. Responses of Plants to Environmental Stress. In Water, Radiation, Salt, and other Stress. New York: New York: Academic Press, 1980.

[2] Sanderson, M. A., D. W. Stair, and M. A. Hussey, 1997: Physiological and morphological response of perennial forages to stress. *Advances in Agron.*, 59: 171-224.

[3] Breshears, D. D., N. S. Cobb, P. M. Rich, K. P. Price, C. D. Allen, R. G. Balice, W. H. Romme, J. H. Kastens,

[4] Eagles, C. F., H. Thomas, F. Volaire, and C. J. Howwarth, 1997: Stress physiology and crop improvement. In: Proc. XVIII Intl. Grassland Cong., B. R. Christie, (ed.). Canada

[5] Palsaniya DR, Singh R, Tewari R K, Yadav R S, and Dhyani S K 2012a. Integrated watershed management for natural resource conservation and livelihood security in Semi-Arid Tropics of India. *Indian J Agric Sci* 82(3): 241-47.

[6] IGFRI Vision 2050. Indian Grassland and Fodder Research Institute, Jhansi (UP).

[7] Palsaniya DR, Singh Ramesh, Tewari RK, Yadav RS, Dwivedi RP, Kumar RV, Venkatesh A, Kareemulla K, Bajpai CK, Singh Rajendra, Yadav SPS, Chaturvedi OP and Dhyani SK 2008. Socio-economic and livelihood analysis of people in Garhkundar-Dabar watershed of central India. *Indian J Agroforestry* 10(1): 65-72.

[8] Ghosh P K and Palsaniya D R 2014a. Crop diversification for sustainable intensification and carbon management. In *National Symposium on agricultural diversification for sustainable livelihood and environmental security*, PAU, Ludhiana, 18-20 November 2014, pp. 3-5.

[9] Vollenweider P, Günthardt-Goerg MS. Diagnosis of abiotic and biotic stress factors using the visible symptoms in foliage. Environ. Pollut. 2005; 137(3):455-465.

[10] Greaves J. The gender trap. Health Informatics J. 1996; 2(4):194-198

[11] Fahad S, Hussain S, Saud S, Khan F, Hassan S, Amanullah, *et al.* exogenously applied plant growth regulators affect heat-stressed rice pollens. J Agron. Crop Sci. 2016b; 202:139-150.

[12] Prasad PVV, Boote KJ, Allen LH. Adverse high temperature effects on pollen viability, seed-set, seed yield, and harvest index of grain-sorghum [*Sorghum bicolor* (L.) Moench] are more severe at elevated carbon dioxide due to higher tissue temperatures. Agric. For. Meteorol. 2006; 139(3-4):237-251.

[13] Maestri E, Klueva NK, Perrotta C, Gulli M, Nguyen H, Marmiroli N. Molecular genetics of heat tolerance and heat shock proteins in cereals.Plant Mol. Biol. 2002; 48:667-681

[14] Dudal R. Inventory of major soils of the world with special reference to mineral stress.-Plant Adaption to Mineral Stress in Problem Soils. Ed. M. J Wright. Cornell Univ. Agric. Exp. Stn. Ithaca, N.Y, 1976, 3-23.

[15] Kramer PJ, Boyer JS. Water relations of plants and soils. Academic Press, San Diego, 1995.

[16] Edmeades GOJ, Bolaoos, Lafitte HR. Progress in selecting for drought tolerance in maize. In D. Wilkinson (ed.), Proc. 47th Annual Corn and Sorghum Research Conference, Chicago, December 9ñ10. ASTA, Washington, 1992, 93n111.

[17] Barnabas B, Jager K, Feher A. The effect of drought and heat stress on reproductive processes in cereals. Plant, Cell & Environment. 2008; 31:11-38.

[18] Anjum SA, Xie XYU, Wang L, Chang Saleem MF, Man C, Lei W. Morphological, physiological and biochemical responses of plants to drought stress. African Journal of Agricultural Research. 2011; 6:2026-2032.

[19] Farooq M, Bramley H, Palta JA, Siddique KHM. Heat stress in wheat during reproductive and grain-filling phases. CRC. Crit. Rev. Plant Sci. 2011; 30(6):491-507.

[20] Demirevska K, Simova-Stoilova L, Fedina I, Georgieva K, Kunert K. Response of oryzacystatin I transformed tobacco plants to drought, heat, and light stress. Journal of Agronomy and Crop Science. 2010; 196:90-99.

[21] Smertenko A, Dráber P, Viklický V, Opatrný Z. Heat stress affects the organization of microtubules and cell division in *Nicotiana tabacum* cells. Plant, Cell Environ. 1997; 20(12):1534-1542.

[22] Porter JR. Rising temperatures are likely to reduce crop yields. Nature. 2005; 436:174.

[23] Rascio N, Navari-Izzo F. Heavy metal hyperaccumulating plants: how and why do they do it? And what makes them so interesting? Plant Sci. 2011; 180(2):169-181

[24] Ali H, Khan E, Sajad MA. Phytoremediation of heavy metals-concepts and applications. Chemosphere. 2013; 91(91):869-881.

[25] Mildvan AS. Metal in enzymes catalysis. Enzymes. 1970; 11(11):445-536.

[26] Yruela I. Copper in plants: acquisition, transport, and interactions. Funct. Plant Biol. 2009; 36(5):409-430.

[27] Grieve CM, Grattan SR, Maas EV. Plant Salt Tolerance. Agricultural Salinity. Assessment and Management (2nd Edition). 2012; 71:405-459.

[28] Tanji KK, Wallender WW. Nature and extent of agricultural salinity and sodicity. In: Wallender WW, Tanji KK (Eds.), Agricultural Salinity Assessment and Management. ASCE Manuals and Reports on Engineering Practice No. 71, second review New York, NY: American Society of Civil Engineers, 2012, 1-25

[29] Shainberg I, Singer MJ. Soil response to saline and sodic conditions. In: Wallender WW, Tanji KK (Eds.), Agricultural Salinity Assessment and Management. ASCE Manual and Reports on Engineering Practice No. 71, second Ed. New York, NY: American Society of Civil Engineers, 2012, 139-167.

[30] Munns R, Fisher DB, Tonnet ML. Na+ and Cl-transport in the phloem from leaves of NaCl-treated barley. Australian Journal of Plant Physiology. 1986; 13:757-766.

[31] Shabala S, Munns R. Salinity stress: Physiological constraints and adaptive mechanisms. In: Shabala S. (Ed.), Plant Stress Physiology. London, UK: CAB International, 2012.

[32] Hall, A. E., Singh, B. B., & Ehlers, J. D. (1997). Cowpea breeding. *Plant Breeding Reviews*, *15*, 215-274

[33] Mengel K, Kirkby EA. Principles of plant nutrition. Bern. Int. Potash Inst, 1978, 593.

[34] Zhang H, Jennings A, Barlow PW, Forde BG. Dual pathways for regulation of root branching by nitrate. Proc. Natl. Acad. Sci. The USA. 1999; 96:6529-6534.

[35] Fredeen AL, Rao IM, Terry N. Influence of phosphorus nutrition on growth and carbon partitioning in *Glycine max*. Plant Physiol. 1989; 89:225-230.

[36] Lynch J, Lauchli A, Epstein E. Vegetative growth of the common bean in response to phosphorus nutrition. Crop Sci. 1991; 31:380-387

[37] Barry DAJ, Miller MH. Phosphorus nutritional requirement of maize seedlings for maximum yield. Agron. J. 1989; 81:95-99.

[38] Lindhauer MG. Influence of K nutrition and drought on water relation and growth of sunflower (*Helianthus annuus* L.). Z. Pflanzenernahr. Bodenk. 1985; 148:654-669.

[39] Karmoker JL, Clarkson DL, Saker LR, Rooney JM, Purves JV. Sulphate deprivation depresses the transport of nitrogen to the xylem and the hydraulic conductivity of barley (*Hordeum vulgare* L.) roots. Planta.1991; 185:269-278

[40] Edelbauer A. Auswirkung von abgestuftem Schwefelmangel auf Wachstum, Substanzbildung und Mineralstoffgehalt von Tomate (*Lycopersicon esculentum* Mill.) In: Nahrlosungskultur. Die Bodenkultur. 1980; 31:229-241.

[41] Willenbrink J. Uber Beziehungen zwischen proteinumsatz und Schwefelver-sorgungder chloroplast. Z. Pflanzenphysiol. 1967; 56:427-438.

[42] Chaves, M. M. (1991). Effects of water deficits on carbon assimilation. J. *Exp. Bot.* 42, 1-16.

[43] Antolin, M. C., and Sinchez-Diaz, M. (1993). Effects of temporary droughts on photosynthesis of alfalfa plants. *J. Exp. Bot.* 44, 1341-1349.

[44] Petit, H. V., Pesant, A. R.. Barnett, G. M., Mason, W. N., and Dionne, J. L. (1992). Quality and morphological characteristics of alfalfa as affected by soil moisture, pH, and phosphorus fertilization. *Can.* J. *Plant Sci.* 72, 147-162.

[45] Sheaffer, C. C., Peterson, P. R., Hall, M. H., and Stordahl, J. B. (1992). Drought effects on yield and quality of perennial grasses in the north central United States. J. *Prod. Agric.* 5,556-561.

[46] Buxton, D. R., and Fales, S. L. (1994). Plant environment and quality. *In* "Forage Quality, Evaluation, and Utilization" (G. C. Fahey, Ed.), pp. 155-199. ASA-CSSA-SSSA, Madison, WI.

[47] Osmond, C. B., Austin. M. P., Berry, J. A., Billings, W. D., Boyer, J. S., Dacey, J. W. H., Nobel, P. S.. and Winner, W. E. (1987). Stress physiology and the distribution of plants. *Bioscience* 37,38-48.

[48] Agarawal, P.K. 2008. Global Climate Change and Indian Agriculture: Impacts, Adaptation, and Mitigation. *Indian J. Environ. Sci.*, 78(11): 911-919.

[49] Srinivasarao, Ch., and K.A. Gopinath. 2016. Resilient rainfed technologies for drought mitigation and sustainable food security. *Mausam* 67 (1), 169-182.

[50] Rosenow, D.T., Quisenberry, J.E., Wendt, C.W., and Clark, L.E. (1983) Drought tolerant sorghum and cotton germplasm. Agr. Water Manage., 7 (1-3), 207-222.

[51] Sanchez, A.C., Subudhi, P.K., Rosenow, D.T., and Nguyen, H.T. (2002) Mapping QTLs associated with drought resistance in sorghum (*Sorghum bicolor* L. Moench). Plant Mol. Biol., 48 (5-6), 713-726.

[52] Borrell, A.K., Hammer, G.L., and Henzell, R.G. (2000) Does maintaining green leaf area in sorghum improve yield under drought? II. Dry matter production and yield. Crop Sci., 40 (4), 1037-1048.

[53] Tuinstra, M.R., Grote, E.M., Goldsbough, P.B., and Ejeta, G. (1996) Identification of quantitative trait loci

associated with pre-flowering drought tolerance in sorghum. Crop Sci., 36 (5), 1337-1344.

[54] Blum, A. (2005) Drought resistance, water-use efficiency, and yield potential: are they compatible, dissonant, or mutually exclusive? Aust. J. Agric. Res., 56 (11), 1159-1168.

[55] Jenks, M.A., and Ashworth, E.N. (1999) Plant epicuticular waxes: function, production, and genetics, in Horticultural Reviews, vol. 23 (ed. J. Janick), John Wiley & Sons, Inc., New York, pp. 1-68.

[56] Jenks, M.A., Rich, P.J., Rhodes, D., Ashworth, E.N., Axtell, J.D., and Ding, C.K. (2000) Leaf-sheath cuticular waxes on bloomless and sparse bloom mutants of *Sorghum bicolor*. Phytochemistry, 54 (6), 577-584

[57] Doggett, H. (1988) Sorghum, 2nd edn, John Wiley & Sons, Inc., New York.

[58] Yu, J. and Tuinstra, M.R. (2001) Genetic analysis of seedling growth under cold temperature stress in grain sorghum. Crop Sci., 41 (5), 1438-1443.

[59] Knoll, J., Gunaratna, G., and Ejeta, G. (2008) QTL analysis of early-season cold tolerance in sorghum. Theor. Appl. Genet.,116 (4), 577-587.

[60] Yadav,O.P.(2004).CZP9802—a new drought tolerant cultivar of pearl millet. *IndianFarming* 54, 15-17.

[61] Yadav,R.S.,Hash,C.T., Bidinger,F.R.,Devos,K.M.,and Howarth,C.J.(2004). Genomic regions associated with grain yield and aspects of post-flowering drought tolerance in pearl millet across stress environments and testers backgrounds. *Euphytica* 136, 265-277. DOI: 10.1023/B: EUPH.0000032711. 34599.3a

[62] Ali, G.M. Khan, N.M., Hazara, R., and McNeilly,T.(2004). Variability in the response of pearl millet(*Pennisetumamericanum* (L.) Leeke)accessions to salinity. *ActaAgron. Hung.* 52, 277-286. DOI: 10.1556/ AAgr.52.2004.3.9

[63] Kumar P, Tarafder JC, Painuli DK, Raina P, Singh MP, Beniwal RK, Soni ML, Kumar M, Santra P, Shamsudin M (2009) Variability in arid soils characteristics. In: Kar A, Garg BK, Singh MP, Kathju S (eds) Trends in arid zone research in India. Central Arid Zone Research Institute, Jodhpur, pp 78-112.

[64] Hash CT, Schaffert RE, Peacock JM (2002) Prospects for using conventional techniques and molecular biological tools to enhance performance of 'orphan' crop plants on soils low in available phosphorus. Plant Soil 245:135-146.

[65] J.A. Cusicanqui, J.G. Lauer, Plant density, and hybrid influence on corn forage yield and quality, Agro. J. 91 (1999) 911-915.

[66] M. Koutsika- Sotiriou, Hybrid seed production in maize, in A.S. Basra (Ed.), Heterosis and hybrid seed production in agronomic crops, Food Products Press, NewYork, 1999, pp. 25-64.

[67] W. Schlenker, M.J. Roberts, Nonlinear temperature effects indicate severe damages to U.S. crop yields under climate change, Proc. Nat. Acad. Sci. USA. 106 (2009) 15594-15598.

[68] J.D. Oster, R. C. Reeve M. Fireman, Salt problems in relation to irrigation, In Irrigation of agricultural lands (ed. by R. M. Hagen), Agronomy J. 11 (1967) 988-1008.

[69] Langyintuo AS, Lowenberg-DeBoer J. Potential regional trade implications of adopting Bt cowpea in West and Central Africa. AgBioForum. 2006;9:111-20.

[70] Watanabe, I., Hakoyama, S., Terao, T., & Singh, B. B. (1997). Evaluation methods for drought tolerance of cowpea. In B. B. Singh, D. R. Mohan Raj, K. E. Dashiell, & L. E. N. Jackai (Eds.), *Advances in cowpea research* (pp. 141-146). Ibadan, Nigeria: IITA. Copublication of International Institute of Tropical Agriculture (IITA) and Japan International Research Center for Agricultural Sciences (JIRCAS).

[71] Singh, B. B., & Matsui, T. (2002). Cowpea varieties for drought tolerance. In C. A. Fatokun, S. A. Tarawali, B. B. Singh, P. M. Kormawa, & M. Tamo (Eds.), *Challenges and opportunities for enhancing sustainable cowpea production. Proceedings of the world cowpea conference III, 4-8 September 2000* (pp. 287-300). Ibadan, Nigeria: International Institute of Tropical Agriculture (IITA).

[72] Iuchi S, Yamaguchi-Shinozaki K, Urao T, Shinozaki K. Novel drought inducible genes in the highly drought-tolerant cowpea: cloning of cDNA and analysis of their gene expression. Plant Cell Physiology. 1996a;37(8):1073-1082. DOI: 10.1093/oxfordjournals.pcp.a029056

[73] D'arcy-Lameta A, Ferrari-Iliou R, Contour-Ansel D, Pham-Thi AT, Zuily-Fodil Y. Isolation and characterization are of four ascorbate peroxidase cDNA responsive to water deficit in cowpea leaves. Annals of Botany. 2006;97(1):133-140. DOI: 10.1093/aob/ mcj010

[74] Helgadottir, A., S. Dalmannsdottir, and R. P. Collins, 2001: Adaptational changes in populations of contrasting white clover cultivars selected under Icelandic conditions. *Annals of Bot.* 88: 771-780.

[75] Vogg, G., S. Fischer, J. Leide, E. Emmanuel, R. Jetter, A. A. Levy, and M. Riederer, 2004: Tomato fruit cuticular waxes and their effects on transpiration barrier properties: functional characterization of a mutant deficient in a very-long-chain fatty acid beta-ketoacyl-CoA synthase. *Journal of Experimental Botany*, 55: 1401-1410.

[76] Humphreys, M. W., P. J. Canter, and H. M. Thomas, 2003: Advances in introgression technologies for precision breeding within the *Lolium-festuca* complex. *Annals of Applied Biology*, 143: 1-10.

[77] Barnes, D. K., and C. C. Sheaffer, 1995: Alfalfa. In: Forages: An Introduction to Grassland Agriculture, R. F. Barnes, D. A. Miller, C. J. Nelson, (eds.). Iowa State University Press, Ames, Iowa, pp. 205-216.

[78] Hall, A. E., 2001: Crop responses to the environment. CRC Press, Boca Raton, FL, USA.

[79] Rahman, M.A., I, Alam, Y.G. Kim, N.Y.Ahn, S.H. Heo, P.G. Lee, G. Lu and B.H. Lee, 2015: Screening for salt responsive proteins in two contrasting alfalfa cultivars using a comparative proteome approach. *Plant Physiol. Biochem.*, 89: 112-122.

[80] Bernard, A., M, Bipfubusa, C, Dhont, F. P, Chalifour, P, Drouin and C. J. Beauchamp. 2016 : Rhizobial strains exert a major effect on the amino acid composition of alfalfa nodules under Nacl stress. *Plant Physiol. Biochem.*, 108 : 344-352.

[81] Pederson, G. A. 1995: White clover and other perennial clovers. In: Forages. I. An Introduction to Grassland Agriculture, R. F. Barnes, D. A. Miller, C. J. Nelson, (eds.). Iowa State University Press, Ames, Iowa, pp 205-216.

[82] Annicchiarico, P. and E. Piano, 2004: Indirect selection for root development of white clover and implications for drought tolerance. *Journal of Agronomy and Crop Science*, 190: 28-34.

[83] Lee, B. R., W. J. Jung, K. Y. Kim, J. C. Avice, A. Ourry, and T. H. Kim. 2005. Transient increase of *de novo* amino acid synthesis and its physiological significance in water-stressed white clover. *Functional Plant Biology*, 32: 831-838.

[84] Chankaew, S., T, Semura, K. Naito, O.E. Tanaka, N. Tomorka, P. Somta, A. Kaga, D.A. Vaughan, and Srinives, 2014: QTL mapping for salt tolerance and domestication related traits in *Vigna marina ssp. oblonga*. *Theor. Appl. Genet.*, 127(3); 691-702.

[85] Choudhary, S. K., M. K. Jat, and A. K. Mathur. 2017: Effect of micronutrient on yield and nutrient uptake in sorghum. *Journal of Pharmacognosy and Phytochemistry*, 6(2): 105-108.

[86] Ali, S., K.A. Riaz, G.M. Mairaj, M. Arif, S. Fida, and Bibi. 2008: Assessment of different crop nutrient management practices for yield improvement. *Australian Journal of Crop Science*. 2(3):150-157.

[87] Prasanna, B. M., S. K. Vasal, B. Kassahun and N. N. Singh. 2001 : Quality protein maize. *Curr Sci.,* 81: 48-53.

[88] Chen, X. P., Z. L. Cui, P. M. Vitousek, K. G. Cassman, and P. A. Matson. 2011: Integrated soil-crop system management for food security. *Proc. Natl. Acad. Sci.*, 108: 6399-6404.

[89] Rehim, A., M. Farooq, F. Ahmad, M. Hussain. 2012: Band placement of phosphorus improves the phosphorus use efficiency and wheat productivity under different irrigation regimes. *Int. J. Agri. Biol*. 14: 727-733.

Chapter 3

Responses of Neotropical Savannah Plant Species to Abiotic Stresses: A Structural and Functional Overview

Silvana Aparecida Barbosa de Castro
and Vinícius Coelho Kuster

Abstract

Plants under field conditions are subject to different types of abiotic stresses such as drought, salinity, and light excess that adversely affect their growth and survival. In addition, several studies have pointed out the effect of climate change such as an increase in the concentration of atmospheric CO_2, as well as an increase in global temperature on the distribution and wealth of plants. Adaptation to abiotic stress and survival occurs on different scales, at the cellular level for each individual, and requires a range of strategies, whether morphological, physiological, molecular or structural. Such strategies may be determinant in the distribution of plant species in natural habitats, depending on ecological adaptations shaped by the evolutionary history of species. In this chapter, we discuss recent information about mechanisms of plant adaptation to abiotic stress in the Neotropical savannah based on the cell and individual scales.

Keywords: ecophysiology, water stress, thermotolerance, high luminosity, soil nutrients, fire, Cerrado

1. Introduction

Plants growing under natural conditions are permanently subject to different types of environmental stresses such as drought, nutritional deficiency, salinity, heat stress and, more recently, the anthropogenic pollutants of ecosystems that negatively affect plant growth and survival [1–5]. Moreover, studies have pointed out the effect of climate changes such as the increase of atmospheric CO_2 concentration and global temperature (2 to 4°C) on the distribution and richness of plant species [6–8]. The survival of plant species under such conditions is constantly threatened, leading them to use different cellular and whole plant mechanisms in order to minimize damage and adjust growth to adverse environmental conditions [9, 10]. Plant responses to abiotic conditions have been widely investigated in different ecosystems worldwide, with this issue gaining great prominence for savannas due to the high complexity of abiotic factors that affect the native species of these systems. Among the world's savannas, the Brazilian one, also called Neotropical savannah or Cerrado *sensu lato*, is one of the most diverse in the world in terms of vegetation and plant species types, recognized as one of the biodiversity hotspots in the world [11, 12]. The Neotropical savannah flora comprises some 12,000 plant species, many of them endemic, two-thirds of which are herbaceous species or small shrubs [13]. The Neotropical savannah covers approximately 22% of the Brazilian territory (**Figure 1**) comprising several physiognomies from open grasslands to woodlands in a mosaic determined by abiotic factors such as edaphic and fire ones [14, 15]. The Campo limpo (grasslands) and Cerradão (tall woodland) are both extreme physiognomies, with great predominance of herbaceous and arboreous strata, respectively [14]. The Campo sujo (shrub savannah), Campo cerrado (wooded savannah), and Cerrado sensu stricto (short woodland) occur between them, with a gradual growth of biomass and larger plant species strata [14]. A mountain physiognomy, called rupestrian fields, also occurs and is restricted to areas of more than 900 m of altitude [16].

The diversity of life-form strategies of the Neotropical savannah plant species is associated with the response to many abiotic factors, especially the climatic seasonality with a well-defined dry and rainy seasons (**Figure 2**). The rainy season normally lasts from October to March, and the dry season from April to September, with an annual rainfall average of about 1,477 mm, a minimum temperature of 16°C and a maximum temperature of 26°C over the past thirty years (**Figure 2**) [17, 18]. High temperatures and irradiance, oligotrophic soils and high content of solubilized Al^{3+} are environmental features of the Neotropical savannah, which select local plant species and induce them to develop physiological and morphological adaptations to this harsh environment [19, 20]. Adaptation to and survival of abiotic stress occur on different scales from the cellular to the individual levels and require a range of strategies, whether morphological, physiological, molecular or structural,

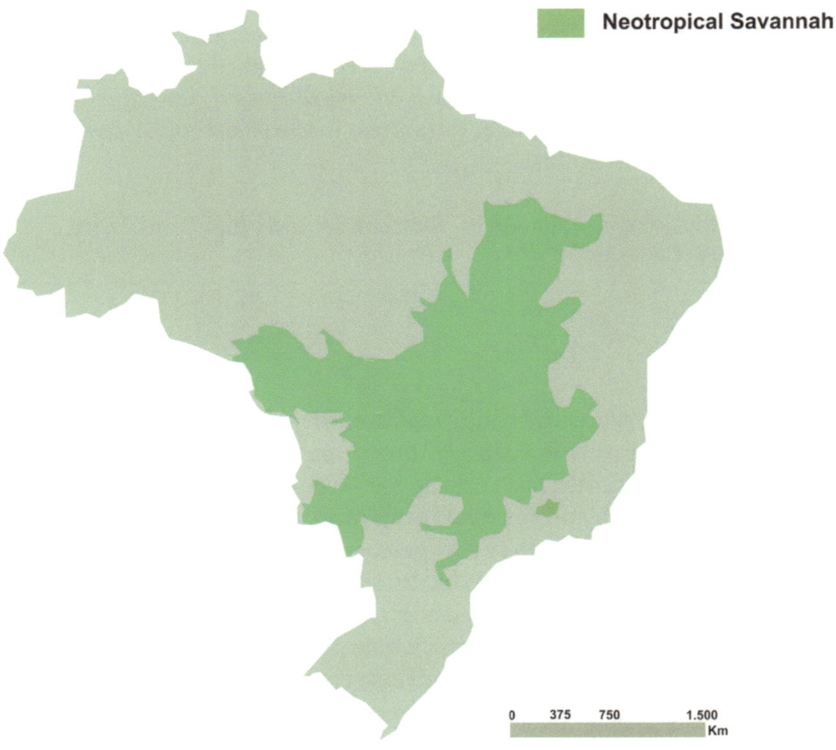

Figure 1.
Map of Brazil showing the occurrence of the Neotropical savannah, which covers approximately 22% of its surface area (Adapted from WWF Brazil, 2020).

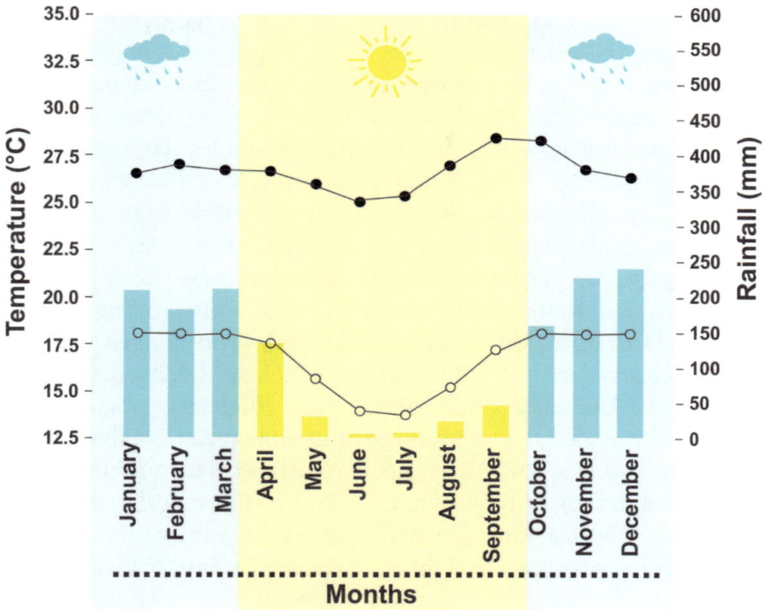

Figure 2.
Rainfall (mm) and temperature (°C) data of the Neotropical savannah during the rainy season (represented in blue area) and the dry season (represented in yellow area), over the past thirty years. Ps.: Black circles indicate the maximum temperature and white circles the minimum temperature (data obtained from Instituto Nacional de Meteorologia - INMET, 2020).

and can be determinant of species distribution in the natural habitats, depending on ecological adaptations shaped by the evolutionary history of the species.

Based on recent studies on Neotropical savannah flora, the focus of this review is on the dynamics and processes related to plant responses to the most prominent abiotic environmental conditions in this ecosystem, mainly highlighting water, temperature, light, nutrient deficiency and fire approaches. We report here information about the different adaptation mechanisms of plant species under the abiotic stress of the Neotropical savannah, from the cell to the individual scales.

2. The water stress seasonality changes the response of plant species during the seasons and along the day

Drought is considered to be one of the major abiotic stresses since it influences the distribution of plant species and limits the productivity of the ecosystems [4]. All aspects of plant development, i.e. germination, cell division, cell elongation, plant growth, and metabolic activities, are affected by a reduction in soil water content, which subsequently affects the physiological, biochemical, morphological and molecular processes of plants [21, 22]. Water flow from the soil to the roots depends of several factors, such as, potential water gradient between the soil and the plant, the hydraulic conductivity of the soil, the soil moisture content, and the atmospheric demand, that together with leaf transpiration create a tension in the xylem contributing to the creation of this water potential gradient [22, 23].

Regarding water stress, physiological and biochemical responses are more immediate in order to avoid cell dehydration, to maximize the efficiency of water use or to tolerate dehydration [1, 24]. The ability to maintain higher values of cellular water content is known as a strategy of plants sensitive to drought conditions. On the other hand, the ability to survive with lower values of cellular water content is called drought tolerance and requires the expression of molecular mechanisms since the initial growth of the seedling [25, 26]. Drought-tolerant plant species can be found in ecological niches with water seasonality, especially in the Neotropical savannah [27]. Drought-sensitive plant species also occur at the same sites as those of drought-tolerant ones; however, they use different molecular, biochemical, and physiological mechanisms to prevent desiccation, and thus deal with the lack of water in the soil. Drought-tolerant plants are able to limit damage to a repairable level in order to maintain physiological integrity due to the occurrence of repair mechanisms triggered by the dehydration/rehydration cycle [27, 28]. During long periods of drought, maintaining cell homeostasis and withstanding dehydration seems impractical for plants, directly impacting their growth rates [4, 27].

The Neotropical savannah abiotic conditions such as high irradiance, elevated air temperature, and low relative humidity impose a high evaporative demand on plants during the dry season while at the same time water is exhausted in the upper layers of the soil [29, 30]. Water use by Neotropical savannah woody plants is limited by a combination of physiological, plant architectural and tree density constraints evidencing that water use by the plant community seems to be more complex than in other savannas [30]. Even in the dry season, the water supply remains constant in the deeper soil layers, allowing access to water for tree species with long roots (in trees and shrubs) and differences between functional groups may be more related to the topography of an environmental continuum in which grass species are most subject to water restriction (**Figure 3**) [30, 31]. The hydrological niche segregation, defined as a fine scale partitioning of space for soil moisture gradients or different strategies of water acquisition, has been pointed out in different plant communities [32, 33] including the Neotropical savannah physiognomy [34, 35]. A marked decrease in water potential during the dry season has been reported for several species of the Neotropical savannah both in predawn and midday measurements [31, 35–37]. However, the water potential can vary according to the microhabitat, with plant species in mesic microhabitats showing smaller drops in the values of water potential even during the dry season (**Figure 3**). Costa et al. [38] also observed a decrease in water potential in *Hymenaea stigonocarpa* (Fabaceae) subjected to water stress, followed by recovery after rehydration. In an experimental study with *Laivosiera campos-portoana* (Melastomataceae), França et al. [39] reported that drought stress triggered a decrease of stomatal conductance and that osmotic adjustment played an important role in keeping high relative water content (RWC) values. These data suggest that the maintenance of high RWC values in plant species, despite a decrease in xylem water potential during the dry season, may be related to adjustment to stress, allowing survival in different microhabitats.

Plant responses to water deficit occur in different forms and scales, e.g. morphological, physiological, biochemical, and molecular ones, which may occur at the same time or at different times. For instance, the synthesis of cellulose microfibrils, hemicelluloses, and pectins in the cell wall may be changed under water stress, especially regarding the cell elongation pattern [40–42]. Strategies for adaptation and survival under water deficit conditions are also associated with the production and accumulation of organic solutes such as glycine-betaine, proline, and soluble sugars, which protect the integrity of the membrane and promote osmotic adjustment [43–47]. In addition, an increased content of several amino acids such as glutamic acid, glutamine, lysine, proline, serine, tryptophan, tyrosine and valine has been reported after water stress imposition in Neotropical savannah species [47, 48]. Proline is an important osmoregulator that protects plants submitted to abiotic stresses, favoring osmotic adjustment and an increase in the other osmolytes [20]. Souza et al. [48] detected increased carbohydrate content in *Hymenae courbaril* (Fabaceae) roots subjected to water deficit, which appeared to be directly linked to

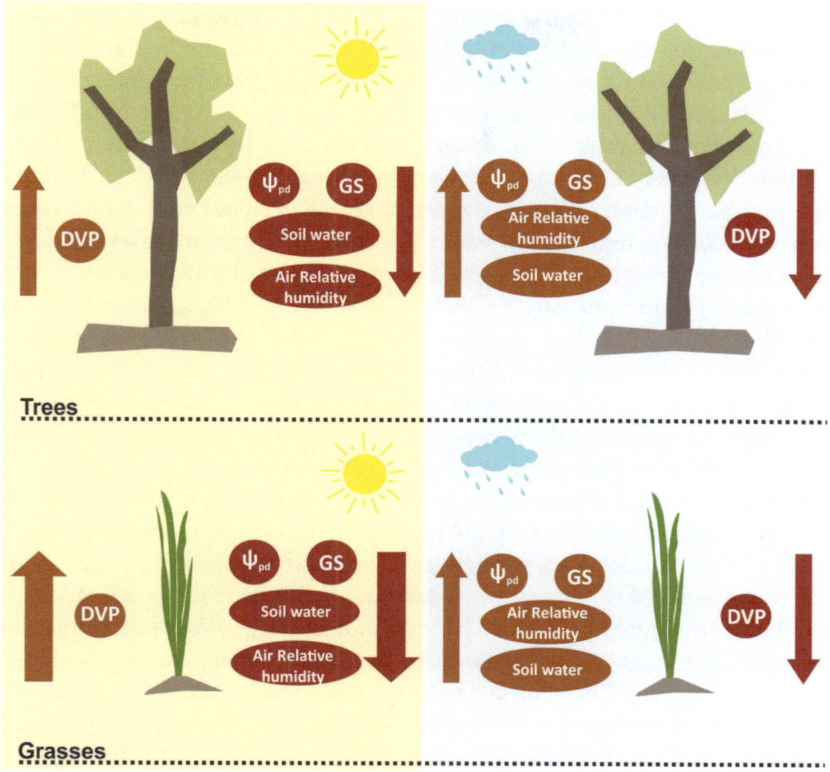

Figure 3.
Schematic presentation of the main water parameters of tree and grass species from the Neotropical savannah during the rainy (represented in blue area) and dry (represented in yellow area) seasons. Trees are able to capture water from deep layers of soil in the dry season, but there is still a certain water restriction that may affect water potential and stomatal conductance in some species. Grasses have lower water potential values and greater restriction of stomatal opening than tree species in the dry season. Brown arrows indicate an increase in rates while the red arrows show a decrease. Abbreviations: ψpd - Predawn leaf water potential; GS - Stomatal conductance; VPD - Relative vapor pressure deficit.

the osmotic adjustment of the cell. Some studies have indicated an accumulation of sucrose, raffinose and arabinose during the dry season, while glucose and fructose levels decreased with the reduction of water availability in the soil, suggesting that these are used for sucrose synthesis [49, 50].

The plant antioxidant system also plays a key role in water stress by responding with the production of enzymes such as superoxide dismutase (SOD), catalase (CAT), ascorbate peroxidase (APX), and glutathione peroxidase (GPX), as well as non-enzymatic components such as ascorbate, glutathione, tocopherols (vitamin E) and carotenoid pigments [51, 52]. Imbalance in the production of reactive oxygen species (ROS) has been demonstrated to affect negatively cell membrane stability, with oxidative damage to carbohydrates, lipids, amino acids, proteins, and nucleic acids [51]. ROS play a key role in the process of plant acclimation to abiotic stress such as signal transduction, which regulates different pathways during plant acclimation to stress [53]. Moreover, many plants may rapidly produce and scavenge different forms of ROS, inducing fast and dynamic changes in ROS level [51]. Under normal growth conditions, ROS are continuously produced and scavenged in organelles such as chloroplasts, mitochondria, peroxisomes and apoplasts [54]. However, the balance between ROS-producing pathways and ROS-scavenging mechanisms may be disrupted when plants experience environmental stress such as drought [52, 55, 56].

Regarding the photosynthetic apparatus, Assaha et al. [57] reported a super excitation of photosynthetic pigments after ROS accumulation in plants under severe water deficit. For instance, increase in oxidative stress (i.e. increase of H_2O_2) and in SOD, CAT, and APX activity in roots was reported in two Neotropical savannah tree species subjected to water stress imposed by competitive species [58]. As well, Vieira et al. [58] also showed an increase in the activity of enzymes of the antioxidant system in *Vatairea macrocarpa* (Fabaceae) subjected to water stress. Besides, ROS detoxification systems, other strategies are essential for plant tolerance to drought, such as a reduction in chlorophyll synthesis and in the assimilatory apparatus components, as well as a balanced sugar pentose-phosphate pathway [59–61].

The seasonality of water availability can also result in dynamic photoinhibition in woody plants [62], but not in herbaceous species [63–65]. A reduction of stomatal conductance and of net CO_2 assimilation has also been reported in *Barbacenia Purpurea* (Velloziaceae) submitted to water stress, but recovery was detected hours after rehydration [50]. Decreased stomatal conductance followed by a reduction in photosynthetic rates in plants under moderate water deficit has been reported [66]. Midday depression is related to excessive sunlight exposure and has been shown for a wide range of plant species of the Neotropical savannah, with impact on photosynthesis and on CO_2 assimilation [62, 67, 68]. The reduction of the Fv/Fm ratio, of maximum variable fluorescence (Fv) and of maximum fluorescence (Fm) during the dry season seems to be a consequence of a combination of strong stomatal control, especially in shrub species, high irradiance, and low soil moisture conditions [35, 67]. Stomatal control may prevent the dissipation of excessive energy and lead to chronic photoinhibition in some Neotropical savannah plant species [68]. Chronic seasonal photoinhibition has been shown for some plant species in the xeric microhabitat of the Neotropical savannah, as reported for Melastomataceae species, although it does not occur in mesic habitats [35, 63, 69]. These data reveal the importance of the specific characteristics of the microhabitat in the physiological studies of plants despite the seasonal variation of water availability. Several factors such as microhabitat, root system, functional group and topography directly influence water use strategies by the plant community, causing the Neotropical savannah to be more complex than other savannas.

3. Thermal stress triggers morphological and physiological damage and acts as selecting factor

Plants have different temperature domains that initially determine their geographic distribution [70] but with the progressive increase in air temperature, the temperature domains become displaced to colder regions, forcing plant species to move away from their current distribution [71, 72]. The increase in temperature may cause several types of damage to plants in different stages of life, from germination to adulthood [73, 74]. The temperature range considered optimal for photosynthesis in C3 plants is between 18°C and 30°C, while temperatures above 30°C are considered to be heating zones [70, 75]. Thermotolerance acquisition occurs when a plant is exposed to low or high temperatures for a short period of time, up to a limit where no fatal injuries occur [8] being thermotolerance fundamental for the plant to deal with greater variation in temperature under certain conditions.

Morphological damage has been reported for vascular plants as a response to heat stress, with emphasis on leaf and branch burn, foliar senescence and abscission, inhibition of shoot and root growth, discoloration, and fruit damage [76]. Associated with the damage triggered by heat stress, the response to rising temperatures varies widely among plant species, functional groups and life strategies [77, 78]. For instance, fast-growing plants from tropical forests in high light environments tend to be more tolerant of high temperatures than slow-growing plant species typical of shaded places in the understory [77]. According to Franco et al. [79], there is controversy over the limiting processes that control photosynthesis at high temperatures, with different responses in the metabolism of C3 and C4 plants. In particular, under high temperatures, the photosynthetic rate (A) can increase up to a point above which it starts to be inhibited and can reach zero [77]. At higher temperatures, plants show elevated stomatal conductance as a way to dissipate excess heat on leaf surfaces. However, in combination with water stress, greater control of stomatal opening can limit heat dissipation with an impact on photosynthesis [60, 80, 81]. The reduction of photosynthesis occurs due to damage to the electron transport chain associated with increased fluidity of the thylakoid membrane and damage to photosystem II (PSII) [82, 83]. PSII is highly sensitive to temperature and damage can be observed after a few minutes to a few hours of exposure to heat [60, 82, 84, 85]. A reduction in the membrane lipid content and an increase in lipid degradation during the late period of stress with formation of ROS exposure has been reported and the balance between ROS production and elimination is rapidly disturbed, leading to an increase in ROS content [86]. Changes in chlorophyll content in plants exposed to high temperatures have also been reported, possibly due to inhibition of chlorophyll biosynthesis for the destruction of numerous enzymes involved in their biosynthesis [87].

A highly consolidated mechanism of heat acclimation and rapid response to high temperature in plants is the production of heat shock proteins (HSPs), which are key for thermotolerance acquisition [88]. HSPs protect intracellular proteins from denaturation and preserve their stability, thus acting as chaperones [88–90]. According to Ohama et al. [90], HSPs include HSP100, HSP90, HSP70, HSP60 and small HSPs that play a key role in protein quality regulation by renaturing a variety of proteins denatured by heat stress. Together, HSPs and ROS scavenging enzymes are the major functional proteins induced by heat stress [90].

Duarte et al. [91] detected increased oxidative stress at high temperatures in seedlings of the Neotropical savannah plant species *Vriesea friburgensis* and *Alcantarea imperialis* (Bromeliaceae) due to the accumulation of H_2O_2, which in turn led to a decrease in the number of leaves and in biomass and to a high mortality rate in *Vriesea friburgensis* when submitted to treatment at 35°C. Under field conditions, Neotropical savannah plant species show photoinhibition during the hottest hours of the day, especially in combination with seasonal water stress [62, 92]. Changes in morphological traits have been observed for other Neotropical savannah plant species, such as germination in *Dipteryx alata* (Fabaceae) in an experiment involving increasing temperatures [93] and in the response to zinc absorption in *Pterogyne nitens* (Fabaceae) [94]. Changes more directly related to photosynthesis, water use efficiency, growth rate, factors that can alter the competitive capacity of species, have been observed [73, 95, 96].

Model-based studies predict changes in plant species distribution and loss in areas with greater biodiversity due to increased worldwide global temperature [97–100], also affecting the Neotropical savannah [100, 101]. Chaves et al. [102], in a study of a restricted plant species from the rupestrian field, showed that although some plants were no longer sensitive to the increase in temperature, they showed lower thermal tolerance and less plasticity compared to plant species of wide distribution, with a possible impact of global climate changes on this species. A recent modeling study evaluating the effect of habitat loss and climate change on 2,354 species of restricted distribution (including the Neotropical savannah) showed that 70 to 85% of them are at high risk of extinction [103]. With the increase in global temperature it is possible that plant species may be following three paths: adapting through phenotypic plasticity, migrating to higher altitude, or becoming extinct [7, 101]. Bueno et al. [104] showed that the greatest expansion of the Neotropical savannah occurred precisely during periods of higher temperatures (Last Interglacial). However, this result was certainly based on climate stability that allowed a large number of plant species to persist over time to the detriment of areas with major climatic changes where species might not have had enough time to adapt or to move to regions with more favorable conditions [105, 106].

Multiple environmental factors vary with increasing altitude, with temperature being perhaps the most important one because it directly influences biochemical processes related to photosynthesis and water status. Therefore, studies that evaluate the physiological responses of plant species at high temperatures are important because they provide data that can generate models of future plant species distribution considering different scenarios of temperature change, and because they serve as the basis for public policies aimed at protecting areas with plant species most vulnerable to global climate change.

4. Neotropical savannah plant species must deal with spatial and temporal light heterogeneity

Light is one of the abiotic factors that influence the survival, development and growth of plants, and several traits are highly plastic in the presence of substantial variations in the amount of light received, especially traits related to light capture and photosynthesis [107–110]. Light also shows spatial and temporal heterogeneity as an important factor underlying niche partitioning among plants [111, 112], which has created plants adapted to a variety of light intensity, such as sun-adapted and shade-tolerant plant species [113]. For photosynthesis, the flux of photons in the 400–700 nm range is the most relevant variable to be considered in studies that evaluate changes in light intensity on the development and growth of plants [110]. Under high-light conditions, plants possess a variety of mechanisms that maintain the operational photosynthetic system, such as carotenoids, violaxanthin, antheraxanthin and zeaxanthin, which play an important role in the dissipation of excitation energy at times when light levels exceed the plant capacity for photosynthetic electron transport and carbon fixation [114, 115]. However, plants exposed to high radiation may suffer increased leaf temperature, resulting in photoinhibition of Photosystem II (PSII) [116, 117].

In tropical regions, the irradiance levels normally exceed those that plants use in photosynthesis, so that many plants have efficient mechanisms preventing or minimizing the damage to the photosynthetic apparatus caused by high light availability [118]. Nevertheless, different light regimes in each physiognomy can promote the establishment of different plant species or functional traits [119]. Regarding the Neotropical savannah, most plant species are well adapted to high light intensity, and at the same time they may have low plasticity for adaptation to and growth in shaded environments [120, 121]. Under contrasting light conditions, some traits such as stomatal conductance, chlorophyll and carotenoid contents have been shown to have higher rates of plasticity in plant species growth and in different seasons (i.e. dry and rainy seasons) [120, 122]. According to Franco et al. [92], a balance between the use of light absorbed for photosynthesis and the dissipation of potentially harmful excess light is essential for Neotropical savannah plant species since they still have to deal with other abiotic factors (i.e., high air temperature, low relative humidity and seasonal drought). Regarding the photosynthesis apparatus, Franco et al. [92] also reported that woody plant species are capable of maintaining the functionality of PSII under conditions of high irradiance and high evaporative demand of the atmosphere during long sunny periods at the end of the dry season. In contrast, in a study of *Hymenaea stigonocarpa* (Fabaceae) under high light intensity, Fv/Fm values decreased to 0.64, indicating the occurrence of photoinhibition as well as a lower water potential value [123]. In a study of *Copaiba langsdorffii* (Fabaceae) growing under full sun and shade conditions, Ronquim et al. [119] observed that juvenile plants under full sun conditions showed higher maximum net photosynthetic rates, light compensation and saturation points than juvenile plants growing under shade conditions. Other leaf traits associated with light have been pointed out, such as low specific leaf area (SLA) values and high carotenoid and chlorophyll content in plant species from open areas, which may represent adaptive mechanisms to high luminosity environments [36, 79, 121].

In another approach, recent studies have considered the impact of plant invasions in open areas of the Neotropical savannah caused by the widespread expansion of forest in the absence of fire, which tend to change the availability of light for native species already adapted to high radiation [121]. In areas with invasion and expansion of forest, changes in leaf area index (LAI) have been reported, as well as changes in the photosynthetically active radiation available to Neotropical savannah plant species [124]. Moreover, Rossatto et al. [121] showed that invasion of originally open savannah areas by forest trees has an impact on the persistence of savannah plant species, which need greater availability of light to maintain a positive carbon balance. Under these conditions, it is expected that grasses and herbaceous species, already adapted to high light availability, will not be able to maintain the CO_2 assimilation necessary for the maintenance of root carbon stock, which is essential for adaptation to fire and seasonal water stress.

5. Neotropical savannah plants live in nutrient-poor soils with aluminum phytotoxicity

The Neotropical savannah soils are considered to be among the most chemically infertile soils in the world [124, 125], being generally dystrophic, acidic and poor in available nutrients [126, 127], as well as having high levels of exchangeable aluminum (Al) [127, 128]. Soils are mainly represented by well-drained oxisols [129], named "latossolos" in the Brazilian soil classification system [130], followed by sandy entisol, ultisol and cambisol soil types [129, 131]. Many relevant plant nutrients such as K, Ca, Mg, N, S, and P, are deficient in the soil of most Neotropical savannah physiognomies due to the nutritional deficiency of the initial matrix rocks and the intense weathering that leads to daily loss of nutrients [127]. Although Neotropical savannah soils are generally defined as dystrophic, mesotrophic physiognomies have also been reported [132, 133], influencing the variation of floristic structure, as reported by Neri et al. [133] when differentiating dystrophic and mesotrophic Cerradão physiognomies.

In the nutritional approach to studies of the Neotropical savannah, aluminum phytotoxicity and the responses of native plants to low nutrient availability, mainly nitrogen and phosphorus, are some of the most highlighted topics, which will be discussed here. Al is one of the main metals present in the crust of the Earth [134], being a limiting factor of plant development [135, 136]. Most Al is available in soils in the form of non-phytotoxic elements such as Al oxides and Al silicates [137]. However, under acid conditions (pH < 5), Al is solubilized to Al^{3+} [138–140], as well as to $Al(OH)^{2+}$, $Al(OH)^{2+}$ and $Al(OH)_6{}^{3+}$ [141], all known as phytotoxic forms. The relationship between Al phytotoxicity and acidic soils stands out mainly in the

Neotropical savannah, where some physiognomies such as Neotropical savannah *sensu stricto* and associated open physiognomies contain acid soils [126]. High Al concentrations, especially Al^{3+}, occur in dystrophic soils, while soils almost free of exchangeable Al occur in mesotrophic physiognomies with pH > 6.0 [127, 133, 142].

Al^{3+} is extremely toxic for the majority of cultivated plants, but, as expected, the native Neotropical savannah plants are highly tolerant to it [133] and may be divided into Al-accumulating and non-accumulating plants [143–145]. Plants that store 1000 mg kg^{-1} Al are defined as Al-accumulating species [127, 144, 146], which can accumulate Al in roots or aerial parts [142]. For instance, *Qualea grandiflora* (Vochysiaceae), *Qualea cordata* (Vochysiaceae) and *Miconia albicans* (Melastomataceae) have been reported to be Al-accumulating species [132, 147]. Moreover, concentrations of about 20,000 mg Al kg^{-1} dry leaves have been observed for some native Al-accumulating plant species of the Neotropical savannah [132, 143] such as noted for *Vochysia tucanorum* (Vochysiaceae) [132]. Many Al-accumulating plant species not only tolerate Al but also need it for their growth and development, as is the case for some plants of the Rubiaceae, Melastomataceae, and Vochysiaceae families [148]. Thus, Al-accumulating plant species, in many cases, reduce their growth rates when subjected to the absence or a low concentration of aluminum [127, 147, 149]. On the other hand, many plant species do not accumulate more than 1000 mg/kg^{-1} Al in their leaves, as reported for 24 out of 32 plants from the central part of the Neotropical savannah, such as *Caryocar brasiliense* (Caryocaraceae), *Vellozia squamata* (Velloziaceae) and *Roupala montana* (Proteaceae) [143].

Reduced developmental rates and physiological changes may be some symptoms related to high Al concentrations in non-Al accumulating Neotropical savannah plants. The inhibition of root growth [103], the lack of lateral roots [150], and a low biomass of stems and leaves [141, 148, 151] have been demonstrated in plants growing in the presence of high Al concentrations. According to Čiamporová [152], the inhibition of root growth may be due to Al impregnation of root cell walls that promotes unequal expansion and changes in tissue organization. This process occurs in many Al-sensitive plants, while in Al-tolerant species the high Al deposition on the cell wall normally does not inhibit root elongation, as reported for soybean seedlings [153]. Organic acid exudation from the roots has been cited as an important strategy for Al exclusion [150, 154], which avoids Al uptake and the associated structural and physiological symptoms. Bittencourt et al. [155] have shown that *Styrax camporum* (Styracaceae) releases organic acids in the roots as a defense mechanism against high Al accumulation. However, this process only works when the plant is exposed to low to moderate Al levels.

Regarding the physiological responses, previous studies have demonstrated that CO_2 assimilation rate and stomatal conductance were reduced in Al non-accumulating species submitted to high Al concentrations [150, 151, 156]. The leaf gas exchange performance of Al-sensitive species has been associated with the lack of lateral roots [150, 157, 158], which are the plant's water uptake site. The lack of a water supply from the roots reduces its amount in the mesophyll, leading to low gas exchange [159]. Associated with reduced gas exchange performance, the CO_2 assimilation rate was also reduced in some plant species submitted to high Al doses [160], as reported for *Citrus* [161, 162] and for two sorghum cultivars [163]. From a photochemical perspective, quantum yield, maximum variable fluorescence ratio (Fv), minimum fluorescence (F_o), and Fv/Fm have been reduced in the presence of high rates of Al [164], an event that seems to be related to the impairment of PSII photochemistry [163] to the reduction of PSII electron transport [165], and to other changes in photochemical processes. However, substantial changes in Fv/Fm values have not been observed in Al accumulating and Al non-accumulating species subjected to high Al concentrations [147, 150].

In contrast to Al, nitrogen (N) and phosphorus (P) are reported as the limiting factors in the Neotropical savannah [166] both due to the old age of the soil and to the recurring effect of fire, which contributes to the volatility of N parts [167, 168]. N is a prevalent element in living systems and is a constituent of phytohormones, amino acids, proteins, nucleic acids, and enzymes [169]. It acts on the initial growth processes, helping with the synthesis of deoxyribonucleic acids and the replication of chromosomes [169]. P is also required for plant development since it is involved in metabolic regulation, such as amino acid synthesis and protein activation [170], and is also a constituent of ATP/ADP and phospholipids [171]. Thus, N and P are nutrients required in large amounts for initial plant development [172]. According to Reis et al. [173], both nutrients are normally present in limited amounts in Neotropical savannah soils since low N availability may be related to the presence of minimal organic matter and to the fact that P is low in soils containing iron and aluminum oxides [174], both typical conditions of most Neotropical savannah soils. N and P are constituents of photosynthetic process-related proteins such as plant pigments [141], thus representing fundamental elements for the maintenance of photosynthetic quantum yield and biochemical steps. For instance, *Eugenia dysenterica* (Myrtaceae) seedlings showed increased photosynthetic activity and growth with N fertilization at rates between 50 and 200 mg dm^{-3}, while P fertilization at rates of 100 mg dm^{-3} and 200 mg dm^{-3} increased growth and photosynthetic activity, respectively [173]. In addition, some findings suggest that the growth of many plant species is reduced in Savannah soils, although the chemical attributes *per se* do not seem to exclude the occurrence of plants in Neotropical savannah physiognomies [175]. Despite the low availability of N and P, plant species growing in Neotropical savannah soils have a variety of mechanisms dealing with shortages of these nutrients. For instance, P is allocated to photosynthetically active mesophyll cells, showing higher photosynthetic P-use efficiency (PPUE), association with mycorrhizal species, different ways of nutrient uptake, and an efficient use of nitrogen [168, 176].

From another perspective, over the last decades, the Neotropical savannah has suffered enormous losses of its original cover (more than 50%) [11, 177]. Moreover, the change in land use, such as the introduction of extensive pecuary and immense extension of soybean crops, has contributed to the growing supply of these nutrients in the Neotropical savannah soils [178]. Studies have shown that the increase in N and P favors the increase in biomass, the decrease in species richness, the establishment of invasive species and changes in plant structure (root/shoot ratio, and root structure) and in competitive relationships [179–181]. It has also been shown that ancient ecosystems with less P availability in the soil are the richest in plant species [182, 183]. Thus, it is expected that an increase in P supply in the Neotropical savannah soils may have an impact on plant species richness in the medium and long-term.

6. Fire as a selective agent for plant communities

Fire has shaped different terrestrial environments since early geological ages and has been part of Earth's system since the origin of land plants in the Silurian period [184, 185]. In the Neotropical savannah, fire is one of the determining factors that select and structure plant communities and that promote the cycling of nutrients, representing a fundamental abiotic agent for the maintenance of open ecosystems [186, 187]. In the phytophysiognomies of the Neotropical savannah, fire normally occurs at the end of the dry season [188] as a result of the highly combustible material of dry herbaceous plant organs (**Figure 4**), while during the rainy season, natural fires can be caused by lightning [187, 189]. Although fire is a natural and periodic component of this environment, anthropic activities have intensified its action and made it more frequent [190], leading to cyclic burning of the herbaceous layer and a marked reduction of the maintenance and renewal of trees and shrubs, progressively decreasing their density [191]. Thus, the main problem with fire is that it is seen as a bad event when its regime (frequency, extension, and intensity) is somehow changed [187, 192].

Given the seasonality of fire in the Neotropical savannah, plants possess physiological, morphological, ecological, chemical, and phenological adaptations to deal with it. These adaptations, as well as adaptation to drought (*see* [193]), can enable plant species to escape from fire, as is the case for plants with a short life cycle, or to resist fire, as is the case for tree species with extra protective bark [194]. As an escape strategy, many herbaceous species synchronize seed production before annual burning periods and enter a type of dormancy that only ends after high-temperature exposure [192], as reported for different plant species. For instance, the dormancy of *Sporobolus cubensis* (Poaceae), a herbaceous species of the Neotropical savannah, was broken by exposure to 200°C, indicating the need of fire temperature for its germination [195]. However, many seeds and underground plant organs may be protected against scorching since the heat from fire events does not reach deeper layers of the soil [196]. In general, a fire event is followed by a high germination rate and seedling development since the supply of nutrients increases from the ashes and more light reaches the plant from the opening of canopies [192, 197].

In a phenological approach, fire can affect vegetative and reproductive phenophases [186], leading to leaf abscission by the heat created from the fire, as well as decreasing the numbers of flowers and fruits [191, 198]. However, Silvério et al. [199] reported that the negative effects of fire on the vegetative phenophases were normally followed by an increase in sprouting and the production of leaves, revealing an epigeous recovery mechanism from fire damage. In a reproductive approach, Pilon et al. [200] reported a similar pattern, with fire triggering the production of flowers and fruits in 79% of 47 plant species such as grasses, forbs, and subshrubs. Moreover, several studies have reported flowering in several plant species after the passage of fire [201, 202] and, more recently, Fidelis et al. [203] reported massive

Figure 4.
Overview of the Neotropical savannah plant species in the pre-fire (blue circle) and post-fire (pink circle) periods, which coincided with the rainy and dry seasons, respectively. At a pre-fire time during the rainy season, the herbaceous stratum exhibits dry and flammable aerial organs, while tree and shrub species have green and photosynthesizes leaves. At a post-fire time during the dry season, the leaves of the herbaceous stratum regrow, while many tree species are still leafless due to leaf loss caused by heat, and with a stem blackened by the action of fire.

flowering in *Bulbostylis paradoxa* (Cyperaceae) within just 24 h after the fire. Flowering after fire certainly gives and advantage, since it allows the plant to use available resources without germination barriers and in new and fresh open areas for its establishment [204].

Many structural plant features have been proposed as important tools for fire event survival. Underground roots and stems with carbohydrate, nutrient and water storage have been reported for herbaceous and sub-shrub Neotropical savannah species, allowing them to regrow after the fire has passed [205]. For instance, diffuse underground systems, rhizophores, and xylopodium have been reported for 7 Asteraceae species from the Neotropical savannah [206], which can support the individual life during the dry season and after fire events. Xilopodium, as well as other underground systems, have been found in many species as reported for the *Brosimum gaudichaudii* (Moraceae) shrub [207], the *Ocimum nudicaule* herb (Lamiaceae) [208] and three *Erythroxylum* species (Erythroxylaceae) [209], and may indicate be a convergent trait of many plant species in this environment, often related to species dominance in physiognomies.

Regrowth after fire events requires carbon [79] for the creation of new tissues and organs; however, burning that reaches the leaves reduces the photosynthetic area of the plants [210], consequently decreasing the capture of atmospheric CO_2. The loss of leaf area can be total in herbaceous and shrub plants and partial in some sub-arboreal plants during fire events, whereas tree branches normally are not burnt but are damaged by the hot airflow [79]. In this respect, underground reserve organs may be a fundamental carbon source, mostly consisting of carbohydrates such as starch, fructan, glucose and sucrose [189, 206]. Complex carbohydrates such as starch are hydrolyzed to soluble sugars by hydrolytic enzymes such as the exoamylase β-amylase, which cleaves α-1,4 linkages, and the debranching enzyme isoamylase, which cleaves α-1,6 linkages [211] and the product used for plant development. After leaf regrowth, the new leaves of some plant species may return with higher photosynthetic rates, stomatal conductance and nitrogen concentration than during the pre-burn period [212], with high vigor and potential for growth [213]. However, these responses did not occur in *Vochysia cinnamomea* (Vochysiaceae), a native species of Neotropical savannah, since fire did not exert a direct influence on its physiological processes [214].

The bark structure and thickness of tree plants has been cited for years as one of the greatest protective traits against fire in the Neotropical savannah, preventing damage to the vascular cambium [215] and deformations of the xylem [216] and phloem. The thick corky bark may protect internal stem tissues [217], reducing mortality and maintaining water and solute transport during and after the fire episodes [218]. On this basis, the bark properties have been studied across 31 species from the Neotropical savannah, providing shreds of evidence that the outer bark acts as mechanical support and defense against pathogen attack and the inner bark as a place related to storage and transport of water [218]. Other less frequent and more specific characteristics have also been discussed regarding fire survival. For instance, *Vellozia variabilis* (Velloziaceae) has resin glands on its leaves that produce flammable substances capable of starting quick-burning on the more protected plant base, preventing damage to more sensitive upper plant portions [219]. Due to these plant specificities, as well as abiotic and biotic conditions, the Neotropical savannah is a potential environment for new structural and ecological discoveries.

Finally, the occurrence of invasive tree species in open areas of the Neotropical savannah has altered the microclimate conditions, affecting grasses and shrubs and consequently the dynamics of fire occurrence in the Neotropical savannah [124, 220]. This cascade of events will certainly have a negative impact on the plant and animal communities of the Neotropical savannah, changing the local landscape.

7. Conclusions

In the current chapter, we report the most common types of abiotic stress in Neotropical savannah plant species, which can simultaneously influence and determine the different functional types of plants. Plant species are affected and respond differently to abiotic stress in the Neotropical savannah, mostly in terms of more specific characteristics such as microhabitat, seasonality, topography and anthropic interference. Moreover, Neotropical savannah plant species have evolved over thousands of years subject to various types of abiotic stress (i.e. water seasonality, high temperature and light availability, soil with a deficiency of essential nutrients, and fire). All of them have been shaping differently the Neotropical savannah physiognomies, and recent changes in abiotic conditions are occurring in a very rapid manner, mainly due to anthropogenic actions over the last decades. Reduction of native vegetation areas, expansion of agriculture and extensive pecuary, changes in the fire regime and the introduction of nutrients, such as nitrogen and phosphorus, are examples that have been reported for Neotropical savannah in recent times and that may potentially have negative effects on biodiversity. In view of this reality, conservation strategies for the remaining area should be discussed with society and prioritized in government policies in order to conserve such areas and to encourage the sustainable use of the Savannah wealth.

Author details

Silvana Aparecida Barbosa de Castro[1]* and Vinícius Coelho Kuster[2]

1 Instituto de Ciências Biológicas, Universidade Federal de Minas Gerais, Belo Horizonte, Minas Gerais, Brazil

2 Unidade Acadêmica de Ciências Biológicas, Universidade Federal de Jataí, Jataí, Goiás, Brazil

*Address all correspondence to: silvana.abcastro@gmail.com

References

[1] Reddy, AR, Chaitanya, KV, Vivekanandan, M. Drought-induced responses of photosynthesis and antioxidant metabolism in higher plants. Journal of Plant Physiology. 2004; 161: 189-1202.

[2] Mittler, R. Abiotic stress, the field environment and stress combination. Trends in Plant Science. 2006; 11: 15-19.

[3] Zavalloni, C, Gielen, B, De Boeck, HJ, Lemmens, CMHM, Ceulemans, R, Nijs, I. Greater impact of extreme drought on photosynthesis of grasslands exposed to a warmer climate in spite of acclimation. Physiologia Plantarum. 2009; 136: 57-72.

[4] Zlatev, Z, Cebola, LF. An overview on drought induced changes in plant growth, water relations and photosynthesis. Emirates Journal of Food and Agriculture. 2012; 24: 57-72.

[5] Nishiyama, Y, Murata, N. Revised scheme for the mechanism of photoinhibition and its application to enhance the abiotic stress tolerance of the photosynthetic machinery. Applied Microbiology Biotechnology. 2014; 98: 8777-8796.

[6] IPCC Climate change 2014. Impacts adaptation and vulnerability. Pat A: Global and sectorial aspects. Contribution of working group II to the fifth assessement report of the intergovernmental panel on climate change. Cambridge: UK. Cambridge University Press. [2014].

[7] Pecl, GT, Araújo, MB, Bell, JD, Blanchard, J, Bonebrake, TC. et al. Biodiversity redistribution under climate change: Impacts on ecosystems and human well-being. Science. 2017; 355: 1-9.

[8] Scarano, FR. Biodiversity Sector: Risks of Temperature Increase to Biodiversity and Ecosystems. In: Nobre C.A.; Marengo J.A.; Soares W.R. [Org.]. Climate Change Risks in Brazil. 1ed.Cham, Switzerland: Springer International Publishing. 2019; pp. 131-141.

[9] Nishiyama R, Watanabe, Y, Leyva-Gonzalez, MA, Van, HC, Fujita, Y, Tanaka, M. et al. *Arabidopsis* AHP2, AHP3, and AHP5 histidine phosphotransfer proteins function as redundant negative regulators of drought stress response. Proc. Natl. Acad. Sci. U.S.A. 2013; 110: 4840-4845.

[10] Ha, CV, Leyva-Gonzalez, MA, Osakabe, Y, Tran, UT, Nishiyama, R, Watanabe, Y. et al. Positive regulatory role of strigolactone in plant responses to drought and salt stress. Proc. Natl. Acad. Sci. U.S.A. 2014; 111: 581-856.

[11] Myers, N, Mittermeier, RA, Mittermeier, CG, Fonseca, GAB, Ken, J. Biodiversity hotspots for conservation priorities. Nature. 2000; 403: 853-858.

[12] Gottsberger, G, Silberbauer-Gottsberger, I. Life in the cerrado: a South American tropical seasonal ecosystem. Vol. I Origin, Structure, Dynamics and Plant use. Ulm: Reta Verlag. 2006. https://doi.org/10.1086/665973

[13] Mendonça, RC, Felfili, JM, Walter, BMT, Silva-Júnior, MC, Rezende, AB. et al. Flora Vascular do Bioma Cerrado: checklist com 12.356 espécies. *In* Cerrado: Ecologia e Flora. Vol. 2. *Edited by* S.M. Sano, S.P. Almeida, and J.F. Ribeiro. Brasília: Embrapa Cerrados. pp. 213-228.

[14] Coutinho, LM. O conceito de Cerrado. Revista Brasileira de Boânica. 1978; 1: 17-2.

[15] Capuzzo JP, Rossatto, DR, Franco, AC. Differences in morphological

and physiological leaf characteristics between *Tabebuia aurea* and *T. impetiginosa* is related to their typical habitats of occurrence. Acta Botanica Brailica. 2012; 26: 519-526.

[16] Silveira, FAO, Negreiros, D, Barbosa, NPU, Buisson, E, Carmo, FF, Carstensen, DW. et al. Ecology and evolution of plant diversity in the endangered campo rupestre: a neglected conservation priority. Plant Soil. 2016; 403: 29-152.

[17] Franco, AC. Seasonal Patterns of Gas Exchange, Water Relations and Growth of Roupala Montana, an Evergreen Savanna Species. Plant Ecology. 1998; 136: 69-76.

[18] Instituto Nacional de Meteorologia. Ministério da Agricultura, Pecuária e Abastecimento. [Internet]. 2020. Available from: https://portal.inmet.gov.br/

[19] Furley, PA, Ratter, JA. Soil resources and plant communities of the central Brazilian Neotropical savannah and their development. Journal of Biogeography. 1988; 15: 97-108.

[20] Vieira, EA, Silva, MG, Moro, CF, Laura, VA. Physiological and biochemical changes attenuate the effects of drought on the Neotropical savannah species *Vatairea macrocarpa* (Benth.) Ducke. Plant Physiology and Biochemistry. 2017; 115: 472-483.

[21] Yordanov, I, Velikova, V, Tsonev, T. Plant responses to drought and stress tolerance. Bulg. Journal of Plant Physiology. 2003; Special issue: 187-206.

[22] Chavarria G, Santos HP. Plant Water Relations: Absorption, Transport and Control Mechanisms. In: Montanaro G, Dichio B (eds.) Advances in Selected Plant Physiology Aspects. Rijeka: InTech; 2012. pp. 105-132.

[23] Silva, EC, Albuquerque, MB, Neto, ADA, Júnior, CDS. Drought and Its Consequences to Plants - From Individual to Ecosystem. 2013; Intechopen. http://dx.doi.org/10.5772/53833

[24] Mullet, JE, Whitsitt, MS. Plant cellular responses to water deficit. Plant Growth Regulation. 1999; 20: 119-124.

[25] Blum, A. Drought resistance, water-use efficiency, and yield potential – are they compatible, dissonant, or mutually exclusive? Australian Journal of Agricultural Research. 2005; 56: 1159-1168.

[26] Moore, JP, Vicré-Gibouin, M, Farrant, JM, Driouich, A. Adaptations of higher plant cell walls to water loss: Drought *vs.* desiccation. Physiologia Plantarum. 2008; 134: 237-245.

[27] Dinakar, C, Djilianov, D, Bartels, D. Photosynthesis in desiccation tolerant plants: Energy metabolism and antioxidative stress defense. Plant Science. 2012; 182: 29-41.

[28] Farrant, JM, Lehner, A, Cooper, K, Wiswedel, S. Desiccation tolerance in the vegetative tissues of the fern *Mohria caffrorum* is seasonally regulated. Plant Journal. 2009; 57: 65-79.

[29] Franco, AC. Ecophysiology of woody plants. In: Oliveira PS, Marquis RJ [eds] The cerrados of Brazil. Columbia University Press, New York. 2002; pp. 178-197.

[30] Goldstein, G, Meinzer, F, Bucci, SJ, Scholz, G, Franco, AC, Hoffmann, WA. Water economy of Neotropical savanna trees: six paradigms revisited. Tree Physiology. 2008; 28: 395-404.

[31] Rossatto, DR, da Silveira Lobo, SL, Franco, AC. The partitioning of water uptake between growth forms in a Neotropical savanna: do herbs exploit a third water source niche? Plant Biology. 2013; 15: 84-92.

[32] Araya, Y, Silvertown, J, Gowing, DJ, McConway, KJ, Linder, HP, Midgley, G. A fundamental, eco-hydrological basis for niche segregation in plant communities. New Phytologist. 2011; 189: 253-258.

[33] Silvertown, J, Araya, Y, Gowing, D. Hydrological niches in terrestrial plant communities: a review. Journal of Ecology. 2015; 103: 93-108.

[34] Aparecido, LMT, Teodoro, GS, Mosquera, G, Brum, M, Barros, FV, Pompeu, PV. et al. 2017. Ecohydrological drivers of Neotropical vegetation in montane ecosystems. Ecohydrology. 2018;e1932. https://doi.org/10.1002/eco.1932

[35] Castro, SAB, Silveira, FAO, Marcato, MS, Lemos-Filho, JP. So close, yet so different: Divergences in resource use may help stabilize coexistence of phylogenetically-related species in a megadiverse grassland. Flora. 2018; 238: 72-78.

[36] Franco AC, Bustamante M, Caldas LS. et al Leaf functional traits of Neotropical savannah trees in relation to seasonal water deficit. Trees. 2005; 19:326-335.

[37] Bucci, SJ, Goldstei, G, Meinzer, FC, Franco, AC, Campanello, P, Schotz, FG. Mechanisms contributing to seasonal homeostasis of minimum leaf water potential and predawn disequilibrium between soil and plant water potential in Neotropica savanna trees. Trees. 2005; 19: 296-304.

[38] Costa, AC, Rezende-Silva, SL, Megguer, CA, Moura, LMF, Rosa, M, Silva, AA. The effect of irradiance and water restriction on photosynthesis in young jatobá-do-Neotropical savannah (*Hymenaea stigonocarpa*) plants. Photosynthetica. 2015; 53: 118-127.

[39] França, MGC, Prados, LMZ, Lemos-Filho, JP, Ranieri, BD, Vale, FHA. Morphophysiological differences in leaves of *Lavoisiera campos-portoana* (Melastomataceae) enhance higher drought tolerance in water shortage events. Journal of Plant Research. 2012; 125: 85-92.

[40] Wolf, S, Greiner, S. Growth control by cell wall pectins. Protoplasma. 2012; 249: 169-175.

[41] Peaucelle, A, Braybrook, S, Höfte, H. Cell wall mechanics and growth control in plants: The role of pectins revisited. Frontier in Plant Science. 2012; 3: 1-6.

[42] Le Gall, H, Phillipe, F, Domon, JM, Gillet, F, Pelloux, J, Rayon, C. Cell Wall Metabolism in Response to Abiotic Stress. Plants. 2015; 4: 112-166.

[43] Filho, JTC, Stacciarini-Seraphin, E. Alteração no potencial osmótico e teor de carboidratos solúveis em plantas jovens de lobeira (*Solanum lycocarpum* St.-Hil.) em resposta ao estresse hídrico. Revista Brasileira de Botânica. 2001; 24: 199-204.

[44] Tamura, T, Hara, K, Yamaguchi, Y, Koizumi, N, Sano, H. Osmotic stress tolerance of transgenic tobacco expressing a gene encoding a membrane-located receptor-like protein from tobacco plants. Plant Physiology. 2003; 131: 454-62.

[45] Chen, TH, Murata, N. Glycinebetaine: an effective protectant against abiotic stress in plants. Trends in Plant Science. 2008; 13: 499-505.

[46] Silva EC, Nogueira RJMC, Vale FHA, Melo NF, Araujo FP. Water relations and organic solutes production in four umbu tree (*Spondias tuberosa*) genotypes under intermittent drought. Brazilian Journal of Plant Physiology. 2009; 21(1): 43-53.

[47] Suguiyama, VF, Silva, EA, Meireles, ST, Centeno, DC, Braga, MR. Leaf

metabolite profile of the Brazilian resurrection plant *Barbacenia purpurea* Hook. (Velloziaceae) shows two time dependent response during desiccation and recovering. Frontiers in Plant Science. 2014; 5:1-13.

[48] Souza, LC, Luz, LM, Martins, JTS, Neto, CFO, Palheta, JG, et al. Osmoregulators in *Hymenaea courbaril* and *Hymenaea stigonocarpa* under water stress and rehydration. Journal of Forestry Research. 2017; https://doi.org/10.1007/s11676-017-0456-x

[49] Farrant, J, Brandt, WF, Lidsey, GG. An overview of the mechanisms of dessication tolerance in selected angiosperm resurrection plants. Plant Stress. 2007; 1:72-84.

[50] Suguiyama, VF, Sanches, RFE, Meirelles, ST, Centeno, DC, Silva, EA, Braga, MR. Physiological responses to water deficit and changes in leaf cell wall composition as modulated by seasonality in the Brazilian resurrection plant *Barbacenia purpurea*. South African Journal of Botany. 2016; 105: 270-278.

[51] Mittler, R, Vanderauwera, S, Suzuki, N, Miller, G, Tognetti, VB. et al. ROS signaling: the new wave? Trends in Plant Science. 2011; 16: 300-309.

[52] Del Río, LA. ROS and RNS in plant physiology: an overview. Journal of Experimental Botany. 2015; 66: 2827-2837.

[53] Turkey, I. ROS and RNS: key signalling molecules in plants. Journal of Experimental Botany. 201; 69:3313-3315.

[54] Dietz, KJ, Turkan, I, Krieger-Liszkay, A. Redox- and reactive oxygen species-dependent signaling in and from the photosynthesizing chloroplast. Plant Physiology. 2016; 171:1541-1550.

[55] Silva, EC, Albuquerque, MB, Neto, ADA, Junior, CDS. Drought and Its Consequences to Plants – From Individual to Ecosystem. In: Responses of Organisms to Water Stress. Intech. pp. 17-47. 2013.

[56] Turkan I. Emerging roles for ROS and RNS—versatile molecules in plants. Journal of Experimental Botany. 2017; 68: 4413-4416.

[57] Assaha, DV, Liu, L, Ueda, A, Nagaoka, T, Saneoka, HEffects of drought stress on growth, solute accumulation and membrane stability of leafy vegetable, huckleberry (Solanum scabrum Mill.). Journal of Environmental Biology. 2016; 37: 107e114.

[58] Vieira, EA, Galvão, FCA, Barros, AL. Influence of water limitation on the competitive interaction between two Neotropical savannah species and the invasive grass *Brachiaria brizantha* cv. Piata. Plant Physiology and Biochemistry. 2019; 135: 206-214.

[59] Mollo, L, Martins, MCM, Oliveira, VF, Nievola, CC, Figueiredo-Ribeiro, RCL. Effects of low temperature on growth and non-structural carbohydrates of the imperial bromeliad *Alcantarea imperialis* cultured *in vitro*. Plant Cell, Tissue Organ Culture. 2011; 107: 141-149.

[60] Mathur S, Agrawal D, Jajoo A. Photosynthesis: Response to high temperature stress. Journal Photochemistry Photobiology B Biol. 2014; 137: 116-126.

[61] Petrov, V, Hille, J, Mueller-Roeber, B, Gechev, TS. ROS-mediated abiotic stress-induced programmed cell death in plants. Frontier Plant Science. 2015; 69: 1-16.

[62] Lemos-Filho, JP. Photoinhibition in three species of the savanna (*Annona crassifolia*, *Eugenia dysenterica* and *Campomanesia adamantium*) in the dry and rainy season. Revista Brasileira de Botânica. 2000; 23: 45-50.

[63] Lüttge, U. Duarte, HM, Scarano, FR, Mattos, EA, Cavalin, PO. et al. Physiological ecology of photosynthesis of five sympatric species of Velloziaceae in the rupestrian fields of Serra do Cipó, Minas Gerais, Brazil. Flora. 2007; 202: 637-646.

[64] Morales, M, Garcia, QS, Silva, AIS, Silva, MC, Bosch, SM. Tocotrienols in *Vellozia gigantea* leaves: occurrence and modulation by seasonal and plant size effects. Planta. 2014; 240: 437-446.

[65] Castro, SAB, Sá, CEM, Mourão, FA, Duarte, HM, Fernandes, GW, Lemos-Filho, JP. Ecophysiological performance of a threatened shrub under restored and natural conditions in harsh tropical mountain environments. Acta Botânica Brasílica. 2016; 30: 17-26.

[66] Grassi, G, Magnani, F. Stomatal, mesophyll conductance and biochemical limitations to photosynthesis as affected by drought and leaf ontogeny in ash and oak trees. Plant, Cell & Environment. 2005; 28: 834-849.

[67] Kuster, VC, Castro, SAB, Vale, FHA. Environmental conditions modulate plasticity in the physiological responses of three plant species of the Neotropical savannah. Acta Physiology Plantarum. 2017; 39: 103.

[68] Franco, AC, Lüttge, U. Midday depression in savanna trees: coordinated adjustments in photochemical efficiency, photorespiration, CO2 assimilation and water use efficiency. Oecologia. 2002; 131:3 56-365.

[69] Morales, M, Garcia, QS, Bosch, SM. Ecophysiological response to season variations in water availability in the arborescent, endemic plant *Vellozia gigantea*. Tree Physiology. 2015; 35: 253-265.

[70] Ducruet, JM, Peeva, V, Havaux, M. Chlorophyll thermofluorescence and thermoluminescence as complementary tools for the study of temperature stress in plants. Photosynthesys Research. 2007; 93: 159-171.

[71] Gian-Reto, W, Post, E, Convey, P, Menzel, A, Parmesan, C, BeeBee, TJC. et al. Ecological responses to recent climate change. Nature. 2002; 416: 389-395.

[72] Corlett RT, Westcott DA. Will plant movements keep up with climate change? Trends in Ecology & Evolution. 2013; 28(8): 482-488.

[73] Melo, NMJ, Rosa, RSEG, Pereira, EG, Souza, JP. Rising (CO2) changes competition relationships between native woody and alien herbaceous Cerrado species. Functional Plant Biology. 2018; 45(8): 854-864.

[74] Souza, JP, Melo, NMJ, Halfeld, AD, Vieira, KIC, Rosa, BL. Elevated atmospheric CO2 concentration improves water use efficiency and growth of a widespread Cerrado tree species even under soil water déficit. Acta Botanica Brasilica. 2019; 33(3): 425-436.

[75] Yamori, W, Hikosaka K, Way, DA. Temperature response of photosynthesis in C3, C4, and CAM plants: temperature acclimation and temperature adaptation. Photosynthesis Research. 2014; 119:101-117.

[76] Wahid A, Gelani S, Ashraf M, Foolad MR. Heat tolerance in plants: An overview. Environmental Experimental Botany. 2007; 61 (3): 199-223.

[77] Slot, M, Winter, K. In situ temperature response of photosynthesis of 42 tree and liana species in the canopy of two Panamanian lowland tropical forests with contrasting rainfall regimes. New Phytologist. 2017; 214: 1103-1117.

[78] Menezes-Silva, PE, Lourenço, LL, Alves, RDFB, Souza, LF, Almeida,

SES, Farnese, FS. Different ways to die in a changing world: Consequences of climate change for tree species performance and survival through an ecophysiological perspective. Ecology and Evolution. 2019; https://doi.org/10.1002/ece3.5663

[79] Franco, AC, Rossatto, DR, Silva, LCR, Ferreira, CS. Cerrado vegetation and global change: the role of functional types, resource availability and disturbance in regulating plant community responses to rising CO2 levels and climate warming. Theoretical Experimental Plant Physiology. 2014; 26: 19-38.

[80] Salvucci, ME, Crafts-Brandner, SJ. Relationship between the heat tolerance of photosynthesis and the thermal stability of rubisco activase in plants from contrasting thermal environments. Plant Physiology. 2004; 134: 1460-1470.

[81] Scafaro, A.P, Gallé, A, Rie AV, Carmo-Silva, E, Salvucci, ME, Atwell, BJ. Heat tolerance in a wild Oryza species is at- tributed to maintenance of Rubisco activation by a thermally stable Rubisco activase ortholog. New Phytologist. 2016; 211: 899-911.

[82] Zhang, M, Chen, Q, Shen, S. Physiological responses of two Jerusalem artichoke cultivars to drought stress induced by polyethylene glycol. Acta Physiologia Plantarum. 2011; 33: 313-318.

[83] Cui, M, Lin, Y, Zu, Y, Efferth, T, Li, D, Tang, Z. Ethylene Increases Accumulation of Compatible Solutes and Decreases Oxidative Stress to Improve Plant Tolerance to Water Stress in Arabidopsis. Journal Plant Biology. 2015; 58: 193-201.

[84] Nievola, CC, Carvalho, CP, Carvalho, V, Rodrigues, E. Rapid responses of plants to temperature changes. Temperature. 2017; 4: 371-405.

[85] Schrader, SM, Wise, RR, Wacholtz, WF, Ort, DR. Thylakoid membrane responses to moderately high leaf temperature in *Pima cotton*. Plant Cell Environmental. 2004; 27: 725-735.

[86] Tang T, Liu P, Zheng G, Li W. Two phases of response to long-term moderate heat: Variation in thermotolerance between *Arabidopsis thaliana* and its relative *Arabis pani*culata. Phytochemistry. 2016;122: 81-90.

[87] Reda, F, Mandoura, HMH. Response of enzymes activities, photosynthetic pigments, proline to low or high temperature stressed wheat plant (*Triticum aestivum* L.) in the presence or absence of exogenous proline or cysteine. International Journal Academic Research. 2011; 3: 108-115.

[88] Vierling, E. The roles of heat shock proteins in plants. Annual Review Plant Physiology and Plant Molecular Biology. 1991; 42: 579-620.

[89] Hemantaranjan, A, Bhanu, AN, Singh, MN, Yadav, DK. et al. Heat Stress Responses and Thermotolerance. Advances in Plants & Agriculture Research. 2014; 1(3): 00012.

[90] Ohama N, Sato H, Shinozaki K, Yamaguchi-Shinozaki K. Transcriptional regulatory network of plant heat stress response. Trends Plant Sci. 2017; 22(1): 53-65.

[91] Duarte, AA, Silva, CJ, Marques, AR, Modolo, Lv, Lemos-Filho, JP. Does oxidative stress determine the thermal limits of the regeneration niche of *Vriesea friburgensis* and *Alcantarea imperialis* (Bromeliaceae) seedlings? Journal of Thermal Biology. 2019; https://doi.org/10.1016/j.jtherbio.2019.02.003

[92] Franco, AC, Matsubara, S, Orthen, B. Photoinhibi-tion, carotenoid composition and the co-regulation of photochemical and non-photochemical quenching in Neotropical savannah

trees. Tree Physiology. 2008; 27: 717-725

[93] Nabout, JC. Global warming decreases the morphological traits of germination and environmental suitability of *Dipteryx alata* (Fabaceae) in Brazilian Cerrado. Acta Botânica Brasilica. 2019; 33: 446-453.

[94] Gomes, MP, Brito, JCM, Silva, JG, Cruz, FVS, Bicalho, EM. et al. Temperature effects on Zn-responses and Zn-reclamation capacity of two native Brazilian plant species: Implications of climate change. Environmental and Experimental Botany. 2018; 155: 589-599.

[95] Drake, JE, Vårhammar, A, Kumarathunge, D, Medlyn, BE, Pfautsch, S, Reich, PB. et al. A common thermal niche among geographically diverse populations of the widely distributed tree species *Eucalyptus tereticornis*: No evidence for adaptation to climate- of- origin. Global Change Biology. 2017; 23: 5069-5082.

[96] Faria, AP, Marabesi, MA, Gaspar, M, França, MGC. The increase of current atmospheric CO2 and temperature can benefit leaf gas exchanges, carbohydrate content and growth in C4 grass invaders of the Cerrado biome. Plant Physiology and Biochemistry. 2018; 127: 608-616.

[97] Leadley, P, Proença, V, Fernández-Manjarrés, Pereira, HM, Alkemade, R. et al. Interacting regional-scale regime shifts for biodiversity and ecosystem services. Bioscience. 2014; 64: 893-905.

[98] Pamesan, C, Hanley, ME. Plants and climate change: complexities and surprises. Annals of Botany. 2015; 116: 849-864.

[99] Oliveira, U, Soares-Filho, BS, Paglia, AP, Brescovit, AD, Carvalho, CJB. et al. Biodiversity conservation gaps in the Brazilian protected areas. Scientific Reports. 2017; 7: 1-9.

[100] Reis, CR, Bueno, ML, Rocha, LF, Santos, LG, Gorgens, EB. Climate Change Influencing the Potential Distribution of a Brazilian Savanna Indicator Species. Floresta e Ambiente. 2019; 26[3]: e20180085.

[101] Barbosa, ILBD, Oliveira, HR, Terribile, LC, Diniz-Filho, JAF. Geographical distribution of *Stryphnodendron adstringens* Mart. Coville [Fabaceae]: modeling effects of climate change on past, present and future. Brazilian Jornal of Botany. 2019; 42: 53-61.

[102] Chaves, CJ, Leal, BSS, Lemos-Filho, JP. How are endemic and widely distributed bromeliads responding to warming temperatures? A case study in a Brazilian hotspot. Flora. 2018; 238: 110-118.

[103] Silva, JMC, Rapini, A, Barbosa, LCF, Torres, RR. Extinction risk of narrowly distributed species of seed plants in Brazil due to habitat loss and climate change. Peer Journal. 2019; https://doi.org/10.7717/peerj.7333

[104] Bueno, ML, Pennington, RT, Dexter, KG, Kamiro, LHY, Pontara, V. et al. Effects of Quaternary Climatic Fluctuations on the Distribution of Neotropical Savanna Tree Species. Ecography. 2017; 40: 403-414.

[105] Keppel, G, Niel, KPV, Wardell-Johnson, GW, Yates, CJ, Byrne, M. et al. Refugia: identifying and understanding safe havens for biodiversity under climate change. Global Ecology Biogeography. 2012; 21: 393-404.

[106] Collevatti, RG, Ribeiro, MSL, Souza Neto, AC, Franco, AA, Oliveira, G, Terribile, LC. Recovering the demographical history of a Brazilian Cerrado tree species *Caryocar brasiliense*: coupling ecological niche modeling

and coalescent analyses. Natureza e Conservação. 2012; 10: 169-176.

[107] Gómez, JM, Valladares, F, Puerta-Pinero, C. Differences between structural and functional environmental heterogeneity caused by seed dispersal. Functional Ecology. 200; 18: 787-792.

[108] Chen, FS, Zeng, DH, Fahey, TJ, Yao, CY, Yu, ZY. Response of leaf anatomy of Chenopodium acuminatum to soil resource availability in a semi-arid grassland. Plant Ecology. 2010; 209: 375-382.

[109] Valladares, F, Gianoli, E, Gómez JM Ecological limits to plant phenotypic plasticity. New Phytologyst. 2007; 176: 749-763.

[110] Poorter, H, Niinemets, U, Ntagkas, N, Siebenkas, A, Mäenpää et al. A meta-analysis of plant responses to light intensity for 70 traits ranging from molecules to whole plant performance. New Phytologist. 2019; 223: 1073-1105.

[111] Barderrama, SIV, Chazdon, RL. Light-dependent seedling survival and growth of four tree species in Costa Rican second-growth rain forests. Journal of Tropical Ecology. 2005; 21: 383-395.

[112] Rodrıguez-Calcerrada, J, Cano, FJ, Valbuen-Carabanã, M, Aranda, LG. Functional performance of oak seedlings naturally regenerated across microhabitats of distinct overstorey canopy closure. New Forest. 2010; 39: 245-259.

[113] Valladares, F, Arrieta, S, Aranda, I, Lorenzao, D, Sanchez, D. et al. Shade tolerance, photoinhibition sensitivity and phenotypic plasticity of *Ilex aquifolium* in continental Mediterranean sites. Tree Physiol. 2005; 25: 1041-1052.

[114] Demmig-Adams B, Adams, WW. The role of xanthophyll cycle carotenoids in the protection of photosynthesis. Trends in Plant Science. 1996; 1: 21-26.

[115] Kouril, R, Wientjes, E, Bultema, JB, Croce, R, Boekema, EJ. High-light vs. lowlight: effect of light acclimation on photosystem II composition and organization in *Arabidopsis thaliana*. Biochim Biophys Acta. 2013; 1827: 411-419.

[116] Cookson, SJ, Granier, C. A dynamic analysis of the shad induced plasticity in *Arabidopsis thaliana* rosette leaf development reveals new components of the shade-adaptative response. Annals of Botany. 2006; 97: 443-452.

[117] Valladares F, Niinemets, Ü. Shade tolerance, a key plant feature of complex nature and consequences. Annual Rewiew of Ecology, Evolution and Systematics. 2008; 39:237-257.

[118] Rooijen, R, Aarts, MGM, Harbinson, J. Natural genetic variation for acclimation of photosynthetic light use efficiency to growth irradiance in Arabidopsis. Plant Physiology. 2015; 167:1412-1429.

[119] Ronquim, CS, Prado, CHBA, Souza, JP. Growth, photosynthesis and leaf water potential in young plants of *Copaifera langsdorffii* Desf. [Caesalpiniaceae] under contrasting irradiances. Brazilian Society of Plant Physiology. 2009; 21: 197-208.

[120] Barros, FV, Goulart, MF, Telles, SBS, Lovato, MB, Valladares, F, Lemos-Filho, JP. Phenotypic plasticity to light of two congeneric trees from contrasting habitats: Brazilian Atlantic Forest versus cerrado [savanna]. Plant Biology. 2010. doi:10.1111/j.1438-8677.2011.00474.x

[121] Rossatto, DV, Araújo, PE, Silva, BHP, Franco AC. Photosynthetic responses of understory savanna plants: Implications for plant persistence in savannas under tree encroachment. Flora. 2018; https://doi.org/10.1016/j.flora.2017.12.009

[122] Goulart, MF, Lovato, MB, Barros, FV, Valladares, F, Lemos-Filho, JP. Which Extent is Plasticity to Light Involved in the Ecotypic Differentiation of a Tree Species from Savanna and Forest? Biotrópica. 2011; DOI 10.1111/j.1744-7429.2011.00760.x

[123] Costa, AC, Rezende-Silva, SL, Megguer, CA, Moura, LMF, Rosa, M, Silva, AA. The effect of irradiance and water restriction on photosynthesis in young jatobá-do-cerrado (*Hymenaea stigonocarpa*) plants. Photosynthetica. 2015; 53(1): 118-127.

[124] Pinheiro, LFS, Kolb, RM, Rossatto, DR. Changes in irradiance and soil properties explain why typical non-arboreal savanna species disappear under tree encroachment. Australian Journal of Botany. 2016; 64: 333-341.

[125] Lepsch, IF. Cerrado: Soils from the. Encyclopedia of Soil Science, Thirt Edition. Taylor & Francis. 2016; p. 1-3. DOI: 10.1081/E-ESS3-120053880

[126] Lopes, AJ, Cox, FR. A survey of the fertility status of surface soils under cerrado vegetation of Brazil. Soil Science Society of America Journal. 1977; 41: 752-757.

[127] Haridasan, M. Nutritional adaptations of native plants of the cerrado biome in acid soils. Brazilian Journal of Plant Physiology. 2008; 20(3): 183-195.

[128] Ribeiro, JF, Walter, BMT. Fitofisionomias do bioma Cerrado. In S. M. Sano, & S. P. Almeida (Eds.) *Cerrado: Ecologia e flora* Planaltina, DF: EMBRAPA-CPAC. p. 151-212. 2008.

[129] FAO. Soil Map of the World: 1: 500.000, Legend. UNESCO: Paris, 1974; 1 pp.

[130] Reatto, A, Correia, JR, Spera, ST. Solos do bioma cerrado: aspectos pedológicos. In: Sano, SM; Almeida, S.P. de, ed. Cerrado: ambiente e flora. Planaltina: EMBRAPA-CPAC, 1998. p. 47-86.

[131] Lambers, H, Costa, PB, Oliveira, RS, Silveira, FAO. Towards more sustainable cropping systems: lessons from native Cerrado species Hans Lambers Theoretical Experimental of Plant Physiology. 2020; https://doi.org/10.1007/s40626-020-00180-z

[132] Araújo, GM, Haridasan, M. A comparison of the nutritional status of two forests on dystrophic and mesotrophic soils in the cerrado region of central Brazil. Communications in Soil Science and Plant Analysis. 1988; 19: 1075-1089.

[133] Neri, AV, Schaefer, CEGR, Silva, AF, Souza, AL, Ferreira-Junior, WG, Meira-Neto, JAA. The influence of soils on the floristic composition and community structure of an area of brazilian cerrado vegetation. Edinburgh Journal of Botany. 2012; 69(1): 1-27.

[134] Delhaize, E, Ryan, PR. Aluminum toxicity and tolerance in plants. Plant Physiology. 1995; 107: 315-321.

[135] Poschenrieder, C, Gunsé, B, Corrales, I, Barceló, J. A glance into aluminum toxicity and resistance in plants. Science Total Environmental. 2008; 400: 356-368.

[136] Brunner, I, Sperisen, C. Aluminum exclusion and aluminum tolerance in woody plants. Frontier Plant Scirence. 2013; 1: 172.

[137] von Uexkull, HR, Mutert, E. Global extent, development and economic impact of acid soils. In: Date RA, et al. (eds) Plant soil interactions at low pH. Kluwer Academic Publishers, Dordrecht. 1995; pp 5-19.

[138] Kinraide, T.B. Reconsidering the rhizotoxicity of hydroxyl, sulphate, and fluoride complexes of Aluminum.

Journal of Experimental Botany. 1997, 48: 1115-1124.

[139] Siqueira-Neto, Piccolo, MC, Scopel, E, Junior, CC, Cerri, CC, Bernoux, M. Carbono total e atributos químicos com diferentes usos do solo no cerrado. Acta Scientiarum Agronomy. 2009; 31: 709-717.

[140] Bojórquez-Quintal, E, Escalante-Magaña, C, Echevarría-Machado, I, Martínez-Estévez, M. Aluminum, a friend or foe of higher plants in acid soils. Frontier Plant Science. 2017; 8:1-18.

[141] Singh, SK, Reddy, VR, Fleisher, DH, Timlin, DJ. Relationship between photosynthetic pigments and chlorophyll fluorescence in soybean under varying phosphorus nutrition at ambient and elevated CO_2. Photosynthetica. 2017; 55: 421-433.

[142] Souza, MC, Habermann, G, do Amaral, CL, Rosa, AL, Pinheiro, MHO, da Costa, FB*Vochysia tucanorum* Mart.: an aluminum-accumulating species evidencing calcifuge behavior. Plant Soil. 2017; 419: 377-389.

[143] Haridasan, M. Aluminum accumulation by some cerrado native species of central Brazil. Plant Soil. 1982; 65: 265-273.

[144] Jansen, S, Broadley, MR, Robbrecht, E, Smets, E. Aluminum hyperaccumulation in angiosperms: a review of its phylogenetic significance. The Botanical Review. 2002; 68: 235-269.

[145] Souza, MC, Bueno, PCP, Morellato, LPC, Habermann, G. Ecological strategies of Al-accumulating and non-accumulating functional groups from cerrado *sensu stricto*. Anais Academia Brasileira Ciências. 2015; 87(2): 813-823.

[146] Kochian, LV, Piñeros, MA, Liu, J, Magalhaes, JV. Plant adaptation to acid soils: the molecular basis for crop Aluminum resistance. Annual Review Plant Biology. 2015; 66: 571-598.

[147] Alvin, MN, França, MGC, Ramos, FT, Nascentes, CC. Slight soil deacidification compromisse the growth and the aluminum accumulation in *Qualea cordata* plants. Applied Ecology and Environmental Research. 2017; 15(4):2023-2034.

[148] Cury, NF, Silva, RCC, Andre, MSF, Fontes, W, Ricart, CAO. et al. Root proteome and metabolome reveal a high nutritional dependency of Aluminum in *Qualea grandiflora* Mart. (Vochysiaceae). Plant Soil. 2020; 446:125-143 https://doi.org/10.1007/s11104-019-04323-3

[149] Andrade, LRM, Barros, LMG, Echevarria, GF, Amaral, LIV, Cotta, MG. et al. Al-hyperaccumulator Vochysiaceae from the Brazilian Cerrado store aluminum in their chloroplasts without apparent damage. Environmental and Experimental Botany. 2011; 70: 37-42.

[150] Banhos, OFAA, Souza, MC, Habermann, G. High aluminum availability may affect *Styrax camporum*, an Al non-accumulating species from the Brazilian savanna. Theoretical and Experimental Plant Physiology. 2016; https://doi 10.1007/s40626-015-0051-5

[151] Bressan, ACG, Silva, GS, Banhos, OFAA, Tanaka, FAO, Haermann, G. Physiological, anatomical and ultrastructural effects of aluminum on *Styrax camporum*, a native Cerrado woody species. Journal of Plant Research. 2020; https://doi.org/10.1007/s10265-020-01210-2.

[152] Čiamporová, M. Morphological and structural responses of plant roots to Aluminum at organ, tissue, and cellular levels. Biologia Plantarum. 2002; 45: 161-171.

[153] Silva, CO, Brito, DS, Silva, AA, Rosa, VR, Santos, MFS. et al. Diferential accumulation of aluminum in root tips of soybean seedlings. Brazilian Journal of Botany. 2020; 43: 99-107.

[154] Matsumoto, H. Cell Biology of Aluminum Toxicity and Tolerance in Higher Plants. International review of cytology. 2000; 200: 1-46.

[155] Bittencourt, BMOC, Silva, CMS, Filho, SZ, Habermann, G. Aluminum (Al)-induced organic acid exudation in an Al-Accumulating species from the Brazilian savanna. Trees. 2019; 34: 155-162.

[156] Mendonça, AMC, Lira, DA, Vilela, ALO, Vieira, DA, Melo, NC, Barbosa, JPRAD. High aluminum concentration and initial establishment of Handroanthus impetiginosus: clues about an Al non-resistant species in Brazilian Cerrado. Journal of Forestry Research. 2019; https://doi.org/10.1007/s11676-019-01033-5

[157] Konrad, MLF, Silva JAB, Furlani, PR, Machado, EC. Gas exchange and chlorophyll fluorescence in six coffee cultivars under aluminum stress. Bragantia. 2005: 64: 339-347.

[158] Silva, S, Pinto, G, Dias, MC, Correia, CM, Moutinho-Pereira, J, et al. Aluminum long-term stress differently affects photosynthesis in rye genotypes. Plant Physiology Biochemistry. 2012; 54: 105-112.

[159] Vitorello, VA, Capaldi, FR, Stefanuto, VA. Recent advances in aluminum toxicity and resistance in higher plants. Brazilian Journal Plant Physiology. 2005; 17: 129-143.

[160] Chen, LS, Qi, YP, Smith, BR, Liu, XH. Aluminum-induced decrease in CO2 assimilation in citrus seedlings is unaccompanied by decreased activities of key enzymes involved in CO2 assimilation. Tree Physiology. 2005; 25: 317-324.

[161] Chen, LS, Qi, YP, Liu, XH. Effects of aluminum on light energy utilization and photoprotective systems in citrus leaves. Annals of Botany. 2005; 96: 35-41.

[162] Guo, P, Yi-Ping, Q, Yan-Tong, C, Tao-Yu, Y, Lin-Tong, Y. et al. Aluminum effects on photosynthesis, reactive oxygen species and methylglyoxal detoxification in two Citrus species differing in Aluminum tolerance. Tree Physiology. 2018; 38: 1548-1565.

[163] Peixoto, PH, Da Matta, FM, Cambraia, J. Responses of the photosynthetic apparatus to aluminum stress in two sorghum cultivars. Journal Plant Nutrition. 2002; 25: 821-832.

[164] Pereira, WE, Siqueira, DL, Martinez, CA, Puiatti, M. Gas exchange and chlorophyll fluorescence in four citrus rootstocks under aluminum stress. Journal Plant Physiology. 2000; 157: 513-520.

[165] Moustakas, M, Ouzounidou, G, Lannoye, R. Aluminum effects on photosynthesis and elemental uptake in an aluminum-tolerant and non-tolerant wheat cultivar. Journal Plant Nutrition. 1995; 18: 669-683.

[166] Bustamante, MMC, Medina E, Asner GP, Nardoto GB, Garcia-Montiel, DC. et al. Nitrogen cycling in tropical and temperate savannas. Biogeochemistry. 2006; 79: 209-237.

[167] Elser, JJ, Bracken, MES, Cleland, EE, Gruner, D, Stanley, W. et al. Global analysis of nitrogen and phosphorus limitation of primary producers in freshwater, marine and terrestrial ecosystems. Ecology Letters. 2007; 10: 1135-1142.

[168] Lambers, H, Raven, J A, Shaver, GR, Smith, SE. Plant nutrient-acquisition strategies change with soil age. Trends Ecology Evolution. 2008; 23: 95-103.

[169] Kamble, BM, Todmal, SM. Effect of different levels and sources of nitrogen on soil properties, nutrient uptake, and yield of wheat grown on saline soil. International Journal of Chemical Studies. 2020; 8(1): 687-693.

[170] Plaxton, WC, Tran, HT. Metabolic adaptations of phosphate-starved plants. Plant Physiol. 2011; 156: 1006-1015.

[171] Havlin, JL, Tisdale, SL, Nelson, WL, Beaton, JD. Phosphorus. In *Soil Fertility and Fertilizers*, 8th ed.; Lawrensen, W., Gohn, J., Eds.; Pearson Inc.: Upper Saddle River, NJ, USA, 2014; pp. 185-221.

[172] Pallardy, SG. Nitrogen Metabolism. In *Physiology of Woody Plants*, 3rd ed.; Pallardy, S.G., Ed.; Academic Press: Missouri, MO, USA, 2008; pp. 233-254.

[173] Reis, DN, Silva, FG, Santana, RC, Oliveira, TC, Freiberger, MB. et al. Growth, Physiology and Nutrient Use Efficiency in *Eugenia dysenterica* DC under Varying Rates of Nitrogen and Phosphorus. Plants. 2020; 9: 722.

[174] Lopes, AS, Guilherme, LRG. A career perspective on soil management in the Cerrado region of Brazil. Advances in Agronomy. 2016; 137: 1-72.

[175] Viani, RAG, Rodrigues, RR, Dawson, TE, Oliveira, RS. Savanna soil fertility limits growth but not survival of tropical forest tree seedlings. Plant Soil. 2011; 349:341-353.

[176] Pereira, CG, Clode, PL, Oliveira, RS, Lambers, H. Eudicots from severely phosphorus-impoverished environments preferentially allocate phosphorus to their mesophyll. New Phytologist. 2017; 218: 959-973.

[177] Machado, RB, Ramos Neto, MB, Pereira, P, Caldas, EF, Gonçalves DA, et al. Estimativas de Perda da Área do Cerrado Brasileiro. Conservation International do Brasil, Brasília. 2004. pp. 25.

[178] Hunke, P, Mueller, EN, Schröder, B, Zeilhofer, P. The Brazilian Cerrado: assessment of water and soil degradation in catchments under intensive agricultural use. Ecohydrology. 2015; 8: 1154-1180.

[179] Jacobson, TKB, Bustamante, MMC, Kozovits, AR. Diversity of shrub tree layer, leaf litter decomposition and N release in a Brazilian Cerrado under N, P and N plus P additions. Environmental Pollution. 2011; 159: 2236-2242.

[180] Bustamante, MMC, Brito, DQ, Kozovitz, AR, Luedemann, G, Mello, TRB. et al. Effects of nutrient additions on plant biomass and diversity of the herbaceous-subshrub layer of a Brazilian savanna (Cerrado). Plant Ecology. 2012; 213: 795-808.

[181] Lannes, LS, Bustamante, MMC, Edwards, PJ, Venterink, HO. Native and alien herbaceous plants in the Brazilian Cerrado are (co-) limited by different nutrients. Plant Soil. 2015; DOI 10.1007/s11104-015-2725-9

[182] Lambers, H, Brundrett, MC, Raven, JA, Hopper SD. Plant mineral nutrition in ancient landscapes: high plant species diversity on infertile soils is linked to functional diversity for nutritional strategies. Plant Soil. 2010; 334: 11-31.

[183] Laliberté, E, Grace, JB, Huston, MA, Lambers, H, Teste, FP. et al. How does pedogenesis drive plant diversity? Trends in Ecology & Evolution. 2013; 28: 331-340.

[184] Pausas, JG, Keeley, JE. A burning story: the role of fire in the history of life. Bioscience. 2009; 59: 593-601.

[185] Bond, WJ, Midgley, JJ. Fire and the angiosperm revolutions. International Journal Plant Science. 2012; 173: 569-583.

[186] Miranda, HS, Bustamante, MM, Miranda, AC. The fire factor. In: Oliveira, P.S.; Marquis, R.J. The Cerrados of Brazil. New York: Columbia University Press, 2002; pp. 51-68.

[187] Fidelis, A. Is fire always the "bad guy"? Flora. 2020; 268: 151611.

[188] Berlinck, CN, Batista, EKL. Good fire, bad fire: It depends on who burns. Flora. 2020; 268: 1-4.

[189] Moraes, MG, Carvalho, MAA, Franco, AC, Pollock, CJ, Figueiredo-Ribeiro, RCL. Fire and Drought: Soluble Carbohydrate Storage and Survival Mechanisms in Herbaceous Plants from the Cerrado. BioScience. 2016; 66: 107-117.

[190] Bowman, DMJS, Balch. J, Artaxo, P, Bond, WJ, Cochrane, M.A. et al. The human dimension of fire regimes on Earth Journal Biogeography. 2011; 38: 2223-2236.

[191] Hoffmann, W.A. Fire and population dynamics of woody plants in a neotropical savanna: matrix model projections. Ecology. 1999; 80: 1354-1369.

[192] Keeley, JE, Pausas, JG, Rundel, PW, Bond, WJ, Bradstock, RA. Fire as an evolutionary pressure shaping plant traits. Trends in Plant Science. 2011; 16: 406-411.

[193] Fahn, A. and Cutler, D. Xerophytes. Gebruder Borntraeger, Berlin. 1992, Pp. 176.

[194] Loram-Lourenço, L, Farnese, FDS, Sousa, LFD, Alves, RDFB, Andrade, MCP. et al. A Structure Shaped by Fire, but Also Water: Ecological Consequences of the Variability in Bark Properties Across 31 Species From the Brazilian Cerrado. Frontier Plant Science. 2020; 10: 1718.

[195] Dairel, M, Fidelis, A. How does fire affect germination of grasses in the Cerrado? Seed Science Research. 2020; https://doi.org/10.1017/S0960258520000094

[196] Coutinho, LM. Ecological effects of fire in Brazilian Cerrado. Berlin, Heidelberg, Springer. 1982; pp. 273-291.

[197] Soares, JJ, Souza, MHAO, Lima, MIS. Twenty years of post-fire plant succession in a "cerrado", São Carlos, SP, Brazil. Brazilian Journal of Biology. 2006; 66

[198] Felfili, JM, Silva-Junior, MC, Dias, BJ, Rezende, AV. Estudo fenológico de *Stryphnodendron adstringens* (Mart.) Coville no cerrado sensu stricto da Fazenda Água Limpa no Distrito Federal, Brasil. Revista Brasileira Botânica. 1999; 22(1): 83-90.

[199] Silvério, DV, Pereira, OR, Mews, HA, Maracahipes-Santos, L. Santos, JO, Lenza, E. Surface fire drives short-term changes in the vegetative phenology of woody species in a Brazilian savanna. Biota Neotropical. vol. 15 no. 3 Campinas. 2015. Epub Sep 18.

[200] Pilon, NAL, Hoffmann, WA, Abreu, RCR, Durigan, G. Quantifying the short-term flowering after fire in some plant communities of a cerrado grassland. Plant Ecology & Diversity. 2018; DOI: 10.1080/17550874.2018.1517396

[201] Munhoz, C, Felfili, JM. Reproductive phenology of an herbaceous sub-shrub layer of a Savannah (Campo Sujo) in the Cerrado Biosphere Reserve I, Brazil. Brazilian Journal of Biology. 2007; 67: 299-307.

[202] Neves, DRM, Damasceno-Junior, GA. Post-fire phenology in a campo sujo vegetation in the Urucum plateau, Mato Grosso do Sul, Brazil. Brazilian Journal Biology. 2011; 71: 881-888.

[203] Fidelis, A, Rosalem, P, Zanzarini, V, Camargos, LS, Martins, AR. From

ashes to flowers: a savanna sedge initiates flowers 24 h after fire. Ecology. 2019; 100: e02648.

[204] Lamont, BB, Downes, KS. Fire-stimulated flowering among resprouters and geophytes in Australia and South Africa. Plant Ecology. 2011; 212: 2111-2125.

[205] Clarke, PJ, Lawes, MJ, Midgley, JJ, Lamont, BB, Ojeda, F, Burrows, GE, Enright, NJ, Knox, KJEResprouting as a key functional trait: How buds, protection and resources drive persistence after fire. New Phytologist. 2013; 197: 19-35.

[206] Appezzato-da-Glória, B, Cury, G, Soares, MKS. Underground systems of Asteraceae species from the Brazilian Cerrado. Journal of the Torrey Botanical Society. 2008; 135(1): 103-113.

[207] Palhares, D, Paula, JE, Pereira, LAR, Silveira, CES. Comparative wood anatomy of stem, root and xylopodium of *Brosimum gaudichaudii* (Moraceae) IAWA Journal, 2007; 28(1): 83-94.

[208] Braga, MR, Carpita, NC, Dietrich, SMC, Figueiredo-Ribeiro, RCL. Changes in pectins of the Xylopodium of *Ocimum nudicaule* from dormancy to sprouting. Brazilian Journal Plant Physiology. 2006; 18: 325-331.

[209] Alonso, AA, Machado, SR. Morphological and developmental investigations of the underground system of Erythroxylum species from Brazilian cerrado. Australian Journal of Botany. 2007: 55: 749-758.

[210] Hoffman, WA, Schroeder W, Jackson, RB. Positive feedbacks of fire, climate, and vegetation and the conversion of tropical savanna. Geophysical Research Letters. 2002; 29:2052. doi:10.1029/2002GL015424

[211] Smith, AM, Zeeman, SC. Annual Review of Plant Biology Starch: A Flexible, Adaptable Carbon Store Coupled to Plant Growth. Annual Review Plant Biology. 2020; 71: 9.1-9.29

[212] Knapp, AK. Effect of fire and drought on the ecophysiology of *Andropogon gerardii* and *Panicum virgatum* in a tallgrass prairie. Ecology. 1985; 66: 1309-1320.

[213] Furley, PA. The nature and diversity of neotropical savanna vegetation with particular reference to the Brazilian cerrados. Global Ecology and Biogeography. 2002; 8: 223-241.

[214] Paula, SM, Naves, ER, Moreira, I, Padua, KJM, Silva, KR, Fernandes, WP. et al. Desempenho fotossintético de folhas jovens e maduras de *Vochysia cinnamomea* (Vochysiaceae) em áreas de cerrado rupestre intactas e pós-queimada. Bioiscience Journal 2015; 31(2): 591-600.

[215] Graves, S.J, Rifai, SW, Putz, FE. Outer bark thickness decreases more with height on stems of fire-resistant than fire-sensitive Floridian oaks (Quercus spp.; Fagaceae). Americam Journal. Botany. 2014; 101: 2183-2188.

[216] West, AG, Nel, JA, Bond, WJ, Midgley, JJ. Experimental evidence for heat plume-induced cavitation and xylem deformation as a mechanism of rapid post-fire tree mortality. New Phytologyst. 2016; 211: 828-838.

[217] Simon, MF, Pennington, T. Evidence for fire adaptation to fire regimes in the tropical savannas of the Brazilian cerrado. International Journal Plant Science. 2012;173: 711-723.

[218] Loram-Lourenço L, Farnese, FDS, Sousa, LFD, Alves, RDFB, Andrade, MCPD. et al. A Structure Shaped by Fire, but Also Water: Ecological Consequences of the Variability in Bark Properties Across 31 Species From the Brazilian Cerrado. Frontie Plant Science. 2020; 10:1718.

[219] Martins, LC, Paiva, EAS. Flammable resin in *Vellozia variabilis* (Velloziaceae): Gland structure and chemical composition. Flora. 2016; 219: 94-100.

[220] Rossatto, DR, Araújo, PE, Silva, BHP, Franco, AC. Photosynthetic responses of understory savanna plants: Implications for plant persistence in savannas under tree encroachment. Flora. 2017; https://doi.org/10.1016/j.flora.2017.12.009

Chapter 4

Influence of Water Stress on Growth, Chlorophyll Contents and Solute Accumulation in Three Accessions of *Vicia faba* L. from Tunisian Arid Region

Hanen Enneb, Leila Ben Yahya, Mohamed Ilyas, Datta Asaram Dhale, Mohamed Bagues and Kamel Nagaz

Abstract

In this study, we aim to investigate the physiological and biochemical adaptations of *Vicia faba* plants to moderate irrigation regime (T1) and describe the effects of water stress on their growth performance and chlorophyll contents. For this reason, three Tunisia accessions (ElHamma, Mareth and Medenine) were studied. An experiment was conducted for one month. Faba bean plants were first grown in a greenhouse and then, exposed to water stress, whereby they were irrigated up to the field capacity (FC) of 0% (control, T0) and 50% of the control (moderate stress, T1). The effect of water stress on physiological parameters showed differences in relation to the accessions studied and the water regime. Relative water content (RWC) of ElHamma accession does not seem to be affected by stress as compared with the control regime. Total chlorophyll content decreases, whereas soluble sugar contents increases for all accessions studied. ElHamma has the highest content. About morphological parameters, bean growth varies according to the ascension and treatment. Hydric stress impedes the growth of the root part and caused a significant reduction in the shoot and root Dry Weight (DW) of the T1-stressed beans, compared to the optimal irrigation (T0).

Keywords: water stress, *Vicia faba* L., RWC, growth, chlorophyll, soluble sugars

1. Introduction

Environmental stresses, especially water stress, seriously limit plant growth as well as plant productivity [1–3]. Thus, we speak of water stress when the demand for water exceeds the available resources. This category includes countries where the availability of water per year and per inhabitant is less than 1700 m^3, particularly in semi-arid regions. Arid zones are characterized by a negative hydric balance, mean annual rainfall of less than 800 mm, with a mean insolation of 2800 h/year and means annual temperatures of 23–27°C. These conditions lead to high evapotranspiration rates, carrying a water deficit in nearly the entire region [4].

Abiotic stresses and inefficient production techniques are the obvious factors limiting crop productivity, especially in the southern parts characterized by severe aridity [5–7].

Faba bean (*Vicia faba* L.) is a crucial grain crop through the word; ordering as the fourth crop legume after dry beans, dry peas and chickpeas. The bean is of great importance for nutrition, it enters in human consumption and also for animal feed [8]. North Africa appears to be one of the best faba bean output global regions and especially, Tunisia.

In Tunisia, faba bean covers more than 68% (70,000 ha) of the total area annually devoted to grain legumes crops and the national average yield is around 50,000 tons [9, 10]. Although the agro-economic importance of the bean is known, its cultivation is in decline, both in some countries of the world and in Tunisia [9]. The areas reserved for legumes in Tunisia in gradual fall [11]. This mainly affected the peas and broad beans. The national average bean yield is low and fluctuates widely from year to year. This is mainly due to climatic variations [6].

The seeds of the bean are very rich in proteins (25% of the dry matter of the seeds) and especially in lysine but poor in amino acid [12]. It is also a source of carbohydrates, mineral salts, fiber, vitamins and little lipids. The seeds are the most consumed parts and like all vegetables; the bean is self-sufficient for nitrogen nutrition. The bean is used as a green manure which serves as well for the improvement of the structure of the soil as for its enrichment in mineral elements. In addition to its nutritional and fertilizing (green manure) utility, the bean also has medicinal interests [13].

The success of agriculture in arid and semi-arid zones worldwide will be tremendously dependent on the ability of agricultural systems and farmers to adjust to climate change [14]. A vast majority of these areas, despite the huge production, suffer from limiting resource of water [15]. As that resource becomes more scarce and availability more difficult to predict, water managers and farmers will be forced to implement new creative solutions to water supply challenges. Expected exposure indicates improvement and better understanding of culture adaptation to water stress in such areas is high-priority to successful development in these regions under a changing climate. This work focuses on understanding the effect of water stress on morphological, physiological and biochemical aspects of *Vicia faba* L. plant grown in Tunisian Arid Region.

2. Concept of water deficit on plants

Water is a vital element for the growth and development of plants. It generally constitutes 85–90% of the total fresh weight of plants. Water is the natural vehicle for all the substances circulating in the plant, the elements absorbed in the soil or the organic substances produced in the leaves circulating in the saps of the xylem and phloem [16]. The need for water for the plant is proven by multiple observations such as wilting and death of plants left without water, hence the importance of irrigation and the relationship between vegetation and rainfall [17]. Plants can get the water they need through their roots. These extract water from the soil to pass it through the plant; part of this water is then released into the atmosphere (a phenomenon called transpiration) [1]. Omprakash et al. [18] defines water deficit or water stress as a period of insufficient water activity in the plant. According to this same author the notion of stress implies on the one hand a more or less abrupt deviation from the normal conditions of the plant and on the other hand a sensitive reaction of the plant in the different aspects of its physiology which changes appreciably with either adaptation to the new situation, ie to the limit degradation leading to a fatal outcome. On the contrary, according to Dodd and Ryan. [19] any water restriction resulting in a drop in the plant's production potential, following a disturbance of its physiological activity caused by a deficit in water consumption is called water stress.

3. Analyses of faba bean responses to water deficit stress

The experiment was carried out during 2019/2020, in a plastic greenhouse at the IRA (Institute of Arid Regions), located 22.5 km north-east of the city of Medenine. This region is located in the lower arid bioclimatic stage characterized by a low annual rainfall of around 150 mm/year; high temperature and frequent winds. *Vicia faba* seeds used in this test is the *Vicia faba* L. originating from Medenine and Gabès (ElHamma and Mareth). The latter ascensions tested were sown in pots filled with soil in February 2019 Three seeds were planted per pot; Three repetitions for each ascension were established. Water used for irrigation is tap water, titrating ±2 g/l and a pH of 7.6. Two water treatments were tested: T0 = 100% CC (capacity in the fields 100%) and T1 = 50% CC. The quantities of water (in ml/kg of soil) added each week (220 ml/kg for T0 and 110 ml/kg for T1). The dry weight of leaves and roots was carried out separately. The above-ground and underground dry matter is obtained after passing in an oven for 48 hours at 75°C. Relative water content of faba bean for each treatment were determined by the following method: Cut the leaf blade into small squares, weigh them to determine the fresh weights (FW) and then place them in a box of petri dish at 5°C. After 24 hours, the pieces of leaves are removed and then weighed at full turgor (TW). They are then placed in an oven at 75°C and then weighed again to determine their dry weight (DW). Chlorophyll contents determined by Gross method [20] and to quantify soluble sugars, the phenol method of Dubois et al. [21] was used.

4. Morphological changes in faba bean under water stress.

4.1 Plant biomass

Three accessions (El Hamma, Mareth and Medenine) of faba bean have been analyzed for dry matter accumulation. Significant effect of treatment ($P < 0.005$) was observed and no difference has been established between ascensions (**Figure 1**). Under moderate water stress treatment (T1), faba bean (i.e., El Hamma, Mareth and Medenine) accessions were significantly decreased their shoot and root DW as compared to control plants. Total leaf and root dry masses were all two to three times greater in the well-watered treatment (T0) than those in the water-limited one (T1). The watering treatment showed significant effects among localities on apparent shoot biomass and apparent root biomass (**Figure 1**). Risk-taking by minimal reduction of shoot DW under water stress was revealed as a physiological indicator for drought resilient at the flowering stage for faba bean [22]. The capacity of populations to maintain shoot biomass in water shortage was displayed to be an important trait for drought tolerance in soybean plants [23]. Under drought faba bean decreased their root DW (**Figure 2**). Root growth reduction was also reported for pea plant [24]. According to Chavoshi et al. [25] severe drought stress reduces biomass and seed yield (by 20–90%).

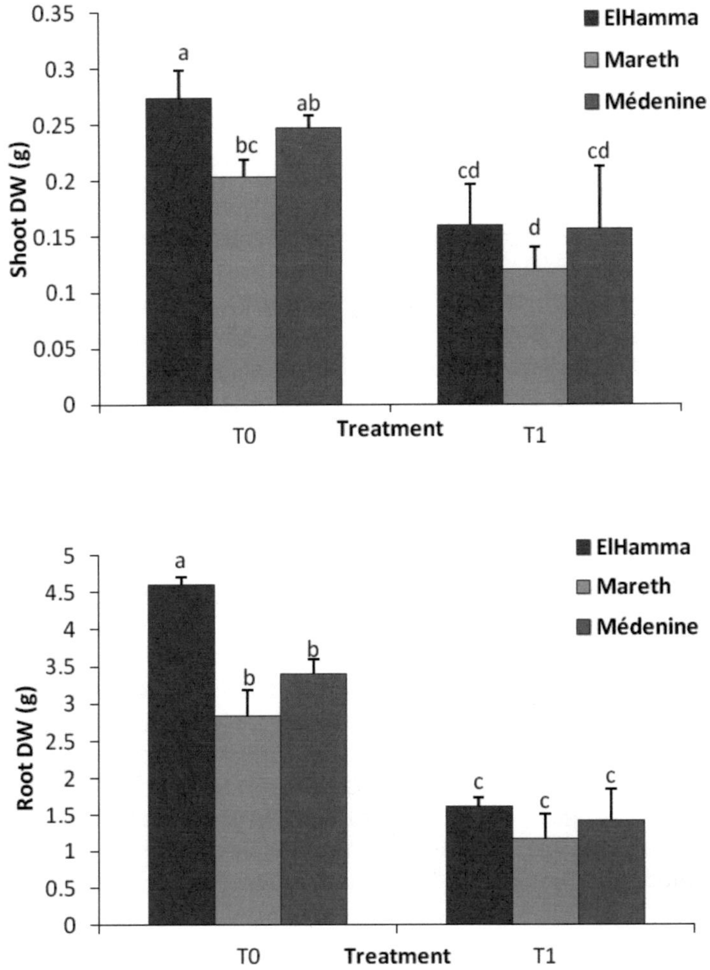

Figure 1.
Changes in (a) Shoot DW, (b) Root DW for Vicia faba *L. accensions in T0 (control) and T1 (moderate deficit irrigation (50%FC)). Means ± S.E. (n = 3). Means with different letters (a, b, c) are significantly different at P < 0.05 and values sharing a common letter are not significantly different at p < 0.05.*

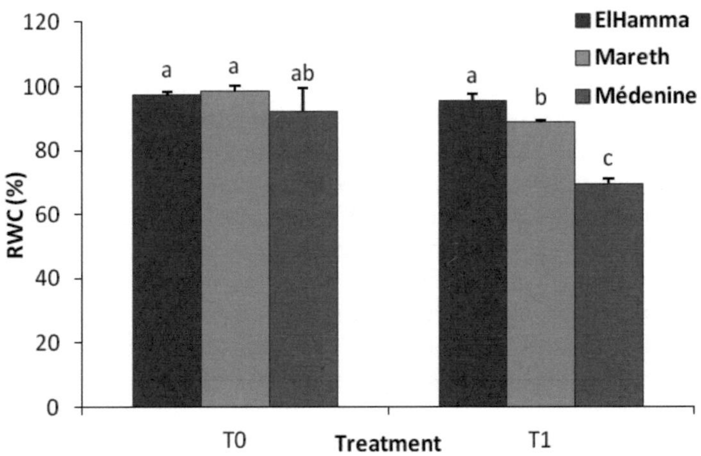

Figure 2.
Changes in relative water content (RWC) for V. faba *leaves in T0 (control) and T1 (moderate deficit irrigation (50% FC)).*

DOI: http://dx.doi.org/10.5772/intechopen.94563

5. Physiological and biochemical responses of faba bean to water stress

5.1 Water status (RWC)

As shown in **Figure 2**, RWC varied significantly among accessions and RWC was significantly higher as expected in non-stress conditions (T0) than in

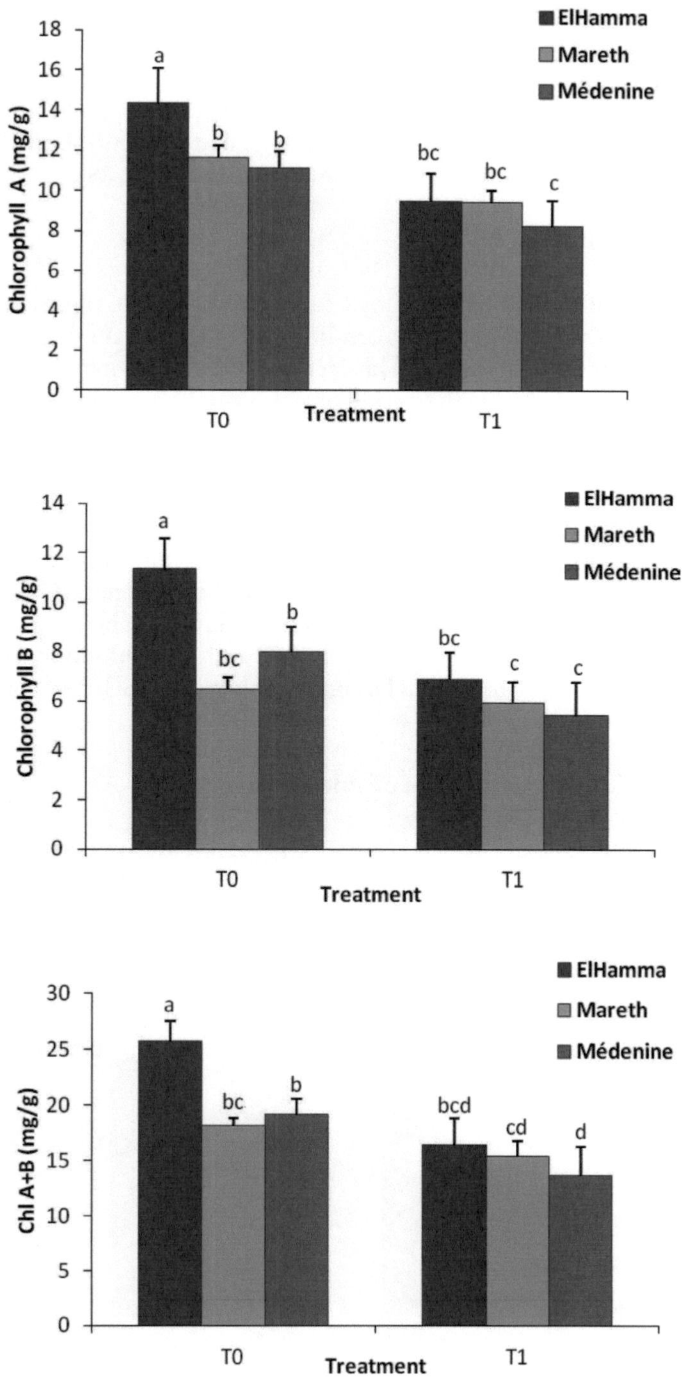

Figure 3.
Changes in Chlorophyll (a), (b) and (a+b) for V. faba *leaves in To (control) and T1 (moderate deficit irrigation (50% FC)).*

stressed plants (T1). This parameter is sufficiently influenced by water restriction ($P < 0.001$). Thus, leaves of the three ascensions studied reacted with a considerable drop in their relative water content (RWC). In El Hamma, the RWC is maintained at high values for irrigated plants and stressed plants, while in Mareth and Medenine, plants cultivated under T1, exhibited a significant decrease compared to the control of about 9.74% and 24.73% respectively. Repeated measures ANOVA, indicated significant interaction between accession and stress deficit ($P < 0.01$) for RWC. The analysis of the relative water content makes it possible to give a clue and describe in a global way the water status of beans in response to water stress. Decrease in RWC in plants under drought stress have been perceived in many plants and may depend on plant vigor. Likely, in this constraint, the ability to achieve good osmoregulation and to preserve cell turgor is reduced [26]. It seems that concentration of organic solutes to maintain membrane is not sufficient in this case [27, 28].

5.2 Chlorophyll contents

Chlorophyll 'A', 'B' and total chlorophyll contents, showed no significant difference among the accessions in the stress conditions, but significant difference under the non-stress conditions (**Figure 3**). The decrease in the chlorophyll contents in the three ascensions studied during the period of stress would be most likely due to the increased in catalytic activity of chlorophylls and degradation of photosynthetic pigments and this process is also the result of not providing the necessary factors for the synthesis of chlorophyll and the destruction of its structure under stress conditions [24]. Zhu [29] has been reported that a decrease in the chlorophyll content will be due to a decrease in the stomata aperture and aimed at limiting water losses by evaporation and by increased resistance to the entry of atmospheric CO2 necessary for photosynthesis [30]. These results are in agreement with Tairo et al. [31] who reported a significant increase in chlorophyll 'A', 'B' and total chlorophyll contents was seen in varieties 1(KAT B9), 4(F8 Drought line) and 5(JESCA) of common bean *Phaseolus vulgaris* (L.).

5.3 Osmoregulation (accumulation of soluble sugars)

Induced drought stress influenced ($P < 0.001$) soluble sugar contents (**Figure 4**). Untreated plants showed the lowest concentration (0.54 mg/g for Mareth accenssion).

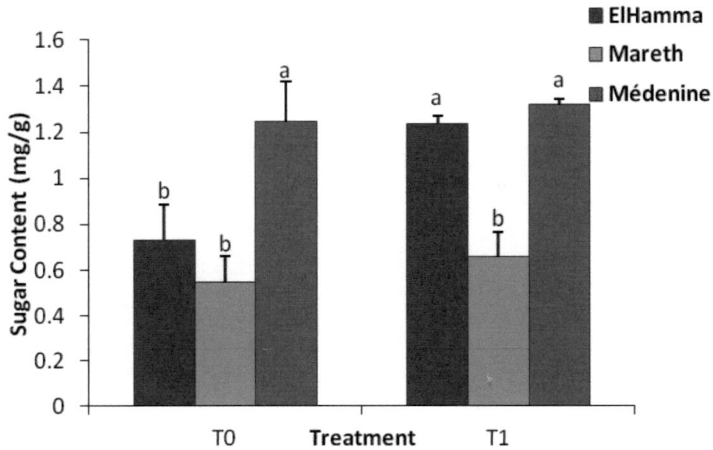

Figure 4.
Effect of water stress on soluble sugar contents of V. faba *L. leaves.*

However, the highest soluble sugar content (1.31 mg/g) was noticed under water deficit condition for ElHamma cultivar. These metabolic adaptations that improve plant tolerance to osmotic or water stress involve an increased synthesis of compatible osmolytes as soluble sugars and proline. This increase in sugar content under stress might contribute for osmotic adjustment and plays an important role in scavenging free radicals and protect DNA from damaging effects of increased reactive oxygen species (ROS) levels due to water stress [32, 33]. Moreover, soluble sugar accumulation was considered as an indicator of plant tolerance to drought conditions. It improves water absorption and reduces water loss when stress leads to slower growth [34, 35].

6. Other environmental stresses effects on *Vicia faba* L.

The effect of sodium chloride (NaCl) concentrations (0.0, 60, 120, 240 mM) on growth and chlorophyll content of (*Vicia faba* L.) seedlings caused an increase for both fresh and dry weights of the shoot. Salinity significantly reduced chlorophyll 'A' content. It also significantly reduced chlorophyll 'B', total chl., and carotenoïds contents after ten days of treatment [36]. High-temperature stress gravely reduced the photosynthesis, stomatal conductanceand water use efficiency of *V. faba* [37–39]. The effects of magnesium (Mg) supplementation on the growth performance as well as, on photosynthetic pigment synthesis, were studied in *Vicia faba* L. plants exposed to heat stress and non-heat-stress conditions. Results revealed that growth attributes and total chlorophyll decreased in plants subjected to heat stress, whereas accumulation of organic solutes had seen. These results suggest that adequate supply of Mg is not only essential for plant growth and development, but also improves plant tolerance to heat stress by suppressing cellular damage induced by ROS through the enhancement of the accumulation of Proline and glycine betaine [40].

7. Improvement of crop production under environmental stresses

Crop production is influenced by multiple environmental abiotic factors including drought, soil salinity, waterlogging and temperature interferes with the whole metabolic activities in plants [41, 42]. Plant adopted different mechanisms and modifications to cope with stresses such as plant hormones that facilitating growth, nutrient allocation and development [43, 44]. Salicylic acid (SA) significantly increased the Mn^{+2}, Co^{+3}, Fe^{+2}, Cu^{+3}, K^{+} and Mg^{+2} while decreased the Na^{+}, Ni^{+3}, Pb^{+4}, Zn^{+2}, and Na^{+}/K^{+} content of roots and soil content under salinity stress. SA (10-5 M) can be implicated to mitigate the adverse effects of salinity on maize plants [45, 46]. Additionally, exogenous application of gibberellic acid (GA3) reduced oxidative stress in *C. capsularis* seedlings. GA3 not only ameliorate photosynthesis, biomass and plant growth, foliar application of GA3 but also, increases metal (Cu) concentration in different organs of the plants when compared to 0 mg/L of GA3. GA3 plays a promising role in reducing ROS generation in the plant tissues [47, 48]. Furthermore, melatonin application (100 mM and 500 mM) maintaining plant growth and relative water content. More than that, improving the photosynthetic characteristics, total carbohydrate, and total phenolic content in leaf tissues of faba bean plants irrigated with saline water. Melatonin might directly eliminate ROS when produced under stressed conditions [49].

Mineral compounds play too a key role in the structural and functional integrity of the plant to better respond to abiotic stresses. Silicon application enhances drought tolerance of *Kentucky bluegrass* by improving plant water relations and morpho-physiological functions [50, 51]. Moreover, the applications of N fertilizer linearly increased kernel weight during both growing seasons [52, 53]. N deficiency reduces photosynthates production in plants while an optimum N fertilization may increase weight and kernel corn quantity resulting in higher crop growth rates [54]. In addition, a combined application of phosphorus (P) and biochar mitigates heat-induced adversities on physiological, agronomical and quality attributes of rice. The application of biochar+P recorded a grain yield of 7% (plant-1) of rice compared with control throughout different temperature treatments and cultivars. The highest grain production and better grain quality in biochar+P treatments might be due to enhanced water use efficiency, photosynthesis and grain size, which offsetting the adversities of high temperature stress [55–57]. So, fertilization enhances heat injury in rice (*Oryza sativa* L.) [58].

8. Conclusions

In conclusion, water stress seems to be the main limitation factor to growth process in *V faba* L. Under moderate stress (50%CC), the tested ascensions responded similarly on how to cope with water stress. But, El Hamma had the highest performance as compared with Medenine and Mareth accessions. *V. faba* maintains their relative water content at high values and accumulates a high concentration of soluble sugars in their tissues. Accordingly, the dehydration avoidance is considered to be an adaptive mechanism for minimizing water loss. The evaluated *V. faba* accessions responded to water stress also, by reducing shoot and root biomass. Besides, morpho-physiological attributes relative to shoots and roots play an important role in the response to water stress and contributed to drought tolerance of plants. Therefore, it is imperative to identify the specific physiological process of drought tolerance in faba bean, and to define as valuable selection criteria under stressful conditions. Thus, nutritional and economic values of *V. faba* L. can be enhanced in Tunisia.

Acknowledgements

The authors are grateful to Mr. Lachiheb Belgacem. lab technician from the Arid Land Institue for their collaboration in the laboratory analysis.

Conflict of interest

The authors declare no conflict of interest.

Author details

Hanen Enneb[1*], Leila Ben Yahya[1], Mohamed Ilyas[2], Datta Asaram Dhale[3], Mohamed Bagues[1] and Kamel Nagaz[1]

1 Arid and Oases Cropping Laboratory, Arid Lands Institute, IRA, Medénine, Tunisia

2 Department of Agriculture, Hazara University, Mansehra, Khyber Pakhtunkhwa, Pakistan

3 Department of Botany, SSVPS'S, L.K. Dr. P.R. Ghogrey Science College, Dhule, M.S., India

*Address all correspondence to: ennebhanen@gmail.com

References

[1] González-Chavira MM, Herrera-Hernández MG, Guzmán-MaldonadoH,Pons-HernándezJL. Controlled water deficit as abiotic stress factor for enhancing the phytochemical content and adding-value of crops. Scientia Horticulturae. 2018;**234**:354-360. DOI: 10.1016/j.scienta.2018.02.049

[2] Fahad S, Chen Y, Saud S, Wang K, Xiong D, Chen C, et al. Ultraviolet radiation effect on photosynthetic pigments, biochemical attributes, antioxidant enzyme activity and hormonal contents of wheat. J Food, Agri Environ. 2013;**11**(3&4):1635-1641

[3] Jan M, Anwar-ul-Haq M, Adnan NS, Muhammad Y, Javaid I, Xiuling L, et al. Modulation in growth, gas exchange, and antioxidant activities of salt-stressed rice (Oryza sativa L.) genotypes by zinc fertilization. Arabian Journal of Geosciences. 2019;**12**:775 https://doi.org/10.1007/s12517-019-4939-2

[4] Thapa S, Rudd JC, Xue QW, Bhandari M, Reddy SK, Jessup KE, et al. Use of NDVI for characterizing winter wheat response to water stress in a semi arid environement. J. Crop. improve. 2019;**33**:633-648. DOI: 10.1080/15427528.2019.1648348

[5] Tariq M, Ahmad S, Fahad S, Abbas G, Hussain S, Fatima Z, et al. The impact of climate warming and crop management on phenology of sunflower-based cropping systems in Punjab, Pakistan. Agri and Forest Met. 2018;**256**:270-282

[6] Kharrat M, Le Guen J, Tivoli B. Genetics of resistance to 3 isolates of Ascochyta fabae on faba bean (Vicia faba L.) in controlled conditions. Euphytica. 2006;**151**:49-61. DOI: 10.1007/s10681-006-9127-2

[7] Saud S, Chen Y, Long B, Fahad S, Sadiq A. The different impact on the growth of cool season turf grass under the various conditions on salinity and drought stress. Int J Agric Sci Res. 2013;**3**:77-84

[8] López-Bellido FJ, López-Bellido L, López-Bellido RJ. Competition, growth and yield of faba bean (Vicia faba L.). European Journal of Agronomy. 2005;**23**:359-378. DOI: 10.1016/j.eja.2005.02.002

[9] Yahia Y, Guetat A, Elfalleh W, Ferchichi A, Yahia H, Loumerem M. Analysis of agromorphological diversity of southern Tunisia faba bean (Vicia faba L.) germplasm. African Journal of Biotechnology. 2012;**11**:11913-11924. DOI: 10.5897/AJB11.3366

[10] Rebaa F, Abid G, Aouida M, Abdelkarim S, Aroua I, Muhovski Y, et al. Genetic variability in Tunisian populations of faba bean (Vicia faba L. var. major) assessed by morphological and SSR markers. Physiology and Molecular Biology of Plants. 2017;**23**:397-409. DOI: 10.1007/s12298-017-0419-x

[11] Hachana A, Hemissi I, Cherif H, Hlel D, Bouraoui M, Abdi N, Maazaoui H, Cherif A, Sifi B. Diagnosis of Biodiversity of Rhizobium Leguminosarum Nodulating Pea (*Pisum sativum* L.) in Differents Soils of Tunisia. In: Kallel A., Ksibi M., Ben Dhia H., Khélifi N. (eds) Recent Advances in Environmental Science from the Euro-Mediterranean and Surrounding Regions. EMCEI 2017. Advances in Science, Technology & Innovation (IEREK Interdisciplinary Series for Sustainable Development). Springer, Cham. DOI: 10.1007/978-3-319-70548-4-365.

[12] Abid G, Mingeot D, Udupa SM, Muhovski Y, Watillon B, Sassi K, et al. Genetic relationship and diversity analysis of Faba Bean (Vicia Faba L.

var. Minor)genetic resources using morphological and microsatellite molecular markers. Plant Mol. Biol. Rep. 2015;**33**:1755-1767. DOI: 10.1007/s11105-015-0871-0

[13] El-Esawi MA. SR analysis of genetic diversity and structure of thegermplasm of faba bean (*Vicia faba* L.). Comptes Rendus Biologies. 2017;**340**:474-480. DOI: 10.1016/j.crvi.2017.09.008

[14] Mubeen M, Ashfaq A, Hafiz MH, Muhammad A, Hafiz UF, Mazhar S, Muhammad Sami ul Din, Asad A, Amjed A, Fahad S, Wajid N (2020) Evaluating the climate change impact on water use efficiency of cotton-wheat in semi-arid conditions using DSSAT model. J Water Climate Change. DOI: 10.2166/wcc.2019.179/622035/jwc2019179.pdf

[15] Saud S, Chen Y, Fahad S, Hussain S, Na L, Xin L, et al. Silicate application increases the photosynthesis and its associated metabolic activities in Kentucky bluegrass under drought stress and post-drought recovery. Environmental Science and Pollution Research. 2016;**23**(17):17647-17655. DOI: 10.1007/s11356-016-6957-x

[16] Saini KS, Brar NS. Crop and Water Productivity of Sugarbeet (Beta vulgaris) Under Different Planting Methods and Irrigation Schedules. Agric. Res. 2018;7:93-97. DOI: 10.1007/s40003-018-0294-x

[17] Hopkins WG, Hüner NPA. Introduction to Plant Physiology. 4th ed. Hoboken: John Wiley & Sons, Inc.; 2009; 503p 978-0-470-24766-2

[18] Omprakash, R. Gobu, Prashant Bisen, Murlimanohar Baghel and Kumar Nishant Chourasia. 2017. Resistance/Tolerance Mechanism under Water Deficit (Drought) Condition in Plants Int. J. Curr.Microbiol.App. Sci. 2017; 6(4): 66-78. DOI: 10.20546/ijcmas.2017.604.009.

[19] Dodd IC, Ryan AC. Whole-Plant Physiological Responses to Water-Deficit Stress. eLS. 2016:1-9. DOI: 10.1002/9780470015902.a0001298.pub3

[20] Gross J. Pigment in vegetables: chlorophylls and carotenoids. New York: Van Nostrand Reinhold; 1991. 351p. DOI: 10.1007/978-1-4615-2033-7.

[21] Dubois M, Gillesk L, Hmilton JK, Reberg PA, Smith F. Colorimetric method for determination of sugars and related substances. Analatycal Chemistry. 1956;**28**(3):350-356. DOI: 10.1021/ac60111a017

[22] Belachew KY, Nagel KA, Poorter H, Stoddard FL. Association of Shoot and Root Responses to Water Deficit in Young Faba Bean (*Vicia faba* L.) Plants. Front. Plant Science. 2019;**10**(1063):2019. DOI: 10.3389/fpls.2019.01063

[23] Fenta BA, Driscoll SP, Kunert KJ, Foyer CH. Characterization of drought-tolerance traits in nodulated soya beans: The importance of maintaining photosynthesis and shoot biomass under drought-induced limitations on nitrogen metabolism. Journal of Agronomy and Crop Science. 2012;**198**:92-103. DOI: 10.1111/j.1439-037X.2011.00491.x

[24] Thakur J, Shinde BP. Effect of water stress and AM fungi on the growth performance of pea plant. International Journal of Applied Biology. 2020;**4**(1). DOI: 10.20956/ijab.v4i1.9446.

[25] Chavoshi S, Nourmohamadi Gh, Madani H, Heidari Sharif Abad H, Alavi Fazel M. The effect of biofertilizers on physiological traits and biomass accumulation of red beans (*Phaseolus vulgaris* cv.Goli) under water stress'. Iranian Journal of Plant Physiology. 2018; 8(4):2555-2562. DOI: 10.22034/ijpp.2018.543427.

[26] El Jaafari S. Durum wheat breeding for abiotic stresses resistance: Defining

physiological traits and criteria. In: Royo C. (ed.), Nachit M. (ed.), Di Fonzo N. (ed.), Araus J.L. (ed.). Durum wheat improvement in the Mediterranean region: New challenges. Zaragoza:CIHEAM,2000.p 251-256. (Options Méditerranéennes: Série A. Séminaires Méditerranéens; n. 40). Seminar on Durum Wheat Improvement in the Mediterranean Region: New Challenges, 2000/04/12-14, Zaragoza (Spain). http://om.ciheam.org/om/pdf/a40/00600038.pdf.

[27] Karimpour M. Effect of Drought Stress on RWC and Chlorophyll Content on Wheat (Triticum durum L.) Genotypes. World. Ess J. 2019;7:52-56

[28] Ilyas M, Mohammad N, Nadeem K, Ali H, Aamir HK, Kashif H, et al. Drought Tolerance Strategies in Plants: A Mechanistic Approach. J Plant Growth Regulation. 2020. DOI: 10.1007/s00344-020-10174-5

[29] Zhu M, Li FH, Shi ZS. Morphological and photosynthetic response of waxy corn inbred line to waterlogging. Photosynthetica. 2016;**54**:636-640. DOI: 10.1007/s11099-016-0203-0

[30] Nemeskéri E, Helyes L. Physiological Responses of Selected Vegetable Crop Species to Water Stress. Agronomy. 2019;**9**:447. DOI: 10.3390/agronomy9080447

[31] Tairo EV, Mtei KM, Ndakidemi PA. Influence of Water Stress and Rhizobial Inoculation on the Accumulation of Chlorophyll in *Phaseolus vulgaris* (L.) Cultivars. International Journal of Plant & Soil Science. 2017;**15**(4):1-13. DOI: 10.9734/IJPSS/2017/32306

[32] Petridis A, Therios I, Samouris G, Tananaki C. Salinity induced changes in phenolic compounds in leaves and roots of four olive cultivars (Olea europaea L.) and their relationship to antioxidant activity. Environmental and Experimental Botany. 2012;**79**:37-43. DOI: 10.1016/j.envexpbot.2012.01.007

[33] Fahad S, Bajwa AA, Nazir U, Anjum SA, Farooq A, Zohaib A, et al. Crop production under drought and heat stress: Plant responses and Management Options. Frontiers in Plant Science. 2017;**8**:1147. DOI: 10.3389/fpls.2017.01147.

[34] Khodadadi S, Chegini MA, Soltani A. Influence of Foliar Applied Humic Acid and Some Key Growth Regulators on Sugar Beet (*Beta vulgaris* L.) Under Drought Stress: Antioxidant Defense System, Photosynthetic Characteristics and Sugar Yield. Sugar Tech. 2020;**22**:765-772. DOI: 10.1007/s12355-020-00839-6

[35] Enneb H, Belkadhi A, Ferchichi A. Physiological adaptations of henna plant (Lawsonia inermis L.) to different irrigation conditions in Tunisian arid region. The. journal of Animal and Plant Science. 2016;**26**(4):1026-1033

[36] Abdul Qados Amira MS. Effect of salt stress on plant growth and metabolism of bean plant *Vicia faba* (L.). Journal of the Saudi Society of Agricultural Sciences. 2011;**10**:7-15. DOI: 10.1016/j.jssas.2010.06.002

[37] Fahad S, Hussain S, Saud S, Hassan S, Ihsan Z, Shah AN, et al. Exogenously applied plant growth regulators enhance the morphophysiological growth and yield of rice under high temperature. Frontiers in Plant Science. 2016c;**7**:1250. DOI: 10.3389/fpls.2016.01250

[38] Fahad S, Adnan M, Hassan S, Saud S, Hussain S, Wu C, et al. Rice responses and tolerance to high temperature. In: Hasanuzzaman M, Fujita M, Nahar K, Biswas JK, editors. Advances in rice research for abiotic stress tolerance. Woodhead publ ltd, Abington hall Abington, Cambridge CB1 6 AH, Cambs, England. 2019b. pp. 201-224

[39] Turan, V, Rahman, MHU. Rice responses and tolerance to metal/ Metalloid toxicity, in: Hasanuzzaman, M and Fujita, M and Nahar, K and Biswas, JK (Ed.), Advances in rice research for abiotic stress tolerance. Woodhead publ ltd, Abington hall Abington, Cambridge CB1 6 AH, Cambs, England, 2019a; pp; 299-312.

[40] Siddiqui MH, Alamri S, Khan MN, Corpas FJ, Al-Amri AA, Alsubaiea QD, et al. Melatonin and calcium function synergistically to promote the resilience through ROS metabolism under arsenic-induced stress. Journal of Hazardous Materials. 2016;**398**(5):122882. DOI: 10.1016/j.jhazmat.2020.122882.

[41] Khan A, Tan DKY, Afridi MZ. Nitrogen fertility and abiotic stresses management in cotton crop: a review. Environmental Science and Pollution Research. 2017;**24**:14551-14566. DOI: 10.1007/s11356-017-8920-x

[42] Yang Z, Zhang Z, Zhang T, Fahad S, Cui K, Nie L, et al. Huang J (2017) The effect of season-long temperature increases on rice cultivars grown in the central and southern regions of China. Frontiers in Plant Science. 2017;**8**:1908. DOI: 10.3389/fpls.2017.01908

[43] Fahad S, Nie L, Chen Y, Wu C, Xiong D, Saud S, et al. Crop plant hormones and environmental stress. Sustain Agric Rev. 2015b;**15**:371-400

[44] Wu C, Tang S, Li G, Wang S, Fahad S, Ding Y. Roles of phytohormone changes in the grain yield of rice plants exposed to heat: a review. 2019. DOI: 10.7717/peerj.7792.

[45] Fahad S, Bano A. Effect of salicylic acid on physiological and biochemical characterization of maize grown in saline area. Pakistan Journal of Botany. 2012;**44**:1433-1438

[46] Fahad S, Hussain S, Matloob A, Khan FA, Khaliq A, Saud S, et al. Phytohormones and plant responses to salinity stress: A review. Plant Growth Regulation. 2014b;**75**(2):391-404. DOI: 10.1007/s10725-014-0013-y

[47] Saleem MH, Fahad S, Adnan M, Mohsin A, Muhammad SR, Muhammad K, et al. Foliar application of gibberellic acid endorsed phytoextraction of copper and alleviates oxidative stress in jute (Corchorus capsularis L.) plant grown in highly copper-contaminated soil of China. Environ Sci Pollution Res. 2020. DOI: 10.1007/s11356-020-09764-3

[48] Saleem MH, Fahad S, Shahid UK, Mairaj D, Abid U, Ayman ELS, et al. Copper-induced oxidative stress, initiation of antioxidants and phytoremediation potential of flax (Linum usitatissimum L.) seedlings grown under the mixing of two different soils of China. Environ Sci Poll Res. 2020c. DOI: 10.1007/s11356-019-07264-7

[49] Ahmad S, Kamran M, Ding R, Meng X, Wang H, Ahmad I, Fahad S, Han Q Exogenous melatonin confers drought stress by promoting plant growth, photosynthetic capacity and antioxidant defense system of maize seedlings. 2019. DOI: 10.7717/peerj.7793.

[50] Saud S, Li X, Chen Y, Zhang L, Fahad S, Hussain S, et al. Silicon application increases drought tolerance of Kentucky bluegrass by improving plant water relations and morph physiological functions. Scientific World Journal. 2014:1-10 https://doi.org/10.1155/2014/368694

[51] Zahida Z, Hafiz FB, Zulfiqar AS, Ghulam MS, Fahad S, Muhammad RA, et al. Effect of water management and silicon on germination, growth, phosphorus and arsenic uptake in rice. Ecotoxicology and Environmental Safety. 2017;**144**:11-18

[52] Hafiz MH, Farhat A, Ashfaq A, Hafiz FB, Wajid F, Carol Jo W, et al.

Predicting Kernel Growth of Maize under Controlled Water and Nitrogen Applications. Int J Plant Prod. 2020a. DOI: 10.1007/s42106-020-00110-8

[53] Hafiz MH, Farhat A, Shafqat S, Fahad S, Artemi C, Wajid F, et al. Offsetting land degradation through nitrogen and water management during maize cultivation under arid conditions. Land Degradation and Development. 2018:1-10. DOI: 10.1002/ldr.2933

[54] Saud S, Fahad S, Cui G, Chen Y, Anwar S. Determining nitrogen isotopes discrimination under drought stress on enzymatic activities, nitrogen isotope abundance and water contents of Kentucky bluegrass. Scientific Reports. 2020;**10**:6415. DOI: 10.1038/s41598-020-63548-w

[55] Fahad S, Hussain S, Saud S, Tanveer M, Bajwa AA, Hassan S, et al. A biochar application protects rice pollen from high-temperature stress. Plant Physiology and Biochemistry. 2015a;**96**:281-287

[56] Fahad S, Hussain S, Saud S, Hassan S, Tanveer M, Ihsan MZ, et al. A combined application of biochar and phosphorus alleviates heat-induced adversities on physiological, agronomical and quality attributes of rice. Plant Physiology and Biochemistry. 2016d;**103**:191-198

[57] Hesham FA and Fahad S. Melatonin application enhances biochar efficiency for drought tolerance in maize varieties: Modifications in physio-biochemical machinery. Agron J 1-22. 2020. DOI: 10.1002/agj2.20263.

[58] Wu C, Kehui C, She T, Ganghua L, Shaohua W, Fahad S, et al. Intensified pollination and fertilization ameliorate heat injury in rice (*Oryza sativa* L.) during the flowering stage. Field Crops Res. 2020;**252**:107795. DOI: 10.1016/j.fcr.2020.107795

Chapter 5

Abiotic Stress Responses in Plants: Current Knowledge and Future Prospects

Deeksha Marothia, Navdeep Kaur and Pratap Kumar Pati

Abstract

Exposure to abiotic stresses has become a major threatening factor that hurdles the sustainable growth in agriculture for fulfilling the growing food demand worldwide. A significant decrease in the production of major food crops including wheat, rice, and maize is predicted in the near future due to the combined effect of abiotic stresses and climate change that will hamper global food security. Thus, desperate efforts are necessary to develop abiotic stress-resilient crops with improved agronomic traits. For this, detailed knowledge of the underlying mechanisms responsible for abiotic stress adaptation in plants is must required. Plants being sessile organisms respond to different stresses through complex and diverse responses that are integrated on various whole plants, cellular, and molecular levels. The advanced genetic and molecular tools have uncovered these complex stress adaptive processes and have provided critical inputs on their regulation. The present chapter focuses on understanding the different responses of the plants involved in abiotic stress adaptation and strategies employed to date for achieving stress resistance in plants.

Keywords: plants, abiotic stress, photosynthesis, reactive oxygen species, ion transport, osmoregulation

1. Introduction

Plants often experience unfavorable environmental conditions such as high salinity, drought, cold, heat, depletion of soil nutrients, and excess of toxic ions, etc. that hamper the plant growth and development [1–3]. These stresses not only play a major role in determining the crop yield and productivity but they also contribute to the differential distribution of plant species across different parts of the earth [4]. About 90% of the arable lands around the globe are susceptible to one or more of the above stresses causing up to 70% annual yield loss of major food crops [5]. The changing climate is further aggravating the impact of abiotic stress factors on the overall growth and development of various crops [6]. It is believed that exposure to salt stress in irrigated lands has been increased by 37% during the last 20 years [7]. Moreover, the occurrence of drought is increased due to alteration in the evapotranspiration and pattern of precipitation caused by global warming [8]. As per a recent meta-analysis study, a further increase of 2.0 to 4.9°C in the average earth temperature by 2100 is speculated which will further impose a huge challenge for sustainable agriculture in the future [9].

Plants respond to different environmental constraints through complex intricate mechanisms [1]. The ability of plants to adjust to different environmental conditions is directly or indirectly related to two major plant strategies - plant stress avoidance and plant stress tolerance. Plant's stress avoidance is a physiologically non- active phase like mature seeds, while stress tolerance is an active reversible adjustment which is generally referred to as acclimation [10]. Acclimation to stress is particularly mediated through profound changes at the level of gene expression which results in changes or modifications in the composition of plant transcriptome, proteome as well as metabolome [11]. During the last few decades, researchers have focused on recognizing and elucidating the different components and molecular partners underlying abiotic stress responses in plants [12]. Several attempts have been made to produce crops/species with improved abiotic stress adaptive traits including drought and salinity. However, one of the massive challenges in modern sustainable agriculture is the development of abiotic stress-resilient crops with new and desired agronomical traits using different approaches. For this purpose, understanding the mechanisms by which plants perceive stress signals and further transmit them to cellular machinery for activating adaptive responses is of huge importance [13–16]. In this context, marrying the various physiological, biochemical, and gene regulatory network knowledge is essential that will aid up in the development of stress-tolerant high-yielding food crop cultivars [17, 18]. Therefore, a holistic understanding of the different responses associated with abiotic stress adaptation by taking advantage of various available high throughput tools like proteomics, metabolomics, and transcriptomics is critical. Hence, the present chapter deals with the various responses associated with abiotic stress stimuli in plants and the current status, and future prospects of different approaches used to date for developing stress-resilient crops.

2. Plant's responses to abiotic stresses

Plants face several types of variations in their physical environment that hampers their growth and development. They respond to these oscillating environmental conditions through a series of external and internal changes [19, 20]. These stress-specific responses are associated with an array of molecular players that modulates the morphology, anatomy, and physiology of plants [12, 13].

2.1 Responses at the level of cellular membranes

Plant cells can sense changing environmental signals leading to significant changes in their physiology, metabolism, and gene expression [12, 13]. The stress stimuli are first perceived at the level of cellular membranes that initiates a cascade of events to transmit the signal to various organelles thus activating the appropriate molecular network [21]. In plants, the primary cell wall is composed of cellulose fibrils connected by hemicellulose tethers embedded in a pectin gel providing mechanical strength for load-bearing. It also contains several structural proteins, phenolics, and calcium [22]. These components are often modified when plants are exposed to abiotic stresses. The overall architecture of the cell wall is affected by exposure to abiotic stress depending upon the species, the stress intensity, plant phenotype, plant genotype as well as the age of plant. It appears to result in both loosening and tightening of the cell wall [23].

The viscoelastic properties of the primary cell wall are improved by elevating the levels of cell wall remodeling and biosynthetic enzymes, and by modulating the other cell wall loosening agents such as pectin, thus contributing to higher hydration status of the plant which aids up in maintaining turgor pressure necessary for growth [23]. The viscoelastic properties are also modulated by reinforcement of the secondary wall with the accumulation of cellulose and non-cellulosic components. In response to abiotic stress stimuli, the biosynthesis of xyloglucan (the most abundant non-cellulosic components of type I primary walls), and cellulose is induced [24, 25]. It is associated with an up-regulation of EXP (expansin), XTH (xyloglucan endo-β-transglucosylases/hydrolases) and Ces A (Cellulose Synthase) encoding genes [25] Moreover, the comparative analysis of changes in the cell wall of two- different drought-resistant varieties of wheat under stress showed an increase in pectin polymers RGI and RGII (rhamnogalacturonan I and II) side chains that probably leads to hydrogel formation of pectin, limiting the damage to the cells [26]. Also, methyl esterification of homogalacturonan (HG) levels regulated by PME (pectin methylesterase) reduces upon exposure to stress stimuli [27]. Such modifications in the cell wall architecture lead to relative maintenance of cell wall extensibility required to cope up with particular abiotic stress. Moreover, the genes encoding for cell wall proteins including arabinogalactan protein (AGP), glycine-rich protein (GRP), and proline-rich protein (PRP) are also induced in response to abiotic stress that could contribute to the cell wall strengthening [23].

One of the alternative responses against abiotic stress stimuli is to decrease the cell wall expansion and cell extensibility that can thus limit the water loss and prevent cell collapse due to dehydration stress [23, 28]. A decrease in cell wall extensibility or turgor pressure is often associated with the rigidification of the secondary cell wall by lignin deposition. As monolignols are the building blocks of lignin, they are synthesized from phenylalanine through the general phenylpropanoid and monolignol-specific pathways in the cytosol. The monolignols are then transported to the cell wall where they are polymerized by apoplastic peroxidase (PRX) and laccases into lignin [23].

A large number of integral plasma membrane proteins are also known to participate in stress perceptions which are the members of different receptor-like kinases RLKs (receptor-like kinases) [29]. Abiotic stresses are often responsible for alterations in wall-associated kinases (WAK) that are required for cell elongation and development [22]. In plants exposed to abiotic stresses, the expression of genes encoding for WAK proteins is up-regulated hinting towards the perception of stress at the cell wall or plasma membrane interface through the detection of released plant cell wall fragments [24, 30]. Thus, it can be concluded that modulation of the cell wall architecture is often a direct response that plays a vital role in the sensitization of the plant against abiotic stress stimuli. However, critical information on understanding this response comes from transcriptomics rather than biochemical analysis [26]. Therefore, a multidisciplinary approach is required for gaining an in-depth knowledge of this complex mechanism in the future.

2.2 Modulation of photosynthetic apparatus and gaseous parameters

Plants suffer numerous physiological reactions on exposure to environmental stress. These responses include alterations in photosynthetic rates, assimilate translocation, nutrient uptake and translocation, changes in water uptake, and evapotranspiration [31]. Among these, photosynthesis is one of the most critical plant processes affected by various abiotic stresses [31, 32]. These stresses negatively influence the photosystems (PS I and PS II) thus reducing the photosynthetic activity along with reduced chlorophyll biosynthesis, and photosynthetic electron transport. They also lead to impaired RuBp (ribulose 1,5-bisphosphate) regeneration that substantially affects the Rubisco activity. Generally, the stress-derived inhibitory effects on photosynthesis in plants may occur due to limitations in CO_2

diffusion factors and/or metabolic factors. Some reports provide evidence that stomatal closure is the key event under stress conditions resulting in a decrease in the sub-stomatal as well as chloroplast CO_2 concentration (Ci and Cc, respectively) thus producing a decline in CO_2 assimilation [32–36].

Under moderate drought stress, decreased stomatal conductance (g_s) is considered as the primary cause of photosynthetic inhibition from reduced supply of CO_2 to the intercellular space. In general, atmospheric CO_2 diffuses to the intercellular space (i.e. stomatal limitation) through stomata and then across the mesophyll (mesophyll limitations) at the carboxylation site [31]. Thus, mesophyll conductance (g_m) and biochemical limitation (b_L) (often termed as non-stomatal limitations to photosynthesis mainly under high water stress) have gained importance in the recent years, however, their relative importance to photosynthesis limitation has been a subject of debate [31, 36, 37]. Although, the function of non-stomatal limitations to photosynthesis is evident, however, controversies still exist because of the error and assumptions in the estimation of g_m and b_L under stress conditions [38].

2.3 Ion stress signaling and homeostasis

Abiotic stresses particularly salt and heavy metal stress are majorly responsible for an imbalance in ionic composition inside the plant cells [10]. For a normal metabolic function of plants, cells need to maintain high K^+ and low Na^+ levels. Thus, systematic exclusion of excess Na^+ ions from the cytoplasm or their accumulation within the vacuoles are the main adaptive mechanisms against ionic stress in plants [21]. This occurs through a highly sophisticated mechanism of ion homeostasis which involves the interplay of different molecular players. Ion homeostasis is maintained by ion pumps like symporters, antiporters, and carrier proteins located on the cell membranes [39]. At the plasma membrane of the cell, the stress signal is perceived by a sensor or a receptor which is generally regulated by the coordination of various ion pumps [40]. Exclusion of ions is typically carried out by transmembrane transport proteins excluding Na^+ from the cytosol, however, compartmentalization is carried out by H^+- pyrophosphatase proteins and vacuolar membrane H^+ -ATPase [12].

Salt Overly Sensitive also known as SOS pathway is an excellent example of intracellular ion management or homeostasis which is turned 'on' after the activation of the receptor in response to stress and transcriptional induction of genes by signaling intermediate compounds along with certain downstream interacting partners which result in the efflux of excess ions [41]. *SOS1, SOS2,* and *SOS3* are the three genes encoding for SOS proteins, which work in a synchronized manner and aids in the transportation of Na^+ ions from the cytoplasm by effluxing excess of Na^+ ions using a plasma membrane Na^+/H^+ antiporter. This pathway is triggered by the high concentrations of Na^+ ions perceived by the intracellular calcium (Ca^{2+}) ion signals. The high concentration of sodium chloride (NaCl) disturbs the intracellular levels of Ca^{2+} via hypothetical plasma membrane sensors. This Ca^{2+} signal is then recognized and interpreted by the SOS3 protein which belongs to the calcineurin B-like protein (CBLs) family which in association with SOS2 activates the SOS1 [42]. SOS1 encodes for a Na^+/H^+ antiporter and various studies have confirmed the functional role of SOS1 in maintaining the homeostatic balance of ions during salt stress adaptation [43]. The vacuolar Na^+/H^+ and $H^{+/}Ca^{2+}$ antiporters are also known to be differentially regulated by SOS2, thus contributing to enhanced Na^+ ions sequestration in vacuole imparting salinity tolerance. Furthermore, the SOS2/SOS3 kinase complex is responsible for the down-regulation of the activity of Na^+ ion transporters, mediating the entry of these ions into the cells of root tissue during salinity. Apart from the well-established function of ion homeostasis, SOS proteins have also been known to play novel functions during stress acclimatization including regulation of cell cytoskeleton dynamics, development of lateral roots via modulation of auxin gradients as well as maxima in roots under moderate salt stress [43].

In plants, potassium (K^+) is one of the most abundant inorganic cations involved in various aspects of plant growth and development including abiotic stress management [44]. Thus, the maintenance of K^+ homeostasis through K^+ ion transporters and channels across the plasma membrane is necessary for the survival of plants, especially during stress conditions [45]. Plants have developed a unique transport system for K^+ acquisition and release using the high-affinity K^+ uptake transporters (HKTs) [46]. There are two sub-groups of these transporters (class I and class II) which have been identified to play a critical role in selective Na^+ ion transport and cationic co-transport of Na^+/K^+, respectively [12]. They also play a significant role in the maintenance and distribution of Na^+ ions between plant shoots and roots [47]. In *Arabidopsis thaliana* (Arabidopsis) knockout mutations in the *AtHKT1* gene along with *AtSOS1* gene {induced either by T-DNA insertion or ethyl methane sulphonate (EMS) treatment} lead to over-deposition of Na^+ ions in leaves due to the decreased amount of Na^+ ions in roots under salt stress [48].

Cl^- is a plant micronutrient which regulates turgor pressure, leaf osmotic potential, and stimulates growth in plants by acting as a critical messenger in plant developmental processes [49]. Cl^- ion signaling and transporters also regulate different pathways conferring abiotic stress tolerance in plants [50]. For instance, as an early salt stress response, the Cl^- ion signal in the soil with elevated salt concentration has been connected to stomatal closure in an ABA dependent manner [21]. However, increased deposition of these ions during ionic stress is detrimental to plant growth and development [51]. Thus, plants tend to decrease the net levels of Cl^- ions during stress through reduced net Cl^- uptake by roots, decreased intracellular compartmentation, reduced net xylem loading of Cl^-, and phloem recirculation and translocation [52]. Also, inside the cytosol, threshold levels of Cl^- ions are maintained primarily through its sequestration with the help of ion transporters and voltage-gated ion channels inside the vacuole [53]. A voltage gradient is maintained between the vacuole and the cytoplasm because of a slightly positive charged vacuole and a negatively charged cytoplasm. Hence, a large number of the Cl^- ions are sequestered through voltage-gated anion channels of the CLC family which are present on the tonoplast. Different CLC proteins function as anion/H^+ exchangers or anion-selective channels. In reports, AtCLCa has been characterized as a two-anion/H^+ exchanger which drives the active uptake of anions inside the vacuoles of Arabidopsis guard cells and mesophyll with higher selectivity for NO_3^- ions over Cl^- ions [54]. Besides, CLCs play a vital role in loading anions in the vacuole of guard cells for stomatal opening in response to light and later releasing them during ABA-induced stomatal closure [55].

2.4 Intracellular osmotic adjustment and osmoprotectants

The intracellular water loss from the cell due to drought and salinity stress results in cellular dehydration thus imposing osmotic stress in plants [56]. To counteract the effects of osmotic stress, plants and bacteria accumulate certain organic solutes like quaternary ammonium compounds, polyamines, fructose, sucrose, sugar alcohols, trehalose, fructans, oxalate, malate, and many others. These metabolites are referred as osmoprotectants or compatible solutes and may accumulate in large quantities without disturbing the intracellular biochemistry [57]. Among these osmoprotectants, quaternary ammonium compounds including proline and glycine betaine (GB) abundantly accumulate in response to abiotic stresses. The imino acid proline is known to be deposited in considerable amounts in plant cells under the influence of drought, salinity, and other stresses [58]. It is synthesized inside the cytoplasm and plastids while it is degraded to glutamate (Glu) in the mitochondria. In addition to its role in osmotic adjustment, proline contributes in the stabilization of the cellular membranes and vital proteins by making clusters with water molecules that later get attached to membranes and proteins, thus, inhibiting their denaturation [59, 60]. Proline also scavenges free radicals to maintain or buffer the redox potential inside the cell under stressful conditions. It alleviates the cytoplasmic acidosis and sustains $NADP^+$/NADPH ratios at required levels for cellular metabolism, hence, supporting redox cycling [60, 61]. Researchers have observed a positive correlation between proline deposition and tolerance against various abiotic stresses in plants [58]. Furthermore, the exogenous application of proline has been used as an effective approach to improve stress tolerance in plants [62].

GB is another critical compound that plays an important role in osmoprotection, stroma adjustment as well as protection of thylakoid membranes for maintaining the photosynthetic activity during stress conditions [63, 64]. It protects the photosystem II (PS-II) complex from the impact of abiotic stresses [65]. GB also possesses a protective role for Rubisco against heat-induced destabilization [65]. The increased accumulation of GB provides abiotic stress resistance in several agronomically important crops including tobacco, potato, tomato, barley, and maize [11, 66, 67]. Moreover, the *Arabidopsis thaliana, Nicotiana tabacum,* and *Brassica napus* plants transformed with bacterial *choline oxidase* cDNA were found to show 5 to 10% increased levels of GB than the naturally found levels of GB in them that moderately improved their tolerance against different abiotic stresses [68].

The content of soluble carbohydrates also varies in response to abiotic stresses in plants. Simple and complex carbohydrates such as sugars, starch, and sugar alcohols accumulate under stress conditions in plants [68]. The major roles of these biomolecules are osmotic adjustment, carbon storage, and free radical scavenging. Their pattern of accumulation in response to stress varies under short- and long-term reactions. In short-term water stress conditions, decreased content of sucrose and starch were observed in the case of *Setaria sphacelata*, which is a naturally adapted C4 grass whereas an increased amount of soluble sugars and decreased amount of starch were reported under long term stress imposition [69]. Trehalose is a rare non-reducing sugar that occurs in some desiccation-tolerant higher plants along with various bacterial and fungal species [70]. It shows significant accumulation in plants in response to various environmental stimuli and acts as an osmolyte thus protecting the plant cells. It also protects the protein functioning by reducing the aggregation of denatured proteins and safeguards the biological molecules from the changing environmental stresses through its reversible water-absorption capacity [68, 71]. The sugar alcohols also show considerable accumulation in response to abiotic stress in plants and help in osmotic adjustment [72]. Mannitol, a sugar alcohol, accumulates upon salt and water stress conditions in plants. Wheat transgenics, expressing the *mtlD* gene (*mannitol-1-phosphate dehydrogenase*) of *Escherichia coli* showed significantly more tolerance towards salt as well as water stress. Upon analysis, increased plant height, biomass, and the number of secondary stems were observed in transgenic wheat [72].

Polyamines are small organic molecules ubiquitously present in all living organisms which play a vital role in diverse cellular processes. They are positively charged at physiological pH and are regarded as growth substances [73–75]. Under stress conditions, different plant species respond differently to polyamines levels. Some of the plants might increase the content of polyamines under stress conditions whereas others decrease their levels of endogenous polyamines when exposed to severe environmental conditions [73]. Exogenous application of polyamine and/or inhibitors of enzymes which are involved in polyamine biosynthesis also hints towards a possible role of such compounds in plant adaptation or defense process in response to environmental stresses [76]. Moreover, studies involving either transgenic overexpression or loss of function mutants support the protective, adaptive, or defensive role of polyamines in plant's response to various abiotic stresses [76, 77].

2.5 Reactive oxygen species (ROS) regulation during stress acclimation

Many evidences suggest that various environmental stresses lead to the generation of ROS in plants. Actually, in plants, each cellular compartment is equipped with its own ROS homeostasis control [78–80]. The ROS signaling is changed depending upon the cell type, developmental stage, and level of stress [81]. Under optimal growth conditions, ROS inside the cell is mainly produced at a low level in organelles like chloroplast, mitochondria, and peroxisomes [82]. It has been estimated that 1–2% of the O_2 consumed by plant tissues, leads to the ROS formation that mainly involves $^1O_2$, H_2O_2, $O^{\bullet-}{}_2$, and OH• [83, 84]. At this low concentration, ROS acts as a signaling molecule that triggers signal transduction pathways involved in growth and development [21, 85]. However, in response to various abiotic stresses, the generation of increased levels of ROS causes irreversible damage to cells through their strong oxidative properties [86]. They possess lethal properties and cause extensive damage to DNA, proteins, and lipids thereby affecting normal cellular functioning [82]. Plants have developed an elaborate and efficient network of ROS generating and scavenging mechanisms to overcome this ROS toxicity. The two systems interplay with each other for maintaining a steady state in plants during stress acclimation [87, 88]. The delicate balance between the generation of ROS and its scavenging is responsible for duality in its function in plants which is orchestrated by a giant network of genes known as 'ROS gene network' [84].

Plant NADPH oxidases also referred as respiratory burst oxidase homologs (RBOHs) are the most studied enzymatic source of ROS in plants [88]. These are superoxide-producing enzymes that are widely involved in various processes including abiotic stress responses in plants [89]. The superoxide radical is a short-lived ROS molecule that is characterized by moderate reactivity and can trigger a series of reactions to produce other ROS species. It is produced inside mitochondria, chloroplasts, endoplasmic reticulum, and peroxisomes as a result of their normal metabolism [90]. The activity of plant NADPH oxidase is regulated by some key regulatory components like Ca^{2+}, calcium-dependent protein kinases (CDPKs), $Ca^{2+/}$CaM-dependent protein kinase, some small GTPases, and others. The production of ROS through NADPH oxidase may result in regulating the acclimation to abiotic stresses in plants. For instance, in barley, NADPH oxidase-mediated apoplastic ROS generation (acting upstream of xylem Na^+ loading) that is linked to ROS-inducible Ca^{2+} uptake systems in the xylem parenchyma tissue is considered as a critical factor contributing to salt stress tolerance in plants [91]. In *Arabidopsis*, the double mutants of *AtRbohD* and *AtRbohF* genes with significantly inhibited ROS generation exhibited less growth and relatively higher cellular Na^+ to K^+ ratios than the wild-type (WT) as well as a single null mutant *ATrbohd* and *ATrbohf* plants under salt stress [92].

Superoxide ions generated by NADPH oxidase are converted to hydrogen peroxide (H_2O_2), catalyzed by the different isoforms of superoxide dismutase (SOD) enzyme [93]. H_2O_2 production in plant cells not only occurs under normal conditions but also by oxidative stress which is caused by different abiotic factors. The major sources of H_2O_2 production in plant cells comprises of the electron transport chain in the chloroplast, endoplasmic reticulum (ER), mitochondria, cell membrane, β-oxidation of fatty acid, and photorespiration along with various other sources including reactions comprising photo-oxidation by NADPH oxidase. The rates of H_2O_2 accumulation in peroxisomes, as well as chloroplasts, may be 30–100 times higher as compared with H_2O_2 generated in the mitochondria. It acts as a systemic signal that alerts various plant tissues to respond and adapt in response to the upcoming stress stimuli [94, 95]. H_2O_2 confer acclamatory stress tolerance by regulating osmotic adjustment, photosynthesis, ROS detoxification, and phytohormones signaling [95]. Studies have suggested that seeds pre-treated with H_2O_2, or together with the application of H_2O_2 and abiotic stress, induce an inductive pulse which aids up in protecting plants under abiotic stresses by the restoration of redox-homeostasis and mitigation of oxidative damage to membranes, lipids, and proteins by modulating the stress signaling pathways [95].

The stress-induced ROS activating responses occur rapidly with the appearance of the stress and it should decay immediately to protect the plants against their toxic effects. For this, plants are equipped with an array of ROS detoxifying proteins that mitigate the toxic effects of ROS generated as a result of different types of stresses [96]. In plants, the redox homeostasis during stressful conditions is maintained by the two arms of the antioxidant machinery—the enzymatic components consisting of the superoxide dismutase (SOD), guaiacol peroxidase (GPX), ascorbate peroxidase (APX), catalase (CAT), glutathione-S-transferase (GST), and the non-enzymatic molecular compounds like reduced glutathione (GSH), ascorbic acid (AA), α-tocopherol, phenolics, carotenoids, flavonoids, and proline. These antioxidant enzymes are situated in different sites of the plant cells and work together to detoxify ROS. The omnipresent behavior of both arms of the antioxidant machinery explains the basic necessity of detoxification of ROS for cell survival [97].

3. Strategies to combat abiotic stresses in plants

Various strategies have been undertaken by the researchers from time to time to improve the abiotic stress tolerance in plants, particularly crop plants [98]. Plant breeding is the most traditional and widely used method for achieving the desired trait in given plants including stress adaptation [99]. However, the success of crop-breeding programs greatly depends on the availability of natural genetic variations among the germplasm resources and tedious selection procedures that are too slow and equally expensive [100]. Moreover, the various environmental factors such as plant developmental stage along with the logistical constraints of physiological screening of large breeding populations on a field-scale can affect the differential selection of a particular stress tolerant plant. Thus, plant breeding is almost always limited by the genetic complexity of the underpinning mechanisms along with the potential interaction among genetic determinants [101]. In this regard, the identification and recognition of discrete chromosomal regions having a major effect on the specific tolerance trait via quantitative trait loci (QTL) mapping and marker-assisted selection remain a valuable option for the success of many breeding programs [102]. Although, QTL mapping holds great promise, but still it remains complicated as the introgression of QTL regions in elite lines is tedious due to linkage drag that may introduce non-target regions. As an alternative, the cellular-based mutant introduction and subsequent selection under controlled *in vitro* conditions offer a method to quickly screen large populations with homogeneous backgrounds for novel fortuitous changes related to tolerance. Subsequent field screening then ensures the adequate performance of the tolerance trait under the external potentially mitigating factors [103].

In the past few decades, the genetic engineering approach has attracted the interest of the research community for producing stress-tolerant elite crops [104].

Genetic transformation with stress-inducible genes has been employed by the researchers to gain an understanding of their functional role in stress tolerance and ultimately to improve the traits in the target genotype [105]. The genetic manipulation techniques including insertional mutagenesis have largely contributed to deciphering the function of genes and thereby identifying the suitable candidates for crop improvement [106]. However, though success has been achieved in introducing desired tolerance traits into various crop varieties from wild relatives like barley and tomato, a restricted success has been reported in achieving abiotic stress tolerance with elite germplasm [107]. Moreover, the integration of transgenes into the host genome is sometimes non-specific and unstable [108]. Recently, the use of targeted genome editing using clustered regularly interspaced short palindromic repeats (CRISPR) and CRISPR-associated protein9 nuclease (Cas9) (CRISPR/Cas) has generated a lot of interest in various fields of plant biology including abiotic stress management [109]. CRISPR/Cas has been adopted in the field of plant developmental biology for characterizing genes as well as to underpin the molecular mechanisms behind various plant traits [110]. It has been used in the model plants such as Arabidopsis and tobacco earlier and likewise, now it is being utilized effectively for crop plants like sorghum, rice, wheat, maize, soybean as well as woody plants. Researchers have worked on the potential use of the CRISPR/Cas9 technique for the production of abiotic stress-tolerant crops by targeting the key sensitivity (*S* genes and *cis*-regulatory sequences) and tolerance (*T* genes) players. In general, *T* genes are deployed to achieve stress tolerance in plants; however, the *S* genes negatively regulates the biological function of the *T* genes. Therefore, the silencing of *S* genes to disturb their functioning can help plants to adjust their physiological and biochemical pathways for providing tolerance in response to abiotic stress [111]. Like *S* genes, various *cis*-regulatory sequences have also been identified that negatively regulates abiotic stress tolerance mechanisms. These sequences are highly conserved and help in the regulation of gene expression by interacting with specific transcription factors [111]. Thus, editing such *cis*-regulatory sequences can also serve as a potential strategy for improving stress tolerance in plants. However, one major limitation of genome editing is the off-target mutations that are caused by Cas9 in transgenic plants. This limitation has been overcome to a considerable extent by the advent of stress-inducible CRISPR/Cas9 technique which reduces the rate of off-target mutations to negligible levels [112]. Thus, we can consider stress-inducible CRISPR/Cas as a promising tool for precise and efficient genome editing in crop plants for numerous traits, including abiotic stress tolerance.

4. Conclusion

In the last few decades, significant progress has been made in our understanding of the complex mechanisms governing abiotic stress tolerance in plants. However, still we are far from pinning the exact battery of gene activation mechanisms responsible for providing tolerance to various abiotic stresses. Our struggle to understand the complex mechanisms is ongoing and recent development of new tools for high-throughput phenotyping and genotyping gives us a new ray of hope. A complete understanding of the physiological, biochemical and molecular mechanisms especially the signaling cascades in response to abiotic stresses in tolerant plants will help to manipulate susceptible crop plants and increase agricultural productivity in the near future. Moreover, advances in genomics strategies including genetic engineering and genome editing have provided new opportunities for crop improvement by employing precise genome engineering for targeted traits in crop plants. However, the selection of the key genes is critical for the success of these approaches.

Acknowledgements

Authors are thankful to the Department of Biotechnology, GOI, and Rashtriya Uchchattar Shiksha Abhiyan (RUSA-II) Program, Ministry of Human Resource Development (MHRD), GOI.

Conflict of interest

The authors declare no conflict of interest.

Author details

Deeksha Marothia[†], Navdeep Kaur[†] and Pratap Kumar Pati*
Department of Biotechnology, Guru Nanak Dev University, Amritsar, Punjab, India

*Address all correspondence to: pkpati@yahoo.com

† These authors contributed equally.

References

[1] Chang YN, Zhu C, Jiang J, Zhang H, Zhu JK, Duan CG. Epigenetic regulation in plant abiotic stress responses. Journal of Integrative Plant Biology. 2020 May;62(5):563-80.

[2] Vaughan MM, Block A, Christensen SA, Allen LH, Schmelz EA. The effects of climate change associated abiotic stresses on maize phytochemical defenses. Phytochemistry Reviews. 2018 Feb 1;17(1):37-49.

[3] Zafar SA, Hameed A, Nawaz MA, Wei MA, Noor MA, Hussain M. Mechanisms and molecular approaches for heat tolerance in rice (*Oryza sativa* L.) under climate change scenario. Journal of Integrative Agriculture. 2018 Apr 1;17(4):726-38.

[4] Kimotho RN, Baillo EH, Zhang Z. Transcription factors involved in abiotic stress responses in Maize (*Zea mays* L.) and their roles in enhanced productivity in the post-genomics era. PeerJ. 2019 Jul 8;7:e7211.

[5] Waqas MA, Kaya C, Riaz A, Li YE. Potential mechanisms of abiotic stress tolerance in crop plants induced by thiourea. Frontiers in plant science. 2019;10:1336.

[6] Fahad S, Bajwa AA, Nazir U, Anjum SA, Farooq A, Zohaib A, Sadia S, Nasim W, Adkins S, Saud S, Ihsan MZ. Crop production under drought and heat stress: plant responses and management options. Frontiers in plant science. 2017 Jun 29;8:1147.

[7] Qadir M, Quillérou E, Nangia V, Murtaza G, Singh M, Thomas RJ, Drechsel P, Noble AD. Economics of salt induced land degradation and restoration, natural resources forum.

[8] Dai A. Drought under global warming: a review. Wiley Interdisciplinary Reviews: Climate Change. 2011 Jan;2(1):45-65.

[9] Raza A, Razzaq A, Mehmood SS, Zou X, Zhang X, Lv Y, Xu J. Impact of climate change on crops adaptation and strategies to tackle its outcome: A review. Plants. 2019 Feb;8(2):34.

[10] Kosová K, Vítámvás P, Prášil IT, Renaut J. Plant proteome changes under abiotic stress—contribution of proteomics studies to understanding plant stress response. Journal of proteomics. 2011 Aug 12;74(8): 1301-22.

[11] Koyro HW, Ahmad P, Geissler N. Abiotic stress responses in plants: an overview. InEnvironmental adaptations and stress tolerance of plants in the era of climate change 2012 (pp. 1-28). Springer, New York, NY.

[12] Haak DC, Fukao T, Grene R, Hua Z, Ivanov R, Perrella G, Li S. Multilevel regulation of abiotic stress responses in plants. Frontiers in plant science. 2017 Sep 20;8:1564.

[13] Mantri N, Patade V, Penna S, Ford R, Pang E. Abiotic stress responses in plants: present and future. InAbiotic stress responses in plants 2012 (pp. 1-19). Springer, New York, NY.

[14] Anwar A, Kim JK. Transgenic Breeding Approaches for Improving Abiotic Stress Tolerance: Recent Progress and Future Perspectives. International journal of molecular sciences. 2020 Jan;21(8):2695.

[15] Le BH, Wagmaister JA, Kawashima T, Bui AQ, Harada JJ, Goldberg RB. Using genomics to study legume seed development. Plant physiology. 2007 Jun 1;144(2):562-74.

[16] Fahad S, Hussain S, Matloob A, Khan FA, Khaliq A, Saud S, Hassan S, Shan D, Khan F, Ullah N, Faiq M. Phytohormones and plant responses to salinity stress: a review.

Plant growth regulation. 2015 Mar 1;75(2):391-404.

[17] Yadav S, Modi P, Dave A, Vijapura A, Patel D, Patel M. Effect of Abiotic Stress on Crops. InSustainable Crop Production 2020 Jun 17. IntechOpen.

[18] Duque AS, de Almeida AM, da Silva AB, da Silva JM, Farinha AP, Santos D, Fevereiro P, de Sousa Araújo S. Abiotic stress responses in plants: unraveling the complexity of genes and networks to survive. Abiotic stress-plant responses and applications in agriculture. 2013 Mar 13:49-101.

[19] Rastogi S, Shah S, Kumar R, Vashisth D, Akhtar MQ, Kumar A, Dwivedi UN, Shasany AK. Ocimum metabolomics in response to abiotic stresses: Cold, flood, drought and salinity. PloS one. 2019 Feb 6;14(2):e0210903.

[20] Hotta CT, Gardner MJ, Hubbard KE, Baek SJ, Dalchau N, Suhita D, Dodd AN, Webb AA. Modulation of environmental responses of plants by circadian clocks. Plant, cell & environment. 2007 Mar;30(3):333-49.

[21] Kaur N, Pati PK. Integrating classical with emerging concepts for better understanding of salinity stress tolerance mechanisms in rice. Frontiers in Environmental Science. 2017 Jul 5;5:42.

[22] Rui Y, Dinneny JR. A wall with integrity: Surveillance and maintenance of the plant cell wall under stress. New Phytologist. 2020 Feb;225(4):1428-39.

[23] Gall, H. L., Philippe, F., Domon, J.-M., Gillet, F., Pelloux, J., and Rayon, C. (2015). Cell wall metabolism in response to abiotic stress. Plants 4, 112-166. doi: 10.3390/plants4010112.

[24] Cosgrove DC. Comparative structure and biomechanics of plant primary and secondary cell walls. Frontiers in plant science. 2012 Aug 22;3:204.

[25] Rao, X., & Dixon, R. A. (2017). Brassinosteroid mediated cell wall remodeling in grasses under abiotic stress. Frontiers in plant science, 8, 806.

[26] Tenhaken R. Cell wall remodeling under abiotic stress. Frontiers in plant science. 2015 Jan 7;5:771.

[27] Wormit A, Usadel B. The multifaceted role of pectin methylesterase inhibitors (PMEIs). International journal of molecular sciences. 2018 Oct;19(10):2878.

[28] Uddin MN, Hanstein S, Faust F, Eitenmüller PT, Pitann B, Schubert S. Diferulic acids in the cell wall may contribute to the suppression of shoot growth in the first phase of salt stress in maize. Phytochemistry. 2014 Jun 1;102:126-36.

[29] Vu MH, Iswanto AB, Lee J, Kim JY. The Role of Plasmodesmata-Associated Receptor in Plant Development and Environmental Response. Plants. 2020 Feb;9(2):216.

[30] Vaahtera L, Schulz J, Hamann T. Cell wall integrity maintenance during plant development and interaction with the environment. Nature plants. 2019 Sep;5(9):924-32.

[31] Morales F, Ancín M, Fakhet D, González-Torralba J, Gámez AL, Seminario A, Soba D, Ben Mariem S, Garriga M, Aranjuelo I. Photosynthetic Metabolism under Stressful Growth Conditions as a Bases for Crop Breeding and Yield Improvement. Plants. 2020 Jan;9(1):88.

[32] Medrano H, Escalona JM, Bota J, Gulías J, Flexas J. Regulation of photosynthesis of C3 plants in response to progressive drought: stomatal conductance as a reference parameter.

Annals of botany. 2002 Jun 15;89(7):895-905.

[33] Sharma A, Kumar V, Shahzad B, Ramakrishnan M, Sidhu GP, Bali AS, Handa N, Kapoor D, Yadav P, Khanna K, Bakshi P. Photosynthetic response of plants under different abiotic stresses: a review. Journal of Plant Growth Regulation. 2019 Aug 19:1-23.

[34] Chen Y, Wang XM, Zhou L, He Y, Wang D, Qi YH, Jiang DA. Rubisco activase is also a multiple responder to abiotic stresses in rice. PLoS one. 2015 Oct 19;10(10):e0140934.

[35] Saud S, Yajun C, Fahad S, Hussain S, Na L, Xin L, Alhussien SA. Silicate application increases the photosynthesis and its associated metabolic activities in Kentucky bluegrass under drought stress and post-drought recovery. Environmental Science and Pollution Research. 2016 Sep 1;23(17):17647-55.

[36] Flexas J, Barón M, Bota J, Ducruet JM, Gallé A, Galmés J, Jiménez M, Pou A, Ribas-Carbó M, Sajnani C, Tomàs M. Photosynthesis limitations during water stress acclimation and recovery in the drought-adapted Vitis hybrid Richter-110 (V. berlandieri× *V. rupestris*). Journal of experimental Botany. 2009 May 1;60(8):2361-77.

[37] Xu Z, Jiang Y, Jia B, Zhou G. Elevated-CO2 response of stomata and its dependence on environmental factors. Frontiers in Plant Science. 2016 May 13;7:657.

[38] Singh SK, Reddy KR. Regulation of photosynthesis, fluorescence, stomatal conductance and water-use efficiency of cowpea (*Vigna unguiculata* [L.] Walp.) under drought. Journal of Photochemistry and Photobiology B: Biology. 2011 Oct 5;105(1):40-50.

[39] Gupta B, Huang B. Mechanism of salinity tolerance in plants: physiological, biochemical, and molecular characterization. International journal of genomics. 2014 Oct;2014.

[40] Van Zelm E, Zhang Y, Testerink C. Salt tolerance mechanisms of plants. Annual Review of Plant Biology. 2020 Mar 13;71.

[41] Soni P, Kumar G, Soda N, Singla-Pareek SL, Pareek A. Salt overly sensitive pathway members are influenced by diurnal rhythm in rice. Plant signaling & behavior. 2013 Jul 1;8(7):e24738.

[42] Zeng H, Xu L, Singh A, Wang H, Du L, Poovaiah BW. Involvement of calmodulin and calmodulin-like proteins in plant responses to abiotic stresses. Frontiers in plant science. 2015 Aug 11;6:600.

[43] Ji H, Pardo JM, Batelli G, Van Oosten MJ, Bressan RA, Li X. The Salt Overly Sensitive (SOS) pathway: established and emerging roles. Molecular plant. 2013 Mar 1;6(2):275-86.

[44] Pandey GK, Mahiwal S. Potassium in Abiotic Stress. In Role of Potassium in Plants 2020 (pp. 45-49). Springer, Cham.

[45] Almeida DM, Oliveira MM, Saibo NJ. Regulation of Na+ and K+ homeostasis in plants: towards improved salt stress tolerance in crop plants. Genetics and molecular biology. 2017;40(1):326-45.

[46] Zhang Y, Lv Y, Jahan N, Chen G, Ren D, Guo L. Sensing of abiotic stress and ionic stress responses in plants. International journal of molecular sciences. 2018 Nov;19(11):3298.

[47] Waters S, Gilliham M, Hrmova M. Plant high-affinity potassium (HKT) transporters involved in salinity tolerance: structural insights to

probe differences in ion selectivity. International journal of molecular sciences. 2013 Apr;14(4):7660-80.

[48] Wang Q, Guan C, Wang P, Ma Q, Bao AK, Zhang JL, Wang SM. The Effect of AtHKT1; 1 or AtSOS1 mutation on the expressions of Na+ or K+ transporter genes and ion homeostasis in *Arabidopsis thaliana* under salt stress. International journal of molecular sciences. 2019 Jan;20(5):1085.

[49] Franco-Navarro JD, Brumós J, Rosales MA, Cubero-Font P, Talón M, Colmenero-Flores JM. Chloride regulates leaf cell size and water relations in tobacco plants. Journal of Experimental Botany. 2016 Feb 1;67(3):873-91.

[50] Brini F, Masmoudi K. Ion transporters and abiotic stress tolerance in plants. ISRN molecular biology. 2012;2012.

[51] White PJ, Broadley MR. Chloride in soils and its uptake and movement within the plant: a review. Annals of Botany. 2001 Dec 1;88(6):967-88.

[52] Wu H, Li Z. The Importance of Cl- Exclusion and Vacuolar Cl- Sequestration: Revisiting the Role of Cl- Transport in Plant Salt Tolerance. Frontiers in Plant Science. 2019;10.

[53] Li B, Tester M, Gilliham M. Chloride on the move. Trends in plant science. 2017 Mar 1;22(3):236-48.

[54] Xing A, Ma Y, Wu Z, Nong S, Zhu J, Sun H, Tao J, Wen B, Zhu X, Fang W, Li X. Genome-wide identification and expression analysis of the CLC superfamily genes in tea plants (*Camellia sinensis*). Functional & Integrative Genomics. 2020 Jan 3:1-2.

[55] Eisenach C, De Angeli A. Ion transport at the vacuole during stomatal movements. Plant physiology. 2017 Jun 1;174(2):520-30.

[56] Xiong L, Zhu JK. Molecular and genetic aspects of plant responses to osmotic stress. Plant, Cell & Environment. 2002 Feb;25(2):131-9.

[57] Slama I, Abdelly C, Bouchereau A, Flowers T, Savoure A. Diversity, distribution and roles of osmoprotective compounds accumulated in halophytes under abiotic stress. Annals of Botany. 2015 Feb 1;115(3):433-47.

[58] KAVI KISHOR PB, Sreenivasulu N. Is proline accumulation per se correlated with stress tolerance or is proline homeostasis a more critical issue?. Plant, cell & environment. 2014 Feb;37(2):300-11.

[59] Ashraf MF, Foolad MR. Roles of glycine betaine and proline in improving plant abiotic stress resistance. Environmental and experimental botany. 2007 Mar 1;59(2):206-16.

[60] Hare PD, Cress WA. Metabolic implications of stress-induced proline accumulation in plants. Plant growth regulation. 1997 Feb 1;21(2):79-102.

[61] Slama I, Ghnaya T, Savouré A, Abdelly C. Combined effects of long-term salinity and soil drying on growth, water relations, nutrient status and proline accumulation of *Sesuvium portulacastrum*. Comptes Rendus Biologies. 2008 Jun 1;331(6):442-51.

[62] Hayat S, Hayat Q, Alyemeni MN, Wani AS, Pichtel J, Ahmad A. Role of proline under changing environments: a review. Plant signaling & behavior. 2012 Nov 1;7(11):1456-66.

[63] Tian F, Wang W, Liang C, Wang X, Wang G, Wang W. Overaccumulation of glycine betaine makes the function of the thylakoid membrane better in wheat under salt stress. The Crop Journal. 2017 Feb 1;5(1):73-82.

[64] Murata N, Mohanty PS, Hayashi H, Papageorgiou GC. Glycinebetaine stabilizes the association of extrinsic proteins with the photosynthetic oxygen-evolving complex. FEBS letters. 1992 Jan 20;296(2):187-9.

[65] Allakhverdiev SI, Los DA, Mohanty P, Nishiyama Y, Murata N. Glycinebetaine alleviates the inhibitory effect of moderate heat stress on the repair of photosystem II during photoinhibition. Biochimica et Biophysica Acta (BBA)-Bioenergetics. 2007 Dec 1;1767(12):1363-71.

[66] Khan MS, Yu X, Kikuchi A, Asahina M, Watanabe KN. Genetic engineering of glycine betaine biosynthesis to enhance abiotic stress tolerance in plants. Plant Biotechnology. 2009 Mar 1;26(1):125-34.

[67] Giri J. Glycinebetaine and abiotic stress tolerance in plants. Plant signaling & behavior. 2011 Nov 1;6(11):1746-51.

[68] Riadh K, Wided M, Hans-Werner K, Chedly A. Responses of halophytes to environmental stresses with special emphasis to salinity. InAdvances in Botanical Research 2010 Jan 1 (Vol. 53, pp. 117-145). Academic Press.

[69] da Silva JM, Arr.abaça MC. Contributions of soluble carbohydrates to the osmotic adjustment in the C4 grass *Setaria sphacelata*: a comparison between rapidly and slowly imposed water stress. Journal of plant physiology. 2004 Jan 1;161(5):551-5.

[70] Iturriaga G, Suárez R, Nova-Franco B. Trehalose metabolism: from osmoprotection to signaling. International journal of molecular sciences. 2009 Sep;10(9):3793-810.

[71] Suprasanna P, Nikalje GC, Rai AN. Osmolyte accumulation and implications in plant abiotic stress tolerance. InOsmolytes and plants acclimation to changing environment: Emerging omics technologies 2016 (pp. 1-12). Springer, New Delhi.

[72] Abebe T, Guenzi AC, Martin B, Cushman JC. Tolerance of mannitol-accumulating transgenic wheat to water stress and salinity. Plant physiology. 2003 Apr 1;131(4):1748-55.

[73] Chen D, Shao Q, Yin L, Younis A, Zheng B. Polyamine function in plants: metabolism, regulation on development, and roles in abiotic stress responses. Frontiers in plant science. 2019 Jan 10;9:1945.

[74] Kusano T, Berberich T, Tateda C, Takahashi Y. Polyamines: essential factors for growth and survival. Planta. 2008 Aug 1;228(3):367-81.

[75] Yu Z, Jia D, Liu T. Polyamine oxidases play various roles in plant development and abiotic stress tolerance. Plants. 2019 Jun;8(6):184.

[76] Groppa MD, Benavides MP. Polyamines and abiotic stress: recent advances. Amino acids. 2008 Jan 1;34(1):35.

[77] Paschalidis K, Tsaniklidis G, Wang BQ, Delis C, Trantas E, Loulakakis K, Makky M, Sarris PF, Ververidis F, Liu JH. The interplay among polyamines and nitrogen in plant stress responses. Plants. 2019 Sep;8(9):315.

[78] Gill SS, Tuteja N. Reactive oxygen species and antioxidant machinery in abiotic stress tolerance in crop plants. Plant physiology and biochemistry. 2010 Dec 1;48(12):909-30.

[79] Nadarajah KK. ROS Homeostasis in Abiotic Stress Tolerance in Plants. International Journal of Molecular Sciences. 2020 Jan;21(15):5208.

[80] Miller GA, Suzuki N, Ciftci-Yilmaz SU, Mittler RO. Reactive oxygen

species homeostasis and signalling during drought and salinity stresses. Plant, cell & environment. 2010 Apr;33(4):453-67.

[81] Apel K, Hirt H. Reactive oxygen species: metabolism, oxidative stress, and signal transduction. Annu. Rev. Plant Biol.. 2004 Jun 2;55:373-99.

[82] Das K, Roychoudhury A. Reactive oxygen species (ROS) and response of antioxidants as ROS-scavengers during environmental stress in plants. Frontiers in environmental science. 2014 Dec 2;2:53.

[83] Thorpe GW, Reodica M, Davies MJ, Heeren G, Jarolim S, Pillay B, Breitenbach M, Higgins VJ, Dawes IW. Superoxide radicals have a protective role during H2O2 stress. Molecular biology of the cell. 2013 Sep 15;24(18):2876-84.

[84] Karuppanapandian T, Moon JC, Kim C, Manoharan K, Kim W. Reactive oxygen species in plants: their generation, signal transduction, and scavenging mechanisms. Australian Journal of Crop Science. 2011 Jun;5(6):709.

[85] Caverzan A, Casassola A, Brammer SP. Antioxidant responses of wheat plants under stress. Genetics and molecular biology. 2016 Mar;39(1):1-6.

[86] Choudhury FK, Rivero RM, Blumwald E, Mittler R. Reactive oxygen species, abiotic stress and stress combination. The Plant Journal. 2017 Jun;90(5):856-67.

[87] Huang H, Ullah F, Zhou DX, Yi M, Zhao Y. Mechanisms of ROS regulation of plant development and stress responses. Frontiers in Plant Science. 2019;10.

[88] Kaur N, Dhawan M, Sharma I, Pati PK. Interdependency of reactive oxygen species generating and scavenging system in salt sensitive and salt tolerant cultivars of rice. BMC plant biology. 2016 Dec;16(1):1-3.

[89] You J, Chan Z. ROS regulation during abiotic stress responses in crop plants. Frontiers in plant science. 2015 Dec 8;6:1092.

[90] Jiménez-Quesada MJ, Traverso JÁ, Alché JD. NADPH oxidase-dependent superoxide production in plant reproductive tissues. Frontiers in Plant Science. 2016 Mar 31;7:359.

[91] Zhu M, Zhou M, Shabala L, Shabala S. Physiological and molecular mechanisms mediating xylem Na+ loading in barley in the context of salinity stress tolerance. Plant, cell & environment. 2017 Jul;40(7):1009-20.

[92] Ma L, Zhang H, Sun L, Jiao Y, Zhang G, Miao C, Hao F. NADPH oxidase AtrbohD and AtrbohF function in ROS-dependent regulation of Na+/K+ homeostasis in Arabidopsis under salt stress. Journal of Experimental Botany. 2012 Jan 1;63(1):305-17.

[93] Ighodaro OM, Akinloye OA. First line defence antioxidants-superoxide dismutase (SOD), catalase (CAT) and glutathione peroxidase (GPX): Their fundamental role in the entire antioxidant defence grid. Alexandria journal of medicine. 2018 Dec 1;54(4):287-93.

[94] Si T, Wang X, Zhao C, Huang M, Cai J, Zhou Q, Dai T, Jiang D. The role of hydrogen peroxide in mediating the mechanical wounding-induced freezing tolerance in wheat. Frontiers in plant science. 2018 Mar 14;9:327

[95] Hossain MA, Bhattacharjee S, Armin SM, Qian P, Xin W, Li HY, Burritt DJ, Fujita M, Tran LS. Hydrogen peroxide priming modulates abiotic oxidative stress tolerance: insights from

ROS detoxification and scavenging. Frontiers in plant science. 2015 Jun 16;6:420.

[96] Sharma P, Jha AB, Dubey RS, Pessarakli M. Reactive oxygen species, oxidative damage, and antioxidative defense mechanism in plants under stressful conditions. Journal of botany. 2012;2012.

[97] Gill, S. S., Khan, N. A., Anjum, N. A., & Tuteja, N. (2011). Amelioration of cadmium stress in crop plants by nutrients management: morphological, physiological and biochemical aspects. Plant Stress, 5(1), 1-23.

[98] Ahmar S, Gill RA, Jung KH, Faheem A, Qasim MU, Mubeen M, Zhou W. Conventional and Molecular Techniques from Simple Breeding to Speed Breeding in Crop Plants: Recent Advances and Future Outlook. International Journal of Molecular Sciences. 2020 Jan;21(7):2590.

[99] Govindaraj M, Vetriventhan M, Srinivasan M. Importance of genetic diversity assessment in crop plants and its recent advances: an overview of its analytical perspectives. Genetics research international. 2015;2015.

[100] Wolter F, Schindele P, Puchta H. Plant breeding at the speed of light: the power of CRISPR/Cas to generate directed genetic diversity at multiple sites. BMC plant biology. 2019 Dec;19(1):1-8.

[101] Nogué F, Mara K, Collonnier C, Casacuberta JM. Genome engineering and plant breeding: impact on trait discovery and development. Plant cell reports. 2016 Jul 1;35(7):1475-86.

[102] Collard BC, Mackill DJ. Marker-assisted selection: an approach for precision plant breeding in the twenty-first century. Philosophical Transactions of the Royal Society B: Biological Sciences. 2008 Feb 12;363(1491):557-72.

[103] Ulukapi K, Nasircilar AG. Induced mutation: creating genetic diversity in plants. InGenetic Diversity in Plant Species-Characterization and Conservation 2018 Nov 5. IntechOpen.

[104] Jain M. Emerging role of metabolic pathways in abiotic stress tolerance. J. Plant Biochem. Physiol. 2013 Jun 15;1(108):10-4172

[105] Parmar N, Singh KH, Sharma D, Singh L, Kumar P, Nanjundan J, Khan YJ, Chauhan DK, Thakur AK. Genetic engineering strategies for biotic and abiotic stress tolerance and quality enhancement in horticultural crops: a comprehensive review. 3 Biotech. 2017 Aug 1;7(4):239.

[106] Vij S, Tyagi AK. Emerging trends in the functional genomics of the abiotic stress response in crop plants. Plant biotechnology journal. 2007 May;5(3):361-80.

[107] Zhang H, Mittal N, Leamy LJ, Barazani O, Song BH. Back into the wild—Apply untapped genetic diversity of wild relatives for crop improvement. Evolutionary Applications. 2017 Jan;10(1):5-24.

[108] Stephens J, Barakate A. Gene editing technologies–ZFNs, TALENs, and CRISPR/Cas9.2017: 157-161.

[109] N, Kaur G, Pati PK. Deciphering Strategies for Salt Stress Tolerance in Rice in the Context of Climate Change. InAdvances in Rice Research for Abiotic Stress Tolerance 2019 Jan 1 (pp. 113-132). Woodhead Publishing.

[110] Li Q, Sapkota M, van der Knaap E. Perspectives of CRISPR/Cas-mediated cis-engineering in horticulture: unlocking the neglected potential for crop improvement. Horticulture Research. 2020 Mar 15;7(1):1-1.

[111] Zafar SA, Zaidi SS, Gaba Y, Singla-Pareek SL, Dhankher OP, Li X,

Mansoor S, Pareek A. Engineering abiotic stress tolerance via CRISPR/Cas-mediated genome editing. Journal of Experimental Botany. 2020 Jan 7;71(2):470-9.

[112] Nandy S, Pathak B, Zhao S, Srivastava V. Heat-shock-inducible CRISPR/Cas9 system generates heritable mutations in rice. Plant direct. 2019 May;3(5):e00145.

Chapter 6

Molecular Abiotic Stress Tolerans Strategies: From Genetic Engineering to Genome Editing Era

Sinan Meriç, Alp Ayan and Çimen Atak

Abstract

In last decades, plants were increasingly subjected to multiple environmental abiotic stress factors as never before due to their stationary nature. Excess urbanization following the intense industrial applications introduced combinations of abiotic stresses as heat, drought, salinity, heavy metals etc. to plants in various intensities. Technological advancements brought novel biotechnological tools to the abiotic stress tolerance area as an alternative to time and money consuming traditional crop breeding activities as well as they brought vast majority of the problem themselves. Discoveries of single gene (as osmoprotectant, detoxyfying enzyme, transporter protein genes etc.) and multi gene (biomolecule synthesis, heat shock protein, regulatory transcription factor and signal transduction genes etc.) targets through functional genomic approaches identified abiotic stress responsive genes through EST based cDNA micro and macro arrays. In nowadays, genetic engineering and genome editing tools are present to transfer genes among different species and modify these target genes in site specific, even single nuclotide specific manner. This present chapter will evaluate genomic engineering approaches and applications targeting these abiotic stress tolerance responsive mechanisms as well as future prospects of genome editing applications in this field.

Keywords: GMO, abiotic stress, genetic engineering, genome editing

1. Introduction

Before the first examples of the crop domestication by late hunter-gatherer man approximately 10.000 years ago, plants evolved most of their traits following the most basic rule of evolution: success in reproduction and producing next generations. To achieve this goal, plants evolved various adaptation mechanisms against harsh environmental conditions described as abiotic stress factors as well as biotic stress factors like diseases and herbivores [1, 2]. The early domesticated agronomical traits desired by the primal farmers were mostly controlled by the limited number of genes. Traits as panicle size (rice, pearl millet), fruit size (tomato, eggplant), seed size (rice, sorghum, maize, common bean), seed dispersal (cereals in general) are easily controlled by small number of genes [3, 4]. Traits as bitter taste which is one of the adaptation mechanism of plants against herbivores can be controlled by single or multiple genes due to the complexity of the biochemical pathway of the compound. In the absence of genetic knowledge, first breeders domesticated almond as its bitterness is controlled by single gene, while the similar bitter taste in oak acorns is controlled my multiple genes and it was never domestication target [5]. Human intervention to natural selection of plants helped plants to achieve their reproduction goal at the cost of losing most of the gene pool diversity including abiotic stress tolerance and disease resistance genes in favor of human desired traits. Until Gregory Mendel's breakthrough findings, plant breeding was based on easily made crosses, observations and mass selections without knowing sexual recombination of alleles recombined in meiosis to produce new traits for hundreds of years. Even after the knowledge of Mendelian segregation of traits, crossing of distantly related plant species for desired traits as disease resistance became problematic due to the abnormal chromosome pairing and infertile hybrids not to mention horizontal gene transfer between different kingdoms as bacteria and viruses. Following the studies on molecular structure and function of genes during 1940s and the discovery of the structure of DNA by Watson and Crick in 1953, technical developments as PCR and use of restriction enzymes led to the first transfer of bacterial antibiotic resistance gene to plant cells in 1983. Foreign bacterial gene introduced to petunia cell cultures and plantlets derived from transformed cells retained antibiotic resistance [6]. Following these developments, tomato was the first genetically engineered plant which was intended to be used in commercial practices in 1982. Commercial use of genetically engineered crops started in 1996 in US, since four genetically engineered plants were allowed by USDA for field testing in 1985. Technical and technological research kept improving on this field. Certain trends in engineered plants divided the use of this technology in three categories. Targeted traits of the first generation genetically engineered plants were mostly oriented according to farmer's benefit as abiotic stress tolerance or herbicide resistance. Second generation of genetically engineered plants emphasized more on commercial benefits as shelf life or nutritional value. The third generation of genetically engineered plants represent the idea of functional foods which were enhanced with pharmaceutical product that are not present in the plant itself [7].

Use of genetic engineering techniques in abiotic stress tolerance of plants requires knowledge on physiology and biochemical processes to identify key gene targets. Functional genomic approaches utilize transcriptome analysis of plants under abiotic environmental stress factor by using techniques as quantitative, real-time PCR, microarray and high-throughput RNAseq. Gene expression profiles of specific tissues under development stage of choice can be searched between previously submitted Expressed Sequence Tags (ESTs) in various cDNA libraries. Abiotic stress responsive genes in particular abiotic stress tolerance mechanisms can be evaluated before progressing further in experimental design [8]. Molecular regulation type of the targeted trait is also important in genetic engineering approach. Abiotic stress tolerance is mostly regulated by multiple genes depending on stress and targeted tolerance mechanism. One of the early effects of the abiotic stresses on plants is disruption of osmotic balance in cells. Therefore, biosynthesis and accumulation of osmoprotectant molecules as proline, glycine betaine, polyamines, mannitol, trehalose can be targeted as they are regulated by single individual genes. Likewise, reactive oxygene species scavenging genes (superoxide dismutase, glutathione reductase, ascorbate peroxidase), late embryogenesis abundant (LEA) proteins, ion transport genes are some single gene targets used in genetic engineering approach for abiotic stress tolerance which will be further detailed in this chapter. Targeting a tolerance mechanism which is regulated by multiple genes is more complicated. Still, there are many successful genetic engineering applications targeting heat shock proteins (HSPs), transcription factors (TFs), signal transduction genes for abiotic stress tolerance [9, 10]. After all of these decision making process which requires intense precision, promoter type is one of the key factors since all transferred genes requires expression regulation. Calliflower Mosaic Virus 35S (CaMV35S), maize ubiquitine-1 (Ubi-1) and rice actin-1 (act-1) are some of the mostly used strongly expressed constitutive promoters which were used in abiotic stress tolerant genetically engineered plant studies. Despite their advantages to plants as strong expression in all developmental stages in all tissues without any stimulation from the environment, they are not economical for plants in contexts of energy and biochemical source consumption. Signal transduction genes as SAPK4, OsITPK2, OsCPKI2, transcription factor genes as JERF3, ZFP182, antioxidant genes as katE, CAT1, GST, OsMIOX, osmotic homeostasis genes as codA, ion homeostasis genes as AgNHX1, nhaA, OshAK5 are some examples which are regulated by CaMV35S promoter in abiotic stress tolerant genetically engineered plants while Ubi-1 promoter is utilized for DSM1, OsCPK21, OsCPK4 signal transduction genes, OsWRKY45-1, OsWRKY45-2, ZmCBF3, OsbZIP46 transcription factor genes, Sod1 antioxidant gene, otsA, otsB, adc osmotic homeostasis genes and OsKAT1 ion homeostasis genes [11]. Constitutive and avoidable use of biochemical sources in stress free periods or environments leads these genetically engineered plants to reduced growth and loss of yield in some cases. Therefore, regulation of abiotic stress induced or tissue/developmental stage specific promoters are more viable options. General aspects of desired stress induced promoters are described as: (i) having basal expression level under stress free conditions, (ii) strongly promoting expression of resistance genes, (iii) having dose dependent and dose sensitive stress induction, (iv) being stress specific and (v) having reversible induction which rapidly reduce to basal level as removal of stress factor. Arabidopsis rd29A promoter is the most frequently used stress inducible promoter in genetically engineered plant studies. Some stress inducible promoters may lack some desired features as stress induced rab16A, LIP9, OsNAC6 which have higher promoting regulation under basal conditions or some may need further characterization and knowledge as HVA1s, Dhn8s and Dhn4s from barley and DGP1 from tobacco [12].

Most of the abiotic stress tolerant genetically engineered plant trials are limited to laboratory experiments and do not have commercial uses due to the regulations or inadequate field performance. Commercial genetically engineered plants need to tolerate or resist to multiple biotic and abiotic stresses in different combination, duration and concentrations. Tolerance or resistance should not be limited only to developmental stages but also reproductive stage which plants are more vulnerable to abiotic stresses.

As an alternative to gene transfer approach, many studies utilize genome editing techniques as they allow researchers to modify the genome even in a few nucleotide level as well as they can be used to alter or replace allels, silence or insert new gene(s) to targeted sites in genome. Genome editing by using site specific endonucleases is not a new concept in developing crop plants by means of biotic and abiotic stress tolerance. Clustered regularly interspaced short palindromic repeats (CRISPR)/CRISPR-associated protein 9 (Cas9) is the most recent site specific genome editing technique which dominated the alternatives as transcription activator-like effector nucleases (TALENs) and zinc-finger nucleases (ZFNs) due to the accuracy, cost and time efficiency and by the means of application advantages as allowing multiple site editing at the same time. Even this new approach evolved in short time due to the occurring limitations during application and new alternatives to Cas9 as Cpf1 are seriously in consideration recently. Also transgene free genome editing applications are introduced lately as the law regulations and public acceptance on this topic are crippling for researchers [13, 14].

Plant	Transformation Event / Developer	Resistance	Expressing	Common Use
Zea mays L.	MON87427 x MON87460 x MON89034 x TC1507 x MON87411 x 59122 / Monsanto	insect resistance, drought and herbicide (glyphosate and spesifically glufosinate ammonium) tolerance	cp4 epsps, cspB, cry2Ab2, cry1A.105, cry1F, pat, cry34Ab1, cry35Ab1, cry3Bb1, dvsnf	Food Feed Biofuel
	MON87460 (Genuity® DroughtGard™) / Monsanto	cold / heat drought tolerance	cspB, nptII	Food Feed Biofuel
	MON87460 x MON88017 / Monsanto	insect resistant (coleopteran and lepidoptera), drought and herbicide (glyphosate) tolerant	cspB, cp4 epsps, cry1A.105, cry2Ab2, cry3Bb1, nptII	Food Feed
	MON87460 x MON89034 x MON88017 / Monsanto	cold / heat, drought, herbicide (glyphosate) tolerance and insect resistance (lepidoptera and coleopteran)	cspB, cry1A.105, cry2Ab2, cp4epsps and cry3Bb1	Food Feed
	MON87460 x MON89034 x NK603 / Monsanto	cold / heat, drought, herbiside (glyphosate) tolerance and insect (lepidoptera) resistance	cspB, cry1A.105, cp4 epsps, cry2Ab2, nptII	Food Feed
	MON87460 x NK603 / Monsanto	drought and herbicide (glyphosate) tolerance	cspB, cp4 epsps nptII	Food Feed
	MON89034 x MON87460 / Monsanto	drought tolerance and insect (lepidoptera) resistance	Cry2Ab2, cry1A.105 and cspB	Food Feed
Glycine max L.	HB4 / Verdeca	drought and hyper salinity tolerance	Hahb-4	Food Feed
	HB4 x GTS 40-3-2 / INDEAR	drought, salinity, herbicide (glyphosate) tolerance	Hahb-4 and cp4 epsps	Food Feed
Saccharum sp.	NXI-1T / PT Perkebunan Nusantara XI (Persero)	drought tolerance	EcBetA, nptII, aph4 (hpt)	Food
	NXI-4T / PT Perkebunan Nusantara XI (Persero)	drought tolerance	RmBetA	Food Feed
	NXI-6T / PT Perkebunan Nusantara XI (Persero)	drought tolerance	RmBetA	Food

Table 1.
Commercial genetically engineered abiotic stress tolerant plant verieties [15, 16].

Consumers and farmers are interested to the traits as taste, productivity, yield, shape of commercial genetically engineered and genome edited varieties more than just survival under abiotic stresses. Therefore, most of the abiotic stress tolerant plant strategies mentioned further in this chapter are laboratory applications. **Table 1** represents successful commercial transformation events of abiotic stress tolerant maize, soybean and sugarcane developed by Monsanto, Verdeca and Persero Companies.

2. Genetic engineering applications in plant abiotic stress tolerance

One of the key points in breeding better crops under various stress conditions is understanding the cellular, biochemical and molecular changes that occur in response to stress [17]. Understanding the mechanisms underlying plants responses to abiotic stress is an important step for genetic engineering that focuses on improving or enhancing tolerance to these stresses (i.e. salinity, cold, dehydration). Identifying key genes that positively affect tolerance within these mechanisms and introduction and overexpression of them allow improvement of genetically modified plants tolerant to abiotic stress. Additionally, understanding the mechanisms by which plants perceive and transmit stress signals is another important point for genetic engineering of crops [18, 19]. Growing number of molecular and biochemical studies on this subject provide an understanding of the signal transduction network involved in the response to abiotic stress and the pathways associated with this network. The development of plants by conventional breeding methods has certain limitations as transfer of a gene requires repeated cycles of crossing and selection. Moreover, the classical breeding process is limited to species with sexual reproduction. Another disadvantage of classical breeding is that genes with undesirable characteristics are transferred together with target genes in this process. With the advent of modern biotechnology tools, it has become possible to modify genetic structure of plants using genes from other living organisms, and recombinant DNA technology has provided an effective alternative to traditional approach. In addition, it has become possible to exchange genetic material between sexually incompatible species. Genes encoding proteins known to play a role in resistance and tolerance to biotic (virus, bacteria, nematode, fungus, herbivores) and abiotic stresses (drought, salinity, temperature, cold) are widely used in the obtaining genetically modified plants. Plants have developed a signal transduction pathway that regulates various stress-response genes such as kinases, molecular chaperones, osmoprotectants, transcription factors, and thus gives an idea of how tolerance to environmental factors is achieved. Since it is known that this signal transduction and associated physiological, biochemical and molecular pathways are governed by more than one gene, it is extremely difficult for only one gene to achieve complete abiotic stress tolerance. Among the molecules known to have protective roles against abiotic stress are proline, which acts as an osmoprotectant, metal chelator, antioxdative defense molecule and signal molecule [20]; trehalose, which acts as an osmoprotectant and is involved in ROS scavenge during abiotic stress [21]; heat shock proteins (HSPs) that serve as molecular chaperones responsible for the folding, assembly, translocation and degradation of proteins [22]; Late embryogenesis abundant (LEA) proteins with antioxidant functions involved in protein protection, membrane protection, and ion binding [23]; aquaporins involved in the transport of water and neutral molecules [24]; calcineurin B-like proteins that act as calcium sensors and play a role in signal transmission. Transcription factors (NAC, WRKY, MYB, bZIP, DREB/CBF), kinases and phosphatases serve a function in stress perception and signal transduction and in the regulation of stress-inducible genes.

In addition to the identification, isolation and characterization of genes involved in all this abiotic stress response, introduction of these genes to plants is an effective molecular approach to understand the roles of genes and products in plant tolerance, behavior and phenotypes.

2.1 Late embryogenesis abundant proteins

Late embryogenesis abundant proteins, which were first discovered in the late developmental stages of plant seeds, are among the proteins most commonly used by plants in their response to abiotic stresses such as drought, high salt, extreme temperature and oxidative stress. Most of the LEA proteins and their mRNAs accumulate in high amounts in the embryo tissues in the final stages of seed development when drought initiates [23]. In addition, LEA proteins also accumulate in plant tissues exposed to dehydration, osmotic and low temperature stress. There is strong evidence that LEA proteins or their genes classified into 6 groups according to their specific domains are correlated with stress resistance [25, 26]. Many studies have reported that stress tolerance occurs as a result of the introduction of LEA genes to various plants such as Arabidopsis, tobacco, rice, barley, and corn by genetic engineering techniques (**Table 2**). Amara [30] reported that the Rab28 LEA gene was overexpressed in maize plants in which they transferred the Rab28 LEA gene with a constitutive maize promoter by particle bombardment method. With the expression of the Rab28 transcripts, the Rab28 protein accumulated in transgenic plants and the transgenic plants continued their growth in the dehydration condition in medium containing polyethyleneglycol (PEG) compared to wild-type controls. These results showed that LEA Rab28 (belonging to the 5th subgroup of the LEA protein family) protein is one of the important candidates that can be used to increase stress tolerance in maize plants. In another transformation study performed with the LEA genes, the barley (HVA1) gene encoding the late embryogenesis abundant protein was transferred to mulberry, which is an important industrial plant used in silk production, by *Agrobacterium tumafaciens*. The HVA1 gene expressing with the rd29a promoter increased the tolerance to cold, drought and salinity in transgenic mulberry [28]. Similarly, overexpression of the ZmLEA3 gene in the tobacco (*Nicotiana tabacum*) plants plays important roles in tolerance to osmotic and oxidative stresses. ZmLEA3 protein plays an active role in protecting plants from damage by preserving the protein structure and holding metals under osmotic and oxidative stress [32]. Late embryogenesis abundant genes have been isolated and characterized from many different plants. Thus, along with transgenic approaches, new gene sources that can be used in abiotic stress tolerance are discovered. The overexpression of the gene responsible for the JcLEA protein (belonging to the 5th subgroup of the LEA protein family) isolated from a tropical plant, *Jatropha curcas*, provided increased resistance to both drought and salt stress compared to the wild-type in transgenic Arabidopsis plants [31]. Transgenic *Arabidopsis thaliana* and *Setaria italica* (foxtail millet) seedlings showed higher tolerance to salt and osmotic stress than the wild-type with overexpression of the SiLEA14 gene [35]. After transformation of plants with LEA genes, chlorophyll content increases, while lipid peroxidation rates, as indicator of ROS decrease. Higher chlorophyll and low lipid peroxidation values obtained by the expression of the AdLEA (belonging to the 5th subgroup of the LEA protein family) gene isolated from the perennial *Arachis diogoi* plant in the tobacco plant (*Nicotiana tabacum*) is a representing study. These transgenic plants have been shown to tolerate dehydration, salinity and oxidative stress. Chlorophyll fluorescence measurements have shown that tobacco plants in which AdLEA is overexpressed can maintain their photosynthetic performance under drought conditions [36].

Gene Action	Gene Source	Promoter	Target Traits	Trans-formed Plant	Transfer Method	Ref.
LEA	ROB5 from Bromegrass, DHN4 from *Hordeum vulgare* SOD3.1 from *Triticum aestivum*	CaMV35S Cor78	Drought Heat Cold	*Solanum tuberosum*	Agrobacterium	[27]
LEA	Hva1 from *Hordeum vulgare*	rd29A	Salinity Drought Cold	*Morus indica*	*Agrobacterium tumefaciens* Ag11	[28]
LEA	OsLEA3-2 from *Oryza sativa*	CaMV35S	Salinity Drought	*Oryza sativa Arabidopsis thaliana*	Agrobacterium EHA105	[29]
LEA	LEA Rab28 gene from *Zea mays*	UBQ	Water stress	*Zea mays*	Biolistic system	[30]
LEA	JcLEA from *Jatropha curcas*	CaMV35S	Salinity	*Arabidopsis thaliana*	*Agrobacterium tumefaciens* LBA4404 floral dip	[31]
LEA	ZmLEA3 from *Zea mays*	CaMV35S	Osmotic, Oxidative stress	*Nicotiana tabacum*	*Agrobacterium tumefaciens* strain LBA4404	[32]
LEA	WCI16 from *Triticum aestivum*	CaMV35S	Freezing	*Arabidopsis thaliana*	*Agrobacterium tumefaciens* floral dip	[33]
LAE	ZmLEA5C from *Zea mays*	CaMV35S	Osmatic stress Cold	*Nicotiana benthamiana*	*Agrobacterium tumefaciens* strain LBA4404	[34]
LAE	SiLAE from *Setaria italica*	Super promoter	Osmotic stress Salinity	*Arabidopsis thaliana*	*Agrobacterium tumefaciens* strain GV1301	[35]
LEA	AdLEA from *Arachis diogoi*	CaMV35S	Salinity Oxidative stress	*Nicotiana tabacum*	*Agrobacterium tumefaciens* LBA4404	[36]

Table 2.
Summary of transgenic plants over-expressing late embryogenesis abundant (LEA) genes for abiotic stress tolerance.

2.2 Aquaporins (AQPs)

Aquaporins (AQPs) are a family of membrane water channel proteins that control the osmatic movement of water and other molecules such as carbon dioxide, glycerol, urea, and ammonia [24]. This system, which has high osmotic water permeability, operates with lower activation energy [37]. Since environmental conditions such as high temperatures and winds encountered by plants living on land cause an increase in the rate of transpiration. These plants have developed various strategies such as controlling transpiration with hormones, cellular signaling mechanisms, ion channels, and transporters to protect hemeostasis [38, 39]. However, the current situation worsens when osmotic pressure increases, water is absent and root hydraulic conductivity (Lpr) decreases [18]. In order to overcome this negative situation, the isolation of water channel genes and their introduction to the desired plant by using genetic transformation emerges as one of the most important solutions. Thus, this situation can be reorganized by transgenic expression of water channels that ensure the continuity of processes such as transcellular water transport, stomatal closure, osmoregulation, vacuolar differentiation and plant cell expansion (**Table 3**). Under saline conditions, plant growth is suppressed due to osmotic stress, and in addition, this suppression is further increased due to reduced photosynthesis and gas exchange due to salt accumulation in the leaves. Aquaporins are thought to be extremely effective in plant growth thanks to their effects on leaf gas exchange as well as root water intake. While overexpression of plasma membrane intrinsic proteins (PIP) causes an increase in leaf gas exchange under normal conditions, it increases assimilation and development under salt stress [43]. Moreover, it has been reported that shoot length and fresh weight significantly increased in PIP introduced plants under salt stress [43]. They indicated that transgenic plants developed by transferring the GmPIP1;6 genes to soybean plant (*Glycine max*) are more tolerant of salt stress than wild-type plants. In another study, overexpression of GhPIP2;7 in Arabidopsis increased the tolerance of the transgenic plant to drought stress [41]. It has also been reported that the introduced plasma membrane intrinsic gene may play a role in leaf development and in absorption in remaining water under drought stress and carry certain substrates instead of water to protect plants from drought. Additionally, seed germination rates increased in most of the transgenic plants produced by introducing plasma membrane intrinsic genes [44]. Transgenic *Arabidopsis thaliana* plants developed with introduction of BvCOLD1 gene from *Beta vulgaris* have been observed to increase in germination and early development rates at 10°C and under different abiotic stress conditions such as NaCl, LiCl, and sorbic acid [44]. Plant aquaporins function by settling in different subcellular locations. Liu [42] reported that one of the most abundant plasma membrane proteins in transgenic rice leaves and roots is OsPIP1 and that its overexpression causes changes in many physiological properties of transgenic plants depending on the stress dose. Rice seed yield, salt resistance, root hydraulic conductivity and seed germination rates increased with the moderate expression of OsPIP1 [42]. Beside plasma membranes, they can be localized in mitochondria, endoplasmic reticulum, chloroplast and vacuole [45].

2.3 Osmoprotectants

Trehalose is a non-reducing disaccharide sugar found in bacteria, fungi, invertebrates, including plants. The accumulation of trehalose, which is an important component of plant growth, affecting sugar metabolism and causes an osmoprotectant effect under abiotic stresses [46]. Trehalose is considered as an osmoprotectant as it counteracts against the effects of dehydration of plants caused by drought, salinity or low temperature [47]. Studies have shown that enzymes, proteins and lipid membranes can be stabilized by trehalose and that biological structures can be protected from damage by trehalose during abiotic stress [18]. Trehalose is an osmolite that both regulates osmosis and protects macromolecules. In higher plants, overexpression of trehalose-6-phosphate synthases (TPS) and/or trehalose-6 phosphate phosphatases genes originated from microorganisms has led to tolerance to various abiotic stresses. In addition, several genes involved in trehalose metabolism have been used together to improve the stress tolerance of some plants. The results show that the increased production of trehalose with overexpression of the bifonional fusion gene (TPSP) of the trehalose-6-phosphate synthase and trehalose-6-phosphate phosphatase genes increases the soluble sugar content and increases the tolerance to drought and salt stress in rice [48]. As a result of the expression of

Gene Action	Gene Source	Promoter	Target Traits	Trans-formed Plant	Transfer Method	Ref.
Aquaporin (plasma membrane intrinsic proteins)	TdPIP1 and TdPIP2 from *Triticum turgidum*	CaMV35S	Osmotic stress Salinity	*Nicotiana tabacum*	*Agrobacterium tumefaciens*	[40]
Aquaporin (plasma membrane intrinsic protein)	GhPIP2;7 from *Gossypium hirsutum*	GhPIP2;7	Drought	*Arabidopsis thaliana*	*Agrobacterium tumefaciens*	[41]
Aquaporin (plasma membrane intrinsic protein)	OsPIP1 from *Oryza sativa*	UBQ	Salinity	*Oryza sativa*	*Agrobacterium tumefaciens* EHA101	[42]
Aquaporin (plasma membrane intrinsic protein)	GmPIP1;6 from *Glycine max*	CaMV35S	Salinity	*Glycine max*	*Agrobacterium tumefaciens*	[43]
Aquaporin (plasma membrane intrinsic protein)	BvCOLD1 from *Beta vulgaris*	CaMV35S	Cold	*Arabidopsis thaliana*	*Agrobacterium tumefaciens* C58C1	[44]

Table 3.
Summary of transgenic plants over-expressing aquaporin genes for abiotic stress tolerance.

the TPSP fusion gene, tomatoes that accumulate high amounts of trehalose in their leaves exhibited a tolerance to drought and salt stress, and showed higher photosynthetic rates under salt stress conditions than wild-type plants [49]. Proline, another osmoprotectant, is an amino acid that plays a role in plant functions exposed to various abiotic stresses. Proline accumulates in response to many environmental stresses such as salinity, drought, heavy metals, and there is a positive correlation between plant stress and proline accumulation. Proline protects proteins and other molecules in the cell against denaturation by stabilizing them. In addition to being an excellent osmolyte, it also functions as a metal chelator, antioxidant defense molecule and signal molecule [20]. In addition, it plays an important role in the protection of subcellular structures, adjustment of cell turgor and osmatic balance, prevention of electrolyte leakage by stabilizing membranes, and balancing the concentration of reactive oxygen species (ROS). Proline synthesis in plants is carried out by two important pathways; glutamate pathway and orinithine pathway. However, among these, the glutamate pathway is the primary pathway of proline accumulation during osmatic stress [50]. This reaction is catalyzed in most plants by the enzyme Δ'-pyrroline-5 carboxylate synthetase (P5CS) encoded by two genes and the enzyme Δ'-pyrroline-5-carboxylate reductase (P5CR) encoded by a gene. In addition to the effects of proline on the physiology and biochemical processes of various plants, a high amount of endogenous proline has been shown to act as a regulator and/or signaling molecule that can alter the transcription levels of stress-related genes such as ascorbate peroxidase (APX), catalase (CAT), superoxide dismutase (SOD) and glutathione reductase (GR) [51, 52]. A P5CS gene that can be transferred in addition to the P5CS genes present in plants by genetic transformation helps plants to adapt to osmatic conditions by regulating proline accumulation, and also supports legume plants for optimum nodule formation and maintenance of nodule function [53]. Ghanti [54] reported that when they transferred the *Vigna aconitifolia* P5CS gene to *Cicer arietinum* plant, more proline accumulation in the leaves and roots of the plants and higher chlorophyll content under salt stress, as well as increased tolerance to salinity. It has also been shown that a single copy of P5CS transferred to plants can be effective in the development of tolerance under stress, regardless of the integration position or copy number [54]. P5CS-transferred transgenic plants, when exposed to high salt stress, grew to maturity by showing good growth and flowering, but control plants died within a certain period of time under the same conditions [55]. Moreover, introduction of a mutagenized version (P5CSF129A) of the *Vigna aconitifolia* P5CS gene to pigeonpea (*Cajanus cajan*), both proline accumulation in plants increased and tolerance to salt stress enhanced. In addition to improved growth performances and higher chlorophyll and relative water content, especially transgenic plants also showed lower levels of lipid peroxidation under 200 mM NaCl.

Plants synthesize and accumulate another osmoprotectant, glycine betaine (GB), in response to environmental stresses. Glycine betaine, which also plays a role in osmoregulation, protects the activity of macromolecules and the integrity of the membranes against stresses and scavenge the ROS. Glysine betaine synthesis reactions are catalyzed in two steps by the enzymes choline monooxygenase (CMO) and NADC-linked betaine aldehyde dehydrogenase. Many studies have been carried out showing that the abiotic stress tolerance of transgenic plants obtained as a result of isolating these genes in the biosynthetic pathway and transferring them to other plants has developed. Several osmoprotectant target for gene transfer approach are listed in **Table 4**.

2.4 Calcineurin B-like proteins (CBLs), calcium-dependent protein kinases (CDPKs)

Plants perceive the stimuli from the external environment through receptors in their membranes. After transmitting this signal, they respond to this stimulus. Ca^{2+} acts as one of the secondary messengers in the ABA signaling pathway, which plays an active role in this process. Calcium acts as a central link in numerous signaling pathways, and thus external signals such as hormone, light, biotic and abiotic stresses cause changes in Ca^{2+} concentration in the cell. The fluctuation in the cytosolic calcium concentration (calcium signatures) initiates various biochemical and physiological processes in the cell. Cellular calcium sensor molecules such as calcineurin B-like proteins (CBLs), calcium-dependent protein kinases (CDPKs) and calcium-dependent protein kinases (CIPKs), calmodulin (CaM) detect Ca^{2+} signals and enable Ca^{2+} ions to be transmitted to downstream pathways. Especially, calcineurin family of B-like proteins (CBLs) interacts and activates calcium-dependent protein kinases (CIPKs). Many studies have shown that this regulatory pathway plays a role in plants in response to environmental stresses (**Table 5**). It has been reported that overexpression of GmCBL1 in Arabidopsis increases tolerance to both high salt and drought stress in transgenic plants. It has also been shown that hypocotyl elongation is promoted under light conditions by overexpression of GmCBL1. Calcineurin B-like proteins (CBLs) can regulate stress tolerance by activating stress-related genes such as DREB1A, DREB2A, RD29A and KIN1, while also controlling hypocotyl development by changing the expression of genes related to gibberellin biosynthesis [68]. BdCIPK31, a CIPK gene belonging to *Brachypodium distachyon*, decreased under polyethylene glycol, NaCl, H_2O_2 and abscisic acid (ABA) treatments. Transgenic tobacco plants overexpressing BdCIPK31 developed tolerance to drought and salinity and lost less water than control plants under dehydration conditions. BdCIPK31 affects the expression of some ion channels and transporter

Gene Action	Gene Source	Promoter	Target Traits	Trans-formed Plant	Transfer Method	Ref.
Pyrroline-5-carboxylate synthetase	P5CS from *Arabidopsis thaliana*	CaMV35S	Salinity	*Solanum tuberosum*	*Agrobacterium tumefaciens* C58C1	[53]
Pyrroline-5-carboxylate synthetase	P5CS from *Vigna aconitifolia*	CaMV35S	Salinity	*Oryza sativa*	*Agrobacterium tumefaciens* strain LBA 4404	[56]
Pyrroline-5-carboxylate synthetase	P5CS from *Vigna aconitifolia*	CaMV35S	Salinity	*Cicer arietinum*	*Agrobacterium tumefaciens* LBA4404	[57]
Pyrroline-5-carboxylate synthetase	VaP5CSF129A from *Vigna aconitifolia*	CaMV35S	Drought	*Swingle citrumelo*	-	[54]
Pyrroline-5-carboxylate synthetase	MtP5CS3 from *Medicago truncatula*	CaMV35S	Salinity	*Medicago truncatula*	-	[58]
Pyrroline-5-carboxylate synthetase	P5CSF129A from *Vigna aconitifolia*	CaMV35S	Salinity	*Cajanus cajan*	*Agrobacterium tumefaciens* LBA4404	[59]
Trehalose-6-phosphate synthase	OsTSP1 from *Oryza sativa*	CaMV35S	Salinity Drought Cold	*Oryza sativa*	*Agrobacterium tumefaciens* AGL0	[60]
Trehalose-6-phosphate synthase/ phosphatase	Recombinant fusion TPSP from *E. coli*	UBQ	Drought Salinity	*Oryza sativa*	-	[51]
Trehalose-6-phosphate synthase	TPS1 from yeast	StDS2	Drought	*Solanum tuberosum*	*Agrobacterium tumefaciens* strain C58C1	[61]
Trehalose-6-phosphate synthase/ phosphatase	Recombinant fusion TPSP from *E. coli*	CaMV35S	Drought Salinity	*Solanum lycopersicum*	*Agrobacterium tumefaciens* LBA4404	[52]
Glycine betaine	codA (Choline oxidase) from *Arthrobacter globiformis*	SWPA2	Cold Salinity	*Solanum tuberosum*	*Agrobacterium tumefaciens* strain AGL0	[62]
Glycine betaine	BADH from *Spinacia oleracea*	CaMV35S	Salinity Cold	*Ipomoea batatas*	*Agrobacterium tumefaciens* LBA4404	[63]
Glycine betaine	betaine aldehyde dehydrogenase gene (AhBADH) from *Atriplex hortensis*,	CaMV35S	Salinity	*Poncirus trifoliata*	*Agrobacterium tumefaciens* LB4404	[64]

Gene Action	Gene Source	Promoter	Target Traits	Trans-formed Plant	Transfer Method	Ref.
Glycine betaine	betA gene from *E. coli*	UBQ	Salinity	*Triticum aestivum*	*Agrobacterium tumefaciens* LBA4404	[65]
Glycine betaine	betA gene from *E. coli*	UBQ	Drought	*Triticum aestivum*	*Agrobacterium tumefaciens* LBA4404	[66]
Glycine betaine	Mpgsmt and Mpsdmt from *Methanohalophilus portucalensis*	CaMV35S	Drought Salinity	*Arabidopsis thaliana*	*Agrobacterium tumefaciens* GV3101	[67]
Glycine betaine	codA from *Arthrobacter globiformis*	-	Heat	*Solanum lycopersicum*	*Agrobacterium tumefaciens* EHA101	[68]
Glycine betain	BADH betaine aldehyde dehydrogenase from *Spinacia oleracea*	CaMV35S	Heat	*Lycopersicon esculentum*	*Agrobacterium tumefaciens* LBA4404	[69]
Glycine betaine	codA from *Arthrobacter globiformis*	CaMV35S	Salinity	Solanum *lycopersicum*	*Agrobacterium tumefaciens* EHA101	[70]

Table 4.
Summary of transgenic plants over-expressing osmoprotectant-related genes for abiotic stress tolerance.

Gene Action	Gene Source	Promoter	Target Traits	Transformed Plant	Transfer Method	Ref.
CIPK	OsCIPK23m from *Oryza sativa*	UBQ	Drought	*Oryza sativa*	*Agrobacterium tumefaciens* EHA105	[71]
CBL-CIPK	BnCBL1-BnCIPK6 from *Brassica napus*	CaMV35S	Salinity	*Arabidopsis thaliana*	*Agrobacterium tumefaciens* floral dip method	[72]
CBL protein	GmCBL1 from *Glycine max*	CaMV35S	Salinity Drought	*Arabidopsis thaliana*	*Agrobacterium tumefaciens* C58C1	[68]
CBL protein	PeCBL10 from *Populus euphratica*	CaMV35S	Salinity Drought Cold	*Arabidopsis thaliana*	*Agrobacterium tumefaciens* GV3101	[73]
CBL protein	PtCBL10A PtCBL10B from *Populus trichocarpa*	CaMV35S	Salinity	*P. davidiana* × *P. bolleana*	*Agrobacterium tumefaciens* EHA105	[74]
CBL protein	NsylCBL10 from *Nicotiana sylvestris*	CaMV35S	Salinity	*Arabidopsis thaliana*	*Agrobacterium tumefaciens* EHA105	[70]
CBL protein	SpCBL6 from *Stipa purpurea*	CaMV35S	Cold	*Arabidopsis thaliana*	*Agrobacterium tumefaciens* strain GV3101	[75]
CIPK	BdCIPK31 from *Brachypodium distachyon*	CaMV35S	Drought Salinity	*Nicotiana tabacum*	*Agrobacterium tumefaciens* EHA105	[69]

Table 5.
Summary of transgenic plants over-expressing calcineurin B-like (CBL) and CBL-interacting protein kinase (CIPK) genes for abiotic stress tolerance.

genes under high salinity stress. Scavenging of reactive oxygen species and osmolite accumulation has been increased with the overexpression of BdCIPK31, thereby mitigate oxidative and osmotic damages [69]. In another transformation study performed with CBL, overexpression of SpCBL6 gene isolated from *Stipa purpurea* has been shown to increase tolerance to cold stress in transgenic *Arabidopsis thaliana*. In further analysis, it was reported that SpCBL6 overexpressing plants had increased water potential and photosynthetic efficiency (Fv/Fm) and decreased ion leakage compared with wild-type plants after cold application. In another study, NsylCBL10, a CBL gene belonging to *Nicotiana sylvestris*, has increased tolerance to high salt stress by maintaining Na^+ balance. Ion content analyzes revealed that transgenic plants maintained the lower Na^+/K^+ ratio in the roots and the higher Na^+/K^+ ratio in the shoots with the overexpression of NsylCBL10 [70]. Similarly, overexpression of PtCBL10A or PtCBL10B in transgenic poplar plants under salinity stress played an important role in tolerance to salt stress by maintaining ion homeostasis in shoot tissues [74]. Overexpression of BnCIPK6, its activated form BnCIPK6M and BnCBL1 CBL genes also increased tolerance to high salinity and low phosphate conditions in transgenic Arabidopsis plants, demonstrating how important the interaction of BnCBL1 and BnCIPK6 is for signaling pathways.

2.5 Heat shock protein

Heat shock proteins (HSPs) were originally identified as proteins that respond to high temperature conditions and involve in eukaryotes and prokaryotes. However, many studies have revealed the relationship of these proteins with various abiotic stresses. Heat shock proteins are powerful chaperones produced in response to many physiological and environmental stresses. In general, heat shock proteins function in the cytoplasm, while they are also located in organelles such as the nucleus, mitochondria, chloroplasts, endoplasmic reticulum [22]. These proteins that are evolutionarily conserved in prokaryotes and eukaryotes are also abundant in plants. HSPs divided into 5 different groups including HSP100, HSP90, HSP70, HSP60 and sHSP (small heat shock protein) according to their molecular weights. Among these groups, small heat shock proteins are also induced by other stresses such as drought, salinity and cold stress, and also play an active role in processes such as seed germination, embryogenesis and fruit development [76]. Transgenic *Arabidopsis thaliana* plants introduced with the chloroplastic sHSP26 gene from wheat (*Triticum aestivum*) show tolerance to higher temperatures than wild-type plants. In addition, these plants show more photosynthetic pigment accumulation and higher biomass and seed yield [76]. Another sHSP26 gene belonging to rice (*Oryza sativa*), which encodes small heat shock protein localized in chloroplast, provided less electrolyte leakage in transgenic plants under heat stress, while it played a role in the accumulation of thiobarbituric acid reactive substances in plants. Transgenic plants showed more photosystem II (PSII) (Fv/Fm) photochemical activity under temperature stress at 42°C than control plants [77]. There is much evidence that small heat shock proteins are also associated with scavenging reactive oxygen species. Thermo tolerance of the plants increased in correlation with ROS scavenging by overexpression of the OsHSP18.6 in transgenic plants. In addition, with the overexpression of OsHSP18.6, malondialdehyde (MDA) levels decreased and CAT and SOD activities increased under heat and drought stress [78]. HSP70, another heat shock protein, stabilizes the proteins and also prevents their denaturation and aggregation. In addition, it maintains protein homeostasis by taking part in processes such as the transport of certain proteins in the cell, folding of newly synthesized proteins, denaturation of unwanted proteins, and the formation and separation of protein complexes [79]. In rice plants produced by introducing anti-apoptotic genes associated with programmed cell death (PCD) such as AtBAG4 (*Arabidopsis thaliana*), Hsp70 (Citrus tristeza virus) and p35 (Baculo virus), the salinity tolerance of rice plants has increased and along with this, ROS production and plant damage have been reduced [79]. In transgenic sugarcane (*Saccharum spp*. Hybrid) plants transformed with the HSP70 gene from *Erianthus arundinaceus*, the expression of stress-related genes increases with the overexpression of the HSP70 gene. These transgenic plants have higher germination ability and higher chlorophyll content under salt stress [80]. HSP90 is one of the proteins that have been well preserved in the evolutionary process and found in large amounts. As one of the important components of the stress response, they are actively involved in various processes such as signal transduction, protein degradation, cell cycle control and protein traffic (**Table 6**). Xu [83] reported that overexpression of HSP90 genes (GmHsp90A2, GmHsp90A4, GmHsp90B1, GmHsp90C1.1 and GmHsp90C2.1) in *Arabidopsis thaliana* plants reduced the damage caused by abiotic stress. In addition, high proline accumulation has been detected in HSP90 transgenic events and it has been reported that HSP90 genes may affect proline synthesis in relation to the AtP5CS1 (pyrroline-5-carboxylate synthases 1) gene responsible for proline synthesis [83]. Polyamine homeostasis is regulated by many different regulatory mechanisms such as transport, turnover and modification of amino groups. At the same time, these molecules affect the extent of heat shock proteins synthesis under increasing temperatures. It has been reported that silencing of HSP90 genes in *Arabidopsis thaliana* plant increased the levels of soluble spermidine (S Spd), acetylated Spd (N8-acetyl-Spd) and acetylated spermine (N1 acetyl-Spm). Moreover,

Gene Action	Gene Source	Promoter	Target Traits	Trans-formed Plant	Transfer Method	Ref.
Heat Shock	GHSP26 from *Gossypium arboreum*	CaMV35S	Drought	*Gossypium hirsutum*	*Agrobacterium tumefaciens* LBA4404	[81]
Heat Shock	OsHSP6 from *Oryza sativa*	CaMV35S	Oxidative stress Heat	*Festuca arundinacea*	*Agrobacterium tumefaciens* EHA105	[77]
Heat shock	TaHSP26 from *Triticum aestivum*	CaMV35S TaHSP26	Heat	*Arabidopsis thaliana*	*Agrobacterium tumefaciens* AGL1	[76]
Heat Shock	ZmHSP16.9 from *Zea mays*	CaMV35S	Heat	*Nicotiana tabacum*	*Agrobacterium tumefaciens* LBA4404	[82]
Heat Shock	GmHsp90s from *Glycine max*	CaMV35S	Osmotic stress Salinity	*Arabidopsis thaliana*	*Agrobacterium tumefaciens* EHA105	[83]
Heat Shock	EaHSP70 from *Erianthus arundinaceus*	Port Ubi 2.3	Salinity Water deficiency	*Saccharum spp. hybrid*	*Agrobacterium tumefaciens* LBA4404	[80]
Heat Shock	HSP70 from *Citrus tristeza* virus (CTV)	UBQ	Salinity	*Oryza sativa*	*Agrobacterium tumefaciens* AGL1	[84]
Heat Shock	sHSP18.6 from *Oryza sativa*	CaMV35S	Heat Drought Salt Cold	*Oryza sativa*	*Agrobacterium tumefaciens* EHA101	[78]

Table 6.
Summary of transgenic plants over-expressing heat shock genes for abiotic stress tolerance.

HSP90s and polyamine (PA) oxidases (PAOs) organize the acetylation, oxidation and PA/H_2O_2 homeostasis of polyamines [85].

2.6 Transcription factors

The characterization and identification of genes involved in the stress response of plants is a prerequisite for producing genetically modified plants that are tolerant of various stresses. The fact that these identified genes are regulatory genes responsible for regulation become prominent as an effective approach in terms of ensuring the efficient control of many other genes involved in stress management. The most important regulatory gene candidates that can be used in the development of stress-tolerant plants are transcription factors (TFs) that also regulate the expression of stress genes. Transcription factors belonging to many different families such as AP2/EREBP, NAC, WRKY, MYB, bZIP are involved in various abiotic and biotic stress processes.

NAC proteins stand out as one of the largest family of transcription factors in plants. In particular, they perform vital tasks in the developmental processes of plants and their response to environmental stresses. They play a role in the regulation of many processes such as cell division, flower development, lateral root development, senescence, phytohormone homeostasis, secondary cell wall formation, abiotic and biotic stress responses. There are many studies that develop tolerance to abiotic stress as a result of transformation of plants with NAC transcription factors. Huang [86] transferred the TaNAC29 gene, which they isolated from wheat

(*Triticum aestivum*), to *Arabidopsis thaliana* plants. It has been reported that plants have developed tolerance to high salt and dehydration as a result of the overexpression of this gene. Transgenic plants accumulated less MDA under stress, but showed higher amounts of SOD and CAT activities. This shows that ABA-signaling pathway and antioxidant enzyme mechanisms are involved in NAC29-mediated stress tolerance. Similarly, overexpression of the TaNAC67 gene isolated from wheat plant increased the expression of many multiple abiotic stress response genes such as DREB1A, DREB2A, RD29A, RD29B, RD22, COR15 and Rab18 simultaneously in transgenic plants. Thus, improved cell membrane stability, higher chlorophyll content and Na^+ efflux rates, and enhanced tolerance to drought, salinity and cold stress were observed in plants [87, 88].

Another important transcription factor family that is widely distributed in plants is MYB TFs. Many MYB transcription factors function in many physiological and biochemical processes, including cell development, cell cycle, hormone cycle, primary and secondary metabolism and signal transduction, biotic and abiotic stresses [89]. There are many studies reporting that transgenic plants enhance tolerance by overexpression of MYB transcription factors (**Table 7**). GmMYBJ1 introduced transgenic Arabidopsis plants have been shown to have an increased tolerance to drought and salinity compared to wild-type plants. At the same time, these plants showed higher plant height and less water loss rate when exposed to dehydration and cold stresses. Moreover, less MDA accumulation was detected in transgenic plants under stress with the overexpression of MYB [115]. TaMYB3R1 (*Triticum aestivum*) transcription factors provided drought tolerance by promoting the closure of stomata and limiting transpiration in Arabidopsis plants under stress [116].

The basic leucine zipper (bZIP) is a class of transcription factors with a conserved bZIP domain consisting of a region for DNA binding and nuclear localization at the N-terminus and a lysine-rich region for dimerization at the C-terminus. Similar to other TFs, it not only plays a role in developmental processes but also plays a regulatory role in the response to various abiotic stresses. In this context, *Nicotiana tabacum* plants in which the LrbZIP (*Nelumbo nucifera*) transcription factor was transferred have developed tolerance to salt stress by showing less electrolyte leakage and higher chlorophyll content under salt stress [137]. Another bZIP, group S transcription factor CaBZ1 (*Capsicum annuum*), reduced the rate of water loss of *Solanum tuberosum* plants under drought stress and increased the expression of ABA and stress-related genes with faster closure [110].

Large regulatory WRKY transcription factors protein family contain two conserved WRKY domains, consisting of approximately 60 amino acids containing the amino site sequence WRKYGQK at the N-terminus and zinc-finger motif at the C-terminus [138]. WRKYs manage many developmental and physiological processes including leaf senescence, regulation of biosynthetic pathways, hormone signaling, embryogenesis, trichome development. Various WRKY transcription factors are also known to be involved in abiotic stress responses. Overexpression of transcription factors, such as ZmWRKY33 (*Zea mays*), that activate various stress-induced genes plays important roles in the acquisition of stress tolerance [131]. Moreover, WRKY TFs increased physiological characteristics such as seed germination, root length and chlorophyll content in transgenic plants under stress conditions [132, 134]. Transgenic *Arobidopsis thaliana* plants developed with the transfer of TaWRKY79 (*Triticum aestivum*) gene have gained tolerance to salinity and ion stress with the improvement in their capacity to elongate their primary roots under salt stress. Transgenic plants developed by the transfer of the GhWRKY34 (*Gossypium hirsutum*) gene have similarly increased tolerance to salt stress, on the other hand, they contain a lower rate of Na^+/K^+ in their leaves and roots [134]. They also act as regulators in the ABA-related pathway [132].

Gene Action	Gene Source	Target Traits	Transformed Plant	Transfer Method	Ref.
AP2-ERFBP	GmERF3 from *Glycine max*	Salinity Drought	*Nicotiana tabacum*	Agrobacterium-mediated transformation	[90]
DREB1	OsDREB1D from *Oryza sativa*	Cold Salinity	*Arabidopsis thaliana*	*Agrobacterium tumefaciens* GV3101	[90]
AP2-ERFBP	SbDREB2 from *Sorghum bicolor*	Drought	*Oryza sativa subsp indica,*	*Agrobacterium tumefaciens* LBA4404	[91]
DREB1A	AtDREB1A from *Arabidopsis thaliana*	Drought	*Glycine max*	Particle-bombardment transformation	[92]
AP2-ERFBP	LcDREB2 from *Leymus chinensis*	Salinity	*Arabidopsis thaliana*	*Agrobacterium tumefaciens* EHA105	[93]
AP2-ERFBP	OsERF4a from *Oryza sativa*	Drought	*Oryza sativa,*	*Agrobacterium tumefaciens* LBA4404	[94]
AP2ERFB	EaDREB2 from *Erianthus arundinaceus*	Drought Salinity	*Saccharum spp. hybrid,*	*Agrobacterium tumefaciens* LBA4404	[95]
DREB1A	AtDREB1A from *Arabidopsis thaliana*	Drought	*Oryza sativa*	*Agrobacterium tumefaciens* LBA4404	[96]
AP2-ERFBP	LcERF054 from *Lotus corniculatus*	Salinity	*Arabidopsis thaliana*	*Agrobacterium tumefaciens* GV3101 floral dip	[97]
AP2-ERFBP	TaPIE1 from *Triticum aestivum*	Freezing	*Triticum aestivum*	Biolistic bombardment	[98]
AP2-ERFBP	VrDREB2A from *Vigna radiata*	DroughtSalinity	*Arabidopsis thaliana*	*Agrobacterium tumefaciens* EHA105 floral dip	[99]
AP2-ERF	OsEREBP1 from *Oryza sativa*	Drought	*Oryza sativa,*	*Agrobacterium tumefaciens* EHA105	[100]
AP2-ERFBP	SsDREB from *Suaeda salsa*	Salt Drought	*Nicotiana tabacum*	*Agrobacterium tumefaciens* EHA105	[101]
bZIP	PtrABF from *Poncirus trifoliata*	Drought	*Nicotiana tabacum*	*Agrobacterium tumefaciens* EHA105	[102]
bZIP	GmbZIP1 from *Glycine max*	DroughtSalinity Cold	*Nicotiana tabacum-Arabidopsis thaliana*	*Agrobacterium tumefaciens* EHA105	[103]
bZIP	OsbZIP39 from *Oryza sativa*	ER stress	*Oryza sativa*	*Agrobacterium tumefaciens* EHA105	[104]
bZIP	MsbZIP from *Medicago sativa*	Drought Salinity	*Nicotiana tabacum*	*Agrobacterium tumefaciens* LBA4404	[105]
bZIP	ZmbZIP72 from *Zea Mays*	Drought Salinity	*Arabidopsis thaliana*	*Agrobacterium tumefaciens* GV3101	[106]
bZIP	LrbZIP from *Nelumbo nucifera*	Salinity	*Nicotiana tabacum*	*Agrobacterium tumefaciens* GV3101 leaf disc method	[107]

Gene Action	Gene Source	Target Traits	Transformed Plant	Transfer Method	Ref.
bZIP	OsbZIP71 from *Oryza sativa*	Drought Salinity	*Oryza sativa*	*Agrobacterium tumefaciens* AGL1	[108]
bZIP	TabZIP60 from *Triticum aestivum*	DroughtSalinity Freezing	*Arabidopsis thaliana*	*Agrobacterium tumefaciens* GV3101 floral dip	[109]
bZIP	CaBZ1 from *Capsicum annuum*	Drought	*Solanum tuberosum*	*Agrobacterium tumefaciens* LBA4404	[110]
bZIP	AtTGA4 from *Arabidopsis thaliana*	Drought	*Arabidopsis thaliana*	*Agrobacterium tumefaciens* floral dip	[111]
MYB	AtMYB15 from *Arabidopsis thaliana*	Drought Salinity	*Arabidopsis thaliana*	Agrobacterium-mediated floral dip	[112]
MYB	TaPIMP1 from *Triticum aestivum*	Drought Salinity	*Nicotiana tabacum*	Agrobacterium-mediated leaf disc method	[113]
MYB	LcMYB1 from *Leymus chinensis*	Salinity	*Arabidopsis thaliana*	*Agrobacterium tumefaciens* EHA105	[114]
MYB	GmMYBJ1 from *Glycine max*	Drought Cold	*Arabidopsis thaliana*	*Agrobacterium tumefaciens* EHA105	[115]
MYB	TaMYB3R1 from *Triticum aestivum*	Drought Salinity	*Arabidopsis thaliana*	*Agrobacterium tumefaciens* EHA105	[116]
MYB	LeAN2 from *Lycopersicum esculentum*	Heat	*Lycopersicum esculentum*	*Agrobacterium tumefaciens* LBA4404 leaf disc method	[117]
MYB	SbMYB2 and SbMYB7 from *Scutellaria baicalensis*	Salinity	*Nicotiana tabacum*	-	[118]
NAC	ONAC063 from *Oryza sativa*	Salinity Osmotic stress	*Arabidopsis thaliana*	*Agrobacterium tumefaciens* GV3101 floral dip	[119]
NAC	GmNAC20 GmNAC11 from *Glycine max*	Salt Freezing	*Arabidopsis thaliana*	Agrobacterium-mediated vacuum infiltration	[120]
NAC	ZmSNAC1 from *Zea may*	Cold Salinity Drought	*Arabidopsis thaliana*	*Agrobacterium tumefaciens* GV3101 floral dip	[121]
NAC	TaNAC2a from *Triticum aestivum*	Drought	*Nicotiana tabacum*	Agrobacterium-mediated transformation	[122]
NAC	AhNAC3 from *Arachis hypogaea*	Drought	*Nicotiana tabacum*	-	[123]
NAC	SNAC1 from *Oryza sativa*	Drought Salinity	*Triticum aestivum*	Bombardment with a biolistic gun	[124]
NAC	OsNAP from *Oryza sativa*	Cold Salinity Drought	*Oryza sativa*	*Agrobacterium tumefaciens* EHA105	[125]

Gene Action	Gene Source	Target Traits	Transformed Plant	Transfer Method	Ref.
NAC	TaNAC67 from *Triticum aestivum*	Cold Salinity Drought	*Arabidopsis thaliana*	*Agrobacterium tumefaciens* floral infiltration	[87]
NAC	TaNAC29 from *Triticum aestivum*	Drought Salinity	*Arabidopsis thaliana*	*Agrobacterium tumefaciens* EHA105	[86]
NAC	MLNAC5 from *Miscanthus lutarioriparius*	Drought Cold	*Arabidopsis thaliana*	*Agrobacterium tumefaciens* EHA105 floral dip	[126]
NAC	TaNAC47 from *Triticum aestivum*	Salt Cold Drought	*Arabidopsis thaliana*	*Agrobacterium tumefaciens* GV3101	[127]
WRKY	OsWRKY45 from *Oryza sativa*	Drought Salinity	*Arabidopsis thaliana*	*Agrobacterium tumefaciens* LBA4404	[128]
WRKY	VvWRKY11 from *Vitis vinifera*	Drought	*Arabidopsis thaliana*	*Agrobacterium tumefaciens* GV3101	[129]
WRKY	AtWRKY28 from *Arabidopsis thaliana*	Salinity	*Arabidopsis thaliana*	*Agrobacterium tumefaciens* EHA105 floral dip	[130]
WRKY	ZmWRKY33 from *Zea may*	Salinity	*Arabidopsis thaliana*	*Agrobacterium tumefaciens* EHA105	[131]
WRKY	TaWRKY79 from *Triticum aestivum*	Drought	*Arabidopsis thaliana*	*Agrobacterium tumefaciens* GV3101 floral dip	[132]
WRKY	BdWRKY36 from *Brachypodium distachyon*	Drought	*Nicotiana tabacum*	*Agrobacterium tumefaciens* EHA105 leaf disc method	[133]
WRKY	GhWRKY34 from *Gossypium hirsutum*	Salinity	*Arabidopsis thaliana*	*Agrobacterium tumefaciens* LBA4404 floral dip	[134]
WRKY	MtWRKY76 from *Medicago truncatula*	Drought Salinity	*Medicago truncatula*	Agrobacterium-mediated transformation	[135]
HomeoBox 4	HaHB4 from *Helianthus annuus*	Drought	*Glycine max-Triticum vulgaris*	-	[136]

Table 7.
Summary of transgenic plants over-expressing transcription factor genes for abiotic stress tolerance.

The Apetala2/Ethylene responsive factor family is characterized by the highly conserved APETALA2 (AP2)/Ethylene Responsive Element Binding Factor (EREB) domain containing 40–70 amino acid sequences involved in DNA binding. AP2/ERFs are divided into four main groups; Apetala2 (AP2), related to abscisic acid intensive 3/Viviparous (RAV), dehydration-responsive element binding protein (DREB), ethylene responsive factor (ERF). These transcription factors play an important role in the regulation of both plant growth and the response to various stresses [139]. The AP2 family of Ethylene Responsive Element Binding Factors supports the survival of plants under stress by activating jasmonate and abscisic acid signaling pathways. The amount of α linolenate, some jasmonate derivatives and abscisic acid increased in rice plants, in which the EREB1 transcription factor gene from AP2/ERF transcription factor family was transferred. The role of this gene family in the response to both biotic and abiotic stress is very important for the regulation of multiple stress tolerance for genetic engineering [100]. Dehydration responsive element-binding proteins (DREB) cooperated with the genes involved in polyamine biosynthesis to enable plants to tolerate salt stress [93]. Thanks to the transfer of another subgroup of ERF transcription factor LcERF054 (*Leymus chinensis*) gene, an increase in relative moisture content, dissolved sugar and proline amount occurred in Arabidopsis plants. Moreover, these plants have developed salinity stress tolerance. At the same time, the expression of the hyperosmatic salinity stress response genes COR15A, LEA4-5, P5CS1 and RD29A increased with the expression of the LcERF054 gene [97]. It has been reported that DREB transcription factor is used effectively in stress tolerance like other transcription factors belonging to the Apetala2/Ethylene responsive factor family [96].

3. Genome editing applications for abiotic stress tolerance in plants on emphasis to recent CRISPR applications

Since the bacterial defense mechanism against biotic agents as viruses based on the detection and elimination of invader nucleic acids suggested as novel tool for site specific genome editing tool by Jinek [140] in 2012, CRISPR/Cas9 is used in developing crop plants for abiotic stress tolerance as well as many other plant biotechnology applications. All of the abiotic stress tolerance mechanisms mentioned earlier in this chapter which were targets for gene transfer applications can be targeted also by CRISPR/Cas9 system as it allows both induction (CRISPRactivation) or repression (CRISPRinterference) of genes. Therefore, it can be used for activation of tolerance genes (T genes) as well as suppressing sensitivity (S genes) genes [141].

Targeting hormonal regulation for abiotic stress tolerance is one of the viable options. Abscisic acid (ABA) regulates physiological processes as seed dormancy, stomatal closure, plant development, as well as plant responses to environmental stimuli and multiple stresses. 9-cis-epoxycarotenoid dioxygenase (NCED), which is responsible from regulation of ABA in rice, targeted for multiple abiotic stress tolerance including salinity, drought and subsequent H_2O_2 stress. *nced3* mutant rice plants were very susceptible to these stresses as they express very low level of ABA. Overexpression of *OsNCEB3* in rice by CRISPRa system increased ABA accumulation and tolerance to salinity and drought stresses [142]. Downstream genes of the ABA signaling pathway in response to drought stress are regulated by ABA-responsive element binding proteins/ABRE binding factors (AREB/ABFs). CRISPRa system, set by CRISPR/dCAS9 fusion with histone acetyltransferase which promotes gene expression, is used to enhance drought resistance of Arabidopsis plants through activation of endogenous promoter of *AREB1* [143].

Transcription factors (TFs) regulate the expression patterns of the genes on promoter regions. Therefore, they are important targets for both genetic engineering and genome editing applications in abiotic stress tolerance of plants. CRISPRi system designed for 696–amino acid B-type response regulator transcription factor encoding *OsRR22* gene successfully improved salinity tolerance of rice plants since its involvement in cytokinin signal transduction and metabolism [144]. In maize, ARGOS genes regulate ethylene signal transduction. It is known to be a negative regulator of ethylene responses. ARGOS genes are reported to increase ethylene sensitivity of Arabidopsis and maize when they were overexpressed through transgenic approach and enhance drought tolerance in plants [145]. Likewise, 400 inbred lines of maize were evaluated for ARGOS8 mRNA expression comparing to the transgenic ARGOS8 events. All inbred lines presented less expression than transgenic events. Therefore, overexpression of ARGOS8 also achieved by using CRISPRa approach and improved traits as drought tolerance and grain yield were generated [146]. Auxin response factors (ARFs) undertake important regulation roles in auxin response gene expression. Differential expression of ARFs were shown to involve in many physiological processes and abiotic stress responses. Downregulation and loss of function applications of *SlARF4* mediated by CRISPR/Cas9 presented salinity and osmotic stress tolerance in tomato [147].

Antioxidant scavenging is also very important in abiotic stress tolerance. Hence, elevation of antioxidant enzyme capacities in plants is a legitimate approach to improve tolerance. There are several successful transgenic application as described before. *Oryza sativa* stress-related RING Finger Protein 1 RING finger (OsSRFP1) which is a E3 ligase is responsive to cold, dehydration, salt, H_2O_2 and abscisic acid. However, overexpression of OsSRFP1 leads increased sensitivity to all of these factors. Transgenic RNAi silenced OsSRFP1 mutant plants obtained enhanced salinity and cold tolerance due to increased antioxidant capacity. It is also proposed as potential candidate target for CRISPRi applications [148, 149].

Ion homeostasis is another key factor in abiotic stress tolerance in plants especially against salinity. In Cucurbitaceae family, salt sensitive cucumber and salt tolerant pumpkin varieties were compared in context of K^+ uptake during salt stress and salt tolerant pumpkins were found superior to cucumber plants for this trait. CRISPRi knocking out of NADPH oxidase (respiratory burst oxidase homolog D; RBOHD), which was previously tested by its inhibitor diphenylene iodonium, presented decrease in salinity tolerance. On the contrary, ectopic expression of pumpkin RBOHD in Arabidopsis enhanced the salinity tolerance [150].

Along with CRISPRa and CRISPRi application to enhance tolerance to abiotic stresses, it can also be used to identify roles of particular genes in abiotic stress tolerance. Nonexpressor of pathogenesis-related gene 1 (NPR1) is well known gene in plant pathogen response. To evaluate its involvement in drought stress CRISPRi derived tomato *slnpr1* mutants presented reduced drought tolerance which presented that gene's role in tolerance [151]. Likewise, CRISPRi application on mitogen-activated protein kinases (MAPKs) presented role of these signal molecules on drought tolerance of tomato plants through generating *slmapk3* mutants with altered drought stress responsive gene expression including SlLOX, SlGST, and SlDREB [152]. Ca^{2+}-dependent phospholipid-binding proteins called annexins were targeted by CRISPRi system. *OsAnn3* knockout rice mutants generated by this method were susceptible to cold stress indicating their role in mechanism [153]. Tomato is also very susceptible plant for cold and chilling stress. C-repeat binding factors (CBFs) were tested their potential role in chilling resistance. slcbf1 mutants generated by CRISPRi system presented symptoms of chilling more severely comparing to the wild-type relatives [154].

Despite all of these present efforts and obviously more to come in near future, there are some bottlenecks on this topic beyond the genome editing technique itself. The topic of law regulations and public opinion on genetically engineered plants still have many uncertainties. In these days, when many countries regulated or starting to regulated their laws for genetically engineered plants, CRISPR/Cas9 derived products started another debate due to the lack of standardized detection methods and difficulty to distinguish these changes from naturally occurring mutations. Court of Justice of the European Union (ECJ) ruled their opinion on genome edited products in 25 July 2018 as they are to be subjected under the same stringent regulations previously determined for genetically engineered products in 2001 directives. On the other hand, United States Department of Agriculture (USDA) announced their opinion as regulation is not needed for genome edited mutations since they can already occur naturally. Likewise, some countries as Brazil, Argentina and Australia shared the same opinion. In 2016, France inquired to ECJ about reviewing 2001 directive in favor of genome editing techniques as many researchers share their opinion as this method should be evaluated similar to use of radiation in mutation breeding allowing random mutations on genome. But the deliberate and intentional nature of the genome editing mutations brings the opposition. Still, Calyxt, an agricultural biotechnology company in Minnesota, announced the first US sale of high-oleic-acid oil product generated from gene edited soybeans in February, 2019. Following this development, Intrexon Company in Maryland announced start of commercial non-browning gene edited lettuce trials [13, 14].

Woo [155] suggested a novel approach to overcome this dilemma as DNA-free genome editing. In contrary to existing CRISPR/Cas9 system which is based on delivery of RNA guided nucleases into plants either by *Agrobacterium tumefaciens* or transfecting plasmids, they suggested delivery of pre-assembled Cas9-gRNA ribonucleoprotein (RNPs) into protoplast cells of *Arabidopsis thaliana*, tobacco, lettuce and rice plants instead of plasmids encoding these required components. They modified six genes in four different plants and these targeted mutations were maintained in plants regenerated from these protoplasts. Since, this system is free from involvement of recombinant DNA, they shared their opinion as this gene editing should be exempt from current genetically engineered plant regulations. This, direct delivery of purified CRISPR/Cas9 RNPs system is also adopted for other plants. MLO-7, a susceptible gene responsible from increasing resistance to powdery mildew disease in grapevine and DIPM-1, DIPM-2, DIPM-4 which are responsible from increased resistance to fire blight disease in apple were edited through DNA-free genome editing approach [156]. Due to the fact that regeneration of whole plant from the protoplast cells can be challenging in some plants especially in monocots as hexaploid bread wheat. Liang [157] reported use of particle bombardment delivery system into immature embryo cells for CRISPR/Cas9 RNPs system. The Cpf1 protein was suggested as an alternative to type II CRISPR-Cas system in 2015 [158]. Unlike Cas9 protein which recognizes proximal 3′-G-rich PAM sequences, Cpf1 recognized 5′-T-rich PAM sequences. Cpf1 is also a ribonuclease which processes precursor crRNAs and does not require trans-acting crRNA for guidance. Merging these two approaches is also achieved as CRISPR/Cpf1-mediated DNA-free plant genome editing. Soybean and tobacco protoplast genomes were successfully edited by this approach. Researchers suggested three potential advantages of this method as: no foreign DNA insertion, larger deletions in target site comparing to the Cas9 and different cleavage pattern of Cpf1 which may assist NHEJ-mediated insertion of donor DNAs [159].

In conclusion, following the footsteps of the data generated by the genetic engineering application for abiotic stress tolerance in crop plants, the most recent developments in genome editing techniques as transgene free approach combined with superior endonucleases provides promising results on this topic from many perspectives as law regulations, public acceptance, technical issues.

Author details

Sinan Meriç*, Alp Ayan and Çimen Atak
Department of Molecular Biology and Genetic, Faculty of Science and Letters, Istanbul Kultur University, Ataköy, Istanbul, Turkey

*Address all correspondence to: s.meric@iku.edu.tr

References

[1] S. Fahad, L. Nie, Y. Chen, C. Wu, D. Xiong, S. Saud, L. Hongyan, K. Cui, J. Huang, Crop Plant Hormones and Environmental Stress, in: Springer, Cham, 2015: pp. 371-400. doi:10.1007/978-3-319-09132-7_10.

[2] S. Fahad, S. Hussain, S. Saud, F. Khan, S. Hassan, Amanullah, W. Nasim, M. Arif, F. Wang, J. Huang, Exogenously Applied Plant Growth Regulators Affect Heat-Stressed Rice Pollens, J. Agron. Crop Sci. 202 (2016) 139-150. doi:10.1111/jac.12148.

[3] A. Frary, S.D.-T.J. of A. and, undefined 2003, Comparative genetics of crop plant domestication and evolution, n.d. https://journals.tubitak.gov.tr/agriculture/abstract.htm?id=6104 (accessed September 29, 2020).

[4] C. Wu, S. Tang, G. Li, S. Wang, S. Fahad, Y. Ding, Roles of phytohormone changes in the grain yield of rice plants exposed to heat: a review, PeerJ. 7 (2019) e7792. doi:10.7717/peerj.7792.

[5] D.J.-N.W.N.& Company, undefined 1997, Guns, germs, and steel: the fates of human societies, (n.d.).

[6] M. Somssich, A Short History of Plant Transformation 5, (n.d.). doi:10.7287/peerj.preprints.27556v2.

[7] J. Fernandez-Cornejo, S. Wechsler, M. Livingston, L. Mitchell, Genetically engineered crops in the United States, in: Genet. Eng. Crop. Am. Anal. Adopt. Trends, Nova Science Publishers, Inc., 2014: pp. 1-74. doi:10.2139/ssrn.2503388.

[8] N. Sreenivasulu, S.K. Sopory, P.B. Kavi Kishor, Deciphering the regulatory mechanisms of abiotic stress tolerance in plants by genomic approaches, Gene. 388 (2007) 1-13. doi:10.1016/j.gene.2006.10.009.

[9] P. Bhatnagar-Mathur, V. Vadez, K.K. Sharma, Transgenic approaches for abiotic stress tolerance in plants: Retrospect and prospects, Plant Cell Rep. 27 (2008) 411-424. doi:10.1007/s00299-007-0474-9.

[10] S. Fahad, S. Hussain, A. Matloob, F.A. Khan, A. Khaliq, S. Saud, S. Hassan, D. Shan, F. Khan, N. Ullah, M. Faiq, M.R. Khan, A.K. Tareen, A. Khan, A. Ullah, N. Ullah, J. Huang, Phytohormones and plant responses to salinity stress: a review, Plant Growth Regul. 75 (2015) 391-404. doi:10.1007/s10725-014-0013-y.

[11] D. Lavania, A.K. Singh, M. Al-Whaibi, Abiotic Stress Tolerant Transgenic Plants and Nanotechnology Deciphering the Drought and Thermo-tolerance Mechanisms in Plants: The Road Ahead of Future Research View project, Springer. (2015) 165-181. doi:10.1007/978-3-319-14502-0_9.

[12] H. Teng, B. Shen, E. Liu, J. Zhang, X. Peng, Responsiveness comparison of three stress inducible promoters in transgenic rice, Acta Physiol. Plant. 40 (2018) 179. doi:10.1007/s11738-018-2753-1.

[13] E.C.- Nature, undefined 2018, CRISPR plants now subject to tough GM laws in European Union, Go.Gale.Com. (n.d.). https://go.gale.com/ps/i.do?id=GALE%7CA572745870&sid=googleScholar&v=2.1&it=r&linkaccess=abs&issn=00280836&p=AONE&sw=w (accessed October 3, 2020).

[14] H. Ledford, CRISPR conundrum: Strict European court ruling leaves food-testing labs without a plan, Nature. 572 (2019) 15. doi:10.1038/d41586-019-02162-x.

[15] Commercial GM Trait:Abiotic Stress Tolerance| GM Approval

Database - ISAAA.org, (n.d.). http://www.isaaa.org/gmapprovaldatabase/commercialtrait/default.asp (accessed October 1, 2020).

[16] Living Modified Organism (LMO) Registry, (n.d.). http://bch.cbd.int/database/lmo-registry/ (accessed October 1, 2020).

[17] M. Ilyas, M. Nisar, N. Khan, A. Hazrat, A.H. Khan, K. Hayat, S. Fahad, A. Khan, A. Ullah, Drought Tolerance Strategies in Plants: A Mechanistic Approach, J. Plant Growth Regul. (2020). doi:10.1007/s00344-020-10174-5.

[18] S.P. dos Reis, D.N. Marques, N.L. Ferreira Barros, C. de Nazaré Monteiro Costa, C.R. Batista de Souza, Genetically engineered food crops to abiotic stress tolerance, in: Genet. Eng. Foods, Elsevier Inc., 2018: pp. 247-279. doi:10.1016/B978-0-12-811519-0.00010-8.

[19] S. Fahad, A.A. Bajwa, U. Nazir, S.A. Anjum, A. Farooq, A. Zohaib, S. Sadia, W. Nasim, S. Adkins, S. Saud, M.Z. Ihsan, H. Alharby, C. Wu, D. Wang, J. Huang, Crop production under drought and heat stress: Plant responses and management options, Front. Plant Sci. 8 (2017) 1147. doi:10.3389/fpls.2017.01147.

[20] S. Hayat, Q. Hayat, M.N. Alyemeni, A.S. Wani, J. Pichtel, A. Ahmad, Role of proline under changing environments: A review, Plant Signal. Behav. 7 (2012). doi:10.4161/psb.21949.

[21] O. Fernandez, L. Béthencourt, A. Quero, R.S. Sangwan, C. Clément Christophe, Trehalose and plant stress responses: Friend or foe?, Trends Plant Sci. 15 (2010) 409-417. doi:10.1016/j.tplants.2010.04.004.

[22] C.-J. Park, Y.-S. Seo, Heat Shock Proteins: A Review of the Molecular Chaperones for Plant Immunity., Plant Pathol. J. 31 (2015) 323-33. doi:10.5423/PPJ.RW.08.2015.0150.

[23] I. Amara, I. Zaidi, K. Masmoudi, ... M.L.-A.J. of, undefined 2014, Insights into late embryogenesis abundant (LEA) proteins in plants: from structure to the functions, Scirp.Org. (n.d.). https://www.scirp.org/html/7-2601719_51811.htm (accessed October 5, 2020).

[24] C. Maurel, Y. Boursiac, D.T. Luu, V. Santoni, Z. Shahzad, L. Verdoucq, Aquaporins in plants, Physiol. Rev. 95 (2015) 1321-1358. doi:10.1152/physrev.00008.2015.

[25] M. der Shih, F.A. Hoekstra, Y.I.C. Hsing, Chapter 4 Late Embryogenesis Abundant Proteins, in: Adv. Bot. Res., Academic Press Inc., 2008: pp. 211-255. doi:10.1016/S0065-2296(08)00404-7.

[26] S.C. Hand, M.A. Menze, M. Toner, L. Boswell, D. Moore, LEA proteins during water stress: Not just for plants anymore, Annu. Rev. Physiol. 73 (2011) 115-134. doi:10.1146/annurev-physiol-012110-142203.

[27] D. Waterer, N.T. Benning, G. Wu, X. Luo, X. Liu, M. Gusta, A. McHughen, L. V. Gusta, Evaluation of abiotic stress tolerance of genetically modified potatoes (solanum tuberosum cv. desiree), Mol. Breed. 25 (2010) 527-540. doi:10.1007/s11032-009-9351-2.

[28] V.G. Checker, A.K. Chhibbar, P. Khurana, Stress-inducible expression of barley Hva1 gene in transgenic mulberry displays enhanced tolerance against drought, salinity and cold stress, Springer. (n.d.). doi:10.1007/s11248-011-9577-8.

[29] J. Duan, W. Cai, OsLEA3-2, an Abiotic Stress Induced Gene of Rice Plays a Key Role in Salt and Drought Tolerance, PLoS One. 7 (2012) e45117. doi:10.1371/journal.pone.0045117.

[30] I. Amara, M. Capellades, M.D. Ludevid, M. Pagès, A. Goday, Enhanced water stress tolerance of transgenic maize plants over-expressing LEA Rab28 gene, J. Plant Physiol. 170 (2013) 864-873. doi:10.1016/j.jplph.2013.01.004.

[31] J. Liang, M. Zhou, X. Zhou, Y. Jin, M. Xu, J. Lin, JcLEA, a novel LEA-like protein from Jatropha curcas, confers a high level of tolerance to dehydration and salinity in Arabidopsis thaliana, PLoS One. 8 (2013). doi:10.1371/journal.pone.0083056.

[32] Y. Liu, L. Wang, X. Xing, L. Sun, ... J.P.-P. and cell, undefined 2013, ZmLEA3, a Multifunctional Group 3 LEA Protein from Maize (Zea mays L.), is Involved in Biotic and Abiotic Stresses, Academic.Oup.Com. (n.d.). https://academic.oup.com/pcp/article-abstract/54/6/944/1838431 (accessed October 5, 2020).

[33] K. Sasaki, N.K. Christov, S. Tsuda, R. Imai, Identification of a novel LEA protein involved in freezing tolerance in wheat, Plant Cell Physiol. 55 (2014) 136-147. doi:10.1093/pcp/pct164.

[34] Y. Liu, L. Wang, S. Jiang, J. Pan, G. Cai, D. Li, Group 5 LEA protein, ZmLEA5C, enhance tolerance to osmotic and low temperature stresses in transgenic tobacco and yeast, Plant Physiol. Biochem. 84 (2014) 22-31. doi:10.1016/j.plaphy.2014.08.016.

[35] M. Wang, P. Li, C. Li, Y. Pan, X. Jiang, D. Zhu, Q. Zhao, J. Yu, SiLEA14, a novel atypical LEA protein, confers abiotic stress resistance in foxtail millet, BMC Plant Biol. 14 (2014) 1-16. doi:10.1186/s12870-014-0290-7.

[36] A. Sharma, D. Kumar, S. Kumar, S. Rampuria, A.R. Reddy, P.B. Kirti, Ectopic expression of an atypical hydrophobic group 5 LEA protein from wild peanut, Arachis diogoi confers abiotic stress tolerance in tobacco, PLoS One. 11 (2016) e0150609. doi:10.1371/journal.pone.0150609.

[37] A. Madeira, T.F. Moura, G. Soveral, Detecting aquaporin function and regulation, Front. Chem. 4 (2016). doi:10.3389/fchem.2016.00003.

[38] S.E. Nilson, S.M. Assmann, The control of transpiration. Insights from arabidopsis, Plant Physiol. 143 (2007) 19-27. doi:10.1104/pp.106.093161.

[39] G. Li, V. Santoni, C. Maurel, Plant aquaporins: Roles in plant physiology, Biochim. Biophys. Acta - Gen. Subj. 1840 (2014) 1574-1582. doi:10.1016/j.bbagen.2013.11.004.

[40] M. Ayadi, D. Cavez, N. Miled, F. Chaumont, K. Masmoudi, Identification and characterization of two plasma membrane aquaporins in durum wheat (Triticum turgidum L. subsp. durum) and their role in abiotic stress tolerance, Plant Physiol. Biochem. 49 (2011) 1029-1039. doi:10.1016/j.plaphy.2011.06.002.

[41] J. Zhang, D. Li, D. Zou, F. Luo, X. Wang, Y. Zheng, X. Li, A cotton gene encoding a plasma membrane aquaporin is involved in seedling development and in response to drought stress, Acta Biochim. Biophys. Sin. (Shanghai). 45 (2013) 104-114. doi:10.1093/abbs/gms096.

[42] C. Liu, T. Fukumoto, T. Matsumoto, P. Gena, D. Frascaria, T. Kaneko, M. Katsuhara, S. Zhong, X. Sun, Y. Zhu, I. Iwasaki, X. Ding, G. Calamita, Y. Kitagawa, Aquaporin OsPIP1;1 promotes rice salt resistance and seed germination, Plant Physiol. Biochem. 63 (2013) 151-158. doi:10.1016/j.plaphy.2012.11.018.

[43] L. Zhou, C. Wang, R. Liu, Q. Han, R.K. Vandeleur, J. Du, S. Tyerman, H. Shou, Constitutive overexpression of soybean plasma membrane intrinsic protein GmPIP1;6 confers salt tolerance, BMC Plant Biol. 14 (2014) 1-13. doi:10.1186/1471-2229-14-181.

[44] R. Porcel, A. Bustamante, R. Ros, R. Serrano, J.M. Mulet Salort, BvCOLD1: A novel aquaporin from sugar beet (*Beta vulgaris* L.) involved in boron homeostasis and abiotic stress, Plant. Cell Environ. 41 (2018) 2844-2857. doi:10.1111/pce.13416.

[45] M.M. Wudick, D.T. Luu, C. Maurel, A look inside: Localization patterns and functions of intracellular plant aquaporins, New Phytol. 184 (2009) 289-302. doi:10.1111/j.1469-8137.2009.02985.x.

[46] I. Delorge, M. Janiak, S. Carpentier, P. Van Dijck, Fine tuning of trehalose biosynthesis and hydrolysis as novel tools for the generation of abiotic stress tolerant plants, Front. Plant Sci. 5 (2014). doi:10.3389/fpls.2014.00147.

[47] L. O'Hara, M. Paul, A.W.-M. plant, undefined 2013, How do sugars regulate plant growth and development? New insight into the role of trehalose-6-phosphate, Elsevier. (n.d.). https://www.sciencedirect.com/science/article/pii/S1674205214600897 (accessed October 5, 2020).

[48] M.C.F.R. Redillas, S.H. Park, J.W. Lee, Y.S. Kim, J.S. Jeong, H. Jung, S.W. Bang, T.R. Hahn, J.K. Kim, Accumulation of trehalose increases soluble sugar contents in rice plants conferring tolerance to drought and salt stress, Plant Biotechnol. Rep. 6 (2012) 89-96. doi:10.1007/s11816-011-0210-3.

[49] J. Il Lyu, S.R. Min, J.H. Lee, Y.H. Lim, J.K. Kim, C.H. Bae, J.R. Liu, Overexpression of a trehalose-6-phosphate synthase/phosphatase fusion gene enhances tolerance and photosynthesis during drought and salt stress without growth aberrations in tomato, Plant Cell. Tissue Organ Cult. 112 (2013) 257-262. doi:10.1007/s11240-012-0225-7.

[50] A. Hmida-Sayari, R. Gargouri-Bouzid, A. Bidani, L. Jaoua, A. Savouré, S. Jaoua, Overexpression of Δ1-pyrroline-5-carboxylate synthetase increases proline production and confers salt tolerance in transgenic potato plants, Plant Sci. 169 (2005) 746-752. doi:10.1016/j.plantsci.2005.05.025.

[51] K. De Carvalho, • Marília, K. Freitas De Campos, • Douglas, S. Domingues, • Luiz, F.P. Pereira, L. Gonzaga, E. Vieira, The accumulation of endogenous proline induces changes in gene expression of several antioxidant enzymes in leaves of transgenic Swingle citrumelo, Springer. (n.d.). doi:10.1007/s11033-012-2402-5.

[52] S. Ahmad, M. Kamran, R. Ding, X. Meng, H. Wang, I. Ahmad, S. Fahad, Q. Han, Exogenous melatonin confers drought stress by promoting plant growth, photosynthetic capacity and antioxidant defense system of maize seedlings, PeerJ. 2019 (2019) e7793. doi:10.7717/peerj.7793.

[53] G.B. Kim, Y.W. Nam, A novel Δ1-pyrroline-5-carboxylate synthetase gene of Medicago truncatula plays a predominant role in stress-induced proline accumulation during symbiotic nitrogen fixation, J. Plant Physiol. 170 (2013) 291-302. doi:10.1016/j.jplph.2012.10.004.

[54] S.K. Kumar Ghanti, K.G. Sujata, B.M.V. Kumar, N.N. Karba, K.J. Reddy, M.S. Rao, P.B.K. Kishor, Heterologous expression of P5CS gene in chickpea enhances salt tolerance without affecting yield, Biol. Plant. 55 (2011) 634-640. doi:10.1007/s10535-011-0161-0.

[55] A. Karthikeyan, S.K. Pandian, M. Ramesh, Transgenic indica rice cv. ADT 43 expressing a Δ1-pyrroline-5-carboxylate synthetase (P5CS) gene from Vigna aconitifolia demonstrates salt tolerance, Plant Cell. Tissue Organ Cult. 107 (2011) 383-395. doi:10.1007/s11240-011-9989-4.

[56] M. Kondrák, F. Marincs, F. Antal, Z. Juhász, Z. Bánfalvi, Effects of yeast trehalose-6-phosphate synthase 1 on gene expression and carbohydrate contents of potato leaves under drought stress conditions, BMC Plant Biol. 12 (2012) 1-12. doi:10.1186/1471-2229-12-74.

[57] H.W. Li, B.S. Zang, X.W. Deng, X.P. Wang, Overexpression of the trehalose-6-phosphate synthase gene OsTPS1 enhances abiotic stress tolerance in rice, Planta. 234 (2011) 1007-1018. doi:10.1007/s00425-011-1458-0.

[58] C. Surekha, K.N. Kumari, L. V. Aruna, G. Suneetha, A. Arundhati, P.B. Kavi Kishor, Expression of the Vigna aconitifolia P5CSF129A gene in transgenic pigeonpea enhances proline accumulation and salt tolerance, Plant Cell. Tissue Organ Cult. 116 (2014) 27-36. doi:10.1007/s11240-013-0378-z.

[59] R. Ahmad, J. Hussain, M. Jamil, M. Duck Kim, S. Kwak, M. Maroof Shah, Glycinebetaine Synthesizing Transgenic Potato Plants Exhibit Enhanced Tolerance To Salt And Cold Stresses, 2014. http://www.pakbs.org/pjbot/PDFs/46(6)/08.pdf (accessed October 5, 2020).

[60] W. Fan, M. Zhang, H. Zhang, P. Zhang, Improved Tolerance to Various Abiotic Stresses in Transgenic Sweet Potato (Ipomoea batatas) Expressing Spinach Betaine Aldehyde Dehydrogenase, PLoS One. 7 (2012) e37344. doi:10.1371/journal.pone.0037344.

[61] X.Z. Fu, E.U. Khan, S.S. Hu, Q.J. Fan, J.H. Liu, Overexpression of the betaine aldehyde dehydrogenase gene from Atriplex hortensis enhances salt tolerance in the transgenic trifoliate orange (Poncirus trifoliata L. Raf.), Environ. Exp. Bot. 74 (2011) 106-113. doi:10.1016/j.envexpbot.2011.05.006.

[62] C. He, A. Yang, W. Zhang, Q. Gao, J. Zhang, Improved salt tolerance of transgenic wheat by introducing betA gene for glycine betaine synthesis, Plant Cell. Tissue Organ Cult. 101 (2010) 65-78. doi:10.1007/s11240-009-9665-0.

[63] C. He, W. Zhang, Q. Gao, A. Yang, X. Hu, J. Zhang, Enhancement of drought resistance and biomass by increasing the amount of glycine betaine in wheat seedlings, Euphytica. 177 (2011) 151-167. doi:10.1007/s10681-010-0263-3.

[64] S.-J. Lai, M.-C. Lai, R.-J. Lee, Y.-H. Chen, · Hungchen, E. Yen, Transgenic Arabidopsis expressing osmolyte glycine betaine synthesizing enzymes from halophilic methanogen promote tolerance to drought and salt stress, Plant Mol Biol. 85 (2014) 429-441. doi:10.1007/s11103-014-0195-8.

[65] S. Li, F. Li, J. Wang, W. Zhang, Q. Meng, T.H.H. Chen, N. Murata, X. Yang, Glycinebetaine enhances the tolerance of tomato plants to high temperature during germination of seeds and growth of seedlings, Plant. Cell Environ. 34 (2011) 1931-1943. doi:10.1111/j.1365-3040.2011.02389.x.

[66] M. Li, Z. Li, S. Li, S. Guo, Q. Meng, G. Li, X. Yang, Genetic Engineering of Glycine Betaine Biosynthesis Reduces Heat-Enhanced Photoinhibition by Enhancing Antioxidative Defense and Alleviating Lipid Peroxidation in Tomato, Springer. (n.d.). doi:10.1007/s11105-013-0594-z.

[67] D. Wei, W. Zhang, C. Wang, Q. Meng, G. Li, T.H.H. Chen, X. Yang, Genetic engineering of the biosynthesis of glycinebetaine leads to alleviate salt-induced potassium efflux and enhances salt tolerance in tomato plants, Plant Sci. 257 (2017) 74-83. doi:10.1016/j.plantsci.2017.01.012.

[68] Z.Y. Li, Z.S. Xu, G.Y. He, G.X. Yang, M. Chen, L.C. Li, Y.Z. Ma, Overexpression of soybean GmCBL1 enhances abiotic stress tolerance

and promotes hypocotyl elongation in Arabidopsis, Biochem. Biophys. Res. Commun. 427 (2012) 731-736. doi:10.1016/j.bbrc.2012.09.128.

[69] Q. Luo, Q. Wei, R. Wang, Y. Zhang, F. Zhang, Y. He, S. Zhou, J. Feng, G. Yang, G. He, BdCIPK31, a Calcineurin B-Like Protein-Interacting Protein Kinase, Regulates Plant Response to Drought and Salt Stress, Front. Plant Sci. 8 (2017) 1184. doi:10.3389/fpls.2017.01184.

[70] L. Dong, Q. Wang, S.M.N. Manik, Y. Song, S. Shi, Y. Su, G. Liu, H. Liu, Nicotiana sylvestris calcineurin B-like protein NsylCBL10 enhances salt tolerance in transgenic Arabidopsis, Plant Cell Rep. 34 (2015) 2053-2063. doi:10.1007/s00299-015-1851-4.

[71] W. Yang, Z. Kong, E. Omo-Ikerodah, W. Xu, Q. Li, Y. Xue, Calcineurin B-like interacting protein kinase OsCIPK23 functions in pollination and drought stress responses in rice (Oryza sativa L.), J. Genet. Genomics. 35 (2008) 531-S2. doi:10.1016/S1673-8527(08)60073-9.

[72] L. Chen, F. Ren, L. Zhou, Q.Q. Wang, H. Zhong, X.B. Li, The Brassica napus Calcineurin B-Like 1/CBL-interacting protein kinase 6 (CBL1/CIPK6) component is involved in the plant response to abiotic stress and ABA signalling, J. Exp. Bot. 63 (2012) 6211-6222. doi:10.1093/jxb/ers273.

[73] D.D. Li, X.L. Xia, W.L. Yin, H.C. Zhang, Two poplar calcineurin B-like proteins confer enhanced tolerance to abiotic stresses in transgenic Arabidopsis thaliana, Biol. Plant. 57 (2013) 70-78. doi:10.1007/s10535-012-0251-7.

[74] R.J. Tang, Y. Yang, L. Yang, H. Liu, C.T. Wang, M.M. Yu, X.S. Gao, H.X. Zhang, Poplar calcineurin B-like proteins PtCBL10A and PtCBL10B regulate shoot salt tolerance through interaction with PtSOS2 in the vacuolar membrane, Plant, Cell Environ. 37 (2014) 573-588. doi:10.1111/pce.12178.

[75] Y. Zhou, Y. Cheng, Y. Yang, X. Li, B. Supriyo, X. Sun, Y. Yang, Overexpression of SpCBL6, a calcineurin B-like protein of Stipa purpurea, enhanced cold tolerance and reduced drought tolerance in transgenic Arabidopsis, Mol. Biol. Rep. 43 (2016) 957-966. doi:10.1007/s11033-016-4036-5.

[76] H. Chauhan, N. Khurana, A. Nijhavan, J.P. Khurana, P. Khurana, The wheat chloroplastic small heat shock protein (sHSP26) is involved in seed maturation and germination and imparts tolerance to heat stress, Plant, Cell Environ. 35 (2012) 1912-1931. doi:10.1111/j.1365-3040.2012.02525.x.

[77] K.H. Kim, I. Alam, Y.G. Kim, S.A. Sharmin, K.W. Lee, S.H. Lee, B.H. Lee, Overexpression of a chloroplast-localized small heat shock protein OsHSP26 confers enhanced tolerance against oxidative and heat stresses in tall fescue, Biotechnol. Lett. 34 (2012) 371-377. doi:10.1007/s10529-011-0769-3.

[78] A. Wang, X. Yu, Y. Mao, Y. Liu, G. Liu, Y. Liu, X. Niu, Overexpression of a small heat-shock-protein gene enhances tolerance to abiotic stresses in rice, Plant Breed. 134 (2015) 384-393. doi:10.1111/pbr.12289.

[79] T.M.L. Hoang, L. Moghaddam, B. Williams, H. Khanna, J. Dale, S.G. Mundree, Development of salinity tolerance in rice by constitutive-overexpression of genes involved in the regulation of programmed cell death, Front. Plant Sci. 6 (2015).

[80] S.M. Augustine, J.A. Narayan, D.P. Syamaladevi, C. Appunu, M. Chakravarthi, V. Ravichandran, N. Subramonian, Erianthus arundinaceus HSP70 (EaHSP70) overexpression increases drought and salinity tolerance in sugarcane (Saccharum spp.

hybrid), Plant Sci. 232 (2015) 23-34. doi:10.1016/j.plantsci.2014.12.012.

[81] A. Maqbool, W. Abbas, A.Q. Rao, M. Irfan, M. Zahur, A. Bakhsh, S. Riazuddin, T. Husnain, *Gossypium arboreum GHSP26* enhances drought tolerance in *Gossypium hirsutum*, Biotechnol. Prog. 26 (2009) NA-NA. doi:10.1002/btpr.306.

[82] L. Sun, Y. Liu, X. Kong, D. Zhang, J. Pan, Y. Zhou, L. Wang, D. Li, X. Yang, ZmHSP16.9, a cytosolic class I small heat shock protein in maize (Zea mays), confers heat tolerance in transgenic tobacco, Plant Cell Rep. 31 (2012) 1473-1484. doi:10.1007/s00299-012-1262-8.

[83] J. Xu, C. Xue, D. Xue, J. Zhao, J. Gai, N. Guo, H. Xing, Overexpression of GmHsp90s, a Heat Shock Protein 90 (Hsp90) Gene Family Cloning from Soybean, Decrease Damage of Abiotic Stresses in Arabidopsis thaliana, PLoS One. 8 (2013) e69810. doi:10.1371/journal.pone.0069810.

[84] T.M.L. Hoang, L. Moghaddam, B. Williams, H. Khanna, J. Dale, S.G. Mundree, Development of salinity tolerance in rice by constitutive-overexpression of genes involved in the regulation of programmed cell death, Front. Plant Sci. 6 (2015) 175. doi:10.3389/fpls.2015.00175.

[85] I. Toumi, M.G. Pagoulatou, T. Margaritopoulou, D. Milioni, K.A. Roubelakis-Angelakis, Genetically Modified Heat Shock Protein90s and Polyamine Oxidases in Arabidopsis Reveal Their Interaction under Heat Stress Affecting Polyamine Acetylation, Oxidation and Homeostasis of Reactive Oxygen Species, Plants. 8 (2019) 323. doi:10.3390/plants8090323.

[86] Q. Huang, Y. Wang, B. Li, J. Chang, M. Chen, K. Li, G. Yang, G. He, TaNAC29, a NAC transcription factor from wheat, enhances salt and drought tolerance in transgenic Arabidopsis, BMC Plant Biol. 15 (2015) 1-15. doi:10.1186/s12870-015-0644-9.

[87] X. Mao, S. Chen, A. Li, C. Zhai, R. Jing, Novel NAC Transcription Factor TaNAC67 Confers Enhanced Multi-Abiotic Stress Tolerances in Arabidopsis, PLoS One. 9 (2014) e84359. doi:10.1371/journal.pone.0084359.

[88] F. Shah, L. Nie, K. Cui, T. Shah, W. Wu, C. Chen, L. Zhu, F. Ali, S. Fahad, J. Huang, Rice grain yield and component responses to near 2°C of warming, F. Crop. Res. 157 (2014) 98-110. doi:10.1016/j.fcr.2013.12.014.

[89] H. Wang, H. Wang, H. Shao, X. Tang, Recent advances in utilizing transcription factors to improve plant abiotic stress tolerance by transgenic technology, Front. Plant Sci. 7 (2016) 67. doi:10.3389/fpls.2016.00067.

[90] G. Zhang, M. Chen, L. Li, Z. Xu, X. Chen, J. Guo, Y. Ma, Overexpression of the soybean GmERF3 gene, an AP2/ERF type transcription factor for increased tolerances to salt, drought, and diseases in transgenic tobacco, J. Exp. Bot. 60 (2009) 3781-3796. doi:10.1093/jxb/erp214.

[91] P. Bihani, B. Char, D.S. Bhargava, Transgenic expression of sorghum DREB2 in rice improves tolerance and yield under water limitation, (2010). doi:10.1017/S0021859610000742.

[92] A.M. Polizel, M.E. Medri, K. Nakashima, N. Yamanaka, J.R.B. Farias, M.C.N. de Oliveira, S.R.R. Marin, R. V. Abdelnoor, F.C. Marcelino-Guimarães, R. Fuganti, F.A. Rodrigues, R. Stolf-Moreira, M.A. Beneventi, A.A.P. Rolla, N. Neumaier, K. Yamaguchi-Shinozaki, J.F.C. Carvalho, A.L. Nepomuceno, Molecular, anatomical and physiological properties of a genetically modified soybean line transformed with rd29A:AtDREB1A for the improvement of drought tolerance, Genet. Mol. Res. 10 (2011) 3641-3656. doi:10.4238/2011.October.21.4.

[93] X. Peng, L. Zhang, L. Zhang, Z. Liu, L. Cheng, Y. Yang, S. Shen, S. Chen, G. Liu, The transcriptional factor LcDREB2 cooperates with LcSAMDC2 to contribute to salt tolerance in Leymus chinensis, Plant Cell. Tissue Organ Cult. 113 (2013) 245-256. doi:10.1007/s11240-012-0264-0.

[94] J. Joo, H.J. Choi, Y.H. Lee, Y.K. Kim, S.I. Song, A transcriptional repressor of the ERF family confers drought tolerance to rice and regulates genes preferentially located on chromosome 11, Planta. 238 (2013) 155-170. doi:10.1007/s00425-013-1880-6.

[95] S.M. Augustine, J. Ashwin Narayan, D.P. Syamaladevi, C. Appunu, M. Chakravarthi, V. Ravichandran, N. Tuteja, N. Subramonian, Overexpression of EaDREB2 and pyramiding of EaDREB2 with the pea DNA helicase gene (PDH45) enhance drought and salinity tolerance in sugarcane (Saccharum spp. hybrid), Plant Cell Rep. 34 (2014) 247-263. doi:10.1007/s00299-014-1704-6.

[96] G. Ravikumar, P. Manimaran, S.R. Voleti, D. Subrahmanyam, R.M. Sundaram, K.C. Bansal, B.C. Viraktamath, S.M. Balachandran, Stress-inducible expression of AtDREB1A transcription factor greatly improves drought stress tolerance in transgenic indica rice, Transgenic Res. 23 (2014) 421-439. doi:10.1007/s11248-013-9776-6.

[97] Z.M. Sun, M.L. Zhou, X.G. Xiao, Y.X. Tang, Y.M. Wu, Genome-wide analysis of AP2/ERF family genes from Lotus corniculatus shows LcERF054 enhances salt tolerance, Funct. Integr. Genomics. 14 (2014) 453-466. doi:10.1007/s10142-014-0372-5.

[98] X. Zhu, L. Qi, X. Liu, S. Cai, H. Xu, R. Huang, J. Li, X. Wei, Z. Zhang, The wheat ethylene response factor transcription factor pathogen-induced ERF1 mediates host responses to both the necrotrophic pathogen rhizoctonia cerealis and freezing stresses, Plant Physiol. 164 (2014) 1499-1514. doi:10.1104/pp.113.229575.

[99] H. Chen, L. Liu, L. Wang, S. Wang, X. Cheng, VrDREB2A, a DREB-binding transcription factor from Vigna radiata, increased drought and high-salt tolerance in transgenic Arabidopsis thaliana, J. Plant Res. 129 (2016) 263-273. doi:10.1007/s10265-015-0773-0.

[100] V. Jisha, L. Dampanaboina, J. Vadassery, A. Mithöfer, S. Kappara, R. Ramanan, Overexpression of an AP2/ERF Type Transcription Factor OsEREBP1 Confers Biotic and Abiotic Stress Tolerance in Rice, PLoS One. 10 (2015) e0127831. doi:10.1371/journal.pone.0127831.

[101] X. Zhang, X. Liu, L. Wu, G. Yu, X. Wang, H. Ma, The SsDREB transcription factor from the succulent halophyte Suaeda salsa enhances abiotic stress tolerance in transgenic tobacco, Int. J. Genomics. 2015 (2015). doi:10.1155/2015/875497.

[102] X.S. Huang, J.H. Liu, X.J. Chen, Overexpression of PtrABF gene, a bZIP transcription factor isolated from Poncirus trifoliata, enhances dehydration and drought tolerance in tobacco via scavenging ROS and modulating expression of stress-responsive genes, BMC Plant Biol. 10 (2010) 1-18. doi:10.1186/1471-2229-10-230.

[103] S.Q. Gao, M. Chen, Z.S. Xu, C.P. Zhao, L. Li, H. jun Xu, Y. miao Tang, X. Zhao, Y.Z. Ma, The soybean GmbZIP1 transcription factor enhances multiple abiotic stress tolerances in transgenic plants, Plant Mol. Biol. 75 (2011) 537-553. doi:10.1007/s11103-011-9738-4.

[104] H. Takahashi, T. Kawakatsu, Y. Wakasa, S. Hayashi, F. Takaiwa, A rice transmembrane bZIP transcription

factor, OsbZIP39, regulates the endoplasmic reticulum stress response, Plant Cell Physiol. 53 (2012) 144-153. doi:10.1093/pcp/pcr157.

[105] Y. Li, Y. Sun, Q. Yang, F. Fang, J. Kang, T. Zhang, Isolation and characterization of a gene from Medicago sativa L., encoding a bZIP transcription factor, Mol. Biol. Rep. 40 (2013) 1227-1239. doi:10.1007/s11033-012-2165-z.

[106] S. Ying, D.F. Zhang, J. Fu, Y.S. Shi, Y.C. Song, T.Y. Wang, Y. Li, Cloning and characterization of a maize bZIP transcription factor, ZmbZIP72, confers drought and salt tolerance in transgenic Arabidopsis, Planta. 235 (2012) 253-266. doi:10.1007/s00425-011-1496-7.

[107] L. Cheng, S. Li, J. Hussain, X. Xu, J. Yin, Y. Zhang, X. Chen, L. Li, Isolation and functional characterization of a salt responsive transcriptional factor, LrbZIP from lotus root (Nelumbo nucifera Gaertn), Mol. Biol. Rep. 40 (2013) 4033-4045. doi:10.1007/s11033-012-2481-3.

[108] C. Liu, B. Mao, S. Ou, W. Wang, L. Liu, Y. Wu, C. Chu, X. Wang, OsbZIP71, a bZIP transcription factor, confers salinity and drought tolerance in rice, Plant Mol. Biol. 84 (2014) 19-36. doi:10.1007/s11103-013-0115-3.

[109] L. Zhang, L. Zhang, C. Xia, G. Zhao, J. Liu, J. Jia, X. Kong, A novel wheat bZIP transcription factor, TabZIP60, confers multiple abiotic stress tolerances in transgenic *Arabidopsis*, Physiol. Plant. 153 (2015) 538-554. doi:10.1111/ppl.12261.

[110] S.J. Moon, S.Y. Han, D.Y. Kim, I.S. Yoon, D. Shin, M.O. Byun, H. Bin Kwon, B.G. Kim, Ectopic expression of a hot pepper bZIP-like transcription factor in potato enhances drought tolerance without decreasing tuber yield, Plant Mol. Biol. 89 (2015) 421-431. doi:10.1007/s11103-015-0378-y.

[111] L. Zhong, D. Chen, D. Min, W. Li, Z. Xu, Y. Zhou, A. Li, M. Chen, Y. Ma, AtTGA4, a bZIP transcription factor, confers drought resistance by enhancing nitrate transport and assimilation in Arabidopsis thaliana, Biochem. Biophys. Res. Commun. 457 (2015) 433-439. doi:10.1016/j.bbrc.2015.01.009.

[112] Z. Ding, S. Li, X. An, X. Liu, H. Qin, D. Wang, Transgenic expression of MYB15 confers enhanced sensitivity to abscisic acid and improved drought tolerance in Arabidopsis thaliana, J. Genet. Genomics. 36 (2009) 17-29. doi:10.1016/S1673-8527(09)60003-5.

[113] H. Liu, X. Zhou, N. Dong, X. Liu, H. Zhang, Z. Zhang, Expression of a wheat MYB gene in transgenic tobacco enhances resistance to Ralstonia solanacearum, and to drought and salt stresses, Funct. Integr. Genomics. 11 (2011) 431-443. doi:10.1007/s10142-011-0228-1.

[114] L. Cheng, X. Li, X. Huang, T. Ma, Y. Liang, X. Ma, X. Peng, J. Jia, S. Chen, Y. Chen, B. Deng, G. Liu, Overexpression of sheepgrass R1-MYB transcription factor LcMYB1 confers salt tolerance in transgenic Arabidopsis, Plant Physiol. Biochem. 70 (2013) 252-260. doi:10.1016/j.plaphy.2013.05.025.

[115] L.T. Su, J.W. Li, D.Q. Liu, Y. Zhai, H.J. Zhang, X.W. Li, Q.L. Zhang, Y. Wang, Q.Y. Wang, A novel MYB transcription factor, GmMYBJ1, from soybean confers drought and cold tolerance in Arabidopsis thaliana, Gene. 538 (2014) 46-55. doi:10.1016/j.gene.2014.01.024.

[116] H. Cai, S. Tian, H. Dong, C. Guo, Pleiotropic effects of TaMYB3R1 on plant development and response to osmotic stress in transgenic Arabidopsis, Gene. 558 (2015) 227-234. doi:10.1016/j.gene.2014.12.066.

[117] X. Meng, J.R. Wang, G.D. Wang, X.Q. Liang, X.D. Li, Q.W. Meng, An R2R3-MYB gene, LeAN2, positively

regulated the thermo-tolerance in transgenic tomato, J. Plant Physiol. 175 (2015) 1-8. doi:10.1016/j.jplph.2014.09.018.

[118] L. Qi, J. Yang, Y. Yuan, L. Huang, P. Chen, Overexpression of two R2R3-MYB genes from Scutellaria baicalensis induces phenylpropanoid accumulation and enhances oxidative stress resistance in transgenic tobacco, Plant Physiol. Biochem. 94 (2015) 235-243. doi:10.1016/j.plaphy.2015.06.007.

[119] N. Yokotani, T. Ichikawa, Y. Kondou, M. Matsui, H. Hirochika, M. Iwabuchi, K. Oda, Tolerance to various environmental stresses conferred by the salt-responsive rice gene ONAC063 in transgenic Arabidopsis, Planta. 229 (2009) 1065-1075. doi:10.1007/s00425-009-0895-5.

[120] Y.J. Hao, W. Wei, Q.X. Song, H.W. Chen, Y.Q. Zhang, F. Wang, H.F. Zou, G. Lei, A.G. Tian, W.K. Zhang, B. Ma, J.S. Zhang, S.Y. Chen, Soybean NAC transcription factors promote abiotic stress tolerance and lateral root formation in transgenic plants, Plant J. 68 (2011) 302-313. doi:10.1111/j.1365-313X.2011.04687.x.

[121] M. Lu, S. Ying, D.F. Zhang, Y.S. Shi, Y.C. Song, T.Y. Wang, Y. Li, A maize stress-responsive NAC transcription factor, ZmSNAC1, confers enhanced tolerance to dehydration in transgenic Arabidopsis, Plant Cell Rep. 31 (2012) 1701-1711. doi:10.1007/s00299-012-1284-2.

[122] Y. Tang, M. Liu, S. Gao, Z. Zhang, X. Zhao, C. Zhao, F. Zhang, X. Chen, Molecular characterization of novel TaNAC genes in wheat and overexpression of TaNAC2a confers drought tolerance in tobacco, Physiol. Plant. 144 (2012) 210-224. doi:10.1111/j.1399-3054.2011.01539.x.

[123] X. Liu, S. Liu, J. Wu, B. Zhang, X. Li, Y. Yan, L. Li, Overexpression of Arachis hypogaea NAC3 in tobacco enhances dehydration and drought tolerance by increasing superoxide scavenging, Plant Physiol. Biochem. 70 (2013) 354-359. doi:10.1016/j.plaphy.2013.05.018.

[124] A.S.I. Saad, X. Li, H.P. Li, T. Huang, C.S. Gao, M.W. Guo, W. Cheng, G.Y. Zhao, Y.C. Liao, A rice stress-responsive NAC gene enhances tolerance of transgenic wheat to drought and salt stresses, Plant Sci. 203-204 (2013) 33-40. doi:10.1016/j.plantsci.2012.12.016.

[125] X. Chen, Y. Wang, B. Lv, J. Li, L. Luo, S. Lu, X. Zhang, H. Ma, F. Ming, The NAC family transcription factor OsNAP confers abiotic stress response through the ABA pathway, Plant Cell Physiol. 55 (2014) 604-619. doi:10.1093/pcp/pct204.

[126] X. Yang, X. Wang, L. Ji, Z. Yi, C. Fu, J. Ran, R. Hu, G. Zhou, Overexpression of a Miscanthus lutarioriparius NAC gene MlNAC5 confers enhanced drought and cold tolerance in Arabidopsis, Plant Cell Rep. 34 (2015) 943-958. doi:10.1007/s00299-015-1756-2.

[127] L. Zhang, L. Zhang, C. Xia, G. Zhao, J. Jia, X. Kong, The Novel Wheat Transcription Factor TaNAC47 Enhances Multiple Abiotic Stress Tolerances in Transgenic Plants, Front. Plant Sci. 6 (2016) 1174. doi:10.3389/fpls.2015.01174.

[128] Y. Qiu, D. Yu, Over-expression of the stress-induced OsWRKY45 enhances disease resistance and drought tolerance in Arabidopsis, Environ. Exp. Bot. 65 (2009) 35-47. doi:10.1016/j.envexpbot.2008.07.002.

[129] H. Liu, W. Yang, D. Liu, Y. Han, A. Zhang, S. Li, Ectopic expression of a grapevine transcription factor VvWRKY11 contributes to osmotic stress tolerance in Arabidopsis,

Mol. Biol. Rep. 38 (2011) 417-427. doi:10.1007/s11033-010-0124-0.

[130] K.C. Babitha, S. V. Ramu, V. Pruthvi, P. Mahesh, K.N. Nataraja, M. Udayakumar, Co-expression of AtbHLH17 and AtWRKY28 confers resistance to abiotic stress in Arabidopsis, Transgenic Res. 22 (2013) 327-341. doi:10.1007/s11248-012-9645-8.

[131] H. Li, Y. Gao, H. Xu, Y. Dai, D. Deng, J. Chen, ZmWRKY33, a WRKY maize transcription factor conferring enhanced salt stress tolerances in Arabidopsis, Plant Growth Regul. 70 (2013) 207-216. doi:10.1007/s10725-013-9792-9.

[132] Y. Qin, Y. Tian, L. Han, X. Yang, Constitutive expression of a salinity-induced wheat WRKY transcription factor enhances salinity and ionic stress tolerance in transgenic Arabidopsis thaliana, Biochem. Biophys. Res. Commun. 441 (2013) 476-481. doi:10.1016/j.bbrc.2013.10.088.

[133] J. Sun, W. Hu, R. Zhou, L. Wang, X. Wang, Q. Wang, Z. Feng, Y. Li, D. Qiu, G. He, G. Yang, The Brachypodium distachyon BdWRKY36 gene confers tolerance to drought stress in transgenic tobacco plants, Plant Cell Rep. 34 (2015) 23-35. doi:10.1007/s00299-014-1684-6.

[134] L. Zhou, N.N. Wang, S.Y. Gong, R. Lu, Y. Li, X.B. Li, Overexpression of a cotton (Gossypium hirsutum) WRKY gene, GhWRKY34, in Arabidopsis enhances salt-tolerance of the transgenic plants, Plant Physiol. Biochem. 96 (2015) 311-320. doi:10.1016/j.plaphy.2015.08.016.

[135] L. Liu, Z. Zhang, J. Dong, T. Wang, Overexpression of MtWRKY76 increases both salt and drought tolerance in Medicago truncatula, Environ. Exp. Bot. 123 (2016) 50-58. doi:10.1016/j.envexpbot.2015.10.007.

[136] F.G. González, N. Rigalli, P.V. Miranda, M. Romagnoli, K.F. Ribichich, F. Trucco, M. Portapila, M.E. Otegui, R.L. Chan, An Interdisciplinary Approach to Study the Performance of Second-generation Genetically Modified Crops in Field Trials: A Case Study With Soybean and Wheat Carrying the Sunflower HaHB4 Transcription Factor, Front. Plant Sci. 11 (2020) 178. doi:10.3389/fpls.2020.00178.

[137] L. Cheng, S. Li, • Javeed Hussain, X. Xu, J. Yin, Y. Zhang, X. Chen, L. Li, Isolation and functional characterization of a salt responsive transcriptional factor, LrbZIP from lotus root (Nelumbo nucifera Gaertn), Springer. (n.d.). doi:10.1007/s11033-012-2481-3.

[138] T. Eulgem, P.J. Rushton, S. Robatzek, I.E. Somssich, The WRKY superfamily of plant transcription factors, Trends Plant Sci. 5 (2000) 199-206. doi:10.1016/S1360-1385(00)01600-9.

[139] A.M. Sharoni, M. Nuruzzaman, K. Satoh, T. Shimizu, H. Kondoh, T. Sasaya, I.R. Choi, T. Omura, S. Kikuchi, Gene structures, classification and expression models of the AP2/EREBP transcription factor family in rice, Plant Cell Physiol. 52 (2011) 344-360. doi:10.1093/pcp/pcq196.

[140] M. Jinek, K. Chylinski, I. Fonfara, ... M.H.-, undefined 2012, A programmable dual-RNA–guided DNA endonuclease in adaptive bacterial immunity, Science.Sciencemag.Org. (n.d.). https://science.sciencemag.org/content/337/6096/816.abstract (accessed October 2, 2020).

[141] S.A. Zafar, S.S.E.A. Zaidi, Y. Gaba, S.L. Singla-Pareek, O.P. Dhankher, X. Li, S. Mansoor, A. Pareek, C. Foyer, Engineering abiotic stress tolerance via CRISPR/Cas-mediated genome editing, J. Exp. Bot. 71 (2020) 470-479. doi:10.1093/jxb/erz476.

[142] Y. Huang, Y. Guo, Y. Liu, F. Zhang, Z. Wang, H. Wang, F. Wang, D. Li, D. Mao, S. Luan, M. Liang, L. Chen, 9-cis-Epoxycarotenoid Dioxygenase 3 Regulates Plant Growth and Enhances Multi-Abiotic Stress Tolerance in Rice, Front. Plant Sci. 9 (2018) 162. doi:10.3389/fpls.2018.00162.

[143] J.F. Roca Paixão, F.X. Gillet, T.P. Ribeiro, C. Bournaud, I.T. Lourenço-Tessutti, D.D. Noriega, B.P. de Melo, J. de Almeida-Engler, M.F. Grossi-de-Sa, Improved drought stress tolerance in Arabidopsis by CRISPR/dCas9 fusion with a Histone AcetylTransferase, Sci. Rep. 9 (2019) 1-9. doi:10.1038/s41598-019-44571-y.

[144] A. Zhang, Y. Liu, F. Wang, T. Li, Z. Chen, D. Kong, J. Bi, F. Zhang, X. Luo, J. Wang, J. Tang, X. Yu, G. Liu, L. Luo, Enhanced rice salinity tolerance via CRISPR/Cas9-targeted mutagenesis of the OsRR22 gene, Mol. Breed. 39 (2019) 1-10. doi:10.1007/s11032-019-0954-y.

[145] J. Shi, J.E. Habben, R.L. Archibald, B.J. Drummond, M.A. Chamberlin, R.W. Williams, H. Renee Lafitte, B.P. Weers, Overexpression of ARGOS genes modifies plant sensitivity to ethylene, leading to improved drought tolerance in both arabidopsis and maize, Plant Physiol. 169 (2015) 266-282. doi:10.1104/pp.15.00780.

[146] J. Shi, H. Gao, H. Wang, H.R. Lafitte, R.L. Archibald, M. Yang, S.M. Hakimi, H. Mo, J.E. Habben, ARGOS8 variants generated by CRISPR-Cas9 improve maize grain yield under field drought stress conditions, Plant Biotechnol. J. 15 (2017) 207-216. doi:10.1111/pbi.12603.

[147] S. Bouzroud, K. Gasparini, G. Hu, M.A.M. Barbosa, B.L. Rosa, M. Fahr, N. Bendaou, M. Bouzayen, A. Zsögön, A. Smouni, M. Zouine, Down regulation and loss of auxin response factor 4 function using CRISPR/Cas9 alters plant growth, stomatal function and improves tomato tolerance to salinity and osmotic stress, Genes (Basel). 11 (2020). doi:10.3390/genes11030272.

[148] H. Fang, Q. Meng, J. Xu, H. Tang, S. Tang, H. Zhang, J. Huang, Knock-down of stress inducible OsSRFP1 encoding an E3 ubiquitin ligase with transcriptional activation activity confers abiotic stress tolerance through enhancing antioxidant protection in rice, Plant Mol Biol. 87 (2015) 441-458. doi:10.1007/s11103-015-0294-1.

[149] H. Fang, Q. Meng, H. Zhang, J. Huang, Knock-down of a RING finger gene confers cold tolerance, Bioengineered. 7 (2016) 39-45. doi:10.1080/21655979.2015.1131368.

[150] Y. Huang, H. Cao, L. Yang, C. Chen, L. Shabala, M. Xiong, M. Niu, J. Liu, Z. Zheng, L. Zhou, Z. Peng, Z. Bie, S. Shabala, C. Foyer, Tissue-specific respiratory burst oxidase homolog-dependent H2O2 signaling to the plasma membrane H+-ATPase confers potassium uptake and salinity tolerance in Cucurbitaceae, J. Exp. Bot. 70 (2019) 5879-5893. doi:10.1093/jxb/erz328.

[151] R. Li, C. Liu, R. Zhao, L. Wang, L. Chen, W. Yu, S. Zhang, J. Sheng, L. Shen, CRISPR/Cas9-Mediated SlNPR1 mutagenesis reduces tomato plant drought tolerance, BMC Plant Biol. 19 (2019) 38. doi:10.1186/s12870-018-1627-4.

[152] L. Wang, L. Chen, R. Li, R. Zhao, M. Yang, J. Sheng, L. Shen, Reduced drought tolerance by CRISPR/Cas9-mediated SlMAPK3 mutagenesis in tomato plants, J. Agric. Food Chem. 65 (2017) 8674-8682. doi:10.1021/acs.jafc.7b02745.

[153] C. Shen, Z. Que, Y. Xia, N. Tang, D. Li, R. He, M. Cao, Knock out of the annexin gene OsAnn3 via CRISPR/Cas9-mediated genome editing decreased cold tolerance in rice, J. Plant Biol. 60 (2017) 539-547. doi:10.1007/s12374-016-0400-1.

[154] R. Li, L. Zhang, L. Wang, L. Chen, R. Zhao, J. Sheng, L. Shen, Reduction of Tomato-Plant Chilling Tolerance by CRISPR-Cas9-Mediated SlCBF1 Mutagenesis, J. Agric. Food Chem. 66 (2018) 9042-9051. doi:10.1021/acs.jafc.8b02177.

[155] J.W. Woo, J. Kim, S. Il Kwon, C. Corvalán, S.W. Cho, H. Kim, S.G. Kim, S.T. Kim, S. Choe, J.S. Kim, DNA-free genome editing in plants with preassembled CRISPR-Cas9 ribonucleoproteins, Nat. Biotechnol. 33 (2015) 1162-1164. doi:10.1038/nbt.3389.

[156] M. Malnoy, R. Viola, M.-H. Jung, O.-J. Koo, S. Kim, J.-S. Kim, R. Velasco, C. Nagamangala Kanchiswamy, DNA-Free Genetically Edited Grapevine and Apple Protoplast Using CRISPR/Cas9 Ribonucleoproteins, Front. Plant Sci. 7 (2016) 1904. doi:10.3389/fpls.2016.01904.

[157] Z. Liang, K. Chen, T. Li, Y. Zhang, Y. Wang, Q. Zhao, J. Liu, H. Zhang, C. Liu, Y. Ran, C. Gao, Efficient DNA-free genome editing of bread wheat using CRISPR/Cas9 ribonucleoprotein complexes, Nat. Commun. 8 (2017) 1-5. doi:10.1038/ncomms14261.

[158] B. Zetsche, J. Gootenberg, O.A.-Cell, undefined 2015, Cpf1 is a single RNA-guided endonuclease of a class 2 CRISPR-Cas system, Elsevier. (n.d.). https://www.sciencedirect.com/science/article/pii/S0092867415012003 (accessed October 3, 2020).

[159] H. Kim, S.T. Kim, J. Ryu, B.C. Kang, J.S. Kim, S.G. Kim, CRISPR/Cpf1-mediated DNA-free plant genome editing, Nat. Commun. 8 (2017) 1-7. doi:10.1038/ncomms14406.

Chapter 7

Elevated CO_2 Concentration Improves Heat-Tolerant Ability in Crops

Ayman EL Sabagh, Akbar Hossain, Mohammad Sohidul Islam, Muhammad Aamir Iqbal, Ali Raza, Çetin Karademir, Emine Karademir, Abdul Rehman, Md Atikur Rahman, Rajesh Kumar Singhal, Analía Llanes, Muhammad Ali Raza, Muhammad Mubeen, Wajid Nasim, Celaleddin Barutçular, Ram Swaroop Meena and Hirofumi Saneoka

Abstract

The rising concentration of atmospheric carbon dioxide (aCO_2) and increasing temperature are the main reasons for climate change, which are significantly affecting crop production systems in this world. However, the elevated carbon dioxide (CO_2) concentration can improve the growth and development of crop plants by increasing photosynthetic rate (higher availability of photoassimilates). The combined effects of elevated CO_2 (eCO_2) and temperature on crop growth and carbon metabolism are not adequately recognized, while both eCO_2 and temperature triggered noteworthy changes in crop production. Therefore, to increase crop yields, it is important to identify the physiological mechanisms and genetic traits of crop plants which play a vital role in stress tolerance under the prevailing conditions. The eCO_2 and temperature stress effects on physiological aspects as well as biochemical profile to characterize genotypes that differ in their response to stress conditions. The aim of this review is directed the open-top cavities to regulate the properties like physiological, biochemical, and yield of crops under increasing aCO_2, and temperature. Overall, the extent of the effect of eCO_2 and temperature response to biochemical components and antioxidants remains unclear, and therefore further studies are required to promote an unperturbed production system.

Keywords: elevated CO_2, heat stress, physio-biochemical mechanisms

1. Introduction

Climate change in the form of increasing temperature and increasingly variable rainfall patterns threatens the production of the crop [1, 2]. Therefore, elevated CO_2 (eCO_2) absorption may indorse plant growth, whereas increased temperature

IntechOpen

is repressive for C_3 plants. Both CO_2 and temperature caused significant changes in crop productivity. The collaborative properties of eCO_2 and increased temperatures on the crop growth and carbon metabolism are not well known.

The fluctuating climatic surroundings are predictable to upsurge the atmospheric CO_2 (aCO_2) meditation, temperatures and modify the rainfall outline. The aCO_2 meditation is prophesied to range 550 ppm by 2050, and possibly surpass 700 ppm by the end of the present century [3]. These fluctuations are expected to disturb the creation and output of cultivated crops, and stimulus the upcoming food safety. The influence analysis of climate alteration on worldwide food construction reveals a 0.5% failure by 2020 and 2.3% by 2050 [4]. The progress of climate arranged germplasm to counterbalance these wounded is of the highest reputation [5].

The eCO_2 is significant abiotic stress and has a noteworthy fertilization encouragement on crops. Widespread preceding educations have described that eCO_2 meaningfully enhanced the water use efficiency (WUE), reduced transpiration frequency, abridged maize growth rate, and augmented plant height, leaf number, leaf area, growth frequency, and overall yield [6]. Furthermore, the cumulative of aCO_2 disturbs precipitation equilibrium, which can alter the periodic precipitation circulation [7]. It has been predicted that this result would carry about a 10% upsurge or decline in water capitals in several areas [8]. The raised temperature, i.e., heat stress (HS) damage growth and physiological ailments, and consequentially reduce yield [9]. Increased temperature due to eCO_2 has a primary effect on the food grain invention reliant on the places. With the rising temperature by 1.0–2.0°C in tropical and subtropical states and the food grain manufacture in India is predictable to decline up to 30% [9].

The C_4 grass maize (*Zea mays* L.) is the third most vital food crop worldwide in the relation of the invention, and its claim is prophesied to rise by 45% from 1997 to 2020 [10]. Educations with maize retort to dual the ambient CO_2 presented variable possessions on growth fluctuating from no inspiration of yield [11] to 50% stimulation [12]. The growth and productivity of maize are expected to be pretentious by raised aCO_2 and temperature. Raised temperature severely disturbs the growth, and yield of maize plants [13]. There is unpredictable information on the properties of eCO_2 on the vintage of maize changing from slight positive consequence [14], no consequence [15] to rice harvest by 50% [16].

Record of the experimentations on the influences of eCO_2 and temperature on the crop yield, though, used measured atmosphere amenities like phytotron and plant growth cavities or crop growth reproduction models [13, 17]. Determination of the impact of raised CO_2 on the photosynthesis tolerance to severe HS is vital to expect the plant replies for universal warming since photosynthesis is sensitive to severe HS and aCO_2 upsurges slightly [18, 19]. Further, flowering is a critical element for plant generative achievement and seed-set. The increase in temperature and eCO_2 is the main climate revolution issues that might influence plant suitability and associated flowering actions. Resolving the influence of these ecological issues on the flowering actions like time of days to anthesis and flowering (duration from germination till flowering) is serious to appreciate the acclimatization of crops in altering climate [20].

2. Interaction of eCO_2 with high-temperature stress and other factors to climate change

The impacts of eCO_2 and stress factors on crops have been made using controlled atmosphere amenities like plant growth cavities or crop growth reproduction models in many studies [13]. The eCO_2 contributes to global warming, causing alterations in the precipitations, water scarcity and changes at temperatures in several regions affecting the growth and development of crop plants [3]. However, the interactive effects of eCO_2 and environmental stress conditions on the crop growth and carbon metabolism are not well predictable. The interactions between eCO_2 and stress factors are critical to photosynthesis performance. It has been reported how stomata react to eCO_2 levels, but the effects on photosynthesis performance of other environmental factors are poorly understood [21].

Numerous studies showed the combining impacts of eCO_2 and drought and revealed that the machines are different from singular eCO_2 and drought. The impact of combined eCO_2 and drought are varying with crop stage such as during vegetative stage, and it restricted the shoot development, decreased leaf area, diminished mobilization of nutrients due to weak root growth, reduced stomatal closure, transpiration, and relative water contents (RWC). However, it enhances the resource use efficiencies of the plant, including WUE, light use efficiency (LUE), and nutrient use efficiency (NUE) at a certain level [22, 23]. Similarly, eCO_2 and drought affect reproductive growth severely such as the impacts on the pollen abortion, pollination, flower formation, panicle length, panicle weight, seed formation, seed size and, yield potential of important agriculture crops [22].

Altering climate, counting eCO_2, increasing temperatures, changing precipitation designs have influenced terrestrial environment assembly and function, carbon and water balance, and finally production of crops [24, 25]. Several experiments have described the biological replies to CO_2 enhancement and their communication with ecological alteration at different levels [26]. The temperature has an important role in plant growth and development and regulates the several functions and enzymatic reactions in plants. Although, the increased value of temperature causes several abnormalities in plants such as reduces the chlorophyll contents, leaf growth, fresh and dry biomass, photosynthesis, and stomata limitations, inhibits the functions of several temperature-sensitive enzymes such as ribulose-1,5-bisphosphate carboxylase/oxygenase (RuBisCO). Impacts of eCO_2 and temperature combined stresses are very destructive at the reproductive phase. They may cause reduced pollination, spike sterility, less filling of grain, reduction of grain size and number, test weight, and yield potential of major crops [22].

It has been reported that eCO_2 promotes an initial stimulation of photosynthesis by an increase of substrate or RuBisCO carboxylation activity and self-consciousness of RuBisCO oxygenation, which might ultimately underwrite for advanced biomass in cereal crops [27, 28]. Nevertheless, growth responses over the long-standing under eCO_2 conditions include a reduction of photosynthesis measurements and several regulatory mechanisms to avoid potential damage by this condition. A regulatory mechanism to maintain the growth and expansion of plants consists of the equilibrium among manufacture and removal of reactive oxygen species (ROS) at the intracellular level. This balance is continued by both enzymatic and non-enzymatic antioxidant defense systems [29–31].

The enzymatic machinery involves numerous antioxidant enzymes, such as superoxide dismutase (SOD), catalase (CAT), glutathione peroxidase (GPX), guaiacol peroxidase (POX), peroxiredoxins (Prxs), and enzymes of the ascorbate-glutathione (AsAGSH) cycle. The AsAGSH includes different enzymes like ascorbate peroxidase (APX), monodehydroascorbate reductase (MDHAR), dehydroascorbate reductase (DHAR), and glutathione reductase (GR) [30, 32]. Modifications of this enzymatic antioxidant component have been reported in studies of cereal crops under eCO_2 conditions with contradictory results. Thus, the response of these antioxidants components is unclear with studies that find to increase [33], decreases [34], or no consistent alterations [35].

Besides, plant cells have non-enzymatic components that involve ascorbate and glutathione (GSH) along with phenolic acids, flavonoids, carotenoids anthocyanins, and phenolic composites. The free hydroxyl groups on the phenolic rings or the chromanol rings of these non-enzymatic compounds are responsible for their antioxidant properties [36]. The ring hydrogen atom can be given to free radicals, dropping and counteracting ROS. The phenolic compounds can lose a hydrogen atom which develops a free radical that is directly non-reactive by character delocalization in the entire ring assembly [28, 37]. Several studies reported only an increase in some of the individual phenolic compounds in cereal crops under high CO_2 conditions [38].

Furthermore, crop growth responses to eCO_2 rely on the tissue category, developmental stage as well as strength and duration of these conditions which also depend on the diversity of apparatuses of construction and purification of ROS, and the result of free radicals on antioxidants [39]. Several studies have reported a high production of hydrogen peroxide (H_2O_2) after exposure of crops to eCO_2, and the concentration of H_2O_2 is dependent on the duration of these conditions. Besides, H_2O_2 production differed among various cellular compartments [40]. The outcome of elevated aCO_2 meditation on growth and various antioxidant actions is superior in C_3 plants to C_4 plants [41, 42]. For example, rice as a C_3 plant is further sensitive to the variations of the aCO_2. However, there are inconsistent studies on the impacts of eCO_2 on the antioxidant responses and yield of rice changing from the reduction of the growth responses [43] to increment of antioxidants components and enhancement of growth [44, 45].

Kumar et al. [46] reported that rice plants under eCO_2 conditions showed modifications in electrolyte leakage, leaf water potential, proline, CAT, and POD activity as compared to ambient CO_2, which assisted the plant to battle contrary effects of stressful environments. Thus, these authors suggest that undesirable possessions on rice yield subsequent from abiotic stress conditions may be moderated by the eCO_2 meditations [46]. In agreement, leaves of a susceptible wheat cultivar (*Triticum aestivum* L.*cv*. Yitpi) infected with Barley yellow dwarf virus-PAV (*PadiAvenae* virus) and grown under eCO_2 presented that the eCO_2 conditions may decrease the oxidative stress caused by virus infection [47]. Nevertheless, more evidence for direct communicating possessions of eCO_2 and biotic and abiotic stress conditions in cereal crops is necessary.

3. eCO_2 mitigates oxidative stress in plants

Various environmental stresses induce the production of ROS, which triggers oxidative stress in plants [30, 48, 49]. The most common ROS are $O_2^{\bullet-}$, $^{\bullet}OH$, and H_2O_2. In response to stressful conditions, H_2O_2 is mainly synthesized by photorespiration, beta (β)-oxidation, or due to the activity of nicotinamide adenine dinucleotide phosphate (NADPH) oxidase [30, 50]. The eCO_2 can efficiently reduce the ROS level by increasing RuBisCO carboxylation along with reducing photorespiratory H_2O_2 production. Several reports have provided indications regarding the eCO_2 influences on the mitigation of abiotic stress in plants [30, 51, 52]. Elevated CO_2 known to be induced plant growth by supplying additional Carbon (C) sources, subsequently alleviates abiotic stress in plants. Although the elevated CO_2 mediated particular physiological and molecular mechanisms related to abiotic stress alleviation are still to be explored. In the physiological aspect, supplying extra C by eCO_2 leads to induce stomatal closing, improves WUE that protect drought stress in plants [53]. However, abiotic stress-induced ROS (e.g., $O_2^{\bullet-}$, $^{\bullet}OH$, and H_2O_2) and cellular oxidative damages (e.g., protein oxidation, lipid peroxidation) are involved in non-stomatal factors with metabolic changes [30, 54].

A systematic study of recently published articles addressed the two major hypotheses such as enhancement of antioxidant (antioxidant hypothesis) and reduction of stress impact (relaxation hypothesis) by reducing ROS generation in the plant under stressful conditions [30, 51, 55]. Several reports have been found in favor of the relaxation hypothesis in plants in response to eCO_2 under stressful conditions. The ROS level was found to be reduced by eCO_2 in plants under drought, and heat stresses through increasing RuBisCO carboxylation as well as reducing the level of photorespiratory H_2O_2 [51]. In the same study, glycine/serine (Gly/Ser) ration, glycolate oxidase (GO), and hydroxypiruvate reductase (HPR) level were evaluated as an indicator of photorespiration, which was found to be decreased in response to eCO_2 under drought and heat stresses. In barley, these all parameters were found at a lower level in response to eCO_2 [56]. Moreover, reduced photorespiration is correlated to the decreased level of NADPH oxidase. Therefore, a combined effect of lower photorespiration and NADPH oxidase responses may lead to reduce H_2O_2 in plants.

According to the antioxidant approach, the availability of additional C by eCO_2 enhances antioxidant molecules, which increase ROS scavenging activity as well as protects grapevine and tomato plants from abiotic stress induced-oxidative damages [57, 58]. More specifically, higher C availability due to eCO_2 may enhance the supply of defense molecules, which improve protection against oxidative injury (antioxidant hypothesis) under stressful conditions in plant cells. However, changes in antioxidant levels are not specific, or C_3 or C_4 based metabolism, or not for a particular group of species. It has been reported in C_4 plants that photorespiration mildly active, in which eCO_2 reduces ROS level as well as oxidative injury without alteration of antioxidants level. It suggests a distinct non-stomatal process that except antioxidant defense or reduces photorespiration processes. Besides, several reports have provided the evidence regarding eCO_2 reduces NADPH oxidase activity and ROS formation in mitochondria and chloroplast in plants [59, 60], but the activity of beta (β)-oxidation is still to be explored.

The ascorbate-glutathione (ASC-GSH) cycle is one of the major mechanisms for stress-induced H_2O_2 regulation. However, only limited reports have been found concerning eCO_2 mediated changes of ASC-GSH cycle components in plants under stress conditions. For example, HS alleviated through enhancement of DHAR, MDHAR, APX, and GR in tomato [58]. Similarly, GR and APX were found to be increased by eCO_2 in wheat under ozone stress [61]. Also, responses of ASC-GSH cycle components varied based on the plant species and experimental set-up. Therefore, additional studies are needed concerning the eCO_2 mediated oxidative stress alleviation as well as enhancement of ASC-GSH cycle components in plants under abiotic stresses.

4. eCO_2 improves photosynthesis under high temperature

Several researchers on the consequences of eCO_2 and stress factors on crops have been made using monitored situation amenities like plant growth chambers, free-air concentration enrichment (FACE) experiment, open-top chamber (OTC) or stimulated crop growth models [13]. The eCO_2 contributes to global warming causing alterations in the precipitations patterns, water scarcity, flood, and changes at extreme temperatures in several regions affecting the growth and development of plants [3]. However, the interactive effects of eCO_2 and environmental stress conditions on the development of crops and metabolism are not well documented. It has been shown the average reduction of stomatal conductance (20–30%), stomatal density (5–7%), stomatal developments, and increment in WUE (8–18%)

under eCO_2 conditions. However, these changes vary with the crop species, developmental stages, nature of stressors and duration, surrounding environments, and plant attributes [62]. Likewise, the interactions between an eCO_2 with stress factors are crucial to understanding the photosynthesis performance. Therefore, there is a considerable deviation in the light-saturated photosynthetic assimilation rate under eCO_2 rely upon the plant type, plant functional traits, micro or surrounding environment, and resource availability. For instance, the stimulation in photosynthetic assimilation rate under eCO_2 is varied from 30 to 80% (strong stimulations in C_3 species as compared to C_4) in FACE experiment or pot conditions but diminished in field conditions because of the integration of other multiple stressors such as drought, heat, flood and nutrient deficiencies.

In this regard, several studies were conducted to observe the combining effects of eCO_2 and drought and reveals that the growth mechanisms are distinct to the singular eCO_2 or drought. The combined stress resulted in the longer retention time of dissolved organic carbon (accumulation of soil organic C), induce invertase and catalase activity in the soil, and ameliorate stress conditions via improving plant physiological traits and activates feedback mechanisms [63]. Further, it limits the activity of some antioxidant enzymes such as proline and MDA content and stimulates others such as SOD, CAT, and GPX [46, 64]. Abscisic acid, calcium-dependent protein kinase and glutathione *S*-transferase (GST) play an important role in the amelioration of drought stress responses by inducing signaling mechanisms under the combined form. Conclusively, it is suggested that drought and HS generate ROS, and affect the antioxidant defense mechanism of plants, which might be ameliorated by the eCO_2 via stimulation of antioxidant defense enzymes [51, 62]. Similarly, eCO_2 combined with drought and HS regulates the sugars (starch, sucrose) and amino acids (alanine, pyruvate, arginine, glutamate) and secondary metabolites (coumaric acid, salicylic acid) metabolism, protective proteins and readjusted the metabolic, redox, and osmotic equilibrium of plants under combined eCO_2 and drought [65, 66].

The impact of combined eCO_2 and drought are varied with crop stage such as during the vegetative phase decreased the shoot elongation and leaf area, diminished mobilization of nutrients due to weak root growth, reduced stomatal conductance by increasing stomatal resistance and stomatal movements, plant hydraulic conductance, aquaporins, and reduce transpiration and RWC. However, at certain levels, it enhances the biomass allocations to the reproductive part, improves resource use efficiencies of plants including WUE, LUE, and NUE [67]. Similarly, eCO_2 and drought affect reproductive growth severely such as its impacts on the assimilate partitioning, pollen abortion, pollination, flower formation, panicle length, panicle weight, productive tiller number, seed formation, seed size, and yield potential of important crops [22, 68]. Besides the yield potential, eCO_2 decreased the grain quality via affecting macro- and micro-nutrients content such as phosphorus (P), Sulphur (S), and Iron (Fe), Zinc (Zn) contents of dryland legumes, which further associated with yield dilutions [23].

Several findings have stated the biological responses to CO_2 enhancement and their communication with conservation alteration at diverse levels [26]. In these aspects, the temperature has an important role in plant growth and development and regulates the several functions and enzymatic reactions in plants. Although, the increased value of temperature cause several abnormalities in plants such as change the emission of volatile organic compounds, reduce nitrogen uptake, chlorophyll contents, leaf growth, fresh and dry biomass, photosynthesis, and stomata limitations (by membrane damage and photosystem II (PSII) activity), enhance the activity of mitochondrial electron transport, stress proteins and plant growth regulators, limitation of several temperature-sensitive enzymes such as RuBisCO.

The combined effects of eCO_2 and temperature stresses are very destructive at the reproductive phase. They may cause changes in flowering time, pollination, spike sterility, less filling of grain, reduction of grain size and number, test weight, grain quality, and yield potential of major crops [20, 22, 69].

Besides the drought and HS (major), other stressors also interact with the eCO_2 under field conditions and influence the plant growth and development. For instance, eCO_2 and salt stress conditions influence the nitrogen metabolism, water balance, photosynthetic inhibitions, nutrient deficiency or toxicity, stomatal conductance, carbohydrate metabolism, phenolic enrichments, and generation of secondary metabolites [63, 70]. Similarly, limited nitrogen supply under eCO_2 modifies the C/N ratio, nitrogen metabolism, protein supply, protein structure, gene expression, sugar metabolism, and decreases antioxidant enzyme, amino acid synthesis, photosynthetic pigments, and elevated ROS, which influence the redox equilibrium and leads to early senescence in plants [71]. Likewise, under N limitations, the photosynthetic rate is more affected in C_3 species (because of more N requirement for RuBisCO synthesis), and eCO_2 could help in mitigation of N limitations by reducing photorespiration, elevating starch level, increase chloroplast size, higher stomatal resistance, mitochondrial respiration, metabolites and dilution of chlorophyll concentrations [62]. Therefore, the eCO_2 and stressors impacts have differed than singular stress, and up to a certain level of eCO_2 try to recover plants via inducing defense machinery, feedback mechanisms, activating secondary messenger signaling, and expression of stress proteins.

5. eCO_2 improves yield under high temperature

By the end of the 21st century, CO_2 is expected to rise from the current level 370 $\mu mol{\cdot}mol^{-1}$ to 540–970 $\mu mol{\cdot}mol^{-1}$, and about to grasp 550 $\mu mol{\cdot}mol^{-1}$ near 2050 and 750 $\mu mol{\cdot}mol^{-1}$ in 2100. In the meantime, Earth's global temperature will rise by about 2–4°C [72]. The increased concentration of CO_2 in the atmosphere would increase the temperature of Earth that is why global heating will develop, the most important aspect of upcoming climate variation. The relations between the temperature and CO_2 will have an intense effect on global agricultural production and the Earth's environment [73]. The rise in the atmospheric temperature and CO_2 would also accelerate the procedure of growth in plants [58, 74].

Climate change impacts on crop growth are becoming global concerns. They are particularly important for food supply and sustainable agricultural development [75, 76]. CO_2 concentration and temperature are two key factors affecting crop growth, development and yield [77]. Combined or individual possessions of temperature increase and eCO_2 meditation change on crop growth and yield during the recent decades have been observed [78]. For instance, the modeled improvements in soybean absorption of CO_2 with an increase in the growing season temperature, and aCO_2 hindered the photorespiration by 23–48%, which depends on the future climatic conditions [79].

The growth and distribution of crops are reduced by environmental factors like CO_2 and temperature. The production of the biomass of modern C_3 plants was decreased by 50% when it was grown at a low concentration of CO_2 (180–220 ppm), while the other conditions were optimal. Crops need the almost dual amount of water at 2°C increase in temperatures at a higher elevation of agricultural plains. Elevated CO_2 concentration increases the yield of the crop once the substrate for the photosynthesis process of leaf and the incline of CO_2 absorption of air increases. C_3 plants are more benefitted at eCO_2 than C_4 plants [80]. However, the doubling of CO_2 does not deteriorate the adverse effects of high temperature on the reproductive growth of crops or fiber quality. Therefore, increased CO_2 concentration is associated with higher temperatures, crop yield, and quality that reduce particularly in areas where current temperatures are near to optimal [81].

Elevated CO_2 resulted in major changes in morpho-physiological restrictions. Besides, eCO_2 along with atmospheric temperatures during the phenological stages of rice cultivars showed contrasting results of the time of flowering and maturation such as eCO_2 in combination with the lower atmospheric temperature that stopped flowering g in the CR-1014 cultivar while with the higher temperature increased grain yield the in Naveen cultivar [82].

Growth and photosynthesis of C_3 crops are enhanced when it is grown at a high level of CO_2, although, the degree of stimulation differs with temperature among cultivars as well as species. The probable decline in the transpiration process due to the partial closure of stomata in the eCO_2 level is largely invalid by the energy balance between the crop and its environment, which could result in total water use in similar climate conditions. The yield of seeds is increased by an increase in the CO_2 under the ideal temperature. On the other hand, at supra-optimal temperature, the yield of seeds is decreased under both raised and ambient CO_2. The yield of kidney bean decreased in that region where temperatures are at or above optimal conditions in combination with increased CO_2 concentration [83].

The effects of HS on the grain and biomass yield of plants depend on the duration and magnitude of HS. HS at the vegetative phase decreased the grain and biomass yield mostly by increasing plant growth and dropping the time obtainable to capture possessions, and also by dropping the rate of photosynthesis [84]. At the anthesis or flowering phase, HS decreases the amount of grain due to pollen abortion. In contrast, at the grain-filling phase, HS decreases the heaviness of grain by restraining the translocation of integrating, and margarine the period of grain-filling [85, 86].

Notably, eCO_2 may lessen the harmful influence of heat stress on the grain and biomass yield by inspiration of photosynthesis, defense of the photosynthetic devices from HS injury, and improvement in the water status of plants owing to reduced transpiration. Moreover, high levels of hexoses and sucrose in plants with eCO_2 are related to increase fertile florets and dry spike mass [87], and osmotic modification [88], which can develop heat stress tolerance [89]. It is hypothesized that HS at anthesis has a drastic impact on the grain yield and plant biomass, but it has a less impact eCO_2 than ambient CO_2 [90].

6. Conclusion

It is very important to know that the impacts of climate change on crop growth are becoming global concerns. Interactive effects of eCO_2 and environmental stress conditions on crop growth and carbon metabolism are not well predictable. The influence of eCO_2 with the connection of temperature is considerable on crops under stress environments. The CO_2 concentration improves the productivity of crops because of improved carbon exchange rates, and superior vegetative and reproductive growth. In contrast, crop productivity is decreased with increased temperatures. Hence, there is a need to develop genotypes that are different intolerant to various environments or to identify genotypes that perform better under predicted climate change. In this review, cereal genotypes have been characterized by differing responses to eCO_2 and HS and identified the mechanisms of tolerance to HS. It can promote the crop potential to assist the breeding program for the development of new genotypes tolerance to HS.

Conflicts of interest

The authors declare no conflicts of interest.

Disclosure statement

Authors declare that no conflict of interest could arise.

Author details

Ayman EL Sabagh[1,2]*, Akbar Hossain[3], Mohammad Sohidul Islam[4], Muhammad Aamir Iqbal[5], Ali Raza[6], Çetin Karademir[2], Emine Karademir[2], Abdul Rehman[7], Md Atikur Rahman[8], Rajesh Kumar Singhal[9], Analía Llanes[10], Muhammad Ali Raza[11], Muhammad Mubeen[12], Wajid Nasim[13], Celaleddin Barutçular[14], Ram Swaroop Meena[15] and Hirofumi Saneoka[16]

1 Department of Agronomy, Faculty of Agriculture, Kafrelsheikh University, Kafrelsheikh, Egypt

2 Department of Field Crops, Faculty of Agriculture, Siirt University, Turkey

3 Bangladesh Wheat and Maize Research Institute, Dinajpur, Bangladesh

4 Department of Agronomy, Hajee Mohammad Danesh Science and Technology University, Dinajpur, Bangladesh

5 Department of Agronomy, University of Poonch Rawalakot Azadkashmir, Pakistan

6 Key Lab of Biology and Genetic Improvement of Oil Crops, Oil Crops Research Institute, Chinese Academy of Agricultural Sciences (CAAS), Wuhan, China

7 Department of Crop Sciences and Biotechnology, Dankook University, 119, Dandae-ro, Dongnam-gu, Cheonan-si, Korea

8 Grassland and Forage Division, National Institute of Animal Science, Rural Development Administration, Cheonan, Korea

9 ICAR-Indian Grassland and Fodder Research Institute, Jhansi, India

10 Plant Physiology Laboratory, Universidad Nacional de Río Cuarto (UNRC)-Instituto Nacional de Investigaciones Agrobiotecnológicas (INIAB), Consejo Nacional de Investigaciones Científicas y Técnicas (CONICET), Río Cuarto, Argentina

11 College of Agronomy, Sichuan Agricultural University, Chengdu, P.R. China

12 Department of Environmental Sciences, COMSATS University Islamabad, Vehari Campus, Pakistan

13 Department of Agronomy, University College of Agriculture and Environmental Sciences, The Islamia University of Bahawalpur (IUB), Bahawalpur, Pakistan

14 Department of Field Crops, Faculty of Agriculture, Cukurova University, Turkey

15 Department of Agronomy, Institute of Agricultural Sciences, Banaras Hindu University, Varanasi, UP, India

16 Graduate School of Integrated Sciences for Life, Hiroshima University, 1-4-4 Kagamiyama, Higashi-Hiroshima, Japan

*Address all correspondence to: aymanelsabagh@agr.kfs.edu.eg

References

[1] Raza A, Razzaq A, Mehmood SS, Zou X, Zhang X, Lv Y, Xu J. Impact of climate change on crops adaptation and strategies to tackle its outcome: A review. Plants. 2019;**8**:34

[2] Hossain A, Farooq M, EL Sabagh A, Hasanuzzaman M, Erman M, Islam T. Morphological, Physiobiochemical and Molecular Adaptability of Legumes of Fabaceae to Drought Stress, with Special Reference to *Medicago sativa* L. In: Hasanuzzaman M, Araújo S, Gill S. (eds) The Plant Family Fabaceae. Springer, Singapore; 2020. pp. 289-317. https://doi.org/10.1007/978-981-15-4752-2_11.

[3] IPCC. Climate Change 2014: Synthesis Report. Contribution of Working Groups I, II and III to the Fifth Assessment Report of the Intergovernmental Panel on Climate Change [Core Writing Team, Pachauri RK, Meyer LA (eds.)]. IPCC, Geneva, Switzerland, 2014. p.151

[4] Calzadilla A, Zhu T, Rehdanz K, Tol RSJ, Ringler C. Economy wide impacts of climate change on agriculture in Sub-Saharan Africa. Ecological Economics.2013;**93**:150-165

[5] Cairns JE, Sonder K, Zaidi PH, Verhulst N, Mahuku G, et al. Maize Production in a Changing Climate: Impacts, Adaptation, and Mitigation Strategies. Advances in Agronomy. 2012;**114**:1-65

[6] Ziska L. Observed changes in soyabean growth and seed yield from *Abutilon theophrasti* competition as a function of carbon dioxide concentration. Weed Research. 2013;53:140-145

[7] Easterling DR, Evans JL, Groisman PY, Karl TR, Kunkel KE, et al. Observed variability and trends in extreme climate events: a brief review. Bulletin of American Meteorological Society. 2000;**81**:417-425

[8] Wallace JS. Increasing agricultural water use efficiency to meet future food production. Agriculture, Ecosystems &Environment. 2000;**82**:105-119

[9] Johkan M, Oda M, Maruo T, Shinohara Y. Crop Production and Global Warming Impacts, Case Studies on the Economy, Human Health, and on Urban and Natural Environments. 2011;**9**:139-152. http://www.intechopen.com.

[10] Young KJ, Long SP. Crop ecosystem responses to climatic change: maize and sorghum. In: Reddy KR, Hodges HF (Eds.). Climate change and global crop productivity, CABI International, Oxon, United Kingdom; 2000. pp. 107-131

[11] Hunt R, Hand D, Hannah M, Neal A. Response to CO_2 enrichment in 27 herbaceous species. Functional Ecology. 1991;**5**:410-421

[12] Rogers HH, Dahlman RC. Crop responses to CO_2 enrichment. CO_2 and biosphere. Advances in vegetation Science. 1993;14:117-131

[13] Pathak H, Aggarwal PK, Singh SD. Climate change impacts, adaptations and mitigation in agriculture: methodology for assessment and application, Indian Agricultural Research Institute, New Delhi, India, 2012;pp. 1-302

[14] Leakey A, Uribelarrea M, Ainsworth E, Naidu S, Rogers A, et al. Photosynthesis, productivity and yield of maize are not affected by open-air elevation of CO_2 concentration in the absence of drought. Plant Physiology. 2006;140:779-790

[15] Kim SH, Gitz DC, Sicher RC, Baker JT, Timlin DJ. Temperature

dependence of growth development, and photosynthesis in maize under elevated CO_2. Environment and Experimental Botany. 2007;**61**:224-236

[16] Vanaja M, Maheswari M, Jyothi Lakshmi N, Sathish P, Yadav SK, Salini K, Vagheera P, Vijay Kumar G, Razak A. Variability in growth and yield response of maize genotypes at eCO_2 concentration. Advances in Plants & Agriculture Research. 2015;**2**:42. DOI: 10.15406. apar.2015.02.00042

[17] Meena RS, Kumar V, Yadav GS, Mitran T. Response and interaction of *Bradyrhizobiumjaponicum*and *Arbuscularmycorrhizal*fungi in the soybean rhizosphere: A review. Plant Growth Regulation. 2018;**84**:207-223

[18] Wang D, Heckathorn SA, Barua D, Joshi P, Hamilton EW, LaCroix JJ. Effects of eCO_2 on the tolerance of photosynthesis to acute heat stress in C_3, C_4, and CAM species. American Journal of Botany. 2008;**95**:165-176. https://doi.org/10.3732/ajb.95.2.165.

[19] Meena RS, Lal R, Yadav GS. Long-term impact of topsoil depth and amendments on carbon and nitrogen budgets in the surface layer of an Alfisol in Central Ohio. Catena, 2020;**194**:104752.https://doi.org/10.1016/j.catena.2020.104752

[20] Jagadish SVK, Bahuguna RN, Djanaguiraman M, Gamuyao R, Prasad PV, Craufurd PQ. Implications of high temperature and elevated CO_2 on flowering time in plants. Frontiersin Plant Science.2016;**7**:11http://dx.doi.org/10.3389/fpls. 2016.00913.

[21] Xu Z, Jiang Y, Jia B, Zhou G. Elevated-CO_2 response of stomata and its dependence on environmental factors. Frontiers in Plant Science. 2016;**7**:657. https://doi.org/10.3389/fpls.2016.00657

[22] Kadam NN, Xiao G, Melgar RJ, Bahuguna RN, Quinones C, Tamilselvan A, Prasad PVV, Jagadish KS. Agronomic and physiological responses to high temperature, drought, and elevated CO_2 interactions in cereals. Advances in Agronomy. 2014;**127**:111-156

[23] Salim N, Raza A. Nutrient use efficiency (NUE) for sustainable wheat production: a review. Journal of Plant Nutrition. 2020;**43**:297-315

[24] Lavania D, Dhingra A, Siddiqui MH, Al-Whaibi MH, Grover A. Current status of the production of high temperature tolerant transgenic crops for cultivation in warmer climates. Plant Physiology and Biochemistry. 2015;**86**:100-108. Doi:10.1016/j.plaphy.2014.11.019

[25] Raza A, Ashraf F, Zou X, Zhang X, Tosif H. Plant Adaptation and Tolerance to Environmental Stresses: Mechanisms and Perspectives. In Plant Ecophysiology and Adaptation under Climate Change: Mechanisms and Perspectives I. Springer, Singapore, 2020;pp. 117-145

[26] Rodrigues WP, Martins MQ, Fortunato AS, Rodrigues AP, Semedo JN, Simões-Costa MC, et al. Long-term elevated air [CO_2] strengthens photosynthetic functioning and mitigates the impact of supra-optimal temperatures in tropical *Coffeaarabica* and *C. canephora* species. Global Change Biology. 2016;**22**:415-431. Doi:10.1111/gcb.13088

[27] Aranjuelo I, Sanz-Sáez Á, Jauregui I, Irigoyen JJ, Araus JL, Sánchez-Díaz M, Erice G. Harvest index, a parameter conditioning responsiveness of wheat plants to elevated CO_2. Journal of Experimental Botany. 2013;**64**:1879-1892

[28] Goufo P, Trindade H. Rice antioxidants: phenolic acids, flavonoids, anthocyanins, proanthocyanidins, tocopherols, tocotrienols, γ-oryzanol,

and phytic acid. Food Science & Nutrition. 2014;**2**:75-104

[29] Horvat D, Šimić G, Drezner G, Lalić A, Ledenčan T, Tucak M, ... Zdunić Z. Phenolic Acid Profiles and Antioxidant Activity of Major Cereal Crops. Antioxidants.2020;**9**:527

[30] Hasanuzzaman M, Bhuyan MHM, Zulfiqar F, Raza A, Mohsin SM, Mahmud JA, ... Fotopoulos V. Reactive Oxygen Species and Antioxidant Defense in Plants under Abiotic Stress: Revisiting the Crucial Role of a Universal Defense Regulator. Antioxidants. 2020;**9**:681

[31] EL Sabagh A, Hossain A, Islam MS, Fahad S, Ratnasekera D, Meena RS, Wasaya A, Yasir TA, Ikram M, Mubeen M, Fatima M, Nasim W, Çığ A, Çığ F, Erman M, Hasanuzzaman M. Nitrogen Fixation of Legumes Under the Family Fabaceae: Adverse Effect of Abiotic Stresses and Mitigation Strategies. In: Hasanuzzaman M, Araújo S, Gill S. (eds) The Plant Family Fabaceae. Springer, Singapore; 2020. pp. 75-111. https://doi.org/10.1007/978-981-15-4752-2_4.

[32] Mittler R, Vanderauwera S, Gollery M, Van Breusegem F. Reactive oxygen gene network of plants. Trends in Plant Science. 2004;**9**:490-498

[33] diToppi LS, Marabottini R, Badiani M, Raschi A. Antioxidant status in herbaceous plants growing under elevated CO_2 in mini-FACE rings. Journal of Plant Physiology.2002;**159**:1005-1013

[34] Pérez-López U, Robredo A, Lacuesta M, Sgherri C, Mena-Petite A, Navari-Izzo F, Muñoz-Rueda A. Lipoic acid and redox status in barley plants subjected to salinity and elevated CO_2. PhysiologiaPlantarum.2010;**139**:256-268

[35] Tausz-Posch S, Borowiak K, Dempsey RW, Norton RM, Seneweera S, Fitzgerald GJ, Tausz M. The effect of elevated CO_2 on photochemistry and antioxidativedefence capacity in wheat depends on environmental growing conditions-A FACE study. Environmental and Experimental Botany. 2013;**88**:81-92

[36] Landete JM. Dietary intake of natural antioxidants: vitamins and polyphenols. Critical Reviews in Food Science and Nutrition. 2013;**53**:706-721

[37] Visioli F, Lastra CADL, Andres-Lacueva C, Aviram M, Calhau C, Cassano A, ...Llorach R. Polyphenols and human health: a prospectus. Critical Reviews in Food Science and Nutrition. 2011;**51**:524-546

[38] Cho JY, Lee HJ, Kim GA, Kim GD, Lee YS, Shin SC, Park KY, Moon JA. Quantitative analyses of individual g-oryzanol (sterylferulates) in conventional and organic brown rice (*Oryza sativa* L.). Journalof Cereal Science. 2012;**55**:337e343

[39] Zinta G, AbdElgawad H, Domagalska MA, Vergauwen L, Knapen D, Nijs I, ... Asard H. Physiological, biochemical, and genome-wide transcriptional analysis reveals that elevated CO_2 mitigates the impact of combined heat wave and drought stress in *Arabidopsis thaliana* at multiple organizational levels. Global Change Biology. 2014;**20**:3670-3685

[40] Slesak I, Libik M, Karpinska B, Karpinski S, Miszalski Z. The role of hydrogen peroxide in regulation of plant metabolism and cellular signalling in response to environmental stresses. Acta Biochimica Polonica. 2007;**54**:39-50

[41] Hymus GJ, Baker NR, Long SP. Growth in elevated CO_2 can both increase and decrease photochemistry and photoinhibition of photosynthesis in a predictable manner. Dactylisglomerata grown in two levels

of nitrogen nutrition. Plant Physiology. 2001;**127**:1204-1211

[42] Cunniff J, Charles M, Jones G, Osborne CP. Reduced plant water status under sub-ambient p CO2 limits plant productivity in the wild progenitors of C_3 and C_4 cereals. Annals of Botany. 2016;**118**:1163-1173

[43] Satapathy BS, Duary B, Saha S, Pun KB, Singh T. Effect of weed management practices on yield and yield attributes of wet direct seeded rice under lowland ecosystem of Assam. ORYZA-An International Journal on Rice. 2017;**54**:29-36

[44] Bhattacharyya P, Roy KS, Neogi S, Dash PK, Nayak AK, Mohanty S, ...Rao KS. Impact of elevated CO_2 and temperature on soil C and N dynamics in relation to CH_4 and N_2O emissions from tropical flooded rice (*Oryza sativa* L.). Science of the Total Environment. 2013;**461**:601-611

[45] Roy S, Banerjee A, Mawkhlieng B, Misra AK, Pattanayak A, Harish GD, ... Bansal KC. Genetic diversity and population structure in aromatic and quality rice (*Oryza sativa* L.) landraces from North-Eastern India. PloS One. 2015;**10**:e0129607

[46] Kumar A, Nayak AK, Sah RP, Sanghamitra P, Das BS. Effects of elevated CO_2 concentration on water productivity and antioxidant enzyme activities of rice (*Oryza sativa* L.) under water deficit stress. Field Crops Research. 2017;**212**:61-72

[47] Vandegeer RK, Powell KS, Tausz M. Barley yellow dwarf virus infection and elevated CO_2 alter the antioxidants ascorbate and glutathione in wheat. Journal of Plant Physiology. 2016;**199**:96-99

[48] Sharma P, Jha AB, Dubey RS, Pessarakli M. Reactive oxygen species, oxidative damage, and antioxidative defense mechanism in plants under stressful conditions. Journal of Botany. 2012;**2012:**217037:http://doi.org/10.1155/2012/217037.

[49] EL Sabagh A, Hossain A, Islam MS, Barutcular C, ... et al. Drought and Heat Stress in Cotton (*Gossypiumhirsutum* L.): Consequences and Their Possible Mitigation Strategies. In: Hasanuzzaman M. (eds) Agronomic Crops. Springer, Singapore. 2020a. https://doi.org/10.1007/978-981-15-0025-1_30

[50] Apel K, Hirt H. Reactive oxygen species: metabolism, oxidative stress, and signal transduction. Annual Review of Plant Biology. 2004;**55**:373-399

[51] AbdElgawad H, Farfan-Vignolo ER, De Vos D, Asard H. Elevated CO_2 mitigates drought and temperature-induced oxidative stress differently in grasses and legumes. Plant Science. 2015;231:1-10. 10.1016/j.plantsci.2014.11.001

[52] AbdElgawad H, Zinta G, Beemster GTS, Janssens IA, Asard H. Future climate CO_2 levels mitigate stress impact on plants: increased defense or decreased challenge? Frontiers in Plant Science. 2016;7:556. Doi: 10.3389/fpls.2016.00556.

[53] Xu Z, Shimizu H, Yagasaki Y, Ito S, Zheng Y, Zhou G. Interactive effects of elevated CO_2, drought, and warming on plants. Journal of Plant Growth Regulation. 2013;**32**:692-707

[54] Ghannoum O. C_4 photosynthesis and water stress. Annals of Botany. 2009;**103**:635-644

[55] Yassin M, El Sabagh A, Mekawy AMM, Islam MS, Hossaın A, Barutcular C, Alharby H, Bamagoos A, Liu L, Ueda A, Saneoka H Comparative performance of two bread wheat (*Triticumaestivum* L.) genotypes under salinity stress. Applied Ecology and

Environmental Research. 2019;**17**:5029-5041. https://doi.org/10.15666/aeer/1702_50295041

[56] Fair P, Tew J, Cresswell C. Enzyme activities associated with carbon dioxide exchange in illuminated leaves of *Hordeum vulgare* L. II. Effects of external concentrations of carbon dioxide and oxygen. Annals of Botany. 1973;**37**:1035-1039

[57] Salazar-Parra C, Aguirreolea J, Sánchez-Díaz M, Irigoyen JJ, Morales F. Climate change (elevated CO_2, elevated temperature and moderate drought) triggers the antioxidant enzymes' response of grapevine cv. Tempranillo, avoiding oxidative damage. Physiologia Plantarum. 2012;**144**:99-110

[58] Li W, Zhu Q, Wang Y, Wang S, Chen X, Zhang D, Wang W. The relationships between physiological and biochemical indexes and the yield characteristics of rice under high temperature stress. 2013;**29**:

[59] Booker FL, Reid CD, Brunschön-Harti S, Fiscus EL, Miller JE. Photosynthesis and photorespiration in soybean [*Glycine max* (L.). Merr.] chronically exposed to elevated carbon dioxide and ozone. Journal of Experimental Botany. 1997;**48**:1843-1852

[60] Lin J-S, Wang G-X. Doubled CO_2 could improve the drought tolerance better in sensitive cultivars than in tolerant cultivars in spring wheat. Plant Science. 2002;**163**:627-637

[61] Rao MV, Hale BA, Ormrod DP. Amelioration of ozone-induced oxidative damage in wheat plants grown under high carbon dioxide (Role of antioxidant enzymes), Plant Physiology. 1995;**109**:421-432

[62] Xu Z, Jiang Y, Zhou G. Response and adaptation of photosynthesis, respiration, and antioxidant systems to elevated CO_2 with environmental stress in plants. Frontiers in Plant Science. 2015;**6**:701

[63] Dietzen CA, Larsen KS, Ambus PL, Michelsen A, Arndal MF, Beier C, Reinsch S, Schmidt IK. Accumulation of soil carbon under elevated CO_2 unaffected by warming and drought. Global Change Biology. 2019;**25**:2970-2977

[64] Li S, Li Y, Gao Y, He X, Zhang D, Liu B, Li Q. Effects of CO_2 enrichment on non-structural carbohydrate metabolism in leaves of cucumber seedlings under salt stress. ScientiaHorticulturae. 2020;**265**:109275

[65] Zinta G, AbdElgawad H, Peshev D, Weedon JT, Van den Ende W, Nijs I, Janssens IA, Beemster GT, Han A. Dynamics of Metabolic Responses to Combined Heat and Drought Spells in *Arabidopsis thaliana* under Ambient and Rising Atmospheric CO_2. Journal of Experimental Botany. 2018;69:2159-2170. doi: 10.1093/jxb/ery055.

[66] Li M, Li Y, Zhang W, Li S, Gao Y, Ai X, Zhang D, Liu B, Li Q. Metabolomics analysis reveals that elevated atmospheric CO_2 alleviates drought stress in cucumber seedling leaves. Analytical Biochemistry. 2018;**559**:71-85.

[67] Oliveira MFD, Marenco RA. Gas exchange, biomass allocation and water-use efficiency in response to elevated CO_2 and drought in andiroba (*Carapasurinamensis*, Meliaceae). iForest-Biogeosciences and Forestry. 2019;**12**:61-68

[68] Mphande W, Nicolas ME, Seneweera S, Bahrami H. Dynamics and contribution of stem water-soluble carbohydrates to grain yield in two wheat lines contrasted under drought and elevated CO2 conditions. Journalof Plant Physiology. 2016;**214**:1037-1058

[69] Vicente R, Bolger AM, Martinez-Carrasco R, Pérez P, Gutiérrez E, Usadel B, Morcuende R. De novo transcriptome analysis of durum wheat flag leaves provides new insights into the regulatory response to elevated CO_2 and high temperature. Frontiers in Plant Science. 2019;**10**:1605.

[70] Sgherri C, Pérez-López U, Micaelli F, Miranda-Apodaca J, Mena-Petite A, Muñoz-Rueda A, Quartacci MF. Elevated CO_2 and salinity are responsible for phenolics-enrichment in two differently pigmented lettuces. Plant Physiology and Biochemistry. 2017;**115**:269-278

[71] Agüera E, De la Haba P. Leaf senescence in response to elevated atmospheric CO_2 concentration and low nitrogen supply. Biologia Plantarum. 2018;**62**:401-408

[72] Collins M, Knutti R, Arblaster J, Dufresne JL, Fichefet T, Friedlingstein P, ...Krinner G. Long-term climate change: projections, commitments and irreversibility. In Climate Change 2013. The Physical Science Basis: Contribution of Working Group I to the Fifth Assessment Report of the Intergovernmental Panel on Climate Change. Cambridge University Press. 2013. pp. 1029-1136

[73] Rosenzweig C, Elliott J, Deryng D, Ruane AC, Müller C, Arneth A, ...Khabarov N. Assessing agricultural risks of climate change in the 21st century in a global gridded crop model intercomparison. Proceedings of the National Academy of Sciences. 2014;**111**:3268-3273

[74] Cai C, Yin X, He S, Jiang W, Si C, Struik PC, . . .Xiong Y. Responses of wheat and rice to factorial combinations of ambient and elevated CO_2 and temperature in FACE experiments. Global Change Biology. 2016;**22**:856-874

[75] Tan K, Zhou G, Lv X, Guo J, Ren S. Combined effects of elevated temperature and CO_2 enhance threat from low temperature hazard to winter wheat growth in North China. Scientific Reports. 2018;**8**:1-9

[76] EL Sabagh A, Hossain A, Barutçular C, Gormus O, Ahmad Z, Hussain S, Islam MS, Alharby H, Bamagoos A, Kumar N, Akdeniz A, Fahad S, Meena RS, Abdelhamid M, Wasaya A, Hasanuzzaman M, Sorour S, Saneoka H. Effects of drought stress on the quality of major oilseed crops: implications and possible mitigation strategies–a review. Appllied Ecology and Environmental Research. 2019;**17**:4019-4043. https://doi.org/10.15666/aeer/1702_40194043

[77] Mubeen M. et al. Evaluating the climate change impact on crop water requirement of cotton- wheat in semi-arid conditions using DSSAT model. Journal of Water and Climate Change. 2019; 201979. doi: 10.2166/wcc.2019.179.

[78] Hatfield JL, Boote KJ, Kimball B, Ziska L, Izaurralde RC, Ort D, . . . Wolfe D. Climate impacts on agriculture: implications for crop production. Agronomy Journal. 2011;**103**:351-370

[79] Walker BJ, VanLoocke A, Bernacchi CJ, Ort DR. The costs of photorespiration to food production now and in the future. Annual Review of Plant Biology. 2016;**67**:107-129.

[80] Madan P, Jagadish S, Craufurd P, Fitzgerald M, Lafarge T, Wheeler T. Effect of elevated CO_2 and high temperature on seed-set and grain quality of rice. Journal of Experimental Botany. 2012;**63**:3843-3852

[81] Reddy KR, Vara Prasad P, Kakani VG. Crop responses to elevated carbon dioxide and interactions with

temperature: cotton. Journal of Crop Improvement. 2005;**13**:157-191

[82] Ziska LH. et al. Food security and climate change: on the potential to adapt global crop production by active selection to rising atmospheric carbon dioxide. Proceedings of the Royal Society B: Biological Sciences. 2012;**279**:4097-4105 http://doi.org/10.1098/rspb.2012.1005

[83] Prasad PV, Boote KJ, Allen Jr LH, Thomas JM. Effects of elevated temperature and carbon dioxide on seed-set and yield of kidney bean (*Phaseolus vulgaris* L.). Global Change Biology. 2002;**8**:710-721

[84] Lobell DB, Gourdji SM. The influence of climate change on global crop productivity. Plant Physiology. 2012;**160**:1686-1697. Doi: https://doi.org/10.1104/pp.112.208298

[85] Farooq M, Bramley H, Palta JA, Siddique KHM. Heat stress in wheat during reproductive and grain-filling phases. Critical Reviewsin Plant Sciences. 2011;**30**:491-507.DOI: 10.1080/07352689.2011.615687

[86] Prasad PVV, Djanaguiraman M. Response of floret fertility and individual grain weight of wheat to high temperature stress: sensitive stages and thresholds for temperature and duration. Functional Plant Biology. 2014;**41**:1261-1269.http://dx.doi.org/10.1071/FP14061

[87] Dreccer MF, Wockner KB, Palta JA, McIntyre CL, Borgognone MG, Bourgault M, Reynolds M, Miralles DJ. More fertile florets and grains per spike can be achieved at higher temperature in wheat lines with high spike biomass and sugar content at booting. Functional Plant Biology 2014;**41**:482-495.http://dx.doi.org/10.1071/FP13232.

[88] Wahid A, Gelani S, Ashraf M, Foolad MR. Heat tolerance in plants: an overview. Environmental and Experimental Botany. 2007;**61**:199-223

[89] Shanmugam S, Kjaer KH, Ottosen CO, Rosenqvist E, Kumari Sharma D, Wollenweber B. The alleviating effect of elevated CO_2 on heat stress susceptibility of two wheat (*Triticumaestivum* L.) cultivars. Journal of Agronomy and Crop Science. 2013;**199**:340-350

[90] Chavan SG, Duursma RA, Tausz M, Ghannoum O. ECO_2alleviates the negative impact of heat stress on wheat physiology but not on grain yield. Journal of Experimental Botany.2019;**70**:6447-6459. doi:10.1093/jxb/erz386

Chapter 8

Ecological and Economic Potential of Major Halophytes and Salt Tolerant Vegetation in India

T.V. Vineeth, Shrvan Kumar, Monika Shukla, Anil Chinchmalatpure and Parbodh Chander Sharma

Abstract

Soil salinization is a global and climatic phenomenon that affects various spheres of life. The present rate of salinization is perilously fast because of global climate change and associated events leading to enhanced land degradation, loss of soil fertility and crop productivity. In this chapter, we tried to focus on the arid and semiarid regions of India along with our coastal zone which are economically fragile regions and need much closer attention. In future, India will face extreme pressure on its land resources in agriculture because of likely rapid degradation of these resources. Thus, salt affected soils must be brought under cultivation by adopting site specific strategies to ensure national food and nutritional security. In this regard, a comprehensive review of the major halophytes of these ecological zones, its mechanism of salt tolerance, ecological and economic potential is done. The potential applications of saline land vegetation including halophytes in climate change mitigation, phytoremediation, desalination, food, secondary metabolite and nutraceutical production, medicine, and saline agriculture have been discussed. Further, we tried to focus on popular farmer adopted halophytic species including edible ones, their uses, products of economic significance etc. which is highly imperative for effective utilization of these saline soils leading to improved livelihood and sustenance of resource poor farmers along with improved ecological balance.

Keywords: coastal soil, dry land, ecology, economic products, halophytes, saline vegetation, salinity stress

1. Introduction

Food security and ecological security are the two pillars of all societies. The incessant growth of human population across the world is evident from the United Nations projection of 9.7 billion people by 2050 [1]. The global food demand is expected to rise by 70–100% and to meet this additional food demand, a predicted 50% increase in yields of the major food crops will be required [2]. Moreover, our global food security is significantly affected by drastic alterations in the climate [3]. This unpredictable change in climate has resulted in to several abiotic stresses such as salinity, drought, heat and low temperature, threatening the agricultural sustenance and productivity [4]. This has also brought about a significant spike in the occurrences of extreme climatic events like flood, cyclones, storms etc. [5]. Land degradation is one of the severe consequences of global climate change and it is happening at a viciously alarming pace because of other reasons too, like vegetation loss, soil erosion, inadequate water management, excessive use of fertilizers, soil and groundwater degradation, urbanization and poor agricultural practices [6]. One of the major causes of land degradation is buildup of soil salinity as a result of natural and anthropogenic factors [7, 8]. It is estimated that agricultural soils are decreasing about 1–2% every year especially in the arid and semiarid regions of the world due to soil salinity [9].

On an average, 831 million hectares (Mha) of global soil is affected by salt (397 and 434 Mha of saline and sodic soils, respectively) which is more than 6% of the world's total land area [10]. Out of total irrigated land (230 Mha), almost 20% (45 Mha) area has already been damaged by excess soil salinity [11], whereas 14% (32 Mha) is affected by varying degrees of salt [12]. In India, the total land degraded due to soil salinity and sodicity is estimated to be 6.74 Mha [13] and it is likely to increase to 16.2 Mha by 2050 [14]. Soil salinization is grouped in to two types*viz.* Primary salinization and secondary salinization. Primary salinization occurs due to the existence of subsoil salts and by other natural processes like weathering of parent material, deposition of sea salt carried by wind and rain, inundation of coastal land by tidal water, etc. Secondary salinization is a consequence of anthropogenic or human induced activities like excessive irrigation, irrigation with salt containing water, poor drainage due to loss of natural water passages, unscientific application of inorganic fertilizers and other soil amendments [14, 15].

The arid and semiarid regions of India along with our coastal zones are economically fragile regions and need much closer attention. Low rainfall and high temperature are two specific features of the arid and semiarid regions of India which are highly conducive to excessive soil salinity [16]. India has a coastline of 7516.6 km viz. 5422.6 km of mainland coastline and 1197 km of Indian islands spread across 9 Indian states and two Union Territories [17]. Soil salinity is estimated to have affected 1.25 Mha soils in these coastal tracts of the nation [13]. The predicted rise in sea level of 1 m due to global warming may aggravate this problem to take on serious dimensions in future [14]. Moreover, there is a predicted increase in water demand by 45% in coastal areas by 2050, which will inevitably result in increased ground water depletion and induce more and more soil salinity [14]. Altogether, in both these ecological zones of the nation, soil salinity has become an important factor limiting the growth of major crops and even some halophytes [18].

Moreover, factors like global climate change, diversion of more and more productive agricultural land towards other nonagricultural purposes as a part of rapid urbanization and the alarming rate of decline in soil organic carbon will lead to further increase in soil salinity [19–24]. So we are running out of options, and the only option left is to bring these saline, sodic and other barren land in to cultivation to meet the growing food demand and more importantly to address the growing climatic and ecological concerns. Major agricultural crops and tress cannot be grown in such soils as they exhibit a low tolerance to salt [25]. So future agricultural production in these saline soils should be based on salt tolerant crops which are edible and/or of economic and ecological significance [26, 27].

Halophytes, salt mangroves, and other salt tolerant plants and trees are the options which can grow and survive in extreme environments. Many halophytes have potential agricultural value and can be grown in these highly saline areas. They not only survive in these conditions, but also produce considerable biomass coupled with various other potential/economic uses in industry, feed, medicine etc. [28]. Indian mangrove vegetation covers about 6749 km^2 along the 7516.6 km long coast line, including island territories. 82 species of mangroves distributed in 52 genera and 36 families have been identified that are spread across 12 habitats of India [29]. They along with many other associate species thrive well in coastal saline soils. Moreover, these saline soils can further be brought in to cultivation through agroforestry, silvipasture, nonconventional crops including medicinal and aromatic plants and other high value crops. These will not only utilize the saline barren land for economic growth but will also help significantly in conserving our biodiversity and improvement of ecological and social environment.

A systematic attempt has been made in bringing together critical aspects of saline agriculture, including major halophytes and other saline vegetation of India, its salt tolerance mechanisms coupled with ecological and economic potential of these plants. This is highly imperative for effective utilization of these saline soils leading to improved livelihood and sustenance of resource poor farmers along with improved ecological balance.

2. Halophytes

Halophytes are plants capable of completing their life cycle under highly saline conditions [30]. To be more scientifically concrete, halophytes are those plant species which can successfully survive, grow and reproduce in soils of salt concentration more than 200 mM of NaCl (Electrical conductivity of soil saturation extract (ECe) of 20 dS m^{-1}) [31]. Some halophytes can even grow well at higher salt concentrations than that of sea water (>500 mM Nacl) [32]).As they prosper well in severely saline soils, they are also considered to be extremophiles [33]. They constitute around 1% of the global flora [15]. The question of how they survive and complete their life cycle under such extreme conditions has led to detailed studies on the various morphological, anatomical, physiological and molecular mechanisms of high salt tolerance. The tolerance of halophytes to salts varies with species and developmental stage. In general, dicot halophytes are reported to be more tolerant (optimal growth in 100–200 mM NaCl) as compared to monocot species (optimal growth in 50–100 mM NaCl) [31].

During 1980s, James Aronson compiled a comprehensive database [34] of 1554 halophyte species, and named it as HALOPH. This database has been converted into an interactive eHALOPH repository [35] that provides online details and bibliography pertaining to halophytes. Till now 2000 to 3000 halophytic plant species are identified [32] in the world mainly belonging to angiosperms. In India, their distribution is mainly confined to arid, semiarid inlands and highly saline wetlands along the tropical and sub-tropical coasts [36]. The major halophytes/saline vegetation found across the various ecological hotspots spread across the country are summarized in **Table 1**.

A summary of classification of halophytes by researchers across the globe based on different aspects is compiled and presented below.

2.1 Based on salt tolerance

i. **Mio-halophytes**: Plants which grows in the habitats of low salinity levels (below 0.5% NaCl) [37].

ii. **Eu-halophytes**: Plants which grow in highly saline habitats. They have been further sub-divided into the following groups:

  a. **Mesohalophytes**: Plants that can tolerate salinity range of 0.5 to 1%.

  b. **Mesoeuhalophytes**: Plants that can tolerate salinity range of 5% and higher.

  c. **Eneuhalophytes**: Plants that can tolerate salinity range of 1% and above.

Scientific name	Common name	Family	Area of occurrence	Plant type
Aeluropus lagopoides	Mangrove grass, usargas, kharoga	Poaceae	Kachchh, Thar Desert	Tufted grass
Atriplex hortensis	French Spinach, Pahari Paleng	Amaranthaceae	Coastal marshes Ladakh, Kashmir	Herb
Atriplex stocksii	Khati palakh, kharo tanko	Amaranthaceae	Kachchh (Gujarat)	Bush
Avicennia marina	Whitemangrove, tavarian, tivar	Acanthaceae	Kachchh, Kerala Maharashtra, Tamilnadu,	Mangrove
Cenchrus biflorus	Indian sandbur, bhurut	Poaceae	Kachchh, Thar Desert	Grass
Chenopodium album	Lamb's quarters, bathua	Amaranthaceae	Northarn India, Kachchh	Herb
Cressa cretica	Rudravanti, machul	Convolvulaceae	Kachchh, Thar Desert	Sub-shrub
Cyperus conglomeratus	—	Cyperaceae	Kachchh, Thar Desert, Kerala	Herb
Cynodon dactylon	Bermuda grass, doob	Poaceae	Throughout India	Grass
Dactyloctenium sindicum	Tantia, ganthio	Poaceae	Kachchh	Grass
Dichanthium annulatum	Marvel grass, karad, bansi	Poaceae	Kachchh, Thar Desert	Grass
Halopyrum mucronatum	Dariyai kans	Poaceae	Kachchh	Grass
Haloxylon salicornicum	Rimth saltbush, khar, lana	Amaranthaceae	Kachchh, Thar Desert	Shrub
Heliotropium bacciferum	—	Boarginaceae	Kachchh, Southern India	Herb
Ipomoea pes-caprae	Atampa, dopatti lata	Convolvulaceae	Kachchh	Herb
Limonium stocksii	Kharia	Plumbaginaceae	Kachchh, North west India	Herb
Portulaca oleracea	Common purslane, lunia, badi-noni	Portulacaceae	Maharastra, Thar desert	Herb
Prosopis juliflora	Mesquite, vilaiti keekar	Fabaceae	Throughout India	Shrub
Salicornia brachiata	Glasswort, umari keerai	Amaranthaceae	Kachchh, Thar desert	Herb
Salsola baryosma	Loonuk, lani	Amaranthaceae	Kachchh, Thar desert	Shrub
Salvadora oleoides	Bada peelu, meethi jaal	Salvadoraceae	Kachchh, Thar desert	Shrub
Salvadora persica	Meswak, piludi, khari jaal		Kachchh, Thar desert	Shrub/Tree
Sesuvium portulacastrum	Sea purslane, lunio, dhapa	Aizoaceae	Kachchh, Rajasthan,	Herb
Sporobolus marginatus	poolongi, khevai	Poaceae	Kachchh	Grass

Scientific name	Common name	Family	Area of occurrence	Plant type
Suaeda fruticosa	lunaki, moras	Amaranthaceae	Kachchh, Thar desert	Shrub
Suaeda maritima	Annual sea blite, alur		Maharastra	
Suaeda nudiflora	Muchole		Kachchh	
Tamarix dioica	Red tamarisk, lai, arseli	Tamaricaceae	Kachchh	Shrub
Urochondra setulosus	Kkariyu	Poaceae	Kachchh	Grass
Ziziphus nummularia	Jhar beri, chanibor		Kachchh, Thar desert	Shrub
Zygophyllum simplex	Bean-Caper, pat lani	Zygophyllaceae	Kachchh, Thar desert	Herb

Table 1.
Major halophytic species and other saline vegetation spread across different ecological zones of India.

2.2 Based on mechanism of tolerance

i. **Salt excluding**: The root architecture of this category of plants is embraced by an ultrafiltration mechanism which leads to establishment of such species in saline conditions. Mangrove vegetation shows such type of tolerance. e.*G. Rhizophora mucronata*, *Bruguiera gymnorrhiza* [38].

ii. **Salt excreting**: This category of plants release excess salts in their internal tissues to outside via. Specialized structures called as salt glands. e.g. *Avicennia officinalis*, *Avicennia alba*, *Avicennia marina*.

iii. **Salt accumulating**: These plants are able to maintain very high levels of salt in their tissues either by virtue of succulence or by compartmentation of excess salts in to comparatively safer cellular locations like vacuole. e.g. *Salvadora persica*, *Sesuvium portulacastrum*, *Suaeda nudiflora*.

2.3 Based on ecological aspect

i. **Obligate halophytes**: They grow only in salty habitats and show satisfactory growth and development under high saline condition. Many plant species belonging to chenopodiceae family comes under this category. e.g. *Salicornia bigelovii* [39].

ii. **Facultative halophytes**: Plants of this group are able to establish themselves on salty soils, but their optimum growth is observed in a salt free or low salt condition. Most poaceae, cyperaceae, and brassicaceae species as well as a large number of dicotyledons like *Aster tripolium*, *Glaux maritima, Plantagomaritima* belong to this group.

iii. **Habitat-indifferent halophytes**: They normally grow on salt free soils but can thrive better than sensitive species under saline conditions. Plants like *Chenopodium glaucum, Myosurus minimus,* and *Potentillaanserina* are categorized as habitat insensitive plants.

Figure 1.
Major halophytes and mangroves found in India. (a) Sporobolus marginatus, *(b)* Salicornia brachiate, *(c)* Chenopodium album, *(d)* Portulaca oleracea, *(e)* Heliotropium curassavicum, *(f)* Haloxylon salicornicum, *(g)* Cresa cretica, *(h)* Aeluropus lagopoides, *(i)* Prosopis juliflora, *(j)* Salvadora oleoides, *(k)* Salvodera persica, *(l)* Sesuvium portulacastrum, *(m)* Suaeda fruticosa, *(n)* Halopyrum mucronatum, *(o)* Tamarix spp, *(p)* Urochondra setulosa, *(q)* Rhizophora mucronata, *(r)* Heritera fomes, *(s)* Avicennia marina, *(t)* Bruguiera cylindrical, *(u)* Bruguiera gymnorrhiza.

2.4 Based on habitat

i. **Hydro-halophytes**: These are halophytic plants which grow in aquatic conditions. Most of the mangroves and salt marsh species along costal lines belong to this group [40].

ii. **Xero-halophytes**: They grow in environment, where the soil is saline and the soil moisture content is very low due to high evaporation. Most plant varieties in desert areas and succulents belong to this group.

There is an enormous diversity among halophytes with regard to its ecological hotspots like coastal saline soils, arid and semiarid saline and sodic soils, soils of mangrove forests, wet lands, marshy lands and even agricultural fields [15]. A picturesque view of the major halophytes and mangroves found in India is given in **Figure 1**.

3. Mechanism of salt tolerance

Any review on halophytes/extremophiles is incomplete, without touching on the mechanisms by which these extremophiles survive under saline conditions. Indian Council of Agricultural Research- Central Soil Salinity Research Institute (ICAR-CSSRI) initiated such basic studies long back to critically understand the mechanism of salt tolerance in the local, native dry land saline vegetation [41]. This precise knowledge is highly imperative in developing other crops by a combination of improved salt tolerance and high yield. Another mandate was to document such species and, to the maximum extend, popularize them among the resource poor dry land farmers for better livelihood and sustenance.

3.1 Salt stress and halophytes

Any plant species initially suffers from osmotic stress as a result of increased soluble salts in the soil solution and later gets subjected to ionic stress due to specific accumulation of toxic ions. The osmotic phase of salt stress is characterized by disruption of water potential gradient and there by leads to reduced water uptake and inhibition of cell expansion [42]. Most plant species tries to adapt to this osmotic stress by accumulation of compatible solutes and thereby lowering water potential of cells, but this process consumes lot of energy and hence growth is heavily compromised.

Another major deleterious effect of salt stress is nutrient imbalance where in which high Na^+ ions in the soil solution reduces the availability of other cations like K^+ Ca^{++} and Mg^{++} [43]. This second phase of salt induced injury (specific ion toxicity) results from very high levels of Na^+ and Cl^- in the plant cells. In normal soils, plants maintain around 100–200 mM of K^+ and 1–10 mM of Na^+ in cellular cytosol for optimum cellular functions. Any salt concentration above this threshold level disrupts enzyme activity, protein synthesis, photosynthesis and other metabolic activities [44]. The light reactions of photosynthesis are comparatively less affected by salt stress as compared to the carboxylation reactions. A summary of salinity mediated effects and responses of plant cells are presented in **Table 2**.

The cell membrane permeability, composition and integrity gets affected as excess Na^+ replaces Ca^{2+} from its surface [46]. A lot of salt tolerant crop varieties have been developed by various research institutes which can survive up to moderate levels of salt stress. Under these circumstances, it is highly imperative to identify and propagate halophytic species in the arid and semiarid, high salt affected regions of the country, so that the vast tract of unproductive land can be put to cultivation for realization of sustainable income to resource poor farmers.

3.2 Salt induced responses and adaptations in halophytes

Halophytes and glycophytes have similar components in the stress tolerance network, but certain additional characteristics help these halophytes tolerate very high levels of salinity. Similar to glycophytes, halophytes also use osmoprotective and

Causes	Effect	Response/adaptation
Osmotic stress	Reduced water uptake, inhibition of cell elongation and expansion and leaf bud development	Compatible solute accumulation (ions/organic compounds)
Ion specific stress (high levels of Na and Cl in plant cell)	Inhibition of enzyme activity, protein synthesis, photosynthesis and leaf senescence and necrosis	Ion homeostasis through ion accumulation or ion exclusion
Imbalanced ion uptake	Nutritional deficiencies, reduced availability of other cations like K, Ca and Mg	Ion reabsorption

[a]*Adopted and modified from [45].*

Table 2.
Salt stress mediated effects and adaptations in plants.[a]

ion-detoxification strategies consisting of Na^+ removal from cytosol, Na^+ transport from root cells to xylem, and ion compartmentation in the vacuoles, involving salt overly sensitive (SOS1), high affinity potassium transporter (HKT1) and Na^+ H^+ antiporter (NHX) ion transporters, respectively. Under salt stress, most halophytes accumulate more Na^+ in their shoots than in their roots while retaining higher levels of K^+ and, thus, a more optimal K^+/Na^+ ratio [47, 48].

Salt stress induced accumulation of specific osmolytes like proline, glycine betaine and sugar alcohols have also been reported in both halophytes and glycophytes [47]. Nevertheless, halophytes exhibit a greater capacity to accumulate very high levels of such osmolytes, even under normal conditions which explains its preparedness to stress [49]. In line with this, over accumulation of proteins involved in carbohydrate metabolism have been reported in the leaves of many halophytes [50]. Specific transporter mediated intracellular compartmentalization of excess ions in to vacuole is another key mechanism used by halophytes to maintain a moderate cytosolic K^+/Na^+ ratio in the cytosol. Thus, membrane ATPases and ion transporters play vital roles in salt tolerance of halophytes [51].

Salvadorapersica, a very common inhabitant of dry saline tracts was grown at different in situ salinities, and the partitioning of sodium and chloride ions in to different plant parts were studied by [52] (**Figure 2**). The results showed very high content of these ions in the bark and senescing leaf tissue and a comparatively low content in photosynthetically active leaves [53]. The capacity of these sink tissues (bark and senescing leaves) to accumulate more and more salts increased with increasing salinity as well as with age of plant, which indicates a well established salt compartmentation mechanism in this halophyte species [41].

However, recent physiological and molecular observations across the globe indicate that halophytes may employ different mechanisms in ion transportation and homeostasis under salt stress. Latest studies showed that over expression of *SOS 1* and *HKT 1* in the halophytes *Eutrema salsugineum* and *Schrenkiella parvula* conferred much stronger salinity tolerance than that of their glycophytic counterpart [54]. Many halophytic species have now been understood to be constitutively higher expressers of the component genes of salt overly sensitive (SOS) pathway [55]. Genome studies of *Schrenkiella parvula* via. next generation sequencing (NGS) platform revealed the presence of three tandem duplicates of *nhx8* and two copies of *hkt1* which clearly explains the higher transcript abundance of these genes as compared to their glycophytic counterpart [56]. Thus genomic variation may be another important factor leading to enhanced salt tolerance in these halophytes which needs further confirmation. Moreover, the presence of specific Na^+/Ca^{2+} converse transportation mechanism also facilitates better adaptation to salt stress in halophytes [57].

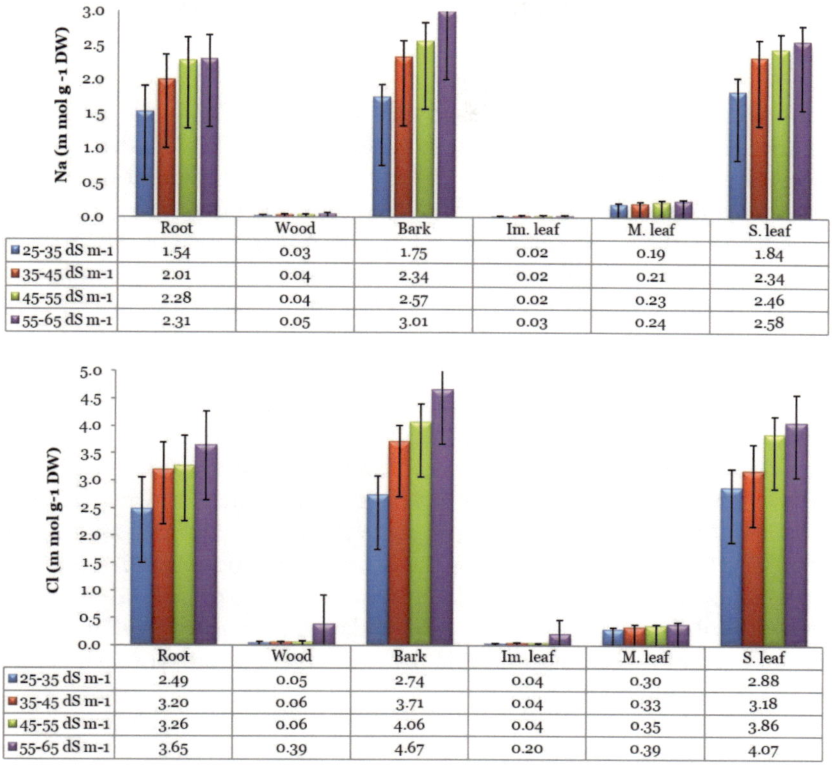

	Root	Wood	Bark	Im. leaf	M. leaf	S. leaf
25-35 dS m-1	1.54	0.03	1.75	0.02	0.19	1.84
35-45 dS m-1	2.01	0.04	2.34	0.02	0.21	2.34
45-55 dS m-1	2.28	0.04	2.57	0.02	0.23	2.46
55-65 dS m-1	2.31	0.05	3.01	0.03	0.24	2.58

	Root	Wood	Bark	Im. leaf	M. leaf	S. leaf
25-35 dS m-1	2.49	0.05	2.74	0.04	0.30	2.88
35-45 dS m-1	3.20	0.06	3.71	0.04	0.33	3.18
45-55 dS m-1	3.26	0.06	4.06	0.04	0.35	3.86
55-65 dS m-1	3.65	0.39	4.67	0.20	0.39	4.07

Figure 2.
Sodium (Na^+) and chloride (Cl^-) ion partitioning in Salvadora persica grown at different in situ salinities. Data represents mean of 2nd, 3rd and 4th years [43]. DW = dry weight, Im. Leaf = immature leaf, M. leaf = mature leaf & S. leaf = senescing leaf.

Epidermal bladder cells (EBCs) is another important feature of certain halophytes by which they secrete excess salt. 50% of the existing halophytes have been reported to contain these EBCs [31]. Removal of these EBCs resulted in enhanced salt sensitivity in a facultative halophyte called quinoa (*Chenopodium quinoa*) [58]. These EBCs are 1000-fold bigger than the general epidermic cells in volume, indicating its enhanced sodium excretion ability [59]. EBCs are similar to the trichome cells of *Arabidopsis* and transcriptomic analysis of quinoa EBCs showed enhanced expression of salt stress responsive genes, genes belonging to transporter family, sugar transporters, while photosynthesis-related genes showed reduced expression [60]. Nevertheless, the genetic determinants of EBCs coupled with the molecular mechanism of sodium pumping in to the EBCs is far from clear except a few studies in *Arabidopsis* [61]. This EBCs mediated strategy to tolerate high salt stress is not present in all halophytes. So certain other mechanisms are adopted by those halophytes which do not rely on EBCs. One category of halophytes called as succulent halophytes, further classified in to stem succulent and leaf succulent, accumulate large quantities of salt in its cells and tissues instead of secretion or compartmentation [40].

Succulence aids them in resisting the salt induced toxicity which can even reach up to 60% of leaf dry weight [62]. *Suaeda* is a classic example of this category of halophytes which can tolerate very high levels of salt without external secretion [31, 63]. Several other succulent halophytes have been reported belonging to amaranthaceae family [26, 64]. *Sesuvium portulacastrum* is another typical example of succulent halophyte often called as euhalophyte which accumulates lot of salt in its succulent leaf tissue [65].

3.3 Photosynthetic adaptation in halophytes

Photosynthesis is the most vital physiological process in plants, which if affected by any kind of stress, leads to significant yield reduction. Under salt stress, stomatal conductance gets reduced in order to save water, so internal CO_2 also gets reduced leading to reduced photosynthesis [66]. This reduced CO_2 limits the dark reactions of photosynthesis and leads to accumulation of reducing powers in grana thylakoid, thereby damaging the photosystem [67]. The differential effect of salt stress on photosystem II activity of halophytes has been studied in *Spartina versicolor* as compared to the glycophyte *Cyperus longus. Cyperus* displayed a significant reduction in the activity of photosystem II (PSII) under salt stress, but on the contrary, *Spartina* did not show distinct reduction [68]. Moreover, a deeper mechanism oriented study showed that halophytes were able to absorb light even under high salt stress, indicating a much stronger PSII complex [69]. Proteomic study showed that the halophytic relative of wheat was able to accumulate more chlorophyll a-b binding protein (CP24) protein under high salt stress, which in turn stabilized the PSII complex [70]. The ability of halophytes to change their carbon assimilation pathways from C3 to C4 and even crassulacean acid metabolism (CAM) according to stress levels is highly unique and vital trait for tolerance to high salt stress [57]. Altogether, the ability to protect PSII from oxidative damage coupled with situation specific shift of carbon assimilation pathways is the photosynthetic adaptive mechanism in halophytes under salt stress.

3.4 Molecular signatures of halophytes

The unique molecular signatures of halophytes place them very high on the salt tolerance hierarchy as compared to glycophytes [71]. A concentrated effort is made to bring together diverse research attempts on the various signaling aspects under salt stress conditions in halophytes, starting from sensing, activation of downstream signaling elements and all other potential candidates in the salt stress signaling network. A much stringent regulation of gene expression has been reported in halophytes as compared to its glycophytic relatives. Research attempts to exploit the rich source of salt responsive genes and promoter regions of diverse halophytes have also yielded significant achievements in the recent past [72]. Moreover, halophytic gene sequences are more complex with the presence of transposons and intergenic sequences [73]. The copy number of genes related to salt tolerance is also high in halophytes as depicted by the presence of three copies of calcineurin B-like 10 (*CBL 10)* in *Thellungiella parvula* as compared to one copy in *Arabidopsis* [74]. *SOS1,NHX1* and many other salt stress related genes have constitutive expression in halophytesas compared to salt induced expression in *Arabidopsis* [75]. Further, the transcript abundance of H^+ ATPase gene was low in the halophyte *Chenopodium quinoa* as compared to *Arabidopsis*, but the activity of the transporter protein was much higher, indicating some kind of post translational modification in halophytes [76].

4. Ecological and economic potential of halophytes and other saline vegetation

Harnessing the huge economic potential of the diverse halophytic plants and other saline vegetation is highly imperative to reduce the damage caused by soil and water salinization with special reference to the poor rural agrarian sector. This section of the chapter deals with the wide range of utilization of halophytes in various ecological zones of India as food, fodder, bio fuel, medicine and industrial raw materials for mass production of various compounds. The commercially untapped economic potential of certain abundantly found halophytes are also discussed as a possible livelihood source to the resource poor farmers of these saline dry lands. The oldest known attempt to use halophytic plants was the utilization of *Alhaji maurorum* as a soil ameliorant. However, Israel has to be credited for the prolific rediscovery and utilization of halophytes in saline tracts [77]. The first half of this section deals with the prime potential of halophytes that has to be harnessed from a national point of view, relating to replanting and ecological recovery of barren saline dry tracts, cheap biomass for renewable energy, climate change mitigation, CO2 sequestration and biological reclamation. The second part consists of the economic potential of halophytes related to use as food, fodder, medicine, chemicals, sea weed, mangrove based aquacultures etc. which can be a possible livelihood source to the resource poor farmers of these saline lands.

4.1 Halophyte mediated climate change mitigation

The climate foot print of present agricultural practices is univocally accepted to be similar to that of fossil fuel burning. The sustainability of current agricultural practices is questioned as they squander rich resources and acts as priming agents to global climate change [78]. So there has to be an inevitable change in the ethos of research and policy making to focus more on crops that grow well on limiting resources. Next generation agriculture should also be based on alternative crops like halophytes and mangroves which are water and nutrient use efficient and have sustainable yields across varying environmental conditions. The climate change mitigation potential of halophytes can be explained by just one example ie. *Suaeda fruticosa* thatcan survive and complete its life cycle under soil salinity of 65 dS m^{-1}, pH of 10.5 and under little or no water [79]. Such halophytes and other saline vegetation has tremendous potential in saving water, preventing soil erosion and replanting barren saline and sodic soils which is highly imperative in the current scenario of global change. Moreover, halophytes are suited to our brackish or saline water resources there by helping in fresh water conservation and replenishment. Scientific management of halophytic fields has tremendous potential to prevent further salinization of aquifers and groundwater of adjacent landscape. Due to climate change, global flood has become quite common and it is accompanied by post flood rise in water table and waterlogging. Under such situation, halophytes possess the ability to reduce saline water table and reclaim the affected land [53].

Mangroves species play a pivotal role in protection of our coastal ecosystem. The abundant aerial roots of these species are home to hundreds of creatures which are mostly endangered. Many species are true to sandy beaches and hence prevent soil erosion. More importantly, the role of mangroves as a livelihood to resource poor farmers cannot be underestimated [28, 80, 81]. As per National cyclone risk mitigation project (NCRMP), 308 cyclones of varying intensity impacted the eastern coast of India between 1981 and 2000. Extreme wave conditions are very common after effect of such cyclones originating in the Bay of Bengal [82]. As a result thousands of fertile agricultural land gets transformed in to wastelands. To cope with this, ICAR-CSSRI, regional research station (RRS), canning town, West Bengal developed a land shaping technology [83]. In this technique, land was converted in to medium ridge and shallow furrows where suitable halophytic tress along with paddy cum fish cultivation was practiced in *kharif* and in *rabi* season, furrows were used for rice cultivation [83]. Deep furrow and high ridge shaping technique was also practiced on coastal land of Sundarbans to overcome post cyclone stress [84]. Altogether, this technology plays a major role in rain water harvest, improved drainage and there by reduction in soil salinity and reduces environmental foot print by increased carbon sequestration.

4.2 Carbon sequestration potential of halophytes

Carbon sequestration is defined as the process of increasing organic carbon reserves by appropriate scientific land management interventions [85]. A much up scaled effort has to be put forth to harness more and more atmospheric CO_2 and store it in our soil. This has been reported to have enormous potential to reduce greenhouse gas (GHG) emissions [86, 87]. Halophytes helps in restoration of barren saline and sodic soils and sequester more and more carbon to enrich the organic carbon status of our infertile soils. ICAR-CSSRI has worked on the carbon sequestration potential of various agroforestry components in saline sodic soils. Results showed that these trees acts as carbon sink by virtue of their high growth rate, attractive wood and bio drainage properties. *Eucalyptus tereticornis* plantation was able to sequester 9.5 to 22.5 Mg ha^{-1} carbon in different spacing and 90.6 Mg ha^{-1} in block plantations along the canal after 6 years of planting in waterlogged saline soils of northwestern India [88–90]. Six years old *Eucalyptus tereticornis* plantation in sodic land showed a cumulative carbon stock (above ground biomass C+ below ground biomass C + soil carbon) of 122.6 Mg ha^{-1} with CO_2 mitigation capability of 369.2 Mg ha^{-1}. Moreover, carbon storage in soil was found to be 44.4% higher in agri-silviculture as compared to rice wheat cropping system in partially saline and sodic soils [91]. Several other studies by ICAR-CSSRI research groups showed very high carbon sequestration potential of tree species like *Eucalyptus tereticornis*, *Syzygium cumini*, *Pongamia pinnata* and *Populus deltoides* [92, 93].

4.3 Source of bioenergy

The dream of producing bioenergy can only be conceptualized if we can identify alternate species that can grow and survive in barren saline and sodic soils and therefore, would not compete with our conventional agricultural components. Halophytes are potential candidates as bioenergy crops as they can be watered even with sea water without any significant reduction in biomass or seed yield. Second generation biofuels are the topic of discussion, where nonfood biomass is used as raw material for biofuel production. Four prominent raw materials used for second generation biofuel production in India are the lignocellulosic biomass of four halophytes namely *Pongamia*, *Jatropha, Panicum virgatum* and *Miscanthus* [94]. Two significant attributes of such fuels are reduced environmental foot print and improvement in soil quality [95]. The salt excluder category of halophytes is better suited to biofuel production as fouling is a common problem in fuel developed from salt accumulating halophytes as salt is noncombustible. The latest review on major halophytes used for second generation biofuel production in China is the benchmark reference [96]. *Salicornia* (glasswort), *Suaeda* (sea-blite), *Atriplex* (saltbush), *Distichlis* (arid salt grass) and *Batis* spp. are another set of promising halophytes rich in lignocelluloses content [26, 97]. Abundantly found halophytic species like *Salicornia*, sea grass along with two mangrove species, *Avicenia berminans* (black mangrove) and *Rhizospora mangle* (red mangrove) has found a place in the green lab of National Aeronautics and Space Administration (NASA), USA as a viable alternative energy resource [98]. Another innovative and attractive concept that has emerged is to develop an integrated seawater energy and agriculture system (ISEAS) where there will be a coupling of biofuel feed stock, aquaculture and mangrove silviculture for the ultimate aim of producing sustainable aviation fuel [99].

4.4 Phytoremediation potential of halophytes

Recently, large tracts of agricultural lands in arid and semiarid regions have been subjected to heavy salinization and heavy metal pollution, arousing serious health and environmental concerns. Phytoremediation is defined as the use of plants to remove pollutants from soil and (or) render them harmless [15, 100, 101]. Deep rooted, high biomass producing accumulator category halophytes are potential candidates for phytoremediation of saline heavy metal contaminated soils [102]. It is cost effective and provides additional output in terms of forage [103]. The added advantage of halophytes in phytoremediation is its high tolerance to heavy metals and increased uptake of these heavy metals [104, 105]. Recent research attempts on halophytes mediated phytoremediation of heavy metal contaminated saline soils across the world are more in number indicating its environmental significance [106, 107]. Functional biology or potential of halophytes based studies also took place but are few in number [105, 108, 109]. Halophytes such as *Atriplex halimus* [110], *Spartina alterniflora* [111], *Sesuviumportulacastrum* [112] and *Tamarix africana* [113] are well proven examples in phytoremediation of heavy metal contaminated saline soils.

4.5 Halophytes: potential source of nutrition and value addition

It is next to impossible to grow conventional crops in the barren salt affected soils spread across the country which develop the gap of demand and supply of quality food in such areas. It ultimately creates barrier to food and nutritional security of the resource poor farmers living in such harsh areas. So halophytes have a huge potential as an edible source of nutrition there by securing livelihood of farmers. Such concept is gaining popularity very quickly and has spread across the world [101]. Among other salt tolerant crops, beetroot (*Beta vulgaris*), date palm (*Phoenix dactylifera*), amla (*Emblica officinalis*), karonda (*Carissa carandas*), guava (*Psidium guajava*), jamun (*Syzygium cumini*), pomegranate (*Punica granatum*), ber (*Ziziphus mauritiana*), bael (*Aegle marmelos*) and clusterbean (*Cyamopsis tetragonoloba)* are well known for their food value, and these are successfully grown in such saline soils using saline water irrigation. Also in India there is an extensive traditional use of non-conventional halophyte crops and naturally occurring halophytic vegetation as a nutrition source and further value addition (**Table 3**).

"Saji" or "barilla" is anindigenous value addition of locally available halophytes. It is the soda ash obtained from the air dried foliage materials of chenopod shrubs like *Haloxylon*, *Salsola* and *Suadea*. It is an essential ingredient of papad, and contributes to organoleptic qualities in terms of crispness and expansion of fried papad. The local Rajasthani *Banwaria* community has expertise in making *Saji*. The *Saji* produced from *Khara lana* (*Haloxylon recurvum*) is of the best quality, whereas*Saji* produced from *Pichki lana* or *luni* (*Suaeda fruticosa*) is of medium quality, and that produced from *lani* (*Salsola baryosma*) is of inferior quality [121]. Another aspect on which ICAR-CSSRI has extended its research is edible cactus (*Opuntiaficus indica*) and its growth and yield potential under saline soils. Certain specific moderately salt tolerant clones have been identified for raised bed planting in saline tracts [122].

Moreover, around 50 species of seed producing halophytes are potential sources of edible oil and proteins. The best part of their use for oilseed is that they generally do not accumulate salts in their seeds. Prominent halophytes such as *Suaeda fruticosa*, *Arthrocnemummacrostachyum*, *Salicornia* spp., *Halogeton glomeratus* and *Haloxylon stocksii* have been reported to produce high grade edible oil with an unsaturation content of 70–80% [123]. Seeds of *Salvadora oleoides* and *S. persica* contain 40–50% fat and are a good source of industrially important lauric acid and myristic acids. Its purified fat is used for soap and candle making. *Terminalia catappa* (seed oil 52%) is another tree species widely used for edible oil production at industrial level. *Salicornia bigelovii* is another halophytic species used for

Plant scientific name/ common name	Plant parts	Nutritional aspects and uses
Anthrocnemum indicum	Phylloclades	Highly nutritious, used in pickles
Amaranthus spinosus (Kanta chaulai, kante bhaji)	Tender shoots, young leaves	High in Ca content, leaves cooked as vegetable, mature stems cooked with small fishes
Aster tripolium (Salt bush)	Leaves	Fresh salads, cooked vegetable
Atriplex hortensis (Pahari paleng)	Leaves	High nutritive value as a green leafy vegetable
Balanites aegyptiaca (Hingor, hingod)	Fruits	Both fresh and dried fruits are edible
Beta maritime (Palak)	Young shoots	Highly nutritious, consumed like spinach
Borassus flabellifer (Palmirah palm)	Edible radicles, fruits	Consumed as toddy, jaggery and vinegar
Capparis decidua (Kair)	Raw Fruits	Used as pickles and high medicinal values
Carissa carandas (Karaunda)	Fruits	Rich in iron, vitamin C, A, Ca and P, immature fruits are pickled and eaten raw.
Centella asiatica (Brahma manduki)	Leaves, young shoots	Widely used in Indian regional cuisines as a culinary vegetable.
Chenopodium album (Bathua)	Leaves, young shoots	High in protein, vitamin A, Ca, P, and K, cooked as vegetable, curries, raita and paratha-stuffed breads in northern India
Corchorus capsularis (Chanachedi, mora-pat)	Leaves, shoots, fruits	Rich in vitamins and minerals, used in salads, leafy vegetable
Eleocharis dulcis (Neerchelli)	Tubers	Rich in vitamin B_6, Mn, and K, tubers cooked
Grewia tenax (Gondni, kanger, kaladi)	Fruit, leaves, seeds	Fe-rich fruit, consumed raw or use as refreshing drink or porridge, leaves eaten as a green vegetable
Haloxylon salicornicum (Lana, khar)	Leaves, stem, seed	Fatty acids, vitamin C & A, used as salad and pickles, seeds mixed with bajra for roti making
Hemidesmus indicus (Anantbel, nannari)	Roots	Roots used for pickle, "nannari sharbat" also made from roots
Leptadenia pyrotechnica (Khimp, jivanti)	Pods	Pods are of medicinal value and used as vegetables in Rajasthan
Morinda citrifolia (Noni)	Fruit	Starvation fruit during famine, consumed as raw, pickled and used for extracting juice
Nypa fruticans (Nipa palm)	Young shoots, Fruits	Eaten as a green vegetable and the immature fruit is used in deserts
Oxalis corniculata (Amrit sak)	Young shoots, Leaves	Rich in vitamin C, Ca, beta-carotene and P, consumed as vegetables
Pisonia alba (Chandu, muruval)	Leaves	Leaves are eaten as a green vegetable
Portulaca oleracea (Lunia, badi-Noni)	Above ground part	Vitamin-rich with high omega-3 fatty acids, consume as a salad or with yoghurt
Prosopis cineraria (Khejri)	Pods	Pod called as Singhri or Sangri used in various types of bhaaji and kadhi in Thar desert of India
Salicornia brachiata (Umari keerai)	Biomass	High protein edible oil similar to safflower, value-added by-products like vegetable salt
Sesuvium portulacastrum (Lunio, dhapa)	Leaves, stems	Rich in Ca, Fe, and carotene, consumed as vegetables by local peoples in arid region

Plant scientific name/ common name	Plant parts	Nutritional aspects and uses
Suaeda fruticosa (Khario luno)	Leaves, shoots	Protein rich, eaten by local people as vegetables
Suaeda maritimum (Luno, lano)	Young leaves	Leafy vegetable, can be used in combination foods as seasoning
Suaeda nudiflora (Morus)	Leaves	Pickles, salad, and as a vegetable
Terminalia catappa (Deshi badam)	Fruits	Sweet fruit and edible kernel
Ziziphus nummularia (Jhar beri, chanibor)	Fruits	Fruits eaten raw

Source: [114–120].

Table 3.
List of halophytes, other salt tolerant crops, their edible portion and nutritional properties.

commercial oil production in the western states of India. The seed (31% protein) is pressed for its high quality edible oil (28%) which is an alternative source of omega-3 polyunsaturated fatty acids [124].

Quinoa (*Chenopodium quinoa*) is another annual facultative halophyte which has every potential to become a highly economic crop in saline areas of India. The exceptional tolerance of this crop to soil salinity has already been reported [125]. Protein content of quinoa ranges from 12 to 17% depending on variety, environment and crop management practices [126]. This is higher than our conventional cereal crops like rice (6–7%), wheat (10.5–14%) and barley (8–14%). More importantly, it is rich in lysine and methionine which are the two amino acids absent in cereals and pulses respectively [127]. ICAR-CSSRI has initiated a network project on morpho physiological characterization and standardization of agronomic practices of quinoa (*Chenopodium quinoa*) for salt affected agro-ecosystems. Nineteen germplasm lines were evaluated under four levels of irrigation water salinity ie. best available water, 8 dS m^{-1}, 16 dS m^{-1} and 24 dS m^{-1}. Germplasm EC 507740 gave maximum grain yield of 9.20 g per plant at highest levels of salinity (24 dS m^{-1}) [128].

"Kharchia 65" is a wheat race native to Pali district of Rajasthan which is commonly called as red wheat. It is universally recognized as highest salt tolerant genotype. This genotype has been extensively used in the development of salt tolerant wheat varieties by ICAR-CSSRI, Karnal, namely KRL1-4, KRL 19 and KRL 39. It is a universal donor to salinity breeding in wheat coupled with high yield and has been registered in ICAR- National Bureau of Plant Genetic Resources (NBPGR), New Delhi (registration number: INGR99020). The main reason for growing this wheat variety in the arid and semiarid regions of the country is that it is highly salt tolerant and requires very less water as compared to other hybrids and can also be grown as rain fed crop purely on conserved soil moisture. Another useful trait of this genotype is that the plants are tall with high straw content which can be effectively used as a fodder to cattle as it is much preferred by animals as compared to other wheat straw [129].

4.6 Silvipastoral system for ecological restoration of saline sodic soils

Biodrainage is a term that is getting more and more popularized which signify the use of salt tolerant trees and grasses for reducing salinity and waterlogging [130].

Apart from reclamation of these saline lands, silvipastural system aids in improved carbon sequestration [131], increased soil rhizosperic activity [20], reduced greenhouse gas emissions [132] and long term adaptation to changing climates [133]. Mesquite (*Prosopis juliflora*) combined with Kallar grass (*Leptochloa fusca*) was reported to be a promising silvipastoral system in sodic soils [134]. Other successful silvipastural combinations *are Acacia nilotica + Desmostachya bipinnata*, *Dalbergia sissoo + Desmostachya bipinnata* and *Prosopis juliflora + Desmostachya bipinnata* [134]. *Tamarix articulate* is another very productive halophytic tree species with a biomass production of 93 Mg ha^{-1} in 7 years [135]. Aromatic grasses like lemon grass (*Cymbopogon flexuosus*) and palmarosa (*Cymbopogon martinii*) were studied by ICAR-CSSRI and found to be suitable to moderate alkali soils up to pH 9.2. Moreover, the most popular aromatic grass called vetiver (*Vetiveria zizanioides*) was found to be dual tolerant to high pH and waterlogged soils [136]. Licorice (*Glycyrrhiza glabra*) was reported by ICAR-CSSRI to be highly tolerant to sodicity level up to 9.8 [81]. Other promising grasses suitable to saline sodic soils are *Aeluropus lagopoides*, *Dichanthium annulatum*, *Chlorisgayana*, *Bothriochloa pertusa, Eragrostis spp.*, *Sporobolus spp.* and *Panicum spp.* [130]. Other farmer preferred halophytes in saline sodic soils are *Ziziphus*, *Atriplex*, *Kochia*, *Suaeda*, *Salsola*, *Haloxylon* and *Salvadora* as they are preferably browsed by camel sheep and goat [137].

Salvadora persica is an oil yielding salt bush that has been extensively studied by ICAR-CSSRI. A *Salvadora persica* based silvipastural system was developed and popularized with forage grasses like *Leptochloa fusca*, *Eragrostis sp.* and *Dichanthium annulatum* on the saline *Vertisols* of Gujarat, India [138]. This model was successful in saline soils of electrical conductivity ranging from 25 to 70 dS m^{-1}.Moreover, high biomass producing halophytic trees like *Acacia tortilis* (hybrid), *Ziziphus mauritiana*, *Pithecellobium dulce*, *Melia azedarach*, *Cassia fistula*, *C. javanica*, *Callistemon lanceolatus*, and *Acacia farnesiana* were popularized among the resource poor farmers of the saline and sodic soil tracts of northern India. The waterlogged soils of semiarid regions of northern India were also subjected to reclamation via. *Eucalyptus* based agroforestry [81, 139]. *Eucalyptus tereticornis* was the preferred species. *Elaeagnus angustifolia* is another tree species recently found to be effective for bio drainage based on its water use efficiency, salt tolerance and growth rate [89].

4.7 *Salicornia*: a case study in western coasts of India

Salicornia is an obligate halophyte belonging to chenopodeaceae family commonly found at the edges of wetlands, marshes, sea shores, and mudflats [140]. It is commonly called as pickleweed, glasswort, sea beans and sea asparagus across India. Some species of *Salicornia* can even tolerate and complete its life cycle under 3% Nacl [141]. In India, this halophyte has been used as an edible crop as well as for non-edible purposes. Use of this plant as a source of soda for glass manufacture is time immemorial. *Salicornia brachiata,* a leafless shrub, was indeed the first source of salt produced from plants in 2003 by Council of Scientific and Industrial Research (CSIR)- Central Salt and Marine Chemicals Research Institute (CSMCRI), Gujarat, India. This vegetable salt, unlike common salt contains salts of potassium, calcium, magnesium and iron. On farm trials have shown that it has the potential to produce 3–4 tons of vegetable salt/hectare which can fetch a market of Rs 10–12 per Kg to the resource poor farmers of these barren saline tracts [142]. Further, recurrent selection mediated germplasm improvement was also carried out for better yield. In order to minimize the cost of cultivation, this species (improved variety: SOS 10) were sown in monsoon along the western coast of India in a large scale. Public private partnership based salt product named “*Saloni*” was also developed on a commercial scale in Gujarat [142].

Low content of seed sodium makes it a very good source for human heart, apart from its other medicinal properties against diabetes, asthma, hepatitis, gastroenteritis and cancer [143]. Moreover, the edible oil from its seeds is rich in poly unsaturated fatty acids and similar to safflower in fatty acid composition. It is also used as a green salad in the western areas of India. Antibacterial, antitubercular and antioxidant activities of *Salicornia brachiata* has been previously reported [144]. It is very popular and commonly used by villagers of western and eastern coast as an animal fodder, herbal salt and as a source of oil, while the ash of the whole plant has been reported to be useful in itch treatment [145, 146]. Most recently, prolific study on its polysaccharides and other phytochemical profile for phenolic compounds, oils, proteins, flavanoids, sterols, saponins, alkaloids and tannins are under way and shows promising results [147]. The oligosaccharide profiling of Indian species *Salicornia brachiata* was performed and the results showed this plant to be rich source of dietary supplements [148]. The ecological benefits of large scale cultivation of this plant along the coasts of India may be summarized as utilization of barren saline lands, upscaling of green belt, coastal development and protection and biodiversity conservation [89]. Finally, the equity benefits are vast export income and private industrial and institutional collaboration. So, this plant is a plant of future which needs special mention.

4.8 Medicinal halophytes: a formidable source of medicine, neutraceuticals and other products

One of the major physiological adaptations of halophytes to saline stress is the production of different biomolecules which possess highly useful biological activities like antioxidant, antimicrobial, anti-inflammatory and antitumoral [149]. So, if introduced in to our diet, it has the potential to prevent a lot of diseases like cancer and cardiac disorders [36]. These biomolecules also enhances the neutracutical value of halophytes as the concentration is very high as compared to their glycophytic counterparts [150]. Moreover, certain biomolecules are specific to halophytes and hence have great agri food, pharmaceutical and cosmetic value [151]. Presence of a wide array of compounds like alkanes, fatty acids, carbohydrates, aminoacids, alcohols, terpenoids, flavanoids etc. have been reported from major halophytes [152]. Halophytes have long been used as folklore medicine by villagers and very little documentation is done in this aspect. A few glimpses off current use of halophytes as medicine in India is discussed here. *Salicornia* sp. is widely used as a folk medicine for constipation, diabetes and cancer [153]. *Suadea* sp. and *Atriplex* sp. are widely known to possess hypoglycaemic and hypolipidaemic activities [154].

Catharanthus withstands EC of 12 dS m^{-1} and produces about 130 *catharanthus* alkaloids compounds, including vinblastine and vincristine, two drugs used to treat cancer. Mangroves are also good source of biomolecules. For instance, the mangrove *Cynometra ramiflora* oil has antibiotic properties and is used in skin diseases [155]. Among other notable use of halophytes as medicine, use of *Pandanus odoratissimus* in leprosy, scabies, heart diseases, *Salvadora persica,* in cough, rheumatism, toothache and piles, *Salsola baryosma, S. kali* as anthelmintic, emmenagogue, diuretic, *Tamarix articulata* in eczema, ulcers, piles, sore throat, diarrhea, liver disorders and*Cress cretica* as tonic, aphordisiac, stomachic deserves special mention [28]. Among other products, biopetrol/diesel from *Jatropha curcas* and *Euphorbia antisyphilitica*, multiple bioactive compounds from *Salsola baryosma,* essential oil from *Pandanus* sp., *Terminalis catappa,* aromatic oil from *Grindelia camporum, Larrea tridentate* and *Simmondsia chinensis,* rubber from *Chrysothamus nauseousus* and *Parthenium argentatum,* multiple beverages from the mangrove palm *Nypa fruiticans,* pulp and fiber from *Phragmitesaustralis, P. karka, Juncus rigidus* and *J. acutus* deserves special mention [28].

4.9 Seaweeds: valuable resource pool

Sea weeds are an integral component of coastal ecosystems that lend invaluable support to the diverse marine life. The economic value of these sea weeds is of significant importance to the resource poor farmers of saline coastal areas. It is used as a food, but more importantly the phycocolloids derived from it is of significant export value [156]. In India, apart from phycocolloid production, sea weeds are being used for the commercial production of crop growth stimulating agents. Moreover, the agar and alginate industry full depends on this valuable coastal resource. Off late, they are extensively being explored for biofuel, neutracuticals, medicines and food additives [157]. In Indian context, carrageenophytes (red sea weeds), *Gracilaria* spp. and *Gelidiella* spp. for agar production, *Sargassum* spp. and *Turbinaria* spp. for alginate production, *Kappaphycus alvarezii*, *Hypnea musciformis* and *Sarconema filiforme* for phycocolloid kappacarrageenan production is of major economic importance [156]. Another matter of pride for Indian Council of Agricultural Research (ICAR) is the endorsement of ICAR-CIFT (Central Institute of Fisheries Technology) by world health organization (WHO) to fight COVID-19 pandemic. The research group proposed that sulphated polysaccharide from seaweed can be a potent molecule to fight against the COVID-19 Pandemic, hence, is a candidate molecule to be studied against SARS-CoV-2 [158].

5. Conclusion and future ideas

Tangible evidences of global climate change and land degradation due to salinization are quite evident and so it is highly imperative to bring more and more salt affected land into cultivation to satisfy the food and nutritional security of our burgeoning population. In this context, halophytes and other saline vegetation has paramount importance to ensure economic returns and maintain ecological balance. India has a very rich source of halophytic, mangrove and other saline vegetation which has huge potential in monetary as well as environmental terms. It can reduce the gap between demand and supply of food and fodder for livestock, besides being a source to numerous other products of economic significance. The role of halophytes in climate change mitigation, carbon foot print reduction, renewable energy source and greening of our barren saline lands has to be conceptualized on field basis. Moreover, deteriorating water resources and lower availability of good quality water for agriculture demands a paramount shift to edible halophytic crops like quinoa, already being termed as super food, highly tolerant to drought and salinity stress.

Identification and documentation of region specific halophytes, access to seeds, establishment of halophytic nurseries, optimizing package of practice and development of processing plants are immediate requirements to spread the concept of halophyte based biosaline agriculture. Long term experiments are the need of the hour to prove the sustainability of halophyte based cultivation and economic security. More importantly, breeding programs should be initiated to improve traits such as yield, taste, biomolecule quantity and quality for faster adoption of halophyte based production. Another major issue to be tackled is the low or nonexistent demand on the market and a lack of value chain. For this consumer awareness need to be created and a value chain which consists of different players need to be established. Let us all hope that the ultimate dream of greening our barren saline lands and replenishing soils with saline water table in to sites of bio saline agriculture gets fulfilled in coming years as a boon to mankind.

Acknowledgements

We acknowledge the funding and support from Indian Council of Agricultural Research (ICAR), Ministry of Agriculture and Farmers Welfare, Government of India. Vineeth TV acknowledges Dr. Ashwani Kumar, ICAR-CSSRI for providing photographs of major halophytes.

Conflict of interest

The authors declare no conflict of interest.

Author details

T.V. Vineeth[1*], Shrvan Kumar[1], Monika Shukla[1], Anil Chinchmalatpure[1] and Parbodh Chander Sharma[2]

1 Indian Council of Agricultural Research-Central Soil Salinity Research Institute (ICAR-CSSRI), Regional Research Station (RRS), Bharuch, Gujarat, India

2 Indian Council of Agricultural Research-Central Soil Salinity Research Institute (ICAR-CSSRI), Karnal, Haryana, India

*Address all correspondence to: vinee2705@gmail.com; vineeth.tv@icar.gov.in

References

[1] United Nation. United Nations, Department of Economic and Social Affairs, Population Division. World Population Prospects 2019 [Internet]. 2019. Available from: https://population.un.org/wpp/Download/Probabilistic/Population/ [Accessed: 2020-08-20]

[2] Godfray HCJ, Beddington JR, Crute IR, Haddad L, Lawrence D, Muir JF, Pretty J, Robinson S, Thomas SM, Toulmin C. Food security: the challenge of feeding 9 billion people. Science. 2010;327:812-818. DOI: 10.1126/science.1185383

[3] Fahad S, Bajwa AA, Nazir U, Anjum SA, Farooq A, Zohaib A, Sadia S, NasimW, Adkins S, Saud S, Ihsan MZ, Alharby H, Wu C, Wang D, Huang J. Crop production under drought and heat stress: Plant responses and management Options. Frontiers in Plant Science. 2017;8:1147-1162. DOI: 10.3389/fpls.2017.01147

[4] Fahad S, Hussain S, Bano A, Saud S, Hassan S, Shan D, Khan FA, Khan F, Chen Y, Wu C, Tabassum MA, Chun MX, Afzal M, Jan A, Jan MT, Huang J. Potential role of phytohormones and plant growth-promoting rhizobacteria in abiotic stresses: consequences for changing environment. Environmental Science and Pollution Research. 2014;22(7):4907-4921. DOI: 10.1007/s11356-014-3754-2

[5] Fahad S, Hussain S, Saud S, Tanveer M, Bajwa AA, Hassan S, Shah AN, Ullah A, Wu C, Khan FA, Shah F, Ullah S, Chen Y, Huang J. A biochar application protects rice pollen from high-temperature stress. Plant Physiology and Biochemistry. 2015;96:281-287. DOI: 10.1016/j.plaphy.2015.08.009

[6] Wingeyer AB, Amado TJC, Pérez-Bidegain M, Studdert GA, Perdomo Varela CH, Garcia FO, Karlen DL. Soil quality impacts of current South American agricultural practices. Sustainability. 2015;7:2213-2242. DOI: 10.3390/su7022213.

[7] Akram R, Turan V, Hammad HM, Ahmad S, Hussain S, Hasnain A, Maqbool MM, Rehmani MIA, Rasool A, Masood N, Mahmood F, Mubeen M, Sultana SR, Fahad S, Amanet K, Saleem M, Abbas Y, Akhtar HM, Waseem F, Murtaza R, Amin A, Zahoor SA, ul Din MS, Nasim W. Fate of organic and inorganic pollutants in paddy soils. In: Hashmi MZ, Varma A, editors. Environmental Pollution of Paddy Soils, Soil Biology. Switzerland: Springer, Cham; 2018. p. 197-214. DOI: 10.1007/978-3-319-93671-0_13

[8] Akram R, Turan V, Wahid A, Ijaz M, Shahid MA, Kaleem S, Hafeez A, Maqbool MM, Chaudhary HJ, Munis, MFH, Mubeen M, Sadiq N, Murtaza R, Kazmi DH, Ali S, Khan N, Sultana SR, Fahad S, Amin A, Nasim. Paddy land pollutants and their role in climate change. In: Hashmi MZ, Varma A, editors. Environmental Pollution of Paddy Soils, Soil Biology. Switzerland: Springer, Cham; 2018. p. 113-124. DOI: 10.1007/978-3-319-93671-0_7

[9] Kafi M, Khan MA, editors. Crop and Forage Production using Saline Waters. 1sted. New Delhi:Daya Publishing House; 2008. 334 p.

[10] FAO. Global Network on Integrated Soil Management for Sustainable Use of Salt-affected Soils. [Internet]. 2000. Available from:http://www.fao.org/soils-portal/soil-management/management-of-some-problem-soils/salt-affected-soils/more-information-on-salt-affected-soils/en/[Accessed: 2020-08-24]

[11] FAO. FAO Soils Portal. [Internet]. 2016. Available from: http://www.fao.

org/soils- portal/soil-management/management-of-some-problem-soils/salt-affected- soils/more-information-on-salt-affected-soils/en/. [Accessed: 2020-07-24]

[12] FAO. FAO Soils Portal. Management. Salt-affected soils [Internet]. 2008. Available from: http://www.fao.org/soils-portal/soil-management/management-of-some-problem-soils/salt-affected-soils/more-information-on-salt-affected-soils/en/ [Accessed: 2020-08-20]

[13] Mandal S, Raju R, Kumar A, Kumar P, Sharma PC. Current status of research, technology response and policy needs of salt-affected soils in India – A Review. Journal of Indian Society of Coastal Agricultural Research. 2018;36(2):40-53. ISSN: 0972-1584

[14] ICAR-CSSRI. ICAR–Central Soil Salinity Research Institute. Vision 2050. Indian Council of Agricultural Research, New Delhi. 2015. p 31.

[15] Hasanuzzaman M, Nahar K, Alam MM, Bhowmik PC, Hossain MA, Rahman MM, Fujita M. Potential use of halophytes to remediate saline soils. BioMed Research International. 2014;8:1-12. DOI: 10.1155/2014/589341

[16] Shrivastava P, Kumar R. Soil salinity: a serious environmental issue and plant growth promoting bacteria as one of the tools for its alleviation. Saudi Journal of Biolgical Sciences. 2015;22:123-131. DOI: 10.1016/j.sjbs.2014.12.001

[17] Sathish S, Kankara RS, Selvan SC, Umamaheswari M, Rasheed K. Wave-beach sediment interaction with shoreline changes along a headland bounded pocket beach, West coast of India. Environmental Earth Sciences. 2018;77:174-185. DOI: 10.1007/s12665-018-7363-0

[18] Sobhanian H, Aghaei K, Komatsu S. Changes in the plant proteome resulting from salt stress: toward the creation of salt-tolerant crops? Journal of Proteomics. 2011;74:1323-1337. DOI: 10.1016/j.jprot.2011.03.018

[19] Lal R. Soil carbon sequestration impacts on global climate change and food security. Science. 2004;304(5677):1623-1627. DOI: 10.1126/science.1097396

[20] Etesami H, Beattie GA. Mining halophytes for plant growth-promoting halotolerant bacteria to enhance the salinity tolerance of non-halophytic crops. Frontiers in Microbiology. 2018;9:148-167. DOI: 10.3389/fmicb.2018.00148

[21] Fahad S, Hussain S, Saud S, Hassan S, Ihsan Z, Shah AN, Wu C, Yousaf M, Nasim W, Alharby H, Alghabari F, Huang J. Exogenously applied plant growth regulators enhance the morphophysiological growth and yield of rice under high temperature. Frontiers in Plant Science. 2016;7:1250. DOI: 10.3389/fpls.2016. 01250

[22] Fahad S, Muhammad ZI, Abdul K, Ihsanullah D, Saud S, Saleh A, Wajid N, Muhammad A, Imtiaz AK, Chao W, Depeng W, Jianliang H. Consequences of high temperature under changing climate optima for rice pollen characteristics-concepts and perspectives. Archives of Agronomy and Soil Science. 2018;64:1473-1488. DOI: 10.1080/03650340.2018.1443213

[23] Farhana G, Ishfaq A, Muhammad A, Dawood J, Fahad S, Xiuling L, Depeng W, Muhammad F, Muhammad F, Syed. Use of crop growth model to simulate the impact of climate change on yield of various wheat cultivars under different agro-environmental conditions in Khyber Pakhtunkhwa, Pakistan. 2020. Arabian Journal of Geosciences. 2020;13:112. DOI: 10.1007/s12517-020-5118-1

[24] Mubeen M, Ashfaq A, Hafiz MH, Muhammad A, Hafiz UF, Mazhar S, Muhammad Sami ul Din, Asad A, Amjed A, Fahad S, Wajid N. Evaluating the climate change impact on water use efficiency of cotton-wheat in semi-arid conditions using DSSAT model. 2020. Journal of Water and Climate Change. DOI: 10.2166/wcc.2019.179/622035/jwc2019179

[25] Glenn EP, Brown JJ, O'Leary JW. Irrigating crops with seawater. Scientific American. 1998;279:76-81. DOI: 10.1038/scientificamerican0898-76

[26] Rozema J, Flowers T. Crops for a salinized world. Science. 2008;322(5907):1478-1480. DOI: 10.1126/science.1168572.

[27] Joshi R, Mangu VR, Bedre R, Sanchez L, Pilcher W, Zandkarimi H,Baisakh N. Salt adaptation mechanisms of halophytes: Improvement of salt tolerance in crop plants. In: Pandey GK, editor. Elucidation of Abiotic Stress Signaling in Plants. 2ndedition. New York: Springer; 2015. p. 243-279. DOI: 10.1007/978-1-4939-2540-7_9

[28] Dagar JC. Biodiversity of Indian saline habitats and management and utilization of high salinity tolerant plants with industrial application for rehabilitation of saline areas. In: Alsharhan AS, Wood WW, Goudie AS, Fowler A, Abdellatif EM, editors. Desertification in the Third Millennium. Lisse: Swets and Zeitlinger Publishers; 2003. p. 151-172. DOI: 10.1201/NOE9058095718.ch17

[29] Mandal RN, Naskar KR. Diversity and classification of Indian mangroves: a review. Tropical Ecology. 2008;49:131-146.

[30] Roy SJ, Tester M, Gaxiola RA, Flowers T J. Plants of saline environments. Access Science. [Internet]. 2018. Available from: https://doi.org/10.1036/1097-8542.525600 [Accessed: 2020-08-20]

[31] Flowers TJ, Colmer TD. Salinity tolerancein halophytes. New Phytologist. 2008;179:945-963. DOI:10.1111/j.1469-8137.2008.02531.x

[32] Sabovljevic M, Sabovljevic A. Contribution to the coastal bryophytes of the northern Mediterranean: Are the halophytes among bryophytes? Phytologia Balcanica. 2007;13(2):131-135.

[33] Kosová K, Vítamvas P, Urban MO, Prasil IT. Plant proteome responses to salinity stress- comparison of glycophytes and halophytes. Functional Plant Biology. 2013;40:775-786. DOI: 10.1071/FP12375

[34] Aronson JA. Haloph: a data base of salt tolerant plants of the world. Whitehead EE, editor. Office of Arid Land Studies. University of Arizona, Tucson, Arizona, USA; 1989. p. 75.

[35] Santos J, Al-Azzawi M, Aronson J, Flowers TJ. eHALOPH a data base of salt-tolerant plants: Helping put halophytes to work. Plant Cell Physiology. 2016;57:e10. DOI:10.1093/pcp/pcv155

[36] Kumari A, Das P, Parida AK, Agarwal PK. Proteomics, metabolomics, and ionomics perspectives of salinity tolerance in halophytes. Frontiers in Plant Science. 2015;6:537-556. DOI: 10.3389/fpls.2015.00537

[37] Chapman VJ. The new perspective in the halophytes. The Quarterly Review of Biology. 1942;17:291-311. DOI:10.1086/394660

[38] Walter H.The adaptations of plants to saline soils. In: Arid Zone Research. XIV. Salinity Problems in the Arid Zones. Proceedings of the Teheran Symposium; 1961; UNESCO. Paris: p.129-134

[39] von Sengbusch P. Halophytes. Botanik Online, University of Hamburg. 2003.

[40] Youssef AM. Salt tolerance mechanisms in some halophytes from Saudi Arabia and Egypt. Research Journal of Agriculture and Biological Sciences. 2009;5:191-206.

[41] Rao GG, Nayak AK, Chinchmalatpure AR, Nath A, Babu VR. Growth and yield of *Salvadora persica*, a facultative halophyte grown on saline black soil (Vertic Haplustept). Arid Land Research and Management. 2004;18:51-61. DOI: 10.1080/15324980490245013

[42] Munns R, Tester M. Mechanisms of salinity tolerance. Annual Review of Plant Biology. 2008;*59*:651-681. DOI: 10.1146/annurev.arplant.59.032607.092911

[43] Epstein E. Mineral Nutrition of Plants. Principles and Perspectives. 1[st]ed. New York: John Wiley and Sons Inc.; 1972. 412 p. DOI: 10.1002/jpln.19721320211

[44] Niu X, Bressan RA, Hasegawa PM, Pardo JM. Homeostasis in NaCl stress environments. Plant Physiology. 1995;109(3):735-742. DOI: 10.1104/pp.109.3.735

[45] Horie T, Karahara I, Katsuhara M. Salinity tolerance mechanisms in glycophytes: An overview with the central focus on rice plants. Rice. 2012;5:11. DOI: 10.1186/1939-8433-5-11

[46] Cramer GR, Läuchli A, Polito VS. Displacement of Ca^{2+} by Na^{+} from the plasmalemma of root cells. A primary response to salt stress? Plant Physiology. 1985;79:207-211. DOI: 10.1104/pp.79.1.207

[47] Flowers TJ, Munns R, Colmer TD. Sodium chloride toxicity and the cellular basis of salt tolerance in halophytes. Annals of Botany. 2015;115:419-431. DOI: 10.1093/aob/mcu217

[48] Orsini F, D'Urzo MP, Inan G, Serra S, Oh DH, Mickelbart MV, Consiglio F, Li X, Jeong JC, Yun DJ, Bohnert HJ, Bressan RA, Maggio A. A comparative study of salt tolerance parameters in 11 wild relatives of *Arabidopsis thaliana*. Journal of Experimental Botany. 2010;61:3787-3798. DOI: 10.1093/jxb/erq188

[49] Gong Q, Li P, Ma S, Indu Rupassara S, Bohnert HJ. Salinity stress adaptation competence in the extremophile *Thellungiella halophila* in comparison with its relative *Arabidopsis thaliana*. ThePlant Journal. 2005;44:826-839. DOI: 10.1111/j.1365-313X.2005.02587.x

[50] van Zelm E,Yanxia Zhang Y, Testerink C. Salt tolerance mechanisms of plants. Annual Review of Plant Biology. 2020;71:403-433. DOI: 10.1146/annurev-arplant-050718-100005

[51] Meng X, Zhou J, Sui N. Mechanisms of salt tolerance in halophytes: current understanding and recent advances. Open Life Sciences. 2018;13:149-154. DOI: 10.1515/biol-2018-0020

[52] Rao GG, Chinchmalatpure AR, Meena RL, Khandelwal MK. Saline agriculture in saline vertisols with halophytic forage grasses. Journal of Soil Salinity and Water Quality. 2011;3:41-48.

[53] Arora S, Rao GG. Bio-amelioration of salt-affected soils through halophyte plant species. In: Arora S, Singh AK, Singh YP, editors. Bioremediation of Salt Affected Soils: An Indian Perspective. Switzerland: Springer Cham; 2017. p. 71-85. DOI: 10.1007/978-3-319-48257-6_4

[54] Ali A, Khan IU, Jan M, Khan HA, Hussain S, Nisar M, Chung WS, Yun DJ. The high-affinity potassium transporter

EpHKT1;2 from the extremophile *Eutrema parvula*mediates salt tolerance. Frontiers in Plant Science. 2018;9:1108. DOI: 10.3389/fpls.2018.01108

[55] Katschnig D, Bliek T, Rozema J, Schat H. Constitutive high-level *SOS1* expression and absence of *HKT1;1* expression in the salt-accumulating halophyte *Salicornia dolichostachya.* Plant Science. 2015;234:144-154. DOI: 10.1016/j.plantsci.2015.02.011

[56] Oh DH, Hong H, Lee SY, Yun DJ, Bohnert HJ, Dassanayake M. Genome structures and transcriptomes signify niche adaptation for the multiple-ion-tolerant extremophyte *Schrenkiella parvula.* Plant Physiology. 2014;164:2123-2138. DOI: 10.1104/pp.113.233551

[57] Xu C, Tang X, Shao H, Wang H. Salinity tolerance mechanism of economic halophytes from physiological to molecular hierarchy for improving food quality. Current Genomics. 2016;17:207-214. DOI: 10.2174/13892029 17666160202215548

[58] Kiani-Pouya A, Roessner U, Jayasinghe NS, Lutz A, Rupasinghe T, Bazihizina N, Bohm J, Alharbi S, Hedrich R, Shabala S. Epidermal bladder cells confer salinity stress tolerance in the halophyte quinoa and *Atriplex* species. Plant, Cell and Environment. 2017;40:1900-1915. DOI: 10.1111/pce.12995

[59] Shabala S, Bose J, Hedrich R. Salt bladders: do they matter? Trends in Plant Science. 2014;19:687-691. DOI: 10.1016/j.tplants.2014.09.001.

[60] Böhm J, Messerer M, Müller HM, Scholz-Starke J, Gradogna A, Scherzer S,Tobias Maierhofer T, Bazihizina N, Zhang H, Stigloher C, Ache P, Al-Rasheid KAS, Mayer KFX, Shabala S, Carpaneto A, Haberer G, Zhu JK, Hedrich R. Understanding the molecular basis of salt sequestration in epidermal bladder cells of *Chenopodium quinoa*. Current Biology. 2018;28:3075-3085. DOI: 10.1016/j.cub.2018.08.004

[61] Glover BJ. Differentiation in plant epidermal cells. Journal of Experimental Botany.2000;51:497-505. DOI: 10.1093/jexbot/51.344.497.

[62] Eshel A. Effects of NaCl and KCl on growth and ionic composition of the halophytic C4 succulent chenopods *Salsola kali*, *Suaeda monoica* and *Suaeda aegyptiaca*. Functional Plant Biology.1985;12:319-328. DOI:10.1071/pp9850319

[63] Song J, Wang B. Using euhalophytes to understand salt tolerance and to develop saline agriculture: *Suaeda salsa* as a promising model. Annals of Botany.2015;115:541-53. DOI: 10.1093/aob/mcu194

[64] Rozema J, Schat H. Salt tolerance of halophytes, research questions reviewed in the perspective of saline agriculture. Environmental and Experimental Botany. 2013;92:83-95. DOI: 10.1016/j.envexpbot.2012.08.004

[65] Yi X, Sun Y, Yang Q, Guo A, Chang L, Wang D, Tong Z, Jin X, Wang L, Yu J. Quantitative proteomics of *Sesuvium portulacastrum* leaves revealed that ion transportation by V-ATPase and sugar accumulation in chloroplast played crucial roles in halophyte salt toleranc*e.* Journal of Proteomics. 2014;99:84-100. DOI: 10.1016/j.jprot.2014.01.017

[66] Ozgur R, Uzilday B, Sekmen AH, Turkan I. Reactive oxygen species regulation and antioxidant defence in halophytes. Functional Plant Biology.2013;40:832-847. DOI: 10.1071/fp12389

[67] Megdiche W, Hessini K, Gharbi F, Jaleel CA, Ksouri R, Abdelly C. Photosynthesis and photosystem 2 efficiency of two salt-adapted halophytic

seashore *Cakile maritima* ecotypes. Photosynthetica.2008;46:410-419. DOI: 10.1007/s11099-008-0073-1

[68] Duarte B, Sleimi N, Caçador I. Biophysical and biochemical constraints imposed by salt stress: learning from halophytes. Frontiers in Plant Science. 2014;5:746. DOI: 10.3389/fpls.2014.00746

[69] Rabhi M, Castagna A, Remorini D, Scattino C, Smaoui A, Ranieri A, Abdelly C. Photosynthetic responses to salinity in two obligate halophytes: *Sesuvium portulacastrum* and *Tecticornia indica.* South African Journal of Botany.2012;79:39-47. DOI:10.1016/j.sajb.2011.11.007

[70] Peng Z, Wang M, Li F, Lv H, Li C, Xia G. A proteomic study of the response to salinity and drought stress in an introgression strain of bread wheat. Molecular and Cellular Proteomics. 2009;8:2676-2686. DOI: 10.1074/mcp.M900052-MCP200

[71] Nikalje GC, Srivastava AK, Pandey GK, Suprasanna P. Halophytes in biosaline agriculture: Mechanism, utilization, and value addition. Land Degradation and Development. 2017;29:1081-1095. DOI: 10.1002/ldr.2819

[72] Mishra A, Tanna B. Halophytes: Potential resources for salt stress tolerance genes and promoters. Frontiers in Plant Science. 2017;8:829. DOI: 10.3389/fpls.2017.00829.

[73] Nah G, Pagliarulo CL, Mohr PG, Luo M, Sisneros N, Yu Y, Collura K, Currie J, Goicoechea JL, Wing RA, Schumaker KS. Comparative sequence analysis of the SALT OVERLY SENSITIVE1 orthologous region in *Thellungiella halophila* and *Arabidopsis thaliana*. Genomics. 2009;94:196-203. DOI: 10.1016/j.ygeno.2009.05.007.

[74] Dassanayake M, Oh DH, Hong H, Bohnert HJ, Cheeseman JM. Transcription strength and halophytic lifestyle. Trends in Plant Science. 2011;16:1-3. DOI: 10.1016/j.tplants.2010.10.006.

[75] Nawaz I, Iqbal M, Bliek M, Schat H. Salt and heavy metal tolerance and expression levels of candidate tolerance genes among four extremophile *Cochlearia* species with contrasting habitat preferences. Science of the Total Environment. 2017;584-585:731-741. DOI: 10.1016/j.scitotenv.2017.01.111.

[76] Bose J, Rodrigo-Moreno A, Lai D, Xie Y, Shen W, Shabala S. Rapid regulation of the plasma membrane H^+-ATPase activity is essential to salinity tolerance in two halophyte species, *Atriplex lentiformis* and *Chenopodium quinoa*. Annals of Botany. 2015;115:481-494. DOI: 10.1093/aob/mcu219.

[77] Boyko H, Boyko E. Principles and experiments regarding irrigation with highly saline and seawater without desalinization. Transactions of the New York Academy Sciences. 1964;26:1087-1102. DOI: 10.1111/j.2164-0947.1964.tb03506.x

[78] Pareek A, Dhankher OP, Foyer CH. Mitigating the impact of climate change on plant productivity and ecosystem sustainability. Journal of Experimental Botany. 2020;71:451-456. DOI: 10.1093/jxb/erz518

[79] Wungrampha S, Joshi R, Rathore RS, Singla-Pareek SL, Govindjee, Pareek A. CO2 uptake and chlorophyll a fluorescence of *Suaeda fruticosa* grown under diurnal rhythm and after transfer to continuous dark. Photosynthesis Research. 2019;142:211. DOI: 10.1007/s11120-019-00659-0

[80] Dagar JC. Indian mangroves: Status, management and their plausible benefits for livelihood security. Journal of Indian Society of Coastal Agricultural Research. 2008;26:121-128.

[81] Dagar JC, Yadav RK, Dar SR, Ahamad S. Liquorice (*Glycyrrhiza glabra*): a potential salt-tolerant, highly remunerative medicinal crop for remediation of alkali soils. Current Science. 2015;108:1683-1687.

[82] Kumar VS, Pathak KC, Pednekar P, Raju NSN, Gowthaman R. Coastal processes along the Indian coastline. Current Science. 2006;91:530-536.

[83] Burman D, Bandyopadhyay BK, Mandal S, Mandal UK, Mahanta KK, Sarangi SK, Maji B, Rout S, Bal AR, Gupta SK, Sharma DK. Land shaping-A unique technology for improving productivity of coastal land. Technical Bulletin: CSSRI/Canning Town/Bulletin/2013/02. Central Soil Salinity Research Institute, Regional Research Station, Canning Town, West Bengal, India; 2013. p. 38. DOI: 10.13140/RG.2.1.1595.2483

[84] Bandyopadhyay BK, Burman B, Bal AR, Sarangi SK, Madal S. Use of available brackish and fresh water in the coastal area for integrated cultivation of crops and fishes. Annual Report. Central Soil Salinity Research Institute, Karnal, Haryana, India; 2008-09. p. 106-108.

[85] Powlson DS, Whitmore AP, Goulding WT. Soil carbon sequestration to mitigate climate change: A critical re-examination to identify the true and the false. European Journal of Soil Science. 2011;62:42-55. DOI: 10.1111/j.1365-2389.2010.01342.x

[86] Wang Q, Li Y, Alva A. Cropping systems to improve carbon sequestration for mitigation of climate change. Journal of Environmental Protection. 2010;1:207-215. DOI: 10.4236/jep.2010.13025

[87] Bhattacharyya T, Chandran P, Ray SK, Mandal C, Tiwary P, Pal DK, Wani SP, Sahrawat KL. Processes determining the sequestration and maintenance of carbon in soils: A synthesis of research from tropical India. Soil Horizons. 2014;55(4):1-16. DOI: 10.2136/sh14-01-0001

[88] Kumar P, Chaudhari SK, Mishra AK, Singh K, Rai P, Singh R, Sharma DK. Labile carbon dynamics and soil amelioration in six-year old *Eucalyptus tereticornis* plantation in sodic soils. Journal of Soil Salinity and Water Quality. 2014;6:91-95.

[89] Dagar JC, Lal L, Jeet R, Kumar M, Chaudhari SK, Yadav RK, Ahamad S, Singh G, Kaur A. Eucalyptus geometry in agroforestry on waterlogged saline soils influences plant and soil traits in North-West India. Agriculture, Ecosystems and Environment. 2016;233:33-42. DOI: 10.1016/j.agee.2016.08.025

[90] Dagar JC, Sharma PC, Chaudhari SK, Jat HS, Ahamad S. Climate change vis-a-vis saline agriculture: Impact and adaptation strategies. In: Dagar JC, Sharma PC, Sharma DK, Singh AK, editors. Innovative Saline Agriculture. 1st ed. India: Springer; 2016. p. 5-53. DOI: 10.1007/978-81-322-2770-0_2

[91] Kumar P, Mishra AK, Kumar M, Chaudhari SK, Singh R, Singh K, Rai P, Sharma DK. Biomass production and carbon sequestration of *Eucalyptus tereticornis* plantation in reclaimed sodic soils of north-west India. Indian Journal of Agricultural Sciences. 2019;89:1091-1095.

[92] Chinchmalatpure AR, Camus D, Shukla M, Kad S. Quasi equilibrium of soil carbon stock in saline Vertisols under different land use systems. ICAR-Central Soil Salinity Research Institute, Karnal, India. Salinity News. 2015; 21:2. p. 4.

[93] Kumar P, Mishra AK, Chaudhari SK, Singh R, Singh K, Rai P, Pandey CB, Sharma DK. Biomass estimation

and carbon sequestration in *Populus deltoides* plantations in India. Journal of Soil Salinity and Water Quality. 2016;8:25-29.

[94] Hadar Y. Sources for ligno cellulosic raw materials for the production of ethanol. In: Faraco V, editor. Lignocellulose Conversion. Berlin: Springer; 2013. p. 21-38. DOI: 10.1007/978-3-642-37861-4_2

[95] Sharma R, Wungrampha S, Singh V, Pareek A, Sharma MK. Halophytes as bioenergy crops. Frontiers in Plant Science. 2016;7:1372. DOI: 10.3389/fpls.2016.01372

[96] Xian-Zhao L, Chun-zhi W, Qing S, Chao-kui L. The potential resource of halophytes for developing bio-energy in China coastal zone. Journal of Agricultural and Food Science Research.2012;1:44-51.

[97] Abideen Z, Ansari R, Khan MA. Halophytes: potential source of ligno-cellulosic biomass for ethanol production. Biomass and Bioenergy. 2011;35:1818-1822. DOI: 10.1016/j.biombioe.2011.01.023

[98] Bomani BMM, Hendricks RC, Elbluk M, Okon M, Lee E, Gigante B. NASA's Green Lab Research Facility-A Guide for a Self-Sustainable Renewable Energy Ecosystem. [Internet]. 2011. Available from: http://ntrs.nasa.gov/archive/nasa/ casi.ntrs.nasa.gov/20120001794.pdf [Accessed: 2020-08-20]

[99] Cornwell A. Etihad and Boeing to Fund UAE Biofuel Farm. [Internet]. 2014. Available from: http://projects.zawya.com/Etihad_and_Boeing_to_fund_UAE_biofuel_farm/ story/GN_22012014_230127/ [Accessed: 2016-08-21]

[100] Ravindran KC, Venkatesan K, Balakrishnan V, Chellappan KP, Balasubramanian T. Restoration of saline land by halophytes for Indian soils. Soil Biology and Biochemistry. 2007;39:2661-2664. DOI: 10.1016/j.soilbio.2007.02.005

[101] Panta S, Flowers T, Lane P, Doyle R, Haros G, Shabala S. Halophyte agriculture: Success stories. Environmental and Experimental Botany. 2014;107:71-83. DOI: 10.1016/j.envexpbot.2014.05.006

[102] Liang L, Liu W, Sun Y, Huo X, Li S, Zhou Q. Phytoremediation of heavy metal contaminated saline soils using halophytes: current progress and future perspectives. Environmental Reviews. 2017;25:269-281. DOI: 10.1139/er-2016-0063

[103] Qadir M, Oster JD, Schubert S, Noble AD, Sahrawat KL. Phytoremediation of sodic and saline-sodic soils. Advances in Agronomy. 2007;96:197-247. DOI: 0.1016/S0065-2113(07)96006-X

[104] Flowers TJ, Galal HK, Bromham L. Evolution of halophytes: multiple origins of salt tolerance in land plants. Functional Plant Biology. 2010;37:604-612. DOI: 10.1071/FP09269.

[105] Wang HL, Tian CY, Jiang L, Wang L. Remediation of heavy metals contaminated saline soils: A halophyte choice? Environmental Science and Technology. 2014;48:21-22. DOI: 10.1021/es405052j

[106] Santos MSS, Pedro CA, Goncalves SC, Ferreira SMF. Phytoremediation of cadmium by the facultative halophyte plant *Bolboschoenus maritimus* (L.) Palla, at different salinities. Environmental Science and Pollution Research. 2015;22:15598-15609. DOI: 10.1007/s11356-015-4750-x

[107] Christofilopoulos S, Syranidou E, Gkavrou G, Manousaki E, Kalogerakis N. The role of halophyte *Juncus acutus* L. in the remediation of mixed contamination

in a hydroponic greenhouse experiment. Journal of Chemical Technology and Biotechnology. 2016;91:1665-1674. DOI: 10.1002/jctb.4939

[108] Lutts S, Lefevre I. How can we take advantage of halophyte properties to cope with heavy metal toxicity in salt-affected areas? Annals of Botany. 2015;115:509-528. DOI: 10.1093/aob/mcu264

[109] Van Oosten MJ, Maggio A. Functional biology of halophytes in the phytoremediation of heavy metal contaminated soils. Environmental and Experimental Botany. 2015;111:135-146. DOI: 10.1016/j.envexpbot.2014.11.010

[110] Manousaki E, Kalogerakis N. Phytoextraction of Pb and Cd by the Mediterranean saltbush (*Atriplex halimus* L.): Metal uptake in relation to salinity. Environmental Science and Pollution Research. 2009;16:844-854. DOI: 10.1007/s11356-009-0224-3

[111] Nalla S, Hardaway CJ, Sneddon J. Phytoextraction of selected metals by the first and second growth seasons of *Spartina alterniflora*. Instrumentation Science and Technology. 2012;40:17-28. DOI: 10.1080/10739149.2011.633143

[112] Ayyappan D, Sathiyaraj G, Ravindran KC. Phytoextraction of heavy metals by *Sesuvium portulacastrum* L. a salt marsh halophyte from tannery effluent. International Journal of Phytoremediation. 2016;18:453-459. DOI: 10.1080/15226514.2015.1109606

[113] Santos ES, Abreu MM, Peres S, Magalhaes MCF, Leitao S, Pereira AS, Cerejeira MJ. Potential of *Tamarix africana* and other halophyte species for phytostabilisation of contaminated salt marsh soils. Journal of Soils and Sediments. 2017;17:1459-1473. DOI: 10.1007/s11368-015-1333-x

[114] Shankar V. Life support species in the Indian Thar desert. In: Proceedings of the International Workshop on Maintenance and Evaluation of Life Support Species in Asia and the Pacific Region; 4-7 April 1987; New Delhi: ICAR; 1988. P. 37-41

[115] Lokhande VH, Nikam TD, Suprasanna P. *Sesuvium portulacastrum* (L.) a promising halophyte: cultivation, utilization and distribution in India. Genetic Resources and Crop Evolution. 2009;56:741-747. DOI: 10.1007/s10722-009-9435-1

[116] Singh JP, Soni ML, Beniwal RK, Mondal BC, Dasora S. Lila Lana-Paschami Rajasthan ki aanban. Parti Bhumi Samachar, 2003;20-22 p.

[117] Dagla HR, Shekhawat NS. Little known use of *Haloxylon* spp. in traditional food. Natural Product Radiance. 2006;5:131-132.

[118] Singh JP, Rathore VS, Beniwal RK. Kheep (*Leptadenia pyrotechnica*): potential rangeland shrub of western Rajasthan. Indian Journal of Plant Genetic Resources. 2007;20:199-203.

[119] Singh JP, Rathore VS, Beniwal RK. Diversity and conservation of shrubs of hot arid region of Rajasthan, India. In: Singh NP, Chhonkar PK, editors. Biodiversity for Sustainable Development. New Delhi: ASSED; 2008. p. 51-71.

[120] Joshi A, Kanthaliya B, Arora J. Halophytes of Thar desert: Potential source of nutrition and feedstuff. International Journal of Bioassays. 2018;8.1:5668-5683. DOI: 10.14303/ijbio.2019.8.1.1

[121] Rathore VS, Singh JP, Soni ML. Saji: An economic plant production from halophytic Chenopode shrubs. In: Yadav OP, Singh JP, Tripathi RS, Saha D, Das T, editors. Desert Environment Newsletter. 19 vol. Jodhpur: ENVIS Centre on Combating Desertification; 2017. p. 10.

[122] Gajender, Singh G, Dagar JC, Lal K, Yadav RK. Performance of edible cactus (*Opuntiaficus indica*) in saline environments. Indian Journal of Agricultural Sciences. 2014;84:509-513.

[123] Weber DJ, Ansari R, Gul B, Khan MA. Potential of halophytes as source of edible oil. Journal of Arid Environments. 2007;68:315-321. DOI:10.1016/j.jaridenv.2006.05.010

[124] Dagar JC, Yadav RK. Climate resilient approaches for enhancing productivity of saline agriculture. Journal of Soil Salinity and Water Quality. 2017;9:9-29.

[125] González JA, Eisa SSS, Hussin SAES, Prado FE. Quinoa: an Incan crop to face global changes in agriculture. In: Murphy KM, Matanguihan J, editors. Quinoa: Improvement and Sustainable Production. 1st ed. Hoboken, NJ: John Wiley and Sons; 2015. p. 1-18. DOI: 10.1002/9781118628041.ch1

[126] Rojas W, Pinto M, Alanoca C, Pando LG, Leónlobos P, Alercia A, Diulgheroff S, Padulosi S, Bazile D. Quinoa genetic resources and ex situ conservation. In: Bazile D, Bertero D, Nieto C, editors. State of the Art Report on Quinoa Around the World in 2013. Rome: FAO/CIRAD; 2015. P. 56-82.

[127] Hussain MI, Al- Dakheel AJ, Reigosa MJ. Genotypic differences in agro-physiological, biochemical and isotopic responses to salinity stress in quinoa (*Chenopodium quinoa* Willd.) plants: Prospects for salinity tolerance and yield stability. Plant Physiology and Biochemistry. 2018;129:411-420. DOI: 10.1016/j.plaphy.2018.06.023

[128] ICAR-CSSRI Annual Report. ICAR-Central Soil Salinity Research Institute, Karnal- 132001, Haryana, India; 2019. p. 223.

[129] Singh D, Chaudhary MK, Meena ML, Laxman, Kumar C. Kharchia wheat: King of saline environment. In: Yadav OP, Singh JP, Tripathi RS, Saha D, Das T, editors. Desert Environment Newsletter. 19 vol. Jodhpur: ENVIS Centre on Combating Desertification; 2017. p. 11-12.

[130] Dagar JC, Sharma PC, Sharma DK, Singh AK, editors. Innovative Saline Agriculture. 1st ed. India: Springer; 2016. 519p. DOI: 10.1007/978-81-322-2770-0

[131] Toderich K, Shuyskaya E, Taha F, Matsuo N, Ismail S, Aralova D, Radjabov T. Integrating agroforestry and pastures for soil salinity management in dryland ecosystems in Aral Sea basin. In: Shahid SA, Abdelfattah MA, Taha FK, editors. Developments in Soil Salinity Assessment and Reclamation-Innovative Thinking and Use of Marginal Soil and Water Resources in Irrigated Agriculture. Dordrecht: Springer; 2013. p. 579-602. DOI: 10.1007/978-94-007-5684-7_38

[132] Wicke B, Smeets E, Dornburg V, Vashev B, Gaiser T, Turkenburg W, Faaij A. The global technical and economic potential of bioenergy from salt affected soils. Energy and Environmental Science. 2011;4:2669-2680. DOI: 10.1039/C1EE01029H

[133] FAO. Global Forest Resources Assessment. Main report [Internet]. 2010. Available from: http://www.fao.org/forest-resources-assessment/past-assessments/fra-2010/en/ [Accessed: 2020-08-20]

[134] Kaur B, Gupta SR, Singh G. Carbon storage and nitrogen cycling in silvi-pastoral systems on a sodic soil in northwestern India. Agroforestry System. 2002;54:21-29. DOI: 10.1023/A:1014269221934

[135] Singh G, Dagar JC. Greening sodic soils: bichhian model, Technical bulletin no. 2/2005. Central Soil Salinity Research Institute, Karnal; 2005. p. 51.

[136] Dagar JC, Tomar OS, Kumar Y, Bhagwan H, Yadav RK, Tyagi NK. Performance of some under explored crops under saline irrigation in a semiarid climate in northwest India. Land Degradation and Development. 2006;17:285-299. DOI: 10.1002/ldr.712

[137] Dagar JC. Greening salty and waterlogged lands through agroforestry systems for livelihood security and better environment. In: Dagar JC, Singh A, Arunachalam A, editors. Agroforestry Systems in India: Livelihood Security & Environmental Services-Advances in Agroforestry. 1st ed. India: Springer; 2014. p. 273-332. DOI: 10.1007/978-81-322-1662-9_9

[138] Rao GG, Nayak AK, Chinchmalatpure AR. *Salvadora persica*: a life support species for salt affected black soils. Technical bulletin 1/2003. CSSRI, Karnal, Haryana, India; 2003. 44 p.

[139] Ram J, Dagar JC, Lal K, Singh G, Toky OP, Tanwar VS, Dar SR, Chauhan MK. Biodrainage to combat waterlogging, increase farm productivity and sequester carbon in central command areas of northwest India. Current Science. 2011;100:1673-1680.

[140] Singh D, Buhmann AK, Flowers TJ, Seal CE, Papenbrock J. *Salicornia* as a crop plant in temperate regions: selection of genetically characterized ecotypes and optimization of their cultivation conditions. AoB Plants. 2014. 6, plu071. DOI:10.1093/aobpla/plu071

[141] Yamamoto K, Oguri S, Chiba S, Momonoki YS. Molecular cloning of acetylcholinesterase gene from *Salicornia europaea* L. Plant Signaling and Behavior. 2009;4:361-366. DOI: 10.4161/psb.4.5.8360

[142] Jha B, Gontia I, Hartmann A. The roots of the halophyte *Salicornia brachiata* are a source of new halotolerant diazotrophic bacteria with plant growth-promoting potential. Plant and Soil. 2012;356:265-277. DOI: 10.1007/s11104-011-0877-9

[143] Essaidi I, Brahmi Z, Snoussi A, Koubaier HBH, Casabianca H, Abe N, Omri AE, Chaabouni MM, Bouzouita N. Phytochemical investigation of Tunisian *Salicornia herbacea* L., antioxidant, antimicrobial and cytochrome P450 (CYPs) inhibitory activities of its methanol extract. Food Control. 2013;32:125-133. DOI: 10.1016/j.foodcont.2012.11.006

[144] Daffodil ED, Rajalakshmi K, Mohan VM. Antioxidant activity, total phenolics and flavonoids of *Salicornia brachiata* Roxb. leaf extracts (Chenopodiaceae). World Journal of Pharmacy and Pharmaceutical Science. 2013;2:352-366.

[145] Ravindran KC, Venkatesan K, Balakrishnan V, Chellappan KP, Balasubramanian, T. Ethnomedicinal studies of Pichavaram mangroves of East coast, Tamil Nadu. Indian Journal of Traditional Knowledge. 2005;4:409-411.

[146] Eganathan P, Subramanian Sr. HM, Latha R, Rao CS. Oil analysis in seeds of *Salicornia brachiata*. Industrial Crop and Products. 2006;23:177-179. DOI: 10.1016/j.indcrop.2005.05.007

[147] Patel S. *Salicornia*: evaluating the halophytic extremophile as a food and a pharmaceutical candidate. 3 Biotech. 2016;6:104. DOI: 10.1007/s13205-016-0418-6

[148] Mishra A, Joshi M, Jha B. Oligosaccharide mass profiling of nutritionally important *Salicornia brachiata*, an extreme halophyte. Carbohydrate Polymers. 2013;92:1942-1945. DOI: 10.1016/j.carbpol.2012.11.055

[149] Ksouri R, Ksouri WM, Jallali I, Debez A, Magné C, Hiroko I, Abdelly C.

Medicinal halophytes: Potent source of health promoting biomolecules with medical, nutraceutical and food applications. Critical Reviews in Biotechnology. 2012;32:289-326. DOI: 10.3109/07388551.2011.630647

[150] Ventura Y, Eshel A, Pasternak D, Sagi M. The development of halophyte-based agriculture: Past and present. Annals of Botany. 2015;115:529-540. DOI: 10.1093/aob/mcu173

[151] Bourgou S, Megdiche W, Ksouri R. The halophytic genus *Zygophyllum* and *Nitraria* from North Africa: A phytochemical and pharmacological overview. In: Neffati M, Najjaa H, Máthé À, editors. Medicinal and Aromatic Plants of the World—Africa volume 3. Medicinal and Aromatic Plants of the World. Vol 3. Dordrecht: Springer; 2017. p. 345-356. DOI: 10.1007/978-94-024-1120-1_13

[152] Faustino MV, Faustino MAF, Pinto DCGA. Halophytic grasses, a new source of nutraceuticals? A review on their secondary metabolites and biological activities. International Journal of Molecular Sciences. 2019;20:1067-1096. DOI: 10.3390/ijms20051067

[153] Bang MA, Kim HA, Cho YJ. Hypoglycemic and antioxidant effect of dietary hamcho powder in streptozotocin-induced diabetic rats. Journal of the Korean Society of Food Science and Nutrition. 2002;31:840-846. DOI: 10.3746/jkfn.2002.31.5.840

[154] Benwahhoud M, Jouad H, Eddouks M, Lyoussi B. Hypoglycemic effect of *Suaeda fructicosa* in streptozotocin induced diabetic rats. Journal of Ethnopharmacology. 2001;76:35-38. DOI: 10.1016/s0378-8741(01)00207-0

[155] Abeysinghe PD. Antibacterial activity of some medicinal mangroves against antibiotic resistant pathogenic bacteria. Indian Journal of Pharmaceutical Sciences. 2010;72:167-172. DOI: 10.4103/0250-474X.65019

[156] Ganesan M, Trivedi N, Gupta V, Madhav S, Venu Radhakrishna Reddy C, Levine IA. Seaweed resources in India – current status of diversity and cultivation: prospects and challenges. Botanica Marina. 2019;62:463-482. DOI: 10.1515/bot-2018-0056

[157] Kim JK, Yarish C,Hwang EK, Park M, Kim Y. Seaweed aquaculture: cultivation technologies, challenges and its ecosystem services. Algae. 2017;32:1-13. DOI: 10.4490/algae.2017.32.3.3

[158] Jha AK, Mathew S, Ravishankar CN. Can sulphated polysaccharides from seaweed provide prophylactic and/or therapeutic solution to COVID-19 pandemic? Current Science. 2020;119:172-174.

Chapter 9

Protagonist of Mineral Nutrients in Drought Stress Tolerance of Field Crops

Malik Ghulam Asghar and Anam Bashir

Abstract

The food demand is increasing hastily, that is inducing continuous pressure on agriculture sector and industries to fulfill rising dietary needs. To meet with increasing demand, the food production must be elevated up to 70% until the year 2050. On the other hand, changing climate is disturbing crop production around the World. Crops grown under field conditions are affected by more than one abiotic stress. It is continuous task and challenge for agronomists to make crops environment hardy to obtain maximum yield. It is considered that different agronomic managements, if done appropriately, could be beneficial for increasing crop production. The optimal provision of plant nutrients can assist the crops to fight in better way with environmental stress like drought; it can help them to continue their normal metabolism even under hostile abiotic circumstances. The regions that have reduced availability of water for crop production, a balanced nutrient management can assist crops to give adequate production. Some of nutrients have potential of not only maintaining plant metabolism but also to enhance the quality of product. This chapter highlights the protagonist of plant nutrients in alleviation of drought stress in field crops.

Keywords: drought, physiology, consequences, alleviation, macronutrients, micronutrients, mechanisms

1. Introduction

Water shortage is an emerging limitation to crop production due to climate change. It critically influences development and growth of crops and results in significant production loss. It is important to recognize morphological, physiological and bio-chemical effects of drought in relation to nutrient uptake in crops [1]. Drought impairs mineral transport and effects stomatal conductance. By considering nutrients role in plants growth, negative consequences of drought can be avoided by management strategies [2, 3]. Previously, many scientists have worked to understand the role of mineral nutrients in alleviation of drought stress, but more is to be done. Among minerals that are essential for plant growth, macronutrients has significant importance because their shortage lead to quick response and plants become more susceptible to other abiotic and biotic stresses. On the other hand, micronutrients deficiency effect at molecular level and results in altered enzymatic activity and blockage in signal transduction pathways [4]. Those plants that have capability to attain and retain water in large amount, as well as better water usage efficiency, are more tolerant to drought stress. Response in the direction of water stress depends upon crop growth stage, intensity and severity of drought [5, 6].

There are many reports available previously that addresses the consequences of drought on different physiological parameters like photosynthesis, respiration, homeostasis and assimilates transportation but very few discourses the drought effects on mineral in crops. Albeit, if crops are grown on mineral-rich soils, water limitations can be the reason of disruption in nutrient uptake. Minerals are taken up by plants in inorganic ionic forms. When a plant is subject to drought, due to low soil moisture, the diffusion of minerals is disrupted and ultimately transport is affected [3, 7, 8].

2. Effect of drought stress in crops

Field crops are simultaneously subjected to more than one abiotic stress during their complete life cycle. Drought and high temperature are the most detrimental abiotic stresses. It is continuous task for scientists to make crops hardy against biotic and more importantly abiotic stresses to increase food productivity. The simulation model predicts that to cope with rising food demand, supply must be increased to 70% till the year 2050 [9–12].

Drought stress influences crops by disturbing their physiological and biochemical functioning [13–16]. Previously, work is done making crops vigorous to deal with climatic challenges [9, 16–18] but more is still to be done.

Early droughts due to changing climate can reduce crop productivity [19]. The struggle of water use among domestic, industrial and agricultural sector is making situation worse for irrigated agriculture [20]. This problematic situation is shifting agriculture from irrigated to rainfed areas where periodic drought events are occurring due to disturbed rainfall pattern [21, 22].

2.1 Impact of drought on morphological traits

Crops when subjected to drought stress show different behavior. Some crops are resistant to drought while others are susceptible [23]. Those crops that have taproot system are more tolerant to short term drought events. They can stand with mild to moderate drought condition. On the other hand, prolonged drought can affect all crops likewise and can cause significant yield loss [24].

2.1.1 Effect of drought on seedling emergence

Seed germination is the most critical stage in complete life cycle; it is influenced by water availability for imbibition [25]. Drought stress at this stage can results in irregular germination and deprived seedlings [26, 27]. In rainfed areas, absence of shower at seedling establishment stage critically reduces field emergence [28, 29].

2.1.2 Growth phase affected by drought

Water shortage at vegetative stage disturbs growth and development through impaired turgor and stomatal conductance [30]. The reduction of water potential inside cytosol increases solute level. This leads to damage of cell structure and functioning. Cell division and expansion is also inhibited [31]. Under drought stress, nutrient uptake is also exaggerated that primes to reduction in leaf area and photosynthesis [32, 33]. Several traits of crops that are affected by drought at vegetative stage include leaf area, assimilation rate, total dry matter and chlorophyll [34, 35]. Root length and dry weight of leaves and stem is also reduced [36].

2.1.3 Effect of drought on crop yield

The loss of crop yield due to drought stress is decided by many factors like intensity, duration and ability of crop to tolerate drought stress. In higher plants, anthesis is the most drought susceptible stage [37]. Water shortage at that stage can results in substantial yield loss [30].

In oilseed crops, almost all yield related traits are affected by drought [38–40]. Severity of drought is also an important aspect; it distresses all growth stages regardless of crop, eventually results in considerable yield loss [41–44].

2.1.4 Effect of drought on crop quality

Among oilseed crops, sunflower has significant importance because it is rich in linoleic acid. Drought stress at reproductive stage reduces oil quality in oilseed crops and deteriorates its texture [45]. Drought stress also reduces quality of end products. It disturbs biochemical enzymes [46] and gene regulation that are responsible for oil constituents in sunflower [47].

2.2 Effect of drought on physio-biochemical traits

Crops are responsive to abiotic stresses from molecular to morphological level. Those crops that are tolerant to drought stress modify their cells at molecular level like increasing concentration of osmolytes in cytosol under harsh environment [48–52]. However, in susceptible crops, drought can affect at biochemical level [53–56].

2.2.1 Water relation disturbance

The key phenotypic adoption in drought tolerant crops is tap root system. They can extract water from deeper soil layer even under severe environment. Those plants that have shallow root system, when subjected to drought, it affects their water potential inside cell [57]. The low water potential leads to turgor loss and interrupted stomatal conductance [36, 41]. Transport of nutrients through xylem is concerned under drought [58, 59].

2.2.2 Photosynthesis reduction

The metabolic process of carbon fixation that occurs in leaves in the presence of light is called as photosynthesis. This is the main energy harvesting phenomenon that is accountable for growth and development. It is affected by different environmental factors like, availability of moisture, sunshine, humidity and temperature [60].

The plants that have C4 carbon fixation pathway are more efficient in carbon harvesting [61], but under drought, they perform in the same way as C_3 plants. Stomatal closure is triggered by water deficit condition that eventually restricts CO_2 diffusion [62], thus diminishes photosynthesis [36]. Ribulose bisphosphate is a vital enzyme in carbon fixation. The activity of RuBP is affected under drought stress. Those crops that can maintain RuBP production are more resistant to drought stress [63–68].

2.2.3 Disrupted uptake of nutrients

Under drought, absorption capacity of roots is affected that condenses nutrient uptake. Nitrogen, being a vital constituent of plants, is required in high quantity. The reduction of soil moisture reduces ability of roots to absorb adequate moisture. Phosphorus uptake, transport and translocation are also affected in drought conditions [6]. It lessens NPK uptake in sunflower [41].

2.2.4 Drought induced oxidative stress

Free radicals of oxygen, that are also known as reactive oxygen species has significant role in cell signaling. Their production remains continue unceasingly inside cell in controlled amount. When a plant is subjected to any environmental stress, its production increases. This augmented concentration induces oxidative stress to crops. They are highly reactive in action; they can cause injury to cellular structure [69]. In oilseeds like sunflower, drought overproduces ROS [70]. Malondialdehyde is an indicator of cell membrane damage in plants. Water deficiency increases MDA production that specifies increment in cellular injury [71–73].

3. Role of nutrients in drought stress alleviation

Optimum nutrient supply not only improves growth of crops but is also helpful for plants under adversative climatic conditions. There are seventeen nutrients that are crucial for plant growth [74]. Upon their requirement, these are grouped as macronutrient and micronutrient. This review deals with role of essential nutrients in drought stress mitigation.

3.1 Macronutrients

3.1.1 Nitrogen

Under dry climatic conditions, water use efficiency and growth of crops is restricted due to less accessibility of water. Efficient nitrogen application can serve the purpose under drought stress [75, 76]. Plants facing drought stress are more susceptible to heat tremors as well. Nitrogen deficiency in drought stress outcomes as biomass reduction in crops [77, 78]. Previous studies have suggested that shoot biomass is more affected under drought-cum-nitrogen stress, while root biomass is not much exaggerated primarily [79]. On the other hand, plants become drought hardy under sufficient soil nitrogen availability [75, 80, 81]. Increasing nitrogen significantly improved crop performance under drought stress. Nitrogen also play significant role in prevention of plasma membrane damage and osmotic adjustment. Application of N under water deficiency also enhances other major nutrient uptake like potassium and calcium [82].

Nitrogen availability diminishes malondialdehyde content that alleviates in drought stress [80]. It recovers photosynthetic contents and improves cell division that lead to leaf area increment [83]. At molecular level, drought stress greatly influences photosystem-II efficiency that is recovered by optimum nitrogen accessibility [51, 84–93].

3.1.2 Phosphorus

Previously, many researchers have testified that phosphorus application under water deficiency in many crops significantly enhance their water usage ability and helps in drought resistance [74, 94, 95]. It is also well known that optimum phosphorus in crops improves root growth and stomatal activity [96, 97]. Phosphorus availability also optimizes leaf area [98], plasma membrane stability and water use efficiency [99–102]. It was observed that phosphorus in leaves was relatively higher under drought condition as compared to optimum water availability which suggests that phosphorus has contribution in drought tolerance [94, 96].

Phosphorus also improves nitrogen mobility under water deficiency [103]. Morphological and physiological parameters were also improved when phosphorus was applied at high rate in drought such as, plant height, leaf area, dry weight and water use efficiency [102, 104]. Application method of phosphorus also influences crop growth in drought, deep phosphorus placement (DPP) method works excellently for drought affected areas that ultimately promotes root growth [101, 105].

3.1.3 Potassium

Potassium is well-known for its osmoregulatory functions in crops. It regulates stomatal conductance and water uptake; the optimum K application increases WUE [106, 107]. Potassium soothes aquaporins and osmotic pressure that regulates water uptake, stomatal regulation, carbon intake, cell elongation and ROS detoxification [108, 109]. In grasses like sorghum, K application under drought improves photosynthesis which leads to growth and yield [106, 110]. In maize, potassium plays role photosynthates assimilation [111]. Potassium availability is correlated with aquaporins activity and stem cell expansion [112].

The hydraulic conductivity of root and anatomical traits has great influence on crop performance. The increment in hydraulic conductivity is associated with drought tolerance [113]. In higher plants, reduction in K influences aforementioned traits, hence compromised yield. Drought simulates ethylene production that in return hinders abscisic acid activity. The starvation of K further worsen the situation, it delays stomatal conductance [109]. Potassium also play role in ROS detoxification and promotes photosynthesis process [114, 115].

3.1.4 Magnesium

Magnesium has central place in chlorophyll molecule, thus has significant importance. It has great role in dry matter partitioning from sink to source. Passable Mg is required at reproductive stage to avoid flower sterility. Foliage application also improves nutrient mobility and helps in growth maintenance under stressful environment [116, 117]. Magnesium is highly mobile nutrient. It has positive correlation with nitrogen and potassium. Adequate magnesium increases their mobility; they are helpful in stress tolerance [118].

Drought stress in field crops affects magnesium uptake from soil. This deficiency can be fulfilled by foliar Mg application [119]. Earlier, it is known that foliage applied Mg can satisfy plant's need [120]. The mechanisms of Mg that are responsible for drought stress induction include growth of root, NPK uptake and improvement of WUE [74].

3.1.5 Calcium

Drought stress leads to overgeneration of ROS that result in cell damage [121–124]. Calcium has its role in detoxification of ROS [125]. It is known that in the activity of aquaporins, pH and calcium are of significance importance [126, 127]. Exogenous application of Ca induces drought resistance in wheat cultivars. Calcium has cell signaling mechanism, which simulates proline accumulation.

Calcium, when it is applied under drought stress, it improves chlorophyll and catalase activity and decreases plasma membrane damage. It also maintains osmolytes like proline and other soluble antioxidants [128, 129]. Foliage applied Ca under drought stress helps to improves drought stress alleviation by refining catalase, peroxidase and superoxide dismutase activity [130].

3.1.6 Sulfur

The role of sulfur application in mitigation of drought stress is very little known previously. It has a substantial role in stress signaling pathway. It improves crop growth, morphological parameters and nutrient contents [131]. In counter stress mechanism, increment in glutathione also has significant importance. It aids in ROS detoxification [132]. The uptake of sulfur in adequate amount helps crops to stand with drought events. Its transport and assimilation is among one of the drought stress responses [133, 134].

3.2 Micronutrients

3.2.1 Zinc

Zinc has role in various physiological processes like activity of catalytic, carboxypeptidase, superoxide dismutase, RNA polymerase and alkaline phosphates [4, 118, 135, 136]. Under water shortage, zinc has been known to improve drought resistance by improving WUE and water activity [4, 137, 138]. The reduction in zinc uptake, that is caused by water shortage, leads plants toward stress condition. Under limited soil moisture, zinc is immobile [118].

In cereals like wheat, when drought is subjected at anthesis and grain filling, it constrains nutrient uptake which become cause of stunted growth [139]. The process of photosynthesis and water activity is affected under zinc-cum-drought stress, however, when zinc is present in optimum amount, it helps crop to stand with drought. It aids in deactivation of ROS [4, 140]. At reproductive stage, plants are highly susceptible to Zn shortage [141]. When plants are subjected to prolonged drought, it impairs activity of different cell metabolic contents like NADPH. Zinc application inhibits photooxidative damage, reduces ROS generation, and promoting osmolytes concentration like SOD [74, 142–145].

3.2.2 Manganese

It is vital micronutrient that has several functions in plants. It assists in activation of various metabolic enzymes of tricarboxylic cycle. It is the part of photosystem-II, also aids in ATP synthesis and RuBP carboxylase activity. It helps to maintain balance among superoxide dismutase activity and chlorophyll contents, even under water stress [130].

The role of manganese is well known for detoxification of ROS like superoxide and hydrogen peroxide [146]. On the other hand, manganese shortage leads to oxidative stress in plants that causes chlorophyll damage thus stunted photosynthetic activity [4]. Water shortage can also be responsible for manganese deficiency. Low soil availability of manganese as it occurs under dry conditions makes it unavailable for plants [147]. The starvation of manganese leads to WUE reduction. In cereals like barley, lower WUE is correlated with abrupt stomatal control during the day and imperfection in stomatal closure during night. This leads to degradation of waxy layer of plasma membrane that is consequence of ROS activity [148].

3.2.3 Iron

It is involved in chlorophyll pigments production. It is the part of enzymes that are involved in transfer of energy, reduction of nitrogen and formation of lignin. It creates compounds along with sulfur that are the catalysts for other vital biochemical procedures in plants. The iron deficiency results in chlorosis which is the consequent of low chlorophyll concentration. Severe deficiency of iron turns leaf color from yellow to white that is sign of leaf death. Under high soil pH, iron uptake is affected. It also has antagonistic effects with phosphorus and manganese [149].

The moisture in soil greatly inhibits iron uptake [150]. The iron has vital protagonist in oxidative damage protection of leaves under stress. Its deficiency is highly dreadful for plants growth [4]. Sufficient iron amount in plant is essential for activities of antioxidants [151].

3.2.4 Boron

Boron is unavailable in soil barring basic pH and low moisture. It is highly immobile in pedosphere as well as plant. The continuous supply of boron can prevent crops from its deficiency and detrimental effects [152].

Low soil moisture greatly hampers boron uptake from rhizosphere. Its uptake via roots involve passive uptake frequently that is maintained by water uptake. As the water decreases in soil, its uptake is compromised [153]. Main function of boron is to take part in synthesis of cell wall and its extension. It also recovers biosynthesis of lignin and differentiation of xylem. It increases photosynthetic activity and plasma-membrane integrity. It facilitates assimilate transportation [4, 74].

It is necessarily required for H-ATPase activity and the coding involved for it. It also influences uptake of other nutrients like K and deteriorate cell expansion [4]. Boron is also involved in lessening of photochemical damage of cell. Among reasons for low photoinhibition, boron deficiency and drought are well known [153].

3.2.5 Copper

Among micronutrients, copper is essential for growth of plants. It has vital role in electron transport chain and cell wall loosening. It also involves in sensing ethylene, metabolism of cell wall and oxidative stress protection [154, 155]. The well-known function of copper is its involvement in formation of pollens and upholding their viability [4, 155].

There are many enzymes in which this metal acts as cofactor like ascorbic oxidase, laccase, amino oxidase and polyphenols. At molecular level, copper is also involved in cell signaling, trafficking of proteins, mobilization of iron and oxidative phosphorylation. The reproductive parts of plants are more susceptible to cooper deficiency [155, 156].

4. Conclusion

The changing climate is making situation worse for field crop production. Abrupt variations in rainfall and temperature is limiting crop yield. Under field condition, more than one abiotic stresses are disturbing plant growth simultaneously. Drought stress is among the major agricultural yield limiting factor worldwide. Different agronomic practices like optimum plant nutrition management are greatly obliging for crops under drought stress. It can alleviate drought consequences affectively. Drought stress greatly inhibits different physiological functions and biochemical processes. It leads to ROS over-generation that significantly damages cell structure. Optimal nutrients supply like NPK and Ca be accommodating for ROS detoxification and maintenance of cell functions. Under drought stress, they also facilitate in antioxidant generation like catalase, superoxide dismutase and peroxidase. They inhibit photooxidation of vital cell molecules and maintain cell membrane integrity. Likewise, micronutrients such as Zn and Mg also play role in antioxidant generation. Other mechanisms that are maintained by nutrients to induce drought stress are water uptake and stomatal conduction regulation. Optimum supply of K and Ca helps to regulate water activity and aquaporin function. In a nutshell, efficient nutrient management will be helpful in mitigation of drought stress in field crops. The best practice should be adopted to increase their availability to plants. Effective nutrient utilization cultivars need to be focused on.

Author details

Malik Ghulam Asghar[1*] and Anam Bashir[2]

1 Department of Agronomy, Faculty of Agricultural Sciences and Technology, Bahauddin Zakariya University, Multan, Pakistan

2 Department of Agronomy, University of Agriculture, Faisalabad, Pakistan

*Address all correspondence to: gh.asghar@hotmail.com

References

[1] Ramegowda V, Senthil-Kumar M. The interactive effects of simultaneous biotic and abiotic stresses on plants: mechanistic understanding from drought and pathogen combination. Journal of plant physiology. 2015;176:47-54. DOI: 10.1016/j.jplph.2014.11.008

[2] Swift J, Adame M, Tranchina D, Henry A, Coruzzi GM. Water impacts nutrient dose responses genome-wide to affect crop production. Nature communications. 2019;10(1):1-9. DOI: 10.1038/s41467-019-09287-7

[3] da Silva EC, Nogueira RJ, da Silva MA, de Albuquerque MB. Drought stress and plant nutrition. Plant stress. 2011;5(1):32-41.

[4] Hajiboland R. Effect of micronutrient deficiencies on plants stress responses. In: Abiotic stress responses in plants 2012. pp. 283-329. Springer, New York, NY. ISBN 978-1-4614-0634-1

[5] Laxa M, Liebthal M, Telman W, Chibani K, Dietz KJ. The role of the plant antioxidant system in drought tolerance. Antioxidants. 2019;8(4):94. DOI: 10.3390/antiox8040094

[6] Taiz L, Zeiger E. Plant Physiology. Sinauer Associates. Inc., Publishers. 2010. pp. 782. Sunderland, MA. DOI: 10.1093/aob/mcg079

[7] Akram R, Turan V, Hammad HM, Ahmad S, Hussain S, Hasnain A, Maqbool MM, Rehmani MI, Rasool A, Masood N, Mahmood F. Fate of organic and inorganic pollutants in paddy soils. In: Environmental pollution of paddy soils 2018. pp. 197-214. Springer, Cham.

[8] Akram R, Turan V, Wahid A, Ijaz M, Shahid MA, Kaleem S, Hafeez A, Maqbool MM, Chaudhary HJ, Munis MF, Mubeen M. Paddy land pollutants and their role in climate change. In: Environmental Pollution of Paddy Soils 2018. pp. 113-124. Springer, Cham.

[9] Wani SH, Sah SK. Biotechnology and abiotic stress tolerance in rice. Journal of Rice Research. 2014;2(2):100e105. DOI: 10.4172/jrr.1000e105

[10] Fahad S, Ihsan MZ, Khaliq A, Daur I, Saud S, Alzamanan S, Nasim W, Abdullah M, Khan IA, Wu C, Wang D. Consequences of high temperature under changing climate optima for rice pollen characteristics-concepts and perspectives. Archives of Agronomy and Soil Science. 2018;64(11):1473-88.DOI: 10.1080/03650340.2018.1443213

[11] Gul F, Ahmed I, Ashfaq M, Jan D, Fahad S, Li X, Wang D, Fahad M, Fayyaz M, Shah SA. Use of crop growth model to simulate the impact of climate change on yield of various wheat cultivars under different agro-environmental conditions in Khyber Pakhtunkhwa, Pakistan. Arabian Journal of Geosciences. 2020;13(3):112. DOI: 10.1007/s12517-020-5118-1

[12] ur Rahman MH, Ahmad A, Wajid A, Hussain M, Rasul F, Ishaque W, Islam MA, Shelia V, Awais M, Ullah A, Wahid A. Application of CSM-CROPGRO-Cotton model for cultivars and optimum planting dates: evaluation in changing semi-arid climate. Field Crops Research. 2019;238:139-52. DOI: 10.1016/j.fcr.2017.07.007

[13] Robert GA, Rajasekar M, Manivannan P. Triazole-induced drought stress amelioration on growth, yield, and pigments composition of *Helianthus annuus* L. (sunflower). International Multidisciplinary Research Journal. 2015;5:6-15.

[14] Ibrahim MF, Faisal A, Shehata SA. Calcium chloride alleviates water

stress in sunflower plants through modifying some physio-biochemical parameters. American-Eurasian Journal of Agricultural and Environmental Science. 2016;16(4):677-693. DOI: 10.5829/idosi.aejaes.2016.16.4.12907

[15] Hasanuzzaman M, Nahar K, Alam M, Roychowdhury R, Fujita M. Physiological, biochemical, and molecular mechanisms of heat stress tolerance in plants. International journal of molecular sciences. 2013;14(5):9643-9684. DOI: 10.3390/ijms14059643

[16] Hasanuzzaman M, Nahar K, Fujita M. Plant response to salt stress and role of exogenous protectants to mitigate salt-induced damages. In: Ecophysiology and responses of plants under salt stress 2013. pp. 25-87. Springer, New York, NY. DOI: 10.1007/978-1-4614-4747-4_2

[17] Manivannan P, Rabert GA, Rajasekar M, Somasundaram R. Drought stress induced modification on growth and Pigments composition in different genotypes of *Helianthus annuus* L. Current Botany. 2014;5:7-13.

[18] Jedmowski C, Ashoub A, Momtaz O, Brüggemann W. Impact of drought, heat, and their combination on chlorophyll fluorescence and yield of wild barley (*Hordeum spontaneum*). Journal of Botany. 2015;120868:1-9. DOI: 10.1155/2015/120868

[19] Debaeke P, Bedoussac L, Bonnet C, Bret-Mestries E, Seassau C, Gavaland A, Raffaillac D, Tribouillois H, Véricel G, Justes E. Sunflower crop: environmental-friendly and agroecological. Oilseeds & fats crops and Lipids. 2017;24(3):D304. DOI: 0.1051/ocl/2017020

[20] Alexandratos N, Bruinsma J. World agriculture towards 2030/2050: the 2012 revision. Food and Agriculture Organization, Rome. ESA Working paper; 2012;12(3):1-12. DOI: 10.22004/ag.econ.288998

[21] Elliott J, Deryng D, Müller C, Frieler K, Konzmann M, Gerten D, Glotter M, Flörke M, Wada Y, Best N, Eisner S. Constraints and potentials of future irrigation water availability on agricultural production under climate change. Proceedings of the National Academy of Sciences. 2014;111(9):3239-3244. DOI: 10.1073/pnas.1222474110

[22] Farooq M, Hussain M, Siddique KH. Drought stress in wheat during flowering and grain-filling periods. Critical Reviews in Plant Sciences. 2014;33(4):331-349. DOI: 10.1080/07352689.2014.875291

[23] Hussain S, Ahmad M, Ahmad S, Iqbal J, Subhani MN, Nadeem SM, Atta S, Ibrahim M. Improvement of drought tolerance in sunflower (*Helianthus annuus* L.) by foliar application of abscisic acid and potassium chloride. Pakistan Journal of Nutrition. 2013;12(4):345. DOI: 10.3923/pjn.2013.345.352

[24] Andrianasolo FN, Casadebaig P, Maza E, Champolivier L, Maury P, Debaeke P. Prediction of sunflower grain oil concentration as a function of variety, crop management and environment using statistical models. European Journal of Agronomy. 2014;54:84-96. DOI: 10.1051/ocl/2016052

[25] Luan Z, Xiao M, Zhou D, Zhang H, Tian Y, Wu Y, Guan B, Song Y. Effects of salinity, temperature, and polyethylene glycol on the seed germination of sunflower (*Helianthus annuus* L.). The Scientific World Journal. 2014;170418:1-9. DOI: 10.1155/2014/170418

[26] Wen B. Effects of high temperature and water stress on seed germination of the invasive species Mexican sunflower. PLoS One. 2015;10(10):e0141567. DOI: 10.1371/journal.pone.0141567

[27] Farooq M, Hussain M, Wahid A, Siddique KH. Drought stress in plants: an overview. In: Plant responses to drought stress 2012.pp. 1-33. Springer, Berlin, Heidelberg. DOI: 10.1007/978-3-642-32653-0

[28] Kaya MD, Okçu G, Atak M, Cıkılı Y, Kolsarıcı Ö. Seed treatments to overcome salt and drought stress during germination in sunflower (*Helianthus annuus* L.). European journal of agronomy. 2006;24(4):291-295. DOI: 10.1016/j.eja.2005.08.001

[29] Mwale SS, Hamusimbi C, Mwansa K. Germination, emergence and growth of sunflower (*Helianthus annuus* L.) in response to osmotic seed priming. Seed Science and Technology. 2003;31(1):199-206. DOI: 10.15258/sst.2003.31.1.21

[30] Benlloch-González M, Quintero JM, García-Mateo MJ, Fournier JM, Benlloch M. Effect of water stress and subsequent re-watering on K+ and water flows in sunflower roots. A possible mechanism to tolerate water stress. Environmental and Experimental Botany. 2015;118:78-84. DOI: 10.1016/j.envexpbot.2015.06.008

[31] Lisar SY, Motafakkerazad R, Hossain MM, Rahman IM. Water stress in plants: Causes, effects and responses. In: Water Stress; Rahman, M., Hasegawa, H., (Eds.). 2012. pp. 1-14. InTech: Rijeka, Croatia. DOI: 10.5772/39363

[32] Fernández-Moroni I, Fraysse M, Presotto A, Cantamutto M. Evaluation of Argentine wild sunflower biotypes for drought stress during reproductive stage. Helia. 2012;35(57):29-36. DOI: 10.2298/HEL1257029F

[33] Hussain S, Saleem MF, Ashraf MY, Cheema MA, Haq MA. Abscisic acid, a stress hormone helps in improving water relations and yield of sunflower (*Helianthus annuus* L.) hybrids under drought. Pakistan Journal of Botany. 2010;42(3):2177-2189. **ISSN** : 0556-3321

[34] Canavar O, Gotz KP, Ellmer F, Chmielewski FM, Kaynak MA. Determination of the relationship between water use efficiency, carbon isotope discrimination and proline in sunflower genotypes under drought stress. Australian Journal of Crop Science. 2014;8(2):232. ISSN: 1835-2693

[35] Hemmati MH, Soleymani A. A study about drought stress effects on grain yield components of three sunflower cultivars. International journal of advanced biological and biomedical research. 2014; 2(3):564-572. DOI: JR_IJABBR-2-3_002

[36] GhobadiM,TaherabadiS,GhobadiME, Mohammadi GR, Jalali-Honarmand S. Antioxidant capacity, photosynthetic characteristics and water relations of sunflower (*Helianthus annuus* L.) cultivars in response to drought stress. Industrial Crops and Products. 2013;50:29-38. DOI: 10.1016/j.indcrop.2013.07.009

[37] Hussain S, Saleem MF, Iqbal J, Ibrahim M, Ahmad M, Nadeem SM, Ali A, Atta S. Abscisic acid mediated biochemical changes in sunflower (*Helianthus annuus* L.) grown under drought and well-watered field conditions. Journal of Animal and Plant Sciences. 2015;25(2):406-416. ISSN : 1018-7081

[38] Buriro M, Sanjrani AS, Chachar QI, Chachar NA, Chachar SD, Buriro B, Gandahi AW, Mangan T. Effect of water stress on growth and yield of sunflower. Journal of Agricultural Technology. 2015;11(7):1547-1563. ISSN 1686-9141

[39] Farzad BA, Mahmoud T, Majid N, Mohamad-Reza S. Effect of drought stress on yield and yield components of some sunflower recombinant inbred lines. International Journal

of Biosciences. 2013;3:50-56. DOI: 10.12692/ijb/3.3.50-56 A

[40] Hussain M, Malik MA, Farooq M, Khan MB, Akram M, Saleem MF. Exogenous glycinebetaine and salicylic acid application improves water relations, allometry and quality of hybrid sunflower under water deficit conditions. Journal of Agronomy and Crop Science. 2009;195(2):98-109. DOI: 10.1111/j.1439-037X.2008.00354.x

[41] Hussain RA, Ahmad R, Nawaz F, Ashraf MY, Waraich EA. Foliar NK application mitigates drought effects in sunflower (*Helianthus annuus* L.). Acta physiologiae plantarum. 2016;38(4):83. DOI: 10.1007/s11738-016-2104-z

[42] Elsheikh ER, Schultz B, Adam HS, Haile AM. Crop water productivity for sunflower under different irrigation regimes and plant spacing in Gezira Scheme, Sudan. Journal of Agriculture and Environment for International Development. 2015;109(2):221-233. DOI: 10.12895/jaeid.20152.346

[43] Danish S, Zafar-ul-Hye M, Fahad S, Saud S, Brtnicky M, Hammerschmiedt T, Datta R. Drought Stress Alleviation by ACC Deaminase Producing Achromobacter xylosoxidans and *Enterobacter cloacae*, with and without Timber Waste Biochar in Maize. Sustainability. 2020;12(15):6286. DOI: 10.3390/su12156286

[44] Ilyas M, Nisar M, Khan N, Hazrat A, Khan AH, Hayat K, Fahad S, Khan A, Ullah A. Drought Tolerance Strategies in Plants: A Mechanistic Approach. Journal of Plant Growth Regulation. 2020:1-9. DOI: 10.1007/s00344-020-10174-5

[45] Hussain S, Saleem MF, Iqbal J, Ibrahim M, Atta S, Ahmed T, Rehmani MI. Exogenous application of abscisic acid may improve the growth and yield of sunflower hybrids under drought. Pakistan Journal of Agricultural Sciences. 2014;51(1):49-58. ISSN: 2076-0906

[46] Schuppert GF, Tang S, Slabaugh MB, Knapp SJ. The sunflower high-oleic mutant Ol carries variable tandem repeats of FAD2-1, a seed-specific oleoyl-phosphatidyl choline desaturase. Molecular Breeding. 2006;17(3):241-256. DOI: 10.1007/s11032-005-5680-y

[47] Anastasi U, Santonoceto C, Giuffrè AM, Sortino O, Gresta F, Abbate V. Yield performance and grain lipid composition of standard and oleic sunflower as affected by water supply. Field Crops Research. 2010;119(1):145-153. DOI: 10.1016/j.fcr.2010.07.001

[48] Fahad S, Hussain S, Bano A, Saud S, Hassan S, Shan D, Khan FA, Khan F, Chen Y, Wu C, Tabassum MA. Potential role of phytohormones and plant growth-promoting rhizobacteria in abiotic stresses: consequences for changing environment. Environmental Science and Pollution Research. 2015 Apr 1;22(7):4907-21. DOI: 10.1007/s11356-014-3754-2

[49] Fahad S, Hussain S, Saud S, Hassan S, Chauhan BS, Khan F, Ihsan MZ, Ullah A, Wu C, Bajwa AA, Alharby H. Responses of rapid viscoanalyzer profile and other rice grain qualities to exogenously applied plant growth regulators under high day and high night temperatures. PLoS One. 2016;11(7):e0159590. DOI: 10.1371/journal.pone.0159590

[50] Fahad S, Hussain S, Saud S, Hassan S, Ihsan Z, Shah AN, Wu C, Yousaf M, Nasim W, Alharby H, Alghabari F. Exogenously applied plant growth regulators enhance the morpho-physiological growth and yield of rice under high temperature. Frontiers in Plant Science. 201;7:1250. DOI: 10.3389/fpls.2016. 01250

[51] Hussain MA, Fahad S, Sharif R, Jan MF, Mujtaba M, Ali Q, Ahmad A, Ahmad H, Amin N, Ajayo BS, Sun C. Multifunctional role of brassinosteroid

and its analogues in plants. Plant Growth Regulation. 2020:1-6. DOI: 10.1007/s10725-020-00647-8

[52] Wu C, Tang S, Li G, Wang S, Fahad S, Ding Y. Roles of phytohormone changes in the grain yield of rice plants exposed to heat: a review. PeerJ. 2019;7:e7792. DOI: 10.7717/peerj.7792

[53] Wang D, Fahad S, Saud S, Kamran M, Khan A, Khan MN, Hammad HM, Nasim W. Morphological acclimation to agronomic manipulation in leaf dispersion and orientation to promote "Ideotype" breeding: Evidence from 3D visual modeling of "super" rice (*Oryza sativa* L.). Plant Physiology and Biochemistry. 2019;135:499-510. DOI: 10.1016/j.plaphy.2018.11.010

[54] Wu C, Cui K, Tang S, Li G, Wang S, Fahad S, Nie L, Huang J, Peng S, Ding Y. Intensified pollination and fertilization ameliorate heat injury in rice (*Oryza sativa* L.) during the flowering stage. Field Crops Research. 2020;252:107795

[55] Yang Z, Zhang Z, Zhang T, Fahad S, Cui K, Nie L, Peng S, Huang J. The effect of season-long temperature increases on rice cultivars grown in the central and southern regions of China. Frontiers in plant science. 2017;8:1908. DOI: 10.3389/fpls.2017.01908

[56] Zafar-ul-Hye M, Naeem M, Danish S, Fahad S, Datta R, Abbas M, Rahi AA, Brtnicky M, Holátko J, Tarar ZH, Nasir M. Alleviation of Cadmium Adverse Effects by Improving Nutrients Uptake in Bitter Gourd through Cadmium Tolerant Rhizobacteria. Environments. 2020;7(8):54. DOI: 10.3390/environments7080054

[57] Kiani SP, Grieu P, Maury P, Hewezi T, Gentzbittel L, Sarrafi A. Genetic variability for physiological traits under drought conditions and differential expression of water stress-associated genes in sunflower (*Helianthus annuus* L.). Theoretical and applied genetics. 2007;114(2):193-207. DOI: 10.1007/s00122-006-0419-7

[58] Ruehr NK, Offermann CA, Gessler A, Winkler JB, Ferrio JP, Buchmann N, Barnard RL. Drought effects on allocation of recent carbon: from beech leaves to soil CO_2 efflux. New Phytologist. 2009;184(4):950-961. DOI: 10.1111/j.1469-8137.2009.03044.x

[59] Fahad S, Hussain S, Matloob A, Khan FA, Khaliq A, Saud S, Hassan S, Shan D, Khan F, Ullah N, Faiq M. Phytohormones and plant responses to salinity stress: a review. Plant growth regulation. 2015;75(2):391-404. DOI: 10.1007/s10725-014-0013-y

[60] Andrianasolo F, Debaeke P, Champolivier L, Maury P. Analysis and modelling of the factors controlling seed oil concentration in sunflower: a review. Oilseeds and fats crops and lipids. 2016;23(2):1-12. DOI: 10.1051 / ocl / 2016004

[61] Killi D, Bussotti F, Raschi A, Haworth M. Adaptation to high temperature mitigates the impact of water deficit during combined heat and drought stress in C_3 sunflower and C_4 maize varieties with contrasting drought tolerance. Physiologia plantarum. 2017;159(2):130-147. DOI: 10.1111/ppl.12490

[62] Flexas J, Bota J, Loreto F, Cornic G, Sharkey TD. Diffusive and metabolic limitations to photosynthesis under drought and salinity in C_3 plants. Plant biology. 2004;6(3):269-279. DOI: 10.1055/s-2004-820867

[63] Markulj Kulundžić A, Kovačević J, Viljevac Vuletić M, Josipović A, Liović I, Mijić A, Lepeduš H, Matoša Kočar M. Impact of abiotic stress on photosynthetic efficiency and leaf temperature in sunflower. Poljoprivreda. 2016;22(2): 17-22. DOI: 10.18047/poljo.22.2.3

[64] Galmés J, Aranjuelo I, Medrano H, Flexas J. Variation in Rubisco content and activity under variable climatic factors. Photosynthesis Research. 2013;117(1-3):73-90. DOI: 10.1007/s11120-013-9861-y

[65] Saud S, Fahad S, Cui G, Yajun C, Anwar S. Determining nitrogen isotopes discrimination under drought stress on enzymatic activities, nitrogen isotope abundance and water contents of Kentucky bluegrass. Scientific Reports. 2020;10(1):1-6. DOI: 10.1038/s41598-020-63548-w

[66] Saud S, Fahad S, Yajun C, Ihsan MZ, Hammad HM, Nasim W, Arif M, Alharby H. Effects of nitrogen supply on water stress and recovery mechanisms in Kentucky bluegrass plants. Frontiers in plant science. 2017;8:983. DOI: 10.3389/fpls.2017.00983

[67] Saud S, Li X, Chen Y, Zhang L, Fahad S, Hussain S, Sadiq A, Chen Y. Silicon application increases drought tolerance of Kentucky bluegrass by improving plant water relations and morphophysiological functions. The Scientific World Journal. 2014;2014.1-10. DOI: 10.1155/2014/ 368694

[68] Saud S, Yajun C, Fahad S, Hussain S, Na L, Xin L, A lhussien SA. Silicate application increases the photosynthesis and its associated metabolic activities in Kentucky bluegrass under drought stress and post-drought recovery. Environmental Science and Pollution Research. 2016;23(17):17647-55. DOI: 10.1007/s11356-016-6957-x

[69] Cechin I, Cardoso GS, Fumis TD, Corniani N. Nitric oxide reduces oxidative damage induced by water stress in sunflower plants. Bragantia. 2015;74(2):200-206. DOI: 10.1590/1678-4499.353

[70] Soleimanzadeh H. Response of sunflower (*Helianthus annuus* L.) to selenium application under water stress. World Applied Sciences Journal. 2012;17(9):1115-1119. ISSN: 1818-4952

[71] Adnan M, Fahad S, Zamin M, Shah S, Mian IA, Danish S, Zafar-ul-Hye M, Battaglia ML, Naz RM, Saeed B, Saud S. Coupling phosphate-solubilizing bacteria with phosphorus supplements improve maize phosphorus acquisition and growth under lime induced salinity stress. Plants. 2020;9(7):900. DOI:10.3390/plants9070900

[72] Ahmad S, Kamran M, Ding R, Meng X, Wang H, Ahmad I, Fahad S, Han Q. Exogenous melatonin confers drought stress by promoting plant growth, photosynthetic capacity and antioxidant defense system of maize seedlings. PeerJ. 2019;7:e7793. DOI: 10.7717/peerj.7793

[73] Izhar Shafi M, Adnan M, Fahad S, Wahid F, Khan A, Yue Z, Danish S, Zafar-ul-Hye M, Brtnicky M, Datta R. Application of Single Superphosphate with Humic Acid Improves the Growth, Yield and Phosphorus Uptake of Wheat (*Triticum aestivum* L.) in Calcareous Soil. Agronomy. 2020;10(9):1224. DOI: 10.3390/agronomy10091224

[74] Waraich EA, Ahmad R, Ashraf MY, Saifullah, Ahmad M. Improving agricultural water use efficiency by nutrient management in crop plants. Acta Agriculturae Scandinavica, Section B-Soil & Plant Science. 2011;61(4):291-304. DOI: 10.1080/09064710.2010.491954

[75] Mahpara S, Shahnawaz M, Rehman K, Ahmad R, Khan FU. Nitrogen fertilization induced drought tolerance in sunflower: a review. Pure and Applied Biology. 2019;8(2):1675-1683. DOI: 10.19045/bspab.2019.80110

[76] Shangguan ZP, Shao MA, Dyckmans J. Nitrogen nutrition and water stress effects on leaf

photosynthetic gas exchange and water use efficiency in winter wheat. Environmental and experimental botany. 2000;44(2):141-149. DOI: 10.1016/S0098-8472(00)00064-2

[77] Gong X, Li J, Ma H, Chen G, Dang K, Yang P, Wang M, Feng B. Nitrogen deficiency induced a decrease in grain yield related to photosynthetic characteristics, carbon–nitrogen balance and nitrogen use efficiency in proso millet (*Panicum miliaceum* L.). Archives of Agronomy and Soil Science. 2020;66(3):398-413. DOI: 10.1080/03650340.2019.1619077

[78] Rahimi A, Sayadi F, Dashti H. Effects of water and nitrogen supply on growth, water-use efficiency and mucilage yield of isabgol (*Plantago ovata Forsk*). Journal of soil science and plant nutrition. 2013;13(2):341-354. DOI: 10.4067/S0718-95162013005000028

[79] Song CJ, Ma KM, Qu LY, Liu Y, Xu XL, Fu BJ, Zhong JF. Interactive effects of water, nitrogen and phosphorus on the growth, biomass partitioning and water-use efficiency of *Bauhinia faberi* seedlings. Journal of arid environments. 2010;74(9):1003-1012. DOI: 10.1016/j.jaridenv.2010.02.003

[80] Saneoka H, Moghaieb RE, Premachandra GS, Fujita K. Nitrogen nutrition and water stress effects on cell membrane stability and leaf water relations in *Agrostis palustris* Huds. Environmental and Experimental Botany. 2004;52(2):131-138. DOI: 10.1016/j.envexpbot.2004.01.011

[81] Halvorson AD, Reule CA. Nitrogen fertilizer requirements in an annual dryland cropping system. Agronomy Journal. 1994;86(2):315-318. DOI: 10.2134/agronj1994.00021962008600020020x

[82] Ahanger MA, Tittal M, Mir RA, Agarwal RM. Alleviation of water and osmotic stress-induced changes in nitrogen metabolizing enzymes in *Triticum aestivum* L. cultivars by potassium. Protoplasma. 2017;254(5):1953-1963. DOI: 10.1007/s00709-017-1086-z

[83] Wu FZ, Bao WK, Li FL, Wu N. Effects of water stress and nitrogen supply on leaf gas exchange and fluorescence parameters of *Sophora davidii* seedlings. Photosynthetica. 2008;46(1):40-48. DOI: 10.1007/s11099-008-0008-x

[84] Fahad S, Adnan M, Hassan S, Saud S, Hussain S, Wu C, Wang D, Hakeem KR, Alharby HF, Turan V, Khan MA. Rice responses and tolerance to high temperature. In: Advances in rice research for abiotic stress tolerance 2019. pp. 201-224. Woodhead Publishing.

[85] Fahad S, Bajwa AA, Nazir U, Anjum SA, Farooq A, Zohaib A, Sadia S, Nasim W, Adkins S, Saud S, Ihsan MZ. Crop production under drought and heat stress: plant responses and management options. Frontiers in plant science. 2017;8:1147. DOI: 10.3389/fpls.2017.01147

[86] Fahad S, Bano A. Effect of salicylic acid on physiological and biochemical characterization of maize grown in saline area. Pakistan Journal of Botany. 2012;44(4):1433-8.

[87] Fahad S, Chen Y, Saud S, Wang K, Xiong D, Chen C, Wu C, Shah F, Nie L, Huang J. Ultraviolet radiation effect on photosynthetic pigments, biochemical attributes, antioxidant enzyme activity and hormonal contents of wheat. Journal of Food, Agriculture and Environment. 2013;11(3&4):1635-41.

[88] Hammad HM, Abbas F, Ahmad A, Bakhat HF, Farhad W, Wilkerson CJ, Fahad S, Hoogenboom G. Predicting Kernel Growth of Maize under Controlled Water and Nitrogen Applications. International Journal

of Plant Production. 2020:1-2. DOI: 10.1007/s42106-020-00110-8

[89] Hammad HM, Abbas F, Saeed S, Fahad S, Cerdà A, Farhad W, Bernardo CC, Nasim W, Mubeen M, Bakhat HF. Offsetting land degradation through nitrogen and water management during maize cultivation under arid conditions. Land degradation & development. 2018;29(5):1366-75. DOI: 10.1002/ldr.2933

[90] Hammad HM, Ashraf M, Abbas F, Bakhat HF, Qaisrani SA, Mubeen M, Fahad S, Awais M. Environmental factors affecting the frequency of road traffic accidents: a case study of sub-urban area of Pakistan. Environmental Science and Pollution Research. 2019;26(12):11674-85. DOI: 10.1007/s11356-019-04752-8

[91] Hammad HM, Farhad W, Abbas F, Fahad S, Saeed S, Nasim W, Bakhat HF. Maize plant nitrogen uptake dynamics at limited irrigation water and nitrogen. Environmental Science and Pollution Research. 2017;24(3):2549-57. DOI: 10.1007/s11356-016-8031-0

[92] Hammad HM, Khaliq A, Abbas F, Farhad W, Fahad S, Aslam M, Shah GM, Nasim W, Mubeen M, Bakhat HF. Comparative Effects of Organic and Inorganic Fertilizers on Soil Organic Carbon and Wheat Productivity under Arid Region. Communications in Soil Science and Plant Analysis. 2020;51(10):1406-22. DOI: 10.1080/00103624.2020.1763385

[93] Hussain S, Mubeen M, Ahmad A, Akram W, Hammad HM, Ali M, Masood N, Amin A, Farid HU, Sultana SR, Fahad S. Using GIS tools to detect the land use/land cover changes during forty years in Lodhran District of Pakistan. Environmental Science and Pollution Research. 2019:1-7. DOI: 10.1007/s11356-019-06072-3

[94] Hansel FD, Amado TJ, Ruiz Diaz DA, Rosso LH, Nicoloso FT, Schorr M. Phosphorus fertilizer placement and tillage affect soybean root growth and drought tolerance. Agronomy Journal. 2017;109(6):2936-2944. DOI: 10.2134/agronj2017.04.0202

[95] Garg BK, Burman U, Kathju S. The influence of phosphorus nutrition on the physiological response of moth bean genotypes to drought. Journal of Plant Nutrition and Soil Science. 2004;167(4):503-508. DOI: 10.1002/jpln.200320368

[96] Singh DK, Sale PW. Phosphorus supply and the growth of frequently defoliated white clover (*Trifolium repens* L.) in dry soil. Plant and Soil. 1998;205(2):155-162. DOI: 10.1023/A:1004316726665

[97] Naeem M, Khan MM. Phosphorus ameliorates crop productivity, photosynthesis, nitrate reductase activity and nutrient accumulation in coffee senna (*Senna occidentalis* L.) under phosphorus-deficient soil. Journal of Plant Interactions. 2009;4(2):145-153. DOI: 10.1080/17429140802193178

[98] Singh SK, Badgujar G, Reddy VR, Fleisher DH, Bunce JA. Carbon dioxide diffusion across stomata and mesophyll and photo-biochemical processes as affected by growth CO2 and phosphorus nutrition in cotton. Journal of Plant Physiology. 2013;170(9):801-813. DOI: 10.1016/j.jplph.2013.01.001

[99] Faustino LI, Bulfe NM, Pinazo MA, Monteoliva SE, Graciano C. Dry weight partitioning and hydraulic traits in young *Pinus taeda* trees fertilized with nitrogen and phosphorus in a subtropical area. Tree physiology. 2013;33(3):241-251. DOI: 10.1093/treephys/tps129

[100] Sawwan J, Shibli RA, Swaidat I, Tahat M. Phosphorus regulates osmotic potential and growth of African violet under in vitro-induced

water deficit. Journal of plant nutrition. 2000;23(6):759-771. DOI: 10.1080/01904160009382057

[101] Kang LY, Yue SC, Li SQ. Effects of phosphorus application in different soil layers on root growth, yield, and water-use efficiency of winter wheat grown under semi-arid conditions. Journal of Integrative Agriculture. 2014;13(9):2028-2039. DOI: 10.1016/S2095-3119(14)60751-6

[102] Singh V, Pallaghy CK, Singh D. Phosphorus nutrition and tolerance of cotton to water stress: I. Seed cotton yield and leaf morphology. Field Crops Research. 2006;96(3):191-198. DOI: 10.1016/j.fcr.2005.06.009

[103] Mariotte P, Cresswell T, Johansen MP, Harrison JJ, Keitel C, Dijkstra FA. Plant uptake of nitrogen and phosphorus among grassland species affected by drought along a soil available phosphorus gradient. Plant and Soil. 2020;448:121-132. DOI: 10.1007/s11104-019-04407-0

[104] Osonubi O. Comparative effects of vesicular-arbuscular mycorrhizal inoculation and phosphorus fertilization on growth and phosphorus uptake of maize (*Zea mays* L.) and sorghum (*Sorghum bicolor* L.) plants under drought-stressed conditions. Biology and fertility of soils. 1994;18(1):55-59. DOI: doi.org/10.1007/BF00336445

[105] Graciano C, Guiamét JJ, Goya JF. Impact of nitrogen and phosphorus fertilization on drought responses in *Eucalyptus grandis* seedlings. Forest Ecology and Management. 2005;212(3):40-49. DOI: 10.1016/j.foreco.2005.02.057

[106] Egilla JN, Davies FT, Drew MC. Effect of potassium on drought resistance of *Hibiscus rosa-sinensis* cv. Leprechaun: Plant growth, leaf macro-and micronutrient content and root longevity. Plant and soil. 2001;229(2):213-224. DOI: doi.org/10.1023/A:1004883032383

[107] Jatav KS, Agarwal RM, Tomar NS, Tyagi SR. Nitrogen metabolism, growth and yield responses of wheat (*Triticum aestivum* L.) to restricted water supply and varying potassium treatments. Journal of Indian Botany and Sociology. 2014;93(4):177-189. ISSN: 0019-4468

[108] Zamani S, Naderi MR, Soleymani A, Nasiri BM. Sunflower (*Helianthus annuus* L.) biochemical properties and seed components affected by potassium fertilization under drought conditions. Ecotoxicology and Environmental Safety. 2020;190:110017. DOI: 10.1016/j.ecoenv.2019.110017

[109] Wang M, Zheng Q, Shen Q, Guo S. The critical role of potassium in plant stress response. International journal of molecular sciences. 2013;14(4):7370-7390. DOI: 10.3390/ijms14047370

[110] Sharma PS, Kumari TS. Effect of potassium under water stress on growth and yield of sorghum in Vertisol. Journal of Potash. Research. 1996;12(3): 319-325. DOI: 10.4236/as.2011.23034

[111] Römheld V, Kirkby EA. Research on potassium in agriculture: needs and prospects. Plant and soil. 2010;335(1-2):155-180. DOI: 10.1007/s11104-010-0520-1

[112] Kanai S, Moghaieb RE, El-Shemy HA, Panigrahi R, Mohapatra PK, Ito J, Nguyen NT, Saneoka H, Fujita K. Potassium deficiency affects water status and photosynthetic rate of the vegetative sink in green house tomato prior to its effects on source activity. Plant science. 2011;180(2):368-374. DOI: 10.1016/j.plantsci.2010.10.011

[113] Benlloch-González M, Arquero O, Fournier JM, Barranco D, Benlloch M. K+ starvation inhibits water-stress-induced stomatal closure. Journal of

plant physiology. 2008;165(6):623-630. DOI: 10.1093/jxb/erp379

[114] Jiang M, Zhang J. Involvement of plasma-membrane NADPH oxidase in abscisic acid- and water stress-induced antioxidant defense in leaves of maize seedlings. Planta. 2002;215(6):1022-1030. DOI:10.1007/s00425-002-0829-y

[115] Cakmak I. The role of potassium in alleviating detrimental effects of abiotic stresses in plants. Journal of Plant Nutrition and Soil Science. 2005;168(4):521-530. DOI: 10.1002/jpln.200420485

[116] Senbayram M, Gransee A, Wahle V, Thiel H. Role of magnesium fertilisers in agriculture: plant–soil continuum. Crop and Pasture Science. 2016;66(12):1219-1229. DOI: 10.1071/CP15104

[117] Ashraf M, Athar HR, Harris PJ, Kwon TR. Some prospective strategies for improving crop salt tolerance. Advances in agronomy. 2008;97:45-110. DOI: 10.1016/S0065-2113(07)00002-8

[118] Marschner H. Mineral Nutrition of Higher Plants. 1995. Academic Press, San Diego. ISBN: 9780080571874

[119] Cakmak I, Kirkby EA. Role of magnesium in carbon partitioning and alleviating photooxidative damage. Physiologia plantarum. 2008;133(4):692-704. DOI: 10.1111/j.1399-3054.2007.01042.x

[120] Thalooth AT, Tawfik MM, Mohamed HM. A comparative study on the effect of foliar application of zinc, potassium and magnesium on growth, yield and some chemical constituents of mungbean plants grown under water stress conditions. World Journal of Agricultural Sciences. 2006;2(1):37-46. ISSN: 1817-3047

[121] Jan M, Anwar-ul-Haq M, Shah AN, Yousaf M, Iqbal J, Li X, Wang D, Fahad S. Modulation in growth, gas exchange, and antioxidant activities of salt-stressed rice (*Oryza sativa* L.) genotypes by zinc fertilization. Arabian Journal of Geosciences. 2019;12(24):775. DOI: 10.1007/s12517-019-4939-2

[122] Kamran M, Cui W, Ahmad I, Meng X, Zhang X, Su W, Chen J, Ahmad S, Fahad S, Han Q, Liu T. Effect of paclobutrazol, a potential growth regulator on stalk mechanical strength, lignin accumulation and its relation with lodging resistance of maize. Plant growth regulation. 2018;84(2):317-32. DOI: 10.1007/ s10725-017-0342-8

[123] Khan A, Tan DK, Afridi MZ, Luo H, Tung SA, Ajab M, Fahad S. Nitrogen fertility and abiotic stresses management in cotton crop: a review. Environmental Science and Pollution Research. 2017;24(17):14551-66. DOI: 10.1007/s11356-017-8920-x

[124] Khan A, Tan DK, Munsif F, Afridi MZ, Shah F, Wei F, Fahad S, Zhou R. Nitrogen nutrition in cotton and control strategies for greenhouse gas emissions: a review. Environmental Science and Pollution Research. 2017;24(30):23471-87. DOI: 10.1007/s11356-017-0131-y

[125] Verma G, Srivastava D, Tiwari P, Chakrabarty D. ROS modulation in crop plants under drought stress. Reactive Oxygen, Nitrogen and Sulfur Species in Plants: Production, Metabolism, Signaling and Defense Mechanisms. 2019:311-336. DOI: 10.1002/9781119468677.ch13

[126] Hu W, Tian SB, Di Q, Duan SH, Dai K. Effects of exogenous calcium on mesophyll cell ultrastructure, gas exchange, and photosystem II in tobacco (*Nicotiana tabacum* Linn.) under drought stress. Photosynthetica. 2018;56(4):1204-1211. DOI: 10.1007/s11099-018-0822-8

[127] Hosseini SA, Réthoré E, Pluchon S, Ali N, Billiot B, Yvin JC.

Calcium application enhances drought stress tolerance in sugar beet and promotes plant biomass and beetroot sucrose concentration. International journal of molecular sciences. 2019;20(15):3777. DOI: 10.3390/ijms20153777

[128] Qiang L, Jianhua C, Longjiang Y, Maoteng L, Jinjing L, Lu G. Effects on physiological characteristics of Honeysuckle ('*Lonicera japonica*' Thunb) and the role of exogenous calcium under drought stress. Plant Omics. 2012;5(1):1-5. ISSN: 1836-0661

[129] Xu WZ, Deng XP, Xu BC. Effects of water stress and fertilization on leaf gas exchange and photosynthetic light-response curves of *Bothriochloa ischaemum* L. Photosynthetica. 2013;51(4):603-612. DOI: 10.1007/s11099-013-0061-y

[130] Upadhyaya H, Panda SK, Dutta BK. CaCl 2 improves post-drought recovery potential in *Camellia sinensis* (L) O. Kuntze. Plant cell reports. 2011;30(4):495-503. DOI: 10.1007/s00299-010-0958-x

[131] Heidari M, Galavi M, Hassani M. Effect of sulfur and iron fertilizers on yield, yield components and nutrient uptake in sesame (*Sesamum indicum* L.) under water stress. African Journal of Biotechnology. 2011;10(44):8816-8822. DOI: 10.5897/AJB11.854

[132] Ahmad MN. Impact of drought stress on the sulfur assimilation pathway in *Zea mays* [Thesis]. The Ruperto-Carola University of Heidelberg; Germany; 2006.

[133] Majeed S, Nawaz F, Naeem M, Ashraf MY. Effect of exogenous nitric oxide on sulfur and nitrate assimilation pathway enzymes in maize (*Zea mays* L.) under drought stress. Acta Physiologiae Plantarum. 2018;40(12):206. DOI: 10.1007/s11738-018-2780-y

[134] Chan KX, Wirtz M, Phua SY, Estavillo GM, Pogson BJ. Balancing metabolites in drought: the sulfur assimilation conundrum. Trends in Plant Science. 2013;18(1):18-29. DOI: 10.1016/j.tplants.2012.07.005

[135] Yuvaraj M, Subramanian KS. Significance of Zinc in Plant Nutrition. Biotica Research Today. 2020;2(8):823-825.

[136] Blasco B, Graham NS, Broadley MR. Antioxidant response and carboxylate metabolism in *Brassica rapa* exposed to different external Zn, Ca, and Mg supply. Journal of plant physiology. 2015;176:16-24. DOI: 10.1016/j.jplph.2014.07.029

[137] Hajiboland R, Amirazad F. Growth, photosynthesis and antioxidant defense system in Zn-deficient red cabbage plants. Plant, Soil and Environment. 2010;56(5):209-217. DOI: 10.17221/207/2009-PSE

[138] Khan HR, McDonald GK, Rengel Z. Zinc fertilization and water stress affects plant water relations, stomatal conductance and osmotic adjustment in chickpea (*Cicer arientinum* L.). Plant and Soil. 2004;267(2):271-284. DOI: 10.1007/s11104-005-0120-7

[139] Karim MR, Zhang YQ, Zhao RR, Chen XP, Zhang FS, Zou CQ. Alleviation of drought stress in winter wheat by late foliar application of zinc, boron, and manganese. Journal of Plant Nutrition and Soil Science. 2012;175(1):142-151. DOI: 10.1002/jpln.201100141

[140] Foroutan L, Solouki M, Abdossi V, Fakheri BA. The effects of zinc oxide nanoparticles on enzymatic and osmoprotectant alternations in different Moringa peregrina populations under drought stress. International Journal of Basic Science Medics. 2018;3(4):178-187. DOI: 10.15171/ijbsm.2018.31

[141] Khurana N, Chatterjee C. Influence of variable zinc on yield, oil content, and physiology of sunflower. Communications in Soil Science and Plant Analysis. 2001;32(19-20):3023-3030. DOI: 10.1081/CSS-120001104

[142] Muhammad B, Adnan M, Munsif F, Fahad S, Saeed M, Wahid F, Arif M, Amanullah J, Wang D, Saud S, Noor M. Substituting urea by organic wastes for improving maize yield in alkaline soil. Journal of Plant Nutrition. 2019;42(19):2423-34. DOI: 10.1080/01904167.2019.1659344

[143] Rehman M, Fahad S, Saleem MH, Hafeez M, Rahman MH, Liu F, Deng G. Red light optimized physiological traits and enhanced the growth of ramie (*Boehmeria nivea* L.). Photosynthetica. 2020;58(4):922-31. DOI: 10.3390/agriculture10080334

[144] Saleem MH, Fahad S, Adnan M, Ali M, Rana MS, Kamran M, Ali Q, Hashem IA, Bhantana P, Ali M, Hussain RM. Foliar application of gibberellic acid endorsed phytoextraction of copper and alleviates oxidative stress in jute (*Corchorus capsularis* L.) plant grown in highly copper-contaminated soil of China. Environmental Science and Pollution Research. 2020:1-3. DOI: 10.1007/s11356-020-09764-3

[145] Saleem MH, Fahad S, Khan SU, Din M, Ullah A, Sabagh AE, Hossain A, Llanes A, Liu L. Copper-induced oxidative stress, initiation of antioxidants and phytoremediation potential of flax (*Linum usitatissimum* L.) seedlings grown under the mixing of two different soils of China. Environmental Science and Pollution Research. 2020;27(5):5211-21. DOI: 10.1007/s11356-019-07264-7

[146] Millaleo R, Reyes-Díaz M, Ivanov AG, Mora ML, Alberdi M. Manganese as essential and toxic element for plants: transport, accumulation and resistance mechanisms. Journal of soil science and plant nutrition. 2010;10(4):470-481. DOI: 10.4067/S0718-95162010000200008

[147] Hu Y, Schmidhalter U. Drought and salinity: a comparison of their effects on mineral nutrition of plants. Journal of Plant Nutrition and Soil Science. 2005;168(4):541-549. DOI: 10.1002/jpln.200420516

[148] Gholamin R, Khayatnezhad M. Effect of different levels of manganese fertilizer and drought stress on yield and agronomic use efficiency of fertilizer in durum wheat in Ardabil. Journal of Food, Agriculture & Environment. 2012;10:1326-1328. ISSN: 1459-0255

[149] Waraich EA, Ahmad R, Ashraf MY. Role of mineral nutrition in alleviation of drought stress in plants. Australian Journal of Crop Science. 2011;5(6):764-777. ISSN: 1835-2707

[150] Sardans J, Penuelas J, Ogaya R. Drought's impact on Ca, Fe, Mg, Mo and S concentration and accumulation patterns in the plants and soil of a Mediterranean evergreen *Quercus ilex* forest. Biogeochemistry. 2008;87(1):49-69. DOI: 10.1007/s10533-007-9167-2

[151] Lombardi L, Sebastiani L, Vitagliano C. Physiological, biochemical, and molecular effects of in vitro induced iron deficiency in peach rootstock Mr. S 2/5. Journal of plant nutrition. 2003;26(11):2149-2163. DOI: 10.1081/PLN-120024271

[152] Taiz L, Zeiger E. Plant Physiology. Sinauer Associates. Inc., Publishers. 2006. pp. 764. Sunderland, MA. DOI: 10.1093/aob/mcg079

[153] Hajiboland R, Farhanghi F. Effect of low boron supply in turnip plants under drought stress. Biologia plantarum. 2011;55(4):775. DOI: 10.1007/s10535-011-0186-4

[154] Printz B, Lutts S, Hausman JF, Sergeant K. Copper trafficking in plants and its implication on cell wall dynamics. Frontiers in plant science. 2016;7:601. DOI: 10.3389/fpls.2016.00601

[155] Yruela I. Copper in plants: acquisition, transport and interactions. Functional Plant Biology. 2009;36(5):409-430. DOI: 10.1071/FP08288

[156] Krämer U, Clemens S. Functions and homeostasis of zinc, copper, and nickel in plants. In: Molecular biology of metal homeostasis and detoxification. 2005. pp. 215-271. Springer, Berlin, Heidelberg. DOI: 10.1007/4735_96

Chapter 10

An Insight into the Responses of Early-Maturing *Brassica napus* to Different Low-Temperature Stresses

Xin He

Abstract

Rapeseed (*Brassica napus* L.) is an important oil crop worldwide, responds to vernalization, and shows an excellent tolerance to cold stresses during vegetative stage. The winter-type and semi-winter-type rapeseed were typical winter biennial plants in Europe and China. In recent years, more and more early-maturing semi-winter rapeseed varieties were planted across China. Unfortunately, the early-maturing rapeseed varieties with low cold tolerance have higher risk of freeze injury in cold winter and spring. The molecular mechanisms for coping with different low-temperature stress conditions in rapeseed recently had gained more attention and development. The present review gives an insight into the responses of early-maturing *B. napus* to different low-temperature stresses (chilling, freezing, cold-acclimation, and vernalization), and the strategies to improve tolerance against low-temperature stresses are also discussed.

Keywords: *Brassica napus*, low-temperature, early-maturing

1. Introduction

Low-temperature is a major environmental stress that adversely affects plant growth and development, limiting the productivity and regional distribution of crops [1, 2]. Rapeseed is an important oil crop worldwide, with planting area of 37.58 million hectares producing 75.00 million tons of oilseeds in 2018 (http://www.fao.org/faostat/). Based on vernalization requirement, rapeseed is divided into three main ecotypes, i.e., winter, semi-winter and spring types [3–5]. The winter type rapeseed is mainly grown in Europe and is sown in late summer, which requires strong vernalization and flowerings in spring, exhibiting a classical winter annual and with excellent cold tolerance during vegetative stage [3, 6]. The semi-winter type rapeseed is mainly grown in China only needs moderate or weak vernalization to promote flowering in spring, and with week cold tolerance [3, 7]. The semi-winter type rapeseed excessive exposure to low temperature stress in winter will lead to plant damage at vegetative stage and finally cause yield loss [8]. Yangtze River basin is the major region for planting semi-winter rapeseed in China, which accounts for at least 90% of the nation's total production [9]. The semi-winter rapeseed is usually sown in late September and early October shortly after the harvest of rice, and harvested in May before the cropping of rice in this area [10].

 IntechOpen

However, in recent years, due to the delay of rice harvest which leads to the postpone of rapeseed sowing until late October or early November, therefore, more and more early-maturing semi-winter rapeseed varieties were planted across Yangtze River basin. Unfortunately, the early-maturing rapeseed varieties with low cold tolerance have higher risk of freeze injury in cold winter and spring [11]. Hence, it is vital to compare early-maturing rapeseed varieties tolerant to cold and evaluate molecular mechanisms that adapt to different low-temperature stress conditions.

2. Morphophysiological mechanism of rapeseed in responses to low-temperature stress

Cold (low-temperature) stress included chilling stress (>0°C) and freezing stress (<0°C) [12]. Chilling stress (0–15°C) causes the membrane to rigidify, destabilizes protein complexes and impairs photosynthesis, eventually made plant stop growing, whereas freezing stress (<0°C) causes intracellular and extracellular ice crystal formation, and results in mechanical injury, and plant death [13–15].

Despite the fact that winter and semi-winter rapeseed is an overwintering oil crop, cold stress can still affect rapeseed development and ultimately lead to a decrease in production [8, 11]. The suitable temperature scope is 10 ~ 20°C for the growth of winter and semi-winter rapeseed. The rapeseed flower number was reduced below 10°C and the rapeseed flowering was arrested when the temperature decreased to 5°C. The rapeseed growth was arrested below 3°C and rapeseed leaves was injured below 0°C [8]. The delay of rapeseed sowing results in poor germination [16], decreased seedling biomass [17, 18], delay of floral initiation and floral bud differentiation processes [17, 19], and decreased flower number, effective pod number, pod length, and seed yield [17, 20, 21] due to low-temperature stress. In January 2008, South China was exposed to an extremely ice-frozen weather, which caused serious injuries to winter rape, affected 77.8% of the overall winter rape area in China and resulted in 10.9% yield losses [22]. Due more and more early-maturing semi-winter rapeseed varieties were planted across Yangtze River basin, rapeseed faces increased risks from continuous low temperature overcast and rainy weather in March. Continuous low temperature overcast and rainy weather during the rapeseed flowering stage or after flowering decreased the ratio of effectual silique, seeds per silique and oil content [23]. In March and April 2010, the middle and lower reaches of the Yangtze River region were exposed to continuous low temperature overcast and rainy weather, which resulted in 10–20% yield losses [23].

To date, many studies have investigated the morphological and physiological changes of low-temperature stressed rapeseeds. Leaves are the main organ to perceive low temperature stress and transmit stress signal in plants [24]. The morphological changes (dehydrated and wilting) of leaves became increasingly evident with the decrease of temperature, due to the total water content in leaves of rapeseed decreased [25, 26].

In winter rapeseed, prolonged cold acclimation led to increased thickness of young leaf blades and leaf cell walls, modified dimensions of mesophyll cells, numerous invaginations of plasma membranes and large phenolic deposits in chloroplasts, large vesicles or cytoplasm/tonoplast interfaces [27, 28]. Unlike cold acclimation, transient freezing treatment reduced the thickness of leaf cell walls and phenolic aggregates, caused reversible disorganization of the cytoplasm and chloroplasts swelling [27, 28]. Obvious gaps existed in the chloroplast grana and starch grains increased in quantity and volume [25]. In general, cold-tolerant winter rapeseed usually grows slowly, having small thick creeping deep-green waxy leaves and large root system.

Low temperature-induced thermodynamic constraints on carbon metabolism was the primary reason for lower photosynthetic activity in plants [24]. Photosynthetic efficiency is a good indicator for Low temperature tolerance in plants [10]. Just like in other crop plants, a marked reduction of photosynthetic activity is observed in rapeseed leaves when treated with low temperature [24, 29]. Tough the photosynthetic activities were reduced both in the cold-stressed leaves of cold-tolerant and cold-sensitive rapeseed cultivars, the chlorophyll a, chlorophyll b and photosynthetic activities in the young leaves of cold-tolerant cultivar all were higher than that in cold-sensitive cultivar [24].

Simultaneously, low-temperature stress caused the overproduction of reactive oxygen species (ROS), elevated H_2O_2 level and increased malondialdehyde (MDA) content in plants, which leads to a necrosis of plants. Plants possess an effective antioxidant system includes superoxide dismutase (SOD), ascorbate peroxidase (APX), peroxidase (POD) and catalase (CAT) enzymes, whose combined activities play an important role in elimination of destructive effects of ROS [24, 30]. Furthermore, under natural cold stress in field, the proline, soluble sugar, soluble protein, MDA contents and SOD, POD, CAT activities changed obviously in functional leaves of rapeseed. CAT and SOD activity reached the highest when temperature dropped to 5 and 3°C, respectively. The proline and soluble sugar contents increased when mean daily temperature decreased to 5°C and reached the maximum when temperature was below 0°C. The contents of soluble protein and MDA showed a trend to decrease at first and then increase when mean daily temperature dropped to 10, 5 and 0°C [30]. The SOD and APX activities were both increased by low temperature in the young leaves of cold-tolerant rapeseed cultivar. However, the APX activity was decreased by low temperature in the young leaves of cold-sensitive rapeseed cultivar. While, in the cold-stressed mature leaves, both cold-tolerant and cold-sensitive rapeseed cultivars represented similar antioxidant capacities [24].

Under chilling and freezing stress, the increment of proline accumulation, soluble sugar and protein contents were enhanced in cold-tolerant cultivar compared with cold-sensitive cultivar [24, 26]. Leaf abscisic acid (ABA) was enhanced in cold-tolerant cultivar under chilling and freezing stress [26].

3. Molecular mechanisms influencing responses to different low-temperature stresses in rapeseed

Plants showed increased freezing tolerance during exposure to chilling and low nonfreezing temperatures in a phenomenon known as cold acclimation [31]. The molecular mechanism of cold acclimation and cold tolerance in *Arabidopsis* and winter cereals has been extensively studied. Cold acclimation is a very complex trait involving an array of physiological and biochemical modifications, and these altered processes involved changes in gene expression patterns via phytohormone and the ICE (Inducer of CBF Expressions)-CBF (C-repeat binding factors)-COR (cold-responsive) signaling pathway [13, 32, 33].

3.1 ICE-CBF-COR signaling

In most plant species, CBF transcription factors could bind directly to the promoters of *COR* genes and induce the expression of *COR* genes [34–36]. The *COR* genes protected plant cells against cold-induced damage, repaired cold-rigidified membranes and stabilized cellular osmotic potential by encoding cryoprotective proteins and key enzymes for osmolyte biosynthesis [37]. In Arabidopsis, the basic helix-loop-helix transcription factor ICE1/2 were induced by cold stress, could bind directly to the promoters of CBF and induced CBF expression under low-temperature stress [38–40].

Similar as other plants, the expression of CBF and COR genes were induced by chilling and freezing stresses in different ecotypes rapeseed with different cold tolerance [11, 36, 41–45]. CBFs (BnaAnng34260D/BnaCnng49280D/BnaC03g71900D/BnaC07g39680D), *Kin1* and COR15 all were upregulated in both winter and semi-winter ecotype rapeseeds after cold stress (4°C for 7 days), while *BnaA08g30910D* (a CBF-like gene) and *BnCOR25* were not varied in expression in any cold stressed rapeseeds [41]. Furthermore, COR15A/B, COR413-PM1 and nine CBF1/2/4 genes all were strongly upregulated in winter and spring ecotype rapeseeds after cold stress (4°C for 12 h) [42]. While *ICE1* and COR413-PM2 was downregulated in both winter and spring ecotypes after cold stress, two *CBF3* genes were not induced by cold stress [42]. Most of *COR15A* and *COR15B* were significantly induced in both cold-tolerant and cold-sensitive semi-winter early-maturing rapeseeds after chilling (4°C for 12 h) and freezing (−4°C for 12 h) stress, regardless of cold acclimation (4°C for 14 days) [11]. Ten of 12 *CBF* genes were strongly induced by freezing stress more than chilling stress, regardless of cold acclimation [11]. Unlike other CBFs, BnaC03g71900D was induced by chilling stress more than freezing stress and BnaA03g13620D was induced by freezing stress but suppressed by cold acclimation [11]. However, two *ICE1* genes were not induced by chilling and freezing stress [11], as opposed to *AtICE1* [38]. Inexplicably, no CBF genes were identified in responsive to freezing stress in freezing-tolerant rapeseed as reported by Pu [25].

BnCOR25 were significantly induced by cold and osmotic stress treatment in rapeseed, overexpression of BnCOR25 in Arabidopsis enhances plant tolerance to cold stress [46]. Overexpression of two rapeseed CBF-like transcription factors BnCBF5 and BnCBF17 in spring rapeseed resulted in increased constitutive freezing tolerance, increased photochemical efficiency and photosynthetic capacity [29]. However, constitutively overexpressing *BNCBF5/17* in rapeseed resulted in various degrees of dwarf habit and longer time to flower [29]. The multi-gene (NCED3, ABAR, CBF3, LOS5, and ICE1) transgenic rapeseed plants exhibited pronounced growth advantage under both normal growth and stress conditions [47].

3.2 ABA signaling

Abscisic acid (ABA) is a vital plant hormone that plays a key role in stress resistance during plant growth and development [48–50]. It was reported that ABA levels are increased after cold stress in plants and exogenous application of ABA can induce plant cold tolerance [11, 51, 52]. OST1/SnRK2E, a serine-threonine protein kinase in ABA core signaling pathway, acted upstream of CBFs to positively regulate freezing tolerance via phosphorylating ICE1 to prevent its 26S proteasome-mediated degradation by HOS1 [53]. OST1 phosphorylated basic transcription factors 3 (BTF3) and BTF3-like factors, and facilitated their interactions with CBFs to promote CBF stability under cold stress [54].

27 ABA biosynthesis genes (nine-*cis*-epoxycarotenoid dioxygenase (NCED3/4/5/9), ABA DEFICIENT 1/2 (ABA1/2), abscisic aldehyde oxidase 1/2/3 (AAO1/2/3) and carotenoid cleavage dioxygenase 1 (CCD1)) were regulated by cold stress in winter and/or spring rapeseed. Additionally, many genes involved in ABA signal transduction, such as ABA INSENSITIVE 1/5 (ABI1/5), ABA-responsive element binding protein 3 (AREB3), ABA responsive element-binding factor 1/2/3 (ABF1/2/3), highly ABA-induced PP2C gene 1/2/3 (HAI1/2/3), OPEN STOMATA 1 (OST1), PYR1-like 4/6/7/10 (PYL4/6/7/10), regulatory component of ABA receptor 1/3 (RCAR1/3), SNF1-related protein kinase 2.2/2.5/2.7/2.10 (SnRK2.2/2.5/2.7/2.10),

all were differentially expressed in winter and/or spring rapeseed after cold stress [42]. Furthermore, in freezing-treated (−2°C) leaves of cold-tolerant winter rapeseed line 2016TS(G)10, one PYL gene and one ABI5-like gene were up-regulated, while 13 PP2C and 4 ABI5-like genes were down-regulated [25]. 72.8% ABA signaling genes (94/129) were regulated by chilling and/or freezing in both cold-tolerant and cold-sensitive rapeseed plants but they presented different expression profiles [11]. The ABA receptors *PYL5/7* genes were both induced by cold and freezing stress, while the *PYL1/9* genes were suppressed. The *PYR1/4* genes were only induced by chilling stress but not by freezing stress. The *PYL6* were induced by chilling stress and freezing stress following cold acclimation. ABA co-receptor *ABI1* and *HAB1* were suppressed by all low-temperature treatments, while *HAB2* was upregulated. The *SnRK2B* and *SnRK2D* were induced by all low-temperature treatments, while *SnRK2C* was suppressed. *SnRK2F* and one ABI5 were both induced only by freezing treatment. The *OST1* was induced only in cold-tolerant but not in cold-sensitive rapeseed [11]. While 13 ABI5-like genes have a complex expression pattern in response to different low-temperature conditions, including up-regulated, down-regulated and no changed. Exogenous application of ABA significantly improved the rapeseed seedlings freezing tolerance [11]. Overexpression of *BnaABI3* leads to improved embryo degreening following frost exposure and enhanced pod strength in rapeseed [55].

3.3 Ca^{2+} signaling

Calcium (Ca^{2+}) is an important second messenger of signal transduction in the plant stress responses, plant growth and development. Ca^{2+} signaling were detected and transmitted by calmodulin/calmodulin-like proteins (CaM/CML), calcium-dependent protein kinase (CDPK) and calcineurin B-like proteins (CBLs) [56, 57]. The level of cytosolic Ca^{2+} was transiently increased in plants under cold stress [57–59]. In rice, COLD1 interacts with the G-protein α subunit and activates the Ca^{2+} channel, results the increment of expression of CBF under low-temperature stress [60]. In *Arabidopsis*, overexpression of CaM3 repressed the expression of *COR* genes (*RD29A*, *KIN1* and *KIN2*) [61]. CaM-binding transcription activators (CAMTAs) bind to the promotor of CBF2, promoting CBF2 expression and plant freezing tolerance [62, 63].

88 of 129 *CaM/CML* genes were regulated by cold stress in semi-winter rapeseed cultivar ZS11 [64]. 91 of 129 *CaM/CML* were regulated by chilling and/or freezing stress in semi-winter early-maturing rapeseed, and most of them were strongly induced by freezing stress [11, 64]. Additionally, 22 rapeseed CDPK genes were up-regulated by freezing treatments [11]. Furthermore, there were 91 genes involved in Ca^{2+} signaling (35 *CDPK*, 38 *CaM*, 16 *CIPK* and 2 Ca^{2+}-ATPase) were differentially expressed in winter rapeseed after cold stress, while 79 genes (44 *CDPK*, 19 *CaM*, 15 *CIPK* and 1 Ca^{2+}-ATPase) were differentially expressed in spring rapeseed after cold stress [42]. In Zhang's report, there were 5 CBL, 7 CIPK and 7 CDPK genes were regulated by cold treatment [65, 66].

4. Future directions

Rapeseed is one of the most important oil crops in the world and China and is affected by chilling and freezing stress. In recently years, several studies have tried to identify the main signaling pathways and genes responsible for low-temperature stress (chilling and/or freezing; cold acclimation and/or cold shock) in different rapeseeds (winter, semi-winter and spring type; cold-sensitive and cold-tolerant;

late maturing and early maturing) based on transcriptomics, metabolomics, lipidomics, and QTL analyses [11, 25, 41, 42, 45, 67–69]. Tough there were so many candidate genes involved in the response to low-temperature stress have been identified, only few genes' functions in cold tolerant have been tested and verified in rapeseed [10, 29, 55, 70, 71]. It is a pity that constitutive overexpression of rapeseed BnCBF5 and BnCBF17 resulted in various degrees of dwarf habit and longer time to flower, tough which resulted in increased freezing tolerance remarkably in spring rapeseed "Westar" [29]. There is still much work to be performed to understand rapeseed plants' responses to low-temperature stress and breed cold-tolerant rapeseed.

Genome editing is an efficient approach for crop improvement either by loss or gain of gene function and several different strategies have been developed [72]. Tough there were a few studies using CRISPR/Cas9 system for editing genes associated with plant/pod development, fatty acid synthesis and biotic stress response [72], no application of CRISPR-Cas9 for editing genes involved in chilling and freezing tolerant in rapeseed. It is expected that the newly emerging genome editing system will make a contribution to future gene function research and molecular design breeding in cold-tolerant rapeseed.

Author details

Xin He[1,2,3]

1 Southern Regional Collaborative Innovation Center for Grain and Oil Crops in China, Hunan Agricultural University, Changsha, Hunan, China

2 Oil Crops Research, Hunan Agricultural University, Changsha, Hunan, China

3 Hunan Branch of National Oilseed Crops Improvement Center, Changsha, Hunan, China

*Address all correspondence to: hexinzhsh@126.com

References

[1] Liu J, Shi Y, Yang S. Insights into the regulation of C-repeat binding factors in plant cold signaling. Journal of Integrative Plant Biology. 2018;**60**(9):780-795

[2] Ding Y, Shi Y, Yang S. Advances and challenges in uncovering cold tolerance regulatory mechanisms in plants. The New Phytologist. 2019;**222**(4):1690-1704

[3] Lu K, Wei L, Li X, Wang Y, Wu J, Liu M, et al. Whole-genome resequencing reveals *Brassica napus* origin and genetic loci involved in its improvement. Nature Communications. 2019;**10**(1):1154

[4] Chalhoub B, Denoeud F, Liu S, Parkin IA, Tang H, Wang X, et al. Plant genetics. Early allopolyploid evolution in the post-Neolithic *Brassica napus* oilseed genome. Science. 2014;**345**(6199):950-953

[5] Wei D, Cui Y, He Y, Xiong Q, Qian L, Tong C, et al. A genome-wide survey with different rapeseed ecotypes uncovers footprints of domestication and breeding. Journal of Experimental Botany. 2017;**68**(17):4791-4801

[6] O'Neill CM, Lu X, Calderwood A, Tudor EH, Robinson P, Wells R, et al. Vernalization and floral transition in autumn drive winter annual life history in oilseed rape. Current Biology: CB. 2019;**29**(24):4300-4306 e4302

[7] Wu D, Liang Z, Yan T, Xu Y, Xuan L, Tang J, et al. Whole-genome resequencing of a worldwide collection of rapeseed accessions reveals the genetic basis of ecotype divergence. Molecular Plant. 2019;**12**(1):30-43

[8] Lei Y, Tariq S, Yong C, Yan LÜ, Xue-Kun Z, Xi-ling Z. Physiological and molecular responses to cold stress in rapeseed (*Brassica napus* L.). Journal of Integrative Agriculture. 2019;**12**:60345-60347

[9] Tian Z, Ji YH, Sun LX, Xu XL, Fan DL, Zhong HL, et al. Changes in production potentials of rapeseed in the Yangtze River Basin of China under climate change: A multi-model ensemble approach. Journal of Geographical Sciences. 2018;**28**(11):1700-1714

[10] Huang Y, Hussain MA, Luo D, Xu H, Zeng C, Havlickova L, et al. A *Brassica napus* reductase gene dissected by associative transcriptomics enhances plant adaption to freezing stress. Frontiers in Plant Science. 2020;**11**:971

[11] Xin H, Xianchao N, Pan X, Wei L, Min Y, Yu K, et al. Comparative transcriptome analyses revealed conserved and novel responses to cold and freezing stress in *Brassica napus* L. G3. 2019;**9**(8):2723-2737

[12] Guy C. Molecular responses of plants to cold shock and cold acclimation. Journal of Molecular Microbiology and Biotechnology. 1999;**1**(2):231-242

[13] Shi Y, Ding Y, Yang S. Molecular regulation of CBF signaling in cold acclimation. Trends in Plant Science. 2018;**23**(7):623-637

[14] Kocsy G, Galiba G, Brunold C. Role of glutathione in adaptation and signalling during chilling and cold acclimation in plants. Physiologia Plantarum. 2001;**113**(2):158-164

[15] Sanghera GS, Wani SH, Hussain W, Singh NB. Engineering cold stress tolerance in crop plants. Current Genomics. 2011;**12**(1):30-43

[16] Luo T, Xian M, Zhang C, Zhang C, Hu L, Xu Z. Associating transcriptional regulation for rapid germination of

rapeseed (*Brassica napus* L.) under low temperature stress through weighted gene co-expression network analysis. Scientific Reports. 2019;**9**(1):55

[17] Zhang Y, Zhang D, Yu H, Lin B, Fu Y, Hua S. Floral initiation in response to planting date reveals the key role of floral meristem differentiation prior to budding in canola (*Brassica napus* L.). Frontiers in Plant Science. 2016;**7**:1369

[18] Zhang SJ, Li L, Zhang CL. Effects of sowing date and planting density on the seed yield and oil content of winter oilseed rape. The Journal of Applied Ecology. 2012;**23**(5):1326-1332

[19] Luo T, Zhang J, Khan MN, Liu JH, Xu ZH, Hu LY. Temperature variation caused by sowing dates significantly affects floral initiation and floral bud differentiation processes in rapeseed (*Brassica napus* L.). Plant Science. 2018;**271**:40-51

[20] Hua SJ, Lin BG, Hussain N, Zhang YF, Yu HS, Ren Y, et al. Delayed planting affects seed yield, biomass production, and carbohydrate allocation in canola (*Brassica napus*). International Journal of Agriculture and Biology. 2014;**16**(4):671-680

[21] Ozer H. Sowing date and nitrogen rate effects on growth, yield and yield components of two summer rapeseed cultivars. European Journal of Agronomy. 2003;**19**(3):453-463

[22] Xue-Kun Z, Chun-Lei Z, Xing L, Han-Zhong W. Investigation on 2008 low temperature and freeze injure on winter rape along Yangtze River. Chinese Journal of Oil Crop Sciences. 2008;**30**:122-126

[23] Tu Y-Q, X-L D. Effects of continuous low temperature overcast and rainy weather on yield and oil content of *Brassica napus* during flowering stage. Chinese Journal of Oil Crop Sciences. 2011;**33**:470-475

[24] Do Ru A, Cak Rlar H. Is leaf age a predictor for cold tolerance in winter oilseed rape plants? Functional Plant Biology: FPB. 2020;**47**(3):250-262

[25] Pu Y, Liu L, Wu J, Zhao Y, Bai J, Ma L, et al. Transcriptome profile analysis of winter rapeseed (*Brassica napus* L.) in response to freezing stress, reveal potentially connected events to freezing stress. International Journal of Molecular Sciences. 2019;**20**(11):2771

[26] Yan L, Shah T, Cheng Y, Lu Y, Zhang XK, Zou XL. Physiological and molecular responses to cold stress in rapeseed (*Brassica napus* L.). Journal of Integrative Agriculture. 2019;**18**(12):2742-2752

[27] Stefanowska M, Kuras M, Kacperska A. Low temperature-induced modifications in cell ultrastructure and localization of phenolics in winter oilseed rape (*Brassica napus* L. *var. oleifera* L.) leaves. Annals of Botany. 2002;**90**(5):637-645

[28] Stefanowska M, Kuraś M, Kubacka-Zebalska M, Kacperska A. Low temperature affects pattern of leaf growth and structure of cell walls in winter oilseed rape (*Brassica napus* L., *var. oleifera* L.). Annals of Botany. 1999;**84**(3):313-319

[29] Savitch LV, Allard G, Seki M, Robert LS, Tinker NA, Huner NP, et al. The effect of overexpression of two *Brassica* CBF/DREB1-like transcription factors on photosynthetic capacity and freezing tolerance in *Brassica napus*. Plant & Cell Physiology. 2005;**46**(9):1525-1539

[30] Xiao-Hong Z, Liang-Jie F, Te-Wu Y, Zheng-Hua X, Li-Yong H. Effects of chilling stress on physiological characteristics of rapeseed seedlings in winter. Plant Physiology. 2015;**51**:737-746

[31] Thomashow MF. PLANT COLD ACCLIMATION: Freezing tolerance

genes and regulatory mechanisms. Annual Review of Plant Physiology and Plant Molecular Biology. 1999;**50**:571-599

[32] Miura K, Furumoto T. Cold signaling and cold response in plants. International Journal of Molecular Sciences. 2013;**14**(3):5312-5337

[33] Ritonga FN, Chen S. Physiological and molecular mechanism involved in cold stress tolerance in plants. Plants. 2020;**9**(5):560

[34] Jaglo-Ottosen KR, Gilmour SJ, Zarka DG, Schabenberger O, Thomashow MF. *Arabidopsis* CBF1 overexpression induces COR genes and enhances freezing tolerance. Science. 1998;**280**(5360):104-106

[35] Liu Q, Kasuga M, Sakuma Y, Abe H, Miura S, Yamaguchi-Shinozaki K, et al. Two transcription factors, DREB1 and DREB2, with an EREBP/AP2 DNA binding domain separate two cellular signal transduction pathways in drought- and low-temperature-responsive gene expression, respectively, in *Arabidopsis*. The Plant Cell. 1998;**10**(8):1391-1406

[36] Jaglo KR, Kleff S, Amundsen KL, Zhang X, Haake V, Zhang JZ, et al. Components of the *Arabidopsis* C-repeat/dehydration-responsive element binding factor cold-response pathway are conserved in *Brassica napus* and other plant species. Plant Physiology. 2001;**127**(3):910-917

[37] Chinnusamy V, Zhu J, Zhu JK. Cold stress regulation of gene expression in plants. Trends in Plant Science. 2007;**12**(10):444-451

[38] Chinnusamy V, Ohta M, Kanrar S, Lee BH, Hong X, Agarwal M, et al. ICE1: A regulator of cold-induced transcriptome and freezing tolerance in *Arabidopsis*. Genes & Development. 2003;**17**(8):1043-1054

[39] Kurbidaeva A, Ezhova T, Novokreshchenova M. *Arabidopsis thaliana* ICE2 gene: Phylogeny, structural evolution and functional diversification from ICE1. Plant Science: An International Journal of Experimental Plant Biology. 2014;**229**:10-22

[40] Kim YS, Lee M, Lee JH, Lee HJ, Park CM. The unified ICE-CBF pathway provides a transcriptional feedback control of freezing tolerance during cold acclimation in *Arabidopsis*. Plant Molecular Biology. 2015;**89**(1-2):187-201

[41] Ke L, Lei W, Yang W, Wang J, Gao J, Cheng J, et al. Genome-wide identification of cold responsive transcription factors in *Brassica napus* L. BMC Plant Biology. 2020;**20**(1):62

[42] Jian H, Xie L, Wang Y, Cao Y, Wan M, Lv D, et al. Characterization of cold stress responses in different rapeseed ecotypes based on metabolomics and transcriptomics analyses. PeerJ. 2020;**8**:e8704

[43] Gao MJ, Allard G, Byass L, Flanagan AM, Singh J. Regulation and characterization of four CBF transcription factors from *Brassica napus*. Plant Molecular Biology. 2002;**49**(5):459-471

[44] Weretilnyk E, Orr W, White TC, Iu B, Singh J. Characterization of three related low-temperature-regulated cDNAs from winter *Brassica napus*. Plant Physiology. 1993;**101**(1):171-177

[45] Du C, Hu K, Xian S, Liu C, Fan J, Tu J, et al. Dynamic transcriptome analysis reveals AP2/ERF transcription factors responsible for cold stress in rapeseed (*Brassica napus* L.). Molecular Genetics and Genomics: MGG. 2016;**291**(3):1053-1067

[46] Chen L, Zhong H, Ren F, Guo QQ, Hu XP, Li XB. A novel cold-regulated gene, COR25, of *Brassica napus* is

involved in plant response and tolerance to cold stress. Plant Cell Reports. 2011;**30**(4):463-471

[47] Wang Z, Yang C, Chen H, Wang P, Wang P, Song C, et al. Multi-gene co-expression can improve comprehensive resistance to multiple abiotic stresses in *Brassica napus* L. Plant Science. 2018;**274**:410-419

[48] Vishwakarma K, Upadhyay N, Kumar N, Yadav G, Singh J, Mishra RK, et al. Abscisic acid signaling and abiotic stress tolerance in plants: A review on current knowledge and future prospects. Frontiers in Plant Science. 2017;**8**:161

[49] Fahad S, Nie L, Chen Y, Wu C, Xiong D, Saud S, et al. Crop plant hormones and environmental stress. In: Lichtfouse E, editor. Sustainable Agriculture Reviews. Vol. 15. Cham: Springer International Publishing; 2015. pp. 371-400

[50] Fahad S, Hussain S, Matloob A, Khan FA, Khaliq A, Saud S, et al. Phytohormones and plant responses to salinity stress: A review. Plant Growth Regulation. 2015;**75**(2):391-404

[51] Shinkawa R, Morishita A, Amikura K, Machida R, Murakawa H, Kuchitsu K, et al. Abscisic acid induced freezing tolerance in chilling-sensitive suspension cultures and seedlings of rice. BMC Research Notes. 2013;**6**(1):351

[52] Huang X, Shi H, Hu Z, Liu A, Amombo E, Chen L, et al. ABA is involved in regulation of cold stress response in Bermudagrass. Frontiers in Plant Science. 2017;**8**:1613

[53] Ding Y, Li H, Zhang X, Xie Q, Gong Z, Yang S. OST1 kinase modulates freezing tolerance by enhancing ICE1 stability in *Arabidopsis*. Developmental Cell. 2015;**32**(3):278-289

[54] Ding Y, Jia Y, Shi Y, Zhang X, Song C, Gong Z, et al. OST1-mediated BTF3L phosphorylation positively regulates CBFs during plant cold responses. The EMBO Journal. 2018;**37**(8):e98228

[55] Perkins M, Skori L, Hickerson NMN, Jamshed M, Samuel MA. Genetic manipulation of ABI3 confers frost-tolerant seed degreening in canola. Plant Biotechnology Journal. 2020;**18**(3):602-604

[56] McAinsh MR, Pittman JK. Shaping the calcium signature. The New Phytologist. 2009;**181**(2):275-294

[57] Dodd AN, Kudla J, Sanders D. The language of calcium signaling. Annual Review of Plant Biology. 2010;**61**:593-620

[58] Knight H, Trewavas AJ, Knight MR. Cold calcium signaling in *Arabidopsis* involves two cellular pools and a change in calcium signature after acclimation. Plant Cell. 1996;**8**(3):489-503

[59] Knight MR, Campbell AK, Smith SM, Trewavas AJ. Transgenic plant aequorin reports the effects of touch and cold-shock and elicitors on cytoplasmic calcium. Nature. 1991;**352**(6335):524-526

[60] Ma Y, Dai X, Xu Y, Luo W, Zheng X, Zeng D, et al. COLD1 confers chilling tolerance in rice. Cell. 2015;**160**(6):1209-1221

[61] Townley HE, Knight MR. Calmodulin as a potential negative regulator of *Arabidopsis* COR gene expression. Plant Physiology. 2002;**128**(4):1169

[62] Doherty CJ, Van Buskirk HA, Myers SJ, Thomashow MF. Roles for *Arabidopsis* CAMTA transcription factors in cold-regulated gene expression and freezing tolerance. Plant Cell. 2009;**21**(3):972-984

[63] Kidokoro S, Yoneda K, Takasaki H, Takahashi F, Shinozaki K, Yamaguchi-Shinozaki K. Different cold-signaling pathways function in the responses to rapid and gradual decreases in temperature. The Plant Cell. 2017;**29**(4):760

[64] He X, Liu W, Li W, Liu Y, Wang W, Pan Xie YK, et al. Genome-wide identification and expression analysis of CaM/CML genes in *Brassica napus* under abiotic stress. Journal of Plant Physiology. 2020:153251. DOI: 10.1016/j.jplph.2020.153251

[65] Zhang H, Yang B, Liu W-Z, Li H, Wang L, Wang B, et al. Identification and characterization of CBL and CIPK gene families in canola (*Brassica napus* L.). BMC Plant Biology. 2014;**14**:8-8

[66] Zhang H, Liu W-Z, Zhang Y, Deng M, Niu F, Yang B, et al. Identification, expression and interaction analyses of calcium-dependent protein kinase (CPK) genes in canola (*Brassica napus* L.). BMC Genomics. 2014;**15**:211-211

[67] Edrisi Maryan K, Samizadeh Lahiji H, Farrokhi N, Hasani Komeleh H. Analysis of *Brassica napus* dehydrins and their co-expression regulatory networks in relation to cold stress. Gene Expression Patterns: GEP. 2019;**31**:7-17

[68] Tasseva G, de Virville JD, Cantrel C, Moreau F, Zachowski A. Changes in the endoplasmic reticulum lipid properties in response to low temperature in *Brassica napus*. Plant Physiology and Biochemistry: PPB. 2004;**42**(10):811-822

[69] Huang Z, Zhao N, Qin M, Xu A. Mapping of quantitative trait loci related to cold resistance in *Brassica napus* L. Journal of Plant Physiology. 2018;**231**:147-154

[70] Nokhrina K, Ray H, Bock C, Georges F. Metabolomic shifts in *Brassica napus* lines with enhanced BnPLC2 expression impact their response to low temperature stress and plant pathogens. GM Crops & Food. 2014;**5**(2):120-131

[71] Dahal K, Gadapati W, Savitch LV, Singh J, Huner NP. Cold acclimation and BnCBF17-over-expression enhance photosynthetic performance and energy conversion efficiency during long-term growth of *Brassica napus* under elevated CO_2 conditions. Planta. 2012;**236**(5):1639-1652

[72] Lohani N, Jain D, Singh MB, Bhalla PL. Engineering multiple abiotic stress tolerance in canola, *Brassica napus*. Frontiers in Plant Science. 2020;**11**:3

Chapter 11

Effects of Salinity on Seed Germination and Early Seedling Stage

Cüneyt Uçarlı

Abstract

Salinity is the major environmental stress source that restricts on agricultural productivity and sustainability in arid and semiarid regions by a reduction in the germination rate and a delay in the initiation of germination and subsequent seedling establishment. Salt negatively effects the crop production worldwide. Because most of the cultivated plants are salt-sensitive glycophytes. Salt stress affects the seed germination and seedling establishment through osmotic stress, ion toxicity, and oxidative stress. Salinity may adversely influence seed germination by decreasing the amounts of seed germination stimulants such as GAs, enhancing ABA amounts, and altering membrane permeability and water behavior in the seed. Rapid seed germination and subsequent seedling establishment are important factors affecting crop production under salinity conditions. Seed priming is one of the useful physiological approaches for adaptation of glycophyte species to saline conditions during germination and subsequent seedling establishment. In seed priming, seeds are exposed to an eliciting solution for a certain period that allows partial hydration without radicle protrusion. Seed priming is a simple, low cost, and powerful biotechnological tool used to overcome the salinity problem in agricultural lands.

Keywords: salinity, germination, glycophyte, halophyte, seed priming, plant hormones

1. Introduction

Seed dormancy and germination are distinct physiological processes, and the transition from dormancy to germination is not only a critical developmental step in the life cycle of higher plants but also determines the failure or success of the subsequent seedling establishment and plant growth [1]. Seed germination begins with the water uptake of dry seed (imbibition) and ends with radicle protrusion. Seed germination is affected by adverse environmental conditions including salinity, high temperature, and drought [2].

It is estimated that about approximately 7% of world land is affected by salinity and approximately 20% of 230 million ha irrigated land is salt-affected [3]. This number could be increased in the future due to increased land salinization as a consequence of contaminated artificial irrigation, climate change, and unsuitable land management. Salinity is a major stress responsible for the inhibition of seed germination or reduction in germination percentage and a delay in germination time in crops. At present, around 30 crop plants provide 90% of plant-based human food and the majority of these crops are not salt tolerant, even salt-sensitive, called glycophytes [4]. There have been high yield losses in these crops under moderate salinity (EC 4–8 dS m^{-1}, approximately 40–80 mM NaCl) [5].

High salinity leads a decrease in osmotic potential of ambient soil water, resulting with a decrease in water uptake by dry seeds (imbibition). Besides, the absorption of excess Na^+ and Cl^- ions from soils creates ionic stress and cause toxicity which contributing to disruption in biochemical processes including nucleic and protein metabolism, energy production, and respiration [6]. Salinity also damages the nutrient and hormone balances, especially gibberellin (GA)/abscisic acid (ABA), during germination. As a result, high salinity level causes a delay in germination, even inhibition of seed germination depending on salt tolerance of plants. Dynamic balance between the generation and scavenging of reactive oxygen species (ROS) such as hydroxyl radicals, superoxide, and hydrogen peroxide could be disturbed by high salinity stress. ROS damage the macromolecules including proteins, carbohydrates, nucleic acids, and lipids, or cellular structures like membranes, resulting with inhibition of seed germination [7].

Germination has been found to be under strict regulation of plant hormones, especially GA and ABA [8]. ABA promotes seed dormancy and inhibits germination of seed, whereas GAs release dormancy and stimulate germination. Plant hormones ethylene (ET), and brassinosteroids (BRs) also have positive effect on seed germination by controlling the inhibitory effects of ABA on germination and rupturing testa and endosperm [9, 10]. The plant hormones widely took part in determining the physiological state of a seed and regulating the germination process by interacting each other [11]. Hormones are regulated by distinct transcription factors and signaling components including NO and H_2O_2, showing the complexity of seed germination regulation. While some plant genes control the activity of plant hormones, and the other plant genes are activated by plant hormones [10]. Signaling molecules, such as NO and H_2O_2, also promotes germination and reduce the dormancy by enhancing ABA catabolism and GA biosynthesis [12].

Rapid seed germination and subsequent seedling establishment are important factors determining crop production and yield under salinity stress. One of the useful physiological approaches for glycophytes to adapt saline condition is seed priming [7]. Seed priming is an easy, low cost and low risk technique. The seeds are hydrated in specific solutions including plant hormones (GA3, ET, auxins, kinetin), antioxidant compounds (ascorbic acid, glutathione, tocopherone) organic solutes (proline, glycine betaine), inorganic salts (KNO_3, $CaCl_2$, and KCl), and particular bacteria and fungi species for a certain time to allow metabolic process of germination, followed by drying the seed to inhibit occurring of radicle protrusion [13].

2. Soil salinity and salinity stress

Plants, being sessile nature, are simultaneously subjected to various adverse conditions including salinity, drought, cold, heat, excess water, and heavy metals, which limit their development and growth. Salinity is the major environmental stress source that restricts on agricultural productivity and sustainability in arid and semiarid regions [14]. Salinity is a global issue that affects about 7% of the world's total land area, including 20% total cultivated lands and 33% of irrigated land, causing estimated yield losses of 20% worldwide [15, 16]. Besides, it is estimated that every year 10 million ha of agricultural land destroyed by salinized soil [17]. This rate can be increased by global climate change, use of contaminated irrigation water, intensive farming and poor drainage [18–56]. Without proper and sustainable control, salinity-affected areas will increase to more than 50% of the world's total arable land by 2050 [15]. This rate can be accelerated by increase in sea water level by climate change, excessive use of groundwater for irrigation, increasing use of low-quality water for irrigation and massive introduction of irrigation associated with intensive farming and poor drainage [57].

Soil salinity is a measure of the concentration of all the soluble salts in soil water, and is usually expressed as electrical conductivity (EC) of the saturation extract (ECe) with units of deci siemens per meter (1 dS m^{-1}) [58]. The soils were classified as saline, sodic or saline-sodic based on the total concentration of salt and the ratio of Na^+ to Ca^{2+} and Mg^{2+} in the saturated extract of the soil [59]. When the ECe exceeds 4 dS m^{-1} (approximately 40 mM/L NaCl) and exchangeable sodium percentage is less than 15 with sodium adsorption ratio (SAR) < 13, the soil is saline. The major problem with saline soils is the presence of soluble salts, primarily Cl^-, SO_4^{2-}, and sometimes NO_3^-. The pH of saline soils is usually below 8.5. Sodic (alkali) soils have an ECe < 4 dS m^{-1}, ESP > 15, and SAR > 13. Therefore, Na^+ is the major problem in these soils. Sodic soils have a pH between 8.5 and 10. Saline-sodic soils have an ECe > 4 dS m^{-1}, SAR > 13, and an ESP > 15. Thus, both soluble salts and exchangeable Na^+ are high in these soils. Saline-sodic soils have similar salt and pH levels as saline soils. USSL Staff [59] has described the general relationship of ECe and plant growth as the following:

- non-saline (ECe ≤ 2 dS m^{-1}): salinity effects mostly negligible;
- very slightly saline (ECe = 2–4 dS m^{-1}): yields of very sensitive crops may be restricted;
- slightly saline (ECe = 4–8 dS m^{-1}): yields of many crops are restricted;
- moderately saline (ECe = 8–16 dS m^{-1}): only salt tolerant crops yield satisfactorily; and
- strongly saline (ECe ≥ 16 dS m^{-1}): only a few very salt tolerant crops yield satisfactorily.

3. Seed germination

Seed germination is a complex multi-stage developmental process and regulated by internal and external factors. Internal factors include proteins, plant hormones (gibberellins/ABA balance, ethylene, and auxin), chromatin-related factors such as methylation, acetylation, histone ubiquitination, related genes (maturating genes and hormonal and epigenetics-regulating genes), non-enzymatic processes, seed age, seed size, and structural components of seed including (endosperm and seed coat). Besides, external factors containing moisture, light, salinity, temperature, acidity, and nutrient also affect the seed germination [60, 61].

Seed germination begins with imbibition, the uptake of water by the dry mature seed, and ends with visible protrusion of radicle through testa [62]. Successful germination requires optimum environmental conditions, including water, oxygen, and temperature to initiate this process. Germination/sprouting is regulated by plant hormones such as gibberellic acid (GA), abscisic acid (ABA), ethylene, auxins, cytokinins, and brassinosteroids [63]. Among them, ABA and GA are two important regulators, which play antagonistic roles in seed dormancy and germination [64].

The process of seed germination can be divided into three phases (**Figure 1**) [65]. Phase I begins with imbibition of dry seeds and ends with the early plateau phase of water uptake. Phase II includes reactivation of metabolisms, significant induction of hormonal and enzyme activity using surviving structures and components in the desiccated cells, genes involved in amino acid and nucleic acid synthesis, restarting of cellular respiration with genesis of mitochondria, mobilization of reserved, RNA and protein synthesis machinery [66, 67]. Phase III is post-germination stage involves establishment of seedling and the induction of genes for photosynthetic metabolism after radicle cells elongate and divide [68].

Gibberellins and ABA are two key phytohormones regulating seed germination and seedling growth [69]. While GA breaks dormancy and enhances the seed germination and seedling, ABA inhibits germination and enhances seed dormancy [10]. However, the ratio of the two hormones, rather than the absolute level of each hormone, plays a key role in regulating the breaking of seed dormancy and the onset of germination [70]. GA/ABA balance determines fate of the seed; germination or dormancy. Gibberellins induce the synthesis and production of α-amylase, proteases, and β-glucanases, resulting in the germination of seeds [71]. GAs also stimulate the genes involved in weakening of endosperm and expansion of embryo cell [10]. On the other hand, ABA suppresses expression of many hydrolytic enzyme genes to prevent viviparous germination and inhibits promoting effect of GA on radicle growth and embryo expansion by inhibiting water uptake and hence cell-wall loosening, which is a key step to start germination [72].

Ethylene is a gaseous hormone involved in various processes, including positive regulation of seed germination. Ethylene breaks the primary and secondary dormancy and promotes seed germination by reducing ABA levels or sensitivity [73]. Brassinosteroids (BRs) and auxin induce the secretion of ethylene which works in conjunction with GAs to induce germination [10]. Auxins reduce seed sensitivity to ABA by overexpressing microRNAs and interacting with GAs to counteract ABA suppression during germination [74, 75].

Low temperature decreases seed dormancy and enhances germination in many species, while high temperature has the negative effect on germination and induces secondary dormancy [70]. High temperature down-regulates the genes involved in synthesis of GA synthesis and deactivation of ABA, whereas genes involved in ABA synthesis are up-regulated by high temperature. Therefore, transcriptional changes in ABA and GA metabolism and signal pathways results with inhibition of germination or a delay in germination [76]. Light has been considered both to stimulate

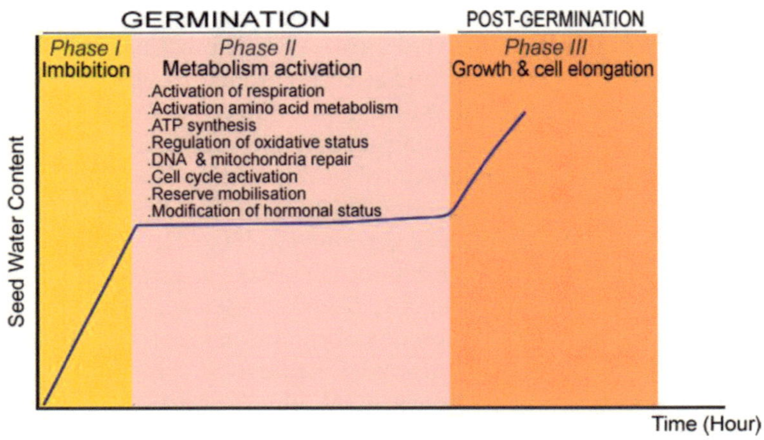

Figure 1.
Major events associated with germination and subsequent post-germinative growth (based on [13, 65]).

germination and to terminate dormancy by increasing the expression of GA anabolic genes, GA3ox1 and GA3ox2, and repressing expression of GA catabolism gene GA2ox2 [77].

In addition to phytohormones, several signal molecules, including as nitric oxide (NO) and reactive oxygen species (ROS), also regulate seed dormancy and germination [68]. ROS is an important regulator during seed germination because of the interaction with lipids, DNA, and protein molecules, as well as phytohormones including ABA and GA in the cell [78]. The biochemical and cellular reactions stimulated by water uptake are accompanied by the generation of ROS [79]. Hydrogen peroxide (H_2O_2) serves as a signaling hub for the regulation of seed dormancy and germination; the accurate regulation of H_2O_2 accumulation by the cell antioxidant mechanism is important to achieve a balance between oxidative signaling that enhances germination and oxidative damage that inhibits germination or delays in germination time [80]. N compounds, including NO, promotes seed germination through increasing amylase activities, adjusting K^+/Na^+ balance, and enhancing seed respiration and ATP production [81].

4. Effect of salinity on seed germination and early seedling stage

Salinity affects seed germination process through osmotic stress, ion-specific effects and oxidative stress, shown by decreasing germination rate and extended germination time [82]. Salinity increases external osmotic potential that reduces water uptake during imbibition [83]. Salinity may affect the germination of seeds by the toxic effects of excess sodium and chloride ions on embryo viability [84, 85]. The toxic effects include disruption to the structure of enzymes and other macromolecules, damage to cell organelles and the plasma membrane, the disruption of respiration, photosynthesis and protein synthesis [85–87].

In general, seed germination progresses in three phases under normal conditions. Seed germination begins with the rapid water uptake by dry seed (imbibition) (Phase I). A plateau phase, known as phase II, follows this phase. The cellular metabolisms are reactivated, and water uptake is restricted in phase II. This is followed by phase III, a post-germination phase, which is characterized by continuous water uptake until germination is complete (**Figure 1**). Based on these three phases, the inhibition of seed germination or delaying in germination time under salinity stress may be generally ascribed to osmotic stress in the phase I and ionic stress in the phase II. Osmotic stress and ionic stress interact together to inhibit or delay germination of seed during the phase III [88].

Salinity may adversely influence seed germination by decreasing the amounts of seed germination stimulants such as GAs, enhancing ABA amounts, and altering membrane permeability and water behavior in the seed [89]. In higher plants, salinity has been demonstrated to change expression profiles of the genes encoding GA metabolic enzymes, including copalyl diphosphate synthase (CPS), ent-kaurene synthase (KS), ent-kaurene oxidase (KO), ent-kaurenoic acid oxidase (KOA), GA 20-oxidase (GA20ox), GA 3-oxidase (GA3ox) and GA 2-oxidase (GA2ox), resulting with change in endogenous GA levels during germination [12].

The germination of seeds is characterized by transcriptional induction of hydrolytic enzymes such as α-amylase [90]. The α-amylase is excreted into the endosperm to break the stored starch to metabolizable sugars that provide ready energy and nutrients for the growing embryo and radicle. Salinity stress may have much effect on delayed germination time than on final germination percentage for most crops. A delay of water uptake and a decrease in the activity of α-amylase with an increase in the concentration of NaCl may be main reasons for delaying of the germination time [91].

The decrease in the α-amylase activity have been reported to be higher in the salt-sensitive genotypes than in the salt-tolerant genotypes. This reduction in the α-amylase activity results with a significant reduction in the translocation of sugars, essential for the developing embryo. Besides, decreasing sugar concentrations also change the osmotic potential of growing cells, resulting in a decrease in water uptake [88].

Both osmotic and ionic effects of salt stress leads to generation of excess reactive oxygen species (ROS) and oxidative damage, which disrupts proteins, lipids, and nucleic acids or the cellular structure including lipid membrane [83].

Plants can be divided into two main groups based on their response to saline stress; salt-tolerant halophytes and salt-sensitive glycophytes (non-halophytes) [6]. The halophytes are plants that are able to grow in the presence of high salt concentrations that generate a low water potential of the soil and kill 99% of other

Plant species	*Maximum salt tolerance	Salt tolerance type	Reference
Salicornia herbacea	1.7 M NaCl	Halophyte	[93]
Suaeda aralocapsica	1.5 M NaCl	Halophyte	[94]
Limonium vulgare	1.5 M NaCl	Halophyte	[95]
Sarcocornia perennis	1.3 M NaCl	Halophyte	[96]
Haloxylon ammodendron	1.3 M NaCl	Halophyte	[97]
Kochia scoparia	1.0 M NaCl	Halophyte	[98]
Kochia prostrata	0.85 M NaCl	Halophyte	[99]
Haloxylon salicornicum	0.8 M NaCl	Halophyte	[100]
Prosopis juliflora	0.6 M NaCl	Halophyte	[100]
Limonium mansanetianum	0.5 M NaCl	Halophyte	[101]
Limonium stocksi	0.4 M NaCl	Halophyte	[102]
Limonium lilacinum	0.3 M NaCl	Halophyte	[103]
Tanacetum cinerariifolium	0.26 M NaCl	Halophyte	[104]
Quinoa (*Chenopodium quinoa* Willd.)	0.3 M NaCl	Halophyte	[105]
Barley (*Hordeum vulgare* L.).	0.25 M NaCl	Glycophyte	[106]
Maize (Zea mays)	0.24 M NaCl	Glycophyte	[107]
Chicory (*Cichorium intybus* L.)	0.21 M NaCl	Glycophyte	[108]
Lentil (*Lens culinaris* Medik.)	0.2 M NaCl	Glycophyte	[14]
Brassica napus	0.2 M NaCl	Glycophyte	[109]
Peanut (*Arachis hypogaea*)	0.2 M NaCl	Glycophyte	[110]
Rice (*Oryza sativa*)	0.16 M NaCl	Glycophyte	[111]
Fig (*Ficus carica* L.)	0.17 M NaCl	Glycophyte	[112]
Button grass (*Dactyloctenium radulans*)	0.1 M NaCl	Glycophyte	[113]
Sorghum (*Sorghum bicolor* Moench)	0.1 M NaCl	Glycophyte	[114]
Ryegrass (*Lolium rigidum*)	0.1 M NaCl	Glycophyte	[115]
Chickpea (*Cicer arietinum* L.)	0.09 M NaCl	Glycophyte	[116]
Tomato (*Solanum lycopersicum*)	0.05 M NaCl	Glycophyte	[117]

**Maximum NaCl concentration at which seed germination percentage reduced to 10–20%.*

Table 1.
Maximum salt tolerance of halophytes and glycophytes at the germination stage.

species. They are adapted to survive and complete their life cycle under saline levels of higher than 200 mM NaCl. However, seed germination was also affected under salt stress and germination percentage was reduced to less than 10% under 1.7 M NaCl [92, 93]. In halophytes, maximum salt tolerance for seed germination has been reported to vary from 1.7 to 0.26 M NaCl depending on halophyte species and other environment conditions such as temperature, moisture, and light (**Table 1**).

A majority of the common crops, such as tomato, bean, rice, corn, etc., are salinity sensitive or even hypersensitive and they are described as glycophytes [5]. The glycophytes contain 99% of the world's flora and are susceptible to even low levels of salinity (ECe < 4 dS m^{-1}, approximately 40 mM NaCl) [92]. Under conditions of moderate salinity (EC 4–8 dS m^{-1}), all important glycophytic crops reduce average yields by 50–80% [118]. Seed germination in glycophytes is severely inhibited under salinity due to both osmotic stress and ionic toxicity stress, while halophytes are less affected by osmotic stress during germination [12].

5. Alleviation salinity stress on germination by seed priming

Most crops are highly susceptible to saline soil, even when soil has electrical conductivity (ECe) as low as 3 dS m^{-1} [119]. Therefore, salinity stress appears to be a major limitation factor for crop productivity. Seed germination and seedling establishments are the two critical stages in plant growth. These stages are the most sensitive to environmental conditions including salinity [120]. Plants are usually seeded within the top layer of the soil which is more saline than lower layers [121]. Salinity stress may delay or prevent germination of germination of high quality seeds, resulting with crop loss. Rapid seed germination and subsequent seedling establishment are important factors affecting crop production under salinity conditions. Therefore, to decrease the negative effects of salinity stress on seed germination, it is important to know to what extent the genotypic variation in the water uptake pattern during these phases is associated with the salt tolerance of genotypes at the germination stage.

Seed priming is one of the useful physiological approaches for adaptation of glycophyte species to saline conditions during germination and subsequent seedling establishment. Seed priming is a simple, low cost and powerful biotechnological tool used to overcome the salinity problem by promoting seed germination and seedling establishment in agricultural lands [122]. Seed are exposed to an eliciting solution for a constant period that allows partial hydration, but radicle emergence does not occur by re-drying of seed. Seed germination occurs three distinct phases: (i) imbibition, (ii) lag phase (reactivation of metabolisms) and (iii) protrusion of the radicle through the testa. The goal of seed priming is to extend the lag phase, which allows pre-germinative physiological and biochemical processes, but prevent the seed transition towards full germination [123]. Enhanced and uniformed germination of primed seeds occurs by reduction in the lag time of imbibition, activation of enzyme involved in seed germination, initiation of biochemical mechanisms of cell repair, increase in the RNA content and DNA replication, decrease in ROS and lipid peroxidation with increased activity of antioxidant enzymes including as superoxide dismutase, catalase, and glutathione reductase, and increase in osmotic adjustment and starch metabolism [124, 125].

Several methods of seed priming have been developed in order to revive seeds under salt stress conditions. Some of these methods are hydro-priming, osmopriming, solid matrix priming, hormonal-priming, bio-priming, chemical priming, and nutripriming [13]. In recent years, many studies have been reported to exhibit the positive effects of seed priming on germination under salinity conditions in many crops (**Table 2**).

Hydro-priming is the simplest and one of the mostly used seed priming method. Hydro-priming depends on seed soaking in pure water without chemical substances for 6–24 h and re-drying to original moisture content prior to sowing without emergence of radicle [144]. This method is a low-cost and environmentally friendly due to no use of additional chemicals. The uncontrolled water uptake by seeds is major disadvantage of this technique. Rapid hydration may cause leakage of

Plant	Treatment	Alleviating effect	Reference
Barley (*Hordeum vulgare* cv. Bülbül 89)	Priming with aqueous solution of 30 μM H_2O_2 for 24 h at room temperature	H_2O_2 increased the germination index from 16.71 to 25.07%, and from 8.19 to 14.65% under 250 mM and 300 mM NaCl, respectively	[126]
Tomato (*Solanum lycopersicum* cv. Hezuo 903)	Priming with 100 μM Epigallocatechin-3-Gallate (EGCG) at 28 ± 3°C	EGCG increased germination rate and index from 84.7 to 97.0%, and from 29.4 to 35.2%, respectively	[127]
Wheat (*Triticum aestivum* cv. Chamran)	Priming with 0.5 mM spermidine for 24 h, 25 mM proline for 2 days, or 1.5 mM silicon (K_2SiO_3) for 6 h	Spermidine, proline, and K_2SiO_3 enhanced the germination rate by 32, 18, and 17%, respectively, under salinity stress (20 dS m^{-1})	[128]
Zea mays, *Pisum sativum*, *Lathyrus sativus*	Priming with 0.2 g/L GA3 solution for 12 h at room temperature without light.	GA3 enhanced germination percentage from 16.67, 26.67, and 50 to 60, 73.3, and 86.67% in *Z. mays*, *P. sativum*, and *L. sativus*, respectively, and resulted in 20% reduction in mean germination time under salinity stress (12 dS m^{-1})	[129]
Pakchoi (*Brassica chinensis* L. cv Tiancuiqing)	Priming with sodium nitroprusside (SNP) for 2 h in dark at 25 ± 1°C	Germination potential, germination index, and vitality index were increased by 7.67%, 14.20% and 74.51% after 10 μM SNP pre-treatment under 100 mM NaCl	[130]
Melilotus officinalis	Soaking with 10 mM Ca^{2+}	Ca^{2+} significantly increased the germination percentage and recovery germination percentage under 200 mM NaCl	[131]
Melon (*Cucumis melo*)	Priming with 10–50 μM melatonin for 6 h	Melatonin increase the germination percentage from 50 to 80% under salinity stress (14 dS m^{-1})	[132]
Wheat (*Triticum aestivum* cv. Khirman)	Priming with 50 mg L^{-1} ascorbate, 50 mM proline, 25 μM triacontanol, or 100 μM indole acetic acid for 12 h	Priming treatments significantly enhanced germination index and final germination percentage, and reduced mean germination time under salinity stress (12 dS m^{-1})	[133]
Grain sorghum (*Sorghum bicolor* Moench)	Priming with 100–500 mg L^{-1} nano-iron oxide (n-Fe_2O_3) for 10 h and soaking with 10 mg L^{-1} n-Fe_2O_3 for 3 days	Treatments improved the speed and percent of germination under 150 mM NaCl	[134]

Plant	Treatment	Alleviating effect	Reference
Lentil (*Lens culinaris* cv. Ncir)	Soaking with 0.5 mM salicylic acid or 0.1 mM H_2O_2 at 25°C in the dark	Salicylic acid and H_2O_2 enhanced the germination percentage from 71 to 86 and 87%, respectively	[135]
Limonium bicolor	Priming with 80 μM salicylic acid (SA)	SA significantly increased germination rate, germination potential, and germination index of the seeds under 200 mM NaCl	[136]
Sweet sorghum (*Sorghum bicolor* cv. Chuntian 1)	Priming with 288 μM Gibberellin (GA3) for 32–48 h	GA3 significantly increased the water uptake, resulting with increased cumulative germination percentage and germination index under 100 mM NaCl	[122]
Maize (*Zea mays*)	Priming with 2 mM silicon (K_2SiO_3) for 7 days at 25°C in the dark	Silicon significantly enhanced the germination rate and percentage, as well as vitality index under 90 mM NaCl	[137]
Oat (*Avena sativa* cv. NDO-2)	Priming with 150 ppm gibberellin (GA3) for 24 h	GA3 enhanced the germination percentage from 56.64 to 76.03% under 100 mM NaCl	[138]
Cucumber (*Cucumis sativus* cv. Jinyou 1)	Priming with 0.3 mM silicon (NaSi) for 36 h	Silicon enhanced the germination percentage and index, and seedling vigor index under 200 mM NaCl	[139]
Limonium bicolor	Priming with 200 μM melatonin	Melatonin significantly increased germination rate, potential and index under 200 mM NaCl	[140].
Ceratoides lanata	Priming with 10 mM ethephon, 5 μM fusicoccin or 50 μM kinetin	Fusicoccin, kinetin, and ethephon increased the germination percentage from 10 to 40, 50, and 84%, respectively under 900 mM NaCl	[141]
Leymus chinensis cv. Jisheng 3	Priming with 200 μM gibberellins (GA4 + 7), 200 μM fluridone (FLU), 200 μM cytokinin (CK), 100 μM sodium nitroprusside (SNP), or 100 μM thiourea (TH) in the dark or light	GA and FLU significantly increased the germination percentage from 7 to 23 and 59% in the light, respectively, while SNP, CK and TH increased the germination percentage from 9 to 54, 55, and 30%, respectively, in the dark under 200 mM NaCl	[142]
Salicornia ramosissima	Inoculation with *Bacillus aryabhattai* SP1016-20	Inoculation with *B. aryabhattai* enhanced the final germination percentage and mean daily germination from 21.3 to 46.7%, and from 1.6 to 4.5%, respectively, under 510 mM NaCl	[143]

Table 2.
The functions of seed priming in plant at the germination stage under salinity condition.

essential nutrients out of the seed during germination, resulting in seed damage in some species [145].

Osmo-priming, also known as osmotic conditioning, involves soaking seeds in aerated low water potential solution including sugar, polyethylene glycol (PEG), glycerol, sorbitol, or mannitol with low water potential instead of pure water, followed by air drying before sowing. Due to low water potential of osmotic solutions, water is absorbed slowly by dry seed, which allows gradual seed imbibition [146]. While osmo-priming promotes activation of early phases of germination, inhibiting radicle emergence. Osmo-priming improves seed germination and enhances general crop performance under salt conditions. Water potential of osmotic agent is critical factor since main purpose is to restrict oxidative damage caused by ROS by inhibiting excess water from entering [147]. If inorganic salts such as NaCl, KCl, KNO_3, K_3PO_4, $MgSO_4$, and $CaCl_2$ are used as an osmo-priming agent, the method is generally referred as halopriming.

In hormonal priming, seed imbibition occurs in the presence of plan hormones such as GA3, ethylene, auxins, and salicylic acid, which can gave effect on seed metabolism. Chemical priming is a promising seed priming technique to enhance germination under high salinity stress. Seeds were pre-treated with different chemical solutions used as priming agents. Chemical agents includes a wide range of both natural and synthetic compounds such as antioxidants (ascorbic acid, glutathione, tocopherol, and melatonin), sodium hydrosulfide, polyamines hydrogen peroxide, sodium nitroprusside, urea, selenium, chitosan, fungicide, etc. [13].

Biopriming involves seed imbibition together with particular bacteria or fungi. These microorganisms are able to create endophytic connections with the plant. As other priming method, this treatment increases rate and uniformity of germination under salt conditions, as well as protects seeds against the soil and seed-borne pathogens [147]. The most frequently used biopriming species are *Bacillus* spp., *Enterobacter* spp., *Pseudomonas* spp., and *Trichoderma* spp. [148].

Seed priming efficiency is influence by many factors and strongly depends on treated plant species and chosen priming technique. Physical and chemical factors including osmotica and water potential, priming agent, duration, temperature, presence or absence of light, aeration, and seed condition also influence priming success and determine germination rate and time, seedling vigor, and further plant development [13, 144].

Conflict of interest

No conflict of interest.

Author details

Cüneyt Uçarlı
Department of Molecular Biology and Genetics, Istanbul University, Istanbul, Turkey

*Address all correspondence to: ucarlicu@istanbul.edu.tr

References

[1] Liu Y, Ye N, Liu R, Chen M, Zhang J. H_2O_2 mediates the regulation of ABA catabolism and GA biosynthesis in Arabidopsis seed dormancy and germination. Journal of Experimental Botany. 2010;**61**:2979-2990

[2] Wahid A, Perveena M, Gelania S, Basra SMA. Pretreatment of seed with H_2O_2 improves salt tolerance of wheat seedlings by alleviation of oxidative damage and expression of stress proteins. Journal of Plant Physiology. 2007;**164**:283-294

[3] Parihar P, Singh S, Singh R, Singh VP, Prasad SM. Effect of salinity stress on plants and its tolerance strategies: A review. Environmental Science and Pollution Research International. 2014;**22**:4056-4075

[4] Zörb C, Geilfus CM, Dietz KJ. Salinity and crop yield. Plant Biology. 2019;**21**:31-38

[5] Koyro HOW, Lieth H, Said S. Salt tolerance of *Chenopodium quinoa* Willd, grains of the Andes, influence of salinity on biomass production, yield, composition of reserves in the seeds, water and solute. In: Lieth H, Garcia Sucre M, Herzog B, editors. Mangroves and Halophytes, Restoration and Utilisation III. Tasks for Vegetation Sciences. Vol. 43. Dordrecht: Springer Netherlands; 2008. pp. 133-145

[6] Mwando E, Han Y, Angessa TT, Zhou G, Hill CB, Zhang XQ, et al. Genome-wide association study of salinity tolerance during germination in barley (*Hordeum vulgare* L.). Frontiers in Plant Science. 2020;**11**:118

[7] Ibrahim EA. Seed priming to alleviate salinity stress in germinating seeds. Journal of Plant Physiology. 2016;**192**:38-46

[8] Han C, Yang PF. Studies on the molecular mechanisms of seed germination. Proteomics. 2015;**15**:1671-1679

[9] Finkelstein R, Reeves W, Ariizumi T, Steber C. Molecular aspects of seed dormancy. Annual Review of Plant Biology. 2008;**59**:387-415

[10] Miransari M, Smith D. Plant hormones and seed germination. Environmental and Experimental Botany. 2014;**99**:110-121

[11] Zhang C, Luo W, Li Y, Zhang X, Bai X, Niu Z, et al. Transcriptomic analysis of seed germination under salt stress in two desert sister species (*Populus euphratica* and *P. pruinosa*). Frontiers in Genetics. 2019;**10**:1-16

[12] Llanes A, Andrade A, Masciarelli O, Alemano S, Luna V. Drought and salinity alter endogenous hormonal profiles at the seed germination phase. Seed Science Research. 2016;**26**(1):1-13

[13] Lutts S, Benincasa P, Wojtyla L, S SK, Pace R, Lechowska K, et al. Seed priming: New comprehensive approaches for an old empirical technique. In: Araujo S, Balestrazzi A, editors. New Challenges in Seed Biology–Basic and Translational Research Driving Seed Technology. Rijeka, Croazia: IntechOpen; 2016. pp. 1-40

[14] Foti C, Khah EM, Pavli OI. Germination profiling of lentil genotypes subjected to salinity stress. Plant Biology. 2019;**21**(3):480-486

[15] Jamil A, Riaz S, Ashraf M, Foolad MR. Gene expression profiling of plants under salt stress. CRC Critical Reviews in Plant Science. 2011;**30**(5):435-458

[16] Ashraf M, Akram NA. Improving salinity tolerance of plants through conventional breeding and

genetic engineering: An analytical comparison. Biotechnology Advances. 2009;**27**:744-752

[17] Pimentel D, Berger B, Filiberto D, Newton M, Wolfe B, Karabinakis E, et al. Water resources: Agricultural and environmental issues. BioScience. 2004;**54**:909-918

[18] Isayenkov SV, Maathuis FJM. Plant salinity stress: Many unanswered questions remain. Frontiers in Plant Science. 2019;**10**:80

[19] Adnan M, Zahir S, Fahad S, Arif M, Mukhtar A, Imtiaz AK, et al. Phosphate-solubilizing bacteria nullify the antagonistic effect of soil calcification on bioavailability of phosphorus in alkaline soils. Scientific Reports. 2018;**8**:4339

[20] Akram R, Turan V, Hammad HM, Ahmad S, Hussain S, Hasnain A, et al. Fate of organic and inorganic pollutants in paddy soils. In: Hashmi MZ, Varma A, editors. Environmental Pollution of Paddy Soils, Soil Biology. Switzerland: Springer; 2018. pp. 197-214

[21] Akram R, Turan V, Wahid A, Ijaz M, Shahid MA, Kaleem S, et al. Paddy land pollutants and their role in climate change. In: Hashmi MZ, Varma A, editors. Environmental Pollution of Paddy Soils, Soil Biology. Switzerland: Springer; 2018. pp. 113-124

[22] Aziz K, Daniel KYT, Fazal M, Muhammad ZA, Farooq S, Fan W, et al. Nitrogen nutrition in cotton and control strategies for greenhouse gas emissions: A review. Environmental Science and Pollution Research. 2017;**24**:23471-23487

[23] Aziz K, Daniel KYT, Muhammad ZA, Honghai L, Shahbaz AT, Mir A, et al. Nitrogen fertility and abiotic stresses management in cotton crop: A review. Environmental Science and Pollution Research. 2017;**24**:14551-14566

[24] Depeng W, Fahad S, Saud S, Muhammad K, Aziz K, Mohammad NK, et al. Morphological acclimation to agronomic manipulation in leaf dispersion and orientation to promote "Ideotype" breeding: Evidence from 3D visual modeling of "super" rice (*Oryza sativa* L.). Plant Physiology and Biochemistry. 2019;**135**:499-510

[25] Fahad S, Bajwa AA, Nazir U, Anjum SA, Farooq A, Zohaib A, et al. Crop production under drought and heat stress: Plant responses and management options. Frontiers in Plant Science. 2017;**8**:1147

[26] Fahad S, Bano A. Effect of salicylic acid on physiological and biochemical characterization of maize grown in saline area. Pakistan Journal of Botany. 2012;**44**:1433-1438

[27] Fahad S, Chen Y, Saud S, Wang K, Xiong D, Chen C, et al. Ultraviolet radiation effect on photosynthetic pigments, biochemical attributes, antioxidant enzyme activity and hormonal contents of wheat. Journal of Food, Agriculture and Environment. 2013;**11**(3&4):1635-1641

[28] Fahad S, Hussain S, Bano A, Saud S, Hassan S, Shan D, et al. Potential role of phytohormones and plant growth-promoting rhizobacteria in abiotic stresses: Consequences for changing environment. Environmental Science and Pollution Research. 2014;**22**(7):4907-4921

[29] Fahad S, Hussain S, Matloob A, Khan FA, Khaliq A, Saud S, et al. Phytohormones and plant responses to salinity stress: A review. Plant Growth Regulation. 2014;**75**(2):391-404

[30] Fahad S, Hussain S, Saud S, Hassan S, Chauhan BS, Khan F, et al. Responses of rapid viscoanalyzer profile and other rice grain qualities to exogenously applied plant growth regulators under high day and high

night temperatures. PLOS One. 2016;**11**(7):e0159590

[31] Fahad S, Hussain S, Saud S, Hassan S, Ihsan Z, Shah AN, et al. Exogenously applied plant growth regulators enhance the morphophysiological growth and yield of rice under high temperature. Frontiers in Plant Science. 2016;**7**:1250

[32] Fahad S, Hussain S, Saud S, Hassan S, Tanveer M, Ihsan MZ, et al. A combined application of biochar and phosphorus alleviates heat-induced adversities on physiological, agronomical and quality attributes of rice. Plant Physiology and Biochemistry. 2016;**103**:191-198

[33] Fahad S, Hussain S, Saud S, Khan F, Hassan S, Amanallah J, et al. Exogenously applied plant growth regulators affect heat-stressed rice pollens. Journal of Agronomy and Crop Science. 2016;**202**:139-150

[34] Fahad S, Hussain S, Saud S, Tanveer M, Bajwa AA, Hassan S, et al. A biochar application protects rice pollen from high-temperature stress. Plant Physiology and Biochemistry. 2015;**96**:281-287

[35] Fahad S, Muhammad ZI, Abdul K, Ihsanullah D, Saud S, Saleh A. Consequences of high temperature under changing climate optima for rice pollen characteristics-concepts and perspectives. Archives of Agronomy and Soil Science. 2018;**64**(11):1473-1488

[36] Fahad S, Nie L, Chen Y, Wu C, Xiong D, Saud S, et al. Crop plant hormones and environmental stress. Sustainable Agriculture Reviews. 2015;**15**:371-400

[37] Fahad S, Adnan M, Hassan S, Saud S, Hussain S, Wu C, et al. Rice responses and tolerance to high temperature. In: Hasanuzzaman M, Fujita M, Nahar K, Biswas JK, editors. Advances in Rice Research for Abiotic Stress Tolerance. Cambridge: Woodhead Publishing; 2019. pp. 201-224

[38] Fahad S, Rehman A, Shahzad B, Tanveer M, Saud S, Kamran M, et al. Rice responses and tolerance to metal/metalloid toxicity. In: Hasanuzzaman M, Fujita M, Nahar K, Biswas JK, editors. Advances in Rice Research for Abiotic Stress Tolerance. Cambridge: Woodhead Publishing; 2019. pp. 299-312

[39] Habib ur R, Ashfaq A, Aftab W, Manzoor H, Fahd R, Wajid I, et al. Application of CSM-CROPGRO-cotton model for cultivars and optimum planting dates: Evaluation in changing semi-arid climate. Field Crops Research. 2019;**238**:139-152

[40] Hafiz MH, Muhammad A, Farhat A, Hafiz FB, Saeed AQ, Muhammad M, et al. Environmental factors affecting the frequency of road traffic accidents: A case study of sub-urban area of Pakistan. Environmental Science and Pollution Research. 2019;**26**:11674-11685

[41] Hafiz MH, Wajid F, Farhat A, Fahad S, Shafqat S, Wajid N, et al. Maize plant nitrogenuptake dynamics at limited irrigation water and nitrogen. Environmental Science and Pollution Research. 2016;**24**(3):2549-2557

[42] Hesham FA, Fahad S. Melatonin application enhances biochar efficiency for drought tolerance in maize varieties: Modifications in physio-biochemical machinery. Agronomy Journal. 2020;**2020**:1-22

[43] Hussain MA, Fahad S, Rahat S, Muhammad FJ, Muhammad M, Qasid A, et al. Multifunctional role of brassinosteroid and its analogues in plants. Plant Growth Regulation. 2020. DOI: 10.1007/s10725-020-00647-8 (In Press)

[44] Kamarn M, Wenwen C, Irshad A, Xiangping M, Xudong Z, Wennan S,

et al. Effect of paclobutrazol, a potential growth regulator on stalk mechanical strength, lignin accumulation and its relation with lodging resistance of maize. Plant Growth Regulation. 2017;**84**:317-332

[45] Muhammad Z, Abdul MK, Abdul MS, Kenneth BM, Muhammad S, Shahen S, et al. Performance of Aeluropus lagopoides (mangrove grass) ecotypes, a potential turfgrass, under high saline conditions. Environmental Science and Pollution Research. 2019;**26**:13410-13421

[46] Qamar-uz Z, Zubair A, Muhammad Y, Muhammad ZI, Abdul K, Fahad S, et al. Zinc biofortification in rice: Leveraging agriculture to moderate hidden hunger in developing countries. Archives of Agronomy and Soil Science. 2017;**64**:147-161

[47] Sajjad H, Muhammad M, Ashfaq A, Waseem A, Hafiz MH, Mazhar A, et al. Using GIS tools to detect the land use/land cover changes during forty years in Lodhran district of Pakistan. Environmental Science and Pollution Research. 2019. DOI: 10.1007/s11356-019-06072-3 (In Press)

[48] Saud S, Chen Y, Fahad S, Hussain S, Na L, Xin L, et al. Silicate application increases the photosynthesis and its associated metabolic activities in Kentucky bluegrass under drought stress and post-drought recovery. Environmental Science and Pollution Research. 2016;**23**(17):17647-17655

[49] Saud S, Chen Y, Long B, Fahad S, Sadiq A. The different impact on the growth of cool season turf grass under the various conditions on salinity and drought stress. International Journal of Agricultural Science and Research. 2013;**3**:77-84

[50] Saud S, Fahad S, Cui G, Chen Y, Anwar S. Determining nitrogen isotopes discrimination under drought stress on enzymatic activities, nitrogen isotope abundance and water contents of Kentucky bluegrass. Scientific Reports. 2020;**10**:6415

[51] Saud S, Fahad S, Yajun C, Ihsan MZ, Hammad HM, Nasim W, et al. Effects of nitrogen supply on water stress and recovery mechanisms in Kentucky bluegrass plants. Frontiers in Plant Science. 2017;**8**:983

[52] Saud S, Li X, Chen Y, Zhang L, Fahad S, Hussain S, et al. Silicon application increases drought tolerance of Kentucky bluegrass by improving plant water relations and morph physiological functions. The Scientific World Journal. 2014;**2014**:10. Article ID: 368694

[53] Shah F, Lixiao N, Kehui C, Tariq S, Wei W, Chang C, et al. Rice grain yield and component responses to near 2°C of warming. Field Crops Research. 2013;**157**:98-110

[54] Wajid N, Ashfaq A, Asad A, Muhammad T, Muhammad A, Muhammad S, et al. Radiation efficiency and nitrogen fertilizer impacts on sunflower crop in contrasting environments of Punjab. Environmental Science and Pollution Research. 2017;**25**:1822-1836

[55] Yang Z, Zhang Z, Zhang T, Fahad S, Cui K, Nie L, et al. Effect of season-long temperature increases on rice cultivars grown in the central and southern regions of China. Frontiers in Plant Science. 2017;**8**:1908

[56] Zahida Z, Hafiz FB, Zulfiqar AS, Ghulam MS, Fahad S, Muhammad RA, et al. Effect of water management and silicon on germination, growth, phosphorus and arsenic uptake in rice. Ecotoxicology and Environmental Safety. 2017;**144**:11-18

[57] Machado R, Serralheiro R. Soil Salinity: Effect on vegetable crop

growth. Management Practices to Prevent and Mitigate Soil Salinization. Horticulturae. 2017;**3**(2):30

[58] Hasanuzzaman M, Nahar K Fujita M. Plant response to salt stress and role of exogenous protectants to mitigate salt-induced damages. In: Ahmad P, Azooz MM, Prasad MNV, editors. Ecophysiology and Responses of Plants under Salt Stress. New York: Springer-Verlag; 2013. pp. 25-87

[59] USSL Staff. Diagnosis and Improvement of Saline and Alkali Soils. USDA Handbook No 60. Washington DC, USA; 1954. p. 160

[60] An Y-Q, Lin L. Transcriptional regulatory programs underlying barley germination and regulatory functions of gibberellin and abscisic acid. BMC Plant Biology. 2011;**11**:105

[61] Finch-Savage WE, Leubner-Metzger G. Seed dormancy and the control of germination. New Phytologist. 2006;**171**(3):501-523

[62] Bewley JD. Seed germination and dormancy. The Plant Cell. 1997;**9**:1055-1066

[63] Nonogaki H, Bassel GW, Bewley JD. Germination - still a mystery. Plant Science. 2010;**179**:574-581

[64] Koornneef M, Bentsink L, Hilhorst H. Seed dormancy and germination. Current Opinion in Plant Biology. 2002;**5**:33-36

[65] Weitbrecht K, Müller K, Leubner-Metzger G. First off the mark: Early seed germination. Journal of Experimental Botany. 2011;**62**(10):3289-3309

[66] Nonogaki H. Seed dormancy and germination-emerging mechanisms and new hypotheses. Frontiers in Plant Science. 2014;**5**:233

[67] Bewley JD, Bradford KJ, Hilhorst HWM, Nonogaki HS. Physiology of Development, Germination and Dormancy. New York: Springer; 2013

[68] Ma Z, Marsolais F, Bykova NV, Igamberdiev AU. Nitric oxide and reactive oxygen species mediate metabolic changes in barley seed embryo during germination. Frontiers in Plant Science. 2016;**7**:138

[69] Bentsink L, Koornneef M. Seed Dormancy and Germination. Arabidopsis Book; 2008;**6**:e0119

[70] Gazzarrini S, Tsai AYL. Hormone cross-talk during seed germination. Essays in Biochemistry. 2015;**58**:151-164

[71] Yamaguchi S. Gibberellin metabolism and its regulation. Annual Review of Plant Biology. 2008;**59**:225-251

[72] Gimeno-Gilles C, Lelievre E, Viau L, Malik-Ghulam M, Ricoult C, Niebel A, et al. ABA-mediated inhibition of germination is related to the inhibition of genes encoding cell-wall biosynthetic and architecture: Modifying enzymes and structural proteins in *Medicago truncatula* embryo axis. Molecular Plant. 2009;**2**(1):108-119

[73] Corbineau F, Xia Q, Bailly C, El-Maarouf-Bouteau H. Ethylene, a key factor in the regulation of seed dormancy. Frontiers in Plant Science. 2014;**5**:539

[74] Liu PP, Montgomery T, Fahlgren N, Kasschau KD, Nonogaki H, Carrington JC. Repression of AUXIN RESPONSE FACTOR10 by microRNA160 is critical for seed germination and post-germination stages. The Plant Journal. 2007;**52**: 133-146

[75] Hentrich M, Bottcher C, Duchting P, Cheng Y, Zhao Y, Berkowitz O, et al. The jasmonic acid signaling pathway is linked to auxin homeostasis through the

modulation of YUCCA8 and YUCCA9 gene expression. Plant Journal. 2013;**74**:626-637

[76] Toh S, Imamura A, Watanabe A, Nakabayashi K, Okamoto M, Jikumaru Y, et al. High temperature-induced abscisic acid biosynthesis and its role in the inhibition of gibberellin action in Arabidopsis seeds. Plant Physiology. 2008;**146**:1368-1385

[77] Cho JN, Ryu JY, Jeong YM, Park J, Song JJ, Amasino RM, et al. Control of seed germination by light- induced histone arginine demethylation activity. Developmental Cell. 2012;**22**: 736-748

[78] Oracz K, Karpiński S. Phytohormones signaling pathways and ROS involvement in seed germination. Frontiers in Plant Science. 2016;**7**:864

[79] Bailly C, El-Maarouf-Bouteau H, Corbineau F. From intracellular signaling networks to cell death: The dual role of reactive oxygen species in seed physiology. Comptes Rendus Biologies. 2008;**331**:806-814

[80] Wojtyla L, Lechowska K, Kubala S, Garnczarska M. Different modes of hydrogen peroxide action during seed germination. Frontiers in Plant Science. 2016;**7**:66

[81] Zheng C, Jiang D, Liub F, Dai T, Liu W, Jing Q, et al. Exogenous nitric oxide improves seed germination in wheat against mitochondrial oxidative damage induced by high salinity. Environmental and Experimental Botany. 2009;**67**:222-227

[82] Munns R. Comparative physiology of salt and water stress. Plant, Cell & Environment. 2002;**25**:239-250

[83] Munns R, Tester M. Mechanisms of salinity tolerance. Annual Review of Plant Biology. 2008;**59**:651-681

[84] Jahromi F, Aroca R, Porcel R, Ruiz-Lozano JM. Influence of salinity on the in vitro development of Glomus intraradices and on the in vivo physiological and molecular responses of mycorrhizal lettuce plants. Microbial Ecology. 2008;**55**:45-53

[85] Daszkowska-Golec A. Arabidopsis seed germination under abiotic stress as a concert of action of phytohormones. OMICS. 2011;**15**:763-774

[86] Parida AK, Das AB. Salt tolerance and salinity effects on plants: A review. Ecotoxicology and Environmental Safety. 2005;**60**(3):324-349

[87] Panda SK, Khan MHG. Growth, oxidative damage and antioxidant responses in Greengram (*Vigna radiata* L.) under short-term salinity stress and its recovery. Journal of Agronomy and Crop Science. 2009;**195**:442-454

[88] El-Hendawy S, Elshafei A, Al-Suhaibani N, Alotabi M, Hassan W, Dewir YH, et al. Assessment of the salt tolerance of wheat genotypes during the germination stage based on germination ability parameters and associated SSR markers. Journal of Plant Interactions. 2019;**14**(1):151-163

[89] Lee SC, Luan S. ABA signal transduction at the crossroad of biotic and abiotic stress responses. Plant, Cell & Environment. 2012;**35**:53-60

[90] Weiss D, Ori N. Mechanisms of cross talk between gibberellin and other hormones. Plant Physiology. 2007;**144**:1240-1246

[91] Kaneko M, Itoh H, Ueguchi-Tanaka M, Ashikari M, Matsuoka M. The α-amylase induction in endosperm during rice seed germination is caused by gibberellin synthesized in epithelium. Plant Physiology. 2002;**128**:1264-1270

[92] Flowers TJ, Colmer TD. Salinity tolerance in halophytes. The New Phytologist. 2008;**179**(4):945-963

[93] Chapman VJ. Salt Marshes and Salt Deserts of the World. New York: Interscience Publishers; 1960

[94] Wang L, Huang Z, Baskin CC, Baskin JM, Dong M. Germination of dimorphic seeds of the desert annual halophyte *Suaeda aralocaspica* (Chenopodiaceae), a C4 plant without kranz anatomy. Annals of Botany. 2008;**102**:757-776

[95] Woodell SRJ. Salinity and seed germination patterns in coastal plants. Vegetation. 1985;**61**:223-229

[96] Redondo S, Rubio-Casal AE, Castillo JM, Luque CJ, Álvarez AA, Luque T, et al. Influences of salinity and light on germination of three Sarcocornia taxa with contrasted habitats. Aquatic Botany. 2004;**78**:255-264

[97] Huang Z, Zhang X, Zheng G, Gutterman Y. Influence of light, temperature, salinity and storage on seed germination of *Haloxylon ammodendron*. Journal of Arid Environments. 2003;**55**:453-464

[98] Khan MA, Gul B, Weber DJ. Effect of salinity and temperature on the germination of Kochia scoparia. Wetlands Ecology and Management. 2001;**9**:483-489

[99] Orlovsky NS, Japakova UN, Shulgina I, Volis S. Comparative study of seed germination and growth of *Kochia prostrata* and *Kochia scoparia* (Chenopodiaceae) under salinity. Journal of Arid Environments. 2011;**75**:532-537

[100] El-Keblawy A, Al-Shamsi N. Effects of salinity, temperature and light on seed germination of *Haloxylon salicornicum*, a common perennial shrub of the Arabian deserts. Seed Science and Technology. 2008;**36**:679-688

[101] Fos M, Alfonso L, Ferrer-Gallego PP, Laguna E. Effect of salinity, temperature and hypersaline conditions on the seed germination in *Limonium mansanetianum* an endemic and threatened Mediterranean species. Plant Biosystems - An International Journal Dealing with all Aspects of Plant Biology. 2020;**0**(0):1-7

[102] Zia S, Khan MA. Effect of light, salinity and temperature on the germination of Limonium stocksii. Canadian Journal of Botany. 2004;**82**:151-157

[103] Yildiz M, Cenkci S, Kargioglu M. Effects of salinity, temperature, and light on seed germination in two Turkish endemic halophytes, *Limonium iconicum* and *L. lilacinum* (Plumbaginaceae). Seed Science and Technology. 2008;**36**:646-656

[104] Li J, Yin LY, Jongsma MA, Wang CY. Effect of light, hydro-priming and abiotic stress on seed germination, and shoot and root growth pyrethrum (*Tanacetum cinerariifolium*). Industrial Crops and Products. 2011;**34**:1543-1549

[105] Causin HF, Bordón DAE, Burrieza H. Salinity tolerance mechanisms during germination and early seedling growth in *Chenopodium quinoa* Wild. genotypes with different sensitivity to saline stress. Environmental and Experimental Botany. 2020;**172**:103995

[106] El Goumi Y, Fakiri M, Lamsaouri O, Benchekroun M. Salt stress effect on seed germination and some physiological traits in three Moroccan barley (*Hordeum vulgare* L.) cultivars. Journal of

Materials and Environmental Science. 2014;**5**(2):625-632

[107] Khodarahmpour Z, Ifar M, Motamedi M. Effects of NaCl salinity on maize (*Zea mays* L.) at germination and early seedling stage. African Journal of Biotechnology. 2012;**11**(2):298-304

[108] Vahabinia F, Pirdashti H, Bakhshandeh E. Environmental factors' effect on seed germination and seedling growth of chicory (*Cichorium intybus* L.) as an important medicinal plant. Acta Physiologiae Plantarum. 2019;**41**(2):1-13

[109] Wu H, Guo J, Wang C, Li K, Zhang X, Yang Z, et al. An effective screening method and a reliable screening trait for salt tolerance of *Brassica napus* at the germination stage. Frontiers in Plant Science. 2019;**10**:530

[110] Chakraborty K, Bishi SK, Goswami N, Singh AL, Bhaduri D, Zala PV. Salinity-induced changes in seed germination and the expression profile of antioxidant enzymes in peanut as early and late responses in emerging radicles. Acta Physiologiae Plantarum. 2019;**41**(8):1-16

[111] Hakim MA, Juraimi AS, Begum M, Hanafi MM, Ismail MR, Selamat A. Effect of salt stress on germination and early seedling growth of rice (*Oryza sativa* L.). African Journal of Biotechnology. 2010;**9**(13):1911-1918

[112] Pakyürek M, Dumanoğlu H. Effects of saline stress on in vitro seed germination and seedling growth of some Turkish fig cultivars (*Ficus carica* L.). Applied Ecology and Environmental Research. 2019;**17**(6):13485-13492

[113] Asaduzzaman M, Koetz E, Rahman A. Factors affecting seed germination and emergence of button grass (*Dactyloctenium radulans*) (R.Br.) P.Beauv. Weed Biology and Management. 2019;**19**(3):85-92

[114] Zhu G, An L, Jiao X, Chen X, Zhou G, McLaughlin N. Effects of gibberellic acid on water uptake and germination of sweet sorghum seeds under salinity stress. Chilean Journal of Agricultural Research. 2019;**79**(3):415-424

[115] Rahman A, Asaduzzaman M. Statistical modelling of seed germination and seedlings root response of annual ryegrass (*Lolium rigidum*) to different stress. Agricultural Research. 2019;**8**(2):262-269

[116] Lavrenko SO, Lavrenko NM, Lykhovyd PV. Effect of degree of salinity on seed germination and initial growth of chickpea (*Cicer arietinum*). Biosystems Diversity. 2019;**27**(2):101-105

[117] Moles TM, Guglielminet L, Reyes TH. Differential effects of sodium chloride on germination and post-germination stages of two tomato genotypes. Scientia Horticulturae. 2019;**257**:108730

[118] Panta S, Flowers T, Lane P, Doyle R, Haros G, Shabala S. Halophyte agriculture: Success stories. Environmental and Experimental Botany. 2014;**107**:71-83

[119] Francois LE, Maas EV. Crop response and management on salt-affected soils. In: Pessarakli M, editor. Handbook of Plant and Crop Stress. New York: Marcel Dekker; 1994. pp. 149-181

[120] Ali E, Hussain N, Shamsi IH, Jabeen Z, Siddiqui MH, Jiang L. Role of jasmonic acid in improving tolerance of rapeseed (*Brassica napus* L.) to Cd toxicity. Journal of Zhejiang University: Science B (Biomedicine & Biotechnology). 2018;**19**(2):130-146

[121] Esechie HA, Al-Saidi A, Al-Khanjari S. Effect of sodium chloride salinity on seedling emergence in chickpea. Journal of Agronomy and Crop Science. 2002;**188**:155-160

[122] Maiti R, Pramanik K. Vegetable seed priming: A low cost, simple and powerful techniques for farmers' livelihood. International Journal of Bio-resource and Stress Management. 2013;**4**:475-481

[123] Hussain T, Koyro HW, Huchzermeyer B, Khan MA. Eco-physiological adaptations of *Panicum antidotale* to hyperosmotic salinity: Water and ion relations and anti-oxidant feedback. Flora. 2015;**212**:30-37

[124] Lemmens E, Deleu LJ, De Brier N, De Man WL, De Proft M, Prinsen E, et al. The impact of hydro-priming and Osmo-priming on seedling characteristics, plant hormone concentrations, activity of selected hydrolytic enzymes, and Cell Wall and Phytate hydrolysis in sprouted wheat (*Triticum aestivum* L.). ACS Omega. 2019;**4**(26):22089-22100

[125] del Carmen Martínez-Ballesta M, Egea-Gilabert C, Conesa E, Ochoa J, Vicente MJ, Franco JA, et al. The importance of ion homeostasis and nutrient status in seed development and germination. Agronomy. 2020;**10**(4):504

[126] Kilic S, Kahraman A. The mitigation effects of exogenous hydrogen peroxide when alleviating seed germination and seedling growth inhibition on salinity-induced stress in barley. Polish Journal of Environmental Studies. 2016;**25**(3):1053-1059

[127] Ahammed GJ, Li Y, Li X, Han WY, Chen S. Epigallocatechin-3-Gallate alleviates salinity-retarded seed germination and oxidative stress in tomato. Journal of Plant Growth Regulation. 2018;**37**(4):1349-1356

[128] Feghhenabi F, Hadi H, Khodaverdiloo H, van Genuchten MT. Seed priming alleviated salinity stress during germination and emergence of wheat (Triticum aestivum L.). Agricultural Water Management. 2020;**231**:106022

[129] Tsegay BA, Andargie M. Seed priming with gibberellic acid (GA 3) alleviates salinity induced inhibition of germination and seedling growth of *Zea mays* L., *Pisum sativum* Var. abyssinicum A. Braun and *Lathyrus sativus* L. Journal of Crop Science and Biotechnology. 2018;**21**(3):261-267

[130] Ren Y, Wang W, He J, Zhang L, Wei Y, Yang M. Nitric oxide alleviates salt stress in seed germination and early seedling growth of pakchoi (*Brassica chinensis* L.) by enhancing physiological and biochemical parameters. Ecotoxicology and Environmental Safety. 2020;**187**:109785

[131] Zhang DW, Vu TS, Huang J, Chi CY, Xing Y, Fu DD, et al. Effects of calcium on germination and seedling growth in *melilotus officinalis* L. (Fabaceae) under salt stress. Pakistan Journal of Botany. 2019;**51**:1-9

[132] Luis Castañares J, Alberto Bouzo C. Effect of exogenous melatonin on seed germination and seedling growth in melon (*Cucumis melo* L.) under salt stress. Horticultural Plant Journal. 2019;**5**(2):79-87

[133] Mahboob W, Khan A. Seed priming modulates germination potential, osmoprotectants accumulation and ionic uptake in wheat seedlings under salt stress screening for abiotic stress view project screening of salt tolerant wheat genotypes view project. International Journal of Agricultural and Biology. 2019;**22**:594-600

[134] Maswada HF, Djanaguiraman M, Prasad PVV. Seed treatment with nano-iron (III) oxide enhances germination, seeding growth and salinity tolerance

of sorghum. Journal of Agronomy and Crop Science. 2018;**204**(6):577-587

[135] Bouallègue A, Souissi F, Nouairi I, Souibgui M, Abbes Z, Mhadhbi H. Physiological and biochemicals changes modulated by seeds' priming of lentil (*Lens culinaris* L.) under salt stress at germination stage. Acta Scientiarum Polonorum Hortorum Cultus. 2019;**18**(5):27-38

[136] Liu J, Li L, Yuan F, Chen M. Exogenous salicylic acid improves the germination of Limonium bicolor seeds under salt stress. Plant Signaling & Behavior. 2019;**14**:10

[137] Khan WUD, Aziz T, Waraich EA, Khalid M. Silicon application improves germination and vegetative growth in maize grown under salt stress. Pakistan Journal of Agricultural Sciences. 2015;**52**(4):937-944

[138] Chauhan A, Abuamarah BA, Kumar A, Verma JS, Ghramh HA, Ali K, et al. Influence of gibberellic acid and different salt concentrations on germination percentage and physiological parameters of oat cultivars. Saudi Journal of Biological Sciences. 2019;**26**(6):1298-1304

[139] Gou T, Chen X, Han R, Liu J, Zhu Y, Gong H. Silicon can improve seed germination and ameliorate oxidative damage of bud seedlings in cucumber under salt stress. Acta Physiologiae Plantarum. 2020;**42**(1):1-11

[140] Li J, Zhao C, Zhang M, Yuan F, Chen M. Exogenous melatonin improves seed germination in Limonium bicolor under salt stress. Plant Signaling & Behavior. 2019;**14**(11):1659705

[141] Khan MA, Gul B, Weber DJ. Action of plant growth regulators and salinity on the seed germination of *Ceratoides lanata*. Canadian Journal of Botany. 2004;**82**:37-42

[142] Wu YP, Chen F, Hu XW, Baskin CC, Baskin JM. Alleviation of salinity stress on germination of *Leymus chinensis* seeds by plant growth regulators and nitrogenous compounds under contrasting light/dark conditions. Grass and Forage Science. 2016;**71**(3):497-506

[143] Figueira C, Ferreira MJ, Silva H, Cunha A. Improved germination efficiency of *Salicornia ramosissima* seeds inoculated with *Bacillus aryabhattai* SP1016-20. The Annals of Applied Biology. 2019;**174**(3):319-328

[144] Waqas M, Korres NE, Khan MD, Nizami A, Deeba F, Ali I, et al. Advances in the concept and methods of seed priming. In: Hasanuzzaman M, Fotopoulos V, editors. Priming and Pretreatment of Seeds and Seedlings. Singapore: Springer; 2019. pp. 11-41

[145] Nawaz J, Hussain M, Jabbar A, Nadeem GA, Sajid M, Subtain M, et al. Seed priming a technique. International Journal of Agriculture and Crop Sciences. 2013;**6**(20):1373-1381

[146] Di Girolamo G, Barbanti L. Treatment conditions and biochemical processes influencing seed priming effectiveness. Italian Journal of Agronomy. 2012;**7**:8-18

[147] Abdelhamid MT, El-masry RR, Darwish DS, Abdalla MMF, Oba S, Ragab R, et al. Mechanisms of seed priming involved in salt stress amelioration. In: Hasanuzzaman M, Fotopoulos V, editors. Priming and Pretreatment of Seeds and Seedlings. Singapore: Springer; 2019. pp. 219-251

[148] Niranjan Raj S, Shetty NP, Shetty HS. Seed bio-priming with Pseudomonas fluorescens isolates enhances growth of pearl millet plants and induces resistance against downy mildew. International Journal of Pest Management. 2004;**50**:41-48

Chapter 12

Abiotic Stress Tolerance in Crop Plants: Role of Phytohormones

Haifa Abdulaziz S. Alhaithloul,
Abdelghafar M. Abu-Elsaoud and Mona H. Soliman

Abstract

Crop plants are encountered by various abiotic pressures which limit their growth and development. Stresses such as drought, heat, pathogen attack, heavy metal, salinity, and radiations impose negative effect on crop plants. The reduction in crop productivity in the current era of climate change is compromising the efforts/strategies used for sustainable agricultural practices. Therefore, plant stress physiologists are engineering plants with suitable exogenous signaling elicitors to engineer tolerance to various stresses. In the present chapter, an appraisal has been made in the "Introduction" section to first assess the damages caused by various abiotic stresses in crop plants. In the second section, we attempt to summarize the role of various plant hormones, namely, salicylic acid (SA), brassinosteroids (BRs), ethylene (ET), and methyl jasmonate (MJ) in enhancing abiotic stress tolerance. The current concept may lead to the development of strategies for unraveling the underlying mechanisms of plant hormone-mediated abiotic stress tolerance in crop plants.

Keywords: abiotic stress, plant hormones, crop plants, stress tolerance

1. Introduction

Due to overexploitation of natural resources for catering to basic human needs coupled with hunger, the aim of Food and Agricultural Organization's [1] of making world agriculture for enhancing 70% food production by 2050 for 2.3 billion newborn individual is facing a lot of challenges. As a result of changing climate and global warming, these threats are now called as "global warning" due to their effects on plant productivity [2]. According to a report published by IPCC [3], global climate change will increase the mean temperature of the Earth by 2–4°C. Due to increased anthropogenic activities, the change in climate will further increase and will definitely affect the agriculture sector as this is being more prone to climate changes [4]. A change in climate scenarios is the main reason of biotic and abiotic pressures of a particular region. Various abiotic stresses all over the world have now been adjudged as the most potential threat for sustainable agricultural productivity [5, 6]. Myriads of anthropogenic emissions in the developmental era have further intensified the abiotic stress-induced negative impacts on agricultural productivity. The major abiotic stresses such as metals/metalloids (hereafter termed as "metal/s"), salt, ozone, UV-B radiation, temperature extremes, nutrient (deficiency and excess), flooding, and water deficit are jeopardizing the agriculture system worldwide [7–19]. These abiotic stresses can modulate almost all plant physiological, biochemical, and molecular processes starting from the seedling to maturity stage and can ultimately cause severe negative impacts on economic yield of various crop plants. Nevertheless, abiotic stresses have been reported to adversely affect 70% yield of crops [20]. The stress of metal/s is known to affect the photosynthetic process to a great extent [19, 21]. According to an estimate of U.S. Environmental Action Group, various metals pose negative effects on health in more than 10 million of human population in eight countries (such as China, India, Russia, Dominican Republic, Kyrgyzstan, Peru, Ukraine, and Zambia) and the effect imposed by them is regarded as one of the major threats to the agriculture [22]. Wang et al. [23] have reports to show that the salinization of arable land could accelerate to a land loss of 30% by the year end of 2028 and up to 50% by 2050.In current times, an increase in atmospheric CO_2 concentration has been reported to reach by 407.4 ppm, and its levels today are more than at any point recorded for at least the past 800,000 years [24]. In addition, the data accessed on NOAA website [25] show that the average global temperature in June 2020 was found to be 1.66 °F (0.92 of a °C) above the twentieth century's average of 59.9 °F (15.5 °C) and the average temperature of June across the U.S. was found to be 70.3 °F (which is 1.8° above average), which ranks it the warmest third month in the history of the 126-year record. Hence, all these abiotic stress factors under the effect of climate change could be expected to significantly pose a potential threat to agriculture [11, 26–30]. Therefore, the knowledge of crop productivity improvement techniques under various abiotic stresses is one of the biggest challenges encountered by the scientific community across the world. Despite much and extensive efforts being utilized in the current field, the underlying quest for obtaining abiotic stress-resistant crops is very low and receives little attention. Various efforts are being utilized to decipher the potential mechanisms behind plant responses to abiotic stress conditions. Plants are sessile and are exposed to various environmental regimes and have to tolerate these abiotic stress-induced adverse impacts for performing their growth and development in an optimal manner. As a result of an exponential increase in the global human population, the demand for enhanced food is also accelerating and in order to meet these challenges under a challenging environment, it has become necessary to increase crop productivity. Nevertheless, the exogenous use of phytohormones in these adaptive mechanisms for engineering abiotic stress tolerance in diverse crop plants can yield better results.

Phytohormones are small chemicals messengers that play important roles in regulating the promotion of growth, development, and metabolism of the plants in a wide range of biotic and abiotic stresses. These are biodegradable and highly promote the growth and crop yield under optimal as well as stressed conditions. The knowledge of engineering in plants could be exploited to impart tolerance against a wide range of abiotic stresses. Thus, the present chapter focuses on the potent role of various phytohormones in enhancing the morphological and physiological properties of crop plants and also sheds light on the alleviative role in providing plant tolerance to various environmental stresses.

2. Regulatory role played by various phytohormones in abiotic stress tolerance in crop plants

In the following sections, we will be discussing the potential role of different phytohormones like SA-, BRs-, ET- and MJ-mediated mechanisms in countering diverse abiotic pressures like salinity, drought, heat, and metal/s stress in diverse crop species. The role of these phytohormones in eliciting various responses in different crop plants has also been represented in **Table 1**.

Name of plant	Abiotic stress, type, and doses	Phytohormone doses	Consequent response	Reference
Mentha arvensis L.	Metal stress Cd (150 mg/Kg)	Methyl jasmonate 1 μM	A significant decrease in length of root and shoot, fresh and dry masses of shoot and leaf area; decreased chlorophyll and leaf gas exchange traits, and leaf N, P, and K content; increase in the activities of APX, GR, SOD, CAT, ROS production, osmolytes; and closure in stomatal movement under Cd. Improvement in these traits and alleviation of stress by MJ application	Zaid and Mohammad [31]
Mentha piperita	Cd 30, 60, and 120 mg/kg	Salicylic acid 10^{-4} M	Decreased growth and photosynthetic traits and enzymes like NR and CA activities; increased phenolics, activities of SOD, POX, and proline content under Cd. SA improved photosynthesis; activity of RuBisCo, CA, and GSH pool, EO content; and yield of peppermint plants	Ahmad et al. [32]
Glycine max L.	Salt 100 mM	24-epibrassinolide 10^{-7} M	EBL application decreased Na^{+} accumulation and promoted the uptake of the mineral nutrients, modulated osmolytes, activities of key antioxidant enzymes, and the levels of nonenzymatic antioxidants thus improves growth and photosynthesis	Alam et al. [33].
Linum usitatissimum	Salt 150 mM	24-epibrassinolide 10^{-8} M	NaCl decreased methylation of CCGG sequences. EBL application induced the total methylation under salinity stress conditions	Amraee et al. [34].
Brassica juncea	Cr 1.2 mM	200 μL/L ethephon application	ET was found to be involved in reversal of Cr-inhibited growth and photosynthesis by increasing PSII efficiency and enhanced N and S assimilation	Asgher et al. [35]
Brassica juncea L.	As 50 and 200 μM	Methyl jasmonate 0.1 and 1.00 μM	MJ application effectively mitigated the adverse effects of As stress by increasing the biomass, chlorophyll fluorescence, and the biosynthesis of secondary metabolites, reducing the oxidative stress biomarkers and maintaining antioxidant system	Farooq et al. [36]
Phaseolus coccineus	Metal stress Cu (50 μM)	Methyl jasmonate 10 μM	The enzymatic activities SOD, CAT, APX, POX, as well as MDA, GSH, and proline increased under Cu stress. MJ also elevated the activities CAT, APX, and POX activities in the roots. Final results showed that MJ was partially effective in modulating the anti-oxidative enzymes' activities and accumulation of metabolites, in both normal and Cu-stressed plants	Hanaka et al. [37]

Name of plant	Abiotic stress, type, and doses	Phytohormone doses	Consequent response	Reference
Mentha piperita	Salt (0, 50, 100, or 150 mM NaCl)	Salicylic acid, gibberellic acid, and triacontanol, each with 10^{-6} M	The salinity stress decreased growth, photosynthesis, activity of CA, contents of N, P, K, peltate glandular trichome density, EO and menthol content, and herbage yield but increased CAT, POX, and SOD activities and proline. The foliar spray of SA surpassed the other two phytohormones and maximally improved all parameters under both salt stress and salt free conditions	Khanam and Mohammad [38]
Brassica juncea	Ni 200 mg/kg	Ethylene 200 μL/L	The results showed that Ni decreased growth and photosynthesis but increased oxidative stress. Supply of ET ameliorated the stress by increasing the proline synthesis, antioxidant defense, and glyoxalase systems to promote the photosynthetic effectiveness under Ni stress	Khan et al. [39]
Artemisia annua	Metalloid stress (Boron) 1.00 and 2.00 mM	Methyl jasmonate 300 μM	B decreased the stem height, fresh and dry masses of the plant, the net photosynthetic rate, stomatal conductance, internal CO_2 concentration, and total chlorophyll content. MJ supply alleviates changes in growth and photosynthetic efficiency by increasing the activities of antioxidant enzymes and yield of secondary metabolites	Aftab et al. [40]

Table 1.
Various studies in diverse crop plants under different abiotic stresses conditions and the consequent response shown by these plants under various phytohormone application.

3. Role of salicylic acid

The name of "salicylic acid SA" is known to be derived from the Latin word *Salix* (willow tree). A German scientist, namely, Johann Andreas Buchner in 1828, isolated a quantity of glucoside of salicylic alcohol "salicin." Later on it was found that salicin, an important salicylate, is present in 36 other plants in addition to willow tree [41]. Chemically, SA is a simple phenolic compound ($C_7H_6O_3$) and consists of an aromatic ring to which one carboxylic and a hydroxyl group are attached. SA participates in several important plant processes either in stress-free or stressful environmental cues. It is now regarded as an important protector and growth regulating molecules in plants [42]. SA has an evidenced role in improving abiotic stress tolerance in various crop plants. Nonetheless, a report by Li et al. [43] suggested that SA acts upstream of NO under high concentration of carbon dioxide (CO_2) to induce enhanced flavonoid biosynthesis in tea plants. The results advocated the cross talk of SA for enhanced biosynthesis of secondary metabolites in tea plants under the era of climate change. In watermelon plants, resistance against root-knot nematode by red light is regulated by the coordination of SA with jasmonate signaling. This shows the potentiality of SA in enhancing tolerance against biotic stress. In Indian mustard plants, SA (10^{-5} M) alleviated the toxicity of Ni (50, 100 and 150 μM) by upregulating the activities of enzymes associated with antioxidant defense and glyoxalase systems [44]. In yet a recent study, Zaid et al. [45] tested the efficacy of foliar supplementation of SA (10^{-6} M) in two menthol mint cultivars, namely, Kosi and Kushal grown under Cd (50 μM) stress conditions. A reduction in growth, photosynthesis, and mineral nutrients with a concomitant increase in electrolyte leakage (EL), malondialdehyde (MDA), and hydrogen peroxide (H_2O_2) contents was observed under Cd stress in both cultivars. Nevertheless, a differential upregulation of proline content and the activities of antioxidant enzymes under Cd stress were also noticed in both cultivars. In contrast, out of three foliar-applied PGRs, the application of SA at three different growth stages proved best in alleviating Cd toxicity that too in both the cultivars but the effect of SA in Kushal was much more pronounced and better than Kosi. Application of SA (10^{-6} and 10^{-5} M) was also found to restrain As toxicity in two varieties of Artemisia annua L., namely, "CIM-Arogya" and "Jeevan Raksha" at 90 and 120 days after plantation [46]. The results suggested that the minimum reduction in growth and yield occurs with enhanced artemisinin production in "CIM Arogya" compared to "Jeevan Raksha" under As toxicity (45 mg kg^{-1} of soil). The application of SA via leaf significantly increased the content and yield of artemisinin in both varieties. El Dakak and Hassan [47] in a study evaluated the underlying physiological and biochemical mechanisms through which SA (1.0 mM) applied as a soil drench treatment alleviated the toxicity of Cd (5 and 10 μM) in maize plants. The results showed that exposure of maize plants to Cd reduced the chlorophyll content, chlorophyll fluorescence traits, photosynthesis, and catalase (CAT) activity, but, however, enhanced non-photochemical quenching (NPQ), MDA content and activities of superoxide dismutase (SOD), ascorbate peroxidase (APX) and guaiacol peroxidase (GPOX). Application of SA through soil drenching had an additive effect on the activities of antioxidant enzymes, but an inverse effect was found in case of MDA content. Es-sbihi et al. [48] studied the effects of SA (0.5 and 1 mM) on physiological traits, distribution of glandular hairs and essential oil (EO) composition in Salvia officinalis L. grown under Cu contamination (40 mM). The Cu stress decreased the growth of stem and root, Ca, P and K content in leaves as well as in roots. Interestingly, Cu increased EO yield. SA spray more prominently at 0.5 mM significantly improved stem and root growth, Ca, P, K and EO content and increased the density of peltate glands in presence of Cu stress. The study concluded with the observation that foliar spray of SA showed the remediating effect on growth traits, biosynthesis, and composition of EO of S. officinalis plants grown under Cu-challenged environment. This shows the key role of SA in enhancing tolerance to metal stress. In case of salt stress, SA also plays a mitigating role in minimizing the salt-induced damages in diverse crop plants. Miao et al. [49] in a hydroponic study applied exogenous SA (0.3 mM) in alleviating salt stress (50 mM NaCl) in cucumber seedlings. The results showed that 50 mM NaCl toxicity reduced relative growth rate and leaf photosynthetic traits like Fv/Fm, ΦII, and Pn. However, plants receiving NaCl + SA treatment showed a significant increment in relative growth rate, leaf photosynthetic traits (mentioned above) and an increase in case of total root length, root surface area, number and total lengths of primary and secondary roots. Both NaCl and SA downregulated and upregulated the expression of *GL2* and *RHD2* gene expression, while the application of SA treatment to NaCl-treated plants showed a notable upregulation in the expression of *NAC1*, *NAC2*, *GL2*, *EXP*, and downregulation of *RHD2* gene expression. Taken together the exogenous application of SA alleviated the NaCl toxicity by enhancing photosynthesis and architecture of root system in cucumber seedlings. Hussain et al. [50, 51] worked out the mechanistic elucidation of SA (0.5 mM) interplay with S (2.0 mM) under salt stress (50 mM NaCl) in mung bean plants. Salt-exposed plants showed an elevation in reactive oxygen species (ROS) content, lipid peroxidation, glucose and enzymatic activities like APX, GR, SOD, and reduced glutathione (GSH) and proline but marked inhibitions in the nitrate reductase (NR) and nitrite reductase (NiR) activities, N content, photosynthesis, photosynthetic N-use-efficiency (NUE), and growth traits. The supplementation of SA and S strengthened the antioxidant machinery, improved NR and NiR activities, N content, GSH, APX, GR and also decrease in ROS, MDA and glucose (a photosynthesis repressor) accumulation. These observations suggested that fine-tuning of SA with S diminished the NaCl-mediated changes in tested plants. Fahad and Bano [52] investigated the effect of foliar spray of SA (10^{-5} M) on maize (Zea mays L.) hybrid cv. 3025 grown in saline environments (pH 8.4 and EC 4.2 ds/m). The salinity treatment was found to significantly increased sugar contents, protein, proline, and activities of SOD, POD, CAT, and APX but decreased the pigment contents like chlorophyll and carotenoid and osmotic potential and membrane stability index. The application of SA to plants grown under salt-stressed plants further incremented the osmolytes, antioxidant enzymes, contents of endogenous abscisic acid (ABA), indole acetic acid (IAA), root length, and fresh and dry weights of roots. The results indicated that foliar application of SA proved to be effective in ameliorating the ill effects of salinity stress on maize plants. Pan et al. [53] worked out the involvement of hydrogen sulfide (H_2S) in SA-induced chilling stress tolerance in cucumber seedlings by using specific scavenger and inhibitor of H_2S. They found that under optimal conditions, the donor of H_2S or removing of endogenous H_2S by its scavenger (hypotaurine) or inhibitor (DL-propargylglycine) showed no impact on the level of endogenous SA, but SA induced H_2S endogenously and triggered the activities and mRNA transcript levels of L-/D-cysteine desulfhydrase (L-/D-CD). In addition, inhibiting endogenous SA with paclobutrazol or 2-aminoindan-2-phosphonic acid inhibited the effects of SA on H_2S. Nevertheless, both SA and H_2S donor NaHS minimized chilling-induced injury by lowering EL, MDA content, and ROS accumulation, improving the activities and mRNA levels of SOD, POD, CAT, APX, and GR, contents of ascorbate (AsA) and GSH contents and enhanced the expression of the chilling responsive genes (*ICE*, *CBF1*, and *COR*). It was postulated from this study that SA acts as an up-streaming signaling molecule under chilling tolerance of cucumber plants via increasing antioxidant defense system and modulating the expression of chilling stress-responsive genes. In a comparative study involving three deficit irrigation regimes, viz., 100, 80, and 60%

of crop evapo-transpiration and four levels of salinity, that is, 2, 4, 8, and 12 dS m^{-1}, the effect of SA in pearl millet and wheat plants was studied [54, 55]. Water and salinity stress caused significant reductions in grain yields, plant height, water and proline content, and Fv/Fm and Na/K^+ ratios in both pearl millet and wheat plants. Supply of SA caused a significant enhancement in plant height and grain yield and significantly improved all physiological, biochemical traits and consequently grain yield in both crop plants. In order to evaluate the participation of NR in SA-mediated improvements in water deficit tolerance in pepper plants, Kaya [56] applied 0.5 mM SA as a foliar spray. The results revealed that water stress caused decrement in total plant dry weight, Fv/Fm, contents of chlorophyll *a* and *b*, relative water content, leaf water potential but increment in MDA, H_2O_2, EL, methylglyoxal (MG), proline, and (nitric oxide) NO, NR and antioxidant enzymes' activities. The supply of SA reduced oxidative stress biomarkers, but triggered a marked increment in antioxidant defense system, enzymes of ascorbate-glutathione (AsA-GSH) cycle and glyoxalase system-glyoxalase I (Gly I) and glyoxalase II (Gly II), growth and photosynthetic traits, activities of NO, NR and proline content. In a recent study, the effect of foliar SA (0, 0.5, and 1 mM) under drought stress (90, 60, and 30% field capacity) in Portulaca oleracea L. was studied [57]. Imposition of drought stress showed a decrement in the contents of photosynthetic pigments, gas exchanges attributes, growth, and biomass production, soluble sugars, total phenolic, flavonoids and unsaturated fatty acids (FA) like oleic, linoleic and linolenic acid, and stearic and behenic acid but increased the contents of H_2O_2, MDA, and palmitic and arachidonic acid, respectively. Application of SA improved the growth, photosynthetic pigment contents, gas exchanges traits, FA contents, compatible solutes, and secondary metabolites, while simultaneously decreasing drought-induced oxidative stress biomarkers. With regard to heat stress, SA is effective in alleviating the damages on plants' physiology. Karpets et al. [58] studied the interplay of hydrogen sulfide (H_2S) in SA-mediated induction of heat resistance via the performance of antioxidant system in wheat plantlets. The data showed that H_2S actively participates under the protective influence of SA application in wheat heat stress tolerance. Khan et al. [59] exposed wheat plants to heat stress (40°C for 6 h) and studied the potential of 0.5 mM SA in alleviating the negative effects of heat stress on photosynthesis. The net photosynthesis (Pn) and activity of ribulose 1,5-bisphosphate carboxylase (RuBisCo) and photosynthetic nitrogen use efficiency (NUE) decreased under heat stress, but metabolism of proline was found to be increased. The application of SA alleviated heat stress by further increasing proline metabolism via the enhancement in the activity of γ-glutamyl kinase (GK) and decrement in proline oxidase (PROX), which maintained optimum osmotic and water potential for optimally maintaining the photosynthesis. Nonetheless, SA restricted the production of stress ethylene under heat-stressed plants to optimal range. The results suggested that SA supplementation alleviates heat stress effects by interacting with proline metabolism and ethylene formation to improve photosynthesis in wheat plants. In yet another related work, Kumar et al. [60] by using MALDI-TOF–TOF/MS analysis showed that spraying 100 mM SA alleviates the heat-induced (38°C) oxidative stress damage in wheat plants via modulation of the expression of heat-stable genes and proteins. Thus, SA application plays a pivotal role in alleviation the damages of myriads of abiotic stress in diverse crop plants.

4. Role of brassinosteroids

Brassinosteroids (BRs) are an important class of plant steroid-based phytohormone that plays multiple roles in regulating plant growth, development, and responses to environmental stresses [7, 8, 50, 51, 61–64]. 28-homobrassinolide (HBL) and 24-epibrassinolide (EBL) are the two isomers of BRs which regulate an array of plant physiological processes [65, 66]. BRs have structural resemblance with androgens, estrogens, corticoids, and ecdysteroids. The discovery of brassinolide (BL) dates back to 1979, and now about 70 compounds which are naturally occurring have been reported as free molecules or form conjugates with glucose and FA [67]. The isomers of BRs play a significant role in imparting abiotic stress tolerance in crop plants. In potato plants, Kolomeichuk et al. [68] studied the EBL-mediated (10^{-10} M) alleviation of NaCl toxicity (root application 100 mM) on photosynthetic processes. Salinity inhibited the contents of photosynthetic pigments, photosynthetic electron transport and photosystem II (PSII) maximal and effective quantum yields, oxidation of the plastoquinone (PQ). The treatment with 24-EBL under salinity stress maintained the efficiency of PSII photochemistry, PQ pool oxidation, and decrease in leaf osmotic potential. Amraee et al. [34] reported the 24-EBL-mediated (10^{-8} M) alteration on the extent and pattern of DNA cytosine methylation in Linum usitatissimum L. plants under salinity stress (150 mM) by using methylation-sensitive amplified polymorphisms (MSAP) technique. In conclusion, they found that the seed priming with 24-EBL play an important regulatory role via the epigenetic modification and induction of methylation in response to salinity toxicity. 24-EBL (10^{-8} mM) also confers tolerance against NaCl (100 mM) stress in soybean plants [33]. Exposure of soybean plants to 100 mM salt stress decreased growth, biomass yield, gas exchange traits, photosynthetic pigment contents, and chlorophyll fluorescence but increased H_2O_2, MDA, EL, and the antioxidant activities like SOD, CAT, accumulation of proline, and glycinebetaine (GB), total phenols, and total flavonoids. Application of 24-EBL (10^{-8} mM) to salt-stressed plants' upregulated the growth, biomass, yield, gas exchange and photosynthetic pigment contents, and chlorophyll fluorescence parameters but concomitantly reduced the accumulation of oxidative stress biomarkers. In a factorial experiment, Kaya et al. [69] studied the effect of 24-EBL (1.5 and 2.0 μM) application in regulating key physio-biochemical traits and oxidative defense system in two cultivars of maize, namely, PR32T83 and PR34N24 under two salinity levels (1.1 and 8.0 dS/m). Salinity levels decreased the fresh and dry weights of plant, yield of PS-II, chlorophyll contents, leaf water potential, and leaf K and Ca levels, but triggered an increase in membrane permeability, lipid peroxidation, H_2O_2, SOD, POD, CAT enzymes' activities, leaf Na and Cl, proline and GB, and leaf sap osmotic pressure. However, 24-EBL supply via both seed treatment and foliar application to the maize cultivars alleviated the detrimental effects of salt stress. In Cucumis sativus L. cv. Jinyan No. 4, plants grown under vermiculite and in hydroponic culture, Ahammed et al. [70] tested the efficacy of 24-EBL in alleviating organic pollutants (OPs), namely, 2,4,6-trichlorophenol, chlorpyrifos and oxytetracycline-retarded root elongation by unraveling the redox homeostasis and secondary metabolism-mediated underlying mechanisms. The results revealed that all the three tested OPs retarded elongation of roots, increased MDA, H_2O_2 and NO, while exogenous 24-EBL markedly decreased the accumulations of H_2O_2, NO and MDA *vis-a-vis* upregulating the expression of genes associated with antioxidant and detoxification systems. In case of pesticide stress, 24-EBL maintained the optimum synthesis of proteins and amino acids in Brassica juncea L. plants [71]. They applied 24-EBL (100 nM) via the seed presoaking treatment to study the changes in the synthesis of protein and amino acid contents grown under im-idacloprid (IMI)-challenged environment. Plants grown under IMI stress (300 mg/kg soil) showed a decrease in protein and 21 amino acid contents. However, application of 24-EBL restored the total protein as well as amino acid contents which was decreased under IMI amended soils. In a recent study laid out in a randomized block design under greenhouse, Jan et al. [72] applied 24-EBL (10^{-7} M) for 8 h as seed priming method to alleviate the toxic effects of Cr (10 mg/kg soil) in tomato plants. The results suggested that that 24-EBL application effectively stimulate growth of tomato plants exposed to Cr stress via the remodulation of the physiological, metabolic, and defense mechanisms and by maintained accumulation of Cr. Alam et al. [33] applied 24-EBL (10^{-7} M) in countering NaCl (100 mM) stress in soybean plants. Salinity decreased growth, gas exchange parameters, pigment contents, and chlorophyll fluorescence, Ca^{2+}, K^{+}, and P and biomass yield but H_2O_2, MDA, EL, MG content, the activities of SOD and CAT, accumulation of proline, GB, total phenols, Na^{+} uptake in root and shoot, and total flavonoids were found to be increased. Overall, 24-EBL supplementation provided tolerance to soybean plants by upregulating the antioxidant system, AsA-GSH cycle, and glyoxalase system. Yusuf et al. [73] under completely randomized block design studied the BRs-mediated protection of growth and N metabolism and the potentiality of antioxidant system under Ni stress (0, 50, 100, or 150 mg/kg) in two cultivars of Vigna radiata, namely, PDM-139 and T-44. The role of two doses (10^{-8} or 10^{-6} M) of HBL in alleviating the Ni-induced damages via the ability of antioxidant enzymes and proline content was investigated. It was found that the Ni imposes toxic impact on the plants in a concentration dependent manner. The supply of HBL was effective in countering Ni stress as its application conferred tolerance up to 100 mg/kg in improving growth, nodulation, photosynthesis and yield attributes consequently. Fariduddin et al. [74] exposed Cucumis sativus L. plants to chilling stress to study the ameliorative role of HBL through studying the growth, photosynthesis, enzymes, and biochemical parameters. Cucumber seedlings were exposed to two chilling regimes, that is, 10/8 and 5/3°C for 18 h. The spray treatment consisted 10^{-8} or 10^{-6} M of HBL applied at 30-day stage to plants. Chilling stress imposes marked reductions in growth, chlorophyll, net photosynthesis, efficiency of photosystem II and alterations in NR and carbonic anhydrase (CA) activities. In contrast, the activities of antioxidant enzymes viz-CAT, POD, and SOD along with the proline content increased in response to chilling stress. HBL supply exerted a protective role via the maintained higher values of antioxidant enzymes and proline content under chilling stress. In order to assess the role of HBL (10^{-8} M) under two Cd doses (3 and 9 mg/kg sand) in Solanum lycopersicum plants, Singh and Prasad [75] designed an experiment in plant growth chamber. Treatment of plants to Cd stress resulted in a significant decrease in enzymes associate with N metabolism. The application of exogenous HBL supply ameliorated the ill effects of Cd stress by triggering an increase in growth, photosynthesis, protein and carbohydrate content and inorganic N and N assimilating enzymes. Kaur et al. [76] proposed the model of HBL-mediated regulation of antioxidant enzyme activities and gene expression in Brassica juncea plants to counteract the combined salt-and temperature-induced oxidative stress. HBL treatments (0, 10^{-6}, 10^{-9}, and 10^{-12} M) was applied as seed priming on plants exposed to the low (4°C) or high (44°C) temperature and salinity (180 mM) stresses. The combined treatments of salt and temperature significantly reduced length but increased H_2O_2 content and the activities of SOD, CAT, APX, GR, DHAR, and MDHAR. Application of HBL improved growth traits, reduced H_2O_2 and the response of plants was dose-dependent.

5. Role of ethylene

Phytohormone "ethylene" (ET) is a small gaseous plant growth regulator which is responsible for ripening of fruits and is known for inducing triple response in plants [77]. The growth-regulating effects of ET were first observed by a Russian physiologist Dimitry K. Neljubov in 1901, however almost 60 years later in 1965 ET was established as a plant hormone [78]. Plant growth and senescence are chiefly regulated by the signaling mechanisms of ET [79–81]. However, it exerts its multifunctional role as one of the diversified signaling molecules to confer acclimatization to crop plants under various abiotic stresses. The signal perception of ET takes place at the membrane of endoplasmic reticulum and is bacterial two-component type which initiates a cascade signaling events through transcriptional regulation in the nucleus, thus eventually controlling a myriad of plant developmental events. Meanwhile, the interplay of ET with other signaling molecules provides adaptation to plants under various abiotic pressures [7, 8, 82–88]. Now, we will focus our attention on various reports on the functions of ET on tolerance toward major abiotic stresses in crop plants. Indian mustard plants were grown under 200 mg Zn/kg soil and the effects of 200 μL/L spray of ethephon at 20 days after sowing (DAS) was tested to alleviate the Zn toxicity. The results after 30 DAS revealed that exposure to Zn stress causes toxicity by inhibiting photosynthesis and growth, while increasing the oxidative stress biomarkers. ET application (as ethephon) to Zn-stressed plants restored photosynthesis and growth of mustard plants by decreasing oxidative stress via augmenting antioxidant activity, the metabolism of proline, glyoxalase system, and optimum nutrient homoeostasis [89]. In tomato plants, Alves et al. [90] provided new insights into the interactive role of ET on selenium-mediated antioxidant enzymes under Cd stress. Overall, the results suggested that cross talk of ET with Se manipulated the ET regulated expression of genes to induce Cd tolerance. The effect of Cr toxicity at 1.2 mM concentration given as $K_2Cr_2O_7$ after 10 DAS through nutrient solution on four cultivars of Brassica juncea, namely, Varuna, Pusa Bold, Rohini and SS2 were studied under ET supplementation [35]. Plants grown under Cr stress showed an enhanced accumulation of H_2O_2, MDA content, and superoxide production and a marked reduction in photosynthesis and growth traits. Ethylene (2-chloroethyl phosphonic acid, 200 μL/L) boosted the activities of enzymes associated with S assimilation, proline accumulation, anti-oxidant defense system, ATP-sulfurylase and serine acetyl transferase, and GSH level and resulted in inhibition of stress ET levels to the optimal range under Cr stress, thus increasing the photosynthesis and growth. The key role of ET was further substantiated by using ET action inhibitor, silver nitrate ($AgNO_3$) in the reversal of Cr-induced changes. In Brassica juncea plants, alleviatory role of ET (200 μL/L as ethephon) in the presence of 200 mg/kg soil each of Ni and Zn stress was investigated [91]. Application of ET reverses photosynthetic inhibition and promotes growth by inducing changes in PS II photochemistry, photosynthetic NUE, and modulation of antioxidant enzymes. The study got further confirmation by using ET action inhibitor-norbornadiene which reversed ET-induced changes. In Arabidopsis thaliana, Hartman et al. [92] studied the changes in NO and oxygen in controlling the stability of TFs in regulating of N-degron pathway and adaptation to flooding-induced hypoxia. They concluded that ET-mediated decrease in NO conferred tolerance to plants against hypoxia stress. The pretreatment of rice plants with the ET precursor under submergence stress plays a key role in the survival of rice plants [93]. ET signaling helped in the survival of rice seedlings under submergence stress by affecting intracellular redox homeostasis and antioxidant systems. Jegadeesan et al. [94] unraveled the proteomics involved in pollen of Solanum lycopersicum thermotolerance mechanisms mediated by ET. The quantitative analysis showed that heat stress-affected protein homeostasis and metabolism of carbohydrate and energy. ET (ethephon) increased pollen quality prior to heat stress exposure. According to gel LC-MS fractionation a total of 1355 proteins were identified. ET application shifted pollen proteome exposed to the heat stress closer to the proteome under non-stressful conditions and caused an upregulation into the protective mechanisms against the heat-induced oxidative stress by causing a significant higher abundance of glutathione-disulfide reductase, glutaredoxin, and protein disulfide isomerase. Iqbal et al. [95] studied the potential of exogenously sourced ET (200 μL/L) under nutrient deficiency (40 mg N/kg soil) at 40 DAS. ET supply increased the stomatal conductance, photosynthesis, and promotes growth under low nutrient availability in mustard plants. Application ET also alleviated salt stress response in various crop plants. In fact, the alteration of ET signaling mechanism significantly affects the response of plants to salt stress [96]. Freitas et al. [97] worked out the capacity of ET in triggering salinity stress tolerance in maize plants. During NaCl exposure, a biphasic ET production (at 5.5 and 12.5 h) and decrease in total polyamine content was noticed in salt-sensitive plants. In the salt-tolerant genotype, there was lack of stress ET biosynthesis and this was attributed to the 1-aminocyclopropane-1-carboxylic acid oxidase (ACO) and *ZmACO5b* gene downregulation activity and expression. The ET was involved in promoting salt stress tolerance by activating pathways of signaling by H_2O_2 that is dependent upon enzymes of polyamine catabolism. Application of ET also improved the tolerance to salt tolerance in Arabidopsis thaliana by maintaining K^+ content in shoots and roots to an appreciable extent [98]. Gharbi et al. [99] showed that inhibition of ET biosynthesis minimizes salt (125 mM) tolerance in Solanum chilense. The involvement of ET was also ascertained in the alleviation of salt-induced photosynthesis and growth reductions in Vigna radiata L. [100]. A central role of ET in BRs-induced alternative respiratory pathway which played an important role in abiotic stress tolerance in Cucumis sativus L. has also been reported [88]. A critical role of ET for the maintenance of root growth and mineral homeostasis under excess Fe stress in Arabidopsis thaliana has also been reported [101].

6. Role of methyl jasmonate

Methyl jasmonate (MJ) is one of the conjugates of jasmonic acids (JAs) and are important as well as ubiquitous phytohormone distributed in plants [102]. MJ is implicated in plant responses to salt, drought, metal/s, and heat stresses. MJ exert its alleviatory role under abiotic stresses by modulating plant morphology and physiology to a great extent. Taheri et al. [103] studied the effects of exogenous MJ-mediated (0, 60, and 120 mM) improvements in physiological and biochemical traits of *Anchusa italic* plants under salinity (5, 10, and 15 dS m^{-1}). Salinity stress significantly decreased growth characters, RWC, chlorophyll a, total chlorophyll, soluble sugars, protein, and K^+ contents but at the same time induced a marked increase in proline, Na^+, and Ca^{2+} contents. The application of MJ increases the aforementioned traits in plants grown under salt stress. In salt-tolerant rice genotypes, a higher concentration of MJ has been reported [104]. In wheat plants, application of 2.0 mM MJ ameliorated salt induced changes by boosting the expression of antioxidant enzymes, such as CAT, SOD and APX, and enhancing the GSH, chlorophyll b, and carotenoids contents [105]. In Hordeum vulgare, MJ-mediated (12 μM) regulation of genes of enzymes such as arginine decarboxylase, apoplastic invertase, and RuBisCo activase were found to be involved in imparting salt stress resistance [106]. The signaling of JAs promotes salt stress tolerance via the wheat allene oxide cyclase gene [107]. The wheat plants were exposed to saline stress conditions (12 dS m^{-1}) and the results showed that height of plant, leaf number per plant, tiller number, fresh and dry weight per plant, spike length, grain number per spike, grain weight per spike, and yield were reduced. MJ application mitigated the ill effects of salt stress by increasing the abovementioned traits in wheat plants [108]. In a polyethylene applied pot study, Tavallali and Karimi [109] studied the effect of foliar application of MJ (0, 0.025, 0.050, and 0.075 mM) in enhancing salinity stress (0, 50, 100, and 150 mM) mitigation of almond root stocks. The MJ-mediated improvement in salt stress tolerance was attributed to the restriction of gene expression and activity of cytokinin oxidase which results in the inhibition of cytokinin decrease. MJ imparted salt tolerance by regulating the endogenous biosynthesis of phytohormones, expression of antioxidant activity, and gas-exchange traits of almond rootstocks. The possible role of MJ (0.25 mM) on the physiological responses and growth of *Fragaria* × *ananassa* cv. "camarosa" under salinity stress (0, 30, and, 60 mM NaCl) was investigated. Application of 0.25 mM MJ boosted the activities of antioxidant enzymes that ameliorated the salt-induced salinity stress effects [110]. Priming rice plants with MJ was reported to ameliorate the polyethylene glycol-induced osmotic stress [111]. In a recent study involving wheat plants, Allagulova et al. [112] applied MJ (0.1 μM) to unravel its alleviatory role under water-induced stress by investigating the potential participation of dehydrins. The stress was imposed by increasing the concentration of mannitol from 3 to 5% in the growth medium. Wheat seedlings pretreated with 0.1 μM MJ found to reduce the level of drought-induced retardations and enhanced the accumulation of the transcripts of *TADHN* dehydrin and proteins as well. In order to evaluate the effects of drought under exogenously applied MJ in wheat cultivars (Yu-Mai-1, Lu-Nong-116, HuaiMai-19, Yan-Nong-19, Bai-Maizi), a simple randomized complete design experiment was executed [113]. The drought was imposed at three stages: jointing, heading, or grain filling. The application of MJ was accomplished one week after the exposure of plants to the imposition of drought stress. The results showed that drought stress substantially decreased the growth and yield of all wheat cultivars reflected in the form of plant height, peduncle length, dry biomass of different plant parts, and also in yield. The application of MJ improved the drought resistance mainly by maintaining growth and yield traits of wheat plants. A pot-culture experiment in order to investigate the role of MJ presoaking (50 μM) for 8 h treatment on maize plants under water stress in the form of water field capacity (WFC) 65, 55, and 45% was setup. MJ application lead to an increase in plant height, fresh and dry weight, the pigment levels total carbohydrates, total soluble sugar, polysaccharides, free amino acids, proline, total protein, activities of CAT, POX and SOD and contents of N, P, and K and endogenous IAA but a decrease in ABA levels of wheat plants under stress condition [114]. In Solanum lycopersicum, the role of JA accumulation in the modification of water deficit-induced stomatal closure was studied. The drought stress-induced the closure of stomata that was correlated with the accumulation of JA, but in grafts of WT and *def-1* mutant which does not accumulate JA, the stomata was found to be insensitive [115]. Improvement of drought stress tolerance by the application of MJ has also been reported in soybean genotypes (Giza 22 and 35) [116], pear [117], sunflower seedlings [118], and *Cistus albidus* L. [119] plants. Nonetheless, application of MJ (10 μM) has been found to reduce chilling injury in cucumber by inhibiting the production of H_2O_2 [120]. MJ-mediated improvements in chilling stress tolerance in cold-stored lemon fruits has also been reported [121]. In case of metal/s stress, MJ application played a significant role in ameliorating the damages in physiology and morphology of diverse crop plants. The effect of exogenous MJ in two Brassica napus L. cultivars (ZS 758—a black seed type, and Zheda 622—a yellow seed type) was investigated against exposure to arsenic toxicity. MJ alleviates As-induced oxidative damages by significantly increasing chlorophyll fluorescence, biomass production and concomitant reduction in MDA and ROS contents. MJ application also maintained the high pools of GSH and AsA along with enhanced transcript expression SOD, APX, CAT, POD, and secondary metabolites (*PAL, PPO, CAD*) and reduction in As uptake [36]. Zaid and Mohammad [31] applied 1 μM MJ to test its alleviating capacity against Cd (150 mg/kg) stress in Mentha arvensis plants. Cd applied through soil imposes restrictions in growth, physio-biochemical traits. The enhanced ROS accumulation and stomatal closure were also observed in plants grown under Cd stress. Application of MJ alleviated the Cd stress by reducing the damages in physio-biochemical traits along with reduced ROS detoxification. Pre-incubating Phaseolus coccineus plants with 10^{-5} M MJ has been found to minimize the Cu toxicity [122]. The identification of differentially expressed protein spots suggested the putative role of MJ in diverse physiological processes including oxidative tolerance and defense-related metabolism [123]. Thus, the above discussion clearly suggests that MJ alleviates the metal/s-induced oxidative stress by modulating the key physio-morphological traits in diverse crop plant to impart stress tolerance.

7. Conclusion and future outlook

Crop plants are exposed to myriads of abiotic stress conditions and undergo a significant retardation in growth and development, thus reducing crop productivity. Stresses are perceived, and modulation in the expression of genes encoding diverse proteins are induced in the signal transduction pathways in plants. Application of discussed phytohormones in the present chapter showed that the cross talk of phytohormones with various signaling agents under different stress conditions in diverse crop plants impart stress tolerance under stress pressures. A schematic representation on phytohormones-mediated abiotic stress tolerance in given in **Figure 1**.

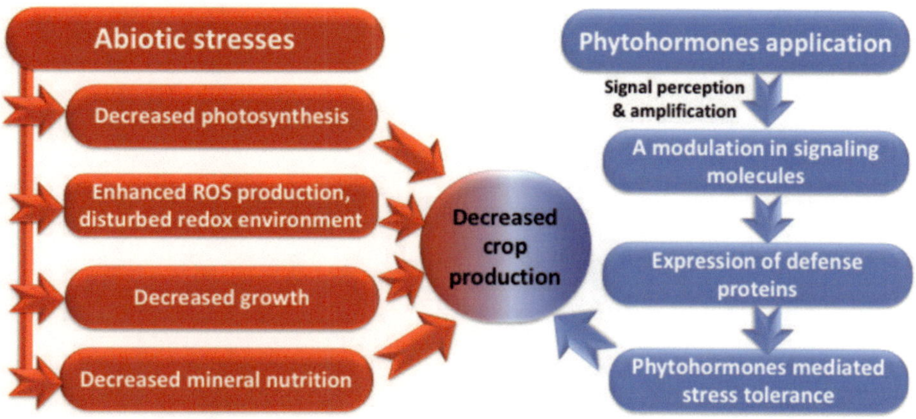

Figure 1.
Abiotic stress and phytohormones' signal transduction pathway in plants.

Author details

Haifa Abdulaziz S. Alhaithloul[1], Abdelghafar M. Abu-Elsaoud[2] and Mona H. Soliman[3*]

1 Biology Department, College of Science, Jouf University, Sakaka, Kingdom of Saudi Arabia

2 Botany Department, Faculty of Science, Suez Canal University, Ismailia, Egypt

3 Botany and Microbiology Department, Faculty of Science, Cairo University, Giza, Egypt

*Address all correspondence to: monahsh1@gmail.com

References

[1] FAO. How to feed the world in 2050. Food and Agriculture Organization, Rome, Italy. 2009. Available from: http://www.fao.org/fileadmin/templates/wsfs/docs/expert_paper/How_to_Feed_the_World_in_2050.pdf

[2] Sadiq Y, Zaid A, Khan MMA. Adaptive physiological responses of plants under abiotic stresses: Role of phytohormones. In: Plant Ecophysiology and Adaptation under Climate Change: Mechanisms and Perspectives I. Singapore: Springer; 2020. pp. 797-824

[3] IPCC Intergovernmental Panel on Climate Change. Climate change 2007: Impacts, adaptation and vulnerability. In: Parry ML, Canziani OF, Palutikof JP, van der Linden PJ, Hanson CE, editors. Contribution of Working Group II to Fourth Assessment Report of the Intergovernmental Panel on Climate Change. Cambridge: Cambridge University Press; 2007. p. 1000

[4] Abbas F. Analysis of a historical (1981-2010) temperature record of the Punjab province of Pakistan. Earth Interactions. 2013;**17**:1-23

[5] Aziz K, Daniel KYT, Muhammad ZA, Honghai L, Shahbaz AT, Mir A, et al. Nitrogen fertility and abiotic stresses management in cotton crop: A review. Environmental Science and Pollution Research. 2017;**24**:14551-14566. DOI: 10.1007/s11356-017-8920-x

[6] Fahad S, Nie L, Chen Y, Wu C, Xiong D, Saud S, et al. Crop plant hormones and environmental stress. Sustainable Agriculture Reviews. 2015;**15**:371-400

[7] Ahammed GJ, Li X, Liu A, Chen S. Brassinosteroids in plant tolerance to abiotic stress. Journal of Plant Growth Regulation. 2020a. DOI: 10.1007/s00344-020-10098-0

[8] Ahammed GJ, Gantait S, Mitra M, Yang Y, Li X. Role of ethylene crosstalk in seed germination and early seedling development: A review. Plant Physiology and Biochemistry. 2020b;**151**:124-131. DOI: 10.1016/j.plaphy.2020.03.016

[9] Ahmad B, Zaid A, Sadiq Y, Bashir S, Wani SH. Role of selective exogenous elicitors in plant responses to abiotic stress tolerance. In: Plant Abiotic Stress Tolerance. Cham: Springer; 2019. pp. 273-290

[10] Anjum NA, Gill SS, Gill R. Plant Adaptation to Environmental Change: Significance of Amino Acids and their Derivatives. 1st ed. Wallingford: CABI; 2014. DOI: 10.1079/9781780642734.0000

[11] Fahad S, Bajwa AA, Nazir U, Anjum SA, Farooq A, Zohaib A, et al. Crop production under drought and heat stress: Plant responses and management options. Frontiers in Plant Science. 2017;**8**:1147. DOI: 10.3389/fpls.2017.01147

[12] Fahad S, Chen Y, Saud S, Wang K, Xiong D, Chen C, et al. Ultraviolet radiation effect on photosynthetic pigments, biochemical attributes, antioxidant enzyme activity and hormonal contents of wheat. Journal of Food, Agriculture and Environment. 2013;**11**(3&4):1635-1641

[13] Fahad S, Hussain S, Bano A, Saud S, Hassan S, Shan D, et al. Potential role of phytohormones and plant growth-promoting rhizobacteria in abiotic stresses: Consequences for changing environment. Environmental Science and Pollution Research. 2014a;**22**(7):4907-4921. DOI: 10.1007/s11356-014-3754-2

[14] Fahad S, Hussain S, Matloob A, Khan FA, Khaliq A, Saud S, et al.

Phytohormones and plant responses to salinity stress: A review. Plant Growth Regulation. 2014b;**75**(2):391-404. DOI: 10.1007/s10725-014-0013-y

[15] Islam S, Zaid A, Mohammad F. Role of triacontanol in counteracting the ill effects of salinity in plants: A review. Journal of Plant Growth Regulation. 2020:1-10

[16] Khan MIR, Fatma M, Per TS, Anjum NA, Khan NA. Salicylic acid-induced abiotic stress tolerance and underlying mechanisms in plants. Frontiers in Plant Science. 2015;**6**:462. DOI: 10.3389/fpls.2015.00462

[17] Zaid A, Ahmad B, Jaleel H, Wani SH, Hasanuzzaman M. A critical review on iron toxicity and tolerance in plants: Role of exogenous Phytoprotectants. In: Aftab T, Hakeem K, editors. Plant Micronutrients. Cham: Springer; 2020c. Available from: http://doi-org-443.webvpn.fjmu.edu.cn/10.1007/978-3-030-49856-6_4

[18] Zaid A, Asgher M, Wani IA, Wani SH. Role of triacontanol in overcoming environmental stresses. In: Protective Chemical Agents in the Amelioration of Plant Abiotic Stress: Biochemical and Molecular Perspectives. USA: John Wiley & Sons Ltd; 2020b. pp. 491-509

[19] Zaid A, Bhat JA, Wani SH. Influence of metalloids and their toxicity impact on photosynthetic parameters of plants. In: Metalloids in Plants: Advances and Future Prospects. 2020a. pp. 113-124

[20] Mantri N, Patade V, Penna S, Ford R. Abiotic stress responses in plants: Present and future. In: Ahmad P, Prasad MNV, editors. Abiotic Stress Responses in Plants. New York, NY: Springer; 2012. pp. 1-19. DOI: 10.1007/978-1-4614- 0634-1_1

[21] Wani W, Masoodi KZ, Zaid A, Wani SH, Shah F, Meena VS, et al. Engineering plants for heavy metal stress tolerance. Rendiconti Lincei. Scienze Matematiche, Fisiche e Naturali. 2018;**29**(3):709-723

[22] ENS. Environment News Service: Report List World's 10 Worst Pollution Spots. 2006. Available from: http://www.google.pt/url?sa=t&rct=j&q=&esrc=s&source=web&cd=2&cad=rja&uact=8&ved=0CCgQFjAB&url=http%3A%2F%2Fwww.chem.unep.ch%2Fpb_and_cd%2FDocuments%2FNews%2F10%2520most%2520polluted%2520cities.pdf&ei=1zZgVb2iOKKR7AaMoIAo&usg=AFQjCNHOAyar_69TBPA143KXTtcAvkP3g&sig2=_lTZpa2YQnsF0slt0Xm7ew&bvm=bv.93990622,d.ZGU

[23] Wang W, Vinocur B, Altman A. Plant responses to drought, salinity and extreme temperatures: Towards genetic engineering for stress tolerance. Planta. 2003;**218**(1):1-14

[24] Lindsey R. Climate Change: Atmospheric Carbon Dioxide. 2020. Available from: https://www.climate.gov/news-features/understanding-climate/climate-change-atmospheric-carbon-dioxide

[25] NOAA, U.S Department of Commerce. 2020. National Oceanic and Atmospheric Administration. Available from: https://www.noaa.gov/news/june-2020-tied-as-earth-s-3rd-hottest-on-record

[26] Fahad S, Hussain S, Saud S, Hassan S, Chauhan BS, Khan F, et al. Responses of rapid viscoanalyzer profile and other rice grain qualities to exogenously applied plant growth regulators under high day and high night temperatures. PLoS One. 2016a;**11**(7):e0159590. DOI: 10.1371/journal.pone.0159590

[27] Fahad S, Hussain S, Saud S, Hassan S, Ihsan Z, Shah AN, et al. Exogenously applied plant

growth regulators enhance the morphophysiological growth and yield of rice under high temperature. Frontiers in Plant Science. 2016b;7:1250. DOI: 10.3389/fpls.2016.01250

[28] Fahad S, Hussain S, Saud S, Khan F, Hassan S, Jr A, et al. Exogenously applied plant growth regulators affect heat-stressed rice pollens. Journal of Agronomy and Crop Science. 2016c;**202**:139-150

[29] Raza A, Ashraf F, Zou X, Zhang X, Tosif H. Plant adaptation and tolerance to environmental stresses: Mechanisms and perspectives. In: Plant Ecophysiology and Adaptation under Climate Change: Mechanisms and Perspectives I. Singapore: Springer; 2020. pp. 117-145

[30] Surówka E, Rapacz M, Janowiak F. Climate change influences the interactive effects of simultaneous impact of abiotic and biotic stresses on plants. In: Plant Ecophysiology and Adaptation under Climate Change: Mechanisms and Perspectives I. Singapore: Springer; 2020. pp. 1-50

[31] Zaid A, Mohammad F. Methyl jasmonate and nitrogen interact to alleviate cadmium stress in *Mentha arvensis* by regulating physio-biochemical damages and ROS detoxification. Journal of Plant Growth Regulation. 2018;**37**(4):1331-1348

[32] Ahmad B, Jaleel H, Sadiq Y, Khan MMA, Shabbir A. Response of exogenous salicylic acid on cadmium induced photosynthetic damage, antioxidant metabolism and essential oil production in peppermint. Plant Growth Regulation. 2018;**86**(2):273-286

[33] Alam P, Albalawi TH, Altalayan FH, Bakht MA, Ahanger MA, Raja V, et al. 24-Epibrassinolide (EBR) confers tolerance against NaCl stress in soybean plants by up-regulating antioxidant system, ascorbate-glutathione cycle, and glyoxalase system. Biomolecules. 2019;**9**(11):640

[34] Amraee L, Rahmani F, Mandoulakani BA. 24-Epibrassinolide alters DNA cytosine methylation of *Linum usitatissimum* L. under salinity stress. Plant Physiology and Biochemistry. 2019;**139**:478-484

[35] Asgher M, Per TS, Verma S, Pandith SA, Masood A, Khan NA. Ethylene supplementation increases PSII efficiency and alleviates chromium-inhibited photosynthesis through increased nitrogen and sulfur assimilation in mustard. Journal of Plant Growth Regulation. 2018;**37**(4):1300-1317

[36] Farooq MA, Gill RA, Islam F, Ali B, Liu H, Xu J, et al. Methyl jasmonate regulates antioxidant defense and suppresses arsenic uptake in *Brassica napus* L. Frontiers in Plant Science. 2016;7:468

[37] Hanaka A, Wójcik M, Dresler S, Mroczek-Zdyrska M, Maksymiec W. Does methyl jasmonate modify the oxidative stress response in *Phaseolus coccineus* treated with Cu? Ecotoxicology and Environmental Safety. 2016;**124**:480-488

[38] Khanam D, Mohammad F. Plant growth regulators ameliorate the ill effect of salt stress through improved growth, photosynthesis, antioxidant system, yield and quality attributes in *Mentha piperita* L. Acta Physiologiae Plantarum. 2018;**40**(11):188

[39] Khan MIR, Jahan B, AlAjmi MF, Rehman MT, Khan NA. Ethephon mitigates nickel stress by modulating antioxidant system, glyoxalase system and proline metabolism in Indian mustard. Physiology and Molecular Biology of Plants. 2020b;**26**(6):1201-1213

[40] Aftab T, Khan MMA, Idrees M, Naeem M, Hashmi N. Methyl jasmonate

counteracts boron toxicity by preventing oxidative stress and regulating antioxidant enzyme activities and artemisinin biosynthesis in *Artemisia annua* L. Protoplasma. 2011;**248**(3):601-612

[41] Raskin I. Role of salicylic acid in plants. Annual Review of Plant Physiology and Plant Molecular Biology. 1992;**43**:439-463

[42] Koo YM, Heo AY, Choi HW. Salicylic acid as a safe plant protector and growth regulator. The Plant Pathology Journal. 2020;**36**(1):1

[43] Li X, Zhang L, Ahammed GJ, Li Y-T, Wei J-P, Yan P. Salicylic acid acts upstream of nitric oxide in elevated carbon dioxide-induced flavonoid biosynthesis in tea plant (*Camellia sinensis* L.). Environmental and Experimental Botany. 2019, 2019;**161**:367-374. DOI: 10.1016/j.envexpbot.2018.11.012

[44] Zaid A, Mohammad F, Wani SH, Siddique KM. Salicylic acid enhances nickel stress tolerance by up-regulating antioxidant defense and glyoxalase systems in mustard plants. Ecotoxicology and Environmental Safety. 2019;**180**:575-587

[45] Zaid A, Mohammad F, Fariduddin Q. Plant growth regulators improve growth, photosynthesis, mineral nutrient and antioxidant system under cadmium stress in menthol mint (*Mentha arvensis* L.). Physiology and Molecular Biology of Plants. 2020d;**26**(1):25-39

[46] Naeem M, Sadiq Y, Jahan A, Nabi A, Aftab T, Khan MMA. Salicylic acid restrains arsenic induced oxidative burst in two varieties of Artemisia annua L. by modulating antioxidant defence system and artemisinin production. Ecotoxicology and Environmental Safety. 2020;**202**:110851

[47] El Dakak RA, Hassan IA. The alleviative effects of salicylic acid on physiological indices and defense mechanisms of maize (*Zea Mays* L. Giza 2) stressed with cadmium. Environmental Processes. 2020;7:873-884. DOI: 10.1007/s40710-020-00448-1

[48] Es-sbihi FZ, Hazzoumi Z, Benhima R, Joutei KA. Effects of salicylic acid on growth, mineral nutrition, glandular hairs distribution and essential oil composition in *Salvia officinalis* L. grown under copper stress. Environmental Sustainability. 2020;**3**:199-208. DOI: 10.1007/s42398-020-00109-x

[49] Miao Y, Luo X, Gao X, Wang W, Li B, Hou L. Exogenous salicylic acid alleviates salt stress by improving leaf photosynthesis and root system architecture in cucumber seedlings. Scientia Horticulturae. 2020;**272**:109577

[50] Hussain MA, Fahad S, Rahat S, Muhammad FJ, Muhammad M, Qasid A, et al. Multifunctional role of brassinosteroid and its analogues in plants. Plant Growth Regulation. 2020a. DOI: 10.1007/s10725-020-00647-8

[51] Hussain SJ, Khan NA, Anjum NA, Masood A, Khan MIR. Mechanistic elucidation of salicylic acid and sulphur-induced defence systems, nitrogen metabolism, photosynthetic, and growth potential of Mungbean (*Vigna radiata*) under salt stress. Journal of Plant Growth Regulation. 2020b. DOI: 10.1007/s00344-020-10159-4

[52] Fahad S, Bano A. Effect of salicylic acid on physiological and biochemical characterization of maize grown in saline area. Pakistan Journal of Botany. 2012;**44**(4):1433-1438

[53] Pan DY, Fu X, Zhang XW, Liu FJ, Bi HG, Ai XZ. Hydrogen sulfide is required for salicylic acid–induced chilling tolerance of cucumber seedlings. Protoplasma. 2020:1-15

[54] Yadav T, Kumar A, Yadav RK, Yadav G, Kumar R, Kushwaha M. Salicylic acid and thiourea mitigate the salinity and drought stress on physiological traits governing yield in pearl millet-wheat. Saudi Journal of Biological Sciences. 2020;**27**(8):2010-2017. DOI: 10.1016/j.sjbs.2020.06.030

[55] Yang YX, Wu C, Ahammed GJ, Wu C, Yang Z, Wan C. Red light-induced systemic resistance against root-knot nematode is mediated by a coordinated regulation of salicylic acid, Jasmonic acid and redox signaling in watermelon. Frontiers in Plant Science. 2020;**9**(899):1-13. DOI: 10.3389%2Ffpls.2018.00899

[56] Kaya C. Nitrate reductase is required for salicylic acid-induced water stress tolerance of pepper by upraising the AsA-GSH pathway and glyoxalase system. Physiologia Plantarum. 2020. DOI: 10.1111/ppl.13153

[57] Saheri F, Barzin G, Pishkar L, Boojar MMA, Babaeekhou L. Foliar spray of salicylic acid induces physiological and biochemical changes in purslane (*Portulaca oleracea* L.) under drought stress. Biologia. 2020:1-12

[58] Karpets YV, Shkliarevskyi MA, Horielova EI, Kolupaev YE. Participation of hydrogen sulfide in induction of antioxidant system in roots of wheat plantlets and their heat resistance by salicylic acid. Applied Biochemistry and Microbiology. 2020;**56**(4):467-472

[59] Khan MIR, Iqbal N, Masood A, Per TS, Khan NA. Salicylic acid alleviates adverse effects of heat stress on photosynthesis through changes in proline production and ethylene formation. Plant Signaling & Behavior. 2013;**8**(11):e26374

[60] Kumar RR, Sharma SK, Goswami S, Verma P, Singh K, Dixit N, et al. Salicylic acid alleviates the heat stress-induced oxidative damage of starch biosynthesis pathway by modulating the expression of heat-stable genes and proteins in wheat (*Triticum aestivum*). Acta Physiologiae Plantarum. 2015;**37**(8):143

[61] Ahanger MA, Ashraf M, Bajguz A, Ahmad P. Brassinosteroids regulate growth in plants under stressful environments and crosstalk with other potential phytohormones. Journal of Plant Growth Regulation. 2018;**37**(4):1007-1024

[62] Li X, Wei JP, Ahammed GJ, Zhang L, Li Y, Yan P. Brassinosteroids attenuate moderate high temperature-caused decline in tea quality by enhancing theanine biosynthesis in *Camellia sinensis* L. Frontiers in Plant Science. 2018;**9**:1016

[63] Wei Z, Li J. Brassinosteroids regulate root growth, development, and symbiosis. Molecular Plant. 2016;**9**(1):86-100

[64] Zhang L, Ahammed GJ, Li X, Wei J-P, Li Y, Yan P, et al. Exogenous brassinosteroid enhances plant defense against colletotrichum gloeosporioides by activating phenylpropanoid pathway in Camellia sinensis L. Journal of Plant Growth Regulation. 2018;**37**(4):1235-1243. DOI: 10.1007/s00344-018-9857-0

[65] Hayat S, Yadav S, Wani AS, Irfan M, Ahmad A. Comparative effect of 28-homobrassinolide and 24-epibrassinolide on the growth, carbonic anhydrase activity and photosynthetic efficiency of *Lycopersicon esculentum*. Photosynthetica. 2011;**49**(3):397

[66] Siddiqui H, Ahmed KBM, Hayat S. Comparative effect of 28-homobrassinolide and 24-epibrassinolide on the performance of different components influencing the photosynthetic machinery in *Brassica juncea* L. Plant Physiology and Biochemistry. 2018;**129**:198-212

[67] Bajguz A, Chmur M, Gruszka D. Comprehensive overview of the Brassinosteroid biosynthesis pathways: Substrates, products, inhibitors, and connections. Frontiers in Plant Science. 2020;**11**:1034

[68] Kolomeichuk LV, Efimova MV, Zlobin IE, Kreslavski VD, Ol'ga KM, Kovtun IS, et al. 24-Epibrassinolide alleviates the toxic effects of NaCl on photosynthetic processes in potato plants. Photosynthesis Research. 2020:1-13

[69] Kaya C, Aydemir S, Akram NA, Ashraf M. Epibrassinolide application regulates some key physio-biochemical attributes as well as oxidative defense system in maize plants grown under saline stress. Journal of Plant Growth Regulation. 2018;**37**(4):1244-1257

[70] Ahammed GJ, He BB, Qian XJ, Zhou YH, Shi K, Zhou J, et al. 24-Epibrassinolide alleviates organic pollutants-retarded root elongation by promoting redox homeostasis and secondary metabolism in *Cucumis sativus* L. Environmental Pollution. 2017;**229**:922-931

[71] Sharma A, Kumar V, Thukral AK, Bhardwaj R. 24-epibrassinolide restores the synthesis of proteins and amino acids in *Brassica juncea* L. leaves under imidacloprid stress. Journal of Horticultural Research. 2017;**25**(2):85-90

[72] Jan S, Noman A, Kaya C, Ashraf M, Alyemeni MN, Ahmad P. 24-Epibrassinolide alleviates the injurious effects of Cr (VI) toxicity in tomato plants: Insights into growth, physio-biochemical attributes, antioxidant activity and regulation of Ascorbate–glutathione and Glyoxalase cycles. Journal of Plant Growth Regulation. 2020:1-18

[73] Yusuf M, Fariduddin Q, Ahmad I, Ahmad A. Brassinosteroid-mediated evaluation of antioxidant system and nitrogen metabolism in two contrasting cultivars of *Vigna radiata* under different levels of nickel. Physiology and Molecular Biology of Plants. 2014;**20**(4):449-460

[74] Fariduddin Q, Yusuf M, Chalkoo S, Hayat S, Ahmad A. 28-homobrassinolide improves growth and photosynthesis in *Cucumis sativus* L. through an enhanced antioxidant system in the presence of chilling stress. Photosynthetica. 2011;**49**(1): 55-64

[75] Singh S, Prasad SM. Effects of 28-homobrassinoloid on key physiological attributes of *Solanum lycopersicum* seedlings under cadmium stress: Photosynthesis and nitrogen metabolism. Plant Growth Regulation. 2017;**82**(1):161-173

[76] Kaur H, Sirhindi G, Bhardwaj R, Alyemeni MN, Siddique KH, Ahmad P. 28-homobrassinolide regulates antioxidant enzyme activities and gene expression in response to salt-and temperature-induced oxidative stress in *Brassica juncea*. Scientific Reports. 2018;**8**(1):1-13

[77] Taiz L, Zeiger E. Stress physiology. In: Plant Physiology. Sunderland, USA: Sinauer Associates. Inc; 2006. pp. 671-702

[78] Neljubow D. Uber die horizontale Nutation der Stengel von Pisum Sativum und einiger anderen Planzen. Beihefte zum Botanischen Centralblatt. 1901;**10**:128-139

[79] Dubois M, Van den Broeck L, Inzé D. The pivotal role of ethylene in plant growth. Trends in Plant Science. 2018;**23**(4):311-323

[80] Iqbal N, Khan NA, Ferrante A, Trivellini A, Francini A, Khan MIR. Ethylene role in plant growth, development and senescence:

Interaction with other phytohormones. Frontiers in Plant Science. 2017;**8**:475

[81] Qin H, He L, Huang R. The coordination of ethylene and other hormones in primary root development. Frontiers in Plant Science. 2019;**10**:874

[82] Husain T, Fatima A, Suhel M, Singh S, Sharma A, Prasad SM, et al. A brief appraisal of ethylene signaling under abiotic stress in plants. Plant Signaling & Behavior. 2020;**15**(9):e1782051-1-7

[83] Khan MIR, Trivellini A, Chhillar H, Chopra P, Ferrante A, Khan NA, et al. The significance and functions of ethylene in flooding stress tolerance in plants. Environmental and Experimental Botany. 2020;**179**:104-188. DOI: 10.1016/j.envexpbot.2020.104188

[84] Kolbert Z, Feigl G, Freschi L, Poór P. Gasotransmitters in action: Nitric oxide-ethylene crosstalk during plant growth and abiotic stress responses. Antioxidants. 2019;**8**(6):167

[85] Sharma A, Kumar V, Sidhu GPS, Kumar R, Kohli SK, Yadav P, et al. Abiotic stress management in plants: Role of ethylene. In: Molecular Plant Abiotic Stress. USA: John Wiley & Sons, Ltd; 2019. pp. 185-208. DOI: 10.1002/9781119463665.ch10.

[86] Sun X, Li X, Zhu J, Huang N, Bian X, Li H, et al. Polyamines and ethylene metabolism during cold acclimation in zoysia grass (*Zoysia Japonica* Steud.). Acta Physiologiae Plantarum. 2020;**42**(8):1-10

[87] Valluru R, Davies WJ, Reynolds MP, Dodd IC. Foliar abscisic acid-to-ethylene accumulation and response regulate shoot growth sensitivity to mild drought in wheat. Frontiers in Plant Science. 2016;**7**:461

[88] Wei L, Deng XG, Zhu T, Zheng T, Li PX, Wu JQ, et al. Ethylene is involved in brassinosteroids induced alternative respiratory pathway in cucumber (*Cucumis sativus* L.) seedlings response to abiotic stress. Frontiers in Plant Science. 2015a;**6**:982

[89] Khan MIR, Jahan B, Alajmi MF, Rehman MT, Khan NA. Exogenously-sourced ethylene modulates defense mechanisms and promotes tolerance to zinc stress in mustard (*Brassica juncea* L.). Plants. 2019;**8**(12):540

[90] Alves LR, dos Reis AR, Prado ER, Lavres J, Pompeu GB, Azevedo RA, et al. New insights into cadmium stressful-conditions: Role of ethylene on selenium-mediated antioxidant enzymes. Ecotoxicology and Environmental Safety. 2019;**186**:109747

[91] Khan MIR, Khan NA. Ethylene reverses photosynthetic inhibition by nickel and zinc in mustard through changes in PS II activity, photosynthetic nitrogen use efficiency, and antioxidant metabolism. Protoplasma. 2014;**251**(5):1007-1019

[92] Hartman S, Liu Z, Van Veen H, Vicente J, Reinen E, Martopawiro S, et al. Ethylene-mediated nitric oxide depletion pre-adapts plants to hypoxia stress. Nature Communications. 2019;**10**(1):1-9

[93] Huang YC, Yeh TH, Yang CY. Ethylene signaling involves in seeds germination upon submergence and antioxidant response elicited confers submergence tolerance to rice seedlings. Rice. 2019;**12**(1):1-8

[94] Jegadeesan S, Chaturvedi P, Ghatak A, Pressman E, Meir S, Faigenboim A, et al. Proteomics of heat-stress and ethylene-mediated thermotolerance mechanisms in tomato pollen grains. Frontiers in Plant Science. 2018;**9**:1558

[95] Iqbal N, Nazar R, Syeed S, Masood A, Khan NA.

Exogenously-sourced ethylene increases stomatal conductance, photosynthesis, and growth under optimal and deficient nitrogen fertilization in mustard. Journal of Experimental Botany. 2011;**62**(14):4955-4963

[96] Cao WH, Liu J, He XJ, Mu RL, Zhou HL, Chen SY, et al. Modulation of ethylene responses affects plant salt-stress responses. Plant Physiology. 2007;**143**(2):707-719

[97] Freitas VS, de Souza Miranda R, Costa JH, de Oliveira DF, de Oliveira Paula S, de Castro Miguel E, et al. Ethylene triggers salt tolerance in maize genotypes by modulating polyamine catabolism enzymes associated with H2O2 production. Environmental and Experimental Botany. 2018;**145**:75-86

[98] Yang L, Zu YG, Tang ZH. Ethylene improves *Arabidopsis* salt tolerance mainly via retaining K+ in shoots and roots rather than decreasing tissue Na+ content. Environmental and Experimental Botany. 2013;**86**:60-69

[99] Gharbi E, Martínez JP, Benahmed H, Lepoint G, Vanpee B, Quinet M, et al. Inhibition of ethylene synthesis reduces salt-tolerance in tomato wild relative species *Solanum chilense*. Journal of Plant Physiology. 2017;**210**:24-37

[100] Khan MIR, Asgher M, Khan NA. Alleviation of salt-induced photosynthesis and growth inhibition by salicylic acid involves glycinebetaine and ethylene in mungbean (*Vigna radiata* L.). Plant Physiology and Biochemistry. 2014;**80**:67-74

[101] Li G, Xu W, Kronzucker HJ, Shi W. Ethylene is critical to the maintenance of primary root growth and Fe homeostasis under Fe stress in *Arabidopsis*. Journal of Experimental Botany. 2015;**66**(7):2041-2054

[102] Ghasemi Pirbalouti A, Sajjadi SE, Parang K. A review (research and patents) on jasmonic acid and its derivatives. Archiv der Pharmazie. 2014;**347**:229-239

[103] Taheri Z, Vatankhah E, Jafarian V. Methyl jasmonate improves physiological and biochemical responses of *Anchusa italica* under salinity stress. South African Journal of Botany. 2020;**130**:375-382

[104] Kang DJ, Seo YJ, Lee JD, Ishii R, Kim KU, Shin DH, et al. Jasmonic acid differentially affects growth, ion uptake and abscisic acid concentration in salt-tolerant and salt-sensitive rice cultivars. Journal of Agronomy and Crop Science. 2005;**191**:273-282

[105] Qiu Z, Guo J, Zhu A, Zhang L, Zhang M. Exogenous jasmonic acid can enhance tolerance of wheat seedlings to salt stress. Ecotoxicology and Environmental Safety. 2014;**104**:202-208

[106] Walia H, Wilson C, Condamine P, Liu X, Ismail AM, Close TJ. Large-scale expression profiling and physiological characterization of jasmonic acid-mediated adaptation of barley to salinity stress. Plant, Cell & Environment. 2007;**30**(4):410-421

[107] Zhao Y, Dong W, Zhang N, Ai X, Wang M, Huang Z, et al. A wheat allene oxide cyclase gene enhances salinity tolerance via jasmonate signaling. Plant Physiology. 2014;**164**(2):1068-1076

[108] Islam MM, Islam S, Parvin S, Haque MA. Mitigation of salt stress by foliar application of Jasmonic acid in wheat. American Journal of Plant Biology. 2019;**4**(1):18-34

[109] Tavallali V, Karimi S. Methyl jasmonate enhances salt tolerance of almond rootstocks by regulating endogenous phytohormones, antioxidant activity and gas-exchange. Journal of Plant Physiology. 2019;**234**:98-105

[110] Faghih S, Ghobadi C, Zarei A. Response of strawberry plant cv.'Camarosa'to salicylic acid and methyl jasmonate application under salt stress condition. Journal of Plant Growth Regulation. 2017;**36**(3):651-659

[111] Sheteiwy MS, Gong D, Gao Y, Pan R, Hu J, Guan Y. Priming with methyl jasmonate alleviates polyethylene glycol-induced osmotic stress in rice seeds by regulating the seed metabolic profile. Environmental and Experimental Botany. 2018;**153**:236-248

[112] Allagulova C, Avalbaev A, Fedorova K, Shakirova F. Methyl jasmonate alleviates water stress induced damages by promoting dehydrins accumulation in wheat plants. Plant Physiology and Biochemistry. 2020;**155**:676-682. DOI: 10.1016/j.plaphy.2020.07.012

[113] Anjum SA, Tanveer M, Hussain S, Tung SA, Samad RA, Wang L, et al. Exogenously applied methyl jasmonate improves the drought tolerance in wheat imposed at early and late developmental stages. Acta Physiologiae Plantarum. 2016;**38**(1):25

[114] Abdelgawad ZA, Khalafaallah AA, Abdallah MM. Impact of methyl jasmonate on antioxidant activity and some biochemical aspects of maize plant grown under water stress condition. Agricultural Sciences. 2014;**5**(12):1077

[115] De Ollas C, Arbona V, Gómez-Cadenas A, Dodd IC. Attenuated accumulation of jasmonates modifies stomatal responses to water deficit. Journal of Experimental Botany. 2018;**69**(8):2103-2116

[116] Mohamed HI, Latif HH. Improvement of drought tolerance of soybean plants by using methyl jasmonate. Physiology and Molecular Biology of Plants. 2017;**23**(3):545-556

[117] Gao XP, Wang XF, Lu YF, Zhang LY, Shen YY, Liang Z, et al. Jasmonic acid is involved in the water-stress-induced betaine accumulation in pear leaves. Plant, Cell & Environment. 2004;**27**(4):497-507

[118] Andrade A, Escalante M, Vigliocco A, del Carmen Tordable M, Alemano S. Involvement of jasmonates in responses of sunflower (*Helianthus annuus*) seedlings to moderate water stress. Plant Growth Regulation. 2017;**83**(3):501-511

[119] Jubany-Marí T, Prinsen E, Munné-Bosch S, Alegre L. The timing of methyl jasmonate, hydrogen peroxide and ascorbate accumulation during water deficit and subsequent recovery in the Mediterranean shrub *Cistus albidus* L. Environmental and Experimental Botany. 2010;**69**(1):47-55

[120] Liu Y, Yang X, Zhu S, Wang Y. Postharvest application of MeJA and NO reduced chilling injury in cucumber (*Cucumis sativus*) through inhibition of H2O2 accumulation. Postharvest Biology and Technology. 2016;**119**:77-83

[121] Siboza XI, Bertling I, Odindo AO. Salicylic acid and methyl jasmonate improve chilling tolerance in cold-stored lemon fruit (*Citrus limon*). Journal of Plant Physiology. 2014;**171**(18):1722-1731

[122] Hanaka A, Maksymiec W, Bednarek W. The effect of methyl jasmonate on selected physiological parameters of copper-treated *Phaseolus coccineus* plants. Plant Growth Regulation. 2015;**77**(2):167-177

[123] Chen Y, Pang Q, Dai S, Wang Y, Chen S, Yan X. Proteomic identification of differentially expressed proteins in *Arabidopsis* in response to methyl jasmonate. Journal of Plant Physiology. 2011;**168**(10):995-1008

Chapter 13

Morphophysiological Traits, Biochemical Characteristic and Productivity of Wheat under Water and Nitrogen-Colimitation: Pathways to Improve Water and N Uptake

Nawab Ali and Mohammad Akmal

Abstract

Drought stress is the most prominent limiting factor and abiotic stress that manipulates the physiological pathway, biochemical traits and hence negatively affects wheat crop productivity. The global nitrogen (N) recovery indicated that about two-fifths of N inputs are lost in the ecosystems through emission, denitrification, gaseous loss, leaching, surface runoff and volatilization etc. Farmers are using higher rates of N to harvest maximum yield but about 50–60% of applied N to crop field is not utilized by the plants and are lost to environment causing environmental pollution. These deleterious environmental consequences need to be reduced by efficient management of N and/or water. N-availability is often regulated by soil water; hence crop is experiencing N- and water-limitation simultaneously. There is great impetus to optimize their uptake through interconnectedness of water and N for yield determination of wheat because of the water scarcity and N losses. It is further advocate that there is need to investigate the intricate role of economizing N rate and water simultaneously for wheat crop growth, yield and backing quality may be beneficial to be investigate.

Keywords: drought, water-N co-limitation, physio biochemical traits of wheat, N losses, NUE and WUE

1. Introduction

Wheat is one of the most important staple food crops supporting the world population. It ranks the second most important cereal crop of the world grown on about 220 million hectares (mha) area with 716 million tons food grain production and 3.2 tons per hectare productivity globally [1]. The current wheat production meets the demand for the existing population, but we will have to increase the production with time keeping in view to achieve the target of 858 million tons by 2050 against the current production globally [2]. Hence about 1.5% annual increase in next three decades should be kept under consideration to feed the overwhelming population. Similarly, about 37% of the wheat is cultivated as rainfed where drought is the most dominant constraint to limit wheat production [3]. Wheat is extensively grown in Asia, where Pakistan ranks eight in term of production (25 million tons) in the world [4]. The productivity of wheat is less than the 2.89 million tons per hectare for population with annual increment of 1.8% [5]. Wheat production affected by various biotic and abiotic stresses. Among the abiotic stress, water stress is affecting the vegetative growth and ultimately affecting yield of the wheat crop [6] (**Figure 1**).

Drought stress is the most prominent factor in crop production that affects the yield stability in the world [6]. Likewise, in rainfed areas the erratic rainfall is also the prime factor that limits crop productivity in in semiarid areas [7]. The occurrence of insufficient rainfall which does not meet the water requirement of the crop life cycle in certain areas occasionally cause the total crop failure [8]. The use of ground water is practice for crop production, and this ground water is declining with time with significant adverse effect on the water table due to the non-judicious use of the ground water resources [9]. Therefore, certain water conservation measures and water saving strategies and techniques should be adopted to cope with this adverse condition thereby conserving the precipitation water in the rhizosphere for the optimum crop production [10, 11].

As an essential macronutrient of plants, N is necessary for the vegetative growth, biomass, protein synthesis and the photosynthetic machinery of the crops [12]. This N is available in various forms in the rhizosphere, i.e. organic, such as in the form of amino acid and proteins, and inorganic forms, urea, nitrate ions ($NO3^-$), ammonium ions (NH4+) and is also found in the form of other nitrogenous compounds [13]. Of them, only NO3- and NH4+ are the forms of soil N that are easily taken by the plants through roots from the rhizosphere [14]. The aforesaid two forms need enough soil moisture for efficient absorption through the plant roots because of their high mobility in water [15]. The soil water is extracted by the plants through its roots and thereby using it for various physiological processes, maintaining internal temperature and for the process of transpiration [16]. Plant uptake various nutrients from soil solution including N for the optimum growth and development. In drought prone areas or areas where there is less and/or no eater availability because of the shortage of water or undependable nature of the

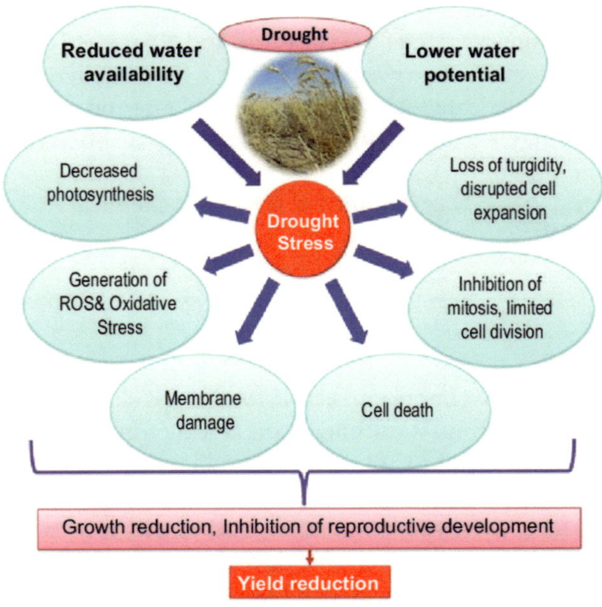

Figure 1.
Schematic representation of the drought effect on physiological processes, growth and yield.

rainfall, plants are with lower N uptake and hence the overall physiological process and activity of N assimilation enzyme is also retarded [17, 18]. Therefore, the plants ability to uptake N from the soil is not only the function of N concentration and availability in soil but also the soil moisture in the root zone [19].

Soil water and Nitrogen (N) are considered as the key limiting factors for maximizing plant growth, yield and quality [20]. The effect of these limiting factors is most prominent in the growing area where water stress is prevalent [21]. All the metabolic processes, biomass assimilation, growth functions and nutrients allocation in the plants are controlled by the interaction of N and water [22]. It is therefore kept in mind that wheat yield might be increased when crop ins not restricted with both N and water supply. For the availability of N optimum moisture in soil is needed. Contrary to this, enough amount of N needs to be in soil to make use of water available. In most areas, wheat yield is affected by the drought induced N deficit. Wheat morphology, physiology and biochemical character is also affected by water stress and nitrogen dynamics in the rhizosphere.

2. Morphophysiological, biochemicals and yield of wheat under water and N-colimitation

Water and N availability, the most limiting crop growth factors affecting adversely the morphological traits, physiology of the crop, alter the biochemical changes in the grains and shoots and final yield of the crop [23]. The resource use efficiency of N and water for the crop is the need of the day. Water stress adversely affects the vegetative growth, physiological processes, biochemical properties of the grains and shoots [24] (**Figure 2**). The amount of N applied to the crop is determined by the soil moisture status, amount and frequency and frequency of precipitation during the crop growth season [25]. The higher photosynthesis, water use efficiency (NUE) and water uptake could be achieved with optimum N supply under optimum water supply [26]. Leaf anatomy, leaf rolling, leaf hairs, stomatal positions and opening/closing, are the traits of wheat crop to withstand with under various environmental stresses especially water

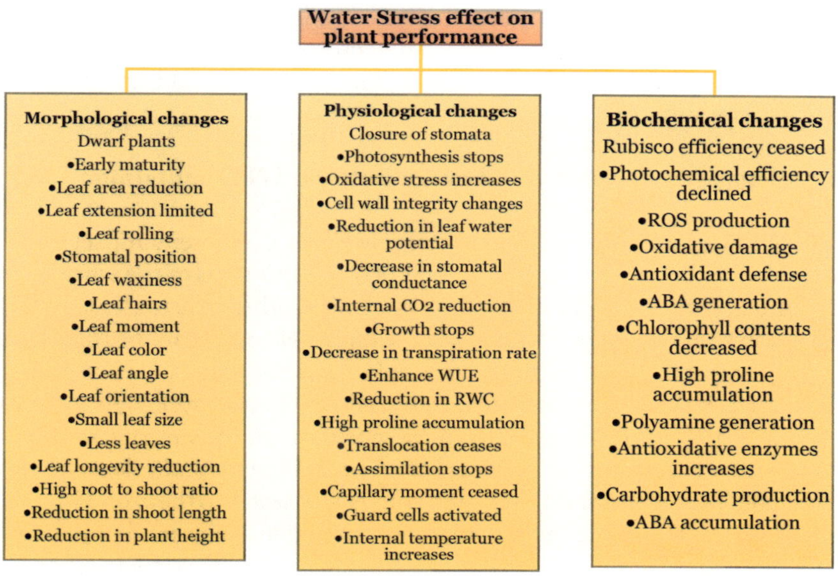

Figure 2.
Morphological, physiological and biochemical dynamics of plants affected by water stress.

deficit [27]. Likewise, the photosynthesis rate, transpiration, osmotic potential, relative water contents, stomatal conductance are also affected by water and N depletion [28]. Rubisco efficiency, photochemical efficiency, reactive oxygen species, antioxidants defense system, abscisic acid generation, proline accumulation, polyamines generation, increase in antioxidative enzymes, carbohydrate and protein dynamics, amylose and amylopectin contents are affected by under water and N colimitation [29].

2.1 Morphological traits of wheat under water and N colimitation

The strong association of WUE and NUE are observed commonly in different agriculture systems including crop husbandry [29]. Certain management practices which are in the way to enhance N and/or WUE at the same time are considered beneficial in term of production cost for the crops especially cereals thereby economizing water and N inputs separately [30]. The morphological attributes of the wheat crop especially the leaf traits i.e. leaf shape, size, expansion, leaf area, senescence, waxiness, leaf hairs and pubescence and the cuticle tolerance of the wheat is affected under water limited conditions [31, 32]. Similarly, the root characteristics i.e. dry weight, density and root length are adversely affected by soil moisture limitation in the rhizosphere [33]. Despite this, the vegetative and reproductive stages of the wheat crop are also affected by water deficit and limited N supply [34]. The better strategies adopted by the plants by modifying its life cycles to theses abiotic stresses (i.e. water and N deficit) are early maturation, small plant size and reduced leaf area for the drought tolerance [35]. It is observed that under water limitation, the flag leaf length is increased, and no significant change was observed in leaf width [36]. Hence, leaf expansion is the function of water extracted from the rhizosphere through rood and its abundance in plant tissues [37]. A decrease in leaf number, size and longevity may be observed under water and N limitation subsequently and their non availability alone as well [37]. In wheat crop, the leaf development is significantly affected by soil moisture depletion thereby decreasing overall biomass [38]. Similarly, N is considered as the integral part of the plant and plays a key role in increasing the leaf area, plant height and finally biomass of the plant under optimum supply [39]. Therefore, both water and N are equally important, their distribution and prevalence should keep in mind as there is a strong linkage between water and N limited supply [40]. Better morphological attributes with less production cost could be achieved under N and water colimitation.

2.2 Physiological response of wheat under water and N colimitation

In regulating plant growth, N plays vital role with a prominent impact on the crop physiology under drought induced N deficit [41]. The biomass of the wheat crop is not only affected under low N availability, but does significantly increased under optimum water supply, mainly because of the higher N content in the rhizosphere which clarifies that with the optimum N supply the presence of adequate availability of soil moisture is therefore essential for biomass, yield and physiological attributes of the wheat crop [42]. The N deficiency the sensitivity of the stomata under water deficit condition in drought prone areas [43]. The leaf photosynthesis rate is significantly affected by the optimum supply of both N and water to the crop [44]. Increase in the photosynthetic activity and biomass accumulation is observed under optimum water and limited N supply [45]. Water use efficiency (WUE) is enhanced through photosynthesis, stomatal conductance and mesophyll conductance in wheat crop and thus ultimately increasing N uptake and nitrogen use efficiency (NUE) [46].

Non availability of soil moisture will not only affect N supply but also have bad effect on N cycling [47]. Thus, water plays vital role in the availability of N in the soil solution to crop and thereby enhancing the photosynthesis process, translocation of assimilates, metabolism and conversion efficiency of the crop [48].

2.3 Biochemical traits of wheat crop under water stress and N limitation

Certain different compounds and metabolites that are accumulated in the crop under water deficit conditions to cope with the drought stress imposition [49]. These metabolites are of key importance because of their peculiar biochemical routes and thereby enhancing the tolerance mechanism in crop plants [50]. A cross talk between various biochemical substances and metabolites are initiated through drought signaling [51]. Various substances are accumulated in the plants body under abiotic stresses especially water stress. Under water limited conditions, proline a key metabolite, that is accumulated in higher concentration under dehydration [52]. The antioxidation enzymes that scavenge the ROS (reactive oxygen species) through various osmo-protection processes that includes stability of the membrane, osmotic adjustment and gene signaling are strongly associated [52]. The vital role in drought tolerance and strong defensive antioxidant mechanism in wheat genotypes are strongly correlated with high proline concentration [53]. The genotypes grown in rainfed areas having high proline concentration compared with the ones grown in irrigated and/or humid areas [54]. Various internal physiological processes are taking place under the crop plants to cope with the drought stress [55]. Plant growth is inhibited, water loss and stomatal closure are regulated, guard cell ion translocation is alternated under water stress through the translocation of ABA from roots to leaves [56]. The wheat crop having less ABA accumulation in leaves are considered to be susceptible to drought stress compared to those which accumulate high proline contents in leaves are drought tolerant [57]. Similarly, the wheat genotypes with high proline concentration in the leaves exhibited better water use efficiency [58]. In wheat crop auxin is considered as negative regulator for drought stress. Similarly, the indole acetic acid (IAA) contents also decreasing under water limitation [59]. The antioxidant enzymes (POD), H2O2, glutathionse (GSH), proline, and malondialdehyde (MDA) concentrations are measured maximum in the wheat plants subjected to water stress [60].

2.4 Yield and yield contributing traits

Most of the yield traits that are associated with the improved productivity in water limited areas are complex and polygenic in nature [61]. Drought stress significantly affected the yield and yield attributes of wheat compared with well water conditions [62]. Thousand kernel weight decreased by 16% under water limited condition thereby affecting overall grain production of wheat crop under drought prone environments [63]. Similarly, the most important yield contributing trait i.e. grain number were also affected by the water stress with less grain number in the crop under drought stress imposition [64]. All the yield components are affected by the drought stress, but the effect was more prominent on grain weight and grain number that contribute to the economic yield of the crop [65]. Water stress causes reduction in grain number to a greater degree that size of the grain, hence grain number exhibiting more plasticity that rest of the components [66]. The narrow variability and low plasticity of the grain size under water stress environments resulting in efficient transport of the pre anthesis assimilates to sink [67]. The reproduction process in the plants is severely affected by water stress, cereal crops exhibits a strong adaptive mechanism up to some extent through the accumulation of the photosynthates in various parts i.e. stem and roots in early growth phase and having the ability to reallocate the assimilates to the sink in reproductive phase of development [68]. The efficient use of the resources like limited water and N rates are the better strategy for wheat production [69]. For primary productivity photosynthetic capacity of the crop plants is the key to higher yield [70]. Water use efficiency provides a simplest mean of assessing whether yield is limited by water supply or other factors and is considered an important component of adaptation to water deficit conditions [71]. As the amount of water used is strongly affected by crop management, a good understanding of crop management effects on WUE may provide researchers opportunities to identify and select appropriate crop management practices for improved water use efficiency [72]. Limited supplemental irrigation and fertilization during the growth season can significantly increase WUE and wheat yield [73].

3. Pathways to improve water and N use efficiency

3.1 Approaches for enhancing water use efficiency (WUE)

Certain agronomic, soil management strategies that may improve the WUE are discussed as under.

3.1.1 Agronomic perspectives

The water use efficiency is the function of the agronomic practices that includes tillage practices, drought resistant and adaptive cultivars, soil fertility and pest management [74]. The crop growth and yield tends to increase under no till system or incorporation of the crop residues which directly influences the availability of soil moisture [75]. Similarly, WUE of the wheat crop enhanced under optimum sowing window, row spacing and seed rate which ultimately increments in wheat production [76]. Decreasing row spacing and increasing seed rate could enhance water soil moisture and thereby increasing water use efficiency [77]. Under optimum row spacing weeds growth is discouraged and hence crop may uptake nutrients and water efficiently [78]. Adjustment in the crop growth season through alteration of the sowing dates keeping in view the seasonal transpiration demand [79]. Early sown crop under conservation agriculture could also increase WUE of grain crops. Drought resistant cultivars resulting in high WUE than the cultivars which are drought susceptible [80]. A well-prepared sowing beds and leveled fields are the prerequisites for water saving and greater yield harvesting. The laser leveling is considered as the effective method responsible for various benefits i.e. germination uniformity, irrigation time reduction, decrease in pest and decreasing production cost [80]. The exogeneous application of certain compounds like silicon, selenium, potassium, hydrogel etc. increase the water use efficiency of the wheat crop under stressful environments, decreasing the adverse effects of biotic and abiotic stresses, enhance bioavailability of nutrients, photosynthetic efficiency and better light harvesting [81].

3.1.2 Soil management perspective

The crop yield and WUE could be significantly increased under better soil management practices through increasing water holding capacity (WHC), ease in soil moisture extraction and by decreasing the water losses through leaching [82]. Increment in WUE could be observed under good soil management practices, that causes soil moisture dynamics by evaporation and infiltration to soil that tends to increase soil water retention capacity and organic matter contents [83]. It is observed over various types and classes of soils, increment in water content at field capacity (FC) is greater than that at permanent wilting point (PWP) when soil organic matter contents are increased [84]. There is also a chance to conserve soil moisture through albedo under high plant density and narrow row spacing [85]. Compared to no tillage, disturbance taken place through tillage practices are more prominent in moisture loss through evaporation. The soil evaporation could be reduced up to 50% by crop residue incorporation [86]. This shows the gaps related to soil management that are responsible and having potential to improve WUE.

3.1.3 Physiological and anatomical perspective

3.1.3.1 Stomatal physiology and biochemistry dynamics

High stomatal conductance in needs more water to fix per unit leaf area for increasing the soil and water use during transpiration [87]. For increasing WUE, during peak hours of the day when transpiration is high by preventing the loss of water from the leaves through prevention by sensitivity of the stomata. Through the process of transpiration water is lost in the form vapor which facilitates CO2 uptake [88]. Thus stomata plays a vital role in increasing WUE, thereby controlling the water loss in the form of vapor and CO2 gain [89]. The opening and closing of stomata is also regulated by the environment determinants [90]. Stomatal moment is controlled by the transport and osmotically active solute accumulation and the guard cells lateral moments [91]. In most of the plant leaves, stomata are separated through an epidermal cell creating space for opening and closing [92]. In certain plant species, the clustering of stomata is also considered a better strategy for lowering evaporation demand and this phenomenon in the crop plants is considered better adaptation in drought prone areas [93].

3.1.3.2 Crop environment, root architecture and harvest index enhancement

Growing of the crops in environments where there is less transpiration will result in higher WUE. The identification and allocation of such areas having less transpiration could be identified through spatial and temporal scale [94]. Enhancement in harvest index increases the water use efficiency, this provides a new research window for increasing WUE in drought prone environments [95]. The harvest index is decreased by the water stress imposition at reproductive stage of the wheat crop, hence certain strategies need to be defined to conserve the soil moisture for the future use in reproductive stage of plant growth and development [96]. Similarly, root canopy and architecture are playing important role in extracting soil moisture [96]. Rood density, root angle, root hairs, root type and root length are the characteristics that determines the water uptake from various depths of the rhizosphere [97]. In wheat crop stele and xylem number are considered to be more responsive in wheat crop in water stress conditions because of the more xylem developmental plasticity in wheat crop [98]. In cereal crops, the metaxylem diameter and density are responsible for high WUE [99].

3.2 Approaches for enhancing nitrogen use efficiency (NUE)

The grain production per unit of soil available N is termed as nitrogen use efficiency. Certain approaches that should be kept in mind while enhancing the NUE of grain crop are described as under.

3.2.1 Agronomic perspective

Nitrogen is accepted universally as an essential component of crop production. The climatic condition is different for every crop. There are some of the main factors i.e. rainfall, solar radiation and temperature that significantly influence the demand for nitrogen, influence nitrogen use efficiency (NUE) and crop growth [100]. The demand of N to plant increases as plant grows showing a sigmoid curve (**Figure 3**). It has generally observed that nitrogen element significantly enhanced the biomass of the plant/crop if it absorbed by plant properly because maximum biomass and plant growth totally determined by the uptake and loss of nitrogen [101]. Yet, due to leaching and volatilization of nitrogen the crop cannot uptake nitrogen easily from the soil because the due to leaching the nitrogen moves down from the below root zone. Thus, efficient nitrogen management could significantly improve the nitrogen use efficiency [102].

Water availability is the essential factor that affected the NUE under limited water availability which is crucial from an environmental and economic standpoint [103]. Maximum nitrogen use efficiency (NUE) in drought tolerant varieties of winter wheat as compared with drought sensitive lines of winter wheat. Drought resistant varieties of wheat crop consistently produced higher yield as compare with drought sensitive varieties [104]. Higher nitrogen use efficiency (NUE) and water use efficiency (WUE) in drought resistant wheat varieties. Nitrogen rate parabolically increased with rainfall, while WUE and NUE were linearly related [105]. Simultaneous enhancement of NUE and WUE were more beneficial than optimization of nitrogen and water individually. It is also stated that low nitrogen application could enhanced the nitrogen use efficiency when supplied it from organic sources rather than commercial. Excess application of nitrogen to the crop could significantly reduce the crop yield and drastically decreased the nitrogen use efficiency. It is therefore, important to obtain the higher nitrogen use efficiency through adaptation of nitrogen management strategies and agronomic practices under deficit irrigation or water stress condition.

3.2.2 Rotation of crops

It is an effective strategy to enhance the nitrogen use efficiency. Effective use of soil resources i.e. nutrient and water and appropriate crop sequences could

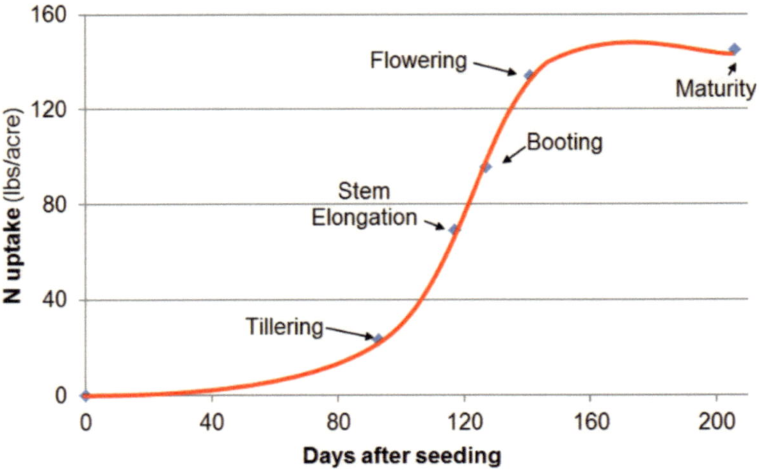

Figure 3.
Graphical representation of N needed for plant growth/N uptake pattern during life cycle the plant.

maximize the crop yield. Increase in crop yield is associated with crop rotation is known as rotation effect while monoculture practice significantly decreased the crop yield [106]. There is an effective management practice of legumes and non-leguminous crop which linearly enhanced the crop yield and other yield components. Monoculture practice in field significantly declined the cop yield and degrade the soil fertility, while crop rotation is an effective strategy to obtain the higher crop yield [107]. Leguminous crops should be included to enhance the crop yield and also suppress the weeds and decreased the risk of disease infestation. There are the following benefits of crop rotation which enhanced the nutrient use efficiency, water use efficiency, reduced allelopathy and improved soil quality [108]. Cereals are the largest consumers of nitrogen use efficiency as compared with leguminous crop. Crop rotation could reduce the immobilization and residual soil nitrate available for leaching. Leguminous crops increased the soil fertility, enhanced the soil microbial activity and improve the soil structure. Leguminous crop also add nitrogen to the soil from the atmosphere due to presence of nodules in the roots [109]. Rotation practices also enhance the nitrogen use efficiency (NUE). As compare to cereals crop, leguminous crops enhanced the soil fertility, improve soil texture and structure and enhance microbial activity. Rotation practice enhance the nitrogen use efficiency (NUE), However, they also stated from their study that nitrogen supplied to the crop following a legume have been inconsistent [110]. Several studies have been documented that NUE could enhance the crop yield when supplied it from organic sources as compared with inorganic sources. Crop rotation practice significantly increased the nitrogen use efficiency (NUE) and other components.

3.2.3 Irrigation management

Irrigation management is an appropriate strategy to improve the water use efficiency and fertilizer use efficiency. Border and furrow irrigation are traditional method used for vegetable and field crops [111]. These two practices are generally associated with higher percolation losses which resulted in low nitrogen use efficiency (NUE) and water use efficiency (WUE). These traditional methods are increasingly becoming difficult in the changing climate scenario. Adoption of water efficient irrigation and other modern irrigation techniques are the best strategies to obtain the higher NUE and WUE. Many researchers confirmed that deficit irrigation strategy enhance the WUE and NUE (water use and nitrogen use efficiency) [112]. According the several authors, maximum recovery efficiency, of nitrogen applied is due to high irrigation of overhead sprinklers. Strategy of water management is linked with the nitrogen nutrition and significantly affects the NUE and WUE [113]. Water application and their adjustment is to match the actual crop demand which is an effective management technique for obtaining higher NUE and WUE.

3.2.4 Nitrogen source and placement method

The uptake efficiency of most of the crops is dependent on the N source and application method. Nitrogen is lost (**Figure 4**) through urease hydrolysis to atmosphere when applied to the soil surface from urea based N source, this is because of the increased soil pH resulting its volatilization [114]. Farmers are increasing the application rate to their field for optimum productivity but in vain because this have a deleterious effect on soil, atmosphere and aquatic bodies as well [115]. Certain slow N release fertilizers are manufactured with the aim to reduce the losses [116]. The NUE could be maximized and losses are reduced under proper N application strategies [117]. Side dressing or band application, mixing in soil, injecting to soil sub surface of N fertilizers are the different strategies that could enhance the N

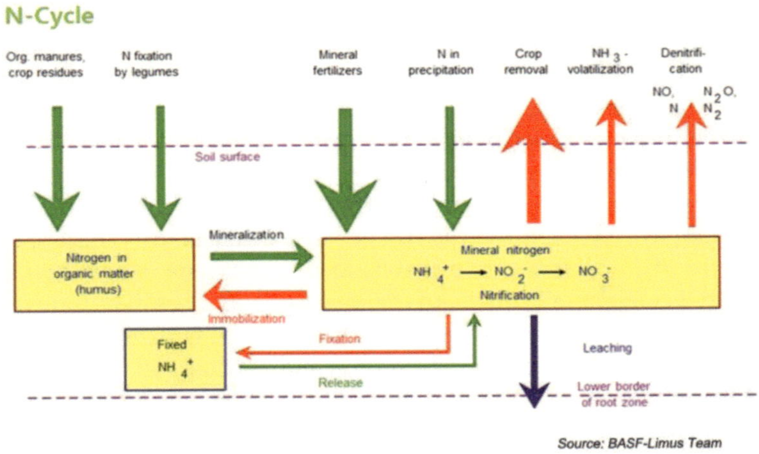

Figure 4.
Schematic diagram showing various processes of N cycle in soil plant atmosphere system.

efficiency than the broadcast application. Surface broadcast N application resulting in greater loss as ammonia-N to atmosphere, this could be minimized if banded, side dressed, placed with or below seed, or injected to soil [118].

3.2.5 Physiological perspective

The vital factor which is responsible for growth and development of the crop is nitrogen, increases various traits of the crop thereby improving final yield [119]. The N availability affects the carbon and N partitioning between roots and shoots of the crop [120]. Under high N availability, the root shoot ratio decreases because of the prominent growth of shoots [121], this varies among varieties and cultivars. The timing of N availability is of utmost importance in wheat crop where it affects all the yield parameters especially tiller density, spike compactness, grain number, kernel weight and grain N ad protein concentration [122]. Wheat crop store and sequester the N in vegetative stage, this stored N is re allocated to the grains during reproductive stage, of them some become the part of cell wall. In grain crops, about 50–70% of the N is translocated and accumulated at maturity [123]. The N harvest index could be improved by the sequestration of N in shoots and grains. Alteration in leaf area index and specific leaf N can enhance the NUE. Understanding these physiological characteristics that regulate plant N uptake, grain N uptake, root N uptake, assimilatory process, redistribution of N within the cell, and leaf structure and architecture can help improve NUE physiologically [124]. N uptake and LAI were found directly proportional for NUE and wheat productivity [125]. Under N- and water-limited conditions the overexpressing a downstream N assimilation gene called alanine amino transferase has increased N uptake and use efficiency [126]. Another way of increasing N uptake in plants is enhancing larger and deeper roots that can capture more available N in the soil profile. Vigorous root and shoot growth are necessary for higher N recovery [127].

4. The intricate relationship between nitrogen and water colimitation

Various literatures are present and research work has been done on the nitrogen and water stress individually but there is need to explore the intricate effect of both

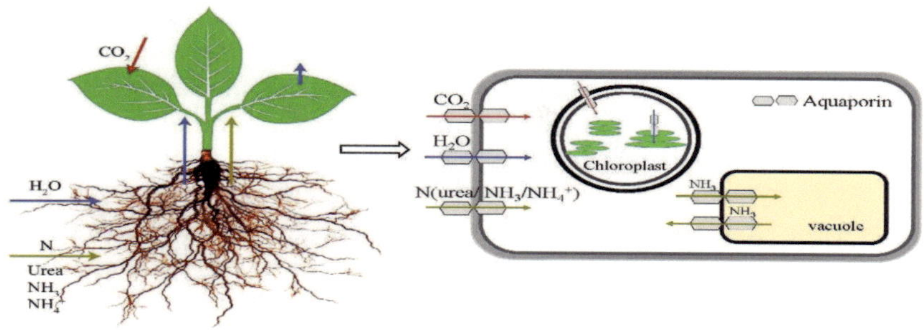

Figure 5.
The intricate relationship of water and N for optimum N uptake affecting plant physiological processes.

N and/or water (**Figure 5**) thereby decreasing cost of production of wheat crop and to enhance the grain and biomass yield [128]. Both the water and N rates need to be economized for optimum wheat production under high WUE and NUE simultaneously [129]. The effect of N and water on the process of photosynthesis is present in literature, claims the positive effect of N rate on photosynthesis. Similarly, with increase in soil moisture the physiological processes are positively influenced and results in optimum wheat production [130]. In fact, there are diminishing return that with the increase of N and water, the photosynthesis rate increases, but the rate of photosynthesis tends to saturate with further increment of water and N resulting in less WUE and NUE for the wheat crop and increasing cost of production [131]. Under certain environments, it is reported that NUE may increase under maximum supply of water and contrary to this N increment may also increase WUE [132]. Therefore, the interaction of N and water is the interplay of management, environment and cultivar collectively [133]. Application of irrigation water at higher rate may cause N losses through leaching resulting in N shortage in the rhizosphere [134], this condition also leads to stomatal closure thereby affecting the crop physiological processes. Under low N availability, photosynthesis process will be less leading to reduction in wheat productivity with less WUE [135]. On the other hand, the crop susceptibility to drought, temperature stress and may increase under high N application. Plant species differs in demand for N and water that affects the photosynthesis process. The studies also suggest that increasing both N and water has positive effect on yield rather than increment in either factor. This increment in both N and water leads to higher production cost, water loss, affecting water quality, aquatic flora and fauna and ultimately causes pollution of water and air as well [136]. It is thought that N input effect will be more efficient at low water input that the higher water supply. Similarly, the NUE decreases with higher N rate and increment in water input may increase NUE at lower N rate.

5. Nitrogen (N)-climate interaction; fix N to fix climate change

The wide spread use of synthetic fertilizers in agriculture [137, 138], fossil fuel use from the industry, leguminous crops cultivation on large scale has therefore increased the reactive nitrogen in the atmosphere in the last five decades [139–141]. Globally, a great threat is there to impair the water and air quality [142], biodiversity and human health due to unpredictable alteration in biogeochemical cycle especially N [143]. The considerable amount of N is lost through leaching in the form of NO3-N [144, 145]. This NO3-N leaching varies due to the N status of the soil and greatly depends on soil properties [146–149]. Thus, the shallow ground water and aquifers are contaminated through the soluble nature of the NO3 in soil water [150, 151].

Worldwide, NO3-N is considered as the major concern as it deteriorates the ground water and surface water quality since few decades [152–154]. Increment in the NO3-N contents in drinking water have bad effect on human and animals health as well [155, 156], also the aquatic bodies are contaminated through its higher concentration accumulation [157–160] thereby favoring the aquatic plants to grow and nourish well [161, 162] by making a surface layer thereby impeding the solar radiation to reach the aquatic fauna [163–166] and ultimately causing the death of the aquatic fauna [167] which creates eutrophication [168]. Besides this, blue baby syndrome in infants (methemoglobinemia) and human birth defects [169] are the major serious problems linked with NO3 contamination [170–171]. Therefore, it is of prime importance to estimate the NO3 leaching from cropland [172] and to evaluate the environmentally compatible and economically suitable agronomic practices [173] that could maintain and/or decrease this phenomenon [174].

The increment in the NO3 leaching loss is strongly correlated with the rainfall occurrence, intensity and irrigation volume and methods [175]. The excessive fertilization increases the risk of N loss through leaching in agriculture system [176, 177]. To feed the overwhelming population of the world, higher fertilizer application to crop field [178] is under practice have rapidly raised the NO2 emission of the atmosphere [179, 180] and is expected to increase linearly in coming decades [188], aggravating the environmental quality [181]. Through the anthropogenic rise in NO2 emission from agriculture sector to atmosphere thereby offset the efforts to decrease the CO2 sully from other sectors [182].

Global warming is directly affected by NO2 emission [183] which is considered as a major greenhouse gas [192], having 300 times more potent than CO2 because of its greater resident time in the atmosphere usually greater than 100 years [184]. The unregulated and increasing trend of the NO2 defines the global warming [185]. This increment in N availability in atmosphere affects the sinks and sources of NO2, CH4 and CO2 thereby potential increase in CO2 sequestration [186].

Climate change will certainly increase the N emission, which in turn will affect the agriculture system, aquatic water bodies, humans and animals' health [187]. Thus, climate N cross talk will expose the crops to elevated O3 [188]. Greater emission of the N will be observed from livestock and agriculture production systems as well [189]. Drought spell, erratic and unpredictable rainfall, temperature events taken place because of climate change will ultimately affects the farmers decisions and practices thereby enhancing C and N release from agriculture ecosystem [190] when higher N rates are practiced keeping in mind leaching losses [191]. N availability to crops impedes and its losses will increase under drought induction, irregularity in rainfall, high temperature events and flooding [192].

Nitrogen, the essential macronutrient responsible for plant growth and development is fundamental for farming [191]. Climate change impact may be taken under consideration for minimizing N losses [193, 194] and to manage N in better way [195–197]. Perturbation to both N and climate needs to be fixed to reduce its adverse effects environmentally compatible and economically suitable agronomic practices [198]. Firstly, irrigation water availability to field crops is a great question and this the situation become worsen in each passing day. Then the application volume to field crops in higher rates decreasing the WUE on one way and depleting the available nutrients in the soil on other hand because farmers are applying higher fertilizers to harvest maximum yield thereby ultimately increasing their cost of production. Therefore, a gate way for future research is to economize the N rates/levels for the crop growth, yield and quality by monitoring the rhizosphere moisture for irrigation water optimization till moisture allowable depletion (MAD) of the respective crops that may not affect its growth and development but also enhance the availability and utilization of the nutrients in soil solution.

6. Conclusion

This chapter provides the information regarding water and N rates on wheat physiological, morphological, biochemical, qualitative changes and yield traits of wheat crop focusing on the NUE and WUE. Drought stress is the most prominent limiting factor that manipulates the physiological pathway, biochemical traits and hence negatively affects wheat crop productivity. The global nitrogen (N) recovery indicated that about two-fifths of N inputs are lost in the ecosystems through leaching, gaseous emission, denitrification, surface runoff and volatilization. About 50–60% of the applied N fertilizer to the field are lost to environment and thus not available to the plants. These deleterious environmental consequences need to be reduced by efficient management of N and/or water focusing on their intricate relationship. N-availability is often regulated by soil water; hence crop is experiencing N- and water-limitation simultaneously. There is great impetus to optimize their uptake through interconnectedness of water and N for yield determination of wheat because of the water scarcity and N losses for enhancing wheat productivity and thereby reducing the production cost. There is need to optimize the N and water colimitation simultaneously to determine the minimum N and water to harvest optimum yield with better baking quality and hence safeguarding the environment, soil micro fauna and water bodies.

Therefore, a gate way for future research is to economize the N rates/levels for the crop growth, yield and quality by monitoring the rhizosphere moisture for irrigation water optimization till moisture allowable depletion (MAD) of the respective crops that may not affect its growth and development but also enhance the availability and utilization of the nutrients in soil solution.

Conflict of interest

The authors declare no conflict of interest.

Author details

Nawab Ali* and Mohammad Akmal
Department of Agronomy, Faculty of Crop Production Sciences, The University of Agriculture Peshawar, Khyber Pakhtunkhwa, Pakistan

*Address all correspondence to: nawab@aup.edu.pk

References

[1] Cakmak I, Kutman UÁ. Agronomic biofortification of cereals with zinc: a review. European Journal of Soil Science. 2018 Jan;69(1): 172-80.

[2] Fischer G. World food and agriculture to 2030/50. Tech. Pap. from Expert Meet. How to Feed World. 2009 Jun 24:24-6.

[3] Sedri MH, Amini A, Golchin A. Evaluation of nitrogen effects on yield and drought tolerance of rainfed wheat using drought stress indices. Journal of Crop Science and Biotechnology. 2019 Sep 1;22(3):235-42.

[4] Younis H, Abbas G, Naz S, Fatima Z, Ali MA, Ahmed M, Khan MA, Ahmad S. Advanced production technologies of wheat. InAgronomic Crops 2019 (pp. 223-236). Springer, Singapore.

[5] Shiferaw B, Smale M, Braun HJ, Duveiller E, Reynolds M, Muricho G. Crops that feed the world 10. Past successes and future challenges to the role played by wheat in global food security. Food Security. 2013 Jun 1;5(3):291-317.

[6] Fahad S, Bajwa AA, Nazir U, Anjum SA, Farooq A, Zohaib A, Sadia S, Nasim W, Adkins S, Saud S, Ihsan MZ. Crop production under drought and heat stress: plant responses and management options. Frontiers in plant science. 2017 Jun 29;8:1147.

[7] Singh HP. Unit-1 Definition and Concept of Rain Fed Farming. IGNOU.

[8] Van Loon AF. Hydrological drought explained. Wiley Interdisciplinary Reviews: Water. 2015 Jul;2(4):359-92.

[9] Sahrawat KL, Wani SP, Pathak P, Rego TJ. Managing natural resources of watersheds in the semi-arid tropics for improved soil and water quality: A review. Agricultural Water Management. 2010 Mar 1;97(3):375-81.

[10] Ullah H, Santiago-Arenas R, Ferdous Z, Attia A, Datta A. Improving water use efficiency, nitrogen use efficiency, and radiation use efficiency in field crops under drought stress: A review. InAdvances in Agronomy 2019 Jan 1 (Vol. 156, pp. 109-157). Academic Press.

[11] Kumar A, Nayak AK, Das BS, Panigrahi N, Dasgupta P, Mohanty S, Kumar U, Panneerselvam P, Pathak H. Effects of water deficit stress on agronomic and physiological responses of rice and greenhouse gas emission from rice soil under elevated atmospheric CO2. Science of the Total Environment. 2019 Feb 10;650:2032-50.

[12] Nowicka B, Ciura J, Szymańska R, Kruk J. Improving photosynthesis, plant productivity and abiotic stress tolerance–current trends and future perspectives. Journal of plant physiology. 2018 Dec 1;231:415-33.

[13] Manikandan A, Subramanian KS. Ability of Urea Impregnated Biochar Fertilizers For Securing the Slow Release of Nitrogen in Soils–Preliminary Study. International Journal of Agriculture Sciences, ISSN. 2015:0975-3710.

[14] Hill PW, Jones DL. Plant-microbe competition: does injection of isotopes of C and N into the rhizosphere effectively characterise plant use of soil N. New Phytol. 2019 Jan 1;221:796-806.

[15] Tardieu F, Simonneau T, Muller B. The physiological basis of drought tolerance in crop plants: a scenario-dependent probabilistic approach. Annual review of plant biology. 2018 Apr 29;69:733-59.

[16] Kathpalia R, Bhatla SC. Plant Mineral Nutrition. InPlant Physiology,

Development and Metabolism 2018 (pp. 37-81). Springer, Singapore.

[17] Kitonyo OM, Sadras VO, Zhou Y, Denton MD. Nitrogen supply and sink demand modulate the patterns of leaf senescence in maize. Field Crops Research. 2018 Aug 1;225:92-103.

[18] Antoniadis V, Levizou E, Shaheen SM, Ok YS, Sebastian A, Baum C, Prasad MN, Wenzel WW, Rinklebe J. Trace elements in the soil-plant interface: Phytoavailability, translocation, and phytoremediation–A review. Earth-Science Reviews. 2017 Aug 1;171:621-45.

[19] Leghari SJ, Wahocho NA, Laghari GM, HafeezLaghari A, MustafaBhabhan G, HussainTalpur K, Bhutto TA, Wahocho SA, Lashari AA. Role of nitrogen for plant growth and development: A review. Advances in Environmental Biology. 2016 Sep 1; 10(9):209-19.

[20] Jaleel CA, Manivannan PA, Wahid A, Farooq M, Al-Juburi HJ, Somasundaram RA, Panneerselvam R. Drought stress in plants: a review on morphological characteristics and pigments composition. Int. J. Agric. Biol. 2009 Jan 1;11(1):100-5.

[21] Hermans C, Hammond JP, White PJ, Verbruggen N. How do plants respond to nutrient shortage by biomass allocation?. Trends in plant science. 2006 Dec 1;11(12):610-7.

[22] Hussain M, Farooq S, Hasan W, Ul-Allah S, Tanveer M, Farooq M, Nawaz A. Drought stress in sunflower: Physiological effects and its management through breeding and agronomic alternatives. Agricultural water management. 2018 Mar 31;201:152-66.

[23] Prasad PV, Staggenborg SA, Ristic Z. Impacts of drought and/or heat stress on physiological, developmental, growth, and yield processes of crop plants. Response of crops to limited water: Understanding and modeling water stress effects on plant growth processes. 2008 Dec 22;1:301-55.

[24] Chakraborty D, Nagarajan S, Aggarwal P, Gupta VK, Tomar RK, Garg RN, Sahoo RN, Sarkar A, Chopra UK, Sarma KS, Kalra N. Effect of mulching on soil and plant water status, and the growth and yield of wheat (*Triticum aestivum* L.) in a semi-arid environment. Agricultural water management. 2008 Dec 1;95(12):1323-34.

[25] Noor MA. Nitrogen management and regulation for optimum NUE in maize–A mini review. Cogent Food & Agriculture. 2017 Jan 1;3(1):1348214.

[26] Tripathi DK, Singh VP, Chauhan DK, Sharma S, Prasad SM, Dubey NK, Ramawat N, editors. Plant Life under Changing Environment: Responses and Management. Academic Press; 2020 Apr 21.

[27] Santander C, Aroca R, Ruiz-Lozano JM, Olave J, Cartes P, Borie F, Cornejo P. Arbuscular mycorrhiza effects on plant performance under osmotic stress. Mycorrhiza. 2017 Oct 1; 27(7):639-57.

[28] Abid M, Tian Z, Zahoor R, Ata-Ul-Karim ST, Daryl C, Snider JL, Dai T. Pre-Drought Priming: A Key Drought Tolerance Engine in Support of Grain Development in Wheat. InAdvances in agronomy 2018 Jan 1 (Vol. 152, pp. 51-85). Academic Press.

[29] Spiertz JH. Nitrogen, sustainable agriculture and food security: a review. InSustainable agriculture 2009 (pp. 635-651). Springer, Dordrecht.

[30] Ahmad Z, Waraich EA, Akhtar S, Anjum S, Ahmad T, Mahboob W, Hafeez OB, Tapera T, Labuschagne M, Rizwan M. Physiological responses

of wheat to drought stress and its mitigation approaches. Acta Physiologiae Plantarum. 2018 Apr 1;40(4):80.

[31] Blum A. Plant water relations, plant stress and plant production. InPlant breeding for water-limited environments 2011 (pp. 11-52). Springer, New York, NY.

[32] Zang U, Goisser M, Häberle KH, Matyssek R, Matzner E, Borken W. Effects of drought stress on photosynthesis, rhizosphere respiration, and fine-root characteristics of beech saplings: A rhizotron field study. Journal of plant nutrition and soil science. 2014 Apr;177(2):168-77.

[33] Zhang X, Wang Y, Sun H, Chen S, Shao L. Optimizing the yield of winter wheat by regulating water consumption during vegetative and reproductive stages under limited water supply. Irrigation Science. 2013 Sep 1;31(5):1103-12.

[34] Mega R, Abe F, Kim JS, Tsuboi Y, Tanaka K, Kobayashi H, Sakata Y, Hanada K, Tsujimoto H, Kikuchi J, Cutler SR. Tuning water-use efficiency and drought tolerance in wheat using abscisic acid receptors. Nature plants. 2019 Feb;5(2):153-9.

[35] Bais HP, Weir TL, Perry LG, Gilroy S, Vivanco JM. The role of root exudates in rhizosphere interactions with plants and other organisms. Annu. Rev. Plant Biol.. 2006 Jun 2;57:233-66.

[36] Ackerly D. Functional strategies of chaparral shrubs in relation to seasonal water deficit and disturbance. Ecological Monographs. 2004 Feb;74(1):25-44.

[37] Dijkstra FA, Cheng W. Moisture modulates rhizosphere effects on C decomposition in two different soil types. Soil Biology and Biochemistry. 2007 Sep 1;39(9):2264-74.

[38] Freschet GT, Swart EM, Cornelissen JH. Integrated plant phenotypic responses to contrasting above-and below-ground resources: Key roles of specific leaf area and root mass fraction. New Phytologist. 2015 Jun;206(4):1247-60.

[39] Li SX, Wang ZH, Hu TT, Gao YJ, Stewart BA. Nitrogen in dryland soils of China and its management. Advances in Agronomy. 2009 Jan 1;101:123-81.

[40] Kamran M, Shahbaz M, Ashraf M, Akram NA. Alleviation of drought-induced adverse effects in spring wheat (*Triticum aestivum* L.) using proline as a pre-sowing seed treatment. Pak. J. Bot. 2009 Apr 1;41(2):621-32.

[41] Hawkesford MJ. Reducing the reliance on nitrogen fertilizer for wheat production. Journal of cereal science. 2014 May 1;59(3):276-83.

[42] Husen A, Iqbal M, Aref IM. Plant growth and foliar characteristics of faba bean (*Vicia faba* L.) as affected by indole-acetic acid under water-sufficient and water-deficient conditions. Journal of Environmental Biology. 2017 Mar 1;38(2):179.

[43] Abid M, Tian Z, Ata-Ul-Karim ST, Cui Y, Liu Y, Zahoor R, Jiang D, Dai T. Nitrogen nutrition improves the potential of wheat (*Triticum aestivum* L.) to alleviate the effects of drought stress during vegetative growth periods. Frontiers in plant science. 2016 Jun 30;7:981.

[44] Reich PB, Hobbie SE, Lee T, Ellsworth DS, West JB, Tilman D, Knops JM, Naeem S, Trost J. Nitrogen limitation constrains sustainability of ecosystem response to CO 2. Nature. 2006 Apr;440(7086):922-5.

[45] Yin L, Xu H, Dong S, Chu J, Dai X, He M. Optimised nitrogen allocation favours improvement in canopy photosynthetic nitrogen-use efficiency:

Evidence from late-sown winter wheat. Environmental and Experimental Botany. 2019 Mar 1;159:75-86.

[46] Moyano FE, Manzoni S, Chenu C. Responses of soil heterotrophic respiration to moisture availability: An exploration of processes and models. Soil Biology and Biochemistry. 2013 Apr 1;59:72-85.

[47] Below FE. Nitrogen metabolism and crop productivity. Handbook of plant and crop physiology. 2002;2:385-406

[48] Ashraf M, Akram NA, Al-Qurainy F, Foolad MR. Drought tolerance: roles of organic osmolytes, growth regulators, and mineral nutrients. InAdvances in Agronomy 2011 Jan 1 (Vol. 111, pp. 249-296). Academic Press.

[49] Kurepin LV, Ivanov AG, Zaman M, Pharis RP, Allakhverdiev SI, Hurry V, Hüner NP. Stress-related hormones and glycinebetaine interplay in protection of photosynthesis under abiotic stress conditions. Photosynthesis research. 2015 Dec 1;126(2-3):221-35.

[50] Reddy AR, Chaitanya KV, Vivekanandan M. Drought-induced responses of photosynthesis and antioxidant metabolism in higher plants. Journal of plant physiology. 2004 Nov 18;161(11):1189-202.

[51] Chen H, Jiang JG. Osmotic adjustment and plant adaptation to environmental changes related to drought and salinity. Environmental Reviews. 2010 Dec;18(NA):309-19.

[52] Hayat S, Hayat Q, Alyemeni MN, Wani AS, Pichtel J, Ahmad A. Role of proline under changing environments: a review. Plant signaling & behavior. 2012 Nov 1;7(11):1456-66.

[53] Wassmann R, Jagadish SV, Heuer S, Ismail A, Redona E, Serraj R, Singh RK, Howell G, Pathak H, Sumfleth K. Climate change affecting rice production: the physiological and agronomic basis for possible adaptation strategies. Advances in agronomy. 2009 Jan 1;101:59-122.

[54] Ullah A, Manghwar H, Shaban M, Khan AH, Akbar A, Ali U, Ali E, Fahad S. Phytohormones enhanced drought tolerance in plants: a coping strategy. Environmental Science and Pollution Research. 2018 Nov 1;25(33):33103-18.

[55] Pei ZM, Kuchitsu K. Early ABA signaling events in guard cells. Journal of Plant Growth Regulation. 2005 Dec 1;24(4):296-307.

[56] Marcińska I, Czyczyło-Mysza I, Skrzypek E, Filek M, Grzesiak S, Grzesiak MT, Janowiak F, Hura T, Dziurka M, Dziurka K, Nowakowska A. Impact of osmotic stress on physiological and biochemical characteristics in drought-susceptible and drought-resistant wheat genotypes. Acta physiologiae plantarum. 2013 Feb 1;35(2):451-61.

[57] Hong-Bo S, Xiao-Yan C, Li-Ye C, Xi-Ning Z, Gang W, Yong-Bing Y, Chang-Xing Z, Zan-Min H. Investigation on the relationship of proline with wheat anti-drought under soil water deficits. Colloids and Surfaces B: Biointerfaces. 2006 Nov 1;53(1):113-9.

[58] Ullah A, Manghwar H, Shaban M, Khan AH, Akbar A, Ali U, Ali E, Fahad S. Phytohormones enhanced drought tolerance in plants: a coping strategy. Environmental Science and Pollution Research. 2018 Nov 1;25(33):33103-18.

[59] Chakraborty U, Pradhan B. Oxidative stress in five wheat varieties (*Triticum aestivum* L.) exposed to water stress and study of their antioxidant enzyme defense system, water stress responsive metabolites and H2O2 accumulation. Brazilian Journal of Plant Physiology. 2012 Jun;24(2):117-30.

[60] Cattivelli L, Rizza F, Badeck FW, Mazzucotelli E, Mastrangelo AM, Francia E, Marè C, Tondelli A, Stanca AM. Drought tolerance improvement in crop plants: an integrated view from breeding to genomics. Field crops research. 2008 Jan 2;105(1-2):1-4.

[61] Wang JY, Xiong YC, Li FM, Siddique KH, Turner NC. Effects of drought stress on morphophysiological traits, biochemical characteristics, yield, and yield components in different ploidy wheat: A meta-analysis. InAdvances in agronomy 2017 Jan 1 (Vol. 143, pp. 139-173). Academic Press.

[62] Mwadzingeni L, Shimelis H, Tesfay S, Tsilo TJ. Screening of bread wheat genotypes for drought tolerance using phenotypic and proline analyses. Frontiers in plant science. 2016 Aug 25;7:1276.

[63] Hafez EM, Gharib HS. Effect of exogenous application of ascorbic acid on physiological and biochemical characteristics of wheat under water stress. Int. J. Plant Prod. 2016 Oct 1;10(4):579-96.

[64] Samarah NH, Alqudah AM, Amayreh JA, McAndrews GM. The effect of late-terminal drought stress on yield components of four barley cultivars. Journal of Agronomy and Crop Science. 2009 Dec;195(6):427-41.

[65] Suneja Y, Gupta AK, Bains NS. Stress adaptive plasticity: *Aegilops tauschii* and *Triticum dicoccoides* as potential donors of drought associated morpho-physiological traits in wheat. Frontiers in Plant Science. 2019 Feb 25;10:211.

[66] Wang JY, Xiong YC, Li FM, Siddique KH, Turner NC. Effects of drought stress on morphophysiological traits, biochemical characteristics, yield, and yield components in different ploidy wheat: A meta-analysis. InAdvances in agronomy 2017 Jan 1 (Vol. 143, pp. 139-173). Academic Press.

[67] Farooq M, Gogoi N, Barthakur S, Baroowa B, Bharadwaj N, Alghamdi SS, Siddique KH. Drought stress in grain legumes during reproduction and grain filling. Journal of Agronomy and Crop Science. 2017 Apr;203(2):81-102.

[68] Mancosu N, Snyder RL, Kyriakakis G, Spano D. Water scarcity and future challenges for food production. Water. 2015 Mar;7(3):975-92.

[69] Driever SM, Lawson T, Andralojc PJ, Raines CA, Parry MA. Natural variation in photosynthetic capacity, growth, and yield in 64 field-grown wheat genotypes. Journal of Experimental Botany. 2014 Sep 1;65(17):4959-73.

[70] Blum A. Drought resistance, water-use efficiency, and yield potential—are they compatible, dissonant, or mutually exclusive?. Australian Journal of Agricultural Research. 2005 Dec 20;56(11):1159-68.

[71] Waraich EA, Ahmad R, Ashraf MY, Saifullah, Ahmad M. Improving agricultural water use efficiency by nutrient management in crop plants. Acta Agriculturae Scandinavica, Section B-Soil & Plant Science. 2011 May 1; 61(4):291-304.

[72] Tavakkoli AR, Oweis TY. The role of supplemental irrigation and nitrogen in producing bread wheat in the highlands of Iran. Agricultural Water Management. 2004 Mar 15;65(3):225-36.

[73] Trethowan RM, Reynolds M, Sayre K, Ortiz-Monasterio I. Adapting wheat cultivars to resource conserving farming practices and human nutritional needs. Annals of applied biology. 2005 Jul;146(4):405-13.

[74] Johnson JF, Allmaras RR, Reicosky DC. Estimating source

carbon from crop residues, roots and rhizodeposits using the national grain-yield database. Agronomy journal. 2006 May;98(3):622-36.

[75] Ali H, Iqbal N, Shahzad AN, Sarwar N, Ahmad S, Mehmood A. Seed priming improves irrigation water use efficiency, yield, and yield components of late-sown wheat under limited water conditions. Turkish Journal of Agriculture and forestry. 2013 Sep 17;37(5):534-44.

[76] Hatfield JL, Sauer TJ, Prueger JH. Managing soils to achieve greater water use efficiency: a review. Agronomy journal. 2001 Mar;93(2):271-80.

[77] Chauhan BS. Weed ecology and weed management strategies for dry-seeded rice in Asia. Weed Technology. 2012 Jan;26(1):1-3.

[78] Debaeke P, Aboudrare A. Adaptation of crop management to water-limited environments. European Journal of Agronomy. 2004 Dec 1;21(4):433-46.

[79] Li Y, Li H, Li Y, Zhang S. Improving water-use efficiency by decreasing stomatal conductance and transpiration rate to maintain higher ear photosynthetic rate in drought-resistant wheat. The Crop Journal. 2017 Jun 1; 5(3):231-9.

[80] Timmusk S, Abd El-Daim IA, Copolovici L, Tanilas T, Kännaste A, Behers L, Nevo E, Seisenbaeva G, Stenström E, Niinemets Ü. Drought-tolerance of wheat improved by rhizosphere bacteria from harsh environments: enhanced biomass production and reduced emissions of stress volatiles. PloS one. 2014 May 8;9(5):e96086.

[81] Abd El-Mageed TA, El-Samnoudi IM, Ibrahim AE, Abd El Tawwab AR. Compost and mulching modulates morphological, physiological responses and water use efficiency in sorghum (bicolor L. Moench) under low moisture regime. Agricultural Water Management. 2018 Sep 30;208:431-9.

[82] Ogle SM, Alsaker C, Baldock J, Bernoux M, Breidt FJ, McConkey B, Regina K, Vazquez-Amabile GG. Climate and soil characteristics determine where no-till management can store carbon in soils and mitigate greenhouse gas emissions. Scientific reports. 2019 Aug 12;9(1):1-8.

[83] Mohanty M, Sinha NK, Painuli DK, Bandyopadhyay KK, Hati KM, Reddy KS, Chaudhary RS. Modelling soil water contents at field capacity and permanent wilting point using artificial neural network for Indian soils. National Academy Science Letters. 2015 Oct 1;38(5):373-7.

[84] He T, Gao F, Liang S, Peng Y. Mapping climatological bare soil albedos over the contiguous United States using MODIS data. Remote Sensing. 2019 Jan;11(6):666.

[85] Li Liu D, Zeleke KT, Wang B, Macadam I, Scott F, Martin RJ. Crop residue incorporation can mitigate negative climate change impacts on crop yield and improve water use efficiency in a semiarid environment. European Journal of Agronomy. 2017 Apr 1;85:51-68.

[86] Patanè C. Leaf area index, leaf transpiration and stomatal conductance as affected by soil water deficit and VPD in processing tomato in semi arid Mediterranean climate. Journal of Agronomy and Crop Science. 2011 Jun;197(3):165-76.

[87] Papanatsiou M, Petersen J, Henderson L, Wang Y, Christie JM, Blatt MR. Optogenetic manipulation of stomatal kinetics improves carbon assimilation, water use, and growth. Science. 2019 Mar 29;363(6434):1456-9.

[88] Bertolino LT, Caine RS, Gray JE. Impact of stomatal density and morphology on water-use efficiency in a changing world. Frontiers in plant science. 2019 Mar 6;10:225.

[89] Casson SA, Hetherington AM. Environmental regulation of stomatal development. Current opinion in plant biology. 2010 Feb 1;13(1):90-5.

[90] Nadeau JA. Stomatal development: new signals and fate determinants. Current opinion in plant biology. 2009 Feb 1;12(1):29-35.

[91] Sibbernsen E, Mott KA. Stomatal responses to flooding of the intercellular air spaces suggest a vapor-phase signal between the mesophyll and the guard cells. Plant physiology. 2010 Jul 1;153(3):1435-42.

[92] Franks PJ, Farquhar GD. The mechanical diversity of stomata and its significance in gas-exchange control. Plant physiology. 2007 Jan 1;143(1):78-87.

[93] Tardieu F. Plant response to environmental conditions: assessing potential production, water demand, and negative effects of water deficit. Frontiers in physiology. 2013 Feb 18;4:17.

[94] Richards RA. Physiological traits used in the breeding of new cultivars for water-scarce environments. Agricultural water management. 2006 Feb 24;80(1-3):197-211.

[95] Wasaya A, Zhang X, Fang Q, Yan Z. Root phenotyping for drought tolerance: a review. Agronomy. 2018 Nov;8(11):241.

[96] Maeght JL, Rewald B, Pierret A. How to study deep roots—and why it matters. Frontiers in plant science. 2013 Aug 13;4:299.

[97] Farooq M, Hussain M, Siddique KH. Drought stress in wheat during flowering and grain-filling periods. Critical Reviews in Plant Sciences. 2014 Jul 4;33(4):331-49.

[98] Schoppach R, Wauthelet D, Jeanguenin L, Sadok W. Conservative water use under high evaporative demand associated with smaller root metaxylem and limited trans-membrane water transport in wheat. Functional Plant Biology. 2014 Mar 4;41(3):257-69.

[99] Brisson N, Mary B, Ripoche D, Jeuffroy MH, Ruget F, Nicoullaud B, Gate P, Devienne-Barret F, Antonioletti R, Durr C, Richard G. STICS: a generic model for the simulation of crops and their water and nitrogen balances. I. Theory and parameterization applied to wheat and corn.

[100] Vangronsveld J, Herzig R, Weyens N, Boulet J, Adriaensen K, Ruttens A, Thewys T, Vassilev A, Meers E, Nehnevajova E, van der Lelie D. Phytoremediation of contaminated soils and groundwater: lessons from the field. Environmental Science and Pollution Research. 2009 Nov 1;16(7):765-94.

[101] Milly PC, Dunne KA, Vecchia AV. Global pattern of trends in streamflow and water availability in a changing climate. Nature. 2005 Nov;438(7066):347-50.

[102] Sylvester-Bradley R, Kindred DR. Analysing nitrogen responses of cereals to prioritize routes to the improvement of nitrogen use efficiency. Journal of Experimental Botany. 2009 May 1;60(7):1939-51.

[103] Quemada M, Gabriel JL. Approaches for increasing nitrogen and water use efficiency simultaneously. Global Food Security. 2016 Jun 1;9:29-35.

[104] Wang X, Dai K, Zhang D, Zhang X, Wang Y, Zhao Q, Cai D,

Hoogmoed WB, Oenema O. Dryland maize yields and water use efficiency in response to tillage/crop stubble and nutrient management practices in China. Field Crops Research. 2011 Jan 14;120(1):47-57.

[105] Hirel B, Tétu T, Lea PJ, Dubois F. Improving nitrogen use efficiency in crops for sustainable agriculture. Sustainability. 2011 Sep;3(9):1452-85.

[106] Bowles TM, Mooshammer M, Socolar Y, Calderón F, Cavigelli MA, Culman SW, Deen W, Drury CF, y Garcia AG, Gaudin AC, Harkcom WS. Long-Term evidence shows that crop-rotation Diversification increases agricultural resilience to Adverse Growing conditions in North America. One Earth. 2020 Mar 6.

[107] Saikia SP, Jain V. Biological nitrogen fixation with non-legumes: An achievable target or a dogma?. Current science. 2007 Feb 10:317-22.

[108] Dawson JC, Huggins DR, Jones SS. Characterizing nitrogen use efficiency in natural and agricultural ecosystems to improve the performance of cereal crops in low-input and organic agricultural systems. Field Crops Research. 2008 May 10;107(2):89-101.

[109] Dawson JC, Huggins DR, Jones SS. Characterizing nitrogen use efficiency in natural and agricultural ecosystems to improve the performance of cereal crops in low-input and organic agricultural systems. Field Crops Research. 2008 May 10;107(2):89-101.

[110] Fang QX, Ma L, Green TR, Yu Q, Wang TD, Ahuja LR. Water resources and water use efficiency in the North China Plain: Current status and agronomic management options. Agricultural Water Management. 2010 Aug 1;97(8):1102-16.

[111] Jat ML, Dagar JC, Sapkota TB, Govaerts B, Ridaura SL, Saharawat YS, Sharma RK, Tetarwal JP, Jat RK, Hobbs H, Stirling C. Climate change and agriculture: adaptation strategies and mitigation opportunities for food security in South Asia and Latin America. InAdvances in agronomy 2016 Jan 1 (Vol. 137, pp. 127-235). Academic Press.

[112] Chai Q, Gan Y, Zhao C, Xu HL, Waskom RM, Niu Y, Siddique KH. Regulated deficit irrigation for crop production under drought stress. A review. Agronomy for sustainable development. 2016 Mar 1;36(1):3.

[113] Rochette P, MacDonald JD, Angers DA, Chantigny MH, Gasser MO, Bertrand N. Banding of urea increased ammonia volatilization in a dry acidic soil. Journal of Environmental Quality. 2009 Jul;38(4):1383-90.

[114] Bukola D, Zaid A, Olalekan EI, Falilu A. Consequences of anthropogenic activities on fish and the aquatic environment. Poultry, Fisheries & Wildlife Sciences. 2015 Aug 28.

[115] Naz MY, Sulaiman SA. Slow release coating remedy for nitrogen loss from conventional urea: a review. Journal of Controlled Release. 2016 Mar 10;225:109-20.

[116] Xu G, Fan X, Miller AJ. Plant nitrogen assimilation and use efficiency. Annual review of plant biology. 2012 Jun 2;63:153-82.

[117] Snyder CS, Bruulsema TW, Jensen TL, Fixen PE. Review of greenhouse gas emissions from crop production systems and fertilizer management effects. Agriculture, Ecosystems & Environment. 2009 Oct 1;133(3-4):247-66.

[118] Fageria NK, Baligar VC, Li YC. The role of nutrient efficient plants in improving crop yields in the twenty first century. Journal of plant nutrition. 2008 May 30;31(6):1121-57.

[119] Andrews M, Raven JA, Sprent JI. Environmental effects on dry matter partitioning between shoot and root of crop plants: relations with growth and shoot protein concentration. Annals of Applied Biology. 2001 Feb;138(1):57-68.

[120] Bonifas KD, Walters DT, Cassman KG, Lindquist JL. Nitrogen supply affects root: shoot ratio in corn and velvetleaf (*Abutilon theophrasti*). Weed Science. 2005 Sep;53(5):670-5.

[121] Feldman M, Levy AA. Origin and evolution of wheat and related Triticeae species. InAlien introgression in Wheat 2015 (pp. 21-76). Springer, Cham.

[122] Jiang L, Dai T, Jiang D, Cao W, Gan X, Wei S. Characterizing physiological N-use efficiency as influenced by nitrogen management in three rice cultivars. Field Crops Research. 2004 Aug 10;88(2-3):239-50.

[123] Ciampitti IA, Vyn TJ. Physiological perspectives of changes over time in maize yield dependency on nitrogen uptake and associated nitrogen efficiencies: A review. Field Crops Research. 2012 Jul 11;133:48-67.

[124] Iqbal A, Qiang D, Alamzeb M, Xiangru W, Huiping G, Hengheng Z, Nianchang P, Xiling Z, Meizhen S. Untangling the molecular mechanisms and functions of nitrate to improve nitrogen use efficiency. Journal of the Science of Food and Agriculture. 2020 Feb;100(3):904-14.

[125] Garnett T, Plett D, Heuer S, Okamoto M. Genetic approaches to enhancing nitrogen-use efficiency (NUE) in cereals: challenges and future directions. Functional Plant Biology. 2015 Oct 6;42(10):921-41.

[126] Ju C, Buresh RJ, Wang Z, Zhang H, Liu L, Yang J, Zhang J. Root and shoot traits for rice varieties with higher grain yield and higher nitrogen use efficiency at lower nitrogen rates application. Field Crops Research. 2015 Apr 1;175:47-55.

[127] Sapkota TB, Majumdar K, Jat ML, Kumar A, Bishnoi DK, McDonald AJ, Pampolino M. Precision nutrient management in conservation agriculture based wheat production of Northwest India: Profitability, nutrient use efficiency and environmental footprint. Field Crops Research. 2014 Jan 1;155:233-44.

[128] Singh Y, Sidhu HS. Management of cereal crop residues for sustainable rice-wheat production system in the Indo-Gangetic plains of India. Proceedings of the Indian National Science Academy. 2014 Mar;80(1):95-114.

[129] Madhu M, Hatfield JL. Interaction of carbon dioxide enrichment and soil moisture on photosynthesis, transpiration, and water use efficiency of soybean. Agricultural Sciences. 2014 Apr 3;2014.

[130] Ye Y, Liang X, Chen Y, Liu J, Gu J, Guo R, Li L. Alternate wetting and drying irrigation and controlled-release nitrogen fertilizer in late-season rice. Effects on dry matter accumulation, yield, water and nitrogen use. Field Crops Research. 2013 Mar 20;144:212-24.

[131] Albrizio R, Todorovic M, Matic T, Stellacci AM. Comparing the interactive effects of water and nitrogen on durum wheat and barley grown in a Mediterranean environment. Field Crops Research. 2010 Jan 20;115(2):179-90.

[132] Kahiluoto H, Kaseva J, Balek J, Olesen JE, Ruiz-Ramos M, Gobin A, Kersebaum KC, Takáč J, Ruget F, Ferrise R, Bezak P. Decline in climate resilience of European wheat. Proceedings of the National Academy of Sciences. 2019 Jan 2;116(1):123-8.

[133] Aulakh MS. Nitrogen losses and fertilizer N use efficiency in

irrigated porous soils. Nutrient cycling in agroecosystems. 1996 Oct 1;47(3):197-212.

[134] Mo X, Liu S, Lin Z, Guo R. Regional crop yield, water consumption and water use efficiency and their responses to climate change in the North China Plain. Agriculture, Ecosystems & Environment. 2009 Nov 1;134(1-2):67-78.

[135] Grizzetti B, Bouraoui F, Billen G, van Grinsven H, Cardoso AC, Thieu V, Garnier J, Curtis C, Howarth RW, Johnes P. Nitrogen as a threat to European water quality.

[136] Adnan M, Zahir S, Fahad S, Arif M, Mukhtar A, Imtiaz AK, Ishaq AM, Abdul B, Hidayat U, Muhammad A, Inayat-Ur R. Phosphate-solubilizing bacteria nullify the antagonistic effect of soil calcification on bioavailability of phosphorus in alkaline soils. Sci Rep 8: 4339.

[137] Ahmad S, Kamran M, Ding R, Meng X, Wang H, Ahmad I, Fahad S, Han Q. Exogenous melatonin confers drought stress by promoting plant growth, photosynthetic capacity and antioxidant defense system of maize seedlings. PeerJ. 2019 Oct 11;7:e7793.

[138] Akram R, Turan V, Hammad HM, Ahmad S, Hussain S, Hasnain A, Maqbool MM, Rehmani MI, Rasool A, Masood N, Mahmood F. Fate of organic and inorganic pollutants in paddy soils. InEnvironmental pollution of paddy soils 2018 (pp. 197-214). Springer, Cham.

[139] Akram R, Turan V, Wahid A, Ijaz M, Shahid MA, Kaleem S, Hafeez A, Maqbool MM, Chaudhary HJ, Munis MF, Mubeen M. Paddy land pollutants and their role in climate change. InEnvironmental Pollution of Paddy Soils 2018 (pp. 113-124). Springer, Cham.

[140] Khan A, Tan DK, Munsif F, Afridi MZ, Shah F, Wei F, Fahad S, Zhou R. Nitrogen nutrition in cotton and control strategies for greenhouse gas emissions: a review. Environmental Science and Pollution Research. 2017 Oct 1;24(30):23471-87.

[141] Khan A, Tan DK, Afridi MZ, Luo H, Tung SA, Ajab M, Fahad S. Nitrogen fertility and abiotic stresses management in cotton crop: a review. Environmental Science and Pollution Research. 2017 Jun 1;24(17):14551-66.

[142] Muhammad B, Adnan M, Munsif F, Fahad S, Saeed M, Wahid F, Arif M, Amanullah J, Wang D, Saud S, Noor M. Substituting urea by organic wastes for improving maize yield in alkaline soil. Journal of Plant Nutrition. 2019 Nov 26;42(19):2423-34.

[143] Wang D, Fahad S, Saud S, Kamran M, Khan A, Khan MN, Hammad HM, Nasim W. Morphological acclimation to agronomic manipulation in leaf dispersion and orientation to promote "Ideotype" breeding: Evidence from 3D visual modeling of "super" rice (*Oryza sativa* L.). Plant Physiology and Biochemistry. 2019 Feb 1;135:499-510.

[144] Adnan M, Fahad S, Zamin M, Shah S, Mian IA, Danish S, Zafar-ul-Hye M, Battaglia ML, Naz RM, Saeed B, Saud S. Coupling phosphate-solubilizing bacteria with phosphorus supplements improve maize phosphorus acquisition and growth under lime induced salinity stress. Plants. 2020 Jul;9(7):900.

[145] Adnan M, Fahad S, Khan IA, Saeed M, Ihsan MZ, Saud S, Riaz M, Wang D, Wu C. Integration of poultry manure and phosphate solubilizing bacteria improved availability of Ca bound P in calcareous soils. 3 Biotech. 2019 Oct 1;9(10):368.

[146] Adnan M, Shah Z, Sharif M, Rahman H. Liming induces carbon dioxide (CO 2) emission in PSB inoculated alkaline soil supplemented with different phosphorus sources.

Environmental Science and Pollution Research. 2018 Apr 1;25(10):9501-9.

[147] Fahad S, Bajwa AA, Nazir U, Anjum SA, Farooq A, Zohaib A, Sadia S, Nasim W, Adkins S, Saud S, Ihsan MZ. Crop production under drought and heat stress: plant responses and management options. Frontiers in plant science. 2017 Jun 29;8:1147.

[148] Fahad S, Bano A (2012) Effect of salicylic acid on physiological and biochemical characterization of maize grown in saline area. Pak J Bot 44:1433-1438

[149] Fahad S, Chen Y, Saud S,Wang K, Xiong D, Chen C,Wu C, Shah F, Nie L, Huang J (2013) Ultraviolet radiation effect on photosynthetic pigments, biochemical attributes, antioxidant enzyme activity and hormonal contents of wheat. J Food, Agri Environ 11(3&4):1635-1641

[150] Fahad S, Hussain S, Bano A, Saud S, Hassan S, Shan D, Khan FA, Khan F, Chen Y, Wu C, Tabassum MA. Potential role of phytohormones and plant growth-promoting rhizobacteria in abiotic stresses: consequences for changing environment. Environmental Science and Pollution Research. 2015 Apr 1;22(7):4907-21.

[151] Fahad S, Hussain S, Matloob A, Khan FA, Khaliq A, Saud S, Hassan S, Shan D, Khan F, Ullah N, Faiq M. Phytohormones and plant responses to salinity stress: a review. Plant growth regulation. 2015 Mar 1;75(2):391-404.

[152] Abbas F, Hammad HM, Ishaq W, Farooque AA, Bakhat HF, Zia Z, Fahad S, Farhad W, Cerdà A. A review of soil carbon dynamics resulting from agricultural practices. Journal of environmental management. 2020 Aug 15;268:110319.

[153] Gul F, Ahmed I, Ashfaq M, Jan D, Fahad S, Li X, Wang D, Fahad M, Fayyaz M, Shah SA. Use of crop growth model to simulate the impact of climate change on yield of various wheat cultivars under different agro-environmental conditions in Khyber Pakhtunkhwa, Pakistan. Arabian Journal of Geosciences. 2020 Feb 1;13(3):112.

[154] Rahman MH, Ahmad A, Wajid A, Hussain M, Rasul F, Ishaque W, Islam MA, Shelia V, Awais M, Ullah A, Wahid A. Application of CSM-CROPGRO-Cotton model for cultivars and optimum planting dates: evaluation in changing semi-arid climate. Field Crops Research. 2019 May 15;238:139-52.

[155] Hammad HM, Khaliq A, Abbas F, Farhad W, Fahad S, Aslam M, Shah GM, Nasim W, Mubeen M, Bakhat HF. Comparative Effects of Organic and Inorganic Fertilizers on Soil Organic Carbon and Wheat Productivity under Arid Region. Communications in Soil Science and Plant Analysis. 2020 May 30;51(10):1406-22.

[156] Hammad HM, Abbas F, Ahmad A, Bakhat HF, Farhad W, Wilkerson CJ, Fahad S, Hoogenboom G. Predicting Kernel Growth of Maize under Controlled Water and Nitrogen Applications. International Journal of Plant Production. 2020 Aug 19:1-2.

[157] Khan MN, Zhang J, Luo T, Liu J, Ni F, Rizwan M, Fahad S, Hu L. Morpho-physiological and biochemical responses of tolerant and sensitive rapeseed cultivars to drought stress during early seedling growth stage. Acta Physiologiae Plantarum. 2019 Feb 1;41(2):25.

[158] Fahad S, Hussain S, Saud S, Hassan S, Chauhan BS, Khan F, Ihsan MZ, Ullah A, Wu C, Bajwa AA, Alharby H. Responses of rapid viscoanalyzer profile and other rice grain qualities to exogenously applied plant growth regulators under high day

and high night temperatures. PLoS One. 2016 Jul 29;11(7):e0159590.

[159] Fahad S, Hussain S, Saud S, Hassan S, Ihsan Z, Shah AN, Wu C, Yousaf M, Nasim W, Alharby H, Alghabari F. Exogenously applied plant growth regulators enhance the morpho-physiological growth and yield of rice under high temperature. Frontiers in Plant Science. 2016 Aug 30; 7:1250.

[160] Fahad S, Hussain S, Saud S, Hassan S, Tanveer M, Ihsan MZ, Shah AN, Ullah A, Khan F, Ullah S, Alharby H. A combined application of biochar and phosphorus alleviates heat-induced adversities on physiological, agronomical and quality attributes of rice. Plant physiology and biochemistry. 2016 Jun 1;103:191-8.

[161] Fahad S, Hussain S, Saud S, Khan F, Hassan S, Nasim W, Arif M, Wang F, Huang J. Exogenously applied plant growth regulators affect heat-stressed rice pollens. Journal of agronomy and crop science. 2016 Apr;202(2):139-50.

[162] Hammad HM, Ashraf M, Abbas F, Bakhat HF, Qaisrani SA, Mubeen M, Fahad S, Awais M. Environmental factors affecting the frequency of road traffic accidents: a case study of sub-urban area of Pakistan. Environmental Science and Pollution Research. 2019 Apr 1;26(12):11674-85.

[163] Hammad HM, Farhad W, Abbas F, Fahad S, Saeed S, Nasim W, Bakhat HF. Maize plant nitrogen uptake dynamics at limited irrigation water and nitrogen. Environmental Science and Pollution Research. 2017 Jan 1;24(3):2549-57.

[164] Alharby HF, Fahad S. Melatonin application enhances biochar efficiency for drought tolerance in maize varieties: Modifications in physio-biochemical machinery. Agronomy Journal. 2020 Jul;112(4):2826-47.

[165] Hussain MA, Fahad S, Sharif R, Jan MF, Mujtaba M, Ali Q, Ahmad A, Ahmad H, Amin N, Ajayo BS, Sun C. Multifunctional role of brassinosteroid and its analogues in plants. Plant Growth Regulation. 2020 Aug 2:1-6.

[166] Ilyas M, Nisar M, Khan N, Hazrat A, Khan AH, Hayat K, Fahad S, Khan A, Ullah A. Drought Tolerance Strategies in Plants: A Mechanistic Approach. Journal of Plant Growth Regulation. 2020 Jun 22:1-9.

[167] Jan M, Anwar-ul-Haq M, Shah AN, Yousaf M, Iqbal J, Li X, Wang D, Fahad S. Modulation in growth, gas exchange, and antioxidant activities of salt-stressed rice (*Oryza sativa* L.) genotypes by zinc fertilization. Arabian Journal of Geosciences. 2019 Dec 1;12(24):775.

[168] Kamran M, Cui W, Ahmad I, Meng X, Zhang X, Su W, Chen J, Ahmad S, Fahad S, Han Q, Liu T. Effect of paclobutrazol, a potential growth regulator on stalk mechanical strength, lignin accumulation and its relation with lodging resistance of maize. Plant growth regulation. 2018 Mar 1;84(2):317-32.

[169] Mubeen M, Ahmad A, Hammad HM, Awais M, Farid HU, Saleem M, Amin A, Ali A, Fahad S, Nasim W. Evaluating the climate change impact on water use efficiency of cotton-wheat in semi-arid conditions using DSSAT model. Journal of Water and Climate Change. 2019 Oct 24.

[170] Fahad S, Hussain S, Saud S, Tanveer M, Bajwa AA, Hassan S, Shah AN, Ullah A, Wu C, Khan FA, Shah F. A biochar application protects rice pollen from high-temperature stress. Plant Physiology and Biochemistry. 2015 Nov 1;96:281-7.

[171] Fahad S, Ihsan MZ, Khaliq A, Daur I, Saud S, Alzamanan S, Nasim W, Abdullah M, Khan IA, Wu C, Wang D. Consequences of high temperature

under changing climate optima for rice pollen characteristics-concepts and perspectives. Archives of Agronomy and Soil Science. 2018 Sep 19;64(11):1473-88.

[172] Fahad S, Nie L, Chen Y, Wu C, Xiong D, Saud S, Hongyan L, Cui K, Huang J. Crop plant hormones and environmental stress. InSustainable Agriculture Reviews 2015 (pp. 371-400). Springer, Cham.

[173] Fahad S, Adnan M, Hassan S, Saud S, Hussain S, Wu C, Wang D, Hakeem KR, Alharby HF, Turan V, Khan MA. Rice responses and tolerance to high temperature. InAdvances in rice research for abiotic stress tolerance 2019 Jan 1 (pp. 201-224). Woodhead Publishing.

[174] Fahad S, Rehman A, Shahzad B, Tanveer M, Saud S, Kamran M, Ihtisham M, Khan SU, Turan V, ur Rahman MH. Rice responses and tolerance to metal/metalloid toxicity. InAdvances in rice research for abiotic stress tolerance 2019 Jan 1 (pp. 299-312). Woodhead Publishing.

[175] Zamin M, Khattak AM, Salim AM, Marcum KB, Shakur M, Shah S, Jan I, Fahad S. Performance of Aeluropus lagopoides (mangrove grass) ecotypes, a potential turfgrass, under high saline conditions. Environmental Science and Pollution Research. 2019 May 1;26(13):13410-21.

[176] Zaman QU, Aslam Z, Yaseen M, Ihsan MZ, Khaliq A, Fahad S, Bashir S, Ramzani PM, Naeem M. Zinc biofortification in rice: leveraging agriculture to moderate hidden hunger in developing countries. Archives of Agronomy and Soil Science. 2018 Jan 28;64(2):147-61.

[177] Rehman M, Fahad S, Saleem MH, Hafeez M, Rahman MH, Liu F, Deng G. Red light optimized physiological traits and enhanced the growth of ramie (*Boehmeria nivea* L.). Photosynthetica. 2020 Jan 1;58(4):922-31.

[178] Hussain S, Mubeen M, Ahmad A, Akram W, Hammad HM, Ali M, Masood N, Amin A, Farid HU, Sultana SR, Fahad S. Using GIS tools to detect the land use/land cover changes during forty years in Lodhran District of Pakistan. Environmental Science and Pollution Research. 2019 Aug 5:1-7.

[179] Saleem MH, Fahad S, Adnan M, Ali M, Rana MS, Kamran M, Ali Q, Hashem IA, Bhantana P, Ali M, Hussain RM. Foliar application of gibberellic acid endorsed phytoextraction of copper and alleviates oxidative stress in jute (*Corchorus capsularis* L.) plant grown in highly copper-contaminated soil of China. Environmental Science and Pollution Research. 2020 Jun 24:1-3.

[180] SALEEM M, REHMAN M, FAHAD S, TUNG S, IQBAL N, HASSAN A, AYUB A, WAHID M, SHAUKAT S, LIU L, DENG G. Leaf gas exchange, oxidative stress, and physiological attributes of rapeseed (*Brassica napus* L.) grown under different light-emitting diodes. PHOTOSYNTHETICA. 2020;58(3):836-45.

[181] Izhar Shafi M, Adnan M, Fahad S, Wahid F, Khan A, Yue Z, Danish S, Zafar-ul-Hye M, Brtnicky M, Datta R. Application of Single Superphosphate with Humic Acid Improves the Growth, Yield and Phosphorus Uptake of Wheat (*Triticum aestivum* L.) in Calcareous Soil. Agronomy. 2020 Sep;10(9):1224.

[182] Saleem, M.H., Fahad, S., Khan, S.U., Din, M., Ullah, A., Sabagh, A.E., Hossain, A., Llanes, A. and Liu, L., 2020. Copper-induced oxidative stress, initiation of antioxidants and phytoremediation potential of flax (*Linum usitatissimum* L.) seedlings grown under the mixing of two

different soils of China. *Environmental Science and Pollution Research*, *27*(5), pp.5211-5221.

[183] Saud S, Yajun C, Fahad S, Hussain S, Na L, Xin L, Alhussien SA. Silicate application increases the photosynthesis and its associated metabolic activities in Kentucky bluegrass under drought stress and post-drought recovery. Environmental Science and Pollution Research. 2016 Sep 1;23(17):17647-55.

[184] Wahid F, Fahad S, Danish S, Adnan M, Yue Z, Saud S, Siddiqui MH, Brtnicky M, Hammerschmiedt T, Datta R. Sustainable Management with Mycorrhizae and Phosphate Solubilizing Bacteria for Enhanced Phosphorus Uptake in Calcareous Soils. Agriculture. 2020 Aug;10(8):334.

[185] Saud SH, Chen Y, Long B, Fahad SH, Sadiq AR. The different impact on the growth of cool season turf grass under the various conditions on salinity and draught stress. Int J Agric Sci Res. 2013;3(4):77-84.

[186] Watanabe S, Iimori M, Chan DV, Hara E, Kitao H, Maehara Y. MDC1 methylation mediated by lysine methyltransferases EHMT1 and EHMT2 regulates active ATM accumulation flanking DNA damage sites. Scientific reports. 2018 Jul 18;8(1):1-0.

[187] SaudS F, YajunC I, HammadHM N, JrA A. AlharbyH (2017) Effects of nitrogen supply on water stress and recovery mechanisms in Kentucky Bluegrass plants. Frontiers of Plant Science8.;983.

[188] Saud S, Li X, Chen Y, Zhang L, Fahad S, Hussain S, Sadiq A, Chen Y. Silicon application increases drought tolerance of Kentucky bluegrass by improving plant water relations and morphophysiological functions. The Scientific World Journal. 2014 Oct;2014.

[189] Shah F, Nie L, Cui K, Shah T, Wu W, Chen C, Zhu L, Ali F, Fahad S, Huang J. Rice grain yield and component responses to near 2 C of warming. Field Crops Research. 2014 Feb 15;157:98-110.

[190] Danish S, Zafar-ul-Hye M, Fahad S, Saud S, Brtnicky M, Hammerschmiedt T, Datta R. Drought Stress Alleviation by ACC Deaminase Producing Achromobacter xylosoxidans and *Enterobacter cloacae*, with and without Timber Waste Biochar in Maize. Sustainability. 2020 Jan;12(15):6286.

[191] Tariq M, Ahmad S, Fahad S, Abbas G, Hussain S, Fatima Z, Nasim W, Mubeen M, ur Rehman MH, Khan MA, Adnan M. The impact of climate warming and crop management on phenology of sunflower-based cropping systems in Punjab, Pakistan. Agricultural and Forest Meteorology. 2018 Jun 15;256:270-82.

[192] Wajid N, Ashfaq A, Asad A. Muhammad Habib ur R, Veysel T, Fahad S, Suad S, Aziz K, Shahzad A (2017) Radiation efficiency and nitrogen fertilizer impacts on sunflower crop in contrasting environments of Punjab. Pakistan Environ Sci Pollut Res 25: 1822-1836.

[193] Wu C, Tang S, Li G, Wang S, Fahad S, Ding Y. Roles of phytohormone changes in the grain yield of rice plants exposed to heat: a review. PeerJ. 2019 Nov 19;7:e7792.

[194] Yang Z, Zhang Z, Zhang T, Fahad S, Cui K, Nie L, Peng S, Huang J. The effect of season-long temperature increases on rice cultivars grown in the central and southern regions of China. Frontiers in plant science. 2017 Nov 6;8:1908.

[195] Zafar-ul-Hye M, Naeem M, Danish S, Fahad S, Datta R, Abbas M, Rahi AA, Brtnicky M, Holátko J, Tarar ZH, Nasir M. Alleviation

of Cadmium Adverse Effects by Improving Nutrients Uptake in Bitter Gourd through Cadmium Tolerant Rhizobacteria. Environments. 2020 Aug;7(8):54.

[196] Zafar-ul-Hye M, Tahzeeb-ul-Hassan M, Abid M, Fahad S, Brtnicky M, Dokulilova T, Datta R, Danish S. Potential role of compost mixed biochar with rhizobacteria in mitigating lead toxicity in spinach. Scientific reports. 2020 Jul 22;10(1):1-2.

[197] Zia Z, Bakhat HF, Saqib ZA, Shah GM, Fahad S, Ashraf MR, Hammad HM, Naseem W, Shahid M. Effect of water management and silicon on germination, growth, phosphorus and arsenic uptake in rice. Ecotoxicology and environmental safety. 2017 Oct 1;144:11-8.

[198] Cousins OH, Garnett TP, Rasmussen A, Mooney SJ, Smernik RJ, Brien CJ, Cavagnaro TR. Variable water cycles have a greater impact on wheat growth and soil nitrogen response than constant watering. Plant Science. 2020 Jan 1;290:110146.

Chapter 14

Adapting Cereal Grain Crops to Drought Stress: 2020 and Beyond

Tinashe Zenda, Songtao Liu and Huijun Duan

Abstract

Among other environmental instabilities, drought stress is the primary limitation to cereal crops growth, development and productivity. In the context of continuing global climate change, breeding of drought resistant crop cultivars is the most economical, effective and sustainable strategy for adapting the crop production system and ensuring food security for the growing human population. Additionally, there is need for improving management practices. Whereas conventional breeding has sustained crop productivity gains in the past century, modern technological advancements have revolutionized our identification of important drought tolerance genes and underlying mechanisms, and accelerated new cultivar development. Large-scale high throughput sequencing, phenotyping, 'omics' and systems biology, as well as marker assisted and quantitative trait loci mapping based breeding approaches have offered significant insights into crop drought stress tolerance and provided some new tools for crop improvement. Despite this significant progress in elucidating the mechanisms underlying drought tolerance, considerable challenges remain and our understanding of the crop drought tolerance mechanisms is still abstract. In this chapter, therefore, we highlight current progress in the identification of drought tolerance genes and underlying mechanisms, as well as their practical applications. We then offer a holistic approach for cereal crops adaptation to future climate change exacerbated drought stress.

Keywords: drought stress, drought tolerance, cereal crops, omics approaches, phenotyping, genetic engineering, climate change, conservation agriculture

1. Introduction

Drought stress is the primary environmental factor influencing the growth, development and productivity of crops and its significance is expected to increase in the wake of global climate change [1–4]. This presents a serious challenge to the food and nutrition security of an ever-rising world human population. Moreover, the current transition from carbon/fossil intensive fuel driven economies to modern climate-smart low-carbon economies further strains our crop production systems [5]. Adapting field crops to drought stress therefore becomes critical for sustainable agricultural production under such climate change scenario [6]. To achieve that goal, breeding drought resilient crop cultivars that maintain yield stability under such conditions befits the most economical, effective and sustainable strategy. This is particularly relevant for cereal grain crops.

Cereal grain crops, chiefly rice (*Oryza sativa* L.), wheat (*Triticum aestivum* L.), maize (*Zea mays* L.), sorghum [*Sorghum bicolor* (L.) Moench], barley (*Hordeum vulgare* L.), and pearl millet (*Pennisetum glaucum* L.) provides above 50% of the total food calorie requirements of billions of people in developing countries across the Sub-Saharan Africa, South Asia and Latin America [7–9]. For instance, the top three cereals; maize, rice, and wheat jointly provide 48% of the total calories and 42% of total protein consumed in developing countries. Additionally, as staple foods, maize and wheat contribute approximately two thirds of the global food energy intake [8]. Further, these cereal crops are important raw materials in the animal feed and bio-fuel manufacturing industries [10]. However, most of these crops are grown across arid and semi-arid regions of the world, where they often endure exposure to recurrent drought episodes throughout their growth cycles. Therefore, understanding how cereal crop plants respond to drought stress is critical for guiding drought tolerance breeding.

In the last century, conventional breeding approach has proven itself capable of sustaining productivity growth in various crops [11]. Meanwhile, modern technological advancements have accelerated the pace and impact of new cultivar development. Such technologies include high throughput omics approaches, identification of quantitative trait loci (QTL) underlying abiotic and biotic stress resistances, marker assisted selection (MAS) and gene cloning [12–15]. Despite this significant progress in elucidating the mechanisms underlying drought tolerance, considerable challenges remain and our understanding of the crop drought tolerance mechanisms is still abstract.

In this chapter, therefore, we look at various aspects of drought stress in major cereal grain crops such as maize, sorghum, wheat, rice and finger millet. We also discuss the current approaches in identifying drought tolerance genes and metabolic pathways. Further, we highlight the progress made to date on elucidation of key drought stress responses, phenotyping and QTL mapping for drought tolerance, genetic engineering of drought tolerant crops and management of crops. We conclude by offering an integrated strategy for adapting cereal grain crops to drought stress in the context of climate change.

2. Drought stress effects in cereal grain crops

A decrease in water inputs into an agro/ecosystem over time that is sufficient to cause soil water deficit (SWD) is often termed drought, and this encompasses various forms such as rainfall anomalies, irrigation failure, seasonal or annual dry spells [16]. In agricultural context, drought signifies a period of below-average precipitation when the available soil water in the plant rhizosphere drops beyond the thresholds for efficient growth and biomass production [17]. The resultant oxidative stress emanating from such SWD is dubbed drought stress.

Numerous research reports have reflected on the effects of drought stress on cereal crops. Drought stress effects span from morphological to molecular levels, and are exhibited at all phenological growth stages at whatever stage the water deficit takes place. Generally, drought stress impairs seed germination resulting in poor crop stand establishment [1, 18]. Drought stress reduces the plant cellular water potential and turgor pressure, thereby increasing the cytosolic and extracellular matrices solute concentrations. Resultantly, cell growth is diminished due to the reduction in turgor pressure [19]. Additionally, abscisic acid (ABA) and compatible osmolytes such as proline are excessively accumulated, causing plant wilting. Simultaneously, reactive oxygen species (ROS) such as H_2O_2 are overly produced. Although they function as signal transduction molecules, over-accumulation of ROS could result in extensive cellular oxidative damage and inhibition of photosynthesis [20].

Moreover, when moisture deficit becomes severe, cell elongation becomes inhibited by the interruption of water flow from the xylem vessels to the surrounding elongating cells [21]. Consequently, vegetative growth, dry matter partitioning, reproductive organ development and reproductive processes, grain filling and grain quality are disrupted [22]. In cereal grain crops, reproductive processes and grain filling are more susceptible to water deficit stress, with optimum and ceiling temperatures that are relatively lower than those for seedling and vegetative growth stages [21, 23–25]. Moisture deficit stress reduces yield by delaying silking, thus increasing the anthesis-to-silking interval [11]. Drought stress at flowering period is critical as it can increase pollen sterility resulting in hampered grain set [26].

Drought stress induced yield reduction in cereal crops depends upon the severity and duration of the stress period. The reduction in photosynthesis, emanating from the decrease in leaf expansion, impaired photosynthesis machinery, premature leaf senescence and related food production decreases, is the major contributing factor on yield reduction [21, 23]. Drought stress induced yield reductions have been reported and reviewed in maize [1, 21, 26–29], wheat [26, 27, 30, 31], rice [1], sorghum [32–33] and pearl millet [25].

3. Plants drought stress responses and resistance mechanisms

Plants have evolved numerous dynamic acclimation and adaptive ways of responding to and surviving short-term and long-term drought stresses [34]. The physiology of plants' drought response at the whole plant level is complicated as it encompasses lethal and adaptive alterations. Moreover, how plants respond to drought stress differ significantly at various organizational levels, and this is generally dependent on plant species; the nature, duration and intensity of the drought stress; plant growth and phenological state at the time of stress exposure [19].

Drought stress triggers a wide range of plant structural changes which are essential for plants to respond to such drought stress conditions. These adjustments include morphological adaptations such as reduced growth rate, deepened rooting system, and root-to-shoot ratio modifications. The increased root-to-shoot ratio under drought stress conditions enables water and nutrient uptake and maintenance of osmotic pressure [19, 26, 35]. Additionally, in their response to moisture fluctuations in the soil rhizosphere, plants alter their physiology, modify their root growth and architecture, and regulate the closure of stomata on their aboveground structures. Such tissue-specific responses adjust the cell signals flux, consequently inducing stunted growth or premature flowering, and generally reduced yield [36]. Thus, drought stress is associated with alterations in leaf anatomy and ultrastructure. Reduced leaf size, decreased number of stomata; thickened leaf cell walls and induced premature senescence are some of those morphological changes [19].

Plants resist drought stress through a combination of strategies, which have been widely classified as drought escape, drought avoidance and drought tolerance [18, 21, 37, 38]. Drought escape is achieved by matching the duration of the crop cycle to water supply through genetic variation in phenology [39]. Plants prioritize early flowering and completing their life cycles before the effects of drought cause harm [21].

Drought avoidance denotes plant's ability to maintain high tissue water potential under drought conditions. Usually, plants achieve drought avoidance through morphological and physiological alterations, including reduced stomatal conductance, decreased leaf area, promotion of extensive rooting systems and increased root to shoot ratios [38]. Drought avoidance mechanisms help in maintaining favorable cellular water balance, by enhancing water absorption, decreasing water loss, or allowing desiccation tolerance at low leaf water potential [34]. Stomatal closure, reduction of leaf growth and increased root length and density all contribute to increased water use efficiency under drought stress conditions. Further, water flux into the plant is reduced or water uptake enhanced to achieve drought avoidance [19].

On the other hand, plants attain a state of drought tolerance by cell and tissue specific physiological, biochemical, and molecular mechanisms. Drought tolerance is a complex trait which refers to the capacity of the plant to be more productive under drought stress [40]. In other words, it denotes the potential of crop plants to maintain their growth and development under drought stress [21]. The main aspects of plant drought tolerance mechanism include homeostasis maintenance, via ionic balance and osmotic adjustment; ROS scavenging and antioxidant enzyme activation; growth regulation and recovery by way of phytohormones; specific gene expression; and accumulation of specific stress responsive proteins [1, 2, 19, 21, 26].

To protect themselves against ROS induced oxidative stress and photo inhibition, plants activate an efficient antioxidant (enzymatic and non-enzymatic) defense system [17, 18, 21, 41]. Enzymatic antioxidant enzymes include superoxide dismutase (SOD), catalase (CAT), ascorbate peroxidase (APX), glutathione peroxidase (GPX) and glutathione S-transferase (GST) among others. SOD acts as the first line of defense by converting singlet oxygen into H_2O_2. APX and GSTs then detoxify H_2O_2 [42]. Non-enzymatic antioxidants including glutathione (GSH), ascorbic acid (AsA), carotenoids, tocopherols and flavonoids are also crucial for ROS homeostasis in plants [43].

At the molecular level, our current knowledge thus far regarding drought stress response pathway suggest that, sequentially, the stress is relayed through the following chain: signal perception, signal transduction, transcriptional control by way of transcription factors (TFs), stress responsive genes activation, and in-turn activation of physiological and metabolic responses [1, 19, 37, 44].

Stress signal perception in plants has revealed the role of plasma membrane in perceiving and transmitting signals to the cell interior, where molecules such as receptor-like kinases and G-proteins function as primary receptor molecules [45]. The roles of Ca^+ and ROS as secondary messengers in stress perception and signaling have also been well acknowledged [4]. These secondary messengers adjust the calcium levels and activate protein phosphorylation. Then, phosphorylation of inactive proteins may be directly involved in cellular protection by protein folding or activation of stress specific genes. The dominant plant signal transduction pathways involve the mitogen activated protein kinase (MAPK) and calcium dependent protein kinases (CDPK) cascades, and their role in abiotic stress response have been well reviewed [38, 42, 46].

Upon drought exposure, the resulting signal transduction triggers the production of several biochemicals, including phytohormones, to respond and adapt to the ensuing drought stress [47]. Phytohormones are molecules produced in low concentrations but are critical in regulating plant growth, development, response to biotic and abiotic stresses, and other physiological processes [48]. These phytohormones include ABA, salicylic acid (SA), ethylene (ET), cytokinins (CKs), gibberellin acid (GA), jasmonic acid (JA) and brassinosteroids (BRs) among others. Among these, ABA is the key and most extensively studied hormone that regulates drought resistance in plants [38, 49]. ABA acts as the second messenger coordinating hormonal cross-talk between several stresses signaling cascades, thereby leading to adaptations to changing physiological and environmental conditions [50]. Additionally, SA, ET, JA, CKs, GA, and BRs play vital roles in regulating various phenomena in plants acclimatization to drought stress [51].

At the molecular level, plants institute stress responsive proteins, TFs and signaling pathways among other strategies. Several studies [52, 53] have identified conserved and species-specific drought responsive genes, including membrane stabilizing proteins and late embryogenic abundant (LEA) proteins, which increase cells' water binding capacity [1, 14]. Several heat shock proteins (HSPs), which play a major role in stabilizing protein structure, were also identified [54, 55]. The HSPs are chiefly involved in unwinding some folded proteins and averting protein denaturation under abiotic stress conditions. The membrane stabilizing proteins, LEA proteins, detoxification enzymes, water channel proteins and ion transporter genes all constitute a group of functional proteins, that is, a group of important enzymes and metabolic proteins which directly function to protect cells from stress [37, 56]. Besides functional genes, regulatory genes are also critical for drought tolerance. These encodes various regulatory proteins such as transcriptional factors (TFs), protein kinases and protein phosphatases – involved in regulating signal transduction and gene expression in response to stress [57, 58]. Several TFs that regulate a large spectrum of downstream stress-responsive genes and provide adaptive response under drought stress have been identified and reviewed, including myeloblastosis (MYB), abscisic acid responsive elements binding factor (ABF), ABRE binding (AREB), dehydration responsive element binding (DREB), C-repeat binding factor (CBF), [NAM, ATAF1/2, and CUC2 containing proteins] (NAC) and WRKYs [59–63].

Additionally, protein kinases and protein phosphatases mediate phosphorylation and dephosphorylation of proteins, respectively. In several signal transduction pathways; they are vital and an effective mechanism for stress signal relaying [19, 56, 63]. At the phosphorylation cascade terminals, protein kinases or phosphatases activate or suppress TFs, respectively. The TFs further specifically bind to cis-elements in the promoters of stress responsive genes, thereby modulating their transcription [64]. The TFs are further subjected to post translational modifications (PTMs), including ubiquitination and sumoylation, thereby forming an intricate regulatory network to modulate stress responsive genes, which consequently trigger the activation of appropriate physiological and metabolic responses [62, 65].

4. Approaches for deciphering drought stress responsive genes, proteins and metabolic pathways: where are we?

The recent convergence of crop physiology, next generation sequencing and molecular biology approaches has offered us convenience in deciphering mechanisms underlying plants' response to various abiotic stresses [13, 15, 66]. Whereas plant physiology enhances our understanding of the complex network of traits related to drought tolerance and improving selection efficiency, genomics and molecular biology methods identify the candidate genes and quantitative trait loci (QTLs) underlying these traits [11, 13]. The classical cDNA and oligonucleotide microarrays have been widely employed to identify candidate genes for drought tolerance in several cereal grain crops including maize [67, 68], rice [69, 70] and barley [71]. Additionally, the use of tilling microarrays has allowed for the identification of differentially expressed DNA sequences at the whole genome level [72]. Other techniques such as differential display; cDNA amplified fragment length polymorphism (cDNA-AFLP); and serial analysis of gene expression (SAGE) have been essential in analyzing global gene expression profiles in functional genomics studies [56, 73–74].

Analysis of large scale, high throughput sequencing data is now facilitating the identification and cloning of important genes at target QTLs. Additionally, the 'omics' analysis approaches are showing monumental capacity to quicken and broaden our understanding of the molecular, genetic and functional basis of crop drought stress tolerance [12, 75]. Encouragingly, some novel insights meant to help us develop new drought tolerant cultivars are being generated [76]. Due to its low cost, high-throughput, and high sensitivity, RNA sequencing (RNA-seq) has offered us breakthrough in performing transcriptome analysis of plants' drought stress responses [77]. Resultantly, we have obtained transcripts from RNA in a tissue- or cell-specific manner, and transcribed at a different developmental stage or functional state; this has been fundamental to fishing out functional genes [78]. Therefore, our knowledge pertaining to gene expression networks modulating drought stress tolerance has been significantly improved. RNA-seq technology has been used in several drought stress response studies in cereal crops and numerous genes have been identified [79–81].

Recently, large scale, high-throughput proteomics has become a very powerful tool for performing comprehensive analysis of crop proteins and identification of stress responsive proteins in comparative abiotic stress studies [82, 83]. Proteomic approaches, particularly gel free methods, that is, those involving digestion of intact proteins into peptides prior to separation, have now become very popular in proteome profiling, comparative expression analysis of two or more protein samples, localization and identification of post translational modifications (PTMs) [14]. For instance, isobaric tags for relative and absolute quantitation (iTRAQ) and isotope-coded affinity tags (ICAT) based methods have become widely used in descriptive and comparative drought stress proteomic studies in cereals [84–87]. The iTRAQ-based method allows for the time-dependent analysis of plant stress responses or biological replicates in a single experiment [88]. Besides, proteomics offers complementarity to genomics; providing clues on the molecular mechanisms underlying plant growth and stress responses, as well as being a crucial link between transcriptomics and metabolomics [82]. Moreover, genomics based methods offer access to agronomically desirable alleles localized at QTLs that affect particular physiological responses. This helps us to effectively improve the drought resilience and yield of crops. Additionally, MAS has aided us in improving drought-related traits [12, 89].

Further, a systems biology approach, premised on the advancement and integration of omics (genomics, transcriptomics, metabolomics, phonomics and proteomics) methods to establish a meaningful relationship between the genotype, phenotype and subsequent abiotic stress tolerance, has also become prominent [76]. Integrated knowledge of the morphological, physiological, biochemical, genetic and molecular events in plants allows for in depth understanding of the complex physiological and cellular processes involved in drought stress adaptation. Thus, meaningful interpretations from complex networks and component integrations can be developed from voluminous omics data, which helps us better decipher the mechanisms underlying cereal crops drought tolerance [63].

Meanwhile, the physiological analysis on contrasting genotypes provides information on the mechanisms underlying drought tolerance and aids as a useful screening strategy for drought tolerance [41]. Therefore, it is important to realize that physiological analysis remains essential in corroborating the molecular analyses in abiotic stress response studies. Thus, it would be essential that, going forward, we build on the progress made to date by harnessing the full potential of genomics-assisted breeding, and integrating our knowledge on the physiological and molecular basis of drought tolerance. This calls for crop physiologists, molecular geneticists, breeders and cytogeneticists to collaborate in a multidisciplinary manner [12].

5. Some identified key drought tolerance mechanisms, genes and metabolic pathways

By applying genetic, biochemical and molecular approaches, we have identified essential genes central in plant responses to drought stress. For instance, several physiological responses contributing to drought tolerance in cereal crops have been identified including thermal dissipation of light energy, stomatal closure, decreased hydraulic conductance, altered source-sink relations and carbon partitioning, ABA biosynthesis, among others (**Table 1**).

No.	Physiological response	Purpose	Reference
1	Thermal dissipation of light energy	Uncoupling of photophosphorylation and electron transport	[41, 90]
2	Activation of photosynthesis (PSII) proteins	Preventing photoinhibition of the PSII and improving leaf photosynthetic capacity	[81, 91]
3	Altered source–sink relations and carbon partitioning	Induction of root growth	[92]
4	Prioritized supply of CHOs to rapidly growing or metabolically hyperactive cells or tissues	Promotion of early seedling/hypoctyle growth	[80]
5	Cell wall biosynthesis	Enhancing cellular contents preservation	[81, 93, 94]
6	Cell wall remodeling	Increasing cell wall elasticity to maintain tissue turgidity	[80]
7	Amino acid biosynthesis	Enhanced protein biogenesis	[80]
8	Osmotic adjustment through increased synthesis of soluble solutes (proline, soluble sugars, etc.)	Increased cellular homeostasis maintenance	[11, 40, 95, 96]
9	Changes in ROS scavenging and enzyme activities	Prevention of cellular oxidative damage	[18, 81]
10	ABA biosynthesis	Stomatal closure regulation and improved stress signaling	[11, 40]
11	Stomatal closure and reduced hydraulic conductance	Prevention of water loss through transpiration	[97]
12	Alteration in root morphology and physiology	Increased water and nutrient absorption under drought conditions	[98, 99]
13	Reduced stomatal density and enhanced control of stomatal opening and closure	Improved cellular water conservation	[100]
14	Alternative oxidase pathway, uncoupling proteins, NADPH dehydrogenases down-regulated	Uncoupling of oxidative phosphorylation and electron transport	[101]

Table 1.
Some key drought stress tolerance mechanisms identified in cereals.

Besides the mechanisms highlighted in **Table 1**, several drought responsive genes have been identified and validated in different crop species. For instance, recent excellent reviews [4, 26, 102–106] provide highly informative details about some crop drought tolerance conferring genes that have been functionally validated to date. Some of these genes are listed in **Table 2**.

Further, several metabolic pathways implicated in drought stress tolerance in cereal grain crops have been identified through comparative physiological and omic analysis approaches. Chief among these pathways are those related to photosynthesis, secondary metabolites biosynthesis, plant hormone signaling, starch and sucrose metabolism, and nitrogen metabolism. Chloroplasts, particularly the thylakoid membranes—PSII reaction centers, are one of the organelles most influenced by drought stress [123, 124]. Photosynthesis (antenna protein) pathway related genes *lhcb5-1* and *lhcb5-2* are part of the light harvesting complexes (LHCs) and the electron transport components of the PSII of the plant photosynthesis machinery, where they participate

No.	Gene name	Donor	Host	Physiological change	References
1	*ZmVPP1*	*Arabidopsis thaliana* L. (Arabidopsis)	Maize	Enhanced photosynthetic efficiency and root development	[81, 107]
2	*ZmPP2C-A10*	Arabidopsis	Maize	ABA signaling	[108]
3	*Zeaxanthin epoxidase (ZEP)*	Arabidopsis	Maize	Improved ABA biosynthesis	[104, 109]
4	Mannitol-1-phosphate dehydrogenase (*mtlD*)	*Escherichia coli*	Wheat	Improved fresh and dry weights, plant height, and flag leaf length	[110]
5	*AtNF-YB1*	Arabidopsis	Maize	Higher photosynthesis capacity	[102, 111]
6	*AtABF3*	Arabidopsis	Rice	Higher Fv/fm	[112]
7	*OsDREB1A*	Arabidopsis	Rice, wheat	Shoot growth retardation at the expense of root growth	[102]
8	*AtHARDY*	Arabidopsis	Rice	Enhanced WUE and photosynthesis efficiency	[113]
9	*OsWRKY11*	Rice	Rice	Sluggish water loss and lessened leaf wilting	[114]
10	*AtLOS5 (LOS5/ ABA3)*	Arabidopsis	Maize	Increased ABA biosynthesis	[115]
11	*HVA1*	Barley	Rice	Higher WUE	[116]
12	*HVA1*	Barley	Wheat	Enhanced biomass accumulation and WUE	[26, 117]
13	*Beta*	Escherichia coli	Wheat	Accumulation of glycine betaine	[118]
14	*Nicotiana protein kinase (NPK1)*	Tobacco	Maize	Preventing dehydration damage to the photosynthesis machinery	[119]
15	*AtSNAC1*	Arabidopsis	Rice	ABA-hypersensitive, stomatal shutdown	[120, 121]
16	*SbER2–1*	Sorghum	Maize	Increased Pn rate and higher WUE	[122]
17	Light harvest complex related genes (*LHCA1, LHCB, LHCA3, LHCA2*)	Maize	Maize	Balancing light capture in the PSII	[91, 93]

Table 2.
Examples of drought tolerance conferring genes that have been functionally validated in cereal crops.

as peripheral antenna systems enabling more efficient absorption of light energy [125, 126]. Further, Lhch5-1 is involved in the intracellular non-photochemical quenching and the cysteine biosynthesis processes [91]. Previously, the photosynthesis pathway has been significantly enriched in drought stress response in maize, with chlorophyll a-b binding proteins being up-regulated in an ABA-dependent manner [87, 91], and pearl millet [127]. This photosynthesis related pathway plays a critical role in balancing light capture and utilization to avoid photoinhibition of the PSII [87].

Phenylpropanoid metabolism is the first step of the secondary metabolites (flavonoids, phenylpropanoids, phenolic compounds and lignin) biosynthesis and phenylpropanoids act as antioxidants to protect plants against oxidative damage [128, 129]. Flavonoids play different molecular functions in plants, including stress protection. All these compounds are widely synthetized in response to several abiotic stresses, including drought [130]. In wheat leaves, an increase in flavonoid and phenolic acids content was shown together with stimulation of genes involved in flavonoid biosynthesis pathway in response to drought stress [131]. Moreover, secondary metabolites biosynthesis related pathways were found to be significantly enriched in response to drought stress in maize [80] and sorghum [37], suggesting their involvement in plant protection. A coordinated reaction of the genes and pathways involved in secondary metabolite biosynthesis is therefore vital for improved drought stress tolerance in plants [80]. Plant hormone signaling pathway participates in drought stress response via either ubiquitin-mediated proteolysis or ABA-mediated response [132], and was observed to be significantly enriched in pearl millet response to drought stress [127].

Sucrose is widely acknowledged to play a crucial role as a key molecule in energy transduction and as a regulator of cellular metabolism [130]. Additionally, sucrose and other sugars are energy and carbon sources required for defense response and are necessary for plant survival under drought stress conditions [133]. Further, sucrose can act as primary messenger controlling the expression of several sugar metabolism related genes [134]. Nitrogen metabolism pathway, being the most basic and central physiological metabolic process during plants' growth period, directly influences the formation of cellular components and regulation of cellular activities, as well as the transformation of photosynthetic products, mineral nutrient absorption and protein synthesis. It therefore follows that the nitrogen pathway is directly affected by drought, hence its significant enrichment under such conditions [80, 127]. These key identified drought responsive pathways should be used as targets for future genetic engineering of drought stress tolerant genotypes.

6. QTL mapping for drought tolerance in cereals

Most yield-related traits in cereal crops are quantitative. Therefore, cloning of the causal genes and deciphering the underlying mechanisms influencing these traits remains critical for continuous genetic improvement [135]. Precisely, drought tolerance is a complex quantitative trait that is multi-genic in its expression and one of the most challenging traits to study and characterize [11]. In comparison to conventional approaches, genomics offers unparalleled opportunities for dissecting quantitative traits into their single genetic determinants, known as QTL, thereby facilitating MAS, gene cloning and their direct manipulation via genetic engineering [12]. Through advances in next generation sequencing, identification of major QTLs regulating specific drought responses has been made possible, via the development of large numbers of genetic markers such as single nucleotide polymorphisms (SNPs) and insertion-deletions (InDels), thereby opening the doors for an efficient way to improving drought tolerance in cereal crops [89]. Additionally, large-scale genome-wide association studies (GWAS) have been conducted to detect genomic regions and candidate genes for various agronomic traits, including drought tolerance in cereals [13, 136, 137]. Resultantly, hundreds of studies reporting thousands of major drought-responsive genes and QTLs in cereal grain crops can be found in the literature, including those for maize [13, 138–140], rice [12, 13, 105, 138, 141, 142], wheat [13, 31, 137, 138, 143], sorghum [138, 144, 145], barley [138, 146], and pearl millet [136, 147, 148]. The high number of studies on QTL mapping suggests that for the past decade, QTL has been the focal target of research to identify the genetic loci regulating the adaptive response of crops to drought stress. Although several QTLs for drought tolerance have already been mapped in these cereal crops, there has been little success in introgression of those QTLs and the number of causal genes that have been confirmed within these QTL regions remains relatively small as compared to Arabidopsis and rice [11, 106].

Going forward, MAS remains a useful tool for major QTL, whereas QTL cloning is increasingly becoming a more routine activity. This has been necessitated by increased use of high-throughput sequencing, precise phenotyping and identification of appropriate candidate genes through omics approaches [89, 136, 137]. Cloned QTL facilitate a more targeted search for novel alleles and will offer novel insights for genetic engineering of drought resilient cereal crops [13]. Moreover, compared to other crops, research in millets is still lagging behind. However, with millets considered predominantly climate resilient crops, millets could serve as valuable source of novel genes, alleles and QTLs for drought tolerance. Therefore, the identification and functional characterization of these genes, alleles and QTLs in millets is critical for their introgression and drought tolerance improvement in cereal grain crops [89].

7. Phenotyping for drought tolerance in cereal grain crops

Phenotyping has become an integral component of the crop improvement programme by contributing towards understanding of the genetics behind crop drought tolerance [105, 149]. Since many component traits of drought tolerance are controlled quantitatively, improving the accuracy of phenotyping has become more important to improve the heritability of the traits, and the target traits would require rapid and precise measurement [106]. High throughput phenotyping now provides an essential link in translating laboratory research to the field. This is vital in developing novel genotypes that incorporate gene(s) expressing promising trait (s) into breeding lines adapted to target field environments [150].

Auspiciously, the recent advances in phenotyping technology and robotics for measuring large number of plants means that large numbers of genotypes could be readily phenotyped [34, 151–153]. More promising approaches that target complex traits tailored to specific requirements at the different main crop growth stages are now available [150]. Precise phenotyping of drought-related physiological traits often requires the utilization of sophisticated and expensive techniques. These include magnetic resonance imaging (MRI) and positron emission tomography (PET), near-infrared (NIR) spectroscopy on agricultural harvesters, canopy spectral reflectance (SR) and infrared thermography (IRT), nuclear magnetic resonance, hyperspectral imaging, laser imaging, 3D imaging and geographical information systems (GIS), among others [34, 138, 154, 155]. For example, 3D visual modeling can be used to determine the plasticity of the canopy architecture, and to evaluate the architectural and physiological characteristics that contribute to the higher productivity of the super rice varieties under drought stress conditions [156]. Though currently expensive, up scaling the use of these phenotyping platforms will eventually enhance our understanding of crop growth kinetics and aid us improve crop models for systems biology and drought tolerance breeding programs.

Selection of primary (grain yield and yield contributing traits) and secondary agronomically important traits (ASI, root architecture, stay green, etc.) is the way to achieve drought tolerance in cereal grain crops [31, 35, 106, 152]. Yield and yield attributing factors are targeted for direct selection whilst secondary traits are vital in conferring drought tolerance and contributing to final yield indirectly [157]. Crucially, considering that under drought stress conditions, the genetic correlation between grain yield and some secondary traits increases meaningfully, and the heritability of some secondary traits remains high, identification and selection of those highly heritable secondary traits that are positively correlated with yield related traits in the target environment, and responsive to high throughput phenotyping, will be critical in achieving the desired drought tolerance goals [11, 152, 158]. Managed stress screening approaches through the utilization of phenomics offers an opportunity to keep heritability high and phenotyping under controlled environments can be helpful in large-scale characterization studies such as trait mapping experiments [13]. However, great caution needs to be taken when phenotyping for drought stress tolerance since controlled environments may fail to mimic the real field conditions, thereby becoming less useful to study the genotype × environment interactions which are very essential to dissect the drought tolerance mechanisms [106].

8. Genetic engineering of drought tolerant cereal crops

The development of tolerant crops by genetic engineering requires the identification of key genetic determinants underlying stress tolerance in plants, and introducing these genes into crops [159, 160]. The momentous progress garnered in abiotic stress response research in the model plant Arabidopsis has created an avenue for the identification of drought tolerance conferring genes and the development of transgenics carrying these genes in other crop species. Therefore, genetic engineering approach has opened the door to the development of new crop genotypes with improved drought tolerance [103]. Over the past decade, numerous articles and reviews on drought stress tolerant transgenic crop species have been gathered [4, 26, 102–106, 160]. A selected list of transgenic cereal grain crops, which includes information on transgenes used for the transformations and the resultant drought stress tolerance mechanisms, has already been provided above in **Table 2.** Despite the availability of numerous reports in the scientific literature of transgenic approaches to improving drought tolerance, restrictions on the establishment of transgenic plants in the field presents a bottleneck in true testing of the effectiveness of transgenic approaches to improve crop drought tolerance [161].

In recent years, transgenesis has taken center stage in our crop improvement efforts. Advances in genome engineering has made it possible to precisely alter DNA sequences in living cells, providing unprecedented control over a plant's genetic material [162]. The genome engineering approaches, also known as gene editing or genome editing techniques, involve the use of programmable site-directed nucleases (SDNs) engineered to modify target genes at desirable locations on the genome [163]. These SDNs cleave the double-stranded DNA at a particular location by means of clustered regularly interspaced short palindromic repeats and CRISPR-associated protein 9 (CRISPR/Cas9), zinc finger nucleases (ZFNs), or transcriptional activator-like effector nucleases (TALENs). The double-stranded DNA break then undergoes natural reparation either via homologous recombination or non-homologous end joining [105, 164]. The restoration of the DNA break can be directed to create a variety of targeted DNA sequence modifications such as DNA deletions or insertions of large arrays of transgenes [162].

Among the several genome editing methods developed to date, CRISPR-Cas9 is the most advanced and has received much attention because of its great accuracy, quickness, adaptability and simplicity [165, 166]. This technique has been successfully used in major food crops [167–169]. Whereas convectional genomics and breeding approaches alone cannot resolve the global food security challenge [170], genetic engineering approaches have great potential to improve crops that feed the burgeoning populations of developing countries [162, 166, 171]. However, the extent of regulation imposed upon crop cultivars generated through genetic engineering will have a huge impact on the cost of their development itself and how rapidly they will be deployed into the food supply chain. Linked to that, the readiness with which the public will accept food products made from genetically engineered crops will also play a role in the extent to which this new technology will be fully utilized for crop improvement, particularly in the developing countries where cereal grain crops are staple diets [162].

9. Field management of crops in the context of climate change

Genetic and management strategies that are aimed at improving grain yields under water constrained environments target three variables, which are the amount of water captured by the plant (W), the efficiency with which that water is converted to biomass (water use efficiency, WUE), and the harvest index (HI) or the proportion of biomass forming grain. Interestingly, each of these variables can be altered [172]. For instance, WUE can be maximized by early planting of crops, and by maintaining healthy leaves with high levels of nutrients [173]. Additionally, extending leaves` longevity through selection for delayed leaf senescence is commonly regarded important for maintaining WUE and root health, as well as increasing the kernel filling duration [172]. Supplementation of irrigation, where available, would be the major means for combating drought stress condition, besides being a prime approach to the intensification of agriculture and the generation of stable income for farmers [174]. However, its uptake will depend on various environmental, economic and social factors on both micro and macro levels [175].

Additionally, agronomic interventions, that is, improved crop management methods can complement the use of drought tolerant cultivars, contributing meaningfully to enhanced and stabilized yields under water constrained environments. Conservation agriculture (CA), a collection of practices embodying the use of reduced tillage and mulch to reduce evaporation of soil water, is an obvious means of increasing water available to the plants [172, 176]. Scaling up CA, which has recently gained wider acceptance in developing countries, offers a great potential of increasing drought resilience and sustainability of cropping systems and ensuring food security. The use of plastic mulch in semi-arid cooler areas on the Loess Plateau of China has significantly increased WUE in maize [177]. Moreover, CA has become the cornerstone of dryland systems in some regions of USA, Canada and Australia [174]. Recently, the application of melatonin with date (*Phoenix dactylifera* L.) residue and wheat straw biochars has enhanced biochar efficiency for drought tolerance in maize cultivars [178]. This can be a game changer in CA farming systems where biochar is a key component for ground cover. Therefore, the management of a sustainable dryland farming system would require that farmers apply good agronomic practices (GAPs) such as improved soil and water conservation and the associated reduced tillage practices, holistic weed and pathogen control, soil fertility management with respect to water regimes, optimized plant population densities and effective control of soil biotic stress factors that may inhibit root development, as well as practicing crop diversification in order to reduce the risks associated with farming in unpredictable environments. Further, governments should increase investments in weather forecasting and cloud seeding.

All these approaches and decision support systems, when integrated, would birth a more holistic strategy for adapting cereal grain crops to future climate change induced drought stress as summarized in **Figure 1.**

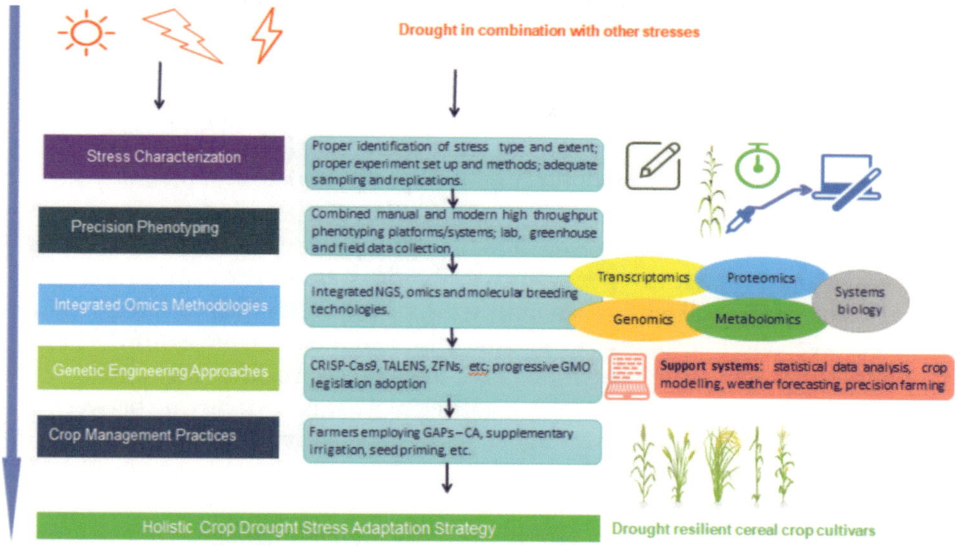

Figure 1.
Proposed holistic approach to cereal crop drought stress adaptation.

10. Future outlook

Currently, crop improvement efforts are shifting focus from solely yield, quality, or abiotic stress resistance to a holistic approach integrating breeding for a combination of these factors. For example, targeting breeding for combined drought and heat stress tolerance [179, 180], drought tolerance and nutritional quality [161] or drought tolerance and disease resistance [3, 181] are being underlined. Drought-tolerant micro-nutrient dense cereal crop cultivars have been developed [8]. In that regard, future-proofing of global food security would call for double-pronged transformation aimed at developing high yielding cultivars possessing both adaptability to abiotic or biotic stresses and higher nutritional quality. Thus, going forward, it is necessary that more breeding programs use high-priority abiotic stresses in their portfolios [11]. In the wake of addressing malnutrition challenges in developing countries, breeding for bio-fortified drought tolerant cultivars should be strengthened [8, 182]. In addition, promising genotypes in trials for fall armyworm (*Spodoptera frugiperda* L.) resistance, once approved, should be screened for drought tolerance and improved cultivars harboring both traits be developed. Though seemingly a daunting task, combined efforts from crop physiologists, molecular geneticists, breeders and pathologists would make this a reality.

In our 'omics research', future proteomics studies should intensify the identification, quantification and analysis of stress-responsive proteins PTMs, since PTMs can affect protein functions, interactions, subcellular targeting and stability [183]. Moreover, breeding for resource use efficiency, for instance, nitrogen use efficiency (NUE) should be integrated in future drought tolerance breeding programs. Previously, researchers have revealed that improving NUE in crops such as paddy rice and cotton will be critical in greenhouse gas emission management [184–188]. Additionally, application of biochar, either singly [189] or in combination with P [190] has been shown to alleviate heat-induced oxidative stress damage on the cellular physiological processes in rice plants. Thus, as we navigate the future, breeding for NUE and its proper management in the field remains essential in helping adapting crop plants to abiotic stresses such as drought.

Another area of focus will be crop physiology. Since photosynthesis is the basis of plant growth, improving photosynthesis can significantly contribute towards greater food security in the future. Multiple targets for manipulation of crop photosynthesis have been extensively reviewed. These include improving Rubisco kinetic properties and improving canopy architecture to enhance light penetration [191]. Therefore, harnessing the benefits of improved photosynthesis for greater yield potential will require that we intensify manipulation of these targets.

Furthermore, the microbes resident in the rhizosphere can potentially improve plant growth and enhance crop resilience to abiotic stresses [192]. For instance, phosphate-solubilizing bacteria can nullify the antagonistic effect of soil calcification on bioavailable phosphorus in alkaline soils, and thus, can be one of the best options for improving soil P nutrition [193]. It would be crucial to investigate, understand and quantify the complex feedback mechanisms occurring between root and microbial responses to drought stress, particularly in cereal crops. Integrating crop ecology, physiology and molecular methodologies in a multi-disciplinary approach would be central [194].

In the short to medium term, exogenous application of plant growth regulators at different crop growth phases would be an important strategy in inducing drought resistance. In a very short term, seed priming will be of value [3]. The crucial roles of plant phytohormones and growth-promoting rhizobacteria in abiotic stress responses have been extensively reviewed [195–197]. Particularly, the exogenously applied plant growth regulators can enhance morpho-physiological, growth and abiotic stress responses of crops such as rice [198–203]. Recently, Saleem et al. [204] have shown that exogenously applied gibberrellic acid (GA3) can reduce metal toxicity induced oxidative stress in jute (*C. capsularis* L.) seedlings. Moreover, exogenous application of salicicic acid (SA) has been shown to ameliorate the adverse effects of salinity on maize plants [205]. Further, silicon (Si) application improved plant water relations, photosynthesis and drought tolerance in Kentucky bluegrass [206, 207], and Si enhanced germination, growth, P and arsenic uptake in rice [208]. Therefore, targeting the modification of hormone biosynthetic pathways may be a gateway to the development of drought tolerant transgenic plants.

Meanwhile, the potential of transgenic technologies across developing countries, particularly in Sub-Saharan Africa, is being hampered by uncoordinated over-regulation by authorities. Unfortunately, the biosafety framework leading to ordered testing and deregulation in such countries is being developed on individual countries basis, instead of a more efficient resource-use regional approach. Moreso, present systems are modeled on overestimated, not science-evidence-based risks [172]. There is need for policy makers to revisit their stance on genetically modified organisms (GMOs) with science guaranteed evidence, not political grand-standing, guiding formulation, adoption and utilization of GMO related polices for food and nutritional security.

11. Conclusion

Though a daunting task, breeding for drought tolerance in cereal crops remains the most economical, effective and sustainable strategy for ensuring food security for the ever increasing human population. While a significant progress has been made to date towards achieving that goal, our understanding of the mechanisms underpinning plant drought stress tolerance remains fragmentary. In the face of global climate change, a multi-disciplinary research strategy becomes obligatory to integrate physiological, genotypic, omics and epigenetics data essential to dissect the complex networks regulating plant drought tolerance, which can then be manipulated through genetic engineering to develop drought resilient crops. Additionally, cereal crop breeding programs should integrate high-priority abiotic stresses, particularly drought and heat, with other goals such as resource use efficiency, biotic resistance, and nutritional improvement. Further, formulation and effective deployment of efficient screening and precise phenotyping approaches at both laboratory and field levels to identify drought tolerant genotypes or mutants remains critical. Promisingly, our new cultivar development thrusts are now endowed with some innovative methodologies that include high-throughput phenotyping, doubled haploidy, mutation and speed breeding, as well as CRISP-Cas 9 technologies. Going forward, we should harness the potential of these technologies. Eventually, our understanding of the crop drought tolerance mechanisms will be quickened and broadened, greatly assisting our development of new drought-resilient cereal crop cultivars. This should be supported by robust, science-evidence-based and progressive policy frameworks that recognize the centrality of GMOs and modern biotechnology in increasing food production. Consequently, this would lead to improved sustainable crop productivity and global food security.

Acknowledgements

We acknowledge the funding received from the National Key Research and Development Project of China (Selection and Efficient Combination Model of Wheat and Maize Water Saving, High Yield and High Quality Varieties; Grant No. 2017YFD0300901).

Conflict of interest

Authors declare that they have no conflict of interests.

Author details

Tinashe Zenda[1,2,3], Songtao Liu[1,2,3] and Huijun Duan[1,2,3]*

1 State Key Laboratory of North China Crop Improvement and Regulation, Hebei Agricultural University, Baoding, China

2 North China Key Laboratory for Crop Germplasm Resources of the Education Ministry, Hebei Agricultural University, Baoding, China

3 Department of Crop Genetics and Breeding, College of Agronomy, Hebei Agricultural University, Baoding, China

*Address all correspondence to: hjduan@hebau.edu.cn

References

[1] Farooq M, Wahid A., Kobayashi N, Fujita D, Basra SMA. Plant drought stress: Effects, mechanisms and management. Agron. Sustain. Dev. 2009;29(1):185-212. DOI: 10.1051/agro:2008021

[2] Feller U, Vaseva II. Extreme climatic events: Impacts of drought and high temperature on physiological processes in agronomically important plants. Front Environ Sci. 2014;2:39. DOI: 10.3389/fenvs.2014.00039.

[3] Fahad S, Bajwa AA, Nazir U, Anjum SA, Farooq A, Zohaib A, Sadia S, et al. Crop production under drought and heat stress: Plant responses and management options. Front. Plant Sci. 2017;8:1147. DOI: 10.3389/fpls.2017.01147

[4] Kumar S, Sachdeva S, Bhat KV, Vats S. Plant responses to drought stress: Physiological, biochemical and molecular basis. In: S Vats (ed.) Biotic and Abiotic Stress Tolerance in Plants. Springer Nature, Singapore Pte Ltd. 2018; pp. 1-25. DOI: 10.1007/978-981-10-9029-5_1

[5] Schils R, Olesen JE, Kersebaum K, Rijk B, Oberforster M, et al. Cereal yield gaps across Europe. Eur J Agron. 2018;101:109-120. DOI: 10.1016/j.eja.2018.09.003

[6] Prasad P VV, Staggenborg SA. Impacts of drought and/or heat stress on physiological, developmental, growth, and yield processes of crop plants. 2008. ASA, CSSA, SSSA, 677 S. Segoe Rd., Madison, WI 53711, USA

[7] Cordain L. Cereal grains: Humanity's double-edged sword. In: Simopoulos AP (ed.) Evolutionary aspects of nutrition and health diet, exercise, genetics and chronic disease. World review of nutrition and dietetics. 1999;19-73. Basel, Switzerland: Karger.

[8] Curiel R, Domínguez L, Donovan M, Doody A, Johnson J, Listman GM, MacNeil M, et al. The cereals imperative of future food systems. Seeds of Change Annual Report 2019. CIMMYT (International Maize and Wheat Improvement Center), Mexico. 2020;pp.18-19.

[9] Shiferaw B, Prasanna BM, Hellin J, Bänziger M. Crops that feed the world 6. Past successes and future challenges to the role played by maize in global food security. Food Sec. 2011;3:307-327. DOI: 10.1007/s12571-011-0140-5.

[10] Ray DK, Mueller ND, West PC, Foley JA. Yield trends are insufficient to double global crop production by 2050. PLoS One. 2013;8: e66428.

[11] Maazou ARS, Tu JL, Qiu J, Liu ZZ. Breeding for drought tolerance in maize (*Zea mays* L.). American Journal of Plant Sciences. 2016;7:1858-1870.

[12] Tuberosa R, Salvi S. Genomics-based approaches to improve drought tolerance of crops. Trends in Plant Science. 2006;11(8):405-412. DOI: https: 10.1016/j.tplants.2006.06.003.

[13] Choudhary M, Wani SH, Kumar P, Bagaria PK, Rakshit S, Roorkiwal M, Varshney RK. QTLian breeding for climate resilience in cereals: progress and prospects. Funct Integr Genomics. 2019;19:685-701. DOI: 10.1007/s10142-019-00684-1

[14] Ghatak A, Chaturvedi P, Weckwerth W. Cereal crop proteomics: Systemic analysis of crop drought stress responses towards marker-assisted selection breeding. Front. Plant Sci. 2017;8:757. DOI: 10.3389/fpls.2017.00757.

[15] Bokszczanin KL, SPOT-ITN Consortium, Fragkostefanakis S. Perspectives on deciphering

mechanisms underlying plant heat stress response and thermotolerance. Front. Plant Sci. 2013;4:315. DOI: 10.3389/fpls.2013.00315

[16] Gilbert ME, Medina V. Drought adaptation mechanisms should guide experimental designs. Trends in Plant Sci. 2016;21:8. DOI:10.1016/j.tplants.2016.03.003

[17] Osmolovskaya N, Shumilina J, Kim A, Didio A, Grishina T, Bilova T, et al. Methodology of Drought Stress Research: Experimental Setup and Physiological Characterization. Int. J. Mol. Sci. 2018;19:4089. DOI: 10.3390/ijms19124089.

[18] Anjum SA, Xie XY, Wang LC, Saleem MF, Man C, Lei W. Morphological, physiological and biochemical responses of plants to drought stress. Afr. J. Agric. Res. 2011;6:2026-2032.

[19] Lisar SYS, Motafakkerazad R, Hossain MM, Rahman IMM. Water stress in plants: causes, effects and responses. Introductory Chapter. In: Rahman IMM, Hasegawa H, editors. Water Stress. IntechOpen. 2012;pp. 1-14. DOI: 10.5772/39363

[20] Choudhury FK, Rivero RM, Blumwald E, Mittler R. Reactive oxygen species, abiotic stress and stress combination. The Plant Journal. 2017;90:856-867. DOI: 10.1111/tpj.13299

[21] Aslam M, Maqbool MA, Cengiz R. Drought stress in maize (*Zea mays* L.): Effects, resistance mechanisms, global achievements and biological strategies for improvement. SpringerBriefs in Agriculture. Springer Cham. 2015. ISBN 978-3-319-25440-1. DOI:10.1007/978-3-319-25442-5

[22] Boyer JS, Westgate M E. Grain yield with limited water. J. Exp. Bot. 2004;55:2385-2394. DOI: 10.1093/jxb/erh219

[23] Sehgal A, Sita K, Siddique KHM, Kumar R, Bhogireddy S, Varshney RK, Hanumantha Rao B, et al. Drought or/and heat-stress effects on seed filling in food crops: Impacts on functional biochemistry, seed yields, and nutritional quality. Front. Plant Sci. 2018;9:1705. DOI: 10.3389/fpls.2018.01705

[24] Reddy AR, Chaitanya KV, Vivekanandan M. Drought-induced responses of photosynthesis and antioxidant metabolism in higher plants. J Plant Physiol. 2004;161:1189-1202

[25] Yadav RS, Hash CT, Bidinger FR, Devos KM, Howarth CJ. Genomic regions associated with grain yield and aspects of post flowering drought tolerance in pearl millet across environments and tester background. Euphytica. 2004;136:265-277

[26] Farooq M, Hussain M, Wahid A, Siddique KHM. Drought stress in plants: An overview. In: R Aroca, editor. Plant responses to drought stress: From morphological to molecular features. Springer-Verlag Berlin Heidelberg. 2012;1-36. DOI: 10.1007/978-3-642-32653-0_1

[27] Daryanto S, Wang L, Jacinthe PA. Global synthesis of drought effects on maize and wheat production. PLoS ONE. 2016;11:e0156362. DOI: 10.1371/journal.pone.0156362.

[28] Lobell DB, Burke MB. On the use of statistical models to predict crop yield responses to climate change. Agricultural and Forestry Meteorology. 2010;150:1443-1452.

[29] Gammans M, Mérel P, Ortiz-Bobea A. Negative impacts of climate change on cereal yields: Statistical evidence from France. Environ. Res. Lett. 2017;12:054007.

[30] You L, Rosegrant MW, Wood S, Sun D. Impact of growing

season temperature on wheat productivity in China. Agric. For. Meteorol. 2009;149:1009-1014.

[31] Khadka K, Earl HJ, Raizada MN, Navabi A. A physio-morphological trait-based approach for breeding drought tolerant wheat. Front. Plant Sci. 2020;11:715. DOI: 10.3389/fpls.2020.00715

[32] Assefa Y, Staggenborg SA, Prasad VPV. Grain sorghum water requirement and responses to drought stress: A review. [Online]. Crop Management. 2010. DOI:10.1094/CM-2010-1109-01-RV

[33] Jabereldar AA, El Naim AM, Abdalla AA, Dagash YM. Effect of water stress on yield and water use efficiency of sorghum (*Sorghum bicolor* L. Moench) in semi-arid environment. International Journal of Agriculture and Forestry. 2017;7(1):1-6. Doi: 10.5923/j.ijaf.20170701.01

[34] Badigannavar A, Teme N, Costa de Oliveira A, Li G, Vaksmann M, Viana VE, et al. Physiological, genetic and molecular basis of drought resilience in sorghum [*Sorghum bicolor* (L.) Moench]. Ind J Plant Physiol. 2018;23(4):670-688. DOI: 10.1007/s40502-018-0416-2

[35] Calleja-Cabrera J, Boter M, Oñate-Sánchez L , Pernas M. Root growth adaptation to climate change in crops. Front. Plant Sci. 2020;11:544. DOI: 10.3389/fpls.2020.00544

[36] Gupta A, Rico-Medina A, Caño-Delgado AI. The physiology of plant responses to drought. Science. 2020;6488:266-269. Doi:10.1126/science.aaz7614

[37] Goche T, Shargie NG, Cummins I, Brown AP, Chivasa S, Ngara R. Comparative physiological and root proteome analyses of two sorghum varieties responding to water limitation. Scientific Reports. 2020;10:11835. DOI: 10.1038/s41598-020-68735-3

[38] Basu S, Ramegowda V, Kumar A, Pereira A. Plant adaptation to drought stress [version 1; referees: 3 approved] F1000 Research. 2016;5 (F1000 Faculty Rev):1554 . DOI: 10.12688/f1000research.7678.1

[39] Shavrukov Y, Kurishbayev A, Jatayev S, Shvidchenko V, Zotova L, Koekemoer F, et al. Early flowering as a drought escape mechanism in plants: How can it aid wheat production? Front. Plant Sci. 2017;8:1950. DOI: 10.3389/fpls.2017.01950

[40] Ribaut JM, Betran J, Monneveux P, Setter T. Drought tolerance in maize. In: Handbook of maize: its biology. Springer, New York. 2009;311-344.

[41] Bhargava S, Sawant K. Drought stress adaptation: metabolic adjustment and regulation of gene expression. Plant Breeding. 2013;132:21-32. DOI: 10.11/pbr.12004.

[42] You J, Chan Z. ROS regulation during abiotic stress responses in crop plants. Front. Plant Sci. 2015;6:1092. DOI: 10.3389/fpls.2015.01092

[43] Gill SS, Tuteja N. Reactive oxygen species and antioxidant machinery in abiotic stress tolerance in crop plants. Plant Physiol. Biochem. 2010; 48:909-930. DOI: 10.1016/j.plaphy.2010.08.016

[44] Upadhyaya H, Sahoo L, Panda SK. Molecular physiology of osmotic stress in plants. In: Rout GR, Das AB. (Eds.). Molecular Stress Physiology of Plants. 2013;179-193. Springer, India. DOI: 10.1007/978-81-322-0807-5_7

[45] Tuteja N, Sopory SK. Plant signaling in stress, G-protein coupled receptors, heterotrimeric G-proteins and signal coupling via phospholipases. Plant Signal Behav. 2008;3(2):79-86.

[46] Zhu JK. Abiotic stress signaling and responses in plants. Cell. 2016;167:313-324. DOI: 10.1016/j.cell.2016.08.029.

[47] Fahad S, Ullah A, Ali U, Ali E, Saud S, Hakeem KR, Alharby H, et al. Drought tolerance in plants role of phytohormones and scavenging system of ROS. In book: Plant Tolerance to Environmental Stress: Role of Phytoprotectants. CRC Press. 2019; 103-113.

[48] Wani SH, Kumar V, Shriram V, Sah SK. Phytohormones and their metabolic engineering for abiotic stress tolerance in crop plants. The Crop Journal 4. 2016;1 6 2 – 1 7 6.

[49] Aimar D., Calafat M., Andrade A.M., Carassay L., Abdala G.I., Molas M.L. Drought tolerance and stress hormones: From model organisms to forage crops. Plants and Environment. 2011;137-164. DOI: 10.5772/24279 ·

[50] Basu S, Rabara R. Abscisic acid — An enigma in the abiotic stress tolerance of crop plants. Plant Gene. 2017;11:90-98.

[51] Ullah A, Manghwar H, Shaban M, Khan AH, Akbar A, Ali U, Ali E, Fahad S. Phytohormones enhanced drought tolerance in plants: a coping strategy. Environ Sci Pollut Res. 2018. Springer-Verlag, part of Springer Nature, GmbH, Germany. DOI: 10.1007/s11356-018-3364-5

[52] Hanin M, Brini F, Ebel C, Toda Y, Takeda S, Masmoudi K. Plant dehydrins and stress tolerance versatile proteins for complex mechanisms. Plant Signal. Behav. 2011;6:1503-1509. DOI: 10.4161/psb.6.10.17088.

[53] Kosova K, Vitamvas P, Prasil IT. Wheat and barley dehydrins under cold, drought, and salinity—What can LEA-II proteins tell us about plant stress response? Front. Plant Sci. 2014;5:343. DOI:10.3389/fpls.2014.00343.

[54] Wang W, Vinocur B, Shoseyov O, Altman A. Role of plant heat-shock proteins and molecular chaperons in the abiotic stress response. Trends Plant Sci. 2004;9:244-252. DOI: 10.1016/j.tplants.2004.03.006.

[55] Zhang X, Lei L, Lai J, Zhao H, Song W. Effects of drought stress and water recovery on physiological responses and gene expression in maize seedlings. BMC Plant Biol. 2018;18:68. DOI:10.1186/s12870-018-1281-x.

[56] Harb A. Identification of candidate genes for drought stress tolerance. In: Hossain MA, Wani SH, Bhattacharjee S, Burritt DJ, Tran L.SP. (Eds). Drought stress tolerance in plants. Springer Cham, Switzerland. 2016;2:385-414. DOI:10.1007/978-3-319-32423-4_14.

[57] Shinozaki K, Yamaguchi-Shinozakiy K, Sekiz M. Regulatory network of gene expression in the drought and cold stress responses. Curr. Opin. Plant Biol. 2003;6:410-417. DOI: 10.1016/S1369-5266(03)00092-X

[58] Ciarmiello LF, Woodrow P, Fuggi A, Pontecorvo G, Carillo P. Plant genes for abiotic stress. 2011;283-308. DOI: 10.5772/22465

[59] Singh D, Laxmi A. Transcriptional regulation of drought response: A tortuous network of transcriptional factors. Front. Plant Sci. 2015;6:895. DOI: 10.3389/fpls.2015.00895.

[60] Mun BG, Lee SU, Park EJ, Kim HH, Hussain A, Imran QM, Lee IJ, Yun BW. Analysis of transcription factors among differentially expressed genes induced by drought stress in Populus davidiana. 3 Biotech. 2017;7:209. DOI: 10.1007/s13205-017-0858-7.

[61] Wang CT, Ru JN, Liu YW, Yang JF, Li M, Xu ZS, Fu JD. The maize WRKY transcription factor ZmWRKY40 confers drought resistance in transgenic Arabidopsis. Int. J. Mol.

Sci. 2018;19:2580. DOI: 10.3390/ijms19092580.

[62] Wang H, Wang H, Shao H, Tang X. Recent advances in utilizing transcription factors to improve plant abiotic stress tolerance by transgenic technology. Front. Plant Sci. 2016;7:67. DOI: 10.3389/fpls.2016.00067.

[63] Jogaiah S, Govind SR, Tran LS. Systems biology-based approaches toward understanding drought tolerance in food crops. Crit. Rev. Biotechnol. 2013;33:23-39. DOI: 10.3109/07388551.2012.659174.

[64] Pérez-Clemente RM, Vives V, Zandalinas SI, López-Climent MF, Muñoz V, Gómez-Cadenas A. Biotechnological approaches to study plant responses to stress. Biomed Res Int. 2013;654120. DOI: 10.1155/2013/654120

[65] Hirayama T, Shinozaki K. Research on plant abiotic stress responses in the post-genome era: past, present and future. Plant J. 2010;61:1041-1052. DOI: 10.1111/j.1365-313X.2010.04124.x

[66] Andorf C, Beavis WD, Hufford M, Smith S, Suza WP, Wang K, Woodhouse M, et al. Technological advances in maize breeding: past, present and future. Theoret. App.Gen. 2019;132:817-849. DOI: .10.1007/s00122-019-03306-3

[67] Zheng J, Fu J, Gou M, Huai J, Liu Y, Jian M, Huang Q, et al. Genome-wide transcriptome analysis of two maize inbred lines under drought stress. Plant Mol. Biol. 2010;72:407-421. DOI: 10.1007/s11103-009-9579-6.

[68] Luo M, Liu J, Lee DR, Scully BT, Guo BZ. Monitoring the expression of maize genes in developing kernels under drought stress using oligo-microarray. J. Integr. Plant Biol. 2010;52:1059-1074. DOI: 10.1111/j.1744-7909.2010.01000.

[69] Rabbani MA, Maruyama K, Abe H, Khan MA, Katsura K, Ito Y, Yoshiwara K, et al. Monitoring expression profiles of rice genes under cold, drought, and high-salinity stresses and abscisic acid application using cDNA microarray and RNA Gel-Blot analyses. Plant Physiol. 2003;133:1755-1767

[70] Wang H, Zhang H, Gao F, Li J, Li Z. Comparison of gene expression between upland and lowland rice cultivars under water stress using cDNA microarray. Thoer Appl Genet. 2007;115:1109-1126

[71] Guo P, Baum M, Grando S, Ceccarelli S, Bai G, Li R, von Korff M, et al. Differentially expressed genes between drought-tolerant and drought-sensitive barley genotypes in response to drought stress during the reproductive stage. J Exp Bot. 2009;60:3531-3544

[72] Matsui A, Ishida J, Morosawa T, Okamoto M, Kim JM, Kurihara Y, et al. Arabidopsis Tiling Array Analysis to Identify the Stress-Responsive Genes. In: Sunkar R, editor. Plant Stress Tolerance. Methods in Molecular Biology (Methods and Protocols). 2010;639:141-155. Humana Press.

[73] Deokar AA, Kondawar V, Jain PK, Karuppayil SM, Raju NL, Vadez V, Varshney RK, Srinivasan R. Comparative analysis of expressed sequence tags (ESTs) between drought-tolerant and -susceptible genotypes of chickpea under terminal drought stress. BMC Plant Biol. 2011;11:70-90

[74] Song Y, Wang Z, Bo W, Ren Y, Zhang Z, Zhang D. Transcriptional profiling by cDNA-AFLP analysis showed differential transcript abundance in response to water stress in Populus hopeiensis. BMC Genom. 2012;13:286-304.

[75] Miao Z, Han Z, Zhang T, Chen S, Ma C. A systems approach to spatio-temporal understanding of the drought stress response in maize.

Sci. Rep. 2017;7:6590. DOI: 10.1038/s41598-017-06929-y.

[76] Mohanta TK, Bashir T, Hashem A, Abd_Allah EF. Systems biology approach in plant abiotic stresses. Plant Physiol. Biochem. 2017;121:58-73. DOI: 10.1016/j.plaphy.2017.10.019.

[77] Kakumanu A, Ambavaram MMR, Klumas C, Krishnan A, Batlang U, Myers E, Grene R, Pereira A. Effects of drought on gene expression in maize reproductive and leaf meristem tissue revealed by RNA-seq. Plant Physiol. 2012;160:846-867.

[78] Luz HP, Juan DR. RNA-seq in kinetoplastids: A powerful tool for the understanding of the biology and host-pathogen interactions. Infect. Genet. Evol. 2017;49:273-282. DOI: 10.1016/j.meegid.2017.02.003.

[79] Shan X, Li Y, Jiang Y, Jiang Z, Hao W, Yuan Y. Transcriptome profile analysis of maize seedlings in response to high-salinity, drought and cold stresses by deep sequencing. Plant Mol. Biol. Rep. 2013;31:1485-1491. DOI: 10.1007/s11105-013-0622-z.

[80] Zenda T, Liu S, Wang X, Liu G, Jin H, Dong A, Yang Y, Duan H. Key maize drought-responsive genes and pathways revealed by comparative transcriptome and physiological analyses of contrasting inbred lines. Int J Mol Sci. 2019;20:(6):1268. DOI: 10.3390/ijms20061268

[81] Zhang Q, Liu H, Wu X, Wang W. Identification of drought tolerant mechanisms in a drought-tolerant maize mutant based on physiological, biochemical and transcriptomic analyses. BMC Plant Biology. 2020; 20:315. DOI: 10.1186/s12870-020-02526-w

[82] Tan CT, Lim YS, Lau SE. Proteomics in commercial crops: an overview. J. Protozool. 2017;169: 176-188.

[83] Komatsu S. Plant Proteomic Research 2.0: Trends and Perspectives. Int. J. Mol. Sci. 2019;20:2495. DOI: 10.3390/ijms20102495.

[84] Jiang Z, Jin F, Shan X, Li Y. iTRAQ-Based Proteomic Analysis Reveals Several Strategies to Cope with Drought Stress in Maize Seedlings. Int J Mol Sci. 2019;20(23):5956. Doi: 10.3390/ijms20235956.

[85] Labuschagne MT. A review of cereal grain proteomics and its potential for sorghum improvement. J. Cereal Sci. 2018;84:151-158. DOI: 10.1016/j.jcs.2018.10.010.

[86] Hu X, Wu L, Zhao F, Zhang D, Li N, Zhu G, Li C, Wang W. Phosphoproteomic analysis of the response of maize leaves to drought, heat and their combination stress. Front. Plant Sci. 2015;6:298. DOI: 10.3389/flps.2015.00298.

[87] Zhao Y, Wang Y, Yang H, Wang W, Wu J, Hu X. Quantitative proteomic analyses identify aba-related proteins and signal pathways in maize leaves under drought conditions. Front. Plant Sci. 2016;7:1-23. DOI: 10.3389/fpls.2016.01827.

[88] Wu X, Wang W. Increasing confidence of proteomics data regarding the identification of stress-responsive proteins in crop plants. Front. Plant Sci. 2016;7:702. DOI: 10.3389/fpls.2016.00702

[89] Kole C, Muthamilarasan M, Henry R, Edwards D, Sharma R, Abberton M, Batley J, et al. Application of genomics-assisted breeding for generation of climate resilient crops: progress and prospects. Front. Plant Sci. 2015;6:563. DOI: 10.3389/fpls.2015.00563

[90] Kopecky J, Azarkovich M, Pfundel EE, Shuvalov VA, Heber U. Thermal dissipation of light energy is

regulated differently and by different mechanisms in lichens and higher plants. Plant Biol. 2005;7:156-167.

[91] Zenda T, Liu S, Wang X, Jin H, Liu G, Duan H. Comparative proteomic and physiological analyses of two divergent maize inbred lines provide more insights into drought stress tolerance mechanisms. Int J Mol Sci. 2018;19:3225. DOI: 10.3390/ijms19103325

[92] Roitsch T. Source-sink regulation by sugar and stress. Curr. Opin. Plant Biol. 1999;2:198-206.

[93] Min H, Chen C, Wei S, Shang X, Sun M, Xia R, Liu X, et al. Identification of drought tolerant mechanisms in maize seedlings based on transcriptome analysis of recombination inbred lines. Front Plant Sci. 2016;7:1080. DOI: 10.3389/fpls.2016.01080.

[94] Marowa P, Ding A, Kong Y. Expansins: roles in plant growth and potential applications in crop improvement. Plant Cell Rep. 2016;35:949-65. DOI: 10.1007/s00299-016-1948-4

[95] Mattioni C, Lacerenza NG, Troccoli A, Leonardis AMD, Fonzo ND. Water and salt stress-induced alterations in proline metabolism of *Triticum durum* seedling. Physiol Plant. 1997;101:787-92. DOI: 10.1111/j.1399-3054. 1997.tb01064.x.

[96] Kumar SG, Reddy AM, Sudhakar C. NaCl effects on proline metabolism in two high yielding genotypes of mulberry (*Morus alba* L.) with contrasting salt tolerance. Plant Sci. 2003;165:1245-51. DOI: 10.1016/s0168- 9452(03)00332-7

[97] Ghannoum, O. C4 photosynthesis and water stress. Ann. Bot. 2009;103:635-644. DOI: 10.1093/aob/mcn093.

[98] Comas LH, Becker SR, Cruz VMV, Byrne PF, Dierig DA. Root traits contributing to plant productivity under drought. Front Plant Sci. 2013;4:442. DOI: 10.3389/fpls.2013.00442.

[99] Chimungu JG, Brown KM, Lynch JP. Large root cortical cell size improves drought tolerance in maize. Plant Physiol. 2014;166:2166-78. DOI: 10.1104/pp.114.250449

[100] Liu J, Zhang F, Zhou J, Chen F, Wang B, Xie X. Phytochrome B control of total leaf area and stomatal density affects drought tolerance in rice. Plant Mol Biol. 2012;78:289-300. DOI: 10.1007/s11103-011-9860-3

[101] Xu F, Yuan S, Lin HH. Response of mitochondrial alternative oxidase (AOX) to light signals. Plant Signal. Behav. 2011;6:55-58.

[102] Yang S, Vanderbeld B, Wan J, Huang Y. Narrowing down the targets: Towards successful genetic engineering of drought-tolerant crops. Molecular Plant. 2010;3(3): 469-490

[103] Todaka D, Shinozaki K, Yamaguchi-Shinozaki K. Recent advances in the dissection of drought-stress regulatory networks and strategies for development of drought-tolerant transgenic rice plants. Front. Plant Sci. 2015;6:84. DOI: 10.3389/fpls.2015.00084

[104] Krannich CT, Maletzki L, Kurowsky C, Horn R. Network candidate genes in breeding for drought tolerant crops. Int. J. Mol. Sci. 2015;16:16378-16400. DOI: 10.3390/ijms160716378

[105] Oladosu Y, Rafii MY, Samuel C, Fatai A, Magaji U, Kareem I, Kamarudin ZS, Muhammad I, Kolapo K. Drought resistance in rice from conventional to molecular breeding: A Review. Int. J. Mol. Sci. 2019;20:3519. DOI: 10.3390/ijms20143519

[106] Nepolean T, Kaul J, Mukri G, Mittal S. Genomics-Enabled

Next-Generation Breeding Approaches for Developing System-Specific Drought Tolerant Hybrids in Maize. Front. Plant Sci. 2018;9:361. DOI: 10.3389/fpls.2018.00361

[107] Wang XL, Wang HW, Liu SX, Ferjani A, Li JS, Yan JB, Yang XH, Qin F. Genetic variation in ZmVPP1 contributes to drought tolerance in maize seedlings. Nat Genet. 2016;48:1233. DOI: 10.1038/ng.3636.

[108] Xiang Y, Shan X, Qin F, Dai M. Deletion of an endoplasmic reticulum stress response element in a ZmPP2C-A gene facilitates drought tolerance of maize seedlings. Mol Plant. 2017;10:456-69. DOI: 10.1016/j.molp. 2016.10.003.

[109] Park HY, Seok HY, Park BK, Kim SH, Goh CH, Lee B-H, Lee et al. Overexpression of Arabidopsis ZEP enhances tolerance to osmotic stress. Biochem. Biophys. Res. Commun. 2008;375:80-85.

[110] Abebe T, Guenzi AC, Martin B, Cushman JC. Tolerance of mannitol-accumulating transgenic wheat to water stress and salinity. Plant Physiol. 2003;131:1748-1755

[111] Nelson DE, Repetti PP, Adams TR, Creelman RA, Wu J, Warner DC, Anstrom DC, Bensen RJ, et al. Plant nuclear factor Y (NF-Y) B subunits confer drought tolerance and lead to improved corn yields on water-limited acres. Proc. Natl Acad. Sci. U S A. 2007;104(42):16450-16455.

[112] Oh SJ, Song SI, Kim YS, Jang HJ, Kim SY, Kim M, Kim YK, Nahm BH, Kim JK. Arabidopsis CBF3/DREB1A and ABF3 in transgenic rice increased tolerance to abiotic stress without stunting growth. Plant Physiol. 2005;138:341-351. DOI:10.1104/pp.104.059147

[113] Karaba A, Dixit S, Greco R, Aharoni A, Trijatmiko KR, MarschMartinez N, Krishnan A, Nataraja KN, Udayakumar M, Pereira A. Improvement of water use efficiency in rice by expression of HARDY, an Arabidopsis drought and salt tolerance gene. Proc. Natl Acad. Sci. U S A. 2007;104: 15270-15275. DOI:10.1073/pnas.0707294104

[114] Wu X, Shiroto Y, Kishitani S, Ito Y, Toriyama K. Enhanced heat and drought tolerance in transgenic rice seedlings overexpressing OsWRKY11 under the control of HSP101 promoter. Plant Cell Rep. 2009;28: 21-30. DOI:10.1007/s00299-008-0614-x

[115] Xiao BZ, Chen X, Xiang CB, Tang N, Zhang QF, Xiong LZ. Evaluation of seven function-known candidate genes for their effects on improving drought resistance of transgenic rice under field conditions. Mol. Plant. 2009;2: 73-83. DOI:10.1093/mp/ssn068

[116] Xu D, Duan X, Wang B, Hong B, Ho THD, Wu R. Expression of a late embryogenesis abundant protein gene, HVA1, from barley confers tolerance to water deficit and salt stress in transgenic rice. Plant Physiol. 1996;110:249-257. DOI:10.1104/pp.110.1.249

[117] Sivamani E, Bahieldin A, Wraith JM, Al-Niemi T, Dyer WE, Ho THD, Qu R. Improved biomass productivity and water use efficiency under water deficit conditions in transgenic wheat constitutively expressing the barley HVA1 gene. Plant Sci 2000;155:1-9. DOI:10.1016/S0168-9452(99)00247-2

[118] He C, Zhang CW, Gao Q, Yang A, Hu X, Zhang J. Enhancement of drought resistance and biomass by increasing the amount of glycine betaine in wheat seedlings. Euphytica 2011; 177: 16-151. DOI:10.1007/s10681-010-0263-3

[119] Shou H, Bordallo P, Wang K. Expression of the Nicotiana protein kinase (NPK1) enhanced drought

tolerance in transgenic maize. J Exper Bot. 2004; 55:1013-1019.DOI:10.1093/jxb / e r h129

[120] Hu H, Dai M, Yao J, Xiao B, Li X, Zhang Q, Xiong, L. Overexpressing a NAM, ATAF, and CUC (NAC) transcription factor enhances drought resistance and salt tolerance in rice. Proc. Natl Acad. Sci. U S A. 2006;103: 12987-12992.DOI:10.1073/pnas.0604882103

[121] Nakashima K, Tran LS, Van Nguyen D, Fujita M, Maruyama K, Todaka D, Ito Y, Hayashi N, et al. Functional analysis of a NAC-type transcription factor OsNAC6 involved in abiotic and biotic stress-responsive gene expression in rice. Plant J. 2007;51: 617-630. DOI:10.1111/j.1365-313X.2007.03168.x

[122] Li H, Han X, Liu X, Zhou M, Ren W, Zhao B, Ju C, Liu Y, Zhao J. A leucine-rich repeat-receptor-like kinase gene SbER2-1 from sorghum (*Sorghum bicolor* L.) confers drought tolerance in maize. . BMC Genomics. 2019; 20:737. DOI:10.1186/s12864-019-6143-x

[123] Tai FJ, Yuan ZL, Wu XL, Zhao PF, Hu XL, Wang W. Identification of membrane proteins in maize leaves, altered in expression under drought stress through polyethylene glycol treatment. Plant Omics J. 2011; 4:250-256, ISSN 1836-3644.DOI:10.1007/s11105-010-0270-5

[124] Zhao F, Zhang D, Zhao Y, Wang W, Yang H, Tai F, Li C, Hu X. The difference of physiological and proteomic changes in maize leaves adaptation to drought, heat, and combined both stresses. Front. Plant Sci. 2016; 7: 1-19. DOI:10.3389/fpls.2016.01471

[125] Murata N, Allakhverdiev SI, Nishiyama Y. The mechanism of photoinhibition in vivo: Re-evaluation of the roles of catalase, α-tocopherol, non-photochemical quenching, and electron transport. Biochim. Biophys. Acta. 2012; 1817: 1127-1133. DOI:10.1016/j.bbabio.2012.02.020.

[126] Melkozernov AN, Blankenship RE. Photosynthetic functions of chlorophylls. In Grimm B, Porra RJ, Rüdiger W, Scheer H. (Eds). Chlorophylls and Bacteriochlorophylls. Springer: Dordrecht. 2006; pp. 397-412.

[127] Dudhate A, Shinde H, Tsugama D, Liu S, Takano T. Transcriptomic analysis reveals the differentially expressed genes and pathways involved in drought tolerance in pearl millet [*Pennisetum glaucum* (L.) R. Br]. PLoS ONE. 2018; 13: e0195908. DOI:10.1371/journal.pone.0195908

[128] Shinde H, Tanaka K, Dudhate A, Tsugama D, Mine Y, Kamiya T, Gupta SK, Liu S, Takano T. Comparative de novo transcriptomic profiling of the salinity stress responsiveness in contrasting pearl millet lines. Environ. Exp. Bot. 2018; 155: 619-627. DOI:10.1016/j.envexpbot.2018.07.008

[129] Ramakrishna A, Ravishankar GA. Influence of abiotic stress signals on secondary metabolites in plants. Plant Signal Behavior. 2011;6:1720-31.

[130] Fracasso A, Trindade LM, Amaducci S. Drought stress tolerance strategies revealed by RNA-Seq in two sorghum genotypes with contrasting WUE. BMC Plant Biology. 2016; 16:115. DOI: 10.1186/s12870-016-0800-x

[131] Ma D, Sun D, Wang C, Li Y, Guo T. Expression of flavonoid biosynthesis genes and accumulation of flavonoid in wheat leaves in response to drought stress. Plant Phys Biochem. 2014;80:60-66. DOI:10.1016/j.plaphy.2014.03.024

[132] Verma V, Ravindran P, Kumar PP. Plant hormone-mediated regulation of stress responses. BMC Plant

Biology. 2016;16:86. DOI: 10.1186/s12870-016-0771

[133] Bray EA. Plant responses to water deficit. Trends Plant Sci. 1997;2:48-54. DOI:10.1016/S1360-1385(97)82562-9

[134] Smeekens S. Sugar-induced signal transduction in plants. Ann Rev Plant Phys Plant Mol Biol. 2000;51:49-81. DOI:10.1146/annurev.arplant.51.1.49

[135] Liu T, Wu L, Gan X, Chen W, Liu B, Fedak G, Cao W, Chi D, Liu D, Zhang H, Zhang B. Mapping Quantitative Trait Loci for 1000-Grain Weight in a Double Haploid Population of Common Wheat. Int. J. Mol. Sci. 2020;21: 3960.

[136] Srivastava RK, Singh RB, Pujarula VL, Bollam S, Pusuluri M, Chellapilla TS, Yadav RS, Gupta R . Genome-Wide Association Studies and Genomic Selection in Pearl Millet: Advances and Prospects. Front. Genet. 2020;10:1389. DOI: 10.3389/fgene.2019.01389

[137] Pang Y, Liu C, Wang D, Amand PS, Bernardo A, Li W, He F, Li L, Wang L, Yuan X, Dong L, Su Y, et al. High-Resolution Genome-Wide Association Study Identifies Genomic Regions and Candidate Genes for Important Agronomic Traits in Wheat. Molecular Plant.2020. DOI:10.1016/j.molp.2020.07.008

[138] Mir RR, Zaman-Allah M, Sreenivasulu N, Trethowan R, Varshney RK. Integrated genomics, physiology and breeding approaches for improving drought tolerance in crops. Theor Appl Genet. 2012; 125:625-645. DOI: 10.1007/s00122-012-1904-9

[139] Messmer R, Fracheboud Y, Bänziger M, Vargas M, Stamp P, Ribaut JM. Drought stress and tropical maize: QTL-by-environment interactions and stability of QTLs across environments for yield components and secondary traits. Theor Appl Genet. 2009; 119: 913-930. DOI:10.1007/s00122-009-1099-x

[140] Rahman H, Pekic S, Lazic-Jancic V, Quarrie SA, Shah SMA, Pervez A, Shar MM. Molecular mapping of quantitative trait loci for drought tolerance in maize plants. Genet Mol Res. 2011;10:889-901. DOI:10.4238/vol10-2gmr1139

[141] Bernier J, Kumar A, Ramaiah V, Spaner D, Atlin G. A large effect QTL for grain yield under reproductive stage drought stress in upland rice. Crop Science. 2007;47:507-516. DOI:0.2135/cropsci2006.07.0495

[142] Kumar A, Dixit S, Ram T, Yadaw RB, Mishra KK, Mandal NP. Breeding high-yielding drought-tolerant rice: genetic variations and conventional and molecular approaches. J Exp Bot. 2014;65:6265-6278. DOI:10.1093/jxb/eru363

[143] Saint Pierre C, Crossa JL, Bonnett D, Yamaguchi-Shinozaki K, Reynolds MP. Phenotyping transgenic wheat for drought resistance. J Exp Bot. 2012;63:1799-1808. DOI:10.1093/jxb/err385

[144] Woldesemayat AA, Van Heusden P, Ndimba BK, Christoffels A. An integrated and comparative approach towards identification, characterization and functional annotation of candidate genes for drought tolerance in sorghum (*Sorghum bicolor* (L.) Moench). BMC Genet. 2017;18:119. DOI:10.1186/s12863-017-0584-5

[145] Harris-Shultz KR, Hayes CM, Knoll JE. Mapping QTLs and Identification of Genes Associated with Drought Resistance in Sorghum. Methods Mol Biol. 2019;1931:11-40. DOI:10.1007/978-1-4939-9039-9_2

[146] Pham AT, Maurer A, Pillen K, Brien C, Dowling K, Berger B, Eglinton JK, March TJ. Genome-wide

association of barley plant growth under drought stress using a nested association mapping population. BMC Plant Biology. 2019;19:134. DOI:10.1186/s12870-019-1723-0

[147] Yadav RS, Sehgal D, Vadez V. Using genetic mapping and genomics approaches in understanding and improving drought tolerance in pearl millet. J Exp Bot. 2011;62:397-408. DOI:10.1093/jxb/erq265

[148] Sehgal D, Rajaram V, Armstead IP, Vadez V, Yadav YP, Hash CT et al. Integration of gene-based markers in a pearl millet genetic map for identification of candidate genes underlying drought tolerance quantitative trait loci. BMC Plant Biol. 2012;12:9. DOI:10.1186/1471-2229-12-9

[149] Großkinsky DK, Svensgaard J, Christensen S, Roitsch T. Plant phenomics and the need for physiological phenotyping across scales to narrow the genotype-to-phenotype knowledge gap. J. Exp. Bot. 2015; 66: 5429-5440. DOI:10.1093/jxb/erv345

[150] Passioura JB. Phenotyping for drought tolerance in grain crops: when is it useful to breeders? Functional Plant Biology. 2012;39: 851-859. DOI:10.1071/FP12079

[151] Monneveux P, Jing R, Misra SC. Phenotyping wheat for adaptation to drought. Drought phenotyping in crops: from theory to practice: CGIAR Generation Challenge Program. 2011.

[152] Araus JL, Sanchez C, Edmeades GO. Phenotyping maize for adaptation to drought. In: Monneveux P, Ribaut JM, editors. Drought phenotyping in crops: from theory to practice CGIAR Generation Challenge Program; 2011. p. 263-283.

[153] Fischer KS, Fukai S, Kumar A, Leung H, Jongdee B. Phenotyping rice for adaptation to drought. In: Monneveux P, Ribaut JM, editors. Drought phenotyping in crops: from theory to practice: CGIAR Generation Challenge Program; 2011. p. 215-243.

[154] Li L, Zhang Q, Huang D. A Review of Imaging Techniques for Plant Phenotyping. Sensors. 2014; 14: 20078-20111. DOI:10.3390/s141120078

[155] Hussain S, Mubeen M, Ahmad A, Akram W, Hammad HM, Ali M, Masood N, et al. Using GIS tools to detect the land use/land cover changes during forty years in Lodhran district of Pakistan. Environ Sci Pollut Res. 2019. https://doi.org/10.1007/s11356-019-06072-3

[156] Wang D, Fahad S, Saud S, Kamran M, Khan A, Khan MN, Hammad HM, Nasim W. Morphological acclimation to agronomic manipulation in leaf dispersion and orientation to promote "Ideotype" breeding: Evidence from 3D visual modeling of "super" rice (*Oryza sativa* L.). Plant Physiology and Biochemistry. 2019;(1)135:499-510.

[157] Nepolean T, Singh I, Hossain F, Pandey N, Gupta HS. Molecular characterization and assessment of genetic diversity of inbred lines showing variability for drought tolerance in maize. J. Plant Biochem. Biotechnol. 2013;22: 71-79. DOI: 10.1007/s13562-012-0112-7

[158] Zaman-Allah M, Zaidi PH, Trachsel S, Cairns JE, Vinayan MT, Seetharam K. Phenotyping for abiotic stress tolerance in maize – Drought stress. A field manual. CIMMYT, Mexico. 2016.

[159] International Service for Acquisition of Agri-biotech Applications (ISAAA). Biotechnology for the development of drought tolerant crops. [Internet]. 2008. Available from: http://www.isaaa.org/resources/

publications/pocketk/32/default.asp [Accessed 2020-07-07].

[160] Khan S, Anwar S, Yu S, Sun M, Yang Z, Gao ZQ. Development of Drought-Tolerant Transgenic Wheat: Achievements and Limitations. Int. J. Mol. Sci. 2019;20: 3350.

[161] Thomas WTB. Drought-resistant cereals: impact on water sustainability and nutritional quality. Conference on 'Carbohydrates in health: friends or foes' Symposium 1: Global challenges and solutions in cereal production. Proceedings of the Nutrition Society. 2015; 74: 191-197. DOI:10.1017/S0029665115000026

[162] Voytas DF, Gao C . Precision genome engineering and agriculture: opportunities and regulatory challenges. PLoS Biol. 2014; 12: e1001877. DOI:10.1371/journal.pbio.1001877

[163] Hilscher J, Bürstmayr H, Stoger E. Targeted modification of plant genomes for precision crop breeding. Biotechnology Journal. 2017;12:1-20. DOI:10.1002/biot.201600173

[164] Miglani GS. Genome editing in crop improvement: Present scenario and future prospects. J. Crop Improv. 2017; 31: 453-559. DOI:10.1080/15427528.2017.1333192

[165] Weeks DP, Spalding MH, Yang B. Use of designer nucleases for targeted gene and genome editing in plants. Plant Biotechnol. J. 2016;14:483-495.

[166] Razzaq A, Saleem F, Kanwal M, Mustafa G, Yousaf S, Imran Arshad HM, Hameed MK, Khan MS, Joyia FA. Modern trends in plant genome editing: An inclusive review of the CRISPR/Cas9 toolbox. Int. J. Mol. Sci. 2019;20:4045. DOI: 10.3390/ijms20164045

[167] Xu R, Yang Y, Qin R, Li H, Qiu C, Li L. Rapid improvement of grain weight via highly efficient CRISPR/Cas9-mediated multiplex genome editing in rice. J. Genet. Genomics. 2016;43:529-532. DOI:10.1016/j.jgg.2016.07.003

[168] Kim D, Kim D, Alptekin B, Budak H. CRISPR/Cas9 genome editing in wheat. Funct. Integr. Genomics. 2017;18:31-41. DOI:10.1007/s10142-017-0572-x

[169] Shi J, Gao H, Wang H, Lafitte HR, Archibald RL, Yang M, Hakimi SM, Mo H, Habben JE. ARGOS8 variants generated by CRISPR-Cas9 improve maize grain yield under field drought stress conditions. Plant Biotechnol. J. 2017;15:207-216. DOI: 10.1111/pbi.12603

[170] Ahmar S, Gill RA, Jung KH, Faheem A, Qasim MU, Mubeen M, Zhou W. Conventional and Molecular Techniques from Simple Breeding to Speed Breeding in Crop Plants: Recent Advances and Future Outlook. Int. J. Mol. Sci. 2020;21:2590.

[171] Baltes NJ, Gil-Humanes J, Voytas DF. Genome Engineering and Agriculture: Opportunities and Challenges. Prog Mol Biol Transl Sci. 2017;149:1-26. DOI: 10.1016/bs.pmbts.2017.03.011

[172] Edmeades GO. Progress in Achieving and Delivering Drought Tolerance in Maize—An Update; ISAA: Ithaca, NY, USA. 2013;pp. 1-39.

[173] Passioura JB, Angus JF. Improving productivity of crops in water-limited environments. Advances in Agronomy. 2010;106:37-75.

[174] Blum A. Mitigation of drought stress by crop management. [Internet]. 2020. Available from: https://plantstress.com/drought-mitigation/. [Accessed 2020-07-06].

[175] Hafiz MH, Wajid F, Farhat A, Fahad S, Shafqat S, Wajid N, Hafiz FB. Maize plant nitrogen uptake dynamics

at limited irrigation water and nitrogen. Environ Sci Pollut Res. 2016;24(3):2549-2557. https://doi.org/10.1007/s11356-016-8031-0

[176] Thierfelder C, Cheesman S, Rusinamhodzi L. A comparative analysis of conservation agriculture systems: benefits and challenges of rotations and intercropping in Zimbabwe. Field Crops Research. 2012;137:237-250.

[177] Fan T, Stewart BA, Payne WA, Wang Y, Song S, Luo J, Robinson CA. Supplemental irrigation and water-yield relationships for plasticulture crops in the Loess Plateau of China. Agronomy Journal. 2005;97:177-188.

[178] Alharby HF, Fahad S. Melatonin application enhances biochar efficiency for drought tolerance in maize varieties: Modifications in physio-biochemical machinery. Agronomy Journal. 2020;112(4):2826-47.

[179] Lamaoui M, Jemo M, Datla R, Bekkaoui F. Heat and Drought Stresses in Crops and Approaches for Their Mitigation. Front. Chem. 2018;6:26. DOI: 10.3389/fchem.2018.00026

[180] Zandalinasa SI, Mittlerb R, Balfagóna D, Arbonaa V, Gómez-Cadenasa A. Plant adaptations to the combination of drought and high temperatures. Physiologia Plantarum. 2018;162: 2-12. DOI:10.1111/ppl.12540

[181] Badu-Apraku B, Yallou CG. Registration of striga-resistant and drought tolerant tropical early maize populations TZE-W Pop DT STR C4 and TZE-Y Pop DT STR C4. J. Plant Regist. 2009;3:86-90. DOI: 10.3198/jpr2008.06.0356crg

[182] Qamar-uz Z, Zubair A, Muhammad Y, Muhammad ZI, Abdul K, Fahad S, Safder B, et al. Zinc biofortification in rice: leveraging agriculture to moderate hidden hunger in developing countries. Arch Agron Soil Sci. 2017;64:147-161. https://doi.org/10.1080/03650340.2017.1338343

[183] Wu S, Ning F, Zhang Q, Wu X, Wang W. Enhancing omics research of crop responses to drought under field conditions. Front. Plant Sci. 2017;8:174. DOI: 10.3389/fpls.2017.00174.

[184] Aziz K, Daniel KYT, Fazal M, Muhammad ZA, Farooq S, FanW, Fahad S, Ruiyang Z. Nitrogen nutrition in cotton and control strategies for greenhouse gas emissions: a review. Environ Sci Pollut Res. 2017a;24:23471-23487. https://doi.org/10.1007/s11356-017-0131-y

[185] Aziz K, Daniel KYT, Muhammad ZA, Honghai L, Shahbaz AT, Mir A, Fahad S. Nitrogen fertility and abiotic stresses management in cotton crop: a review. Environ Sci Pollut Res. 2017b; 24:14551-14566. https://doi.org/10.1007/s11356-017-8920-x

[186] Ding W, Xu X, He P, Ullah S, Zhang J, Cui Z, Zhou W. Improving yield and nitrogen use efficiency through alternative fertilization options for rice in China: A meta-analysis. Field Crops Res. 2018;227:11-8.

[187] Akram R, Turan V, Hammad HM, Ahmad S, Hussain S, Hasnain A, Maqbool MM, et al. Fate of organic and inorganic pollutants in paddy soils. In: Hashmi MZ, Varma A. (Ed.). Environmental Pollution of Paddy Soils, Soil Biology. Springer International Publishing AG, Gewerbestrasse 11, CHAM, CH-6330, Switzerland. 2018a; pp. 197-214

[188] Akram R, Turan V, Wahid A, Ijaz M, Shahid MA, Kaleem S, Hafeez A, et al. Paddy land pollutants and their role in climate change. In: Hashmi MZ, Varma A. (Ed.). Environmental Pollution of Paddy Soils, Soil Biology. Springer International Publishing AG, Gewerbestrasse 11, CHAM, CH-6330, Switzerland. 2018b; pp. 113-124.

[189] Fahad S, Hussain S, Saud S, Tanveer M, Bajwa AA, Hassan S, Shah AN, et al. A biochar application protects rice pollen from high-temperature stress. Plant Physiol Biochem. 2015a;96:281-287

[190] Fahad S, Hussain S, Saud S, Hassan S, Tanveer M, Ihsan MZ, Shah AN, et al. A combined application of biochar and phosphorus alleviates heat-induced adversities on physiological, agronomical and quality attributes of rice. Plant Physiol Biochem. 2016d;103:191-198

[191] Evans JR. Improving photosynthesis. Plant Physiology. 2013;162:1780-1793. DOI:10.1104/pp.113.219006

[192] Hartman K, Tringe SG. Interactions between plants and soil shaping the root microbiome under abiotic stress. Biochemical Journal. 2019;476:2705-2724. DOI: 10.1042/BCJ20180615

[193] Adnan M, Zahir S, Fahad S, Arif M, Mukhtar A, Imtiaz AK, Ishaq AM, Abdul B, et al. Phosphate-solubilizing bacteria nullify the antagonistic effect of soil calcification on bioavailability of phosphorus in alkaline soils. Sci Rep. 2018;8:4339. https://doi.org/10.1038/s41598-018-22653-7

[194] de Vries FT, Griffiths RI, Knight CG, Nicolitch O, Williams A. Harnessing rhizosphere microbiomes for drought-resilient crop production. Science. 2020; 368:270-274.

[195] Fahad S, Hussain S, Bano A, Saud S, Hassan S, Shan D, Khan FA, Khan F, et al. Potential role of phytohormones and plant growth-promoting rhizobacteria in abiotic stresses: consequences for changing environment. Environ Sci Pollut Res. 2014a;22(7):4907– 4921. https://doi.org/10.1007/s11356-014-3754-2

[196] Fahad S, Hussain S, Matloob A, Khan FA, Khaliq A, Saud S, Hassan S, Shan D, et al. Phytohormones and plant responses to salinity stress: A review. Plant Growth Regul. 2014b;75(2):391–404. https://doi.org/10.1007/s10725-014-0013-y

[197] Fahad S, Nie L, Chen Y, Wu C, Xiong D, Saud S, Hongyan L, et al. Crop plant hormones and environmental stress. Sustain Agric Rev. 2015b;15:371-400

[198] Fahad S, Hussain S, Saud S, Hassan S, Chauhan BS, Khan F, et al. Responses of rapid viscoanalyzer profile and other rice grain qualities to exogenously applied plant growth regulators under high day and high night temperatures. PLoS One. 2016a;11(7):e0159590. https://doi.org/10.1371/journal.pone.0159590

[199] Fahad S, Hussain S, Saud S, Khan F, Hassan S, Jr A, Nasim W, et al. Exogenously applied plant growth regulators affect heat-stressed rice pollens. J Agron Crop Sci. 2016b;202:139– 150

[200] Fahad S, Hussain S, Saud S, Hassan S, Ihsan Z, Shah AN,Wu C, et al. Exogenously applied plant growth regulators enhance the morphophysiological growth and yield of rice under high temperature. Front Plant Sci. 2016c;7:1250. https://doi.org/10.3389/fpls.2016. 01250

[201] Kamarn M, Wenwen C, Irshad A, Xiangping M, Xudong Z, Wennan S, Junzhi C, et al. Effect of paclobutrazol, a potential growth regulator on stalk mechanical strength, lignin accumulation and its relation with lodging resistance of maize. Plant Growth Regul. 2017;84:317-332. https://doi.org/10.1007/ s10725-017-0342-8

[202] Fahad S, Rehman A, Shahzad B, Tanveer M, Saud S, Kamran M, Ihtisham M, et al. Rice responses and tolerance to metal/metalloid toxicity. In: Hasanuzzaman M, Fujita M, Nahar K,

Biswas JK (Ed.). Advances in Rice Research for Abiotic Stress Tolerance. Woodhead Publ Ltd, Abington Hall Abington, Cambridge CB1 6AH, CAMBS, England. 2019a; pp. 299-312.

[203] Fahad S, Adnan M, Hassan S, Saud S, Hussain S, Wu C, Wang D, et al. Rice responses and tolerance to high temperature. In: Hasanuzzaman M, Fujita M, Nahar K, Biswas JK (Ed.). Advances in Rice Research for Abiotic Stress Tolerance. Woodhead Publ Ltd, Abington Hall Abington, Cambridge CB1 6AH, CAMBS, England. 2019b; pp.201-224.

[204] Saleem MH, Fahad S, Adnan M, Ali M, Rana MS, Kamran M, Ali Q, et al. Foliar application of gibberellic acid endorsed phytoextraction of copper and alleviates oxidative stress in jute (*Corchorus capsularis* L.) plant grown in highly copper-contaminated soil of China. Environmental Science and Pollution Research. 2020;1-13. https://doi.org/10.1007/s11356-020-09764-3

[205] Fahad S, Bano A. Effect of salicylic acid on physiological and biochemical characterization of maize grown in saline area. Pak J Bot. 2012;44:1433-1438

[206] Saud S, Li X, Chen Y, Zhang L, Fahad S, Hussain S, Sadiq A, Chen. Silicon application increases drought tolerance of Kentucky bluegrass by improving plant water relations and morph physiological functions. Sci World J. 2014; 2014:1-10. https://doi.org/10.1155/2014/ 368694

[207] Saud S, Chen Y, Fahad S, Hussain S, Na L, Xin L, Alhussien SA. Silicate application increases the photosynthesis and its associated metabolic activities in Kentucky bluegrass under drought stress and post-drought recovery. Environ Sci Pollut Res. 2016;23(17):17647– 17655. https://doi.org/10.1007/s11356-016-6957-x

[208] Zahida Z, Hafiz FB, Zulfiqar AS, Ghulam MS, Fahad S, Muhammad RA, Hafiz MH, et al. Effect of water management and silicon on germination, growth, phosphorus and arsenic uptake in rice. Ecotoxicol Environ Saf. 2017;144:11-18

Chapter 15

Abiotic Stress-Induced Molecular and Physiological Changes and Adaptive Mechanisms in Plants

Sivaji Mathivanan

Abstract

Abiotic stress is the primary cause of crop loss worldwide, reducing average yields for most major crop plants by more than 50%. Among abiotic stress, drought, salinity, high temperature, and cold are major adverse environmental factors that limit the crop production and productivity by inhibiting the genetic potential of the plant. So, it leads to complete change of morphological, physiological, biochemical, and molecular behavior of the plants and modifies regular metabolism of life, thereby adversely affecting plant productivity. Major effects of the drought, salinity, extreme temperatures, and cold stress are often interconnected and form similar cellular damage. To adopt plants with various abiotic stresses, plants can initiate a number of molecular, cellular, and physiological changes in its system. Sensors are molecules that perceive the initial stress signal from the outside of the plant system and initiate a signaling cascade to transmit the signal and activate nuclear transcription factors to induce the expression of specific sets of genes. Understanding this molecular and physiological basis of plant responses produced because of abiotic stress will help in molecular and modern breeding applications toward developing improved stress-tolerant crops. This review presents an overview and implications of physiological and molecular aspects of main abiotic stress, i.e., drought, heat, salt, and cold. Potential strategies to improve abiotic tolerance in crops are discussed.

Keywords: abiotic stress, signal transduction, stress-inducible genes, gene expression, mitigation process, genetic engineering and genome editing

1. Introduction

Plants live in constantly changing environments that are often unfavorable or stressful for growth and development. These unfavorable environmental conditions for plant growth are drought or water stress, high temperature or heat stress, low temperature or cold stress, excessive salt or salinity stress, and heavy metals toxicity like aluminum, arsenate, and cadmium in the soil. These adverse abiotic stresses are major threat that limits agriculture production and productivity, thereby creating great food insecurity. In the near future, it is predicted that because of climate change, abiotic stresses may become more intense and frequent. Drought and salinity are becoming drastically increased in many regions and may cause serious salinization of more than 50% of all arable lands by the year 2050. Consequently, because of rising temperatures and frequent flooding events for several decades,

IntechOpen

fertile agricultural land and crop yields may decrease rapidly, especially in the mid-latitudes [1, 2]. In addition to these factors, anthropogenic activities may lead to an increased abundance of soil, water, and air pollutants, factors that plants must cope with. Moderate estimates propose that more than 90% of the land in rural areas is affected by abiotic stress factors at some point during the growing season [3]. On the other hand, the population explosion has resulted in a higher demand for food and other natural resources.

Thus, understanding stress responses is essential when attempting to develop stress-resistant cultivars that can withstand abiotic stressors and in order to feed the growing population. Plants, which undergo various abiotic stresses, sense the enormous stress signal in order to respond to the stress condition. The primary signal caused by drought is hyperosmotic stress, which is often referred to simply as osmotic stress, and salt stress has both osmotic and ion toxicity effects in cells. The secondary effects of drought and salt stresses are complex and include oxidative stress; damage to cellular components such as membrane lipids, proteins, and nucleic acids; and metabolic dysfunction. Thus, drought and salt have unique and overlapping signals. Salt and drought stress disrupts homeostasis in water potential (osmotic homeostasis) and ion distribution (ionic homeostasis). This disruption of homeostasis occurs at both the cellular and the whole plant levels. Drastic changes in ion and water homeostasis lead to molecular damage, growth arrest, and even death. While some cellular responses are induced from primary stress signals, others arise mainly from secondary signals. An important feature of drought and salt stress is that the hyperosmotic signal causes the accumulation of the phytohormone abscisic acid (ABA), which in turn elicits many adaptive responses in plants [4]. Cold or chilling stress affects plant growth and development, by changing the cell structure. First, symptom of cold stress is changing the cell membrane structure in plants; this event initiates primary cold stress responses in plants [5]. Second, chilling stress disturbs the stability of proteins or protein complexes and reduces the activities of enzymes such as ROS scavenging enzymes. These processes result in photo-inhibition and impaired photosynthesis, as well as considerable membrane damage [6, 7]. Third, chilling stress affects gene expression and protein synthesis, as it favors the formation of secondary structures in RNA [8].

To achieve stress tolerance, three interconnected aspects of plant activities are important. First, damage must be prevented or alleviated. Second, homeostatic conditions must be reestablished in the new, stressful environment. Third, growth must resume, even though at a reduced rate [9]. Significant progress has been made in understanding the physiological, cellular, and molecular mechanisms of plant responses to environmental stress factors [10]. The detection of a stressful condition results in variations in gene expression, causing changes in the composition of plant transcriptome, proteome, and metabolome. Responses to stress are not linear pathways, but are complicated integrated circuits involving multiple pathways and specific cellular compartments, tissues, and the interaction of additional cofactors and/or signaling molecules to coordinate a specified response to a given stimulus [11]. With advancement of omic technologies, i.e., genomics, transcriptomics, proteomics, and metabolomics, now it is possible to analyze and identify the most complicated interlink between various stress response, signal transduction, gene expression, and metabolites production in plants with respective to the abiotic stress [12, 13].

2. Crop plants and abiotic stresses

Generally, many stress factors act at same time, such as the frequently combined, heat, water, and high-light stress [14]. Abiotic stress changes the expression pattern of the various genes in crop plants. So, this modification affects the regular function of plant metabolism, and source-sink relationship in turn reduces the growth, production, and productivity.

2.1 Drought

Distribution of rainfall is uneven due to the change in climate, which acts as an important stress as drought. Drought is the main abiotic stressor around the world and drastically reduces grain yields. It devastatingly influences the capability to meet the food demands of an ever-increasing global population. Drought stress is associated with water deficit and cellular dehydration. Plant adaptation to drought is a trait involving morphological, physiological, and biochemical changes. Plants reduce their growth of shoots under drought conditions and reduce their metabolic demands. Reduction in yield by as much as 40% was observed for maize and 21% for wheat at approximately a 40% water reduction [15]. In the case of cowpea, yield reduction can vary between 34 and 68% depending on the developmental timing of the drought stress [16].

2.2 Heat or higher temperature

Plants are more sensitive to the temperature conditions; in extreme cases, the unfavorable temperature condition leads to plant death. Normally plant growth and function would be better at optimum temperature level; both conditions below and higher temperature than optimum temperature severely affect the plant growth and production. The rate of most biochemical, enzymatic reactions rises two-fold for every 10°C increase between 20 and 30°C. Temperatures outside this range reduce the reaction rate because enzymes become either inactivated gradually or denatured.

Change of a few degrees considerably affects the plants' growth and development, especially reproduction. Abiotic stresses, specifically high and low temperatures, have a harmful effect on the early stage of male gametophyte in several agricultural crops such as rice, wheat, maize, barley, sorghum, and chickpea [17]. Male sterility and abnormalities in the spikelets' production were induced by heat stress in rice and wheat [18]. In both wheat and rice, heat and cold stresses caused tapetum degradation, microspore callose wall and exine formation, and changes in carbohydrate metabolism, eventually resulting in male sterility [19]. By contrast, temperature stress has no negative effect on female gametophyte development [20].

2.3 Salt

Soil salinization is a major threat to agriculture in arid and semi-arid regions, where water scarcity and inadequate drainage of irrigated lands severely reduce crop yield. More than 6% of the world's total land area and out of 230 M ha of irrigated land 45 M ha (19.5%) is already affected by salt [21]. Salt accumulation inhibits plant growth and reduces the ability to uptake water and nutrients, leading to osmotic or water-deficit stress. Salt stress tolerance level varies from one species to another. For cereal crops, barley (*Hordeum vulgare*), the most tolerant cereal, can tolerate up to 250 mM NaCl (equivalent to 50% seawater) and bread wheat is a moderately salt-tolerant crop, whereas rice, durum wheat (*Triticum turgidum* ssp.), maize (*Zea mays*), and sorghum (*Sorghum bicolor*) are less tolerant to salinity [22]. The reduction in plant growth following salt exposure is due to two phases, osmotic stress and ionic toxicity [23].

2.4 Cold

Cold stress has proved to be the main abiotic stress that decreases productivity of agricultural crops by affecting the quality of crops and their postharvest life. Cold stress, including chilling (0–15°C) and freezing (<0°C), is an abiotic stress that adversely affects the growth and agricultural productivity of plants [24, 25]. Freezing stress is highly detrimental to plants when compared with chilling stress. Usually, freezing damage will start with formation of ice nucleation in between the cell, then slowly grow and form ice crystals, and induce water leakage, leading to cell dehydration [26, 27]. However, many important crops are still incompetent to the process of cold acclimation. Rice (*Oryza sativa*), maize (*Zea mays*), tomato (*Solanum lycopersicum*), soybean (*Glycine max*), and cotton (*Gossypium hirsutum*) lack the ability to acclimate to cold temperatures and can only grow in tropical or subtropical regions [28]. Thus, cold stress adversely affects plant growth and development, limits the geographical distribution of plant species, and decreases crop yields worldwide [26].

3. Stress and crops

Stress refers to any substance or stimulus that restricts plant metabolism, growth, development, and crop productivity, including biotic and abiotic stresses [29]. Once this threshold is surpassed, an organism is stressed and mechanisms are activated at molecular, biochemical, physiological, and morphological levels. The activation of the mechanisms can result in the establishment of a new physiological state and homeostasis is reestablished [30]. Stress-related alterations in plant development, growth, and productivity reduce yield and cause unacceptable economic losses in agriculture. It has been assessed that abiotic stresses may reduce up to 70% of crop production of many economically important crops and perform at only 30% of their genetic potential with respect to yield [31].

4. Abiotic stress sensing and responding mechanism in plants

Sensors are biological molecules that recognize the adverse environmental modification and evoke the immediate response to the particular environment change by initiating the signal molecules in the system. Drought, salt, and cold stresses are inducing more amount of Ca^{2+} entry into the cell cytoplasm from internal stores or apoplastic source. Passages controlling Ca^{2+} entry are considered as one type of sensor for the stress signals [32–34]. Other than Ca^{2+}, ROS and nitric oxide (NO) are other messenger molecules involved in inducing plant response to cold stress. Reactive oxygen species (ROS) like superoxide ($O_2^{\cdot -}$), hydroxyl radicals (OH^-), and hydrogen peroxide (H_2O_2) are produced in plants in order to face various stresses [35]. Receptor-like kinases (RLKs) have an extracellular domain in which ligand is binding or protein-protein interaction will occur, a transmembrane domain, and an intracellular kinase domain. When the ligand or signal binds extracellular domain, histidine residue present in the intracellular kinase domain is auto-phosphorylated and the phosphoryl moiety is received by aspartate receiver part of the sensor protein or a separate protein. Then, the activated sensor protein(s) may induce cellular responses specific to signal through the mitogen-activated protein kinase (MAPK) cascade or directly phosphorylate specific targets. Intracellular signaling mode, i.e., protein

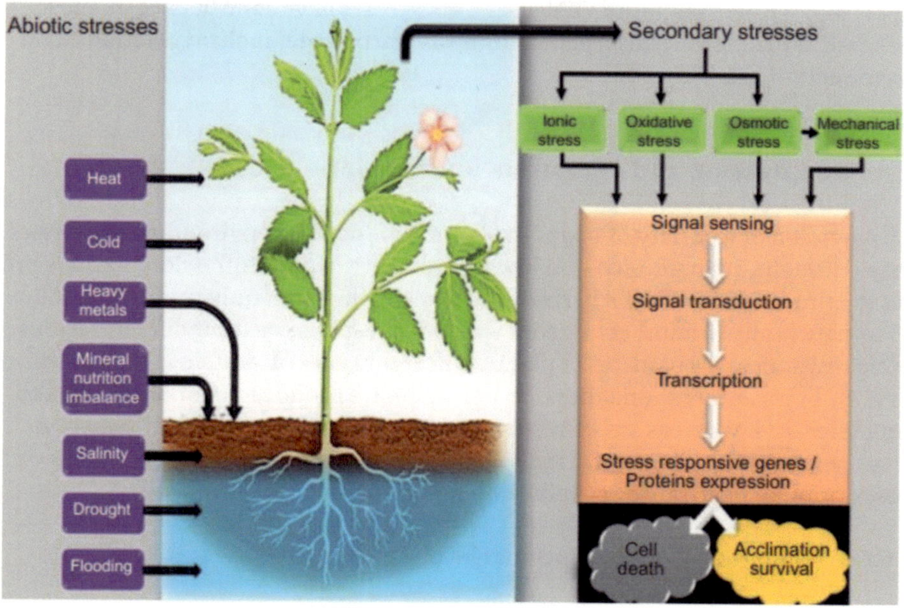

Figure 1.
Plant responses to abiotic stress.

phosphorylation and dephosphorylation regulate a wide range of cellular processes such as enzyme activation, assembly of macromolecules, protein localization, and degradation [36]. Upon sensing of abiotic stress by plants, signaling cascades are induced that activate ion channels, kinase cascades, assembly of reactive oxygen species (ROS), and accumulation of plant hormones leads to induce expression of specific subsets of genes that responsible to combat the abiotic stress (**Figure 1**) [37].

5. Similar and variable features of drought, salt, and cold stress

All the three stresses such as drought, salt, and cold stresses cause a primary loss of cell water, which leads to decrease in cell osmotic potential but the reason of cell water loss varies among stresses: (i) the decrease of the cell water content under drought stress is due to water shortage in soil or/and in the atmosphere. (ii) In salt stress, osmotic or water potential of surrounding root zone is decreased by Na^+ and Cl^- solutes, which in turn create more difficulty in uptake by roots and water translocation to metabolically active cells; (iii) osmotic stress is created in cold stress mainly because of inability to transport the water available from the soil to the living cell of leaf mesophyll. This condition is called as physiological drought. Anyhow, water loss in cell increases abscisic acid (ABA) biosynthesis, and it is well-known fact that it is involved in activation of various drought, salt, and cold stresses, responsive genes in plant system [38]. In plant system losing of cell water and increasing of solute concentration (especially Na^+) by these three stress cause lower osmatic potential and it creates harmful effects to the protein and enzymes. This effect can be avoided by producing more amounts of low molecular osmolytes (carbohydrates [39], betaine [40], and proline [41]) that can counteract cellular dehydration and turgor loss [42]. Production of low molecular osmolytes in higher quantity is a common stress alleviating process for drought, cold, and salt stress. While main cause of drought stress is osmotic, Na^+, ion toxicity, for salt stress, and physiological drought for cold stress, all stresses have an influence on most biochemical reactions such as photosynthesis, carbon metabolism reactions, and enzyme activities.

6. Gene expression and regulation under abiotic stress

Expression of a variety of genes in plants is induced by environmental stresses such as drought, high salinity, and low temperature. Upon expression, various proteins are produced in various parts of the plants, which not only protect the cell but also initiate humpty number of genes which are responsible for inducing various abiotic resistance mechanisms in plants. Different types of proteins, i.e., chaperones or late embryogenesis abundant (LEA) proteins are produced mainly involved to create tolerance, whereas stress-responsive genes are all involved in to generate stress response [43]. The regulation of stress-responsive plant genes at three levels: transcriptional, posttranscriptional, and posttranslational.

6.1 Gene regulation at transcriptional level

Transcriptional regulation involves (i) chromatin and its alteration and remodeling; (ii) cis-regulatory elements such as enhancers and promoters, which are often binding sites, located upstream and downstream the coding region; and trans-regulatory elements, usually transcription factors. Different environmental stresses create altered methylation pattern of DNA and changing histones protein in order to suppress or increase the transcription of the gene.

Promoters are specific sequences involved in regulatory function, where they bind RNA polymerase and different transcription factors to start the transcription [44]. Dehydration-responsive element-binding (DREB) or C-repeat binding factor (CBF), MYB, basic-leucine zipper (bZIP), and zinc-finger families are some of the trans-regulatory elements involved in the regulation of plant defense and stress-responsive genes upon binding in cis element of the respective gene promoters [45]. Overexpression of the *Oryza sativa* WRKY11 (transcription factor or trans-regulatory elements) under the control of heat shock protein 101 (HSP101) promoter led to enhanced drought tolerance [46, 47]. The important discovery of a novel *cis*-acting element, C-repeat/dehydration response element (CRT/DRE), is responsive to drought, cold, and high-salt stress [48]. Since this discovery, CBF proteins have been isolated sequentially by screening for DNA-binding proteins that bind to the CRT/DRE motif [49, 50]. *Arabidopsis* contains three cold-induced *CBF* genes, *CBF1–3* (*CBF1/DREB1B*, *CBF2/DREB1C*, and *CBF3/DREB1A*), which are arranged in tandem on chromosome IV. CBF 1–3 are APETALA2/ETHYLENE-RESPONSIVE (AP2/ERF1)-type transcription factors that directly bind to the conserved CRT/DRE motifs in the promoters of *COR* genes (known as CBF regulons) and activate their expression under cold conditions [50–52]. Transgenic *Arabidopsis* plants overexpressing *CBF1* display increased *COR* expression and enhanced freezing tolerance [53]. *CBF* orthologs have been isolated in many plant species, including rice, tomato, wheat (*Triticum aestivum*), barley (*Hordeum vulgare*), and maize [54]. Heterologous expression of *Arabidopsis CBFs* enhances freezing tolerance in various species, and heterologous expression of *CBFs* from other plant species enhances freezing tolerance in *Arabidopsis* [55–57]. Cold-sensitive tomato (*Lycopersicon esculentum*) become freezing tolerant upon overexpression of its own *CBF* genes, i.e., *LeCBF1*; however, overexpression of cold-tolerant *Arabidopsis CBF3* in tomato plants do not exhibit freezing tolerance; this proves that there are different CBF regulons in tomato and *Arabidopsis* [56]. It also indicates that the biological function of *CBF1–3* in modulating freezing tolerance is not only highly conserved among plants but also species specific.

6.2 Gene regulation at posttranscriptional level

Regulation that occurs in the stage of pre-mRNA till translation of mRNA is called posttranscriptional gene regulation. It occurs in four stages: (i) pre-messenger (mRNA) processing (capping, splicing, and polyadenylation), (ii) mRNA nucleocytoplasmic trafficking, (iii) mRNA turnover and stability, and (iv) mRNA translation [58]. One more strategy, i.e., alternative splicing (AS) is regulating the gene under cold and heat stress. STABILIZED1 (STA1), a gene coding for a nuclear pre-mRNA, is one of the best examples for alternate splicing factor which is involved in cold stress resistance in *A. thaliana* [28, 59]. Posttranscriptional regulation also is important for *COR* gene function. *REGULATOR OF CBF GENE EXPRESSION1* (*RCF1*), encoding a DEAD-box RNA helicase, helps ensure the proper pre-mRNA splicing of many *COR* genes under cold stress [60]. *STABILIZED1* (*STA1*) encodes a pre-mRNA splicing factor that controls the pre-mRNA splicing and mRNA turnover of *COR* genes [59]. Genome-wide AS profiling analysis revealed that hundreds of genes such as *RCF1* and *STA1* that have highly altered AS in the first few hours of cold treatment. This study showed that plant using AS pathway to change the gene expression in order to respond the temperature stress [61, 62]. Small RNAs (20–25 nucleotides) are processed from noncoding double-stranded RNA precursors by RNAses of the DICER-LIKE (DCL) family and mediate a series of gene silencing mechanisms at posttranscriptional level. One of these mechanisms cleaves mRNAs or prevents their translation through the mediation of 21 nucleotide microRNAs [63–65].

6.3 Posttranslational level regulation

Phosphorylation, sumoylation, and ubiquitination of proteins are posttranslational-level processes that play vital roles in the changing of plant response to various abiotic stresses. Mitogen-activated protein kinases (MAPKs) and SNF-1-relatedprotein kinases (SnRKs) are formed numerous signal transduction cascades, induced by dehydration and osmotic stress through the phosphorylation of specificresidues [66]. XERICO controls the level of ABA by enhancing the transcription of the key ABA biosynthetic gene AtNCED3. SnRK2 proteins and XERICO gene, encoding a H2-type zinc-finger E3 ubiquitin ligase, are involved in ABA-dependent responses to water deficit, like stomata closure [67–69].

Posttranslational histone modifications, along with DNA methylation, are associated with gene expression levels in response to cold stress. Histone acetylation/deacetylation catalyzed by histone acetyltransferases (HATs) and histone deacetylases (HDAs) plays a role in cold responses in plants [70]. *Arabidopsis HISTONE DEACETYLASE6* (*HDA6*) is upregulated by cold stress and positively regulates freezing tolerance [71]. *HDAs* also are upregulated by cold stress in maize, leading to global deacetylation at H3 and H4. Under cold stress, HDAs appear to directly activate maize *DREB1* (*ZmDREB1*) expression and histone hyperacetylation. Histone acetylation of *OsDREB1b* in rice and *ZmDREB1A* and *ZmCOR413* in maize is induced by cold stress [72, 73]. RNA-DIRECTED METHYLATION4 (RDM4) protein was reported to function in RNA-directed DNA methylation (RdDM) by working with RNA polymerases Pol V and Pol II in *Arabidopsis* [74]. Under cold stress, RDM4 is important for Pol II occupancy at the promoters of *CBF2* and *CBF3* genes [75].

7. Important signal transduction pathway for drought cold and salt

Abiotic stresses such as cold, drought, and salt inducing signal transduction networks are divided into three types: (I) osmotic/oxidative stress signaling that uses MAPK modules to generate ROS scavenging enzymes, antioxidant compounds, and osmolytes; (II) Ca^{2+}-dependent signaling helps to activate late embryogenesis abundant (LEA)-type genes (such as the DRE/CRT class of genes), and (III) Ca^{2+}-dependent salt overlay sensitive (SOS) signaling that regulates ion homeostasis (**Figure 2**) [14, 76].

7.1 Oxidative or osmatic stress signaling

Formation of reactive oxygen species (ROS) such as superoxide, hydrogen peroxide, and hydroxyl radicals is common for all the stresses, particularly drought, heat, salt, cold, and oxidative stress. Reactive oxygen species which are produced by the environmental stresses cause major plant damage (oxidative stress) [77]. High amount of ROS acts as a signal, and synthesizing ROS scavengers is one of the protective mechanisms in plants. Osmotic stress stimulates many protein kinases; among them, one is mitogen-activated kinases, which are involved in

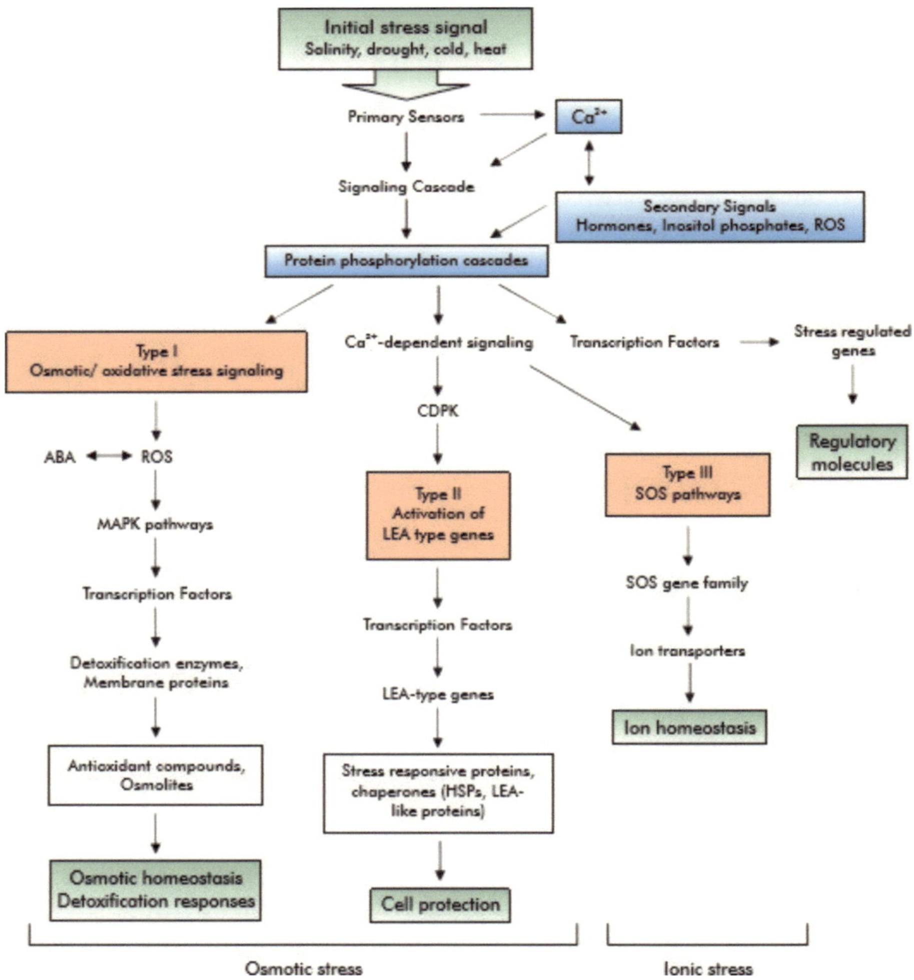

Figure 2.
Schematic pathway for the transduction of osmotic and ionic stress in plants.

reestablishing the osmotic homeostasis. Plant cells which undergo osmatic stress should produce more osmolytes in order to mitigate the negative effect of the ROS and maintain the osmotic homeostasis. So, the osmotic stress activate the receptors/sensors proteins such as protein tyrosine kinases, G-protein-coupled receptors, and two-component histidine kinases; this will trigger MAPK pathway and signal cascade, which is responsible for production of more amount of osmolytes that are necessary for osmotic adjustment. The main purpose of osmolytes is to maintain cell turgor, thus the driving gradient for water uptake. Compatible solutes such as amino acids (e.g., proline), quaternary amines (e.g., glycine betaine and dimethyl sulfoniopropionate), and polyol/sugars (e.g., mannitol and trehalose) will act as free radical scavengers or chemical chaperones by directly stabilizing membranes and proteins [78].

The MAP kinase pathways are intracellular signal modules that mediate signal transduction from the cell surface to the nucleus. The core MAPK cascades consist of three kinases that are activated sequentially by an upstream kinase. The MAP kinase kinase kinase (MAPKKK), upon activation, phosphorylates a MAP kinase kinase (MAPKK) on serine and threonine residues. This dual-specificity MAPKK in turn phosphorylates a MAP kinase (MAPK) on conserved tyrosine and threonine residues. The stimulated MAPK either travels to the nucleus to stimulate the transcription factor directly, or activates additional signal components to regulate gene expression, cytoskeleton-associated proteins or enzyme activities, or targets certain signal proteins for degradation [14].

7.2 Ca^{2+}-dependent signaling to activate late embryogenesis abundant (LEA) genes

Abiotic stress induces more Ca^{2+} entry into the cell cytoplasm; channels which control the Ca^{2+} entry act as a sensor for these abiotic stress signals. Ca^{2+} activates the calcium-dependent protein kinases (CDPKs), and CDPKs are serine/threonine protein kinases with a C-terminal calmodulin-like domain with four EF-hand motifs that can directly bind Ca^{2+}. CDPKs are encoded by multigene families, and the expression levels of these genes are spatially and temporally controlled throughout development. CDPK pathway is involved in production of high amount of anti-desiccation protection protein (LEA proteins) by activation of LEA-type genes including the dehydration-responsive element (DRE)/C-repeat (CRT) class of stress-responsive genes. Activation of LEA-type genes actually represents damage repair pathways but this is completely different from the pathways regulating osmolyte production [79, 80]. At the time of seed maturation, naturally it undergoes desiccation; to avoid desiccation shock during seed germination, seeds accumulate more transcripts and relatively high concentration proteins; for this reason, these proteins were named as late embryogenesis abundant (LEA) proteins [81]. Water deficit, high osmolarity, and low temperature stress induce the accumulation of an LEA protein in crop plants. Such proteins are used to prevent protein denaturation or renaturing unfolded proteins, realm protein structure and membrane integrity, and sequestering ions in stressed tissues. Many scientific reports suggest that LEA proteins and chaperones are involved in protecting macromolecules such as enzymes, lipids, and mRNAs from dehydration, and these proteins have been grouped into at least six families on the basis of sequence similarity [12, 82, 83]. LEA proteins are specialized in desiccation protection of membranes; and antioxidant enzymes and molecules are involved to achieve desiccation tolerance. Both osmolytes and LEA proteins work together in stabilization of membrane and protein structures by conferring preferential hydration at moderate desiccation and replacing water at extreme desiccation [84].

7.3 Ca^{2+}-dependent salt overlay sensitive (SOS) signaling

Calcium-dependent SOS signaling regulates ion homeostasis relatively specific to the salt stress. Ion transporters are main target for this type of signaling that controls ion homeostasis under salt stress. Excess extracellular or intracellular Na^+ acts as an input for the SOS pathway and mainly increases a cytoplasmic Ca^{2+} signal, and this signal changes expression and activity of transporters for ions such as Na^+, K^+, and H^+. The input for osmotic stress signaling is like a change in turgor. Salt stress signal transduction comprises of ionic and osmotic homeostasis signaling pathways, detoxification (e.g., damage control and repair) response pathways, and pathways for growth regulation [66]. This signaling pathway mediates salt induction of the SOS1 gene in *Arabidopsis*. In addition, the SOS2-SOS3 kinase directly phosphorylates and activates the SOS1 transporter [14, 85]. Studies comparing the growth of wild-type and mutant plants in response to NaCl and sequence analysis of the predicted SOS1 protein suggested that SOS1 encodes an Na^+/H^+ exchanger (antiporter) on the plasma membrane [86]. Because the SOS pathway operates during ionic stress, it is thought that homologs of SOS3 and SOS2 may also function in the transduction of other stress or hormonal signals [87]. Transient increases in cytosolic Ca^{2+} are perceived by various Ca^{2+}-binding proteins. In the case of abiotic stress signaling, evidence suggests that Ca^{2+}-dependent protein kinases (CDPKs) and the SOS3 family of Ca^{2+} sensors are major players in coupling this universal inorganic signal to specific protein phosphorylation cascades. It seems that calcium signaling is crucial for salt tolerance in plants [14].

Second messengers can control intracellular Ca^{2+} levels, often initiating a protein phosphorylation cascade that finally targets proteins directly involved in cellular protection or transcription factors controlling specific sets of stress-regulated genes. The products of stress-regulated genes like the plant hormones ABA, ethylene, and salicylic acid (SA) are mainly involved in the induction of stress-tolerant mechanism in plants. Salt and water-deficit stress participates in the production and activation of regulatory molecules, and to some extent, cold stress causes an increased biosynthesis and accumulation of ABA by activating genes coding for ABA biosynthetic enzymes, which can be quickly catabolized after the stress. Most of the abiotic stress-responsive genes are upregulated by ABA [88].

8. Managing abiotic stress by genetic engineering or genetic manipulation or genome editing

All over the world, abiotic stress significantly affects the production and yield potential of the crops and creates a major challenge for crop improvement sectors such as plant breeders and biotechnologists. Genetic manipulation of crops can be done to generate desirable character by so many ways. Among them, transgenic technology or genetic engineering and genome editing are the best strategies to develop abiotic resistant crops. Abiotic stress resistance in crops is possible by transgenic approach through boosting endogenous defense mechanisms by overexpressing of genes, which normally involves the synthesis of compatible osmolytes, antioxidants, polyamines, maintenance of hormone homeostasis, and modification of transporters and/or regulatory proteins, including transcription factors and alternative splicing events. Sometimes overexpression of some genes and thereby synthesizing of specific protein and metabolites will affect the normal metabolism and reducing the yield. Transgenic crop with abiotic tolerance but diminished yield potential and reduced growth is undesirable. So, it is important to analyze the functions of stress-inducible genes not only to understand the molecular mechanisms of stress tolerance and responses of higher plants but also to improve the stress tolerance without reducing the yield potential of crops by gene manipulation. Hundreds of genes are thought to be involved in abiotic stress responses [89].

8.1 Transgenic crop technology for abiotic tolerance

Transgenic crop or genetic engineering technology has ample opportunity in development of crops with specific objectives by overexpression of responsible genes or suppression of undesirable genes. Present engineering approaches rely on the transfer of one or several genes that are either involved in signaling and regulatory pathways or that encode enzymes present in pathways leading to the synthesis of functional and structural protectants, such as osmolytes and antioxidants, or that encode stress-tolerance-conferring proteins [78]. Phytohormones such as ABA are major targets for genetic manipulation to obtain abiotic stress-tolerant crops. Overexpression of ABA biosynthetic pathway-related TFs imparts an ABA-hypersensitive response and also improves the osmotic stress tolerance in transgenic plants [90, 91]. Under moderate drought stress, during the flowering period, the yields of transgenic canola overexpressing a farnesyltransferase protein were significantly higher comparatively to the control [92]. The overexpression of TFs that control root architecture induced drought tolerance in rice and transgenic *Arabidopsis* plants by promoting root growth and thus enhancing WUE [93, 94]. Other TFs linked to WUE, such as those stimulating wax deposition in cuticle and suberin deposition [95]. Many scientific researches revealed that glyoxalase pathway is involved in enhancing tolerance to abiotic stress; so, overexpression of glyoxalase I and glyoxalase II genes enhances the various abiotic stress tolerance in plants [96–103]. Transgenic rice with overexpression of choline oxidase (codA), D-pyrroline-5-corboxylate synthase (P5CS), LEA protein group 3 (HVA1), alcohol dehydrogenase (ADH), and pyruvate decarboxylase (PDC) genes have shown improved tolerance to abiotic stress [104, 105]. Usually rice does not accumulate glycine betaine but transgenic rice with overexpression of codA gene in the chloroplast and the cytosol accumulate more amount of glycine betaine, which recovered to normal growth at a faster rate give comparatively more yield under salt and cold stress [106]. Overexpression of *Escherichia coli* trehalose biosynthetic genes (otsA and otsB) with stress-inducible promoter enhances abiotic stress tolerance in rice [107–109]. Transgenic tobacco plants overexpressing chloroplastic Cu/Zn-SOD showed increased resistance to oxidative stress caused by high light and low temperatures. Transgenic tobacco plants expressing alfalfa aldose aldehyde reductase, a stress-activated enzyme, showed reduced damage when exposed to oxidative stress and increased tolerance to heavy metals, salt and dehydration stress. Targeting detoxification pathways are an appropriate approach for obtaining plants with multiple stress-tolerant traits [26, 110–113].

Water stress increases the formation of ROS through membrane perturbation of electron transport chains. The loss of catalase and gain of the glutathioneS-transferase/peroxidase functions in plants associate defenses against oxidative damage,which are more important in plant salt tolerance [79]. Loss of osmotic homeostasis is the important process in the abiotic stress, which affects the cell ion concentration; so, to achieve abiotic tolerance especially salt, plants should re-establish homeostasis (ionic and osmotic homeostasis) under stressful conditions. More amount of Na^+ within the cell inhibits enzyme activity and is harmful to the proteins; so, either compartmentalization of Na^+ in the vacuole or controlling of Na^+ accumulation in the cytoplasm is more important [9]. To revert ionic and osmotic homeostasis, many ion transporters act as terminal determinants. Protein AtNHX1 of *Arabidopsis thaliana*, one type of transporter, is located in cell vacuolar membranes involving transport of excessive Na^+ ions from cytoplasm to vacuole; thereby protecting the plant from the drying effect of salt. Overexpression of the AtNHX1 gene in transgenic tomato increased the salt tolerance and also produced good quality fruit containing less Na^+ ions because the plants store the sodium in the leaf vacuoles [114]. Plant salt tolerance genes include defenses against osmotic and oxidative stresses. The genetic analysis indicates that enzymes involved in osmolyte synthesis, osmoprotecting LEA proteins, and antioxidant enzymes such as catalases and glutathione S-transferase/peroxidase are important for plant salt tolerance [79].

8.2 Genome editing technology

Despite the benefits of commercial genetically engineered plants [115] and successfully addressing abiotic stresses, still this technology is not accepted unanimously because of the negative perception of the public; so, it limits the usage of this technique to develop the abiotic resistant crop varieties. The major concern in transgenic technology might be in many cases; the source of the gene to generate transgenic crop is taken from non-related organism, i.e., microorganism and non-related plant and animal; so, this issue can be addressed in a better manner by genome editing technology. In this technique, genetic modifications are accomplished by creating minor genome changes that are comparable with changes generated through mutation breeding (conventional crop improvement method) by using chemical and physical mutagenic agents. Mutational breeding produce changes in the genome at random manner; so, the success rate or chance of obtaining desirable genotype is quite low. But in genome editing, modification is done at targeted site, by application of sequence-specific nucleases that create double-stranded breaks in the target genomic loci selected for editing; so, the success rate of obtaining desirable genotype is very high. The major genome editing tools are zinc finger (ZF) nucleases, transcription activator-like effector nucleases (TALENs), and clustered regularly interspaced short palindromic repeats (CRISPR) [116]. Many strategies or options are there to improve agronomic traits by using genome editing tools. Introduction of a premature stop codon to terminate the functional protein production or modify to a gene promoter motif to control gene expression is one of the best examples of genome editing. Genome editing techniques such as TALEN and CRISPR-Cas9 are used to introduce targeted mutations in MILDEW-RESISTANCE LOCUS (MLO) proteins in hexaploid bread wheat [117]. On the other hand, Piatek et al. used synthetic transcriptional repressor and activator to increase gene expression [118]. C-repeat binding factors (CBFs) are responsible for cold acclimation in plants. Since *CBF1–3* loci all are on the same chromosome, by traditional genetic crossing it is highly difficult to generate *cbf1,2,3* triple mutant lines. So by using genome editing tool CRISPR/Cas9, generating single, double, and triple mutants of *CBF* genes was achieved successfully. The *cbfs* triple mutants are the most sensitive to freezing stress of these different mutants under cold-acclimation treatment. RNA-seq analysis of the triple mutants revealed that the expression of *c*. 10–20% of *COR* genes is CBF dependent [119, 120]. These findings support the notion that CBFs are key regulators that play redundant roles in cold acclimation in plants.

9. Conclusion

Plants are developed with inherent adaptive mechanisms to cope up with varied and composite abiotic stresses. Now it is possible with the help of science and technological advancement to understand function of gene, gene manipulation strategy, and plant traits development to overcome the abiotic stress. Signaling pathways have to be regarded as complex networks. Molecular analyses of the signaling factors provide a better understanding of the signal-transduction cascades during abiotic stress. A notable improvement in crop genome characterization and functional annotation of the gene will advance our knowledge in genetic manipulation (transgenic technology) and optimization of genome editing technology toward development of abiotic stress-tolerant crops. In due course, to accomplish the desired improved varieties, genome editing and transgenic approaches both should be combined with conventional and marker-assisted breeding activities. Further, identification of new adapted germplasm is also most important to guide breeding programs in target trait identification for changing scenario of climate. These combined efforts will make notable progress to face effect of climate change, especially stress such as drought, heat, and cold stress and will contribute to enhanced crop production, productivity, and thereby food security.

Author details

Sivaji Mathivanan
Agricultural College and Research Institute, Tamil Nadu Agricultural University, Vazhavachanur, Tiruvannamalai, India

*Address all correspondence to: sivajibt@gmail.com

References

[1] Fedoroff NV, Battisti DS, Beachy RN, Cooper PJ, Fischhoff DA, Hodges CN, et al. Radically rethinking agriculture for the 21st century. Science. 2010;**327**:833-834

[2] Kundzewicz Z, Ulbrich U, Brucher T, Graczyk D, Kruger A, Leckebusch GC. Summer floods in Central Europe - Climate change track? Natural Hazards. 2005;**36**:165-189

[3] Cramer GR, Urano K, Delrot S, Pezzotti M, Shinozaki K. Effects of abiotic stress on plants: A systems biology perspective. BMC Plant Biology. 2011;**11**:163-176

[4] Zhu L, Chen S, Alvarez S, Asirvatham VS, Schachtman DP, Wu Y. Cell wall proteome in the maize primary root elongation zone. I. Extraction and identification of water soluble and lightly ionically bound proteins. Plant Physiology. 2006;**140**:311-325

[5] Orvar BL, Sangwan V, Omann F, Dhindsa RS. Early steps in cold sensing by plant cells: The role of actin cytoskeleton and membrane fluidity. The Plant Journal. 2000;**23**:785-794

[6] Siddiqui KS, Cavicchioli R. Cold-adapted enzymes. Annual Review of Biochemistry. 2006;**75**:403-433

[7] Ruelland E, Vaultier MN, Zachowski A, Hurry V. Cold signalling and cold acclimation in plants. Advances in Botanical Research. 2009;**49**:35-150

[8] Rajkowitsch L, Chen D, Stampfl S, Semrad K, Waldsich C, Mayer O, et al. RNA chaperones, RNA annealers and RNA helicases. RNA Biology. 2007;**4**:118-130

[9] Fahad S, Muhammad ZI, Abdul K, Ihsanullah D, Saud S, Saleh A, et al. Consequences of high temperature under changing climate optima for rice pollen characteristics - Concepts and perspectives. Archives of Agronomy and Soil Science. 2018;**64**(11):1473-1488. DOI: 10.1080/03650340.2018.1443213

[10] Hadiarto T, Tran LS. Progress studies of drought-responsive genes in rice. Plant Cell Reports. 2011;**30**:297-310

[11] Dombrowski JE. Salt stress activation of wound-related genes in tomato plants. Plant Physiology. 2003;**132**:2098-2107

[12] Yamaguchi-Shinozaki K, Kasuga M, Liu Q, Nakashima K, Sakuma Y, Abe H. Biological mechanisms of drought stress response. JIRCAS Working Report. 2002:1-8

[13] Shinozaki K, Yamaguchi SK, Seki M. Regulatory network of gene expression in the drought and cold stress responses. Current Opinion in Plant Biology. 2003;**6**:410-417

[14] Xiong L, Schumaker KS, Zhu JK. Cell signaling during cold, drought, and salt stress. The Plant Cell. 2002: 165-183

[15] Daryanto S, Wang L, Jacinthe PA. Global synthesis of drought effects on maize and wheat production. PLoS One. 2016;**11**:e0156362. DOI: 10.1371/journal.pone.0156362

[16] Farooq M, Gogoi N, Barthakur S, Baroowa B, Bharadwaj N, Alghamdi SS. Drought stress in grain legumes during reproduction and grain filling. Journal of Agronomy and Crop Science. 2017;**203**:81-102. DOI: 10.1111/jac.12169

[17] Boyer JS, Westgate ME. Grain yields with limited water. Journal of Experimental Botany. 2004;**55**:2385-2394

[18] Fahad S, Bajwa AA, Nazir U, Anjum SA, Farooq A, Zohaib A,

et al. Crop production under drought and heat stress: Plant responses and management options. Frontiers in Plant Science. 2017;**8**:1147. DOI: 10.3389/fpls.2017.01147

[19] Mamun EA, Alfred S, Cantrill LC, Overall RL, Sutton BG. Effects of chilling on male gametophyte development in rice. Cell Biology International. 2006;**30**:583-591

[20] Fahad S, Chen Y, Saud S, Wang K, Xiong D, Chen C, et al. Ultraviolet radiation effect on photosynthetic pigments, biochemical attributes, antioxidant enzyme activity and hormonal contents of wheat. Journal of Food, Agriculture and Environment. 2013;**11**(3&4):1635-1641

[21] FAO. The State of Food and Agriculture - Climate Change, Agriculture and Food Security. Rome, Italy: Food and Agricultural Organization of the United Nations; 2016

[22] Maas EV, Hoffman GJ. Crop salt tolerance - Current assessment. Journal of the Irrigation and Drainage Division. 1977;**103**:115-134

[23] Munns R, Tester M. Mechanisms of salinity tolerance. Annual Review of Plant Biology. 2008;**59**:651-681

[24] Guo XY, Liu DF, Chong K. Cold signaling in plants: Insights into mechanisms and regulation. Journal of Integrative Plant Biology. 2018;**60**:745-756

[25] Liu H, Yu C, Li H, Ouyang B, Wang T, Zhang J, et al. Overexpression of *ShDHN*, a dehydrin gene from *Solanum habrochaites* enhances tolerance to multiple abiotic stresses in tomato. Plant Science. 2015;**231**:198-211

[26] Pearce RS. Plant freezing and damage. Annals of Botany. 2001;**87**:417-424

[27] Dowgert MF, Steponkus PL. Behavior of the plasma membrane of isolated protoplasts during a freeze-thaw cycle. Plant Physiology. 1984;**75**:1139-1151

[28] Chinnusamy V, Zhu J, Zhu JK. Cold stress regulation of gene expression in plants. Trends in Plant Science. 2007;**12**(10):444-451

[29] Lichtenthaler HK. The stress concept in plants: An introduction. Annals of the New York Academy of Sciences. 1998;**851**:187-198

[30] Jogaiah S, Govind SR, Tran LSP. Systems biology-based approaches toward understanding drought tolerance in food crops. Critical Reviews in Biotechnology. 2013;**33**:23-29

[31] Boyer JS. Plant productivity and environment. Science. 1982;**218**:443-448

[32] Knight H, Trewavas AJ, Knight MR. Cold calcium signaling in *Arabidopsis* involves two cellular pools and a change in calcium signature after acclimation. Plant Cell. 1996;**8**:489-503

[33] Plieth C, Hansen UP, Knight H, Knight MR. Temperature sensing by plants: The primary characteristics of signal perception and calcium response. The Plant Journal. 1999;**18**:491-497

[34] Knight MR, Knight H. Low-temperature perception leading to gene expression and cold tolerance in higher plants. The New Phytologist. 2012;**195**:737-751

[35] Tyystjarvi E. Photoinhibition of photosystem II. International Review of Cell and Molecular Biology. 2013;**300**:243-303

[36] Xiong L, Zhu JK. Abiotic stress signal transduction in plants: Molecular and genetic perspectives. Physiologia Plantarum. 2001;**112**:152-166

[37] Sha Valli Khan PS, Nagamallaiah GV, Dhanunjay Rao M, Sergeant K, Hausman JF. Abiotic stress tolerance in plants insights from proteomics. In: Ahmad P, editor. Emerging Technologies and Management of Crop Stress Tolerance. Elsevier Academic press; Vol. 2. 2014. pp. 23-68

[38] Boudsocq M, Laurière C. Osmotic signaling in plants: Multiple pathways mediated by emerging kinase families. Plant Physiology. 2005;**138**(3):1185-1194

[39] Marques da Silva J, Arrabaça MC. Contributions of soluble carbohydrates to the osmotic adjustment in the C4 grass *Setaria sphacelata*: A comparison between rapidly and slowly imposed water stress. Journal of Plant Physiology. 2004;**161**(5):551-555

[40] Aziz K, Daniel KYT, Muhammad ZA, Honghai L, Shahbaz AT, Mir A, et al. Nitrogen fertility and abiotic stresses management in cotton crop: A review. Environmental Science and Pollution Research. 2017b;**24**:14551-14566. DOI: 10.1007/s11356-017-8920-x

[41] Hesham FA, Fahad S. Melatonin application enhances biochar efficiency for drought tolerance in maize varieties: Modifications in physio-biochemical machinery. Agronomy Journal. 2020:1-22

[42] Beck EH, Fettig S, Knake C, Hartig K, Bhattarai T. Specific and unspecific responses of plants to cold and drought stress. Journal of Biosciences. 2007;**32**(3):501-510

[43] Kazuko YS, Shinozaki K. Transcriptional regulatory networks in cellular responses and tolerance to dehydration and cold stresses. Annual Review of Plant Biology. 2006;**57**:781-803

[44] Juven-Gershon T, Hsu JY, Theisen JW, Kadonaga JT. The RNA polymerase II core promoter—The gateway to transcription. Current Opinion on Cell Biology. 2008;**20**(3):253-259

[45] Hu H, Dai M, Yao J, Xiao B, Li X, Zhang Q, et al. Overexpressing a NAM, ATAF, and CUC (NAC) transcription factor enhances drought resistance and salt tolerance in rice. Proceedings of the National Academy of Sciences of the United States of America. 2006;**103**(35):12987-12992

[46] Chen L, Song Y, Li S, Zhang L, Zou C, Yu D. The role of WRKY transcription factors in plant abiotic stresses. Biochimica et Biophysica Acta. 2012;**1819**(2):120-128

[47] Wu X, Shiroto Y, Kishitani S, Ito Y, Toriyama K. Enhanced heat and drought tolerance in transgenic rice seedlings overexpressing OsWRKY11 under the control of HSP101 promoter. Plant Cell Reports. 2009;**28**(1):21-30

[48] Yamaguchi-Shinozaki K, Shinozaki K. A novel cis-acting element in an *Arabidopsis* gene is involved in responsiveness to drought, low-temperature, or high-salt stress. The Plant Cell. 1994;**6**:251-264

[49] Stockinger EJ, Gilmour SJ, Thomashow MF. *Arabidopsis thaliana CBF1* encodes an AP2 domain-containing transcriptional activator that binds to the C-repeat/DRE, a cis-acting DNA regulatory element that stimulates transcription in response to low temperature and water deficit. Proceedings of the National Academy of Sciences of the United States of America. 1997;**94**:1035-1040

[50] Liu Q, Kasuga M, Sakuma Y, Abe H, Miura S, Yamaguchi-Shinozaki K, et al. Two transcription factors, DREB1 and DREB2, with an EREBP/AP2 DNA binding domain separate two cellular signal transduction pathways in drought- and

low-temperature-responsive gene expression, respectively, in *Arabidopsis*. The Plant Cell. 1998;**10**:1391-1406

[51] Gilmour SJ, Zarka DG, Stockinger EJ, Salazar MP, Houghton JM, Thomashow MF. Low temperature regulation of the *Arabidopsis* CBF family of AP2 transcriptional activators as an early step in cold-induced *COR* gene expression. The Plant Journal. 1998;**16**:433-442

[52] Medina J, Bargues M, Terol J, Perez-Alonso M, Salinas J. The *Arabidopsis CBF* gene family is composed of three genes encoding AP2 domain-containing proteins whose expression is regulated by low temperature but not by abscisic acid or dehydration. Plant Physiology. 1999;**119**:463-470

[53] Jaglo-Ottosen KR, Gilmour SJ, Zarka DG, Schabenberger O, Thomashow MF. *Arabidopsis CBF1* overexpression induces *COR* genes and enhances freezing tolerance. Science. 1998;**280**:104-106

[54] Shi Y, Ding Y, Yang S. Molecular regulation of CBF signaling in cold acclimation. Trends in Plant Science. 2018;**23**:623-637

[55] Gilmour SJ, Sebolt AM, Salazar MP, Everard JD, Thomashow MF. Overexpression of the *Arabidopsis CBF3* transcriptional activator mimics multiple biochemical changes associated with cold acclimation. Plant Physiology. 2000;**124**:1854-1865

[56] Zhang X, Fowler SG, Cheng HM, Lou YG, Rhee SY, Stockinger EJ, et al. Freezing-sensitive tomato has a functional CBF cold response pathway, but a CBF regulon that differs from that of freezing-tolerant *Arabidopsis*. The Plant Journal. 2004;**39**:905-919

[57] Savitch LV, Allard G, Seki M, Robert LS, Tinker NA, Huner NPA, et al. The effect of overexpression of two *Brassica* CBF/DREB1-like transcription factors on photosynthetic capacity and freezing tolerance in *Brassica napus*. Plant & Cell Physiology. 2005;**46**:1525-1539

[58] Floris M, Mahgoub H, Lanet E, Robaglia C, Menand B. Post-transcriptional regulation of gene expression in plants during abiotic stress. International Journal of Molecular Sciences. 2009;**10**(7):3168-3185

[59] Lee BH, Kapoor A, Zhu J, Zhu JK. STABILIZED1, a stress-upregulated nuclear protein, is required for pre-mRNA splicing, mRNA turnover, and stress tolerance in *Arabidopsis*. The Plant Cell. 2006;**18**(7):1736-1749

[60] Guan Q, Wu J, Zhang Y, Jiang C, Liu R, Chai C, et al. A DEAD box RNA helicase is critical for pre-mRNA splicing, cold-responsive gene regulation, and cold tolerance in *Arabidopsis*. The Plant Cell. 2013;**25**:342-356

[61] James AB, Syed NH, Bordage S, Marshall J, Nimmo GA, Jenkins GI, et al. Alternative splicing mediates responses of the *Arabidopsis* circadian clock to temperature changes. The Plant Cell. 2012;**24**:961-981

[62] Calixto CPG, Guo WB, James AB, Tzioutziou NA, Entizne JC, Panter PE, et al. Rapid and dynamic alternative splicing impacts the *Arabidopsis* cold response transcriptome. The Plant Cell. 2018;**30**:1424-1444

[63] Sunkar R, Zhu JK. Novel and stress-regulated microRNAs and other small RNAs from *Arabidopsis*. The Plant Cell. 2004;**16**(8):2001-2019

[64] Bari R, Datt PB, Stitt M, Scheible WR. PHO2, microRNA399, and PHR1 define a phosphate-signaling pathway in plants. Plant Physiology. 2006;**141**(3):988-999

[65] Sunkar R, Kapoor A, Zhu JK. Posttranscriptional induction of two Cu/Zn superoxide dismutase genes in *Arabidopsis* is mediated by downregulation of miR398 and important for oxidative stress tolerance. The Plant Cell. 2006;**18**(8):2051-2065

[66] Zhu JK. Salt and drought stress signal transduction in plants. Annual Review of Plant Biology. 2002;**53**:247-273

[67] Yoshida R, Umezawa T, Mizoguchi T, Takahashi S, Takahashi F, Shinozaki K. The regulatory domain of SRK2E/OST1/SnRK2.6 interacts with ABI1 and integrates abscisic acid (ABA) and osmotic stress signals controlling stomatal closure in *Arabidopsis*. The Journal of Biological Chemistry. 2006;**281**(8):5310-5318

[68] Ko JH, Yang SH, Han KH. Upregulation of an *Arabidopsis* RING-H2 gene, XERICO, confers drought tolerance through increased abscisic acid biosynthesis. The Plant Journal. 2006;**47**(3):343-355

[69] Mazzucotelli E, Mastrangelo AM, Crosatti C, Guerra D, Stanca AM, Cattivelli L. Abiotic stress response in plants: When post-transcriptional and post-translational regulations control transcription. Plant Science. 2008;**174**(4):420-431

[70] Kim JM, Sasaki T, Ueda M, Sako K, Seki M. Chromatin changes in response to drought, salinity, heat, and cold stresses in plants. Frontiers in Plant Science. 2015;**6**:114

[71] To TK, Nakaminami K, Kim JM, Morosawa T, Ishida J, Tanaka M, et al. *Arabidopsis* HDA6 is required for freezing tolerance. Biochemical and Biophysical Research Communications. 2011;**406**:414-419

[72] Hu Y, Zhang L, Zhao L, Li J, He SB, Zhou K, et al. Trichostatin A selectively suppresses the cold-induced transcription of the *ZmDREB1* gene in maize. PLoS One. 2011;**6**:e22132

[73] Roy D, Paul A, Roy A, Ghosh R, Ganguly P, Chaudhuri S. Differential acetylation of histone H3 at the regulatory region of OsDREB1b promoter facilitates chromatin remodelling and transcription activation during cold stress. PLoS One. 2014;**9**:e100343

[74] He XJ, Hsu YF, Zhu SH, Liu HL, Pontes O, Zhu JH, et al. A conserved transcriptional regulator is required for RNA-directed DNA methylation and plant development. Genes & Development. 2009;**23**:2717-2722

[75] Chan ZL, Wang YP, Cao MJ, Gong YH, Mu ZX, Wang HQ, et al. RDM4 modulates cold stress resistance in *Arabidopsis* partially through the CBF-mediated pathway. The New Phytologist. 2016;**209**:1527-1539

[76] Rodríguez M, Canales E, Borras-Hidalgo O. Molecular aspects of abiotic stress in plants. Biotecnologia Aplicada. 2005;**22**:1-10

[77] Sunkar R, Bartels D, Kirch HH. Overexpression of a stress-inducible dehydrogenase gene from *Arabidopsis thaliana* in transgenic plants improves stress tolerance. The Plant Journal. 2003:35

[78] Wang W, Vinocur B, Altman A. Plant responses to drought, salinity and extreme temperatures: Towards genetic engineering for stress tolerance. Planta. 2003;**218**:1-14

[79] Serrano R, Gaxiola R, Rios G, Forment J, Vicente O, Ros R. Salt stress proteins identified by a functional approach in yeast. Monatshefte für Chemie. 2003;**134**:1445-1464

[80] Fahad S, Nie L, Chen Y, Wu C, Xiong D, Saud S, et al. Crop plant

hormones and environmental stress. Sustainable Agriculture Reviews. 2015b;**15**:371-400

[81] Garay-Arroyo A, Colmenero-Flores JM, Garciarrubio A, Covarrubias AA. Highly hydrophilic proteins in prokaryotes and eukaryotes are common during conditions of water deficit. The Journal of Biological Chemistry. 2000;**275**(8):5668-5674

[82] Fahad S, Bano A. Effect of salicylic acid on physiological and biochemical characterization of maize grown in saline area. Pakistan Journal of Botany. 2012;**44**:1433-1438

[83] Eimer M. Transgenic drought- and salt-tolerant plants. Genetic Engineering Newsletter Special Issue. 2004;**15**:1-14

[84] Serrano R, Montesinos C. Molecular bases of desiccation tolerance in plant cells and potential applications in food dehydration. Food Science and Technology International. 2003;**9**(3):157-161

[85] Fahad S, Adnan M, Hassan S, Saud S, Hussain S, Wu C, et al. Rice responses and tolerance to high temperature. In: Hasanuzzaman M, Fujita M, Nahar K, Biswas JK, editors. Advances in Rice Research for Abiotic Stress Tolerance. Cambridge, England: Woodhead Publ Ltd; 2019b. pp. 201-224

[86] Shi H, Ishitani M, Kim C, Zhu JK. The *Arabidopsis thaliana* salt tolerance gene SOS1 encodes a putative Na^+/H^+ antiporter. Proceedings of the National Academy of Sciences of the United States of America. 2000;**97**:6896-6901

[87] Guo Y, Halfter U, Ishitani M, Zhu JK. Molecular characterization of functional domains in the protein kinase SOS2 that is required for plant salt tolerance. The Plant Cell. 2001;**13**:1383-1400

[88] Saud S, Li X, Chen Y, Zhang L, Fahad S, Hussain S, et al. Silicon application increases drought tolerance of *Kentucky bluegrass* by improving plant water relations and morph physiological functions. Scientific World Journal. 2014:1-10. DOI: 10.1155/2014/368694

[89] Seki M, Kamei A, Yamaguchi-Shinozaki K, Shinozaki K. Molecular responses to drought, salinity and frost: Common and different paths for plant protection. Current Opinion in Biotechnology. 2003;**14**:194-199

[90] Fahad S, Hussain S, Bano A, Saud S, Hassan S, Shan D, et al. Potential role of phytohormones and plant growth-promoting rhizobacteria in abiotic stresses: Consequences for changing environment. Environmental Science and Pollution Research. 2014a;**22**(7):4907-4921. DOI: 10.1007/s11356-014-3754-2

[91] Gao JJ, Zhang Z, Peng RH, Xiong AS, Xu J, Zhu B. Forced expression of Mdmyb10, a myb transcription factor gene from apple, enhances tolerance to osmotic stress in transgenic *Arabidopsis*. Molecular Biology Reports. 2011;**38**:205-211. DOI: 10.1007/s11033-010-0096-0

[92] Wang Y, Ying J, Kuzma M, Chalifoux M, Sample A, McArthur C. Molecular tailoring of farnesylation for plant drought tolerance and yield protection. The Plant Journal. 2005;**43**:413-424. DOI: 10.1111/j.1365-313X.2005. 02463

[93] Redillas MC, Jeong JS, Kim YS, Jung H, Bang SW, Choi YD. The overexpression of OsNAC9 alters the root architecture of rice plants enhancing drought resistance and grain yield under field conditions. Plant Biotechnology Journal. 2012;**10**:792-805. DOI: 10.1111/j.1467-7652.2012.00697

[94] He GH, Xu YJ, Wang XY, Liu MJ, Li SP, Chen M. Drought-responsive

WRKY transcription factor genes TaWRKY1 and TaWRKY33 from wheat confer drought and/or heat resistance in *Arabidopsis*. BMC Plant Biology. 2016;**16**:116. DOI: 10.1186/s12870-016-0806-4

[95] Legay S, Guerriero G, Andre C, Guignard C, Cocco E, Charton S. MdMyb93 is a regulator of suberin deposition in russeted apple fruit skins. The New Phytologist. 2016;**212**:977-991. DOI: 10.1111/nph.14170

[96] Fahad S, Hussain S, Matloob A, Khan FA, Khaliq A, Saud S, et al. Phytohormones and plant responses to salinity stress: A review. Plant Growth Regulation. 2014b;**75**(2):391-404. DOI: 10.1007/s10725-014-0013-y

[97] Singla-Pareek SL, Yadav SK, Pareek A, Reddy MK, Sopory SK. Enhancing salt tolerance in a crop plant by overexpression of glyoxalase II. Transgenic Research. 2008;**17**:171-180. DOI: 10.1007/s11248-007-9082-2

[98] Bhomkar P, Upadhyay CP, Saxena M, Muthusamy A, Prakash NS, Pooggin M. Salt stress alleviation in transgenic *Vigna mungo* L. Hepper (blackgram) by overexpression of the glyoxalase I gene using a novel cestrum yellow leaf curling virus (CmYLCV) promoter. Molecular Breeding. 2008;**22**:169-181. DOI: 10.1007/s11032-008-9164-8

[99] Lin F, Xu J, Shi J, Li H, Li B. Molecular cloning and characterization of a novel glyoxalase I gene TaGly I in wheat (*Triticum aestivum* L.). Molecular Biology Reports. 2010;**37**:729-735. DOI: 10.1007/s11033-009-9578-3

[100] Tuomainen M, Ahonen V, Kärenlampi SO, Schat H, Paasela T, Svanys A. Characterization of the glyoxalase 1 gene TcGLX1 in the metal hyperaccumulator plant *Thlaspi caerulescens*. Planta. 2011;**233**:1173-1184. DOI: 10.1007/s00425-011-1370-7

[101] Wu C, Ma C, Pan Y, Gong S, Zhao C, Chen S. Sugar beet M14 glyoxalase I gene can enhance plant tolerance to abiotic stresses. Journal of Plant Research. 2013;**126**:415-425. DOI: 10.1007/s10265-012-0532-4

[102] Alvarez-Gerding X, Cortes-Bullemore R, Medina C, Romero-Romero JL, Inostroza-Blancheteau C, Aquea F. Improved salinity tolerance in Carrizo citrange rootstock through overexpression of glyoxalase system genes. BioMed Research International. 2015:1-7. DOI: 10.1155/2015/827951

[103] Hasanuzzaman M, Nahar K, Hossain MS, Mahmud JA, Rahman A, Inafuku M. Coordinated actions of glyoxalase and antioxidant defense systems in conferring abiotic stress tolerance in plants. International Journal of Molecular Sciences. 2017;**18**:200. DOI: 10.3390/ijms18010200

[104] Adnan M, Zahir S, Fahad S, Arif M, Mukhtar A, Imtiaz AK, et al. Phosphate-solubilizing bacteria nullify the antagonistic effect of soil calcification on bioavailability of phosphorus in alkaline soils. Scientific Reports. 2018;**8**:4339. DOI: 10.1038/s41598-018-22653-7

[105] Datta SK. Recent developments in transgenics for abiotic stress tolerance in rice. JIRCAS Working Report. 2002:43-53

[106] Sakamoto A, Alia HH, Murata N. Metabolic engineering of rice leading to biosynthesis of glycine betaine and tolerance to salt and cold. Plant Molecular Biology. 1998;**38**:1011-1019

[107] Fahad S, Hussain S, Saud S, Hassan S, Ihsan Z, Shah AN, et al. Exogenously applied plant growth regulators enhance the morphophysiological growth and yield of rice under high temperature. Frontiers in Plant Science. 2016c;**7**:1250. DOI: 10.3389/fpls.2016. 01250

[108] Penna S et al. Building stress tolerance through over-producing trehalose in transgenic plants. Trends in Plant Science. 2003;**8**:355-357

[109] Jang IC, Se-Jun O, Ju-Seok S, WonBin C, Song SI. Expression of a bifunctional fusion of the *Escherichia coli* genes for trehalose-6-phosphate synthase and trehalose-6-phosphate. Phosphatase in transgenic rice plants increases trehalose accumulation and abiotic stress tolerance without stunting growth. Plant Physiology. 2003;**131**:516-524

[110] Wang WX, Barak T, Vinocur B, Shoseyov O, Altman A. Abiotic resistance and chaperones: Possible physiological role of SP1, a stable and stabilizing protein from *Populus*. In: Vasil IK, editor. Plant Biotechnology 2000. Dordrecht: Kluwer; 2003. pp. 439-443

[111] Davison PA, Hunter CN, Horton P. Overexpression of b-carotene hydroxylase enhances stress tolerance in *Arabidopsis*. Nature. 2002;**418**:203-206

[112] Oberschall A, Deak M, Torok K, Sass L, Vass I, Kovacs I. A novel aldose/aldehyde reductase protects transgenic plants against lipid peroxidation under chemical and drought stresses. The Plant Journal. 2000;**24**:437-446

[113] Bartels D. Targeting detoxification pathways: An efficient approach to obtain plants with multiple stress tolerance. Trends in Plant Science. 2001;**6**:284-286

[114] Zhang HX, Blumwald E. Transgenic salt-tolerant tomato plants accumulate salt in foliage but not in fruit. Nature Biotechnology. 2001;**19**

[115] National Academies of Sciences, Engineering, and Medicine. Genetically Engineered Crops: Experiences and Prospects. Washington, DC: National Academies Press; 2017. DOI: 10.17226/23395

[116] Voytas DF. Plant genome engineering with sequence specific nucleases. Annual Review of Plant Biology. 2013;**64**:327-350. DOI: 10.1146/annurev-arplant-042811-105552

[117] Wang Y, Cheng X, Shan Q, Zhang Y, Liu J, Gao C. Simultaneous editing of three homoeoalleles in hexaploid bread wheat confers heritable resistance to powdery mildew. Nature Biotechnology. 2014;**32**:947-951. DOI: 10.1038/nbt.2969

[118] Piatek A, Ali Z, Baazim H, Li L, Abulfaraj A, Al-Shareef S. RNA-guided transcriptional regulation in planta via synthetic dCas9-based transcription factors. Plant Biotechnology Journal. 2015;**13**:578-589. DOI: 10.1111/pbi.12284

[119] Jia YX, Ding YL, Shi YT, Zhang XY, Gong ZZ, Yang SH. The *cbfs* triple mutants reveal the essential functions of CBFs in cold acclimation and allow the definition of CBF regulons in *Arabidopsis*. The New Phytologist. 2016;**212**:345-353

[120] Zhao CZ, Zhang ZJ, Xie SJ, Si T, Li YY, Zhu JK. Mutational evidence for the critical role of CBF transcription factors in cold acclimation in *Arabidopsis*. Plant Physiology. 2016;**171**:2744-2759

Chapter 16

HSPs under Abiotic Stresses

Noor ul Haq and Samina N. Shakeel

Abstract

Different organisms respond to the altered environmental conditions by different ways. Heat shock proteins' (HSPs) production is one among the different defense mechanisms which defend the photosystem II and thylokoid membrane in plants. There are different types of HSPs based on their size, that is, high molecular weight (60–100 kDa) and low molecular weight heat shock proteins (15–30 kDa). Small HSPs are further classified based on their localization and role in different sub-cellular organelles. Cp-sHSPs are the chloroplast-specific small HSPs that protect the photosystem II and thylokoid membrane. A model to control the Cp-sHSPs in *Chenopodium album* has been put forward in this chapter. According to this model, Cp-sHSPs of *Chenopodium album* are created in cytoplasm and are moved toward chloroplast. The transit peptide is removed on reaching to the target sub-cellular organelle, that is, chloroplast and the premature Cp-sHSPs are converted into mature ones which have multiple roles under different abiotic stress conditions.

Keywords: plants HSPs, abiotic stresses, HSPs model, *Chenopodium album*

1. Introduction

Organisms respond to the changed growth conditions through heat shock proteins' (HSPs) production [1] and that is the way of survival for the cell which responds differentially [2]. Different environmental conditions including abiotic and biotic stress conditions influence the plants' development and production [3]. Different stress conditions like heat, salt, and low water conditions may majorly influence the plants' physiology and production [4–8], but plants response to the changed environmental conditions may vary depending upon duration, intensity, and combination of different environmental growth conditions [9]. Different processes in the plants including biochemistry, development, and physiology may affected by stress conditions and so the expression of different genes may be turned off or on in response to the changed environmental conditions, which may lead to the creation of different proteins and metabolites that protect the cells against such conditions [10].

2. Stress types

Stresses due to living and non-living things can affect the plants' development and production. Different organisms like viruses, bacteria, and fungi may cause stress conditions for the plants [8] which may activate different defense pathways of the plants [9]. There are reports that abiotic stress conditions are responsible to make mostly changes in plant biochemistry and physiology [10]. So plant growth may negatively be affected by abiotic stress conditions also known as non-living factors [6], and any kind of change in environmental conditions may lead the plants toward adaptation under altered growth conditions [11]. Below are the details of different abiotic stress conditions which may affect the plants.

2.1 Types of abiotic stresses and their effects on plants

2.1.1 High temperature or heat stress

Heat stress is the main factor among abiotic stress conditions that affects the plants yield [12] and so different factors in the plants like metabolite concentration, osmolytes, membrane fluidity, proteins structure, and nucleic acids are seriously changed by temperature [13]. Additionally, high-temperature stress affects the chloroplast photochemical activity [14]. Photosystem II is considered as the most sensitive part of thylokoid membrane [15] and heat stress conditions may influence the photosystem II (PS II) reaction center and the light harvesting complexes [16].

Plants adapt their system to the changed growth conditions through complex mechanisms [17]. Thus, different processes at cellular level are reprogrammed under high- and low-temperature growth conditions and many changes in transcription may happen in different parts of the plants, that is, seedlings, roots, pollens, and leaves [18, 19]. Effect on plants may vary with intensity and duration of temperature [20]. One of the plants responses is the reactive oxygen species (ROS) production which is increased by low- and high-temperature stress conditions, while oxidative damage and cell death have also been reported as a result of high-temperature stress conditions [21]. Photosynthesis inhibition has also been reported by researchers under high-temperature conditions [17], additional to the damage of the oxygen evolving complex (OEC) of photosystem II caused by heat stress [22].

Plants adapt to the high-temperature conditions through heat shock proteins (HSPs) production, which are found to be produced in all organisms from prokaryotes to eukaryotes and have role in cell protection under harsh conditions [2]. Establishment of defense mechanism under high-temperature growth conditions is necessary for cells survival which is not specifically occurred only under high temperature but it is also significant under different stress conditions [23].

2.1.2 Low temperature or cold stress

Low temperature represses the plants development without stopping the cell functions and may cause problems to different processes at cellular level [3]. Temperature is the main factor to control the growth changes from vegetative till reproductive level [24]. Low temperature may increase the ROS production additional to the reduction of cellular respiration [25] as well as damages the cell membrane [26].

Low-temperature stress conditions may reduce photosystem I and this effect has been reported to be increased under low light conditions [27]. The same effects have also been observed by different researchers in different plants like winter rye and barley [28, 29].

Expression of different genes and proteins has been reported to be up- or down-regulated by low-temperature stress conditions [30]. Researchers have reported the up-regulation of the defensive genes under cold stress [24]. For example, almost 300 genes have been reported to get up-regulated under cold stress conditions, while 88 genes (27%) were down-regulated in *Arabidopsis thaliana* [31].

2.1.3 Metal stress

Development of the plants is badly affected by heavy metals [32] and roots are usually damaged by heavy metals which lead to build up different defensive mechanisms for normal growth [33]. Membrane potential and permeability are changed by interactions of heavy metals with membrane components [32]. Plants take up the heavy metals as essential nutrients and are passed to the upper parts of the plants following the pathways of the essential elements transport [34].

Plants respond differentially to the heavy metal toxicity [35] and that is the reason that some plants do not show any phytotoxicity symptoms on heavy metals accumulation [36]. But heavy metals restrict the plants growth and cause cell death due to interruption in different physiological and biochemical pathways [37]. Different essential ions are replaced by heavy metals, for example, Ni replaces Mg ion that results in the changed activity of ribulose-1,5-biphosphate carboxylated oxygenase [38]. Chlorophyll activity is altered [39], while heavy metals break the disulfide bridges of the proteins, which leads to the destabilization of proteins [37]. Besides the formerly mentioned adverse roles in plants, heavy metals interact with the hydroxyl and carboxyl groups of proteins and thus interrupt in the proteins functions [40].

Plants adopt different defense mechanisms while get exposure to heavy metals. These mechanisms include the synthesis of cystein-rich polypeptides phytochelatins and metallothioneins [32]. Researchers have also reported the up-regulation of HSP70 gene and chaperonin 60 family members under different heavy metals, that is, Cd and Ni [41, 42]. Additional to the former HSP families, chloroplast small heat shock proteins (Cp-sHSPs) are also reported to be up-regulated by heavy metals [43].

2.1.4 Salt stress

Based on the response to salt stress, plants may be two types either glycophytes or halophytes. The former kind of plants has no tolerance to the saline environment, while the latter group plants covering are natively grown in saline environment [44]. Halophytes cover almost 1% of the world flora [45]. Salt stress adversely affects the plants growth and productivity by different ways; for example, sodium chloride salt can cause the ionic toxicity and osmotic stress to the plants [46]. Researchers have also reported the adverse effect of salt on growth and photosynthesis of the plants [47] by lowering the intra-cellular CO_2 availability [48] or by changed photosynthetic metabolism [49].

2.1.5 Drought stress

Crops yield and quality are adversely affected by drought conditions. Drought conditions may affect the macro- and micromolecules in a cell including minerals, lipids, proteins, hormones, carbohydrates, or even DNA or RNA [50]. The combination of drought with salt, high- or low-temperature stress conditions becomes more severe for the plants, which affects the plants' growth, development, and signal transduction [51, 52]. Besides the abovementioned macro–/micromolecules, photosynthesis that needs water is adversely affected by environmental stress conditions [53, 54]. Additional to the above, drought conditions may affect the metabolism of the plants because catabolism is enhanced due to hydrolytic enzymatic activity while anabolism is decreased due to lowering synthase activity [52]. In short, drought stress conditions adversely affect the photosynthesis in the chloroplast by decreased nutrient uptake and ion transport [55, 56].

2.2 Effect of stress conditions on gene expression

Stress conditions may activate the defense mechanism of the plants and result the change in different gene expression. The expression of heat shock proteins has been reported to be changed due to heat stress [57]. Heat shock proteins function as chaperones and safeguard the heat sensitive organelles and intra-cellular processes [2].

Proteins other than HSPs have also been reported to get produced and their expression is regulated differentially under heat stress conditions [58]. Besides the HSPs expression under heat stress conditions, these proteins have also been up-regulated under different stress conditions including heavy metal, cold, salt, drought, and oxidative stress conditions [43, 59–62].

3. Heat shock proteins (HSPs)

Heat shock response has been characterized in salivary glands of *Drosophila* [63]. Heat shock proteins have been studied in the result of transcription and translation in chromosomal puffs with active sites [64]. HSPs are produced in all organisms, that is, from bacteria to humans under changed environmental conditions [2] and have chaperone activity that protects the proteins from damage [65].

3.1 Role of heat shock proteins

Genes encoding HSPs respond to abiotic stress factors like high temperature, drought, salt, and low-temperature stress conditions [66]. HSPs having low expression under normal environmental conditions may have different function like chaperone function, prevention of proteins aggregation and folding, as well as to target the miss-folded proteins toward the specific pathways or for degradation [67]. Additional to the HSPs expression under abiotic stress conditions, these proteins have differential expression in different tissues and organelles. Taking all together, HSPs production is to protect the metabolic apparatus for adaptation under different environmental conditions and survival [68].

3.2 Types of HSPs

HSPs are divided into two classes based on their molecular weight, that is, high molecular weight heat shock proteins (HSP100, HSP90, HSP70, HSP60, and HSP40) and low molecular weight heat shock proteins (sHSPs), the weight of which is ranging from 15 to 30 kDa [69].

3.2.1 High molecular weight heat shock proteins

High molecular weight heat shock proteins are further divided into different classes based on molecular weight, that is, HSP100, HSP90, HSP70, and HSP60, the details of which are as below.

3.2.1.1 HSP100

HSP100 (protein family), found in all organisms from prokaryotes to eukaryotes [70], possess two subunits and are reported primarily in prokaryotes, that is, bacteria: (1) large-subunit (ClpA) which is ATP-dependent unfoldase and (2) protease which is a small-subunit ClpP [71]. Nucleotide-binding domain 1 & 2 (NBD1 & NBD2), carboxyl domains, middle domain, and amino and are the five parts of HSP100 proteins family members [72].

HSP100 genes have been reported to be up-regulated under heat stress conditions while the same pattern of expression has not been observed [73] but earlier than these findings, researchers have reported the expression of a member of HSP100 family under abscisic acid (ABA), cold and salt stresses additional to the high-temperature stress conditions [74]. Differential expression of one gene or this family member has been suggested under different abiotic stress conditions [75]. HSP100 family members have been reported with up-regulation under heat stress conditions in different plants like wheat and tobacco [75], rice [74], *Arabidopsis thaliana* [76], soybean [77] and maize [78]. Besides the above, HSP100 family members have also been reported with differential expression at different developmental stages [79] which may be the reason that HSP100 family members have been reported with high concentration in mature seeds of different plants [80].

3.2.1.2 HSP90

All organisms from prokaryotes to eukaryotes have HSP90 [81] and are involved to activate the component proteins involved in proteins transportation, assembling, folding and signal transduction [82]. Seven different isoforms of HSP90 have been identified in *Arabidopsis* and are classified based on sub-cellular localization, that is, three have been reported to be localized in endoplasmic reticulum, chloroplast, and mitochondria while the remaining four are localized in cytosole [83]. Three among the four cytosolic isoforms are expressed constitutively while fourth one is expressed under heat stress conditions [84].

3.2.1.3 HSP70

HSP70 are expressed under normal conditions in plants so these are also named as heat shock cognates [85]. HSP70 are having important role under different environmental conditions including heat stress [86, 87]. This class of proteins may function to stabilize the unstable proteins [82] additional to the proteins transport among sub-cellular compartments and proteins folding [88].

HSP70 family proteins may be classified into four classes based on the sub-cellular localization and thus are localized in four different compartments (cytosol, mitochondria, plastids, and endoplasmic reticulum) of the cell [89].

3.2.1.4 HSP60

HSP60 family members encoded by nuclear DNA [90] are present in prokaryotes to eukaryotes and have function in cells under stress and normal conditions [91]. Bacterial HSP60 plays role in proteins assembling to form complexes (oligomeric) and movement through cell membrane [91] but the same family proteins are involved in organelle (chloroplast and mitochondria)-specific proteins folding [91].

3.2.2 Small heat shock proteins (sHSPs)

Plants' small heat shock proteins having molecular weight from 15 to 30 kDa are encoded by nuclear DNA and are classified into further six classes based on sub-cellular localization [92]. Researchers have classified the abovementioned proteins as per the localization in different cellular organelles, that is, first two are localized in cytosol and the next three classes (III, IV, and V) are localized in endoplasmic reticulum, mitochondria, and plastids, respectively [93]. Additional to the above, class VI has been reported to be localized in endoplasmic reticulum [94].

C-terminal region, N-terminal region, and α-crystallin domain are the three main parts of small heat shock proteins. Small HSPs are characterized by 100 amino acids sequence having α-crystalline domain [95] as well as N-terminal region on one side and C-terminal region on the other side of the formerly mentioned domain [96]. The abovementioned three domains are the conserved regions of small heat shock proteins [97].

Small HSPs expression has been reported in different plants, for example, *Chenopodium album* [43, 62], carrot [98], sugarcane [99], Agave [100], *Arabidopsis* [101], cotton [102], tomato [18], maize [103], tobacco [104], etc. The abovementioned studies of sHSPs in different plants show the importance of this class of HSPs in adaptation under different environmental conditions [92].

3.2.3 Chloroplast small heat shock proteins (Cp-sHSPs) and their role

Cp-sHSPs are produced in cytoplasm followed by its import toward chloroplast [105]. As the name shows, these kinds of proteins are located in chloroplast and have consensus-III or methionine rich region at the N-terminal region additional to the other sHSPs-specific regions [106].

These proteins protect photosynthesis of the plants under heat and oxidative stress conditions [107]. There are different mechanisms to protect photosynthesis, for example, chloroplast membrane stabilization or avoiding everlasting proteins aggregation [108] but the role of Cp-sHSPs is very important in this case [109]. Different researchers have shown the relation of sHSPs with the adaptation of the plants under environmental stress conditions [43, 60–62, 109, 110].

It has been established in vitro by researchers that these chloroplast-specific proteins may protect photosynthetic electron transport under high-temperature stress conditions [59]. Cp-sHSPs associate with photosystem II (PS II) through oxygen-evolving complex (OEC) proteins under high-temperature conditions. It has been confirmed by researchers that these proteins protect PS II from inactivation under heat stress conditions by the protection of oxygen evolution and OEC proteins but have no capability to repair inactivated PS II [107].

4. HSP gene expression and promoters

Promoters regulate gene expression quantitatively and qualitatively [111]. There are three types of promoters that regulate the gene expression, that is, inducible, spatiotemporal, and constitutive promoters. Constitutive promoters promote the gene expression throughout the tissues irrespective to the environmental and developmental conditions, while spatiotemporal promoters direct the target gene expression in specific tissues, but inducible promoters are independent of the endogenous factors but dependent upon the external stimuli and environmental conditions [112]. Almost all kinds of promoters have the same core sequence with TATA-box, initiator, and the TF binding-specific cis-acting motifs specific to the target genes [113].

There are very less reports about the regulation of organelle-localized sHSPs under specific stress conditions or even under combination of stresses though it has been known that these genes are mainly regulated at transcriptional level. Researchers have reported the use of soybean promoter (GmHSP17.3B) to induce the sHSPs expression in *Physcomitrella patens* [114]. Additional to the above, researchers have also reported the rice promoter (Oshsp16.9A) to induce the expression of sHSPs under high-temperature stress conditions [115]. Small heat shock proteins have also been reported to get expressed under different abiotic stress conditions additional to the sHSPs expression at different developmental stages [43, 61, 62, 110, 116].

Heat shock transcription factors (HSFs) and heat shock elements (HSEs) may control the HSPs expression in the result of complex network of interaction [117]. HSFs (more than 20 in number) [118] may control the heat shock response both in vitro and in vivo [119]. Thermotolerance is increased in the result of higher expression of HSPs that is resulted by binding the HSFs to HSEs [120, 121]. Differential expression of HSPs is resulted by the variations in HSEs of HSPs. These HSEs have difference in the location and arrangements of its basic units (nGAAn), for example, AtHsp90–1 gene promoter has heat shock element 1 (HSE1) (tGAAgcTTCtgGAAt), heat shock element 2 (HSE2) (agTCtcGAAacGAAaaGAActTTCtgGAAt), and heat shock element 3 (HSE3) (gGAAgaaTCcaGAAt) [122]. Additional to the above elements, other motifs to regulate HSPs (gap-type 1, gap-type 2, and gap-type 3 with the sequences nTTCnnGAAn[5bp]nGAAn, nTTCn[1bp]nGAAn[5bp]nGAAn and nTTCn[2bp]nGAAn[5bp]nGAAn respectively) have also been reported. Researchers have also reported TTC-rich type regulatory elements with 2–4 units of nTTCn with 0–8 bp gap {e.g., TTC-rich 1 (nTTCn[1bp]nTTCn[6bp]nTTCn) and 3 (nTTCnnTTCn[8bp]nTTCn[1bp]nTTCn)} that have binding capability with HsfA1a of *Arabidopsis*. But some TTC-rich regions are also present with no binding potential with HsfA1a, for example, TTC-rich 2 (nTTCn[5bp]nTTCn[4bp]nTTCn) and TTC-rich 4 (nTTCn[3bp]nTTCn) [119]. Besides the above, other cis-regulatory elements are also present in HSPs promoter to regulate their expression under different growth conditions, for example, stress response elements (STREs), metal response elements (MREs), and CAAT boxes C/EBP [123–125]. Metalothionein gene of animals and plants has also been reported to get activated by heavy metal stress conditions because of the presence of MRE in promoter region of this gene [126–128]. Similarly, another stress-related element, that is, STRE (AGGGG) is also regulated by different abiotic stress conditions in yeast [129].

5. Model to express the Cp-sHSPs under different environmental conditions

There is no model put forward by researchers to control the expression of chloroplast-specific small heat shock proteins (Cp-sHSPs), but a model (**Figure 1**) to control the formerly mentioned genes has been proposed by Haq et al. [62]. According to this model, the presence of different cis-regulatory elements in Cp-sHSPs promoter shows the role of Cp-sHSPs under different abiotic stress

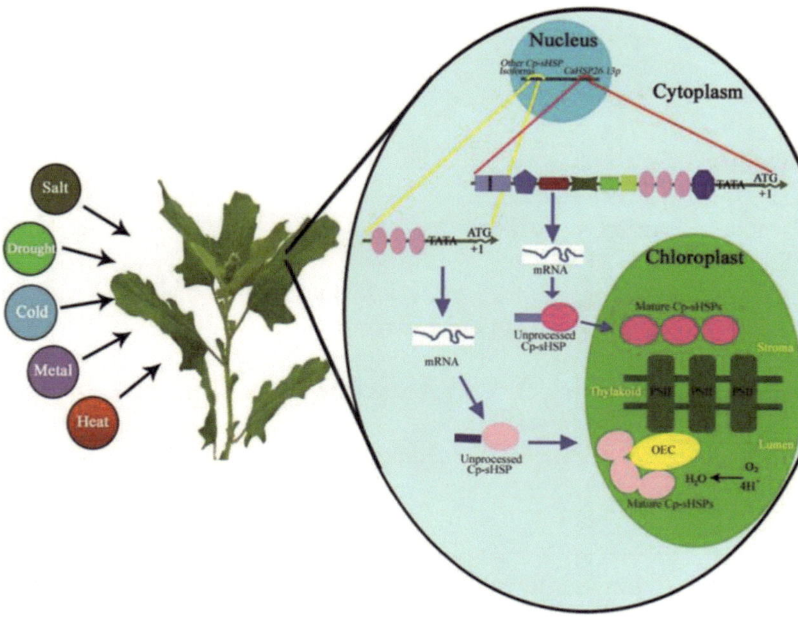

Figure 1.
Proposed model of expression and role of Cp-sHSPs [65].

conditions, that is, salt, drought, cold, metal, and high-temperature stress conditions. Cp-sHSPs in *Chenopodium album* have been shown to protect thylakoid membranes and photosystem II under different abiotic stress conditions. Different abiotic stress conditions, that is, heat, cold, heavy metal, drought, and salt stress conditions may regulate the single Cp-sHSP transcript in *C. album* which produces the precursor proteins that have transit peptide which directs that toward chloroplast. The transit peptide is detached from the proteins while reaching toward chloroplast in the result of which these proteins are matured that have the function in chloroplast. As per this proposed model, differential regulation of the same Cp-sHSP family member in *C. album* makes it able to play multiple roles under different abiotic stress conditions, that is, salt, drought, heavy metal, cold, and heat stress conditions [62].

Author details

Noor ul Haq[1*] and Samina N. Shakeel[2]

1 Department of Computer Science and Bioinformatics, Khushal Khan Khattak University, Karak, Khyber-Pakhtunkhwa, Pakistan

2 Department of Biochemistry, Quaid-i-Azam University, Islamabad, Pakistan

*Address all correspondence to: noorqu@gmail.com; noorulhaq@kkkuk.edu.pk

References

[1] Parsell DA, Lindquist S. The function of heat-shock proteins in stress tolerance: Degradation and reactivation of damaged proteins. Annual Review of Genetics. 1993;**27**:437-496

[2] Lindquist S. The heat-shock response. Annual Review of Biochemistry. 1986;**55**:1151-1191

[3] Balestrasse KB, Tomaro ML, Batlle A, Noriega GO. The role of 5-aminolevulinic acid in the response to cold stress in soybean plants. Phytochemistry. 2010;**71**:2038-2045

[4] Zhang H, Mao X, Wang C, Jing R. Overexpression of a common wheat gene TaSnRK2.8 enhances tolerance to drought, salt and low temperature in *Arabidopsis*. PLoS One. 2010;**5**:e16041

[5] Zhang M, Li G, Huang W, Bi T, Chen G, Tang Z, et al. Proteomic study of *Carissa spinarum* in response to combined heat and drought stress. Proteomics. 2010;**10**:3117-3129

[6] Khan A, Tan DKY, Afridi MZ, Luo H, Tung SA, Ajab M, et al. Nitrogen fertility and abiotic stresses management in cotton crop: A review. Environmental Science and Pollution Research International. 2017;**24**:14551-14566

[7] Fahad S, Bajwa AA, Nazir U, Anjum SA, Farooq A, Zohaib A, et al. Crop production under drought and heat stress: Plant responses and management options. Frontiers in Plant Science. 2017;**8**:1147

[8] Fahad S, Bano A. Effect of salicylic acid on physiological and biochemical characterization of maize grown in saline area. Pakistan Journal of Botany. 2012;**44**:1433-1438

[9] Lefebvre V, Kiani SP, Durand-Tardif M. A focus on natural variation for abiotic constraints response in the model species *Arabidopsis thaliana*. International Journal of Molecular Sciences. 2009;**10**:3547-3582

[10] Tuteja N, Ahmad P, Panda BB, Tuteja R. Genotoxic stress in plants: Shedding light on DNA damage, repair and DNA repair helicases. Mutation Research. 2009;**681**:134-149

[11] Dangl JL, Jones JD. Plant pathogens and integrated defence responses to infection. Nature. 2001;**411**:826-833

[12] Collinge M, Boller T. Differential induction of two potato genes, Stprx2 and StNAC, in response to infection by *Phytophthora infestans* and to wounding. Plant Molecular Biology. 2001;**46**:521-529

[13] Agarwal PK, Agarwal P, Reddy MK, Sopory SK. Role of DREB transcription factors in abiotic and biotic stress tolerance in plants. Plant Cell Reports. 2006;**25**:1263-1274

[14] Kvaalen H, Johnsen O. Timing of bud set in *Picea abies* is regulated by a memory of temperature during zygotic and somatic embryogenesis. The New Phytologist. 2008;**177**:49-59

[15] Boyer JS. Plant productivity and environment. Science. 1982;**218**:443-448

[16] Chinnusamy V, Zhu J, Zhu JK. Cold stress regulation of gene expression in plants. Trends in Plant Science. 2007;**12**:444-451

[17] Ilik P, Kouril R, Kruk J, Mysliwa-Kurdziel B, Popelkova H, Strzalka K, et al. Origin of chlorophyll fluorescence in plants at 55-75°C. Photochemistry and Photobiology. 2003;**77**:68-76

[18] Lichtenthaler HK. The stress concept in plants: An introduction. Annals of the New York Academy of Sciences. 1998;**851**:187-198

[19] Keren N, Berg A, van Kan PJ, Levanon H, Ohad I. Mechanism of photosystem II photoinactivation and D1 protein degradation at low light: The role of back electron flow. Proceedings of the National Academy of Sciences of the United States of America. 1997;**94**:1579-1584

[20] Zinn KE, Tunc-Ozdemir M, Harper JF. Temperature stress and plant sexual reproduction: Uncovering the weakest links. Journal of Experimental Botany. 2010;**61**:1959-1968

[21] Frank G, Pressman E, Ophir R, Althan L, Shaked R, Freedman M, et al. Transcriptional profiling of maturing tomato (*Solanum lycopersicum* L.) microspores reveals the involvement of heat shock proteins, ROS scavengers, hormones, and sugars in the heat stress response. Journal of Experimental Botany. 2009;**60**:3891-3908

[22] Kreps JA, Wu Y, Chang HS, Zhu T, Wang X, Harper JF. Transcriptome changes for *Arabidopsis* in response to salt, osmotic, and cold stress. Plant Physiology. 2002;**130**:2129-2141

[23] Thakur P, Kumar S, Malik JA, Berger JD, Nayyar H. Cold stress effects on reproductive development in grain crops: An overview. Environmental and Experimental Botany. 2010;**67**:429-443

[24] Apel K, Hirt H. Reactive oxygen species: Metabolism, oxidative stress, and signal transduction. Annual Review of Plant Biology. 2004;**55**:373-399

[25] Strasser BJ. Donor side capacity of photosystem II probed by chlorophyll a fluorescence transients. Photosynthesis Research. 1997;**52**:147-155

[26] Walther W, Stein U. Heat-responsive gene expression for gene therapy. Advanced Drug Delivery Reviews. 2009;**61**:641-649

[27] Winfield MO, Lu C, Wilson ID, Coghill JA, Edwards KJ. Plant responses to cold: Transcriptome analysis of wheat. Plant Biotechnology Journal. 2010;**8**:749-771

[28] Lee TM, Lur HS. Role of abscisic acid in cold tolerance of rice (*Oryza sativa* L.) seedlings. II. Modulation of free polyamine levels. Plant Science. 1997;**126**:1-10

[29] Xing W, Rajashekar CB. Glycine betaine involvement in freezing tolerance and water stress in *Arabidopsis thaliana*. Environmental and Experimental Botany. 2001;**46**:21-28

[30] Li XG, Duan W, Meng QW, Zou Q, Zhao SJ. The function of chloroplastic NAD(P)H dehydrogenase in tobacco during chilling stress under low irradiance. Plant & Cell Physiology. 2004;**45**:103-108

[31] Ivanov AG, Morgan RM, Gray GR, Velitchkova MY, Huner NP. Temperature/light dependent development of selective resistance to photoinhibition of photosystem I. FEBS Letters. 1998;**430**:288-292

[32] Teicher HB, Moller BL, Scheller HV. Photoinhibition of photosystem I in field-grown barley (*Hordeum vulgare* L.): Induction, recovery and acclimation. Photosynthesis Research. 2000;**64**:53-61

[33] Thomashow MF. Plant cold acclimation: Freezing tolerance genes and regulatory mechanisms. Annual Review of Plant Physiology and Plant Molecular Biology. 1999;**50**:571-599

[34] Fowler S, Thomashow MF. *Arabidopsis* transcriptome profiling indicates that multiple regulatory pathways are activated during cold acclimation in addition to the CBF cold response pathway. The Plant Cell. 2002;**14**:1675-1690

[35] Bekesiova B, Hraska S, Libantova J, Moravcikova J, Matusikova I.

Heavy-metal stress induced accumulation of chitinase isoforms in plants. Molecular Biology Reports. 2008;**35**:579-588

[36] Pavlikova D, Pavlik M, Staszkova L, Motyka V, Szakova J, Tlustos P, et al. Glutamate kinase as a potential biomarker of heavy metal stress in plants. Ecotoxicology and Environmental Safety. 2008;**70**:223-230

[37] Ma JF, Yamaji N, Mitani N, Xu XY, Su YH, McGrath SP, et al. Transporters of arsenite in rice and their role in arsenic accumulation in rice grain. Proceedings of the National Academy of Sciences of the United States of America. 2008;**105**:9931-9935

[38] Metwally A, Safronova VI, Belimov AA, Dietz KJ. Genotypic variation of the response to cadmium toxicity in *Pisum sativum* L. Journal of Experimental Botany. 2005;**56**:167-178

[39] Hart JJ, Welch RM, Norvell WA, Kochian LV. Characterization of cadmium uptake, translocation and storage in near-isogenic lines of durum wheat that differ in grain cadmium concentration. The New Phytologist. 2006;**172**:261-271

[40] Ahsan N, Renaut J, Komatsu S. Recent developments in the application of proteomics to the analysis of plant responses to heavy metals. Proteomics. 2009;**9**:2602-2621

[41] Van Assche F, Clijsters H. Inhibition of photosynthesis in *Phaseolus valgaris* by treatment with toxic concentration of zinc: Effects on electron transport and photophosphorylation. Physiologia Plantarum. 1986;**66**:717-721

[42] Kupper H, Kupper F, Spiller M. Environmental relevance of heavy metal substituted chlorophylls using the example of water plants. Journal of Experimental Botany. 1996;**47**:259-266

[43] Sahr T, Voigt G, Paretzke HG, Schramel P, Ernst D. Caesium-affected gene expression in *Arabidopsis thaliana*. The New Phytologist. 2005;**165**:747-754

[44] Ingle RA, Smith JA, Sweetlove LJ. Responses to nickel in the proteome of the hyperaccumulator plant *Alyssum lesbiacum*. Biometals. 2005;**18**:627-641

[45] Kieffer P, Dommes J, Hoffmann L, Hausman JF, Renaut J. Quantitative changes in protein expression of cadmium-exposed poplar plants. Proteomics. 2008;**8**:2514-2530

[46] Haq NU, Raza S, Luthe DS, Heckathorn SA, Shakeel SN. A dual role for the chloroplast small heat shock protein of *Chenopodium album* including protection from both heat and metal stress. Plant Molecular Biology Reporter. 2013;**31**:398-408

[47] Kader MA, Lindberg S. Uptake of sodium in protoplasts of salt-sensitive and salt-tolerant cultivars of rice, *Oryza sativa* L. determined by the fluorescent dye SBFI. Journal of Experimental Botany. 2005;**56**:3149-3158

[48] Flowers TJ, Colmer TD. Salinity tolerance in halophytes. The New Phytologist. 2008;**179**:945-963

[49] Kader MA, Lindberg S. Cytosolic calcium and pH signaling in plants under salinity stress. Plant Signaling & Behavior. 2010;**5**:233-238

[50] Munns R, James RA, Lauchli A. Approaches to increasing the salt tolerance of wheat and other cereals. Journal of Experimental Botany. 2006;**57**:1025-1043

[51] Flexas J, Diaz-Espejo A, Galmes J, Kaldenhoff R, Medrano H, Ribas-Carbo M. Rapid variations of mesophyll conductance in response to changes in CO_2 concentration around leaves. Plant, Cell & Environment. 2007;**30**:1284-1298

[52] Lawlor DW, Cornic G. Photosynthetic carbon assimilation and associated metabolism in relation to water deficits in higher plants. Plant, Cell & Environment. 2002;**25**:275-294

[53] Chae L, Sudat S, Dudoit S, Zhu T, Luan S. Diverse transcriptional programs associated with environmental stress and hormones in the *Arabidopsis* receptor-like kinase gene family. Molecular Plant. 2009;**2**:84-107

[54] Halliwell B. Reactive species and antioxidants. Redox biology is a fundamental theme of aerobic life. Plant Physiology. 2006;**141**:312-322

[55] Ni FT, Chu LY, Shao HB, Liu ZH. Gene expression and regulation of higher plants under soil water stress. Current Genomics. 2009;**10**:269-280

[56] Bray EA. Genes commonly regulated by water-deficit stress in *Arabidopsis thaliana*. Journal of Experimental Botany. 2004;**55**:2331-2341

[57] Andjelkovic V, Thompson R. Changes in gene expression in maize kernel in response to water and salt stress. Plant Cell Reports. 2006;**25**:71-79

[58] Boudsocq M, Lauriere C. Osmotic signaling in plants: Multiple pathways mediated by emerging kinase families. Plant Physiology. 2005;**138**:1185-1194

[59] Brooker RW. Plant-plant interactions and environmental change. The New Phytologist. 2006;**171**:271-284

[60] Feder ME, Hofmann GE. Heat-shock proteins, molecular chaperones, and the stress response: Evolutionary and ecological physiology. Annual Review of Physiology. 1999;**61**:243-282

[61] Wahid A, Gelani S, Ashraf M, Foolad MR. Heat tolerance in plants: An overview. Environmental and Experimental Botany. 2007;**61**:199-223

[62] Heckathorn SA, Downs CA, Sharkey TD, Coleman JS. The small, methionine-rich chloroplast heat-shock protein protects photosystem II electron transport during heat stress. Plant Physiology. 1998;**116**:439-444

[63] Heckathorn SA, Ryan SL, Baylis JA, Wang D, Hamilton IEW, Cundiff L, et al. In vivo evidence from an *Agrostis stolonifera* selection genotype that chloroplast small heat-shock proteins can protect photosystem II during heat stress. Functional Plant Biology. 2002;**29**:933-944

[64] Shakeel S, Haq NU, Heckathorn SA, Hamilton EW, Luthe DS. Ecotypic variation in chloroplast small heat-shock proteins and related thermotolerance in *Chenopodium album*. Plant Physiology and Biochemistry. 2011;**49**:898-908

[65] Haq NU, Ammar M, Bano A, Luthe DS, Heckathorn SA, Shakeel SN. Molecular characterization of *Chenopodium album* chloroplast small heat shock protein and its expression in response to different abiotic stresses. Plant Molecular Biology Reporter. 2013;**31**:1230-1241

[66] Ritossa FM. A new puffing pattern induced by a temperature shock and DNP in *Drosophila*. Experientia. 1962;**18**:571-573

[67] Ashburner M, Bonner JJ. The induction of gene activity in *Drosophila* by heat shock. Cell. 1979;**17**:241-254

[68] Taylor RP, Benjamin IJ. Small heat shock proteins: A new classification scheme in mammals. Journal of Molecular and Cellular Cardiology. 2005;**38**:433-444

[69] Cho EK, Hong CB. Molecular cloning and expression pattern analyses of heat shock protein 70 genes from *Nicotiana tabacum*. Journal of Plant Biology. 2004;**47**:149-159

[70] Kalmar B, Greensmith L. Induction of heat shock proteins for protection against oxidative stress. Advanced Drug Delivery Reviews. 2009;**61**:310-318

[71] Huang B, Xu C. Identification and characterization of proteins associated with plant tolerance to heat stress. Journal of Integrative Plant Biology. 2008;**50**:1230-1237

[72] Tower J. Hsps and aging. Trends in Endocrinology and Metabolism. 2009;**20**:216-222

[73] Agarwal M, Katiyar-Agarwal S, Sahi C, Gallie DR, Grover A. *Arabidopsis thaliana* Hsp100 proteins: Kith and kin. Cell Stress & Chaperones. 2001;**6**:219-224

[74] Gottesman S, Squires C, Pichersky E, Carrington M, Hobbs M, Mattick JS, et al. Conservation of the regulatory subunit for the Clp ATP-dependent protease in prokaryotes and eukaryotes. Proceedings of the National Academy of Sciences of the United States of America. 1990;**87**:3513-3517

[75] Batra G, Chauhan VS, Singh A, Sarkar NK, Grover A. Complexity of rice Hsp100 gene family: Lessons from rice genome sequence data. Journal of Biosciences. 2007;**32**:611-619

[76] Agarwal M, Katiyar-Agarwal S, Grover A. Plant Hsp100 proteins: Structure, function and regulation. Plant Science. 2002;**163**:397-405

[77] Pareek A, Singla SL, Grover A. Immunological evidence for accumulation of two high-molecular-weight (104 and 90 kDa) HSPs in response to different stresses in rice and in response to high temperature stress in diverse plant genera. Plant Molecular Biology. 1995;**29**:293-301

[78] Campbell JL, Klueva NY, Zheng HG, Nieto-Sotelo J, Ho TD, Nguyen HT. Cloning of new members of heat shock protein HSP101 gene family in wheat (*Triticum aestivum* (L.) Moench) inducible by heat, dehydration, and ABA(1). Biochimica et Biophysica Acta. 2001;**1517**:270-277

[79] Schirmer EC, Lindquist S, Vierling E. An *Arabidopsis* heat shock protein complements a thermotolerance defect in yeast. The Plant Cell. 1994;**6**:1899-1909

[80] Lee YR, Nagao RT, Key JL. A soybean 101-kD heat shock protein complements a yeast HSP104 deletion mutant in acquiring thermotolerance. The Plant Cell. 1994;**6**:1889-1897

[81] Nieto-Sotelo J, Kannan KB, Martinez LM, Segal C. Characterization of a maize heat-shock protein 101 gene, HSP101, encoding a ClpB/Hsp100 protein homologue. Gene. 1999;**230**:187-195

[82] Queitsch C, Hong SW, Vierling E, Lindquist S. Heat shock protein 101 plays a crucial role in thermotolerance in *Arabidopsis*. The Plant Cell. 2000;**12**:479-492

[83] Singla SL, Pareek A, Grover A. Plant Hsp 100 family with special reference to rice. Journal of Biosciences. 1998;**23**:337-345

[84] Wegele H, Muller L, Buchner J. Hsp70 and Hsp90--a relay team for protein folding. Reviews of Physiology, Biochemistry and Pharmacology. 2004;**151**:1-44

[85] Garavaglia BS, Garafalo CG, Orellano EG, Ottado J. Hsp70 and Hsp90 expression in citrus and pepper plants in response to *Xanthomonas axonopodis* pv.citri. European Journal of Plant Pathology. 2009;**123**:91-97

[86] Krishna P, Gloor G. The Hsp90 family of proteins in *Arabidopsis thaliana*. Cell Stress & Chaperones. 2001;**6**:238-246

[87] Yabe N, Takahashi T, Komeda Y. Analysis of tissue-specific expression of *Arabidopsis thaliana* HSP90-family gene HSP81. Plant & Cell Physiology. 1994;**35**:1207-1219

[88] Sung DY, Kaplan F, Guy CL. Plant Hsp70 molecular chaperones: Protein structure, gene family, expression and function. Physiologia Plantarum. 2001;**113**:443-451

[89] Wu CH, Caspar T, Browse J, Lindquist S, Somerville C. Characterization of an HSP70 cognate gene family in *Arabidopsis*. Plant Physiology. 1988;**88**:731-740

[90] Miersch J, Grancharov K. Cadmium and heat response of the fungus *Heliscus lugdunensis* isolated from highly polluted and unpolluted areas. Amino Acids. 2008;**34**:271-277

[91] Fink AL. Chaperone-mediated protein folding. Physiological Reviews. 1999;**79**:425-449

[92] Nikolaidis N, Nei M. Concerted and nonconcerted evolution of the Hsp70 gene superfamily in two sibling species of nematodes. Molecular Biology and Evolution. 2004;**21**:498-505

[93] Reading DS, Hallberg RL, Myers AM. Characterization of the yeast HSP60 gene coding for a mitochondrial assembly factor. Nature. 1989;**337**:655-659

[94] Craig EA, Gambill BD, Nelson RJ. Heat shock proteins: Molecular chaperones of protein biogenesis. Microbiological Reviews. 1993;**57**:402-414

[95] Gao C, Jiang B, Wang Y, Liu G, Yang C. Overexpression of a heat shock protein (ThHSP18.3) from *Tamarix hispida* confers stress tolerance to yeast. Molecular Biology Reports. 2011;**39**:4889-4897

[96] Heckathorn SA, Downs CA, Coleman JS. Small heat shock proteins protect electron transport in chloroplasts and mitochondria during stress. American Zoologist. 1999;**39**:865-876

[97] LaFayette PR, Nagao RT, O'Grady K, Vierling E, Key JL. Molecular characterization of cDNAs encoding low-molecular-weight heat shock proteins of soybean. Plant Molecular Biology. 1996;**30**:159-169

[98] Kappe G, Leunissen JA, de Jong WW. Evolution and diversity of prokaryotic small heat shock proteins. Progress in Molecular and Subcellular Biology. 2002;**28**:1-17

[99] Nakamoto H, Vigh L. The small heat shock proteins and their clients. Cellular and Molecular Life Sciences. 2007;**64**:294-306

[100] Laksanalamai P, Robb FT. Small heat shock proteins from extremophiles: A review. Extremophiles. 2004;**8**:1-11

[101] Malik MK, Slovin JP, Hwang CH, Zimmerman JL. Modified expression of a carrot small heat shock protein gene, hsp17. 7, results in increased or decreased thermotolerance double dagger. The Plant Journal. 1999;**20**:89-99

[102] Tiroli AO, Ramos CH. Biochemical and biophysical characterization of small heat shock proteins from sugarcane. Involvement of a specific region located at the N-terminus with substrate specificity. The International Journal of Biochemistry & Cell Biology. 2007;**39**:818-831

[103] Lujan R, Lledias F, Martinez LM, Barreto R, Cassab GI, Nieto-Sotelo J. Small heat-shock proteins and leaf cooling capacity account for the unusual heat tolerance of the central spike leaves in *Agave tequilana* var.weber. Plant, Cell & Environment. 2009;**32**:1791-1803

[104] Dafny-Yelin M, Tzfira T, Vainstein A, Adam Z. Nonredundant

functions of sHSP-CIs in acquired thermotolerance and their role in early seed development in *Arabidopsis*. Plant Molecular Biology. 2008;**67**:363-373

[105] Maqbool A, Abbas W, Rao AQ, Irfan M, Zahur M, Bakhsh A, et al. *Gossypium arboreum* GHSP26 enhances drought tolerance in *Gossypium hirsutum*. Biotechnology Progress. 2010;**26**:21-25

[106] Cao Z, Jia Z, Liu Y, Wang M, Zhao J, Zheng J, et al. Constitutive expression of ZmsHSP in *Arabidopsis* enhances their cytokinin sensitivity. Molecular Biology Reports. 2010;**37**:1089-1097

[107] Hamilton Iii EW, Coleman JS. Heat-shock proteins are induced in unstressed leaves of *Nicotiana attenuata* (Solanaceae) when distant leaves are stressed. American Journal of Botany. 2001;**88**:950-955

[108] Vierling E. The roles of heat shock proteins in plants. Annual Review of Plant Physiology and Plant Molecular Biology. 1991;**42**:579-620

[109] Chen Q, Vierling E. Analysis of conserved domains identifies a unique structural feature of a chloroplast heat shock protein. Molecular & General Genetics. 1991;**226**:425-431

[110] Nakamoto H, Suzuki N, Roy SK. Constitutive expression of a small heat-shock protein confers cellular thermotolerance and thermal protection to the photosynthetic apparatus in cyanobacteria. FEBS Letters. 2000;**483**:169-174

[111] Torok Z, Goloubinoff P, Horvath I, Tsvetkova NM, Glatz A, Balogh G, et al. Synechocystis HSP17 is an amphitropic protein that stabilizes heat-stressed membranes and binds denatured proteins for subsequent chaperone-mediated refolding. Proceedings of the National Academy of Sciences of the United States of America. 2001;**98**:3098-3103

[112] Hamilton EW, Heckathorn SA. Mitochondrial adaptations to NaCl. Complex I is protected by anti-oxidants and small heat shock proteins, whereas complex II is protected by proline and betaine. Plant Physiology. 2001;**126**:1266-1274

[113] Shakeel SN, Haq NU, Heckathorn S, Luthe DS. Analysis of gene sequences indicates that quantity not quality of chloroplast small HSPs improves thermotolerance in C4 and CAM plants. Plant Cell Reports. 2012;**31**:1943-1957

[114] Kang TJ, Kwon TH, Kim TG, Loc NH, Yang MS. Comparing constitutive promoters using CAT activity in transgenic tobacco plants. Molecules and Cells. 2003;**16**: 117-122

[115] Potenza C, Aleman L, Sengupta-Gopalan C. Targeting transgene expression in research, agricultural, and environmental applications: Promoters used in plant transformation. In Vitro Cellular & Developmental Biology. Plant. 2004;**40**:1-22

[116] Peremarti A, Twyman RM, Gomez-Galera S, Naqvi S, Farre G, Sabalza M, et al. Promoter diversity in multigene transformation. Plant Molecular Biology. 2010;**73**:363-378

[117] Saidi Y, Domini M, Choy F, Zryd JP, Schwitzguebel JP, Goloubinoff P. Activation of the heat shock response in plants by chlorophenols: Transgenic *Physcomitrella patens* as a sensitive biosensor for organic pollutants. Plant, Cell & Environment. 2007;**30**:753-763

[118] Guan JC, Jinn TL, Yeh CH, Feng SP, Chen YM, Lin CY. Characterization of the genomic structures and selective expression profiles of nine class I small heat shock protein genes clustered on two chromosomes in rice (*Oryza sativa* L.). Plant Molecular Biology. 2004;**56**:795-809

[119] Sun W, Van Montagu M, Verbruggen N. Small heat shock proteins and stress tolerance in plants. Biochimica et Biophysica Acta. 2002;**1577**:1-9

[120] Guan JC, Yeh CH, Lin YP, Ke YT, Chen MT, You JW, et al. A 9 bp cis-element in the promoters of class I small heat shock protein genes on chromosome 3 in rice mediates L-azetidine-2-carboxylic acid and heat shock responses. Journal of Experimental Botany. 2010;**61**:4249-4261

[121] Nover L, Bharti K, Doring P, Mishra SK, Ganguli A, Scharf KD. *Arabidopsis* and the heat stress transcription factor world: How many heat stress transcription factors do we need? Cell Stress and Chaperones. 2001;**6**:177-189

[122] Guo L, Chen S, Liu K, Liu Y, Ni L, Zhang K, et al. Isolation of heat shock factor HsfA1a-binding sites in vivo revealed variations of heat shock elements in *Arabidopsis thaliana*. Plant and Cell Physiology. 2008;**49**:1306-1315

[123] Lee JH, Hubel A, Schoffl F. Derepression of the activity of genetically engineered heat shock factor causes constitutive synthesis of heat shock proteins and increased thermotolerance in transgenic *Arabidopsis*. The Plant Journal. 1995;**8**:603-612

[124] Wunderlich M, Werr W, Schoffl F. Generation of dominant-negative effects on the heat shock response in *Arabidopsis thaliana* by transgenic expression of a chimaeric HSF1 protein fusion construct. The Plant Journal. 2003;**35**:442-451

[125] Haralampidis K, Milioni D, Rigas S, Hatzopoulos P. Combinatorial interaction of cis elements specifies the expression of the *Arabidopsis* AtHsp90-1 gene. Plant Physiology. 2002;**129**:1138-1149

[126] Amin J, Ananthan J, Voellmy R. Key features of heat shock regulatory elements. Molecular and Cellular Biology. 1988;**8**:3761-37619

[127] Xiao H, Lis JT. Germline transformation used to define key features of heat-shock response elements. Science. 1988;**239**:1139-1142

[128] Schoffl F, Prandl R, Reindl A. Regulation of the heat-shock response. Plant Physiology. 1998;**117**:1135-1141

[129] Karin M, Haslinger A, Heguy A, Dietlin T, Cooke T. Metal-responsive elements act as positive modulators of human metallothionein-IIA enhancer activity. Molecular and Cellular Biology. 1987;**7**:606-613

Chapter 17

Plant Growth and Morphophysiological Modifications in Perennial Ryegrass under Environmental Stress

Fuchun Xie, Rahul Datta and Dong Qin

Abstract

Perennial ryegrass (*Lolium perenne* L.) is a popular and important cool-season turfgrass used in parks, landscapes, sports fields, and golf courses, and it has significant ecological, environmental, and economic values. It is also widely used as forage and pasture grass for animals around the world. However, the growth of perennial ryegrass is often affected by various abiotic stresses, which cause declines in turf quality and forage production. Among abiotic stresses, drought, salinity, temperature, and heavy metal are the most detrimental factors for perennial ryegrass growth in different regions, which result in growth inhibition, cell structure damage, and metabolic dysfunction. Many researches have revealed a lot useful information for understanding the mechanism of tolerance to adverse stresses at morphophysiological level. In this chapter, we will give a systematic literature review about morphological and physiological changes of perennial ryegrass in response to main stress factors and provide detail aspects of improving perennial ryegrass resistance based on research progress. Understanding morphophysiological response in perennial ryegrass under stress will contribute to improving further insights on fundamental mechanisms of perennial ryegrass stress tolerance and providing valuable information for breeding resistance cultivars of perennial ryegrass.

Keywords: perennial ryegrass, morphology, physiology, abiotic stress, stress resistance

1. Introduction

Urban green areas have important various functions contributing to the quality of human health. Well-kept lawns enhance the esthetic value of the entire city and are involved in phytoremediation, leading to an improvement in the quality of the air and soil [1–4]. Perennial ryegrass (*Lolium perenne* L.) is an important and widespread perennial cool-season grass cultivated in temperate climates, originating in Europe, temperate Asia, and North Africa [5]. Perennial ryegrass is commonly used in home lawns, sport fields, and parks with rapid growth and establishing rate, and other elements for ecosystem service due to its massive root system, superior regeneration, and tillering ability. It is also widely used as nutritive forage and pasture grass for animals around the world [6–8]. Moreover, numerous perennial ryegrass genotypes and hybrids are now released by commercial utilities [9, 10].

In fields, the growth and development process of plants needs to counteract various environmental stresses such as salinity, drought, cold, heat, and heavy metal [11–13]. Harsh environmental conditions may result in growth inhibition, cell structure damage, and metabolic dysfunction [14–20]. Moreover, stresses will further be intensified for the potential impact of climate change in future. Thus, maintaining proper growth of turfgrass with minimal inputs under abiotic stress conditions is a great challenge for turfgrass industry. This challenge could be addressed through improving the stress tolerance of turfgrass [14, 21]. Understanding morphological and physiological mechanisms of turfgrass adaptation to various abiotic stresses is a key step for the development of stress-tolerant ability and cost-effective and efficient management practices [13]. Morphophysiological mechanisms of turfgrass in abiotic stresses tolerance involve phenotypic changes, multiple physiological and biochemical response, and complex metabolic processes, such as water and nutrient relations, carbohydrate metabolism, protein metabolism, hormone metabolism, as well as antioxidant defenses [22, 23]. Current studies on morphophysiological mechanism controlling turfgrass adaptations to various growth conditions have provided important information for production of abiotic stress-tolerant germplasms and the further understanding of regulation mechanism of turfgrass response to abiotic stresses [13, 24, 25]. However, the mechanisms of the adaptive responses are integrated but are not necessarily the same [14]; thus, studies on how perennial ryegrass adapts to stress conditions will become more important with the increasing pressure of utilizing both ecological and economical strategies in the turf management. Furthermore, insights into mechanisms of stress resistance in perennial ryegrass will aid in identifying important characteristics for selecting the criteria of improving stress tolerance and will ultimately lead to better selection of new cultivars adapted to adverse environments. This chapter, therefore, focuses on an extensive overview of the current understanding of changes in physiology and growth/development of perennial ryegrass under various abiotic stresses. In addition, strategies for improving the stress tolerance of perennial ryegrass are also presented. This review can contribute to the better understanding of the mechanisms of perennial ryegrass response to environmental stresses and can provide valuable information for improving resistance characteristics of perennial ryegrass by breeding. Moreover, enhancing our understanding of physiological effects of abiotic stresses can provide guidelines for the practical management strategies of the maintenance of high-quality turf under limited resource availability.

2. Abiotic stresses

Abiotic stresses are major environmental conditions that reduce plant growth, productivity, and quality. Plants have evolved mechanisms to perceive these environmental challenges, transmit the stress signals within cells as well as between cells and tissues, and make appropriate adjustments in their growth and development for survive and reproduce [26–29]. The morphological and physiological changes of perennial ryegrass under abiotic stress will be discussed in this chapter.

2.1 Responses of drought stress in perennial ryegrass

Growth and development processes are inhibited when plant is exposed to drought stress [30–36]. Morphological adjustments, such as biomass allocation and leave changes, have been proposed as the key mechanisms used by turfgrass to enhance survival under drought [37]. There is a series of morphology changes in perennial ryegrass under drought stress. Drought stress reduced the turf quality (TQ), number of live tillers and dry-matter yield [38–40]. Moreover, drought significantly enhanced root to shoot ratio (R/S) in perennial ryegrass to an less extents, depending on the intensity, the reason may be that perennial ryegrass in drought stress develop a large R/S to maintain water and nutrient uptake [39]. The leaves of perennial ryegrass under drought stress were also dramatically different from that of nonstressed perennial ryegrass, for example, under drought stress, the diurnal variation in the rate of leaf extension was smaller but the leaves tended to grow faster at night compared to normal irrigation controls; however, water stress ultimately reduced the rate of leaf extension and leaf area in perennial ryegrass [40]. Furthermore, the leaves' epidermis of perennial ryegrass under drought reduced the stomatal size and increased the numbers per unit leaf area. Drought also resulted deeper ridging on leaf ad-axial surface, smaller epidermal cells and bigger ridge angle [40]. Under drought stress, leave stomata of perennial ryegrass began to close to reduce their evapotranspiration rate (ET), at leaf water potentials below—13 bars [40].

Drought stress causes significant physiological changes, including photosynthesis, osmotic adjustment substances, proteins, and antioxidant metabolism, in perennial ryegrass. For instance, the content of leaf total nitrogen and leaf relative water content (RWC) were tested to decrease, on the contrary, antioxidant activity including ascorbate peroxidase (APX), superoxide dismutase (SOD), catalase (CAT), glutathione reductase (GR), glutathione S-transferase (GST), amino acids such as aspartic acid, threonine, serine, glutamic acid, abscisic acid (ABA) concentration, and proline content increased under drought stress [41]. Photosynthesis is the primary process controlling plant growth and adaption to drought stress [42]. The canopy photosynthesis of perennial ryegrass at saturating light intensity was reduced by about half in the stressed field swards and by more than 80% in the stressed simulated swards [38]. Drought stress inhibits photosynthesis, which may be the result of low CO_2 availability caused by stomatal closure and/or the inhibition of photochemical reactions and carbon assimilation metabolism [43]. In addition to the photosynthesis, starchis also considered as a buffer for imbalance between acquisition by photosynthesis and C-sink activities such as growth and respiration resulted from drought, also stress due to the excessive use of inorganic fertilizer [44]. However, use of green manure also has risk of xenobiotic contamination [45–47], and the soluble sugars, including sucrose, fructose and glucose, are involved in multiple physiological functions such as respiration, turgor maintenance, signaling and defense. Under drought conditions, starch of perennial ryegrass significantly decreased in shoots, but did not change in roots, which indicated that perennial ryegrass in drought condition preferentially allocates carbon not only to root growth, but also to root storage, while soluble sugars were enhanced in both shoots and roots. Accumulation of soluble sugars has been widely reported for plants upon water stress as a means to provide osmotic protection [39], which suggested that increasing of soluble sugars was benefit to plants to maintain growth and active metabolic activities under water deficit.

It is generally accepted that there is a noticeable genotypic variation in perennial ryegrass for drought stress responses. The research showed that one self-pollinating genotype "S10" showed higher RWC, shoot dry weight (SDW), proline, ABA, nitrogen and amino acid contents, and antioxidant enzymes activities in comparison with two commercial genotypes of "Vigor" and "Speedy" [41]. Proteins involved in carbon and energy metabolism, photosynthesis, tricarboxylic acid cycle (TCA) cycle, redox, and transport categories were upregulated in the two commercial genotypes of "Vigo" and "Speedy," while the protein profile of the "S10" changed slightly under drought stress, and the reason may be that self-pollination in the genetic background of the "S10" genotype may have a lower variation in response to drought stress conditions [41]. Additionally, other research indicated that tetraploid perennial ryegrass exhibited a greater biomass under severe drought, whereas diploids had a greater biomass under the current rainfall [48, 49]. Moreover, tetraploid perennial ryegrass populations were able to develop more shoot and root dry matter than diploid populations in following the application of drought stress [50].

The above researches showed that drought stress caused significantly physiological and morphological changes in perennial ryegrass (**Table 1**) [55, 56]. Thus, the growth of perennial ryegrass is severely restricted by soil water deficits [57]. Increasing drought tolerance of perennial ryegrass via strategies is importance for both water conservation and maintaining growth in water limiting environments. For example, the grass-Epichloë endophytic improved water utilization and drought tolerance in perennial ryegrass [58]. Moreover, arbuscular mycorrhizal fungi (AMF) + Epichloë treatments increased phosphorus (P) uptake, net photosynthetic rate (Pn), root activity, soluble sugar concentration, peroxidase (POD) activity, and decreased malonyldialdehyde (MDA) concentration in perennial ryegrass under drought stress, the reason may be that Plant-AMF-Epichloë symbiosis alleviated the damage caused by drought stress by promoting P uptake, photosynthesis, and the accumulation of osmoregulatory substances [59]. Additionally, application of plant growth regulators (PGRs) have been reported to be a promising way of reducing drought stress impacts [60]. The study manifested that trinexapac ethyl (TE) treatment increased chlorophyll content, proline content, the RWC, soluble sugar content, antioxidant enzymes activities, decreased MDA and hydrogen peroxide (H_2O_2) contents in perennial ryegrass under drought stress, while Paclobutrazol (PAC)- and ABA-treated perennial ryegrasses were all effective in mitigating physiological damages resulting from drought stress [52]. Furthermore, overexpression of some drought-related genes has been shown to effectively improve drought tolerance of plants [61]. According to Patel et al. [53], overexpression of *LpHUB1* gene conferred drought tolerance in perennial ryegrass.

2.2 Responses of temperature stress in perennial ryegrass

Perennial ryegrass can grow throughout the year, and the major constraint on growth is temperature [51, 62]. Perennial ryegrass has an optimal growth temperature of about 20°C, and it is sensitive to high (30–40°C) and low (−20 to 0°C) temperatures [63, 64]. Common perennial ryegrass germinates quickly and can be used as a temporary ground cover while the slower growing bluegrass plants take

Morphological responses	Physiological responses	Strategies
• Decreased turf quality • Enhanced R/S • Decreased leaf area • Reduce the number of live tillers • Had smaller stomata and epidermal cells • Had bigger ridge angle • Controlled stomatal opening [38–40]	• Decline biomass • Decline photosynthetic rate • Increased osmotic adjustment substances • Increased antioxidant activity • Increased amino acid content [39, 41, 51]	• Application of plant growth regulators (PGRs) • Selected drought resistance cultivars from different cultivars • Using endophytes • Using transgenic technology [52–54]

Table 1.
Morphophysiological response of perennial ryegrass under drought stress.

hold in cool temperate region. In warm climates, it is used as an overseed to maintain winter green in the lawn after the warm season grasses go dormant. However, populations of perennial ryegrass will not survive the summer heat. Severe heat stress (40/35°C day/night) caused significant physiological damages, including declining in TQ, RWC, CAT activity, and enhancing in electrolyte leakage (EL) of leaves and MDA content, in perennial ryegrass [65]. Moreover, heat stress decreased plant height (HT), leaf fresh weight (LFW) and leaf fresh dry (LFD), and increased cytokinin and auxin at 35/30°C (day/night) of temperature [66]. Moreover, low temperature is one of the main factors that limit the persistence of perennial ryegrass-dominated grasslands in northern regions. Cold stress decreased TQ, regrowth, dry weight, and tiller density in perennial ryegrass when the winters were mild with short (2–6 weeks) periods of lower than −10°C temperatures and no permanent snow cover [67]. Furthermore, cold stress decreased RWC and increase EL in leaves and roots when perennial ryegrass was exposed to −15 or −25°C [68]. To resolve these problems and maintain high visual quality of perennial ryegrass through the year, it is important to found new cultivars to adapt to temperature stress. Variations of heat- or cold- resistance were also found among different perennial ryegrass cultivars. Thus, selection cold- or heat-tolerant cultivars during perennial ryegrass genotypes can be an effective method for temperature tolerance improvement in perennial ryegrass. The research tested the heat tolerance of 58 cultivars collected from seed companies and research centers in U.S.A., New Zealand, and Europe, the result showed that distinct heat tolerance was found among the cultivars at all the temperature regimes, and the least and most tolerant cultivars were "JPR005" and "JPR178," respectively [69]. The other research indicated that changes of morphology and physiology were different for heat-tolerant accession "PI265351" and sensitive accession "PI225825" [66]. Similarly, the heat-tolerant populations of perennial grass showed significantly lower degree damage in efficiency of photosystem II and cell membrane stability than the sensitive ones at different levels of stress [70]. Additionally, the study showed that 21 accessions sampled from a larger set of 300 accessions with known winter hardiness, the result showed that the degree of semi-lethal temperature in 21 ryegrass varieties varies from −10.31 to −13.95°C, with 3 accessions possessing significantly greater freezing tolerance than the most freeze-tolerant check "NK200" [71]. Moreover, tetraploid genotypes of perennial ryegrass demonstrated higher tolerance to cold stress conditions, better spring growth, and regrowth after cuts, and higher dry matter yield compared to diploid genotypes [67].

The studies indicate that temperature stress caused the morphological and physiological damage in plant, and the response of genotypes to temperature stress was different [72–80]. Therefore, founding some strategies which could improve cold- or heat-tolerant of perennial ryegrass is important. It was also reported that 24-epibrassinolide promoted carbohydrates accumulation in crowns of perennial ryegrass during cold acclimation by regulation of gene expression and enzyme activities, and which resulted in increased frost tolerance [81]. Moreover, drought preconditioning increased in crown fructans, proline, and total soluble protein content for "Buccaneer" and "Sunkissed" during cold acclimation, which suggested a synergistic effect between drought exposure and low temperature, and drought preconditioning resulted in an improvement in freezing tolerance of perennial ryegrass [82]. Additionally, previous studies have shown that the enzyme activity level and gene expression of antioxidants are associated with cold and heat tolerance in a cool-season perennial grass species [83, 84]. For instance, *LpHOX21* was positively associated with heat tolerance of perennial ryegrass [85]. Similarly, P450 gene (*LpCYP72A161*) showed remarkable upregulation in perennial ryegrass under heat and cold treatment. Therefore, transferring key genes into perennial ryegrass will be beneficial to improve heat or freezing tolerance [86].

2.3 Responses of salt stress in perennial ryegrass

Salinity stress has become a more significant problem in turfgrass management in many areas [13]. Responses of plants to salinity stress occur mainly through two distinct phases over time: osmotic-changing and ion specific phases [87–89]. Like other turfgrasses, salt stress caused morphology, physiology, molecular changes in growth and development of perennial ryegrass, such as TQ LFW, LED, and RWC of perennial ryegrass decreasing after exposure to salinity [89, 90]. The alterations of morphological characteristics of turfgrass under salt stress are derived from the changes of physiological traits such as cell membrane stability [14]. It was reported that MDA content and EL enhanced by NaCl concentration in perennial ryegrass [54]. Simultaneously, superoxide radical (O^{2-}), H_2O_2, and singlet oxygen (O_2) concentration increased observably in perennial ryegrass after salt stress treatment [54, 91]. To scavenge reactive oxygen species (ROS), salt-stressed leaves of perennial ryegrass exhibited greater activities of SOD, APX, and CAT at the initial stage of salt stress, but lower levels of enzyme with the extension of salt stress [89]. Salt stress also negatively affected on the total chlorophyll (Chl), Chl a and Chl b, in perennial ryegrass [89], which showed that salt stress induced Chl decomposition in leaves. Moreover, a further research of PSII changes in perennial ryegrass discovered that quantum yields, efficiencies, and energy fluxes were impacted after salt stress treatment [92, 93]. Additionally, a vast amount of Na^+ accumulated in plants could induce ionic imbalance in the cells. It was reported that Na^+ concentration accumulated rapidly and other ion concentrations including K^+, Ca^{2+} and Mg^{2+} were decreased in response to salt stress in perennial ryegrass [89].

Salt stress causes dramatically changes in morphology and physiology of perennial ryegrass as showed above and summaries in **Figure 1**. However, these responses varied greatly among different genotypes. The research compared the salt tolerance in 10 accessions of perennial ryegrass, and determined that "PI275660"

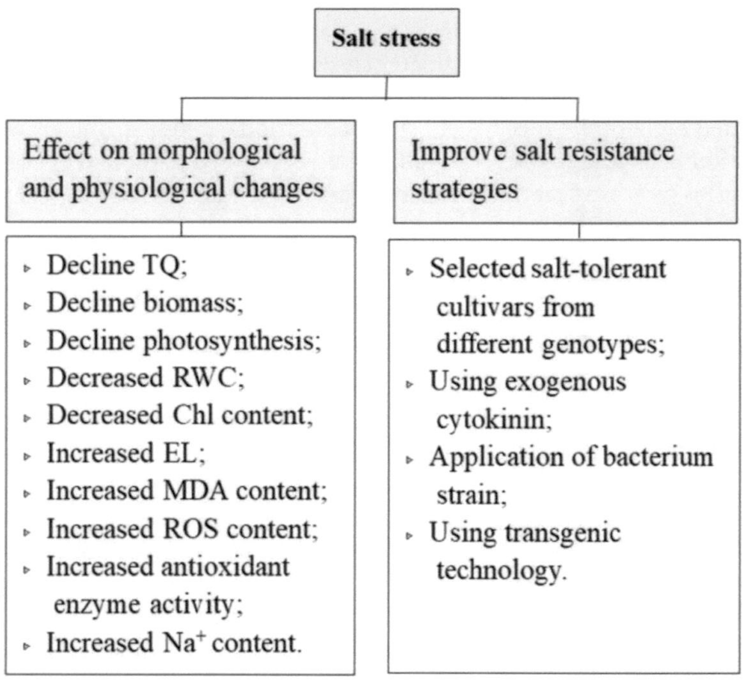

Figure 1.
Morphophysiological response and strategies for salt stress in perennial ryegrass.

and "BrightStar" showed the best tolerance to salt stress, while "PI231595" and "PI251141" were the most sensitive accessions [5]. The other research reported that the effect on parameters of photosynthetic efficiency in perennial ryegrass "Roadrunner" was less than that in "Nira" under salt stress condition [6]. Moreover, the highest salt tolerance accessions were from the European group, wild accessions and exhibited more variation in functional traits and salt tolerance than commercial cultivars [90]. Some other strategies can also improve the salt-tolerance in perennial ryegrass. Salt-tolerant transgenic perennial ryegrass could be obtained by *Agrobacterium tumefaciens*-mediated transformation of the vacuolar Na^+/H^+ antiporter gene [94]. Additionally, exogenous cytokinin applications alleviated salt-induced leaf senescence in perennial ryegrass [8]. Furthermore, salt tolerance of perennial ryegrass can increase by a novel bacterium strain from the rhizosphere of a desert shrub *Haloxylon ammodendron* induced [95].

2.4 Responses of heavy metals stress in perennial ryegrass

The continuing industrialization has led to extensive environmental problems worldwide [96–98]. Heavy metals produced from industry are released to soil. Thus, high accumulation of heavy metal in soil can induce environmental stress on plants [14]. Research on the response of perennial ryegrass to heavy metal stress has also progressed in recent years. It has been proved that heavy metals can induce damage and affect metabolic processes in perennial ryegrass [98–100]. For example, perennial ryegrass had characters in yield reduction and visible symptoms of phytotoxicity under cadmium (Cd) and zinc (Zn) stress [98]. Moreover, the cellular membrane system was damaged because of elevated MDA and EL contents when perennial ryegrass was exposed to salt condition [101]. According to studies, a dramatic inhibition of root and shoot growth was detected in perennial ryegrass after heavy metals treatment [101–103]. Moreover, the composition of the leaves of perennial ryegrass, including apparently opposite effects on the calcium (Ca), potassium (K) and P levels, was changed under the aluminum (Al) stress [104]. Additionally, ROS bursts occurred in perennial ryegrass under heavy mental stress conditions. For instance, H_2O_2 and O^{2-} were significantly accumulated in perennial ryegrass under Cd stress [105]. Hence, the protection mechanisms in perennial ryegrass such as the antioxidant system were triggered under heavy stress, resulting in the increase of SOD, CAT, and POD activities and their corresponding genes [106]. Moreover, content of fructan, sugar, and starch showed an increasing trend in perennial ryegrass after heavy metal stress [98]. However, certain concentrations of heavy metal were beneficial for the growth of perennial ryegrass [107]. Heavy metal stresses not only induce physiological damage, but also inhibit germination and growth of perennial ryegrass [108].

To improve the heavy metal stress tolerance of perennial ryegrass, several investigations were conducted in recent years. It was reported that signal messengers such as nitricoxide (NO) and glycine betaine (GB) play crucial roles in alleviating Cd and Cu-induced damages in perennial ryegrass [109, 110]. Moreover, the exogenous P was testified to improve the Cd tolerance of perennial ryegrass, the reason may be that exogenous P facilitates chelation-mediated Cd detoxification processes [105]. Similarly, a high dose of P amendment alleviated Mn-toxicity in Mn-sensitive genotype in perennial ryegrass [102]. Furthermore, the addition of biochar to a contaminated mine soil improved the nutrient status of this mine soil and contributed to a better establishment of perennial ryegrass [100]. Additionally, AMF enhance both absorption and stabilization of Cd by perennial ryegrass in a Cd-contaminated acidic soil [96], and ethylene diamine tetra acetate (EDTA) enhanced phytoremediation of heavy metals from municipal waste compost and sludge soil by perennial ryegrass [99, 111, 112].

3. Conclusions and future research perspectives

Significant progress has been made in the understanding of morphological and physiological mechanisms associated with perennial ryegrass tolerance to drought, salinity, temperature, and heavy mental stresses. Harsh stress conditions inhibit the growth and development and decrease TQ, root length, and dry weight in perennial ryegrass. Moreover, physiological response to abiotic stress in perennial ryegrass displays changes of the cell membrane, photosystem, metabolites, and antioxidant system. The contents of MDA and EL are increased, while Chl content and photosynthesis are decreased under stress conditions. To regulate the osmotic potential of the cell after stress treatment, some metabolites such as proline, soluble sugars, and proteins accumulate. Meanwhile, antioxidant enzymes' activities increase in perennial ryegrass for scavenging ROS. Perennial ryegrass has protective responses against unfavorable conditions, but there is a threshold to these physiological changes. To understand the response to abiotic stress and resistance attributes in perennial ryegrass will be beneficial to breeding in future.

For improving the stress tolerance of perennial ryegrass, some practical strategies are exploited currently, such as application of phytohormones, endophytes, and chemical compounds. Further research on increasing perennial ryegrass stress tolerance should pay more attention to transgenic technology to identify effective genes for modifying stress-tolerance ability.

Author details

Fuchun Xie[1], Rahul Datta[2] and Dong Qin[1*]

1 College of Horticulture and Landscape Architecture, Northeast Agricultural University, Harbin, China

2 Department of Agrochemistry, Soil Science, Microbiology and Plant Nutrition, Faculty of AgriSciences, Mendel University in Brno, Brno, Czech Republic

*Address all correspondence to: dongq9876@126.com

References

[1] Dabrowski P, Kalaji MH, Baczewska AH, Pawluśkiewicz B, Mastalerczuk G, Borawska-Jarmulowicz B, et al. Delayed chlorophyll a fluorescence, MR 820, and gas exchange changes in perennial ryegrass under salt stress. Journal of Luminescence. 2017;**183**:322-333. DOI: 10.1016/j.jlumin.2016.11.031

[2] Hafiz MH, Muhammad A, Farhat A, Hafiz FB, Saeed AQ, Muhammad M, et al. Environmental factors affecting the frequency of road traffic accidents: A case study of sub-urban area of Pakistan. Environmental Science and Pollution Research. 2019. DOI: 10.1007/s11356-019-04752-8

[3] Sajjad H, Muhammad M, Ashfaq A, Waseem A, Hafiz MH, Mazhar A, et al. Using GIS tools to detect the land use/land cover changes during forty years in Lodhran district of Pakistan. Environmental Science and Pollution Research. 2019. DOI: 10.1007/s11356-019-06072-3

[4] Adnan M, Zahir S, Fahad S, Arif M, Mukhtar A, Imtiaz AK, et al. Phosphate-solubilizing bacteria nullify the antagonistic effect of soil calcification on bioavailability of phosphorus in alkaline soils. Scientific Reports. 2018;**8**:4339. DOI: 10.1038/s41598-018-22653-7

[5] Tang J, Camberato JJ, Yu X, Luo N, Bian S, Jiang Y. Growth response, carbohydrate and ion accumulation of diverse perennial ryegrass accessions to increasing salinity. Scientia Horticulturae. 2013;**154**:73-81. DOI: 10.1016/j.scienta.2013.02.021

[6] Yu X, Bai G, Liu S, Luo N, Wang Y, Richmond DS, et al. Association of candidate genes with drought tolerance traits in diverse perennial ryegrass accessions. Journal of Experimental Botany. 2013;**64**:1537-1551. DOI: 10.1093/jxb/ert018

[7] Yu X, Pijut PM, Byrne S, Asp T, Bai G, Jiang Y. Candidate gene association mapping for winter survival and spring regrowth in perennial ryegrass. Plant Science. 2015;**235**:37-45. DOI: 10.1016/j.plantsci.2015.03.003

[8] Ma X, Zhang J, Huang B. Cytokinin-mitigation of salt-induced leaf senescence in perennial ryegrass involving the activation of antioxidant systems and ionic balance. Environmental and Experimental Botany. 2016;**125**:1-11. DOI: 10.1016/j.envexpbot.2016.01.002

[9] Dabrowski P, Pawluśkiewicz B, Kalaji HM, Baczewska AH. The effect of light availability on leaf area index, biomass production and plant species composition of park grasslands in Warsaw. Plant, Soil and Environment. 2013;**59**:543-548. DOI: 10.17221/140/2013-PSE

[10] Dabrowski P, Pawluśkiewicz B, Baczewska AH, Oglecki P, Kalaji H. Chlorophyll a fluorescence of perennial ryegrass (*Lolium perenne* L.) varieties under long term exposure to shade. Zemdirbyste-Agriculture. 2015;**102**: 305-312. DOI: 10.13080/z-a.2015.102.039

[11] Shah S, Li X, Chen YJ, Zhang L, Fahad S, Hussain S, et al. Silicon application increases drought tolerance of Kentucky bluegrass by improving plant water relations and morphophysiological functions. The Scientific World Journal. 2014. DOI: 10.1155/2014/368694

[12] Muhammad Z, Abdul MK, Abdul MS, Kenneth BM, Muhammad S, Shahen S, et al. Performance of *Aeluropus lagopoides* (mangrove grass) ecotypes, a potential turfgrass, under high saline conditions. Environmental Science and Pollution Research. 2019. DOI: 10.1007/s11356-019-04838-3

[13] Huang B, DaCosta M, Jiang Y. Research advances in mechanisms of turfgrass tolerance to abiotic stresses: From physiology to molecular biology. Critical Reviews in Plant Sciences. 2014;**33**:141-189. DOI: 10.1080/07352689.2014.870411

[14] Fan J, Zhang W, Amombo E, Hu L, Kjorven JO, Chen L. Mechanisms of environmental stress tolerance in turfgrass. Agronomy. 2020;**10**:522. DOI: 10.3390/agronomy10040522

[15] Hafiz MH, Wajid F, Farhat A, Fahad S, Shafqat S, Wajid N, et al. Maize plant nitrogen uptake dynamics at limited irrigation water and nitrogen. Environmental Science and Pollution Research. 2016;**24**(3):2549-2557. DOI: 10.1007/s11356-016-8031-0

[16] Nasim W, Ahmad A, Amin A, Tariq M, Awais M, Saqib M, et al. Radiation efficiency and nitrogen fertilizer impacts on sunflower crop in contrasting environments of Punjab. Pakistan Environmental Science and Pollution Research. 2017;**25**:1822-1836. DOI: 10.1007/s11356-017-0592-z

[17] Aziz K, Daniel KYT, Fazal M, Muhammad ZA, Farooq S, Fan W, et al. Nitrogen nutrition in cotton and control strategies for greenhouse gas emissions: A review. Environmental Science and Pollution Research. 2017;**24**:23471-23487. DOI: 10.1007/s11356-017-0131-y

[18] Aziz K, Daniel KYT, Muhammad ZA, Honghai L, Shahbaz AT, Mir A, et al. Nitrogen fertility and abiotic stresses management in cotton crop: A review. Environmental Science and Pollution Research. 2017;**24**:14551-14566. DOI: 10.1007/s11356-017-8920-x

[19] Hussain MA, Fahad S, Rahat S, Muhammad FJ, Muhammad M, Qasid A, et al. Multifunctional role of brassinosteroid and its analogues in plants. Plant Growth Regulation. 2020. DOI: 10.1007/s10725-020-00647-8

[20] Kamarn M, Wenwen C, Irshad A, Xiangping M, Xudong Z, Wennan S, et al. Effect of paclobutrazol, a potential growth regulator on stalk mechanical strength, lignin accumulation and its relation with lodging resistance of maize. Plant Growth Regulation. 2017;**84**:317-332. DOI: 10.1007/s10725-017-0342-8

[21] Huang S, Jiang S, Liang J, Chen M, Shi Y. Current knowledge of bermudagrass responses to abiotic stresses. Breeding Science. 2019;**69**: 215-226. DOI: 10.1270/jsbbs.18164

[22] Saud S, Chen Y, Fahad S, Hussain S, Na L, Xin L, et al. Silicate application increases the photosynthesis and its associated metabolic activities in Kentucky bluegrass under drought stress and post-drought recovery. Environmental Science and Pollution Research. 2016;**23**(17):17647-17655. DOI: 10.1007/s11356-016-6957-x

[23] Chen Y, Pettersen T, Kvalbein A, Aamlid TS. Playing quality, growth rate, thatch accumulation and tolerance to moss and annual bluegrass invasion as influenced by irrigation strategies on red fescue putting greens. Journal of Agronomy and Crop Science. 2018;**204**:185-195. DOI: 10.1111/jac.12246

[24] Saud S, Chen Y, Long B, Fahad S, Sadiq A. The different impact on the growth of cool season turf grass under the various conditions on salinity and drought stress. International Journal of Agricultural Science and Research. 2013;**3**:77-84

[25] Saud S, Fahad S, Cui G, Chen Y, Anwar S. Determining nitrogen isotopes discrimination under drought stress on enzymatic activities, nitrogen isotope abundance and water contents of Kentucky bluegrass. Scientific

Reports. 2020;**10**:6415. DOI: 10.1038/s41598-020-63548-w

[26] Gong Z, Xiong L, Shi H, Yang S, Herrera-Estrella LR, Xu G, et al. Plant abiotic stress response and nutrient use efficiency. Science China. Life Sciences. 2020;**63**:635-674. DOI: 10.1007/s11427-020-1683-x

[27] Akram R, Turan V, Hammad HM, Ahmad S, Hussain S, Hasnain A, et al. . In: Hashmi MZ, Varma A, editors. Fate of Organic and Inorganic Pollutants in Paddy Soils. Cham, Switzerland; 2018. pp. 197-214

[28] Akram R, Turan V, Wahid A, Ijaz M, Shahid MA, Kaleem S, et al. Paddy land pollutants and their role in climate change. In: Hashmi MZ, Varma A, editors. Environmental Pollution of Paddy Soils, Soil Biology. Cham, Switzerland: Springer International Publishing AG; 2018. pp. 113-124

[29] Depeng W, Fahad S, Saud S, Muhammad K, Aziz K, Mohammad NK, et al. Morphological acclimation to agronomic manipulation in leaf dispersion and orientation to promote "Ideotype" breeding: Evidence from 3D visual modeling of "super" rice (*Oryza sativa* L.). Plant Physiology and Biochemistry. 2018. DOI: 10.1016/j.plaphy.2018.11.010

[30] Fahad S, Chen Y, Saud S, Wang K, Xiong D, Chen C, et al. Ultraviolet radiation effect on photosynthetic pigments, biochemical attributes, antioxidant enzyme activity and hormonal contents of wheat. Journal of Food, Agriculture and Environment. 2013;**11**(3&4):1635-1641

[31] Fahad S, Hussain S, Bano A, Saud S, Hassan S, Shan D, et al. Potential role of phytohormones and plant growth-promoting rhizobacteria in abiotic stresses: Consequences for changing environment. Environmental Science and Pollution Research. 2014;**22**(7):4907-4921. DOI: 10.1007/s11356-014-3754-2

[32] Fahad S, Nie L, Chen Y, Wu C, Xiong D, Saud S, et al. Crop plant hormones and environmental stress. Sustainable Agriculture Reviews. 2015b;**15**:371-400

[33] Zahida Z, Hafiz FB, Zulfiqar AS, Ghulam MS, Fahad S, Muhammad RA, et al. Effect of water management and silicon on germination, growth, phosphorus and arsenic uptake in rice. Ecotoxicology and Environmental Safety. 2017;**144**:11-18

[34] Hesham FA, Fahad S. Melatonin application enhances biochar efficiency for drought tolerance in maize varieties: Modifications in physio-biochemical machinery. Agronomy Journal. 2020:1-22

[35] Habib ur Rahman M, Ashfaq A, Aftab W, Manzoor H, Fahd R, Wajid I, et al. Application of CSM-CROPGRO-Cotton model for cultivars and optimum planting dates: Evaluation in changing semi-arid climate. Field Crops Research. 2017. DOI: 10.1016/j.fcr.2017.07.007

[36] Fahad S, Bajwa AA, Nazir U, Anjum SA, Farooq A, Zohaib A, et al. Crop production under drought and heat stress: Plant responses and management options. Frontiers in Plant Science. 2017;**8**:1147. DOI: 10.3389/fpls.2017.01147

[37] Saud S, Fahad S, Yajun C, Ihsan MZ, Hammad HM, Nasim W, et al. Effects of nitrogen supply on water stress and recovery mechanisms in Kentucky bluegrass plants. Frontiers in Plant Science. 2017;**8**:983. DOI: 10.3389/fpls.2017.00983

[38] Jones MB, Leafe EL, Stiles W. Water stress in field-grown perennial ryegrass. I. Its effect on growth, canopy

photosynthesis and transpiration. Association of Applied Biologists. 1980a;**96**:87-101. DOI: 10.1111/j.1744-7348.1980.Tb04772.x

[39] Guo T, Tian C, Chen C, Duan Z, Zhu Q, Sun L. Growth and carbohydrate dynamic of perennial ryegrass seedlings during PEG-simulated drought and subsequent recovery. Plant Physiology and Biochemistry. 2020;**154**:85-93. DOI: 10.1016/j.plaphy.2020.06.008

[40] Jones MB, Leafe EL, Stiles W. Water stress in field-grown perennial ryegrass. II. Its effect on leaf water status, stomatal resistance and leaf morphology. Association of Applied Biologists. 1980b;**96**:103-110. DOI: 10.1111/j.1744-7348.1980.tb04773.x

[41] Vanani FR, Shabani L, Sabzalian MR, Dehghanian F, Winner L. Comparative physiological and proteomic analysis indicates lower shock response to drought stress conditions in a self-pollinating perennial ryegrass. PLoS One. 2020;**15**(6):e0234317. DOI: 10.1371/journal.pone.0234317

[42] Bray EA. Physiology of plants under stress: Abiotic factors. Field Crops Research. 1998;**55**:192-193. DOI: 10.1016/s0378-4290(97)00069-5

[43] Lawlor DW, Cornic G. Photosynthetic carbon assimilation and associated metabolism in relation to water deficits in higher plants. Plant, Cell & Environment. 2002;**25**:275-294. DOI: 10.1046/j.0016-8025.2001.00814.x

[44] Meena R, Kumar S, Datta R, Lal R, Vijayakumar V, Brtnicky M, et al. Impact of agrochemicals on soil microbiota and management: A review. Land. 2020;**9**(2):34. DOI: 10.3390/land9020034

[45] Molaei A, Lakzian A, Haghnia G, Astaraei A, Rasouli-Sadaghiani M, Teresa Ceccherini M, et al. Assessment of some cultural experimental methods to study the effects of antibiotics on microbial activities in a soil: An incubation study. PLoS One. 2017;**12**(7):e0180663

[46] Datta R, Baraniya D, Wang Y-F, Kelkar A, Meena RS, Yadav GS, et al. Amino acid: Its dual role as nutrient and scavenger of free radicals in soil. Sustainability. 2017;**9**(8):1402

[47] Molaei A, Lakzian A, Datta R, Haghnia G, Astaraei A, Rasouli-Sadaghiani M, et al. Impact of chlortetracycline and sulfapyridine antibiotics on soil enzyme activities. International Agrophysics. 2017;**31**(4):499

[48] Datta R, Baraniya D, Wang YF, Kelkar A, Moulick A, Meena RS, et al. Multi-function role as nutrient and scavenger off reeradical in soil. Sustainability. 2017;**9**:402

[49] Lee MA, Howard-Andrews V, Chester M. Resistance of multiple diploid and tetraploid perennial ryegrass (*Lolium perenne* L.) varieties to three projected drought scenarios for the UK in 2080. Agronomy. 2019;**9**:159. DOI: 10.3390/agronomy9030159

[50] Bothe A, Westermeier P, Wosnitza A, Willner E, Schum A, Dehmer KJ, et al. Drought tolerance in perennial ryegrass (*Lolium perenne* L.) as assessed by two contrasting phenotyping systems. Journal of Agronomy and Crop Science. 2018:1-15. DOI: 10.1111/jac.12269

[51] Thomas H, James AR. Freezing tolerance and solute changes in contrasting genotypes of *Lolium perenne* L. acclimated to cold and drought. Annals of Botany. 1993;**72**:249-254. DOI: 10.1006/anbo.1993.1105

[52] Mohammadi MHS, Etemadi N, Arab MM, Aalifar M, Arab M, Pessarakli M. Molecular and physiological responses of Iranian

perennial ryegrass as affected by trinexapac ethyl, paclobutrazol and abscisic acid under drought stress. Plant Physiology and Biochemistry. 2017;**111**:129-143. DOI: 10.1016/j.plaphy.2016.11.014

[53] Patel M, Milla-Lewis S, Zhang W, Templeton K, Reynolds WC. Over expression of ubiquitin-like *LpHUB1* gene confers drought tolerance in perennial ryegrass. Plant Biotechnology Journal. 2015;**13**:689-699. DOI: 10.1111/pbi.12291

[54] Hu L, Li H, Pang H, Fu J. Responses of antioxidant gene, protein and enzymes to salinity stress in two genotypes of perennial ryegrass (*Lolium perenne*) differing in salt tolerance. Journal of Plant Physiology. 2012;**169**:146-156. DOI: 10.1016/j.jplph.2011.08.020

[55] Norris I. Relationships between growth and measured weather factors among contrasting varieties of *Lolium*, *Dactylis* and *Festuca* species. Grass and Forage Science. 1985;**40**:151-159. DOI: 10.1111/j.1365-2494.1985.tb01732.x

[56] Turner LR, Holloway-Phillips MM, Rawnsley RP, Donaghy DJ, Pembleton KG. The morphological and physiological responses of perennial ryegrass (*Lolium perenne* L.), cocksfoot (*Dactylis glomerata* L.) and tall fescue (*Festuca arundinacea* Schreb.; syn. *Schedonorus phoenix* Scop.) to variable water availability. Grass and Forage Science. 2012;**67**(4):507-518. DOI: 10.1111/j.1365-2494.2012.00866.x

[57] Stiles W, Williams TE. The response of a rye-grass/white clover sward to various irrigation regimes. Journal of Agricultural Science. 1965;**65**:351-364. DOI: 10.1017/S0021859600048917

[58] Van Heeswijck R, McDonald G. *Acremonium endophytes* in perennial ryegrass and other pasture grasses in Australia and New Zealand. Australian Journal of Agricultural Research. 1992;**43**:1683-1709. DOI: 10.1071/ar9921683

[59] Li F, Deng J, Nzabanita C, Li Y, Duan T. Growth and physiological responses of perennial ryegrass to an AMF and an Epichloë endophyte under different soil water contents. Symbiosis. 2019;**75**:151-161. DOI: 10.1007/s13199-019-00633-3

[60] Chen Y, Chen Y, Shi Z, Jin Y, Sun H, Xie FC, et al. Biosynthesis and signal transduction of ABA, JA, and BRs in response to drought stress of Kentucky bluegrass. International Journal of Molecular Sciences. 2019;**20**:1289. DOI: 10.3390/ijms 20061289

[61] Bhatnagar-Mathur P, Vadez V, Sharma KK. Transgenic approaches for abiotic stress tolerance in plants: Retrospect and prospects. Plant Cell Reports. 2008;**27**:411-424. DOI: 10.1007/s00299-007-0474-9

[62] Wehner DJ, Watschke TL. Heat tolerance of Kentucky bluegrasses, perennial ryegrasses, and annual bluegrass. Agronomy Journal. 1981;**73**:79-84. DOI: 10.2134/agronj1981.00021962007300010019x

[63] Harrison J, Tonkinson C, Eagles C, Foyer C. Acclimation to freezing temperatures in perennial ryegrass (*Lolium perenne*). Acta Physiologiae Plantarum. 1997;**19**:505-515. DOI: 10.1007/s11738-997-0047-0

[64] Minner DD, Dernoeden PH, Wehner DJ, Mcintosh MS. Heat tolerance screening of field-grown cultivars of Kentucky bluegrass and perennial ryegrass. Agronomy Journal. 1983;**75**:772-775. DOI: 10.2134/agronj1983.00021962007500050012x

[65] Yang Z, Miao Y, Yu J, Liu J, Huang B. Differential growth and physiological responses to heat stress between two annual and two perennial cool-season

turfgrasses. Scientia Horticulturae. 2014;**170**:75-81. DOI: 10.1016/j.scienta.2014.02.005

[66] Li M, Jannasch AH, Jiang Y. Growth and hormone alterations in response to heat stress in perennial ryegrass accessions differing in heat tolerance. Journal of Plant Growth Regulation. 2019. DOI: 10.1007/s00344-019-10043-w

[67] Kemesyte V, Statkeviciute G, Brazauskas G. Perennial ryegrass yield performance under abiotic stress. Crop Science. 2017;**57**:1935-1940. DOI: 10.2135/cropsci 2016.10.0864

[68] Iwaya-Inoue M, Matsui R, Fukuyama M. Cold- or heat-tolerance of leaves and roots in perennial ryegrass determined by H-NMR. Plant Production Science. 2004;7(2):118-128. DOI: 10.1626/pps.7.118

[69] Imada R, Razmjoo K, Hirano J, Kaneko IR. Response of perennial ryegrass (*Lolium perenne* L.) cultivars to heat stress. Japanese Journal of Grassland Science. 1993;**39**(2):225-235

[70] Soliman WS, Fujimori M, Tase K, Sugiyama S. Oxidative stress and physiological damage under prolonged heat stress in C3 grass *Lolium perenne*. Grassland Science. 2011;**57**:101-106. DOI: 10.1111/j.1744-697X.2011.00214.x

[71] Hulke BS, Watkins E, Wyse DL, Ehlke NJ. Freezing tolerance of selected perennial ryegrass (*Lolium perenne* L.) accessions and its association with field winterhardiness and turf traits. Euphytica. 2007;**163**(1):131-141. DOI: 10.1007/s1 0681-007-9631-z

[72] Yang Z, Zhang Z, Zhang T, Fahad S, Cui K, Nie L, et al. The effect of season-long temperature increases on rice cultivars grown in the central and southern regions of China. Frontiers in Plant Science. 2017;**8**:1908. DOI: 10.3389/fpls.2017.01908

[73] Shah F, Lixiao N, Kehui C, Tariq S, Wei W, Chang C, et al. Rice grain yield and component responses to near 2°C of warming. Field Crops Research. 2013;**157**:98-110

[74] Fahad S, Hussain S, Saud S, Hassan S, Chauhan BS, Khan F, et al. Responses of rapid viscoanalyzer profile and other rice grain qualities to exogenously applied plant growth regulators under high day and high night temperatures. PLoS One. 2016;**11**(7):e0159590. DOI: 10.1371/journal.pone.0159590

[75] Fahad S, Hussain S, Saud S, Hassan S, Ihsan Z, Shah AN, et al. Exogenously applied plant growth regulators enhance the morphophysiological growth and yield of rice under high temperature. Frontiers in Plant Science. 2016;**7**:1250. DOI: 10.3389/fpls.2016.01250

[76] Fahad S, Hussain S, Saud S, Hassan S, Tanveer M, Ihsan MZ, et al. A combined application of biochar and phosphorus alleviates heat-induced adversities on physiological, agronomical and quality attributes of rice. Plant Physiology and Biochemistry. 2016;**103**:191-198

[77] Fahad S, Hussain S, Saud S, Khan F, Hassan S, Jr A, et al. Exogenously applied plant growth regulators affect heat-stressed rice pollens. Journal of Agronomy and Crop Science. 2016;**202**:139-150

[78] Fahad S, Hussain S, Saud S, Tanveer M, Bajwa AA, Hassan S, et al. A biochar application protects rice pollen from high-temperature stress. Plant Physiology and Biochemistry. 2015;**96**:281-287

[79] Fahad S, Muhammad ZI, Abdul K, Ihsanullah D, Saud S, Saleh A, et al. Consequences of high temperature under changing climate optima for rice pollen characteristics-concepts

and perspectives. Archives of Agronomy and Soil Science. 2018. DOI: 10.1080/03650340.2018.1443213

[80] Fahad S, Adnan M, Hassan S, Saud S, Hussain S, Wu C, et al. Rice responses and tolerance to high temperature. In: Hasanuzzaman M, Fujita M, Nahar K, Biswas JK, editors. Advances in Rice Research for Abiotic Stress Tolerance. Cambridge, England: Woodhead Publ Ltd; 2019. pp. 201-224

[81] Pociecha E, Jurczyk B, Dziurka M, Paczyński R, Oklešťková J, Janeczko A. 24-epibrassinolide promotes carbohydrates accumulation in crowns of perennial ryegrass during cold acclimation by regulation of gene expression and enzyme activities which results in increased frost tolerance. Procedia Environmental Sciences. 2015;**29**:234-235. DOI: 10.1016/j.proenv.2015.07.289

[82] Hoffman L, DaCostaa M, Ebdon JS, Zhao J. Effects of drought preconditioning on freezing tolerance of perennial ryegrass. Environmental and Experimental Botany. 2012;**79**:11-20. DOI: 10.1016/j.envexpbot.2012.01.002

[83] Krasensky J, Jonak C. Drought, salt, and temperature stress-induced metabolic rearrangements and regulatory networks. Journal of Experimental Botany. 2012;**63**:1593-1608. DOI: 10.1093/jxb/err460

[84] Du H, Zhou P, Huang B. Antioxidant enzymatic activities and gene expression associated with heat tolerance in a cool-season perennial grass species. Environmental and Experimental Botany. 2013;**87**:159-166. DOI: 10.1016/j.envexpbot.2012.09.009

[85] Wang J, Zhuang L, Zhang J, Yu J, Yang Z, Huang B. Identification and characterization of novel homeodomain leucine zipper (HD-Zip) transcription factors associated with heat tolerance in perennial ryegrass. Environmental and Experimental Botany. 2019;**160**:1-11. DOI: 10.1016/j.envexpbot.2018.12.023

[86] Dai Y, Mao P, Tao X, Wang Y, Wei C, Ma X. Flexible shift on gene body methylation and transcription of *LpCYP72A161* exposed to temperature stress in perennial ryegrass. Environmental and Experimental Botany. 2017;**143**:29-37. DOI: 10.1016/j.envexpbot.2017.08.002

[87] Fahad S, Bano A. Effect of salicylic acid on physiological and biochemical characterization of maize grown in saline area. Pakistan Journal of Botany. 2012;**44**:1433-1438

[88] Fahad S, Hussain S, Matloob A, Khan FA, Khaliq A, Saud S, et al. Phytohormones and plant responses to salinity stress: A review. Plant Growth Regulation. 2014b;**75**(2):391-404. DOI: 10.1007/s10725-014-0013-y

[89] Hu T, Li H, Zhang X, Luo H, Fu J. Toxic effect of NaCl on ion metabolism, antioxidative enzymes and gene expression of perennial ryegrass. Ecotoxicology and Environmental Safety. 2011;**74**:2050-2056. DOI: 10.1016/j.ecoenv.2011.07.013

[90] Hu T, Zhang XZ, Sun JM, Li HY, Fu JM. Leaf functional trait variation associated with salt tolerance in perennial ryegrass. Plant Biology. 2014;**16**:107-116. DOI: 10.1111/plb.12012

[91] Hu T, Jin Y, Li H, Amombo E, Fu J. Stress memory induced transcriptional and metabolic changes of perennial ryegrass (*Lolium perenne*) in response to salt stress. Physiologia Plantarum. 2016;**156**:54-69. DOI: 10.1111/ppl.12342

[92] Pompeiano A, Patrizio ED, Volterrani M, Scartazza A, Guglielminetti L. Growth responses and physiological traits of seashore subjected to short-term salinity stress and recovery. Agricultural Water Management. 2016;**163**:57-65. DOI: 10.1016/j.agwat.2015.09.004

[93] Li X, Han S, Wang G, Liu X, Amombo E, Xie Y, et al. The fungus *Aspergillus aculeatus* enhances salt-stress tolerance, metabolite accumulation, and improves forage quality in perennial ryegrass. Frontiers in Microbiology. 2017;**8**:1664. DOI: 10.3389/fmicb.2017.01664

[94] Wu Y, Chen Q, Chen M, Chen J, Wang X. Salt-tolerant transgenic perennial ryegrass (*Lolium perenne* L.) obtained by *Agrobacterium tumefaciens*-mediated transformation of the vacuolar Na^+/H^+ antiporter gene. Plant Science. 2005;**169**(1):65-73. DOI: 10.1016/j.plantsci.2005.02.030

[95] He A, Niu S, Zhao Q, Li Y, Gou J, Gao H. Induced salt tolerance of perennial ryegrass by a novel bacterium strain from the rhizosphere of a desert shrub *Haloxylon ammodendron*. International Journal of Molecular Sciences. 2018;**19**:469. DOI: 10.3390/ijms19020469

[96] Fahad S, Rehman A, Shahzad B, Tanveer M, Saud S, Kamran M, et al. Rice responses and tolerance to metal/metalloid toxicity. In: Hasanuzzaman M, Fujita M, Nahar K, Biswas JK, editors. Advances in Rice Research for Abiotic Stress Tolerance. Cambridge, England: Woodhead Publ Ltd; 2019. pp. 299-312

[97] Qamar-uz Z, Zubair A, Muhammad Y, Muhammad ZI, Abdul K, Fahad S, et al. Zinc biofortification in rice: Leveraging agriculture to moderate hidden hunger in developing countries. Archives of Agronomy and Soil Science. 2017;**64**:147-161. DOI: 10.1080/03650340.2017.1338343

[98] Frossard R, Stadelmann FX, Niederhauser J. Effects of different heavy metals on fructan, sugar and starch content of ryegrass. Journal of Plant Physiology. 1989;**134**:180-185. DOI: 10.1016/s0176-1617(89)80052-5

[99] Li F, Qiu Y, Xu X, Yang F, Wang Z, Feng J, et al. EDTA-enhanced phytoremediation of heavy metals from sludge soil by Italian ryegrass (*Lolium perenne* L.). Ecotoxicology and Environmental Safety. 2020;**191**:110185. DOI: 10.1016/j.ecoenv.2020.110185

[100] Norini M, Thouin H, Miard F, Battaglia-Brunet F, Gautret P, Guégan R, et al. Mobility of Pb, Zn, Ba, As and Cd toward soil pore water and plants (willow and ryegrass) from a mine soil amended with biochar. Journal of Environmental Management. 2019;**232**:117-130. DOI: 10.1016/j.jenvman.2018.11.021

[101] Zhao S, Liu Q, Qi Y, Duo L. Responses of root growth and protective enzymes to copper stress in turfgrass. Acta Biologica Cracoviensia Series Botanica. 2010;**52**:7-11. DOI: 10.2478/v10182-010-0017-5

[102] Berríos GA, Escobar AL, Alberdi MR, Nunes-Nesi A, Reyes-Díaz MM. Manganese toxicity amelioration by phosphorus supply in contrasting Mn resistant genotypes of ryegrass. Plant Physiology and Biochemistry. 2019;**144**:144-156. DOI: 10.1016/j.plaphy.2019.09.034

[103] Hu Y, Habibul N, Hu Y, Meng F, Zhang X, Sheng G. Mixture toxicity and uptake of 1-butyl-3-methylimidazolium bromide and cadmium co-contaminants in water by perennial ryegrass (*Lolium perenne* L.). Journal of Hazardous Materials. 2020;**386**:121972. DOI: 10.1016/j.jhazmat.2019.121972

[104] Bennet J, Stewart A. The aluminium response network in perennial ryegrass (*Lolium perenne*): II. Water fluxes and ion transport. South African Journal of Plant and Soil. 1999;**16**(1):1-9. DOI: 10.1080/02571862.1999.10634837

[105] Jia H, Hou D, Connor DO, Pan S, Zhu J, Bolan NS, et al. Exogenous

phosphorus treatment facilitates chelation-mediated cadmium detoxification in perennial ryegrass (*Lolium perenne* L.). Journal of Hazardous Materials. 2020;**389**:121849. DOI: 10.1016/j.jhazmat.2019.121849

[106] Luo H, Li H, Zhang X, Fu J. Antioxidant responses and gene expression in perennial ryegrass (*Lolium perenne* L.) under cadmium stress. Ecotoxicology. 2011;**20**:770-778. DOI: 10.1007/s10646-011-0628-y

[107] Hua L, Wang Y, Wu W, McBride MB, Chen Y. Biomass and Cu and Zn uptake of two turfgrass species grown in sludge compost–soil mixtures. Water, Air, and Soil Pollution. 2008;**188**:225-234. DOI: 10.1007/s11270-007-9539-1

[108] Taghizadeh M, Solgi E. Impact of heavy metal stress on in vitro seed germination and seedling growth indices of two turfgrass species. Journal of Rangeland Science. 2017;**7**:220-231

[109] Lou Y, Yang Y, Hu L, Liu H, Xu Q. Exogenous glycinebetaine alleviates the detrimental effect of Cd stress on perennial ryegrass. Ecotoxicology. 2015;**24**:1330-1340. DOI: 10.1007/s10646-015-1508-7

[110] Dong Y, Xu L, Wang Q, Fan Z, Kong J, Bai X. Effects of exogenous nitric oxide on photosynthesis, antioxidative ability, and mineral element contents of perennial ryegrass under copper stress. Journal of Plant Interactions. 2014;**9**:402-411. DOI: 10.1080/17429145.2013.845917

[111] Duo LA, Lian F, Zhao SL. Enhanced uptake of heavy metals in municipal solid waste compost by turfgrass following the application of EDTA. Environmental Monitoring and Assessment. 2010;**165**:377-387. DOI: 10.1007/s10661-009-0953-2

[112] Zhao S, Lian F, Duo L. EDTA-assisted phytoextraction of heavy metals by turfgrass from municipal solid waste compost using permeable barriers and associated potential leaching risk. Bioresource Technology. 2011;**102**:621-626. DOI: 10.1016/j.biortech.2010.08.006

Chapter 18

Advances in Developing Multigene Abiotic and Biotic Stress-Tolerant Rice Varieties

Nitika Sandhu, Shailesh Yadav and Arvind Kumar

Abstract

Increasing incidences of multiple abiotic stresses together with increasing population are the major constraints to attain the global food security. Rice, the major staple food crop is very much prone to various abiotic and biotic stresses, which can occur one at a time or two or more together in a single crop growing season and adversely affects the rice production and productivity. The devastating effect of multiple stresses on rice crop is much more erratic and complex leading to higher losses in the crop grain yield. The concurrent occurrence of multiple streeses can destroy rice production in many of the rainfed areas of South and Southeast-Asia. Genomics-assisted breeding strategies have been instrumental in introgression of various major effect QTLs/genes into rice mega varieties and have proven successful in achieving the desired level of tolerance/resistance to various abiotic stresses in diffferent crop species. Keeping the present scenario of changing climate in mind, the chapter discusses the recent past success in combining tolerance to two or more abiotic stresses in mega rice varieties applying genomics-assisted breeding and development of high-yielding climate resilient rice through stacking of multiple genes/QTLs, which can withstand in a cascade of multiple stresses occurring regularly in rainfed environments.

Keywords: abiotic stress, biotic stress, genomic-assisted breeding, pyramiding, QTLs, rice, yield

1. Introduction

Global warming and the changing climatic conditions lead to the concurrence of multiple abiotic and biotic stresses individually/or in combination [1, 2] thus adversely affecting the rice crop growth and yield [3]. The changing climate, more and more extreme weather events are increasing the probability of simultaneous multiple abiotic stresses, including extra pressure from biotic stresses. Abiotic and biotic stresses reported to have significant negative impact on rice crop survival, growth, development and yield in most parts of the world, especially the Asia and Africa [4, 5]. The abiotic stresses such as drought, salinity, cold, high temperature and heavy metals are known to influence the occurrence of biotic stresses [6–8]. The combined effect of multiple stresses may resulted the minor pests to become the potential threats in the coming future [1, 9, 10].

The rice farming is practiced in various diverse ecological zones. The rice cultivation system in diiferent growing areas are mainly depends upon various factors such as available water, soil type, and the prevailing monsoon. Rice production faces various constraints in various ecology of rice cultivation (**Table 1**). Rice crop faces multiple stresses during different stages of its growth and development and around 70% reduction in yield was reported due to the occurance of abiotic stresses at different stages of growth and development [5]. Similarly, the major biotic stresses such as bacterial leaf blight, blast, brown plant hopper, brown spot, sheath blight and gall midge reported to impart severe crop yield losses or even complete crop failure during infestation [4]. The growth of rice yield has deteriorated from 2.3% per year during 1970s–1980s to about 1.5% during 1990s, and to <1% in the first decads of this present century [11]. Although the rice production has improved considerably over time but it is not sufficient to cope with the increasing demand globally [12]. The annual shortage of rice is expected to rise from 400,000 tons in 2016 to around 800,000 tons by 2030 [13].

As crop plants are immobile, they have to respond to the different abiotic and biotic stresses in the field itself. Breeding efforts in developing tolerance for single stress such as drought, heat, salinity, cold, insect and pathogen or a single stress type viz. abiotic or biotic may be tricky because plants may respond differentially to different or simultaneous occurance of stresses. The increase in resistance/tolerance to one type of stress may be at the cost of resistance/tolerance to another stress [14]. Breeding of high yielding multiple stress tolerant/resistant rice varieties with better grain quality is the urgent need of the hour since many decades [15]. Improvement of germplasm involving improved donors free from undersirable linkage, identification and introgression of genomic regions after validation involving recent advances in genomics-assisted breeding has provided opportunity to combat the challenges arising due the occurance of multiple stresses [16]. An integrated genomics-assisted breeding approach to introgress desirable genes/QTLs conferring tolerance/resistance to major abiotic and biotic stresses in addition to improved yield and quality will help to combat the present situation [16–20]. The commercial use of QTLs/genes-conferred multiple stress tolerant/resistant rice varieties provides an effective, economical and environment friendly approach to protect the crop yield and productivity. In past few years, the identification of genomic regions associated with drought, submergence and heat tolerance and introgression and pyramiding of these regions applying markers assisted selection/backcross approach have successfully led to the development of

Ecosystem	Source of water	Constraints
Upland	Rainfall	Drought, blast, weeds, low soil fertility, Fe toxicity, soil nematode problem, lodging
Rainfed shallow lowland	Rainfall, water table	Lack of assurred irrigation, frequent drought, blast, bacterial leaf blight
Rainfed medium lowland	Rainfall, water table	Lack of assurred irrigation, drought, flood, drought and flood in same or different season, bacterial leaf blight, brown plant hopper, gall midge
Rainfed deep lowland	Rainfall, water table, flood water	Lack of assured irrigation, fragile and low productivity, Prevailing abiotic stresses such as flood, salinity, Biotic stresses such as bacterial leaf blight, gall midge, brown plant hopper
Irrigated	Irrigation	Salinity, bacterial leaf blight, brown plant hopper

Table 1.
Rice production constraints in various ecologies of rice cultivation.

drought or flood tolerant version of some of the mega varieties such as Swarna, IR64 and Sambha Mahsuri. Some of these developed genomics-assisted derived breeding lines have been released as varieties in various countries of South Asia and South East Asia for cultivation.

2. Biotic stresses

2.1 Bacterial blight

Rice crop is most vulnerable to bacterial blight (BLB) caused by *Xanthomonas oryzae* pv. *Oryzae* (*Xoo*). The incidences of BLB reported yield losses of 20–30% and as high 80–100% in moderate and severe infection conditions, respectively [21, 22]. This might be due to the partial grain filling resulted from low photosynthetic activity [23]. Out of already identified 45 BLB resistant (R) genes, 11 genes have been fine mapped and cloned till date applying modern biotechnological [24–27]. The marker-assisted pyramiding approach was applied to pyramid four BLB resistant genes; *Xa4, xa5, xa13* and *Xa21* in background of high yielding rice varieties to achieve wider and durable resistance [28–31]. The combination of these four BLB genes reported as most stable and showed resistance to most of the pathogen isolates [32, 33].

2.2 Blast

Rice blast (*Magnaporthe oryzae*) is another crucial threat to the rice production caused by fungus *Pyricularia oryzae*. It is affecting leaves, collar, nodes, panicles and panicle neck during vegetative to reproductive stage causing 10–30% under mild infestation [34–36] to 70–80% [37] to 100% [38] yield losses under severe infestation conditions. Till now, more than 100 blast resistant genes have been identified, however only 30 of them has been cloned and functionally characterized [17, 39]. The identified linked markers can be used effectively to provide resistance against this devastating fungal disease. The broad-spectrum blast resistant genes *viz. Pi9* which encodes the NBS–LRR gene clusters and *Pita2* which was mapped on the short arm of chromosome 12 confer resistant to many of the blast races in different countries [40, 41]. Transgenic rice lines carrying *Pi-d2* blast resistance gene transformed involving vectors pCB6.3 kb, pZH01–2.72 kb, and pCB5.3 kb showed various levels of resistance (~92%) against 39 strains of rice blast [42]. The durability of rice blast resistance can be further improved by the hybridization of rice varieties carrying complementary genes to attain multi-genic resistance against broad spectrum pathogen races [43], thereby reducing the selection pressure on a single isolate.

2.3 Brown plant hopper

Brown plant hopper (*Nilaparvata lugens*) is one of the most notorious insect-pest of rice causing large scale destructions across Asia amounting to around 60% crop loss [44]. In addition, it is responsible for the transmission of the virus diseases such as rice grassy stunt virus and rice ragged stunt (RRSV) viruses [45, 46]. To date, 37 BPH resistant genes on six of the total twelve rice chromosomes have been reported from cultivated rice and wild *Oryza* species [17, 47, 48]. Out of these 34 genes, 20 genes were fine mapped and only 8 genes (*Bph3*, *Bph14, Bph9, Bph17, Bph26, Bph18, Bph29* and *Bph32*) have been cloned and functionally characterized [49–56]. These resistance genes encode NBS-LRR protein family that are being widely used in marker-assisted breeding programs to develop rice varieties resistance to BPH. However, notable achievements have been made in identification and introgression of BPH resistance genes, rapid evolution of the virulent populations of BPH poses a prime concern. The Srilankan rice cultivar Rathu Heenati was first reported as the potential donor providing resistance against four BPH biotypes [57]. Even after 30 years of deployment in the Philippines, Rathu Heenati still reported to provide resistance to BPH [58]. The cloning and functional characterization of BPH resistant genes provides unique opportunity to effectively use these genes in marker-assisted gene introgression program [53].

2.4 Gall midge

The Asian rice gall midge (*Orseolia oryzae*) is another serious insect pest of rice prevailing mainly in wet season in the South-East Asia, China, and India, while *Orseolia oryzivora*, a closely related species is prevalent in the Africa. The infestation of gall midge (GM) amounting an annual yield loss of $550 million in different countries in Asia [59]. Till date, various genetic studies have identified 11 major resistance (R) genes providing resistance to 7 biotypes of the gall midge of rice which are prevailing mostly in South Asian countries [60–63]. Out of 11, eight of GM resistance genes (*Gm1, Gm2, Gm4, gm3, Gm6, Gm8, Gm7, and Gm11*) have been mapped successfully [64, 65]. Interestingly, none of the identified GM resistant genes confers resistance to all the biotypes of gall midge, while none of the GM biotype is virulent against all the identified resistance genes. Four gall midge resistance genes designated as *Gm1, gm3(NB-ARC), Gm2(NB-ARC),* and *Gm4 (NB-LRR)* have been functionally validated and linked markers can be used for the marker-assisted introgression program [63, 66–68]. Marker-assisted introgression/pyramiding of gall midge resistant genes (*Gm1* + *Gm4*) and (*Gm4* + *Gm8*) in background of improved Samba Mahsuri and an elite rice hybrid DRRH3 respectively was attempted by Divya et al. [62] and Kumar et al. [69]. Further, marker assisted pyramiding of multi-genes conferring to bacterial blight, gall midge, blast along with Saltol QTLs for salinity tolerance was reported [18, 33].

3. Abiotic stresses

3.1 Drought

Among the abiotic stresses, drought is one of the most disruptive, and risky events of the ongoing climate change that affect millions of people every year across the world. Depending upon the intensity and pattern of rainfall, drought can occur from few days to few months or even to years [70]. The development of high yielding drought-tolerant rice varieties is the final goal of rice breeders to reduce the yield losses due to drought and to ensure the projected world food production. However, the development of drought-tolerant rice varieties is immensely tough due to the complex quantitative nature of trait [71, 72]. The selection of lines under differential level of drought and due to the occurrence of drought at different stages [73–75] is again not an easy task. In addition, the strong GxE interactions and low heritability of traits such as grain yield also add to the difficulty of the task [76]. Cost-effective modified breeding strategy involving combined phenotyping and genotyping selection approaches in the development and screening of large segregating populations covering high genetic variation have led to the successful identification of 12 major effect QTLs ($qDTY_{1.1}$ on chromosome 1; $qDTY_{2.1}$, $qDTY_{2.2}$ and $qDTY_{2.3}$ on chromosome 2; $qDTY_{3.1}$ and $qDTY_{3.2}$ on chromosome 3, $qDTY_{4.1}$ on chromosome 4; $qDTY_{6.1}$ and $qDTY_{6.2}$ on chromosome 6, $qDTY_{9.1}$ on chromosome 9, $qDTY_{10.1}$ on chromosome 10, and $qDTY_{12.1}$ on chromosome 12) with consistent effects in background of widely cultivated, popular, high-yielding but drought-susceptible rice varieties, such as IR64, Swarna, Sabitri, MTU1010, Vandana and TDK1 [77–85].

3.2 Salinity

Salinization of soil is an another important crises the world is facing nowadays. Salty soil which is widely distributed across the world is major factor of rice yield reduction. The salt affected land in India accounts for 6.73 mha (million heactare) which is predicted to increase to 16.2 mha by 2050 [86, 87]. The complexity of salt tolerance mechanisms limits the development of high yielding salt tolerance rice varieties [88]. Salinity stress reported to affect rice grain yield from 20 to 100% depends on the severity of the stress and the duration of stress exposed to the rice crop [89]. Fortunately, the exiting wide genetic variability in rice germplasm in response to soil salinity stress makes possible to develop salt tolerant rice varieties [90–92]. The identification and introgression of the trait (s)/genomic regions of interest are the well-known approaches for the development of salinity tolerant varieties [93, 94]. Marker assisted breeding approaches have been proven successful in developing new improved, high yielding salt tolerance rice varieties [95–100].

3.3 High and low temperature

Global changes in the climate conditions and increasing greenhouse gas emission led to a rise in earth's surface temperature in some past decades, and the temperature is predicted to rise by 2 to 4°C by 2050 [101]. The high temperature duration of 3–5 days, 5–7 days and above 8 days is generally considered as mild, moderate and severe heat injury, respectively [102] while low temperature ranged from 0 to 15°C and <0°C categorized as chilling and freezing stress, respectively. Over the past few decades, extensive efforts have been made in identification of genes/QTLs improving heat [103–105] and cold tolerance [106, 107] in rice, which are very complex trait.

4. Marker-assisted pyramiding of multiple QTLs/genes for abiotic/biotic stresses

The challenges from the climate change scenario require the development of climate-adapted rice varieties that combine the tolerance of various abiotic and biotic stresses to better sustain yield losses from unpredicted climate-related events. Recent developments in the identification of major QTLs/genes for drought, submergence, salinity, bacterial blight, brown plant hopper, gall midge, and blast and the successful introgression of identified QTLs to develop improved varieties tolerant of different individual stresses indicate that, with the advent of new marker technology, the development of varieties that combine tolerance of the various abiotic and biotic stresses prevalent in any region is feasible. Such varieties once developed can help farmers overcome yield losses and better farm income under the changed climatic conditions.

The identification of major effect QTLs for the grain yield under drought $qDTY_{12.1}$ [77], $qDTY_{3.1}$ [80], and $qDTY_{1.1}$ [78]; *Sub1*, the gene for submergence [108, 109]; and Saltol, the QTL for salinity [110, 111] using modern breeding tools has provided novel opportunities to the breeders to develop the rice varieties

Variety	QTLs/gene combinations	Targeted trait	Targeted country	Year of release
DRR dhan-42	$qDTY_{2.2}$ + $qDTY_{4.1}$	Drought	India	2014
Yaenelo 4	$qDTY_{2.2}$ + $qDTY_{4.1}$	Drought	Myanmar	2015
Yaenelo 5	$qDTY_{2.2}$ + $qDTY_{4.1}$	Drought	Myanmar	2016
Yaenelo 7	$qDTY_{2.2}$ + $qDTY_{4.1}$	Drought	Myanmar	2016
CR dhan-801	$qDTY_{1.1}$ + $qDTY_{2.1}$ + $qDTY_{3.1}$ + *Sub 1*	Drought + flood	India	2017
Bahuguni dhan-2	$qDTY_{3.1}$ + *Sub 1*	Drought + flood	Nepal	2017
Bahuguni dhan-1	$qDTY_{1.1}$ + $qDTY_{3.1}$ + *Sub 1*	Drought + flood	Nepal	2017

Table 2.
Marker-assisted derived variety released in various countries of South Asia for tolerance to drought and flood in rice.

tolerant of abiotic stresses. Marker-assisted backcrossing of *Sub1* to improve popular variety Swarna had been successfully demonstrated to provide a yield advantage under submerged conditions for up to 18 days [112]. Similarly, the development of drought-tolerant versions of IR64 and Vandna [84, 113] at IRRI that are shown to possess a 1.0 and 0.5 t/ha yield advantage under drought over IR64 and Vandna, respectively, and successful introgression of Saltol into BR28 [114] are some of the recent successes using marker-assisted introgression.

Marker assisted gene pyramiding is an effective breeding strategy to transfer more than one tolerance/resistance genes into a single rice line in order to achieve durable and broader resistance level which can prevent the breakdown of tolerance/resistance against specific races/pathogens [115]. Pyramiding of BLB resistant genes such as *xa5* + *xa13* + *Xa21* [30, 31], *Xa4* + *xa5* + *Xa21* [29] and *Xa4* + *xa5* + *xa13* + *Xa21* [70] had been reported to provide durable resistance in rice against bacterial blight disease. Pyramiding of blast resistance genes *Pi9* and *Pita* has proven effective to combat the blast incidence and increase the durability of blast resistance genes [116].

Rice lines pyramided with multiple disease resistance genes (*Xa4, Xa21, xa5, Bph18* and *Pi40,*) has conferred resistance against BLB, blast, and BPH disease [29]. Among abiotic stresses, recently, drought and flood tolerance were combined using marker assisted pyramiding of the drought QTLs ($qDTY_{1.1}$ + $qDTY_{2.1}$ + $qDTY_{3.1}$) and submergence gene (*Sub1)* together in a popular rice variety, Swarna [19]. The marker-assisted derived rice varieties have been released in different countries (**Table 2**).

5. QTLs/gene pyramiding through multiple parents crossing

To tackle the multiple problems of rice cultivation under ongoing climate change, a high yielding climate smart new rice lines with superior grain quality is the urgent need to intensify the sustainable rice production. Genomics-assisted breeding (GAB) was attempted to introgress and assemble multiple QTL/genes- $qDTY_{1.1}$, $qDTY_{2.1}$, $qDTY_{3.1}$, $qDTY_{12.1}$, *Sub1, Gm4, Pi9, Pita2, Bph3, Bph17, Xa4, xa5, xa13, Xa21* and *Xa23* into the background of a high yielding breeding line suited for lowland ecosystem of rice under Stress Tolerant Rice for Africa and South

Asia(STRASA) project at IRRI. Developed climate resilient rice lines carrying 6–10 QTLs/genes combinations for tolerance to multiple biotic (BLB-*Xa4, xa5, Xa21, xa13* and *Xa23*; Blast-*Pita2*, *Pi9*; BPH-*Bph3 & Bph17* and gall midge-*Gm4*) and abiotic stresses (yield under drought stress- $qDTY_{1.1}$, $qDTY_{2.1}$, $qDTY_{3.1}$, $qDTY_{12.1}$ and submergence- *Sub1*) as well as superior grain quality traits are free from undesirable linkage drags and can be released as varieties on different countries after evaluation in the national system or can be used as an elite parental lines for making diverse crosses targeted to achieve high genetic gain [117]. Recently, introgression lines with 7 to 10 QTLs/genes for multiple-biotic stresses (blast, BLB, BPH and GM) with drought QTLs in background of Swarna has been reported [14]. A MAGIC (multi-parent advanced generation intercross) population developmental strategy has also been proposed to examine the effect of multiple alleles to provide high grain yield, better grain quality, and tolerance to a wide range of multiple biotic and abiotic stresses [118].

An increase in rice productivity through introgression of multiple traits which can improve rice adaptability under dry direct seeded (DSR) and additionally carrying traits for abiotic/biotic stresses looks a promising breeding strategy to adapt with changing climate, limited water and labor resources and increase rice yield under mechanized DSR conditions. QTLs for traits that increase adaptability to direct seeded rice conditions such as root traits [nodal root number ($qNR_{4.1}$, $qNR_{5.1}$) and root hair density ($qRHD_{1.1}$, $qRHD_{5.1}$, $qRHD_{8.1}$)], early vegetative vigor ($qEVV_{9.1}$), early uniform emergence ($qEUE_{1.1}$, $qEUE_{11.1}$), grain yield under DSR conditions ($qGY_{1.1}$, $qGY_{8.1}$, $qGY_{10.1}$), and lodging resistance ($qLDG_{4.1}$) had been pyramided with abiotic stresses (drought QTLs-$qDTY_{1.1}$, $qDTY_{2.1}$, $qDTY_{3.1}$ and $qDTY_{12.1}$) as well as biotic resistance (gall midge-*Gm4*, blast- *Pi9*, *Pita 2*, bacterial leaf blight- *Xa4*, *Xa21, xa5*, *xa13*, and brown plant hoppers-*Bph3* and *Bph17*) using MAS approach [119, 120].

5.1 Steps in multiple-trait breeding

Three steps involved in multiple traits introgression scheme, (a) assemble first (b) line fixation and (c) line evaluation. In assemble first step, a complex crossing scheme utilized in transferring the desirable alleles/traits from all the targeted parents aimed to accumulate one copy of all targeted genes/QTLs in a single genotype. In line fixation step, gene based/SSRs/linked markers were utilized in each generation from F_2 to F_6 generation for tracking the presence of desirable alleles of targeted QTLs/in order to find homozygous plants carrying all the targeted QTLs/genes. Phenotyping of the homozygous lines for the targeted traits were performed in line evaluation step and proceed further for multilocation testing of promising lines in the targeted environments. The detailed description on traits, donors, QTLs/genes and markers associated that were used in genomic-assisted breeding program for the development of climate resilient lines at IRRI, Philippines (**Table 3**).

5.2 Challenges in multiple-trait breeding

The complex breeding program that targets combining tolerance of various abiotic and biotic stresses together in the various genetic backgrounds is unpredictable and more research is needed as such, little is known about the effect that each gene/QTL on the others. However, it is highly assumed that most of the QTLs/genes should work in an additive manner, as far as the targeted QTLs/genes are either located on different chromosomes or in different regions of the same chromosomes. Genomic interactions play a significant role in deciding the performance

Trait	Donor	QTLs/genes	Markers (SNPs/Indels/SSRs/gene based markers)	Reference
Biotic stress				
Blast	WHD-1S-75-1-127, Tadukan, IRBL9	*Pi9*, *Pita2*	*Pi9: Pi9*STS2, Pi9-659T, Pi9-1477G, MSU7_6_10381500 (M492 + M493), M891 (C), Pi9-659T, Pi9-1477G *Pita2:* MSU7_12_9177624 (M535 + M536), SnpOS00488(G), YL155/YL87, YL153/YL154	[40, 117, 121]
Bacterial leaf blight	IRBB60,	*Xa4, xa5, xa13, Xa21, Xa23*	*Xa4:* snpOS0054 (AG), RM224, MP1 + MP2 *xa5: xa5S, xa5R, xa5*DRR *xa13: xa13*-promoter (M478Lm + M479Lm + M480Lm), *xa13*F_130-147/*xa13* R_1678-1662 *Xa21:* Xa21s_exon (M769 + M770), snpOS0061 (C), U1/I1, M1207 (T), pTA248	[117, 122]
Brown plant hopper	Rathu Heenati	*Bph3, Bph17*	RM589,RM586,RM190, RM8213, RM16556, RM586, RM589, RM190, RM7639, RM19311(linked markers)	[14, 117, 123]
Gall midge	Abhaya	*Gm4*	GM4_LRR-del_F, GM4_LRR-del_R	[49, 117, 124]
Abiotic stress				
Drought + submergence	IR96322-34-223-B	$qDTY_{1.1}$ + $qDTY_{2.1}$ + $qDTY_{3.1}$ + *Sub1*	$qDTY_{1.1:}$ RM431, RM11943, RM12233(linked markers), snpOS0071 (A), snpOS0074 (G) $qDTY_{2.1:}$ RM324, RM3549, RM12868, RM12987, RM12995(linked markers), snpOS0078 (A), snpOS0079 (A) $qDTY_{3.1:}$ RM520, RM16030, RM416(linked markers), snpOS0085 (G), snpOS0089 (C) *Sub1:* ART5, snpOS0040 (T)	[19, 117]
Drought	IR74371-46-1-1	$qDTY_{12.1}$	RM28099, RM28166, Indel 8, SnpOS00483(G), SnpOS00484(A)	[77, 117]
Cold	IR 83222-8-1-1-1-1-1-1, IR 66160-121-4-4-2, HGKN	*qCTS4a, qCTS11.1*	*qCTS4a:* RM349, RM17604, RM17623, RM3648, RM2799 *qCTS11:* RM26889, RM21, RM206	[125, 126]
Heat	N22/IR64	*qHTSF4.1, qHTSF4.2*,	id4005120, id4011562	[104]
Salinity	Pokkali/IR29	*Saltol*	G11A, AP3206, RM3412, RM493	[110, 127]

Table 3.
List of traits, donors, QTLs/genes, and markers associated used in crossing program for the development of climate resilient lines at IRRI, Philippines.

of introgression lines pyramided with various drought grain yield QTLs in rice [128–131]. In some cases, epistatic interactions between different loci can enhance or reduce the effect of some of the genes/QTLs. Under such a situation, as an alternative strategy of identifying and advancing lines with different combinations of genes/QTLs- six, seven, or eight for different stresses and showing higher grain yield under nonstress conditions will also be selected and the best combinations that will show tolerance of a maximum number of abiotic and biotic stresses and the highest yield advantage will be advanced for testing. Plants carrying maximum number QTLs/gene but having negative interaction with grain yield and showing inferior plant type can be rejected for further advancement.

Maintenance of larger population size could also be a feasible strategy which can allow to select rare recombinants having maximum number of targeted QTLs/genes and also free from undesirable linkages. In previous studies, drought tolerant rice lines were developed through successful breakage of the linkages between loci for tolerance to drought and undesirable traits by fine mapping and maintain huge population size [132].

6. Role of genomic selection in multiple-trait breeding

Genomic selection (GS) in crop plants facilitates the rapid selection of superior accessions/genotypes and accelerate the breeding cycle targeted for higher genetic gain. It aims to use the genome-wide markers to predict the effects of all associated loci. The developed best prediction model is applied to the tested breeding material which has been charactized only genotypically but not phenotypically. The breeding estimated value called as GEBV (genomic estimated breeding value). The parental lines with higher GEBV can be selected as the candidate lines for future breeding programs. Most of the previous studied in cereal crops has shown great potential for GS to enhance the selection for grain yield and yield related traits [133, 134]. Multi trait genomic selection can be also implemented on phenotypic data of multiple traits *viz.* grain quality traits, grain yield and yield components, and reaction to the biotic and abiotic stresses, however it is important that a favorable genetic correlation exists between traits to implement genomic prediction model effectively [135, 136].

7. Conclusions

To solve the global issue of food security in the era of changing climate, novel approaches involving successful stacking of multiple genes/QTLs in a single rice line utilizing strategic phenotypic-genotypic selection could provide opportunity targeting genetic gain in rice. New advances in hybridization stratgies, genomics, marker development, and sequencing permitted the opportunity to create muti-gene carrying high-yielding rice varieties to combat multiple stresses. The development of rice varieties carrying multiple QTLs/genes in homozygous conditions can address the production constraints faced due to both biotic and abiotic stresses simultaneously. These stress-tolerant rice varieties with desired grain quality can greatly help farmers in improving productivity under multiple stress conditions.

Conflict of interest

The authors declare that they have no conflict of interest.

Author details

Nitika Sandhu[1,2], Shailesh Yadav[2,3] and Arvind Kumar[2,3]*

1 Punjab Agricultural University, Ludhiana, Punjab, India

2 International Rice Research Institute, Philippines

3 IRRI South Asia Regional Centre (ISARC), Varanasi, Uttar Pradesh, India

*Address all correspondence to: a.kumar@irri.org

References

[1] Prasad PV, Pisipati SR, Momčilović I, Ristic Z. Independent and combined effects of high temperature and drought stress during grain filling on plant yield and chloroplast EF-Tu expression in spring wheat. Journal of Agronomy and Crop Science. 2011;**197**(6):430-441. DOI: 10.1111/j.1439-037X.2011.00477.x

[2] Pandey P, Ramegowda V, Senthil-Kumar M. Shared and unique responses of plants to multiple individual stresses and stress combinations: Physiological and molecular mechanisms. Frontiers in Plant Science. 2015;**6**:723. DOI: 10.3389/fpls.2015.00723

[3] Narsai R, Wang C, Chen J, Wu J, Shou H, Whelan J. Antagonistic, overlapping and distinct responses to biotic stress in rice (*Oryza sativa*) and interactions with abiotic stress. BMC Genomics. 2013;**14**(1):93. DOI: 10.1186/1471-2164-14-93

[4] Hasan MM, Rafii MY, Ismail MR, Mahmood M, Rahim HA, Alam MA, et al. Marker-assisted backcrossing: A useful method for rice improvement. Biotechnology and Biotechnological Equipment. 2015;**29**(2):237-254. DOI: 10.1080/13102818.2014.995920

[5] Akram R, Fahad S, Masood N, Rasool A, Ijaz M, Ihsan MZ, et al. Plant growth and morphological changes in rice under abiotic stress. In: Advances in Rice Research for Abiotic Stress Tolerance. Utah, UT: Woodhead Publishing, Elsevier Inc; 2019. pp. 69-85

[6] Ziska LH, Tomecek MB, Gealy DR. Competitive interactions between cultivated and red rice as a function of recent and projected increases in atmospheric carbon dioxide. Agronomy Journal. 2010;**102**(1):118-123. DOI: 10.2134/agronj2009.0205

[7] Peters K, Breitsameter L, Gerowitt B. Impact of climate change on weeds in agriculture: A review. Agronomy for Sustainable Development. 2014;**34**:707-721. DOI: 10.1007/s13593-014-0245-2

[8] McDonald A, Riha S, DiTommaso A, DeGaetano A. Climate change and the geography of weed damage: Analysis of US maize systems suggests the potential for significant range transformations. Agriculture, Ecosystems and Environment. 2009;**130**(3-4):131-140. DOI: 10.1016/j.agee.2008.12.007

[9] Jedmowski C, Ashoub A, Momtaz O, Brüggemann W. Impact of drought, heat, and their combination on chlorophyll fluorescence and yield of wild barley (*Hordeum spontaneum*). Journal of Botany. 2015;**9**:120868. DOI: 10.1155/2015/120868

[10] Duveiller E, Singh RP, Nicol JM. The challenges of maintaining wheat productivity: Pests, diseases, and potential epidemics. Euphytica. 2007;**157**(3):417-430. DOI: 10.1007/s10681-007-9380-z

[11] Khush GS. Strategies for increasing the yield potential of cereals: Case of rice as an example. Plant Breeding. 2013;**132**(5):433-436. DOI: 10.1111/pbr.1991

[12] Sasaki T, Burr B. International Rice genome sequencing project: The effort to completely sequence the rice genome. Current Opinion in Plant Biology. 2000;**3**(2):138-142. DOI: 10.1016/S1369-5266(99)00047-3

[13] Thirze H. Modelling grain surplus and deficit in Cameroon for 2030 [Master's thesis]. Lund, Sweden: Lund University; 2016. p. 59

[14] Dixit S, Singh UM, Singh AK, Alam S, Venkateshwarlu C, Nachimuthu VV, et al. Marker assisted forward breeding to combine multiple biotic-abiotic stress resistance/tolerance

in Rice. Rice. 2020;**13**(1):1-15. DOI: 10.1186/s12284-020-00391-7

[15] Khush GS. What it will take to feed 5.0 billion rice consumers in 2030. Plant Molecular Biology. 2005;**59**:1-6. DOI: 10.1007/s11103-005-2159-5

[16] Kumar A, Sandhu N, Dixit S, Yadav S, Swamy BPM, Shamsudin NAA. Marker-assisted selection strategy to pyramid two or more QTLs for quantitative trait-grain yield under drought. Rice. 2018;**11**:35. DOI: 10.1186/s12284-018-0227-0

[17] Wang Y, Jiang W, Liu H, Zeng Y, Du B, Zhu L, et al. Marker assisted pyramiding of *Bph6* and *Bph9* into elite restorer line 93-11 and development of functional marker for *Bph9*. Rice. 2017;**10**:51. DOI: 10.1186/s12284-017-0194-x

[18] Das G, Rao GJ, Varier M, Prakash A, Prasad D. Improved Tapaswini having four BB resistance genes pyramided with six genes/QTLs, resistance/tolerance to biotic and abiotic stresses in rice. Scientific Reports. 2018;**8**:2413. DOI: 10.1038/s41598-018-20495-x

[19] Sandhu N, Dixit S, Swamy BP, Raman A, Kumar S, Singh SP, et al. Marker assisted breeding to develop multiple stress tolerant varieties for flood and drought prone areas. Rice. 2019;**12**:8. DOI: 10.1186/s12284-019-0269-y

[20] Muthu V, Abbai R, Nallathambi J, Rahman H, Ramasamy S, Kambale R, et al. Pyramiding QTLs controlling tolerance against drought, salinity, and submergence in rice through marker assisted breeding. PLoS One. 2020;**15**:1-18. DOI: 10.1371/journal.pone.0227421

[21] Mew TW, Vera Cruz CM, Medalla ES. Changes in race frequency of Xanthomonas oryzae pv. Oryzae in response to rice cultivars planted in the Philippines. Plant Disease. 1992;**76**(10):1029-1032

[22] Noh TH, Lee DK, Park JC, Shim HK, Choi MY, Kang MH, et al. Effects of bacterial leaf blight occurrence on rice yield and grain quality in different rice growth stage. Research in Plant Disease. 2007;**13**(1):20-23

[23] Ou SH. Rice Diseases. 2nd ed. Kew, Surrey (GB): Commonwealth Mycological Institute; 1985

[24] Neelam K, Mahajan R, Gupta V, Bhatia D, Gill BK, Komal R, et al. High-resolution genetic mapping of a novel bacterial blight resistance gene xa-45 (t) identified from *Oryza glaberrima* and transferred to *Oryza sativa*. Theoretical and Applied Genetics. 2020;**133**(3):689-705. DOI: 10.1007/s00122-019-03501-2

[25] Cheema KK, Grewal NK, Vikal Y, Sharma R, Lore JS, Das A, et al. A novel bacterial blight resistance gene from *Oryza nivara* mapped to 38 kb region on chromosome 4L and transferred to *Oryza sativa* L. Genetics Research. 2008;**90**(5):397-407. DOI: 10.1017/S0016672308009786

[26] Sun X, Yang Z, Wang S, Zhang Q. Identification of a 47-kb DNA fragment containing Xa4, a locus for bacterial blight resistance in rice. Theoretical and Applied Genetics. 2003;**106**(4):683-687. DOI: 10.1007/s00122-002-1117-8

[27] Gu K, Yang B, Tian D, Wu L, Wang D, Sreekala C, et al. R gene expression induced by a type-III effector triggers disease resistance in rice. Nature. 2005;**435**(7045):1122-1125. DOI: 10.1038/nature03630

[28] Sundaram RM, Vishnupriya MR, Biradar SK, Laha GS, Reddy GA, Rani NS, et al. Marker assisted introgression of bacterial blight resistance in samba Mahsuri, an elite indica rice variety. Euphytica.

2008;**160**(3):411-422. DOI: 10.1007/s10681-007-9564-6

[29] Suh JP, Jeung JU, Noh TH, Cho YC, Park SH, Park HS, et al. Development of breeding lines with three pyramided resistance genes that confer broad-spectrum bacterial blight resistance and their molecular analysis in rice. Rice. 2013;**6**(1):5. DOI: 10.1186/1939-8433-6-5

[30] Dokku P, Das KM, Rao GJ. Pyramiding of four resistance genes of bacterial blight in Tapaswini, an elite rice cultivar, through marker-assisted selection. Euphytica. 2013;**192**(1):87-96. DOI: 10.1007/s10681-013-0878-2

[31] Pradhan SK, Nayak DK, Mohanty S, Behera L, Barik SR, Pandit E, et al. Pyramiding of three bacterial blight resistance genes for broad-spectrum resistance in Deepwater rice variety. Jalmagna. Rice. 2015;**8**(1):19. DOI: 10.1186/s12284-015-0051-8

[32] Shanti ML, Shenoy VV. Evaluation of resistance genes and their pyramids against rice bacterial leaf blight pathogen Xanthomonas oryzae pv. Oryzae. Oryza. 2005;**42**(3):169-173

[33] Das G, Rao GJ. Molecular marker assisted gene stacking for biotic and abiotic stress resistance genes in an elite rice cultivar. Frontiers in Plant Science. 2015;**6**:698. DOI: 10.3389/s00698

[34] Wilson RA, Talbot NJ. Under pressure: Investigating the biology of plant infection by Magnaporthe oryzae. Nature Reviews. Microbiology. 2009;7(3):185-195. DOI: 10.1038/nrmicro2032

[35] Ashkani S, Yusop MR, Shabanimofrad M, Harun AR, Sahebi M, Latif MA. Genetic analysis of resistance to rice blast: A study on the inheritance of resistance to the blast disease pathogen in an F3 population of rice. Journal of Phytopathology. 2015;**163**(4):300-309. DOI: 10.1111/jph.12323

[36] Sakulkoo W, Osés-Ruiz M, Garcia EO, Soanes DM, Littlejohn GR, Hacker C, et al. A single fungal MAP kinase controls plant cell-to-cell invasion by the rice blast fungus. Science. 2018;**359**(6382):1399-1403. DOI: 10.1126/science.aaq0892

[37] Zhu YY, Fang H, Wang YY, Fan JX, Yang SS, Mew TW, et al. Panicle blast and canopy moisture in rice cultivar mixtures. Phytopathology. 2005;**95**(4):433-438. DOI: 10.1094/PHYTO-95-0433

[38] Musiime O, Tenywa MM, Majaliwa MJ, Lufafa A, Nanfumba D, Wasige JE, et al. Constraints to rice production in Bugiri district. African Crop Science Conference Proceedings. 2005;7(03):1495-1499

[39] Zhao H, Wang X, Jia Y, Minkenberg B, Wheatley M, Fan J, et al. The rice blast resistance gene Ptr encodes an atypical protein required for broad-spectrum disease resistance. Nature Communications. 2018;**9**(1):1-2. DOI: 10.1038/s41467-018-04369-4

[40] Qu S, Liu G, Zhou B, Bellizzi M, Zeng L, Dai L, et al. The broad-spectrum blast resistance gene Pi9 encodes a nucleotide-binding site–leucine-rich repeat protein and is a member of a multigene family in rice. Genetics. 2006;**172**(3):1901-1914. DOI: 10.1534/genetics.105.044891

[41] Meng X, Xiao G, Telebanco-Yanoria MJ, Siazon PM, Padilla J, Opulencia R, et al. The broad-spectrum rice blast resistance (R) gene Pita2 encodes a novel R protein unique from Pita. Rice. 2020;**13**(1):1-5. DOI: 10.1186/s12284-020-00377-5

[42] Chen DX, Chen XW, Wang YP, Zhu LH, Li SG. Genetic transformation

of rice with Pi-d2 gene enhances resistance to rice blast fungus *Magnaporthe Oryzae*. 2010;**17**(1):19-27. DOI: 10.1016/S1672-6308(08)60100-6

[43] Hittalmani S, Parco A, Mew TV, Zeigler RS, Huang N. Fine mapping and DNA marker-assisted pyramiding of the three major genes for blast resistance in rice. Theoretical and Applied Genetics. 2000;**100**(7):1121-1128

[44] Srivastava C, Chander S, Sinha SR, Palta RK. Toxicity of various insecticides against Delhi and Palla populations of brown planthopper (*Nilaparvata lugens*). Indian Journal of Agricultural Sciences. 2009;**79**:1003-1006

[45] Normile D. Reinventing rice to feed the world. Science. 2008;**321**(5887):330-333. DOI: 10.1126/science.321.5887.330

[46] Cabauatan PQ, Cabunagan RC, Choi IR. Rice viruses transmitted by the brown planthopper Nilaparvata lugens Stål. In: Heong KL, Hardy B, editors. Planthoppers: New Threats to the Sustainability of Intensive Rice Production Systems in Asia. Los Baños: International Rice Research Institute; 2009. pp. 357-368

[47] Li Z, Xue Y, Zhou H, Li Y, Usman B, Jiao X, et al. High-resolution mapping and breeding application of a novel brown planthopper resistance gene derived from wild rice (*Oryza. rufipogon* Griff). Rice. 2019;**12**(1):41. DOI: 10.1186/s12284-019-0289-7

[48] Yang M, Cheng L, Yan L, Shu W, Wang X, Qiu Y. Mapping and characterization of a quantitative trait locus resistance to the brown planthopper in the rice variety IR64. Hereditas. 2019;**156**(1):22. DOI: 10.1186/s41065-019-0098-4

[49] Sun L, Su C, Wang C, Zhai H, Wan J. Mapping of a major resistance gene to the brown planthopper in the rice cultivar Rathu Heenati. Breeding Science. 2005;**55**(4):391-396

[50] Qiu Y, Guo J, Jing S, Zhu L, He G. Development and characterization of japonica rice lines carrying the brown planthopper-resistance genes *BPH12* and *BPH6*. Theoretical and Applied Genetics. 2012;**124**(3):485-494. DOI: 10.1007/s00122-011-1722-5

[51] Lv W, Du B, Shangguan X, Zhao Y, Pan Y, Zhu L, et al. BAC and RNA sequencing reveal the brown planthopper resistance gene BPH15 in a recombination cold spot that mediates a unique defense mechanism. BMC Genomics. 2014;**15**(1):674. DOI: 10.1186/1471-2164-15-674

[52] Wang Y, Cao L, Zhang Y, Cao C, Liu F, Huang F, et al. Map-based cloning and characterization of BPH29, a B3 domain-containing recessive gene conferring brown planthopper resistance in rice. Journal of Experimental Botany. 2015;**66**(19):6035-6045. DOI: 10.1093/jxb/erv318

[53] Liu Y, Wu H, Chen H, Liu Y, He J, Kang H, et al. A gene cluster encoding lectin receptor kinases confers broad-spectrum and durable insect resistance in rice. Nature Biotechnology. 2015;**33**(3):301-305. DOI: 10.1038/nbt.3069

[54] Ren J, Gao F, Wu X, Lu X, Zeng L, Lv J, et al. Bph32, a novel gene encoding an unknown SCR domain-containing protein, confers resistance against the brown planthopper in rice. Scientific Reports. 2016;**6**:1-14. DOI: 10.1038/srep37645

[55] Zhao Y, Huang J, Wang Z, Jing S, Wang Y, Ouyang Y, et al. Allelic diversity in an NLR gene *BPH9* enables rice to combat planthopper variation. Proceedings of the National Academy of Sciences. 2016;**113**(45):12850-12855. DOI: 10.1073/pnas.1614862113

[56] Guo J, Xu C, Wu D, Zhao Y, Qiu Y, Wang X, et al. Bph6 encodes an exocyst-localized protein and confers broad resistance to planthoppers in rice. Nature Genetics. 2018;**50**(2):297-306. DOI: 10.1038/s41588-018-0039-6

[57] Lakshminarayana A, Khush GS. New genes for resistance to the brown planthopper in rice. Crop Science. 1977;**17**(1):96-100

[58] Cruz PA, Arida A, Heong KL, Horgan FG. Aspects of brown planthopper adaptation to resistant rice varieties with the Bph3 gene. Entomologia Experimentalis et Applicata. 2011;**141**(3):245-257. DOI: 10.1111/j.1570-7458.2011.01193.x

[59] Biradar SK, Sundaram RM, Thirumurugan T, Bentur JS, Amudhan S, Shenoy VV, et al. Identification of flanking SSR markers for a major rice gall midge resistance gene Gm1 and their validation. Theoretical and Applied Genetics. 2004;**109**(7):1468-1473. DOI: 10.1007/s00122-004-1763-0

[60] Kumar A, Jain A, Sahu RK, Shrivastava MN, Nair S, Mohan M. Genetic analysis of resistance genes for the rice gall midge in two rice genotypes. Crop Science. 2005;**45**(4):1631-1635. DOI: 10.2135/cropsci2004.0406

[61] Himabindu K, Suneetha K, Sama VS, Bentur JS. A new rice gall midge resistance gene in the breeding line CR57-MR1523, mapping with flanking markers and development of NILs. Euphytica. 2010;**174**(2):179-187. DOI: 10.1007/s10681-009-0106-2

[62] Divya D, Himabindu K, Nair S, Bentur JS. Cloning of a gene encoding LRR protein and its validation as candidate gall midge resistance gene, *Gm4*, in rice. Euphytica. 2015;**203**(1):185-195. DOI: 10.1007/s10681-014-1302-2

[63] Bentur JS, Rawat N, Divya D, Sinha DK, Agarrwal R, Atray I, et al. Rice–gall midge interactions: Battle for survival. Journal of Insect Physiology. 2016;**84**:40-49. DOI: 10.1016/j.jinsphys.2015.09.008

[64] Yasala AK, Rawat N, Sama VK, Himabindu K, Sundaram RM, Bentur JS. In silico analysis for gene content in rice genomic regions mapped for the gall midge resistance genes. Plant Omics. 2012;**5**(4):405-413

[65] Sama VS, Rawat N, Sundaram RM, Himabindu K, Naik BS, Viraktamath BC, et al. A putative candidate for the recessive gall midge resistance gene gm3 in rice identified and validated. Theoretical and Applied Genetics. 2014;**127**(1):113-124. DOI: 10.1007/s00122-013-2205-7

[66] Sundaram RM. Fine Mapping of Rice Gall Midge Resistance Genes Gm1 and Gm2 and Validation of the Linked Markers Dissertation. Hyderabad: University of Hyderabad; 2007

[67] Suvendhu DS, Divya D, Rani CV, Reddy TD, Visalakshmi V, Cheralu C, et al. Characterization of gall midge resistant rice genotypes using resistance gene specific markers. Journal of Experimental Biology and Agricultural Sciences. 2014;**2**(4):439-446

[68] Venkanna V, Hari Y, Rukminidevi K, Chandra BS, Raju J, Malathi S, et al. Markers assisted selection for pyramiding of gallmidge resistance genes in Kavya, a popular rice variety. International Journal of Current Microbiology and Applied Sciences. 2018;**7**(4):745-753. DOI: 10.20546/ijcmas.2018.704.083

[69] Kumar VA, Balachiranjeevi CH, Naik SB, Rekha G, Rambabu R, Harika G, et al. Marker-assisted pyramiding of bacterial blight and gall midge resistance genes into RPHR-1005, the restorer line of the

popular rice hybrid DRRH-3. Molecular Breeding. 2017;**37**(7):86. DOI: 10.1007/s11032-017-0687-8

[70] Stahle DW, Cook ER, Burnette DJ, Villanueva J, Cerano J, Burns JN, et al. The Mexican drought atlas: Tree-ring reconstructions of the soil moisture balance during the late pre-Hispanic, colonial, and modern eras. Quaternary Science Reviews. 2016;**149**:34-60

[71] Reynolds M, Tuberosa R. Translational research impacting on crop productivity in drought-prone environments. Current Opinion in Plant Biology. 2008;**11**(2):171-179. DOI: 10.1016/j.pbi.2008.02.005

[72] Anami S, De Block M, Machuka J, Van Lijsebettens M. Molecular improvement of tropical maize for drought stress tolerance in sub-Saharan Africa. Critical Reviews in Plant Sciences. 2009;**28**(1-2):16-35. DOI: 10.1080/07352680802665305

[73] Kumar A, Verulkar S, Dixit S, Chauhan B, Bernier J, Venuprasad R, et al. Yield and yield-attributing traits of rice (*Oryza sativa* L.) under lowland drought and suitability of early vigor as a selection criterion. Field Crops Research. 2009;**114**(1):99-107. DOI: 10.1016/j.fcr.2009.07.010

[74] Sabetfar S, Ashouri M, Amiri E, Babazadeh S. Effect of drought stress at different growth stages on yield and yield component of rice plant. Persian Gulf Crop Protection. 2013;**2**(2):14-18

[75] Farooq M, Gogoi N, Barthakur S, Baroowa B, Bharadwaj N, Alghamdi SS, et al. Drought stress in grain legumes during reproduction and grain filling. Journal of Agronomy and Crop Science. 2017;**203**(2):81-102. DOI: 10.1111/jac.12169

[76] Fleury D, Jefferies S, Kuchel H, Langridge P. Genetic and genomic tools to improve drought tolerance in wheat. Journal of Experimental Botany. 2010;**61**(12):3211-3222. DOI: 10.1093/jxb/erq152

[77] Bernier J, Kumar A, Ramaiah V, Spaner D, Atlin G. A large-effect QTL for grain yield under reproductive-stage drought stress in upland rice. Crop Science. 2007;**47**(2):507-516. DOI: 10.2135/cropsci2006.07.0495

[78] Vikram P, Swamy BM, Dixit S, Ahmed HU, Cruz MT, Singh AK, et al. $qDTY_{1.1}$, a major QTL for rice grain yield under reproductive-stage drought stress with a consistent effect in multiple elite genetic backgrounds. BMC Genetics. 2011;**12**(1):89. DOI: 10.1186/1471-2156-12-89

[79] Ghimire KH, Quiatchon LA, Vikram P, Swamy BM, Dixit S, Ahmed H, et al. Identification and mapping of a QTL ($qDTY_{1.1}$) with a consistent effect on grain yield under drought. Field Crops Research. 2012;**131**:88-96. DOI: 10.1016/j.fcr.2012.02.028

[80] Venuprasad R, Dalid CO, Del Valle M, Zhao D, Espiritu M, Cruz MS, et al. Identification and characterization of large-effect quantitative trait loci for grain yield under lowland drought stress in rice using bulk-segregant analysis. Theoretical and Applied Genetics. 2009;**120**(1):177-190. DOI: 10.1007/s00122-009-1168-1

[81] Yadaw RB, Dixit S, Raman A, Mishra KK, Vikram P, Swamy BPM, et al. A QTL for high grain yield under lowland drought in the background of popular rice variety Sabitri from Nepal. Field Crops Research. 2013;**144**:281-287. DOI: 10.1016/j.fcr.2013.01.019

[82] Mishra KK, Vikram P, Yadaw RB, Swamy BM, Dixit S, Cruz MT, et al. $qDTY_{12.1}$: A locus with a consistent effect on grain yield under drought in rice. BMC Genetics. 2013;**14**(1):12. DOI: 10.1186/1471-2156-14-12

[83] Dixit S, Singh A, Cruz MT, Maturan PT, Amante M, Kumar A. Multiple major QTL lead to stable yield performance of rice cultivars across varying drought intensities. BMC Genetics. 2014;**15**(1):16. DOI: 10.1186/1471-2156-15-16

[84] Swamy BPM, Ahmed HU, Henry A, Mauleon R, Dixit S, Vikram P, et al. Genetic, physiological, and gene expression analyses reveal that multiple QTL enhance yield of rice mega-variety IR64 under drought. PLoS One. 2013;**8**(5):e62795. DOI: 10.1371/journal.pone.0062795

[85] Sandhu N, Singh A, Dixit S, Cruz MT, Maturan PC, Jain RK, et al. Identification and mapping of stable QTL with main and epistasis effect on rice grain yield under upland drought stress. BMC Genetics. 2014;**15**(1):63. DOI: 10.1186/1471-2156-15-63

[86] CSSRI Vision 2050. Director. Karnal, Haryana, India: ICAR-Central Soil Salinity Research Institute; 2015

[87] Singh G. Climate change and sustainable management of salinity in agriculture. Research in Medical & Engineering Sciences. 2018;**6**(2):1-7. DOI: 10.31031/RMES.2018.06.000635

[88] Yeo AR, Flowers TJ. Salinity resistance in rice (Oryza sativa L.) and a pyramiding approach to breeding varieties for saline soils. Functional Plant Biology. 1986;**13**(1):161-173. DOI: 10.1071/PP9860161

[89] Krishnamurthy SL, Sharma SK, Kumar V, Tiwari S, Singh NK. Analysis of genomic region spanning Saltol using SSR markers in rice genotypes showing differential seedlings stage salt tolerance. Journal of Plant Biochemistry and Biotechnology. 2016;**25**(3):331-336. DOI: 10.1007/s13562-015-0335-5

[90] Krishnamurthy SL, Sharma PC, Sharma DK, Ravikiran KT, Singh YP, Mishra VK, et al. Identification of mega-environments and rice genotypes for general and specific adaptation to saline and alkaline stresses in India. Scientific Reports. 2017;**7**(1):1-14. DOI: 10.1038/s41598-017-08532-7

[91] Krishnamurthy SL, Sharma PC, Gautam RK, Singh RK, Singh YP, Mishra VK, et al. Notification of crop varieties and registration of Germplasm: Variety CSR60. Indian Journal of Genetics and Plant Breeding. 2019;**79**(2):513-514

[92] Krishnamurthy SL, Sharma PC, Gautam RK, Singh RK, Singh YP, Mishra VK, et al. Notification of crop varieties and registration of Germplasm: Variety CSR56 (IET 24537). Indian Journal of Genetics and Plant Breeding. 2019;**79**(2):512-513

[93] Im CH, Kim MK, Kim KH, Cho SJ, Lee JJ, Joung WK, et al. Breeding of Pleurotus eryngii with a high temperature tolerance trait. Journal of Mushroom. 2014;**12**(3):187-192. DOI: 10.4489/KJM.2013.41.2.91

[94] Kumar M, Kim SR, Sharma PC, Pareek A. Simple and efficient way to detect small polymorphic bands in plants. Genomics Data. 2015;**5**:218-222. DOI: 10.1016/j.gdata.2015.06.006

[95] Luu TNH, Luu MC, Abdelbagi MI, Le HH. Introgression the salinity tolerance QTLs Saltol into AS996, the elite rice variety of Vietnam. American Journal of Plant Sciences. 2012;**2012**(3):981-987. DOI: 10.4236/ajps.2012.37116

[96] Usatov AV, Alabushev AV, Kostylev PI, Azarin KV, Makarenko MS, Usatova OA. Introgression the saltol QTL into the elite rice variety of Russia by marker-assisted selection. American Journal of Agricultural and Biological Sciences. 2015;**10**(4):165-169. DOI: 10.3844/ajabssp.2015.165.169

[97] Singh R, Singh Y, Xalaxo S, Verulkar S, Yadav N, Singh S, et al. From

QTL to variety-harnessing the benefits of QTLs for drought, flood and salt tolerance in mega rice varieties of India through a multi-institutional network. Plant Science. 2016;**242**:278-287. DOI: 10.1016/j.plantsci.2015.08.008

[98] Singh VK, Singh BD, Kumar A, Maurya S, Krishnan SG, Vinod KK, et al. Marker-assisted introgression of Saltol QTL enhances seedling stage salt tolerance in the rice variety "Pusa Basmati 1". International Journal of Genomics. 2018;**2018**:1-12. DOI: 10.1155/2018/8319879

[99] Babu NN, Vinod KK, Krishnamurthy SL, Krishnan SG, Yadav A, Bhowmick PK, et al. Microsatellite based linkage disequilibrium analyses reveal Saltol haplotype fragmentation and identify novel QTLs for seedling stage salinity tolerance in rice (*Oryza sativa* L.). Journal of Plant Biochemistry and Biotechnology. 2017;**26**(3):310-320. DOI: 10.1007/s13562-016-0393-3

[100] Bhandari A, Jayaswal P, Yadav N, Singh R, Singh Y, Singh B, et al. Genomics-assisted backcross breeding for infusing climate resilience in high-yielding green revolution varieties of rice. Indian Journal of Genetics and Plant Breeding. 2019;**79**(1):160-170. DOI: 10.31742/IJGPB.79S.1.5

[101] Stocker TF, Qin D, Plattner GK, Tignor M, Allen SK, Boschung J, et al. Climate change 2013: The physical science basis. In: Contribution of Working Group I to the Fifth Assessment Report of the Intergovernmental Panel on Climate Change. 2013. p. 1535

[102] Wei Y, Yang TM, Chen JH. Analysis on law of high temperature in summer and its influence on single-season rice planting in Anhui Province. Journal of Anhui Agricultural Sciences. 2008;**36**:15811-15813

[103] Ye C, Argayoso MA, Redoña ED, Sierra SN, Laza MA, Dilla CJ, et al. Mapping QTL for heat tolerance at flowering stage in rice using SNP markers. Plant Breeding. 2012;**131**(1):33-41. DOI: 10.1111/j.1439-0523.2011.01924.x

[104] Ye C, Tenorio FA, Redoña ED, Morales-Cortezano PS, Cabrega GA, Jagadish KS, et al. Fine-mapping and validating qHTSF4. 1 to increase spikelet fertility under heat stress at flowering in rice. Theoretical and Applied Genetics. 2015;**128**(8):1507-1517. DOI: 10.1007/s00122-015-2526-9

[105] Shanmugavadivel PS, Sv AM, Prakash C, Ramkumar MK, Tiwari R, Mohapatra T, et al. High resolution mapping of QTLs for heat tolerance in rice using a 5K SNP array. Rice. 2017;**10**(1):28. DOI: 10.1186/s12284-017-0167-0

[106] Andaya VC, Tai TH. Fine mapping of the qCTS12 locus, a major QTL for seedling cold tolerance in rice. Theoretical and Applied Genetics. 2006;**113**(3):467-475. DOI: 10.1007/s00122-006-0311-5

[107] Andaya VC, Tai TH. Fine mapping of the qCTS4 locus associated with seedling cold tolerance in rice (*Oryza sativa* L.). Molecular Breeding. 2007;**20**(4):349-358. DOI: 10.1007/s11032-007-9096-8

[108] Xu K, Mackill DJ. A major locus for submergence tolerance mapped on rice chromosome 9. Molecular Breeding. 1996;**2**(3):219-224. DOI: 10.1007/BF00564199

[109] Neeraja CN, Maghirang-Rodriguez R, Pamplona A, Heuer S, Collard BC, Septiningsih EM, et al. A marker-assisted backcross approach for developing submergence-tolerant rice cultivars. Theoretical and Applied Genetics. 2007;**115**(6):767-776. DOI: 10.1007/s00122-007-0607-0

[110] Bonilla P, Dvorak J, Mackill D, Deal K, Gregorio G. RLFP and SSLP mapping of salinity tolerance genes in chromosome 1 of rice (*Oryza sativa* L.) using recombinant inbred lines. Philippine Agricultural Scientist. 2002;**85**:68-76

[111] Gregorio GB, Senadhira D, Mendoza RD, Manigbas NL, Roxas JP, Guerta CQ. Progress in breeding for salinity tolerance and associated abiotic stresses in rice. Field Crops Research. 2002;**76**(2-3):91-101. DOI: 10.1016/S0378-4290(02)00031-X

[112] Septiningsih EM, Pamplona AM, Sanchez DL, Neeraja CN, Vergara GV, Heuer S, et al. Development of submergence-tolerant rice cultivars: The Sub1 locus and beyond. Annals of Botany. 2009;**103**(2):151-160. DOI: 10.1093/aob/mcn206

[113] Kumar A, Dixit S, Ram T, Yadaw RB, Mishra KK, Mandal NP. Breeding high yielding drought-tolerant rice: Genetic variations and conventional and molecular approaches. Journal of Experimental Botany. 2014;**65**(21):6265-6278. DOI: 10.1093/jxb/eru363

[114] Thomson MJ, Ocampo D, Egdane J, Katimbang M, Singh R, Gregorio G, et al. QTL mapping and marker-assisted backcrossing for improved salinity tolerance in rice. BioAsia. 2007;(Supplement Papers):6-12

[115] Collard BCY, Mackill DJ. Marker-assisted selection: An approach for precision plant breeding in the 21st century. Philosophical Transactions of the Royal Society, B: Biological Sciences. 2008;**363**(1491):557-572. DOI: 10.1098/rstb.2007.2170

[116] Khanna A, Sharma V, Ellur RK, Shikari AB, Krishnan SG, Singh UD, et al. Development and evaluation of near-isogenic lines for major blast resistance gene (s) in basmati rice. Theoretical and Applied Genetics. 2015;**128**(7):1243-1259. DOI: 10.1007/s00122-015-2502-4

[117] Yadav S, Sandhu N, Dixit S, Singh VK, Catolos M, Mazumder RR, Rahman MA, Kumar A. 2020. Genomics-Assisted Breeding Enables Successful Development of Multiple Stress Tolerant Climate Smart Rice for South and South East Asia. Communicated

[118] Bandillo N, Raghavan C, Muyco PA, Sevilla MA, Lobina IT, Dilla-Ermita CJ, et al. Multi-parent advanced generation inter-cross (MAGIC) populations in rice: Progress and potential for genetics research and breeding. Rice. 2013;**6**(1):11. DOI: 10.1186/1939-8433-6-11

[119] Sandhu N, Yadav S, Catolos M, Cruz MTS, Kumar A. Developing Climate-Resilient, Direct-Seeded Adapted Multiple-Stress Tolerant Rice Applying Genomic Assisted Breeding. 2020

[120] Sagare DB, Abbai R, Jain A, Kj P, Dixit S, Singh AK, et al. More and more of less and less: Is genomics based-breeding of dry direct seeded rice (DDSR) varieties the need of hour? Plant Biotechnology Journal. 2020;**18**:2173-2186. DOI: 10.1111/pbi.13454

[121] Koide Y, Kobayashi N, Xu D, Fukuta Y. Resistance genes and selection DNA markers for blast disease in rice (*Oryza sativa* L.). Japan Agricultural Research Quarterly. 2009;**43**(4):255-280

[122] Song WY, Pi LY, Wang GL, Gardner J, Holsten T, Ronald PC. Evolution of the rice *Xa21* disease resistance gene family. The Plant Cell. 1997;**9**(8):1279-1287

[123] Jairin J, Phengrat K, Teangdeerith S, Vanavichit A, Toojinda T. Mapping of a broad-spectrum brown planthopper

resistance gene, *Bph3*, on rice chromosome 6. Molecular Breeding. 2007;**19**(1):35-44

[124] Nair S, Kumar A, Srivastava MN, Mohan M. PCR-based DNA markers linked to a gall midge resistance gene, Gm4t, has potential for marker-aided selection in rice. Theoretical and Applied Genetics. 1996;**92**(6):660-665

[125] Suh JP, Lee CK, Lee JH, Kim JJ, Kim SM, Cho YC, et al. Identification of quantitative trait loci for seedling cold tolerance using RILs derived from a cross between japonica and tropical japonica rice cultivars. Euphytica. 2012;**184**(1):101-108

[126] Ranawake AL, Manangkil OE, Yoshida S, Ishii T, Mori N, Nakamura C. Mapping QTLs for cold tolerance at germination and the early seedling stage in rice (*Oryza sativa* L.). Biotechnology and Biotechnological Equipment. 2014;**28**(6):989-998

[127] Krishnamurthy SL, Pundir P, Warraich AS, Rathor S, Lokeshkumar BM, Singh NK, et al. Introgressed saltol QTL lines improves the salinity tolerance in rice at seedling stage. Frontiers in Plant Science. 2020;**11**:833

[128] Dixit S, Swamy BPM, Vikram P, Bernier J, Sta Cruz MT, Amante M, et al. Increased drought tolerance and wider adaptability of *qDTY12.1* conferred by its interaction with *qDTY2.3* and *qDTY3.2*. Molecular Breeding. 2012;**30**:1767-1779. DOI: 10.1007/s11032-012-9760-5

[129] Shamsudin NAA, Swamy BPM, Ratnam W, Cruz MTS, Sandhu N, Raman AK, et al. Pyramiding of drought yield QTLs into a high-quality Malaysian rice cultivar MRQ74 improves yield under reproductive stage drought. Rice. 2016;**9**(1):21. DOI: 10.1186/s12284-016-0093-6

[130] Sandhu N, Dixit S, Swamy BM, Vikram P, Venkateshwarlu C, Catolos M, et al. Positive interactions of major-effect QTLs with genetic background that enhances rice yield under drought. Scientific Reports. 2018;**8**:1626. DOI: 10.1038/s41598-018-20116-7

[131] Yadav S, Sandhu N, Majumder RR, Dixit S, Kumar S, Singh SP, et al. Epistatic interactions of major effect drought QTLs with genetic background loci determine grain yield of rice under drought stress. Scientific Reports. 2019;**9**(1):1-3. DOI: 10.1038/s41598-019-39084-7

[132] Vikram P, Swamy BM, Dixit S, Singh R, Singh BP, Miro B, et al. Drought susceptibility of modern rice varieties: An effect of linkage of drought tolerance with undesirable traits. Scientific Reports. 2015;**5**:14799. DOI: 10.1038/srep14799

[133] He S, Schulthess AW, Mirdita V, Zhao Y, Korzun V, Bothe R, et al. Genomic selection in a commercial winter wheat population. Theoretical and Applied Genetics. 2016;**129**(3):641-651. DOI: 10.1007/s00122-015-2655-1

[134] Michel S, Ametz C, Gungor H, Epure D, Grausgruber H, Löschenberger F, et al. Genomic selection across multiple breeding cycles in applied bread wheat breeding. Theoretical and Applied Genetics. 2016;**129**(6):1179-1189. DOI: 10.1007/s00122-016-2694-2

[135] Calus MP, Veerkamp RF. Accuracy of multi-trait genomic selection using different methods. Genetics Selection Evolution. 2011;**43**(1):26-32. DOI: 10.1186/1297-9686-43-26

[136] Schulthess AW, Wang Y, Miedaner T, Wilde P, Reif JC, Zhao Y. Multiple-trait-and selection indices-genomic predictions for grain yield and protein content in rye for feeding purposes. Theoretical and Applied Genetics. 2016;**129**(2):273-287. DOI: 10.1007/s00122-015-2626-6

Chapter 19

Salt Stress in Plants and Amelioration Strategies: A Critical Review

Sajal Roy and Nasrin Chowdhury

Abstract

High salt concentration in soil is a major abiotic stress, which adversely influences the growth, overall development, and productivity of crops. More than 20% of the land of the world used for crop production is adversely affected by high salt concentration. The problem of salt stress becomes a major concern when previously fertile, productive agricultural lands are salinized more profoundly as a result of anthropogenic activities along with natural causes. Therefore, this review is focused on various aspects of salt-affected soils (SAS), their effects on plants, and different approaches for reclamation of SAS to enhance the potentiality for crop production. Salt-affected soils are categorized into saline, saline-sodic, and sodic soils based on the amount of total soluble salts as expressed by electrical conductivity (EC), sodium adsorption ratio (SAR), exchangeable sodium percentage (ESP), and soil pH. The inhibition of plant growth in saline soils is mainly induced by osmotic stress; reduced uptake of essential macro- and micronutrients, including nitrogen (N), phosphorus (P), potassium (K), calcium (Ca), magnesium (Mg), iron (Fe), manganese (Mn), zinc (Zn), and copper (Cu); and specific toxicities of sodium (Na) and chloride (Cl). Sodic soils adversely affect the plant through high soil pH and poor physical condition resulting from an excessive amount of exchangeable Na. Different plants respond to salt stress in different extents. Salt-affected soils must be reclaimed to restore their productivity for increasing food production. The approaches for the management of SAS include leaching, incorporation of different organic and inorganic amendments, mulching, and development of salt-tolerant crops. The suitability of approaches depends on several considerations such as cost of reclamation, the time required, the extent of the salt stress, soil properties, availability of technology, and other environmental factors. Among different strategies, the incorporation of organic amendments is beneficial, cost-effective, environment friendly, and sustainable for amelioration of salt stress and enhancement of crop production due to the extensive roles of organic amendments in improving the soil's physical (structural stability, porosity, and permeability), chemical [pH, EC, ESP, organic matter, cation exchange capacity (CEC), and Na leaching], and biological and/or biochemical (microbial abundance, microbial activity, biomass carbon, and enzymatic activities) properties.

Keywords: abiotic stress, nutrient uptake, organic amendments, reclamation, salinity, sodicity

1. Introduction

The term stress in plants is defined as the environmental constraint that leads to the inhibition of morphological, physiological, and biochemical functioning of plants adversely affecting their growth and development [1–4]. The stresses may be biotic (pest, diseases, weed, etc.) or abiotic (soil salinity, radiation, water lodging, drought, extreme temperature, organic and inorganic pollutants, etc.) [4–12] that may act alone or in combinations, limiting the productivity of crops and food security worldwide. Salt-affected soils (SAS) cause greatest environmental abiotic stresses to plants [13–15] and cover more than 20% of the cultivated lands throughout the world [16, 17].

Salt-affected soils are grouped into saline, sodic, and saline-sodic and exhibit stresses to plants differently through various mechanisms. The principle mechanisms of salt stress in plants include osmotic effect, ionic toxicity, and nutritional imbalances [4, 15]. The increase in the uptake of Na and Cl caused by the salt constraint is attributed to the reduction in N and P concentrations that may be due to the antagonistic relations of Na and Cl with ammonium (NH_4^+), nitrate (NO_3^-), and phosphate ($H_2PO_4^-$) [18, 19]. The salt stress adversely affects plant growth by inhibiting various steps of N metabolism such as uptake, assimilation, and amino acid and protein synthesis [20]. High concentrations of Na ions at the root surface have detrimental effects on the uptake of K [21]. Because of the similar chemical nature of Na and K ions, Na has an adverse effect on K uptake by the root specifically through high-affinity potassium transporters (HKTs) and nonselective cation channels (NSCCs) [22, 23]. Similar to K, the uptake and transport of Ca and Mg can also be adversely affected by the high concentrations of Na commonly found in saline soils, resulting in lower Ca:Na and Mg:Na ratios in plants [24]. The concentration of micronutrients in plants may be increased, decreased, or remain unaffected under salt stress depending on the plant type, tolerance of plants to salinity, macro- and micronutrient concentrations in soil, pH of the soil solution, the adsorption phenomena on the surface complexes of mineral and organic particles, and different environmental conditions [1, 25].

The approaches for the management of SAS include removal of salts from the root zone through leaching, incorporation of organic and inorganic amendments, mulching, maintaining groundwater table, and cultivating crops tolerant to salt stress [26, 27]. In salt-affected areas, different methods are applied to remove excess soluble salts from the root zone of plants to improve crop growth and production. Scraping, flushing, and leaching are most commonly used, but these methods are found to be very expensive [28]. The application of organic amendments is effectively being practiced to ameliorate physical, chemical, and microbial complications associated with SAS. Organic amendments help to flocculate mineral particles to organic polymers because of their bonding or adhesion properties [29], resulting in a good structural stability, which is a precondition to maintain an appropriate soil structure. The application of organic matter to SAS can improve the aggregate stability and porosity, resulting in increased Na leaching and decreased exchangeable sodium percentage (ESP) and electrical conductivity (EC) values [30]. Several studies reported a significant increase in the soil microbial and enzymatic activities in SAS as a result of organic matter incorporation [31–33]. Salt-affected areas are increasing globally with the intervention of human activities and natural events. With the increase in salt-affected areas and population worldwide, the net cultivable land is decreasing, pushing enormous pressure on food security. It is pertinent to study how SAS harms plant growth and productivity as well as the strategies to ameliorate SAS by improving their physical, chemical, and biological conditions for the sustainable production of crops. Therefore, this chapter is arranged by compiling the existing knowledge in the literature in such a way that will demonstrate plant responses to salt stress, especially the uptake and accumulation of essential nutrient elements. Besides, different approaches that can be practiced for the amelioration of SAS have been discussed with the aim to improve soil health and boost up crop production in sustainable ways for ensuring food security worldwide.

2. Salt-affected soils: a brief preview

Understanding the differences in properties among SAS is important for proper reclamation and management. Salt-affected soils are categorized into saline, sodic, and saline-sodic groups based on the amount of total soluble salts (TSS) (measured by EC), sodium adsorption ratio (SAR; the ratio of Na^+ to Ca^{2+} and Mg^{2+} on the exchange sites of soil), exchangeable sodium percentage [ESP; the relative amount of the Na^+ ion expressed as a percentage (%) to the cation exchange capacity (CEC) or the sum of exchangeable bases], and soil pH. Based on these criteria, the classification of SAS is given in **Table 1**.

The pH of sodic soils is usually greater than 8.5, which may rise as high as 10.5. The high pH of sodic soils may be due to a greater extent of hydrolysis of exchangeable Na compared to more strongly held ions such as Ca and Mg. Upon hydrolysis, exchangeable Na contributes to high soil pH according to the following phenomenon [26]:

$$\mathrm{Na-clay-Na+H-OH} \leftrightarrow \mathrm{Na-clay-H+Na++OH-} \quad (1)$$

The increase in the concentration of OH^- ion contributes to the increase in soil pH. The limited hydrolysis of $CaCO_3$ and $MgCO_3$ causes low solubility of these salts, resulting in soil pH no higher than 8.5, while soils containing Na_2CO_3 have a pH of more than 8.5 or even 10 to 10.5 due to their higher solubility [26, 34, 35].

The processes by which saline soils are formed are known as salinization, whereas the processes that are responsible for the formation of sodic soils are called as sodification. The causes of salinization and sodification are multifactorial, and one factor that affects the salinization and/or sodification may influence the others [36]. Soil salinization and sodification are interrelated with such factors as soil characteristics; the amount and composition of salts in the soil; and the quantity, quality, and methods of irrigation [37]. Salt-affected soils are formed either by natural processes or anthropogenic activities. Natural processes are known as primary salinization/sodification, whereas human-induced processes are known as secondary salinization/sodification [38]. The main sources of soluble salts under natural and anthropogenic factors involving the development of SAS are summarized in **Figure 1**.

Types of SAS	pH	EC (dS/m)	SAR	ESP	Major cations and anions and their relative concentrations
Saline	<8.5	>4.0	<13	<15	$Cl^- > SO_4^{2-} > HCO_3^- > CO_3^{2-}$ $Na^+ > Ca^{2+} + g^{2+} > K^+$
Saline-sodic	<8.5	>4.0	>13	>15	High content of Na^+ on exchange sites as well as in soil solution
Sodic	<8.5	>4.0	>13	>15	High Na^+ on exchange site of soil particles with little amount in soil solution

Table 1.
Classes of SAS.

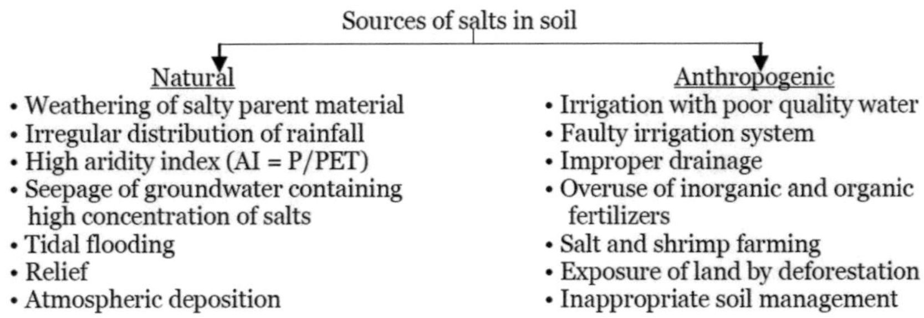

Figure 1.
Natural and anthropogenic sources of salts in soils.

Soil salinity and sodicity have contributed to the changes in land use/land cover features over the years, which are directly related to land degradation and results in many changes in the environment [39]. More than 20% of the total agricultural lands are affected by high salinity [38, 40–42], which accounts for more than 7% of the world's total land area [23, 43]. It has been reported that the majority of the SAS occur mainly in the arid and semiarid regions of Asia, Australia, and South America, covering an estimated 1 billion ha [36]. Data summarized from Abrol et al. [34] and Szabolcs [44] give an account that globally about 932 million ha areas are occupied by SAS. There are sporadic studies on the estimates in the global distribution of salt-affected areas in recent years. For the establishment of better management strategies of SAS, it is important to identify their extent and distribution in systematic ways. However, different countries assess the extent and severity of their SAS at national levels.

Soil salinity can be determined by measuring either the total soluble salts (TSS) by evaporation of a soil-water extract or by determining the EC of a soil-water suspension. In the laboratory, EC as an indicator of soil salinity can be measured by analyzing the soil suspension either in a 1:5 soil:distilled water dilution ($EC_{1:5}$) or in a saturated paste extract (EC_e). The measurement of $EC_{1:5}$ is most commonly employed from an unfiltered 1:5 soil-water suspension prepared by taking a unit of 2-mm sieved soil (usually, 5 g) and 5 units of distilled water (25 ml). The suspension is shaken for 30 min to dissolve the soluble salts and left for 15 min to settle down the soil particles, followed by the EC measurement using EC meter after necessary calibration [45].

3. Mechanisms of salt stress in plants

Though all soils invariably contain soluble salts, under saline and sodic conditions, excess salts in the root zone deteriorate the physical, chemical, and biological properties of soils to such an extent that the crop growth is adversely affected [46]. The adverse effects of salinity on plant growth depend on plant factors such as growth stage, species, and variety; soil factors such as temperature and moisture, degree of salinity, and presence of heavy metals; and environmental factors such as growing season, temperature, humidity, light, pollutants in the atmosphere, etc. [47–52]. Based on the degree of salinity and associated effects on the plant growth, the saline soils can be classified from nonsaline to very strongly saline (**Table 2**).

The mechanisms of how the SAS distress plant growth are shown in **Figure 2**. Under saline environment, the inhibition of growth is mainly induced by the physiological and biochemical disturbances resulting from osmotic stress; changes in the uptake of mineral nutrients such as N, P, K, Ca, and Mg; and specific ion toxicities

Classes of saline soils	EC_e (dS/m)	Salinity effects on plant	Yield loss (%)
Nonsaline	0–2	Salt effects are negligible	0
Slightly saline	2–4	Yields of only sensitive crops may be restricted, and the yields of most of the crops are not likely to be affected	20–30
Moderately saline	4–8	The yields of most of the crops are likely to be hampered	30–60
Strongly saline	8–16	Most of the crops are likely to be affected, while the yields of only tolerant crops are satisfactorily	60–100
Very strongly saline	>16	Only limited tolerant plants can sustain	100

Adapted and modified from Shin et al. [53].

Table 2.
Soil salinity classes and associated effects on plants.

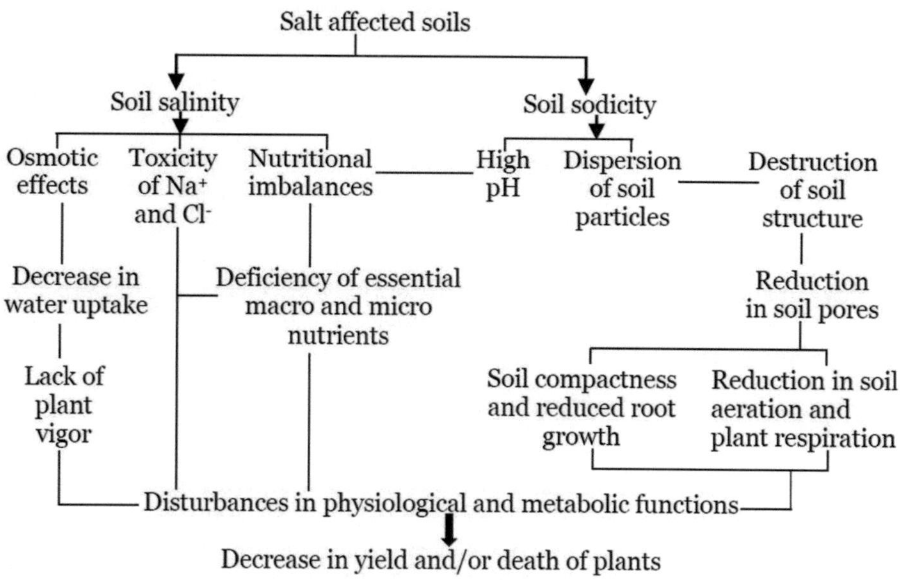

Figure 2.
Mechanism of salt stress in plants.

such as Na and Cl [54–57]. Plants containing approximately 0.25–0.5% Na in leaf tissues on dry weight basis may exhibit toxicity symptoms such as necrosis and burns of leaf edges [1, 58]. When the salt concentration in the soil is equal to that of the plant, there is no net movement of water. But when the salt concentration in the soil solution is greater than that of the plant, water moves from the plant into the soil causing physiological drought [59, 60]. To counteract the low osmotic potential of the soil solution, plants accumulate organic and inorganic solutes to reduce the osmotic potential inside the cell. Maintenance of such osmotic adjustment requires a considerable expenditure of energy, resulting in reduced growth [35]. Besides direct effects of SAS through osmotic stress and nutritional imbalances, the growth of plants may be adversely affected by a high content of exchangeable Na relative to Ca and Mg on soil particles which leads to the breakdown of soil structure, resulting in decreased porosity, permeability, hydraulic conductivity, and aeration in the vicinity of plant root [35, 41].

The adverse effects of salts are first expressed by the inhibition of growth and other physiological symptoms. Salt stress has been found to negatively affect the biomass of cabbage, *Brassica oleracea* [19]; cotton, *Gossypium hirsutum* [61]; Jerusalem artichoke, *Helianthus tuberosus* [62]; pistachio, *Pistacia vera* [63]; poacea, *Catapodium rigidum* [15]; rice, *Oryza sativa* [64]; saltmarsh grass, *Spartina alterniflora* [65]; strawberry, *Fragaria ananassa* [66]; tomato, *Lycopersicon esculentum* [64]; and wheat, *Triticum aestivum* [67].

The capacity of roots to transport nutrients and water to shoots is restricted due to abiotic stress, adversely affecting the functional balance between roots and shoots, which results in lower shoot:root ratio [68]. The leaves of plants are the most sensitive organs and greatly affected, while the roots are reasonably less affected when exposed to high salinity [69]. However, all plants do not respond to salinity to the same extent. Different crops exhibit a broad spectrum of responses to salt stress.

4. Salt stress on nutrient uptake of plants

Though the most common cations found in SAS are Na^+, Ca^{2+}, and Mg^{2+}, accompanied by anions Cl^-, SO_4^{2-}, and HCO_3^-, these soils are mainly dominated by Na^+ and Cl^- [35, 45, 70]. Nitrogen, P, K, Ca, and Mg play important physiological functions in plants, and their replacement by Na and Cl may lead to nutritional imbalances [47]. For the optimum growth and yield of crops, an adequate and balanced supply of mineral nutrients is essential [42]. High contents of Na and Cl in the rhizosphere can interfere with the uptake of essential elements, leading to their deficiencies or imbalances [71, 72] through the processes of fixation, adsorption, and transformation in soil [18]. The effects of high concentrations of Na and Cl on the uptake of essential elements by plants are outlined in the following subsections.

4.1 Nitrogen

Nitrogen is a major constituent of the nucleotides and proteins of all micro- and macroorganisms [9, 73]. It is considered as one of the limiting factors and is required in large quantities compared to other essential nutrients for the growth, yield, and quality of crops [74–77]. In general, the total N content in the plow layer of mineral soils ranges from 0.05 to 0.2% corresponding to approximately 1750 to 7000 kg N/ha, and in plants, it ranges from 1 to 4% on dry matter basis [77]. Nitrogen rarely exists in the lithosphere in rock deposits, and more than 90% of the total N comes from the soil organic matter [35]. It is principally taken up by plants in the forms of NO_3^- and NH_4^+ [61]. As a consequence of high salinity, the uptake and assimilation of NH_4^+ and NO_3^- are inhabited by excessive concentrations of Na and Cl. The uptake of other essential nutrients may be disturbed due to the inhibition of N metabolism as the presence of N in the growth medium stimulates the uptake and assimilation of other essential nutrients [77]. Decreases in N uptake may be a physiological response of plants under salt stress [78].

Salt stress was found to strongly inhibit the uptake and accumulation of N in different plant parts of green chireta, *Andrographis paniculata* [14]; cabbage, *Brassica oleracea* [19]; canola, *Brassica napus* [78]; cotton, *Gossypium hirsutum* [61]; gray poplar, *Populus × canescens* [20]; saltmarsh grass, *Spartina alterniflora* [65]; sesame, *Sesamum indicum* [79]; and wheat, *Triticum aestivum* [67]. However, the uptake and accumulation of N in response to salt stress depend on different plant and soil factors [80]. Kanagaraj and Desingh [79] observed variations in foliar N among different varieties of sesame, where the maximum reduction was found at the highest dose of NaCl.

The decline in the uptake of N by plants could be due to the increased uptake and accumulation of Cl. The antagonistic relationship between the uptake of NO_3^- and Cl^- was found in rice, *Oryza sativa* [64]; cabbage, *Brassica oleracea* [19]; and barley, *Hordeum vulgare* [81, 82]. High soil salinity was found to increase the concentration of Cl in the different plant parts of cotton, *Gossypium hirsutum* [47]; cowpea, *Vigna unguiculata* [72]; and poacea, *Catapodium rigidum* [15]. However, the accumulation of Cl is regulated by the duration of salt stress as well as the phase of the plant life cycle [83]. As a consequence of high salinity, the uptake of NH_4^+ can also be reduced. It was observed that the decreased uptake of NH_4^+ in plants was accompanied by an increased Na^+:NH_4^+ ratio in the growth medium [20].

Several studies reported inhibited accumulation of nutrients with decreasing water uptake under salt stress [15, 64]. The effect of salt stress on N metabolism varies depending on the forms of N applied [20]. Botella et al. [67] reported that the increased concentration of NaCl in the root medium decreased the net uptake of N more profoundly in NO_3^- compared to the NH_4^+ form when their compounds were incorporated as the source of N, which was assumed to be a reason for the greater affinity for NH_4^+ compared to NO_3^- under the saline environment. Similarly, Saud et al. [84] found higher N content, N isotope abundance, and relative water content in both roots and leaves of Kentucky bluegrass, *Poa pratensis*, in $^{15}NH_4^+$ compared to the $^{15}NO_3$ treatment under abiotic stress. By contrast, Dluzniewska et al. [20] observed that the net uptake of NH_4^+ decreased in poplars when exposed to increased NaCl concentrations, resulting in decreased whole-plant N content in comparison to control. Hofman and Cleemput [77] stated that NH_4^+ is more preferably taken up by plants in comparison to other forms as it does not require to be reduced before incorporation into plant compounds. On the other hand, Dai et al. [61] found better root growth and low Na content in NO_3-fed compared to NH_4-fed cotton seedlings, and they reported the superiority of NO_3-N relative to the NH_4-N source in the uptake of N under salt stress. The high salt content in soil may also interrupt the synthesis of protein in plants. Chakraborty et al. [85] observed that the uptake of N of the *Brassica* spp. decreased due to the high salt content, which reflected in the low protein levels in seeds. The physiological drought of plants, which is caused by the low osmotic potential of soil solution, is attributed to the reduced metabolism of N [18] that may result in the low protein content under saline condition.

4.2 Phosphorus

Phosphorus is the second most essential nutrient requiring 0.3–0.5% of the dry matter for the optimal growth of plants [86]. It is an integral part of nucleic acids and membrane lipids [73]. In contrast to N, the main sources of P in the lithosphere are the rock deposits [77]. Phosphorus exists in the soil solution as orthophosphate ions such as $H_2PO_4^-$, HPO_4^{2-}, and PO_4^{3-} [87] and plants take up P from the soil solution in the form of $H_2PO_4^-$ and HPO_4^{2-}, although $H_2PO_4^-$ is taken up to a greater extent [88, 89]. Exposure to high salinity resulted in significant decrease in P levels in green chireta, *Andrographis paniculata* [14]; cabbage, *Brassica oleracea* [19]; canola, *Brassica napus* [78]; pistachio, *Pistacia vera* [63]; saltmarsh grass, *Spartina alterniflora* [65]; and spinach, *Ipomoea aquatica* [90]. The salt stress had been found to magnify the adverse effects on the uptake of P by plants in several studies when plants were exposed simultaneously to both salinity and drought [19, 65]. The reduction in P concentration under high salinity may be due to the competition between $H_2PO_4^-$ and Cl^- ions [18]. The accumulation of P in plants exposed to salt stress varies with the plant organs. Silva et al. [72] found a greater P content in leaves compared to roots. Conversely, Shahriaripour et al. [63] reported that the translocation of P from root to shoot was inhibited under a high saline environment.

The evidence of increased P under high salinity had also been reported in several studies. The presence of NaCl in nutrient solution resulted in an increased concentration of P content in leaves [72, 83]. Zribi et al. [15], however, demonstrated that salinity had no significant effect on both P concentration and acquisition efficiency (PAE), and they also found a higher concentration of Cl in the shoot when subjected to both salt stress and P treatment compared to salt stress only.

The uptake of P by plants under salt stress is influenced by the plant species, plant stage, degree and extent of stress, temperature, moisture, soil pH, and the prevailing soil P level [55, 83, 89]. The solubility of P may also be influenced by strong sorption-desorption processes with divalent cations such as Ca^{2+}, Mg^{2+}, and Fe^{3+} at different pH ranges because of their high sorption capacity regulating the uptake by plants [5, 87, 91].

4.3 Potassium

Potassium is an important and abundant mineral nutrient that comprises 2.6% of the earth's crust [92] and 1–10% of the plant tissue on a dry weight basis [92, 93]. It contributes to important physiological, biochemical, and biophysical roles such as photosynthesis, osmotic adjustment, and turgor maintenance [94, 95], regulating the growth and development of plants. The concentration of K in the growth medium determines the net uptake of K [13]. The seldom deficiency of K in most soils is believed due to comparatively high concentration and greater mobility [96, 97]. However, in the condition of salt stress, the plants may suffer from K deficiency as a result of Na toxicity. Both Na and K are monovalent cations with similar physicochemical properties. The hydrated ion radius of Na ion is 0.358 nm, while that of K ion is 0.331 nm [23]. Due to similar ionic radius and cationic competition for entry into the plant cells, high concentrations of Na can adversely affect the uptake and accumulation of K in plants.

Salt-affected soils led to a significant influence on K in plant tissues, either increasing or decreasing its concentration. The reduction in K content in plant tissues is one of the primary responses of plants to high Na, which ultimately causes nutrient imbalances [98]. High salt stress with increased concentrations of NaCl was found to decrease the total K and cause an increase in the Na content in *Aloe vera*, *Aloe vera* [99]; barley, *Hordeum vulgare* [100]; bean, *Phaseolus vulgaris* and *P. acutifolius* [54]; cabbage, *Brassica oleracea* [19]; cotton, *Gossypium hirsutum* [47]; green chireta, *Andrographis paniculata* [14]; maize, *Zea mays* [101]; pistachio, *Pistacia vera* [63]; poacea, *Catapodium rigidum* [15]; rice, *Oryza sativa* [102]; saltmarsh grass, *Spartina alterniflora* [65]; strawberry, *Fragaria ananassa* [66]; and tomato, *Lycopersicon esculentum* [13, 103, 104]. A positive relation of salt-tolerance capacity with Na and a negative relationship with that of K was found in saltbush, *Atriplex canescens* [105]. Besides, there is evidence that both the contents of Na and K in plants increased with the increase of salinity. The concentrations of Na and K in the above-ground and below-ground portions of Jerusalem artichoke, *Helianthus tuberosus*, were found to increase with the increasing amount of seawater application in a greenhouse experiment [62]. Similarly, the concentrations of Na and K in the leaves of cowpea, *Vigna unguiculata*, increased in salt-stressed condition and the responses differed depending on the duration of stress and leaf age [83]. Al-karaki [13] found higher translocation of K from root to shoot in tomato plants to a greater extent in the saline environment in comparison to nonsaline conditions, with the increase of K concentration in medium.

The higher concentration of Na restricts the transport of K by impairing the route AKT1 (hyperpolarization-activated inward-rectifying K channel) used for the uptake of K and, in this way, reduce the uptake of K [106]. The inhibitory effect of Na on the transport of K through channels in the membranes is probably more important in the phase of uptake of K from the soil solution than in the phase of K transport to the xylem [107]. Besides, the inhibitory effect of Na on K translocation depends on the concentration of K in solution. The effect is lower with low Na and high K levels in the medium [13, 108, 109]. However, the effects of Na on the uptake and accumulation of K can vary among species and even between varieties within such same species of *Aloe vera*, *Aloe vera* [99]; bean, *Phaseolus vulgaris* [110]; cotton, *Gossypium hirsutum* [111]; green chireta, *Andrographis paniculata* [14]; mustard, *Brassica nigra* [112]; tomato, *Lycopersicon esculentum* [104, 113]; and wheat, *Triticum aestivum* [109]. Salt-tolerant species were found to maintain a high concentration of K and low concentration of Na. High concentration of K and low concentration of Na in plants under saline environment can be considered as a good indicator for salinity tolerance [114]. The plant parts also differently respond to salt stress. Generally, a higher concentration of Na in roots relative to leaves in various plant species was explained by the high tolerance in roots in addition to reduced translocation of Na from root to leaf [1]. Different authors reported lower content of K as a result of salt stress to a greater extent in stems in comparison to leaves and roots [99, 110]. A high concentration of K in leaves is important to maintain constant photosynthesis and leaf stomatal conductance [54].

4.4 Calcium and magnesium

Calcium is the fifth most plentiful element, whereas Na is the sixth most abundant element comprising, respectively, about 3 and 2.6% of the earth's crust. Because of the greater charge density of Ca^{2+} compared to Na^{+}, it attracts more water, resulting in a greater hydrated ionic radius of 0.44 nm [48]. Calcium is an essential inorganic nutrient and plays a vital role in maintaining the structural and functional integrity of cell wall and membrane because of its ability to form intermolecular linkages [98, 113]. Considering the importance of external Ca in enduring K transport and K:Na selectivity in plants under Na stress condition, a large number of research was carried out on Na:Ca interactions. The deficiency of Ca in plants is a common indication of Na toxicity [115]. The Na:Ca ratio of plants is a good indication used to express the extent to which a plant can survive under salt stress [69]. A high concentration of Na can decrease the uptake of Ca and Mg by inhabiting their influx through roots, decreasing the extracellular binding sites of Ca and Mg in the plasma membranes [69, 116], decreasing the osmotic potential [19] under saline condition.

Increased level of Na was found to significantly decrease the concentrations of Ca and Mg in saltmarsh grass, *Spartina alterniflora* [65]. Significant negative correlations were also found between salt stress and the contents of Ca and Mg, where the ratios of Ca:Na and Mg:Na in the leaves and roots of cabbage, *Brassica oleracea*, decreased with the increasing salinity levels [19]. Similar results were reported by other authors where the salt stress resulted in an increased accumulation of Na and Cl and decreased the contents of Ca and Mg in plants [18, 117]. On the other hand, Chen et al. [47] found both higher Na and Ca in cotton leaf with increasing salt stress. Rahimi and Biglarifard [66] observed that the Ca and Mg concentrations decreased in the shoot and increased in the root of strawberry, *Fragaria ananassa*, with the increase in salinity. There were evidences from several studies that the supplemental addition of Ca increased the concentration of Ca in rapeseed, *Brassica napus* [114], and in tomato, *Lycopersicon esculentum* [113] under high salinity levels. However, the effects of salinity had been found more detrimental in conditions having different levels of drought stress [19, 65, 118].

The influx of Ca and Mg in the root and transport to the shoot can vary with the amount of Na as well as genotypes. The uptake of Ca had been found to be higher in a salt-sensitive variety of wheat, *Triticum aestivum*, at a low level of Na and the transport became lower as the concentration increased [115]. Furthermore, tolerant varieties of green chireta, *Andrographis paniculata*, contained high concentrations of Ca and Mg and low concentration of Na in comparison to sensitive accessions under high salinity [14].

4.5 Micronutrients

Micronutrients are involved in various important physiological and biochemical functioning of plants, including enzyme activation, chlorophyll formation, protein synthesis, carbohydrates, and lipids and nucleic acids metabolism [119]. Different authors reported differently in the uptake and accumulation of micronutrients such as Fe, Mn, Zn, and Cu in salt-stressed condition. Though the exact mechanism in the literature is scarce, different soil, plant, and environmental factors are believed to influence the uptake and accumulation of micronutrients in different plant parts under salt stress. The response of micronutrient concentrations in plants under salt stress was found to be variable in previous studies. The plants grown in saline soils often showed the deficiency of micronutrients such as Fe, Zn, Mn, and Cu, which was assumed to be the result of their low solubility and availability [120]. A negative relationship was found between the salt content and Cu, Mn, and Fe concentration in tissues of *Avicennia marina* (Forssk.) Vierh [49]. Chakraborty et al. [85] found a reduced accumulation of Fe, Mn, and Zn in different parts of *Brassica* spp. at the flowering and post-flowering stages when exposed to high salt concentration. Similarly, salt stress had been found to decrease the concentrations of Fe and Zn in the roots and leaves of cabbage, *Brassica oleracea* [19]. On the other hand, the concentrations of Fe, Zn, and Cu in green chireta, *Andrographis paniculata*, significantly increased though the Mn content decreased under high salinity [14]. The concentrations of Fe, Mn, and Zn increased in the above-ground part of strawberry, *Fragaria ananassa*, while their contents did not change in the root when plants were exposed to salt stress. On the other hand, while Cu content did not change in the aerial part, the concentration increased in the root as a result of salt stress [121]. Significant reduction in the Fe content in both root and shoot of strawberry was observed, while Zn, Cu, and Mn concentrations remained unaffected under salt stress [66]. The change in concentrations of Fe, Zn, and Mn was not found as a limiting factor for the growth of wheat, *Triticum aestivum*, and their contents in plants were not much affected by salt stress [25]. The response of micronutrient concentrations under salt stress differed among varieties in green chireta, *Andrographis paniculata* [14]; *Brassica* spp. [85]; and strawberry, *Fragaria ananassa* [121].

5. Reclamation of salt-affected soils

To make SAS productive, integrative soil, water and agronomic reclamation, and management approaches can be practiced. Saline and sodic soils differ in their reclamation and management practices because of their unlike complications. The problem of saline soils is mainly oriented with high soluble salt content, whereas sodic soils are associated with high exchangeable Na. The suitability of approaches depends on such considerations as cost of reclamation, the time required, the extent of the salinity or sodicity problem, soil properties, availability of technology, and other environmental factors [26]. This section will mainly focus on the existing information in the literature on various aspects, including the effectiveness and downsides of the leaching approach for removing salts as well as the application of organic and inorganic amendments as ameliorative for the restoration of SAS.

5.1 Removal of salts

Removing salts from the root zone is the first requirement to restore productivity of saline areas. However, saline-sodic and sodic soils cannot be reclaimed only by the leaching approach. Thus, to improve the growth and yield performance of crops in saline soils, the harmful concentration of salts must first be washed out from the root zone, which can be achieved by leaching, the most effective procedure for removing excess soluble salts. The efficiency of leaching to remove the salts from the soil profile depends on such factors as the initial salt content, nature of soluble salts, desired EC of soil after leaching, properties of soil, quality of water to be used for leaching, etc. [34]. The key to leaching of soluble salts is to provide an appropriate amount of water at the proper time with adequate drainage. A reliable estimate of the favorable soil moisture content is essential to alleviate the harmful effects of salinity by leaching. In general, depth of water equal to the depth of soil removes 70–80% of the soluble salts for continuous ponding, that is, 15 cm of water is required to reduce the salt content by 70–80% in the upper 15 cm of soil. However, as continuous ponding leads to reduced soil aeration and quick loss of water in coarse-textured soils, intermediate ponding or sprinkler irrigation is preferred for more efficient leaching of salts through increasing contact time of salts with water [122]. On the other hand, prolonged drying may increase the concentration of salts in soil solution, resulting in lower osmotic potential of the soil solution. Due to high permeability and less workability, a large volume of water can be leached over a shorter period in coarse-textured soils. Fine-textured soils having high CEC and organic matter require more water to remove the salts from the soil profile [123]. The desalinization of soils through leaching also depends upon the drainage condition of the soil. The application of leaching with surface drainage at shallow groundwater levels may further exacerbate salinity problems, while the subsurface drainage can sustain the groundwater depth and prevent additional salinization [124]. The role of organic amendments on the reclamation of SAS through improvements in physical properties found in the literature has been presented later in the organic amendment section. Soils having a good structure and internal drainage favor the leaching process [125]. Therefore, a judicious application of water having proper drainage at the right time is important to prevent irrigated lands from becoming saline. The leaching process is accomplished by either natural precipitation and/or artificial irrigation water containing minimum salt content and allowing the water to drain out. The salinity problem of the agricultural land may become worse if saline water is used for leaching purpose because of considerable quantities of such cations as Na^+, Ca^{2+}, Mg^{2+}, and K^+ and such anions as Cl^-, HCO_3^-, and SO_4^{2-} in saline water [126, 127].

5.2 Reclamation with organic amendments

Appropriate nutritional management is often considered feasible and cost-effective, which can lead to the better performance of plants grown in contrasting environments by reducing the adverse effects [128–130]. The application of organic manures as amendments to reclaim SAS is considered highly sustainable and commonly practiced over the last few years [29, 131]. Due to the high cost and quick release of nutrients involved in chemical fertilizers, the application of organic amendments as the substitute of chemical fertilizers or in a combination with chemical fertilizer has gained worldwide acceptance from the farmers [132]. Besides, the extensive application of inorganic fertilizer contributes to groundwater pollution due to leaching loss, and the global climate change is caused due to the emission of potent greenhouse gas such as nitrous oxide from agricultural lands [133, 134]. Though inorganic fertilizers provides easily available nutrients for plant, these are quickly lost from the soil. On the other hand, organic manures contribute to the physical and biological improvements of soil with a gradual release of nutrients. The incorporation of organic manures along with the chemical fertilizers is considered as an effective and sustainable approach for enhancing the resistance of crops to abiotic stress [135–138]. Different sources of organic materials originating from plant and animal residues such as green manure, cattle manure, poultry manure, food processing wastes, etc. are used to augment the organic matter content and nutrient status. The role of organic amendments on physical, chemical, and biological properties of soils having salinity and sodicity problems is shown in **Table 3**.

Sources of organic amendments	Rate	Soil condition (pH, EC, and ESP/ SAR)	Effects on soil properties	Reference
Farmyard manure, and sewage sludge	5 and 10 t/ha	Different sites having variable PH and EC	Decreased bulk density, increased available water and hydraulic conductivity, decrease in pH, and increased organic matter and available macronutrients	[139]
Biochar, green waste compost (GWC), and municipal sewage sludge	1.0, 2.5 and 5.0%	pH: 4.04 $EC_{1:10}$: 4.91 ESP: 67.62	Increase in pH, total organic carbon, and CEC; decrease in EC, ESP, and SAR; and increase in catalase activity and acid-and alkaline phosphatase activity	[140]
Vermicompost and compost	5 and 10 Mg/ha	pH: 7.3 EC_e: 4.26	Decrease in pH and EC, exchangeable Na^+; increase in organic carbon, CEC, exchangeable K^+, Ca^{2+} and Mg^{2+}; and increased microbial C and N, and basal soil respiration	[32]
Municipal solid waste and palm waste	50, 100, and 150 t/ha	pH: 7.97 ECe: 5.13	Increased organic carbon and N; increased microbial biomass and enzymatic activities	[141]
Cow dung, and paddy husk	1 g/kg	pH: 7.86 EC_e: 24.35 SAR: 26.53	Decreased bulk density; decreased EC and SAR; and increased exchangeable Ca^{2+}	[142]
Cotton gin crushed compost and poultry manure	5 and 10 t/ha	pH: 8.0 $EC_{1:5}$: 9.1 ESP: 15.7	Decreased bulk density and ESP; increased soil microbial biomass, respiration, carbohydrate, and enzymatic activities	[33]
Green waste compost (GWC), sedge peat (SP), furfural residue (FR), and a mixture of GWC, SP, and FR (GSF) (1:1:1 by volume).	4.5 kg/m^3	pH: 7.75 $EC_{1:5}$: 3.69 ESP: 15.8	Decreased bulk density; increased total porosity; increased CEC, organic carbon, and available nutrients (N, P, and K); and decreased EC and ESP	[30]

Sources of organic amendments	Rate	Soil condition (pH, EC, and ESP/SAR)	Effects on soil properties	Reference
Cattle dung, vermicompost, biofertilizer, and their combinations	1.8, 3.7, and 5 t/ha	pH: 7.39 $EC_{1:5}$: 7.44	Decreased EC; increased organic matter and available nutrients; and increased microbial biomass carbon	[143]
Pig manure, cattle dung, chicken manure, rapeseed meal, and biochar	50 g/kg	pH: 8.29 $EC_{1:5}$: 19.35 ESP: 67.62	Decrease in pH; increase in organic carbon, K, Ca, and Mg; and increased enzyme activities such as catalase, urease, alkaline phosphatase, and saccharase	[144]
Poultry manure, commercially available organic fertilizer, reed straw, and fresh reed straw with green leaves	15 g/kg	pH: 8.44 $EC_{1:5}$: 11.61	Decreased bulk density; increased available N, water-soluble organic carbon; decreased SAR; and increased soil respiration	[145]
Straw, composted straw, fresh reed, and chicken manure	15 g/kg	pH: 7.82 $EC_{1:5}$: 6.59 ESP: 36.17	Decreased pH, ESP; increased organic carbon, CEC, macronutrient concentration; and increased soil respiration	[146]

Table 3.
Effects of organic amendments on the physical, chemical, and biological properties of SAS.

Organic amendments have profound influences on soil's physical properties. Several studies revealed that the application of organic manures decreased the bulk density [33, 147–149] and penetration resistance [149], whereas it increased the aggregate stability [150–152], total porosity [152, 153], hydraulic conductivity, and permeability [150, 154]. Soil organic matter is an important attribute of soil quality and aggregate stability, which is influenced by the inherent properties of soil such as soil type and texture [155, 156] as well as agronomic factors such as management, inputs, and nature of the organic matter [157]. Improvement in the aggregate stability is related to increased soil porosity and decreased bulk density. As a consequence of good aggregation with a concomitant increase in porosity and decrease in bulk density, soil water infiltration is facilitated causing the soluble salts to leach down, and an adequate oxygen supply is maintained in the root zone which is necessary for crop production in SAS.

The incorporation of organic amendments had also been found to improve chemical properties such as decrease in pH [27, 147, 158, 159], EC [27, 158, 160], ESP [33, 158], and SAR [27], while there is an increase in the soil organic matter [153, 161], organic carbon [132, 147, 151, 152, 159, 162–164], CEC [152, 162], total nutrients [32, 151, 161], and available nutrients [32, 132, 147, 149, 153, 159, 162, 164, 165].

The decline in soil pH resulting from the incorporation of organic amendments reported in several studies can be explained by the acidic nature of the amendments. Microbial activities resulting from the incorporation of organic materials release carbon dioxide that reacts with water to form carbonic acid, and thus contribute to lowering of the soil pH [166]. The pH of soils was also found to be increased by the addition of organic amendments in several studies [30, 33, 141, 143, 167]. Roy and Kashem [132] found that the addition of organic amendments slightly increased the soil pH initially which thereafter declined significantly with time. However, the change in pH with the addition of organic amendments depends on the initial pH of the original soil, nature of organic materials, and rate of organic matter application [32, 141]. The increase in pH is explained by the mineralization of carbon (C) and the subsequent production of OH^- ions by ligand exchange and release of such basic ions such as K, Ca, and Mg [167]. The decrease in EC due to the application of organic amendments might be a result of Na displacement from the exchange sites and washing out with soluble salts through the leaching process. Organic amendments substantially contain Ca, Mg, and K [30, 131]. The presence of Ca can contribute to the low ESP of SAS due to increased exchange of Na by Ca at the cation exchange sites of soil particles, allowing greater leaching of exchanged Na with percolating water [32, 168]. Moreover, the presence of soluble and exchangeable K can limit the entry of Na into the exchange complex due to a similar ionic balance resulting in lower ESP [169]. Furthermore, Ca improves the aggregation of soil by cationic bridges between the soil organic matter and clay particles, and thus increases the soil porosity. The greater the total porosity, the greater the leaching of the soluble and exchangeable Na ions, and the greater the subsequent reduction in soil sodicity and salinity as expressed by the ESP and EC values, respectively [30]. Salt adsorbing ability of organic amendments is also known to be considered in lowering the EC of soils [140]. The incorporation of different organic amendments in soils had often contributed to a slight increase in EC depending on the rate and duration of incorporation [139, 146]. The increase in EC after the application of organic manures could be attributed to the presence of high amounts of K and Ca [141]. The role of organic matter in increasing the CEC has already been established. Hue [170] reported the decrease in Na in soil solution due to greater CEC resulting from incorporation of organic matter. The removal of organic matter had also been found to increase the CEC of a specific soil, which could be due to the exposure of the permanent charge of the montmorillonitic clay that was blocked by the interaction of the organic matter with the clay [171].

Several studies reported a significant increase in the soil microbial and enzymatic activities as a result of organic matter incorporation. The incorporation of organic amendments resulted in enhanced enzymatic activities [31, 33, 140, 161, 164, 172], microbial biomass C [32, 143, 161, 164, 173], microbial biomass N [32, 161], soil microbial activity as expressed by basal respiration [32, 174], and nematode abundance [143]. However, the response of microbial and enzymatic activities to organic amendments differs depending on the kinds of amendments, rates of incorporation, the types of crops grown, etc. The rate of microbial respiration was found to be highest in poultry manure-amended soils compared to reed-, composted straw-, and straw-treated soils, and all the amended soils resulted in a rapid increase in respiration rate in the beginning of incorporation that decreased gradually with time [163]. Tejada et al. [33] also observed the highest soil microbial biomass C and cumulative C-CO_2 in soils amended with a maximum dose of poultry manure that was 37% higher compared to cotton gin crushed compost-amended soils. Similarly, the urease, protease, b-glucosidase, phosphatase, arylsulfatase, and dehydrogenase activities were found to be increased by 34, 18, 37, 39, 40, and 30%, respectively, in poultry manure-amended soils compared to the cotton gin crushed compost-amended soils. In another study, Liang et al. [31] found that the urease activity increased by 62.3 and 117.4%, respectively, in pig manure- and rice straw plus pig manure-amended soils in comparison to rice straw treatment alone. The author also observed that the addition of rice straw, poultry manure, and their combination increased the urease activity by 21, 96, 163%, respectively, in rice, whereas by 57.4, 93.1, and 152.5%, respectively, in barley compared to the control. On the other hand, Zhang et al. [164] found dual effects where the application of vermicompost increased the activities of dehydrogenase, urease, and phosphatases by 37–68%, 22–107%, and 3.4–56%, respectively, while vermicompost addition decreased the activities of β-1,4-glucosidase and β-1,4-N-acetylglucosaminidase by 17–53% and 24–42%, respectively. Easily decomposable organic substances such as biosolids, swine manure, and chicken manure may likely be retained in the soil over short periods, resulting in an intense and short effect, while recalcitrant, lignin-rich amendments such as woody biomass have a smaller but long-lasting effect on these soil properties [131, 175].

Several studies reported the beneficial effects of phytohormones and plant growth-promoting rhizobacteria in modifying the physiological and metabolic responses of plants to salt stress, enhancing their tolerance as well as growth and yield [2, 176–178]. Egamberdieva [179] found that the indoleacetic acid producing bacterial strains significantly increased the seedling root growth of wheat up to 52% compared to control under conditions of soil salinity. However, in different studies, the morphological and physiological growth and yield attributes of crops were found to be increased under abiotic stress when several plant growth regulators were applied in combination compared to their single dose [180, 181].

The beneficial effects of the organic amendments on physical, chemical, and biological properties of SAS greatly influence the growth, nutrient uptake, and accumulation of plants under salt stress. The application of organic amendments in SAS is considered a useful and effective way to increase soil fertility and enhance crop growth [31, 157]. The application of organic amendments in SAS increased the biomass yield of alfalfa, *Medicago sativa* [182]; barley, *Hordeum vulgare* [31]; cotton, *Gossypium hirsutum* [143]; maize, *Zea mays* [32, 159, 183]; onion, *Allium cepa* [142]; rice, *Oryza sativa* [31, 153, 184]; seepweed, *Suaeda salsa* [185]; sweet fennel, *Foeniculum vulgare* [186]; tomato, *Solanum lycopersicum* [187]; and wheat, *Triticum aestivum* [139]. The quantitative and qualitative improvements in the growth and yield attributes of crops as affected by abiotic stresses in the presence of different additives might be due to the enhanced photosynthesis, cholorophyll contents, stomatal conductance, water-use efficiency, and synthesis of metabolites [137, 188–191].

Organic manure incorporation into the SAS also increased the N, P, and K contents in rice, *Oryza sativa*, barley, *Hordeum vulgare* [31], and sweet fennel, *Foeniculum vulgare* [186]; K content in rice, *Oryza sativa* [184, 192]; and K and Ca contents in tomato, *Solanum lycopersicum* [187], while it decreased the Na uptake [31, 184, 186, 187]. Improved soil physical conditions, availability of macro- and micronutrients, and enhanced microbial activities in soil resulting from the incorporation of organic amendments lead to better growth and yield of crops under salt stress [30, 175]. Maintaining a high K:Na ratio as a result of organic manure incorporation is an important mechanism of plants to resist the harmful effects of salts and perform better growth [31, 186]. Decreased uptake of Na may be due to the action of organic matter which acts as salt-ion chelating agents detoxifying the toxic ions, especially Na and Cl [184]. The C:N ratio also determines the growth of plants by influencing the availability of nutrients, especially N. Incorporation of organic amendments having a lower C:N ratio attributes to higher N availability [147].

5.3 Reclamation with inorganic amendments

As saline soils are usually good in structure, removal of excess salts can be obtained merely by the leaching process, and in most cases, the application of inorganic amendments is unnecessary. However, in the case of saline-sodic and sodic soils, exchangeable Na must first be removed from the exchange sites of soil particles and then leached to wash out from the root zone. As sodic soils are characterized by poor soil structure and limited infiltration rate, in addition to organic materials, various inorganic amendments are used to improve the soil structure and facilitate the leaching process. The application of Ca containing salt especially gypsum along with the organic amendments in SAS, in order to replace exchangeable Na, improves the physical condition of the soil, facilitates leaching of salts, and increases crop yield, which has been previously been reported in several studies [142, 184]. The application of gypsum followed by leaching of soils enhanced the reclamation and decreased the salinity as well as the sodicity levels [193]. Khattak et al. [194] observed a decrease in the pH, EC, and SAR of leached soils and an increase in the yield of rice and wheat by 9.8–25.3% and 10–80%, respectively, in salt-affected soil as a result of gypsum application. Khosla et al. [195] reported that the use of additional quantities of water can be minimized by the application of gypsum to achieve a reduced SAR value to a greater extent. The amount of gypsum required to reclaim saline-sodic and sodic soils is based on the amount of exchangeable Na, soil texture, leaching rate, crop to be sown, solubility, and reaction rates of the amendments [34, 125]. Abdel-Fattah et al. [193] studied the effects of different size fractions of gypsum (<0.5, 0.5–1, and 1.0–2.0 mm) on the efficiency of the reclamation of SAS and found that the salinity and sodicity decreased with the increasing fineness of gypsum. Gypsum is usually required to spread uniformly in the field and is incorporated into the upper 10–15 cm of soil by 2–3 shallow plowings at least 10–15 days before planting [34].

Zeolite ($CaAl_2Si_4O_{12}{\cdot}nH_2O$), an aluminosilicate, had also been studied as an inorganic amendment with the aims to reclaim SAS and enhance plant growth. Zeolite can enhance plant growth and nutrient uptake by mitigating salt stress. In an experiment, Al-Busaidi [196] studied 1 and 5% rates of zeolite and found that the application of zeolite resulted in a significant increase in the biomass of barley, *Hordeum vulgare*, and increased concentration of Ca, Mg, and K in postharvest soils. Application of zeolite in the soil also increased fresh and dry weights of shoots and roots, fruit weight, and the number of achenes of strawberry, *Fragaria ananassa*, as well as the available N, P, K, Ca, and Mg of the medium [197]. Milosevic and Milosevic [198] found higher amounts of humus, total N, and available P and K in soil along with a significant increase in the shoot length and trunk cross-sectional area by the application of zeolite in combination with cattle manure and inorganic fertilizer. Zeolite is characterized by large sorption and ion-exchange capacity. As a sorbent, it has an important effect on the mobilization of heavy metals as well as micronutrients and macronutrients [199]. In the structure of zeolite, the negative charges developed through the replacement of quadri-charged silicon cations by triply charged aluminum can be balanced by the adsorption of Na under salt stress conditions. Besides, the three-dimensional framework of zeolite is made up of $[SiO_4]^{4-}$ and $[AlO_4]^{5-}$ tetrahedra, which are bonded together by sharing the oxygen atoms located at the corner of each tetrahedron in such a way that the framework develops voids or pores in the form of cages and channels between the tetrahedra [200]. The incorporation of zeolite into SAS thus can lead to the adsorption of Na on the surfaces or entrapment in the void spaces, resulting in decreased uptake by plants. Moreover, zeolite plays an important source of Ca (CaO, 16.0%) to the soil-plant system. The release of Ca in the root media from the Ca-type zeolite can maintain a high Ca:Na ratio in the shoot and root by decreasing the Na while increasing the Ca uptake [201]. While using zeolite as an amendment to reclaim SAS, the concentration of Na should be taken into consideration. The application of zeolite contributed to a substantial increase of Na in soil and plant [196]. Besides, in soils having low pH (below 4.2), decomposition of zeolite and concomitant placement of Al^{3+} and Mn^{2+} ions in the sorption complex may lead to increased leaching of Mg and Ca, root damage, deficiency of Mg and P, toxicity of Mn and Fe, and restricted plant growth [199].

6. Conclusion

Plants are subjected to various abiotic and biotic stresses due to natural or human interferences. Among the abiotic factors, the problems of soil salinity and sodicity occurring in arid, humid, coastal, and even in irrigated agricultural lands possess great threats to sustainable food security worldwide. Due to differences in properties of saline, sodic, and saline-sodic soils, their ways of stresses to plants as well as amelioration approaches are different. The adverse effects of salt stress on the uptake of essential nutrients by plants vary depending on the genotype, growth stage, concentration of salts in medium, etc. While salinity can be reclaimed by leaching of salts through good quality water together with a proper drainage system, saline-sodic and sodic soils cannot be reclaimed merely by leaching. The application of inorganic and organic amendments is often required to reclaim the saline-sodic and sodic soils. The incorporation of organic amendments is beneficial to reclaim saline as well as saline-sodic and sodic soils. Organic amendments contribute to the physical, chemical, and biological improvements of saline, saline-sodic, and sodic soils, enhancing the magnitude of their reclamation. Besides, organic amendments act as important sources of essential plant nutrients. Therefore, application of organic amendments in combination with the judicious application of inorganic amendments can be a better approach to improve the properties of SAS and the plant's response to salt stress for sustainable crop production and food security.

Conflict of interest

The authors declare no conflict of interest.

Author details

Sajal Roy* and Nasrin Chowdhury
Department of Soil Science, Faculty of Biological Sciences, University of Chittagong, Chittagong, Bangladesh

*Address all correspondence to: sajal.roy@cu.ac.bd

References

[1] Alam SM. Nutrient uptake by plants under stress conditions. In: Pessarakli M, editor. Handbook of Plant and Crop Stress. 2nd ed. New York, USA: Marcel Dekker, Inc; 1999. pp. 285-313

[2] Fahad S, Hussain S, Bano A, Saud S, Hassan S, Shan D, et al. Potential role of phytohormones and plant growth-promoting rhizobacteria in abiotic stresses: Consequences for changing environment. Environmental Science and Pollution Research. 2015;**22**:4907-4921. DOI: 10.1007/s11356-014-3754-2

[3] Fahad S, Bajwa AA, Nazir U, Anjum SA, Farooq A, Zohaib A, et al. Crop production under drought and heat stress: Plant responses and management options. Frontiers in Plant Science. 2017;**8**:1147. DOI: 10.3389/fpls.2017.01147

[4] Gull A, Lone AA, Wani NUI. Biotic and abiotic stresses in plants. In: de Oliveira AB, editor. Abiotic and Biotic Stress in Plants. London, UK: IntechOpen; 2019. pp. 1-6. DOI: 10.5772/intechopen.85832

[5] Akram R, Turan V, Hammad HM, Ahmad S, Hussain S, Hasnain A, et al. Fate of organic and inorganic pollutants in paddy soils. In: Hashmi MZ, Varma A, editors. Environmental Pollution of Paddy Soils. Cham: Springer; 2018. pp. 197-214. DOI: 10.1007/978-3-319-93671-0_13

[6] Fahad S, Chen Y, Saud S, Wang K, Xiong D, Chen C, et al. Ultraviolet radiation effect on photosynthetic pigments, biochemical attributes, antioxidant enzyme activity and hormonal contents of wheat. Journal of Food, Agriculture and Environment. 2013;**11**(3&4):1635-1641. DOI: 10.1234/4.2013.4911

[7] Fahad S, Ihsan MZ, Khaliq A, Daur I, Saud S, Alzamanan S, et al. Consequences of high temperature under changing climate optima for rice pollen characteristics - Concepts and perspectives. Archives of Agronomy and Soil Science. 2018;**64**(11):1473-1488. DOI: 10.1080/03650340.2018.1443213

[8] Fahad S, Rehman A, Shahzad B, Tanveer M, Saud S, Kamran M, et al. Rice responses and tolerance to metal/metalloid toxicity. In: Hasanuzzaman M, Fujita M, Nahar K, Biswas JK, editors. Advances in Rice Research for Abiotic Stress Tolerance. Woodhead Publishing; 2019. pp. 299-312. DOI: 10.1016/B978-0-12-814332-2.00014-9

[9] Khan A, Tan DKY, Afridi MZ, Luo H, Tung SA, Ajab M, et al. Nitrogen fertility and abiotic stresses management in cotton crop: A review. Environmental Science and Pollution Research. 2017;**24**:14551-14566. DOI: 10.1007/s11356-017-8920-x

[10] Shah F, Nie L, Cui K, Shah T, Wu W, Chen C, et al. Rice grain yield and component responses to near 2°C of warming. Field Crops Research. 2014;**157**:98-110. DOI: 10.1016/j.fcr.2013.12.014

[11] Yang Z, Zhang Z, Zhang T, Fahad S, Cui K, Nie L, et al. The effect of season-long temperature increases on rice cultivars grown in the central and southern regions of China. Frontiers in Plant Science. 2017;**8**:1908. DOI: 10.3389/fpls.2017.01908

[12] Zia Z, Bakhat HF, Saqib ZA, Shah GM, Fahad S, Ashraf MR, et al. Effect of water management and silicon on germination, growth, phosphorus and arsenic uptake in rice. Ecotoxicology and Environmental Safety. 2017;**144**:11-18. DOI: 10.1016/j.ecoenv.2017.06.004

[13] Al-Karaki GN. Growth, sodium, and potassium uptake and translocation

in salt stressed tomato. Journal of Plant Nutrition. 2000;**23**(3):369-379. DOI: 10.1080/01904160009382023

[14] Talei D, Kadir MA, Yusop MK, Valdiani A, Abdullah MP. Salinity effects on macro and micronutrients uptake in medicinal plant king of bitters (*Andrographis paniculata* Nees.). Plant Omics Journal. 2012;**5**(3):271-278

[15] Zribi OT, Slama I, Trabelsi N, Hamdi A, Smaoui A, Abdelly C. Combined effects of salinity and phosphorus availability on growth, gas exchange, and nutrient status of *Catapodium rigidum*. Arid Land Research and Management. 2018;**32**(3):277-290. DOI: 10.1080/15324982.2018.1427640

[16] Etesami H, Noori F. Soil salinity as a challenge for sustainable agriculture and bacterial-mediated alleviation of salinity stress in crop plants. In: Manoj Kumar M, Etesami H, Kumar V, editors. Saline Soil-Based Agriculture by Halotolerant Microorganisms. Singapore: Springer; 2019. pp. 1-22. DOI: 10.1007/978-981-13-8335-9

[17] Khan MA, Duke NC. Halophytes - A resource for the future. Wetlands Ecology and Management. 2001;**6**:455-456. DOI: 10.1023/A:1012211726748

[18] Maksimovic I, Ilin Ž. Effects of salinity on vegetable growth and nutrients uptake. In: Lee TS, editor. Irrigation Systems and Practices in Challenging Environments. London, UK: IntechOpen; 2012. pp. 169-190. DOI: 10.5772/29976

[19] Sahin U, Ekincib M, Orsa S, Turanc M, Yildizb S, Yildirimb E. Effects of individual and combined effects of salinity and drought on physiological, nutritional and biochemical properties of cabbage (*Brassica oleracea* var. capitata). Scientia Horticulturae. 2018;**240**:196-204. DOI: 10.1016/j.scienta.2018.06.016

[20] Dluzniewska P, Gessler A, Dietrich H, Schnitzler JP, Teuber M, Rennenberg H. Nitrogen uptake and metabolism in *Populus* x *canescens* as affected by salinity. The New Phytologist. 2007;**173**(2):279-293. DOI: 10.1111/j.1469-8137.2006.01908.x

[21] Maathuis FJM, Amtmann A. K^+ nutrition and Na^+ toxicity: The basis of cellular K^+/Na^+ ratios. Annals of Botany. 1999;**84**(2):123-133. DOI: 10.1006/anbo.1999.0912

[22] Maria SG, Rubio F, Dubcovsky J, Rodriguez-Navarro A. The HAK1 gene of barley is a member of a large gene family and encodes a high-affinity potassium transporter. The Plant Cell. 1997;**9**:2281-2289. DOI: 10.1105/tpc.9.12.2281

[23] Wakeel A. Potassium-sodium interactions in soil and plant under saline-sodic conditions. Journal of Plant Nutrition and Soil Science. 2013;**176**: 344-354. DOI: 10.1002/jpln.201200417

[24] Hadi MR, Khiyam-Nekoie SM, Khavarinejad R, Sima NAKK, Yavari P. Accumulation and role of ions (Ca^{2+}, Mg^{2+}, SO_4^{2-}) on salt tolerance in *Triticum turgidum* L. Journal of Biological Sciences. 2008;**8**(1):143-148. DOI: 10.3923/jbs.2008.143.148

[25] Hu Y, Schmidhalter U. Effects of salinity and macronutrient levels on micronutrients in wheat. Journal of Plant Nutrition. 2001;**24**(2):273-281. DOI: 10.1081/PLN-100001387

[26] Osman KT. Management of Soil Problems. 1st ed. Switzerland: Springer International Publishing AG; 2018. p. 474. DOI: 10.1007/978-3-319-75527-4

[27] Shaaban MM, El-Fouly MM. Nutrient content and salt removal potential of wild plants grown in salt affected soils. Acta Horticulturae. 2002;**573**:377-385. DOI: 10.17660/ActaHortic.2002.573.45

[28] Jouyban Z. The effects of salt stress on plant growth. Technical Journal of Engineering and Applied Sciences. 2012;**2**(1):7-10

[29] Diacono M, Montemurro F. Effectiveness of organic wastes as fertilizers and amendments in salt-affected soils. Agriculture. 2015;**5**:221-230. DOI: 10.3390/agriculture5020221

[30] Wang L, Sun X, Li S, Zhang T, Zhang W, Zhai P. Application of organic amendments to a coastal saline soil in North China: Effects on soil physical and chemical properties and tree growth. PLoS One. 2014;**9**(2):e89185. DOI: 10.1371/journal.pone.0089185

[31] Liang YC, Chen Q, Liu Q, Zhang WH, Ding RX. Exogenous silicon (Si) increases antioxidant enzyme activity and reduces lipid peroxidation in roots of salt-stressed barley (*Hordeum vulgare* L.). Journal of Plant Physiology. 2003;**160**(10):1157-1164. DOI: 10.1078/0176-1617-01065

[32] Oo AN, Iwai CB, Saenjan P. Soil properties and maize growth in saline and nonsaline soils using cassava-industrial waste compost and vermicompost with or without earthworms. Land Degradation and Development. 2015;**26**(3):300-310. DOI: 10.1002/ldr.2208

[33] Tejada M, Garcia C, Gonzalez JL, Hernandez MT. Use of organic amendment as a strategy for saline soil remediation: Influence on the physical, chemical and biological properties of soil. Soil Biology and Biochemistry. 2006;**38**(6):1413-1421. DOI: 10.1016/j.soilbio.2005.10.017

[34] Abrol IP, Yadav JSP, Massoud FI. Salt Affected Soils and Their Management. FAO Soils Bulletin 39. Rome: Food and Agriculture Organization of the United Nations; 1988

[35] Brady NC, Weil RR. The Nature and Properties of Soils. 13th ed. Delhi, India: Pearson-Prentice Hall; 2005. p. 960

[36] Doula MK, Sarris A. Soil environment. In: Poulopoulos SG, Inglezakis VJ, editors. Environment and Development: Basic Principles, Human Activities, and Environmental Implications. UK: Academic Press; 2016. pp. 213-286. DOI: 10.1016/C2012-0-07133-5

[37] Endo T, Yamamoto S, Larrinaga JA, Fujiyama H, Honna T. Status and causes of soil salinization of irrigated agricultural lands in Southern Baja California, Mexico. Applied and Environmental Soil Science. 2011;**2011**:1-12. DOI: 10.1155/2011/873625

[38] Ghassemi F, Jakeman AJ, Nix HA. Salinization of Land and Water Resources: Human Causes, Extent, Management and Case Studies. Wallingford: CABI Publishing; 1995. p. 526

[39] Hussain S, Mubeen M, Ahmad A, Akram W, Hammad HM, Ali M, et al. Using GIS tools to detect the land use/land cover changes during forty years in Lodhran District of Pakistan. Environmental Science and Pollution Research. 2019. DOI: 10.1007/s11356-019-06072-3

[40] Jamil A, Riaz S, Ashraf M, Foolad MR. Gene expression profiling of plants under salt stress. Critical Reviews in Plant Sciences. 2011;**30**(5):435-458. DOI: 10.1080/07352689.2011.605739

[41] Machado RMA, Serralheiro RP. Soil salinity: Effect on vegetable crop growth. Management practices to prevent and mitigate soil salinization. Horticulturae. 2017;**3**:30. DOI: 10.3390/horticulturae3020030

[42] Sairam RK, Tyagi A. Physiology and molecular biology of salinity stress tolerance in plants. Current Science. 2004;**86**:407-421. DOI: 10.1007/1-4020-4225-6

[43] Szabolcs I. Salt affected soils as the ecosystem for halophytes. In: Squires VR,

Ayoub AT, editors. Halophytes as a Resource for Livestock and for Rehabilitation of Degraded Lands. 1st ed. Netherlands: Springer; 1994. pp. 19-24. DOI: 10.1007/978-94-011-0818-8

[44] Szabolcs I. Salt-Affected Soil in Europe. 1st ed. Netherlands: Springer; 1974. p. 57

[45] Hardie M, Doyle R. Measuring soil salinity. In: Shabala S, Cuin TA, editors. Plant Salt Tolerance: Methods and Protocols, Methods in Molecular Biology. Totowa, NJ: Humana Press; 2012. pp. 415-425. DOI: 10.1007/978-1-61779-986-0_28

[46] Rengasamy P. World salinization with emphasis on Australia. Journal of Experimental Botany. 2006;**57**(5):1017-1023. DOI: 10.1093/jxb/erj108

[47] Chen W, Hou Z, Wu L, Liang Y, Wei C. Effects of salinity and nitrogen on cotton growth in arid environment. Plant and Soil. 2010;**326**:61-73. DOI: 10.1007/s11104-008-9881-0

[48] Cramer GR. Sodium-calcium interactions under salinity stress. In: Läuchli A, Lüttge U, editors. Salinity: Environment-Plants-Molecules. 1st ed. Netherlands: Springer; 2002. pp. 205-227. DOI: 10.1007/0-306-48155-3

[49] Patel NT, Gupta A, Pandey AN. Salinity tolerance of *Avicennia marina* (Forssk.) Vierh. from Gujarat coasts of India. Aquatic Botany. 2010;**93**(1):9-16. DOI: 10.1016/j.aquabot.2010.02.002

[50] Rahman MHU, Ahmad A, Wajid A, Hussain M, Rasul F, Ishaque W, et al. Application of CSM-CROPGRO-cotton model for cultivars and optimum planting dates: Evaluation in changing semi-arid climate. Field Crops Research. 2019;**238**:139-152. DOI: 10.1016/j.fcr.2017.07.007

[51] Shannon MC, Grieve CM, Francois LE. Whole-plant response to salinity. In: Wilkinson RE, editor. Plant-Environment Interactions. New York: Marcel Dekker; 1994. pp. 199-244

[52] Zamin M, Khattak AM, Salim AM, Marcum KB, Shakur M, Shah S, et al. Performance of *Aeluropus lagopoide*s (mangrove grass) ecotypes, a potential turfgrass, under high saline conditions. Environmental Science and Pollution Research. 2019;**26**:13410-13421. DOI: 10.1007/s11356-019-04838-3

[53] Shin W, Siddikee MA, Joe MM, Benson A, Kim K, Selvakumar G, et al. Halotolerant plant growth promoting bacteria mediated salinity stress amelioration in plants. Korean Journal of Soil Science and Fertilizer. 2016;**49**(4):355-367. DOI: 10.7745/KJSSF.2016.49.4.355

[54] Bayuelo-Jiménez JS, Debouck DG, Lynch JP. Growth, gas exchange, water relations, and ion composition of *Phaseolus* species grown under saline conditions. Field Crops Research. 2003;**80**(3):207-222. DOI: 10.1016/S0378-4290(02)00179-X

[55] Grattan SR, Grieves CM. Salinity-mineral nutrient relations in horticultural crops. Scientia Horticulturae. 1999;**78**:127-157. DOI: 10.1016/S0304-4238(98)00192-7

[56] Hauser F, Horie T. A conserved primary salt tolerance mechanism mediated by HKT transporters: A mechanism for sodium exclusion and maintenance of high K^+/Na^+ ratio in leaves during salinity stress. Plant, Cell & Environment. 2010;**33**:552-565. DOI: 10.1111/j.1365-3040.2009.02056.x

[57] Romero-Aranda R, Soria T, Cuartero J. Tomato plant-water uptake and plant-water relationships under saline growth conditions. Plant Science. 2001;**160**:265-272. DOI: 10.1016/s0168-9452(00)00388-5

[58] Ayers RS, Westcot DW. Water Quality for Agriculture. FAO Irrigation

and Drainage Paper. 29 Rev. 1. Food and Agriculture Organization of the United Nations, Rome; 1985

[59] Munns R, Tester M. Mechanisms of salinity tolerance. Annual Review of Plant Biology. 2008;**59**:651-681. DOI: 10.1146/annurev.arplant.59.032607.092911

[60] Seelig BD. Salinity and Sodicity in North Dakota Soils. Fargo, North Dakota: NDSU Extension Service, North Dakota State University; 2000

[61] Dai J, Duan L, Dong H. Comparative effect of nitrogen forms on nitrogen uptake and cotton growth under salinity stress. Journal of Plant Nutrition. 2015;**38**(10):1530-1543. DOI: 10.1080/01904167.2014.983126

[62] Long XH, Chi JH, Liu L, Li Q, Liu ZP. Effect of seawater stress on physiological and biochemical responses of five Jerusalem artichoke ecotypes. Pedosphere. 2009;**19**(2):208-216. DOI: 10.1016/S1002-0160(09)60110-7

[63] Shahriaripour R, Pour AT, Mozaffari V. Effects of salinity and soil phosphorus application on growth and chemical composition of pistachio seedlings. Communications in Soil Science and Plant Analysis. 2011;**42**:144-158. DOI: 10.1080/00103624.2011.535065

[64] Abdelgadir EM, Oka M, Fujiyama H. Characteristics of nitrate uptake by plants under salinity. Journal of Plant Nutrition. 2005;**28**(1):33-46. DOI: 10.1081/PLN-200042156

[65] Brown CE, Pezeshki SR, DeLaune RD. The effects of salinity and soil drying on nutrient uptake and growth of *Spartina alterniflora* in a simulated tidal system. Environmental and Experimental Botany. 2006;**58**:140-148. DOI: 10.1016/j.envexpbot.2005.07.006

[66] Rahimi A, Biglarifard A. Influence of NaCl salinity and different substrates on plant growth, mineral nutrient assimilation and fruit yield of strawberry. Notulae Botanicae Horti Agrobotanici Cluj-Napoca. 2011;**39**(2):219-226

[67] Botella MA, Martínez V, Nieves M, Cerdá A. Effect of salinity on the growth and nitrogen uptake by wheat seedlings. Journal of Plant Nutrition. 1997;**20**(6):793-804. DOI: 10.1080/01904169709365295

[68] Engels C, Marschner H. Root to shoot translocation of macronutrients in relation to shoot demand in maize (*Zea mays* L.) grown at different root zone temperature. Zeitschrift für Pflanzenernährung und Bodenkunde. 1992;**155**:121-128. DOI: 10.1002/jpln.19921550209

[69] Hadi MR, Karimi N. The role of calcium in plants' salt tolerance. Journal of Plant Nutrition. 2012;**35**(13):2037-2054. DOI: 10.1080/01904167.2012.717158

[70] Qadir M, Schubert S, Steffens D. Phytotoxic substances in soils. In: Hillel D, editor. Encyclopedia of Soils in the Environment. UK: Academic Press; 2005. pp. 216-222

[71] Hu Y, Schmidhalter U. Drought and salinity: A comparison of their effects on mineral nutrition of plants. Journal of Plant Nutrition and Soil Science. 2005;**168**:541-549. DOI: 10.1002/jpln.200420516 541

[72] Silva JV, Lacerda CF, Costa PHA, Enéas Filho J, Filho EG, Prisco JT. Physiological responses of NaCl stressed cowpea plants grown in nutrient solution supplemented with $CaCl_2$. Brazilian Journal of Plant Physiology. 2003;**15**:99-105. DOI: 10.1590/S1677-04202003000200005

[73] Elser JJ, Fagan WF, Denno RF, Dobberfuhl DR, Folarin A, Huberty A, et al. Nutritional constraints in terrestrial and freshwater food webs.

Nature. 2000;**408**:578-580. DOI: 10.1038/35046058

[74] Blumenthal JM, Baltensperger DD, Cassman KG, Mason SC, Pavlista AD. Importance and effect of nitrogen on crop quality and health. In: Hatfield JL, Follett RF, editors. Nitrogen in the Environment: Sources, Problems, and Management. 2nd ed. Elsevier Inc.; 2008. pp. 51-70. DOI: 10.1016/B978-0-12-374347-3.00003-2

[75] Chen B, Yang H, Song W, Liu C, Xu J, Zhao W, et al. Effect of N fertilization rate on soil alkali-hydrolyzable N, subtending leaf N concentration, fiber yield, and quality of cotton. The Crop Journal. 2016;**4**(4):423-430. DOI: 10.1016/j.cj.2016.03.006

[76] Hammad HM, Farhad W, Abbas F, Fahad S, Saeed S, Nasim W, et al. Maize plant nitrogen uptake dynamics at limited irrigation water and nitrogen. Environmental Science and Pollution Research. 2017;**24**:2549-2557. DOI: 10.1007/s11356-016-8031-0

[77] Hofman G, Cleemput OV. Soil and Plant Nitrogen. 1st ed. Paris, France: International Fertilizer Industry Association; 2004. p. 48

[78] Bybordi A, Ebrahimian E. Effect of salinity stress on activity of enzymes involved in nitrogen and phosphorous metabolism case study: Canola (*Brassica napus* L.). Asian Journal of Agricultural Research. 2011;**5**:208-214. DOI: 10.3923/ajar.2011.208.214

[79] Kanagaraj A, Desingh R. Salinity influences physiological traits of seven sesame (*Sesamum indicum* L.) varieties. Journal of Scientific Agriculture. 2017;**1**:188-196. DOI: 10.25081/jsa.2017.v1.59

[80] Ashraf M, Shahzad SM, Imtiaz M, Rizwan MS. Salinity effects on nitrogen metabolism in plants - Focusing on the activities of nitrogen metabolizing enzymes: A review. Journal of Plant Nutrition. 2018;**41**(8):1065-1081. DOI: 10.1080/01904167.2018.1431670

[81] Aslam M, Huffaker RC, Rains DW. Early effects of salinity on nitrate assimilation in barley seedlings. Plant Physiology. 1984;**76**(2):321-325. DOI: 10.1104/pp.76.2.321

[82] Ward MR, Aslam M, Huffaker RC. Enhancement of nitrate uptake and growth of barley seedlings by calcium under saline conditions. Plant Physiology. 1986;**80**:520-523. DOI: 10.1104/pp.80.2.520

[83] Lacerda CF, Assis Júnior JO, Filho LCAL, de Oliveira TS, Guimarães FVA, Gomes-Filho E, et al. Morpho-physiological responses of cowpea leaves to salt stress. Brazilian Journal of Plant Physiology. 2006;**18**(4):455-465. DOI: 10.1590/S1677-04202006000400003

[84] Saud S, Fahad S, Cui G, Yajun C, Anwar S. Determining nitrogen isotopes discrimination under drought stress on enzymatic activities, nitrogen isotope abundance and water contents of Kentucky bluegrass. Scientific Reports. 2020;**10**:6415. DOI: 10.1038/s41598-020-63548-w

[85] Chakraborty K, Sairam RK, Bhaduri D. Effects of different levels of soil salinity on yield attributes, accumulation of nitrogen, and micronutrients in *Brassica* spp. Journal of Plant Nutrition. 2015;**39**(7):1026-1037. DOI: 10.1080/01904167.2015.1109105

[86] Pessarakli M, Haghighi M, Sheibanirad A. Plant responses under environmental stress conditions. Advances in Plants & Agriculture Research. 2015;**2**(6):276-286. DOI: 10.15406/apar.2015.02.00073

[87] Hutchins DA, Fu FX. Linking the oceanic biogeochemistry of iron and phosphorus with the marine nitrogen

Cycle. In: Capone DG, Bronk DA, Mulholland MR, Carpenter EJ, editors. Nitrogen in the Marine Environment. 2nd ed. Academic Press, UK; 2008. pp. 1627-1666. DOI: 10.1016/B978-0-12-372522-6.00038-4

[88] Menzies N. The Science of Phosphorus Nutrition: Forms in the Soil, Plant Uptake, and Plant Response [Internet]. 2009. Available from: https://grdc.com.au/resources-and-publications/grdc-update-papers/tab-content/grdc-update-papers/2009/02/the-science-of-phosphorus-nutrition-forms-in-the-soil-plant-uptake-and-plant-response [Accessed: 08 June 2020]

[89] Syers JK, Johnston AE, Curtin D. Efficiency of Soil and Fertilizer Phosphorus Use: Reconciling Changing Concepts of Soil Phosphorus Behaviour with Agronomic Information. FAO Fertilizer and Plant Nutrition Bulletin 18. Rome: Food and Agriculture Organization of the United Nations; 2008

[90] Khan MZ, Islam MA, Azom MG, Amin MS. Short term influence of salinity on uptake of phosphorus by *Ipomoea aquatica*. International Journal of Plant & Soil Science. 2018;**25**(2):1-9. DOI: 10.9734/IJPSS/2018/44822

[91] Marschner H. Mineral Nutrition of Higher Plants. 2nd ed. UK: Academic Press; 1995. p. 889

[92] Maathuis FJM, Podar D. Uptake, distribution, and physiological functions of potassium, calcium, and magnesium. In: Hawkesford MJ, Barraclough P, editors. The Molecular and Physiological Basis of Nutrient Use Efficiency in Crops. 1st ed. UK: John Wiley & Sons Inc; 2011. pp. 265-293. DOI: 10.1002/9780470960707

[93] Epstein E, Bloom AJ. Mineral Nutrition of Plants: Principles and Perspectives. 2nd ed. Sunderland, Massachusetts: Sinauer Associates Inc.; 2004. p. 380

[94] Greenway H, Munns R. Mechanisms of salt tolerance in nonhalophytes. Annual Review of Plant Physiology. 1980;**31**:149-190. DOI: 10.1146/annurev.pp.31.060180.001053

[95] Reggiani R, Bonzo S, Bertani A. The effect of salinity on early seedling growth of seeds of three wheat (*Triticum aestivum* L.) cultivars. Canadian Journal of Plant Science. 1995;**75**:175-177. DOI: 10.4141/cjps95-029

[96] Huq SMI, JUM S. The Soils of Bangladesh. 1st ed. Netherlands: Springer; 2013. p. 165. DOI: 10.1007/978-94-007-1128-0

[97] Mahmud MS, Roy S. Physico-chemical properties of soils collected from different agricultural lands of Chittagong district of Bangladesh. Academia Journal of Environmental Science. 2020;**8**(2):22-29. DOI: 10.15413/ajes.2019.0315

[98] Rameeh V. Ions uptake, yield and yield attributes of rapeseed exposed to salinity stress. Journal of Soil Science and Plant Nutrition. 2012;**12**(4):851-861. DOI: 10.4067/S0718-95162012005000037

[99] Jin ZM, Wang CH, Liu ZP, Gong WJ. Physiological and ecological characters studies on *Aloe vera* under soil salinity and seawater irrigation. Process Biochemistry. 2007;**42**(4):710-714. DOI: 10.1016/j.procbio.2006.11.002

[100] Jiang Q, Roche D, Monaco TA, Durham S. Gas exchange, chlorophyll fluorescence parameters and carbon isotope discrimination of 14 barley genetic lines in response to salinity. Field Crops Research. 2006;**96**:269-278. DOI: 10.1016/j.fcr.2005.07.010

[101] Fahad S, Bano A. Effect of salicylic acid on physiological and biochemical characterization of maize grown in saline area. Pakistan Journal of Botany. 2012;**4**(4):1433-1438

[102] Saleque MA, Choudhury NN, Rezaul-Karim SM, Panaullah GM. Mineral nutrient and yield of four rice genotypes in the farmer's fields of salt affected soils. Journal of Plant Nutrition. 2005;**28**(5):865-875. DOI: 10.1081/PLN-200055560

[103] Adams P, Ho LC. Effects of constant and fluctuating salinity in the yield, quality and calcium status of tomatoes. Journal of Horticulture Science. 1989;**64**:725-732. DOI: 10.1080/14620316.1989.11516015

[104] Kaya C, Kirnak H, Higgs D. Enhancement of growth and normal growth parameters by foliar application of potassium and phosphorus in tomato cultivars grown at high (NaCl) salinity. Journal of Plant Nutrition. 2001;**24**:357-367. DOI: 10.1081/PLN-100001394

[105] Glenn E, Pfister R, Brown J, Thompson T, O'Leary J. Na and K accumulation and salt tolerance of *Atriplex canescens* (Chenopodiaceae) genotypes. American Journal of Botany. 1996;**83**(8):997-1005. DOI: 10.2307/2445988

[106] Fuchs I, Stölzle S, Ivashikina N, Hedrich R. Rice K^+ uptake channel OsAKT1 is sensitive to salt stress. Planta. 2005;**221**(2):212-221. DOI: 10.1007/s00425-004-1437-9

[107] Qi Z, Spalding EP. Protection of plasma membrane K^+ transport by the salt overly sensitive Na^+-H^+ antiporter during salinity stress. Plant Physiology. 2004;**136**:2548-2555. DOI: 10.1104/pp.104.049213

[108] Idowu MK, Aduayi EA. Effects of sodium and potassium application on water content and yield of tomato in southwestern Nigeria. Journal of Plant Nutrition. 2006;**29**(12):2131-2145. DOI: 10.1080/01904160600972761

[109] Krishnasamy K, Bell R, Ma Q. Wheat responses to sodium vary with potassium use efficiency of cultivars. Frontiers in Plant Science. 2014;**5**:631. DOI: 10.3389/fpls.2014.00631

[110] Bayuelo-Jiménez JS, Jasso-Plata N, Ochoa I. Growth and physiological responses of *Phaseolus* species to salinity stress. International Journal of Agronomy. 2012;**2012**:1-13. DOI: 10.1155/2012/527673

[111] Ali L, Rahmatullah MMA, Kanwal S, Ashraf M, Hannan A. Potassium substitution by sodium in root medium influencing growth behavior and potassium efficiency in cotton genotypes. Journal of Plant Nutrition. 2009;**32**(10):1657-1673. DOI: 10.1080/01904160903150917

[112] Purty RS, Kumar G, Singla-Pareek SL, Pareek A. Towards salinity tolerance in *Brassica*: An overview. Physiology and Molecular Biology of Plants. 2008;**14**:39-49. DOI: 10.1007/s12298-008-0004-4

[113] Tuna AL, Kaya C, Ashraf M, Altunlu H, Yokas I, Yagmur B. The effects of calcium sulfate on growth, membrane stability and nutrient uptake of tomato plants grown under salt stress. Environmental and Experimental Botany. 2007;**59**(2):173-178. DOI: 10.1016/j.envexpbot.2005.12.007

[114] Rameeh V, Rezai A, Saeidi G. Study of salinity tolerance in rapeseed. Communications in Soil Science and Plant Analysis. 2004;**35**:2849-2866. DOI: 10.1081/CSS-200036472

[115] Davenport RJ, Reid RJ, Smith FA. Sodium-calcium interactions in two wheat species differing in salinity tolerance. Physiologia Plantarum. 1997;**99**:323-327. DOI: 10.1111/j.1399-3054.1997.tb05419.x

[116] Lynch J, Läuchli A. Salinity affects intracellular calcium in corn root protoplasts. Plant Physiology. 1988;**87**:351-356. DOI: 10.1104/pp.87.2.351

[117] Parida AK, Das AB. Salt tolerance and salinity effects on plants: A review. Ecotoxicology and Environmental Safety. 2005;**60**(3):324-349. DOI: 10.1016/j.ecoenv.2004.06.010

[118] Saud S, Yajun C, Long B, Fahad S, Sadiq A. The different impact on the growth of cool season turf grass under the various conditions on salinity and draught stress. International Journal of Agricultural Science and Research. 2013;**3**(4):77-84

[119] Zaman QU, Aslam Z, Yaseen M, Ihsan MZ, Khaliq A, Fahad S, et al. Zinc biofortification in rice: Leveraging agriculture to moderate hidden hunger in developing countries. Archives of Agronomy and Soil Science. 2018;**64**(2):147-161. DOI: 10.1080/03650340.2017.1338343

[120] Chen WP, Chang AC, Page AL. Deficiencies and toxicities of trace elements. In: Wallender WW, Tanji KK, editors. Agricultural Salinity Assessment and Management. 2nd ed. Virginia: American Society of Civil Engineers; 2012. pp. 207-233. DOI: 10.1061/9780784411698.ch07

[121] Turhan E, Eris A. Changes of micronutrients, dry weight, and chlorophyll contents in strawberry plants under salt stress conditions. Communications in Soil Science and Plant Analysis. 2005;**36**:1021-1028. DOI: 10.1081/CSS-200050418

[122] Sommerfeldt TG, Rapp E, Chang C, Janzen HH. Management of Saline Soils under Irrigation. Publication 1624/E. Ottawa, ON, Canada: Agriculture Canada; 1988

[123] Gaines TP, Gaines ST. Soil texture effect on nitrate leaching in soil percolates. Communications in Soil Science and Plant Analysis. 1994;**25**(13-14):2561-2570. DOI: 10.1080/00103629409369207

[124] Gelaye KK, Zehetner F, Loiskandl W, Klik A. Effects of soil texture and groundwater level on leaching of salt from saline fields in Kesem irrigation scheme, Ethiopia. Soil and Water Research. 2019;**14**:221-228. DOI: 10.17221/137/2018-SWR

[125] Provin T, Pitt JL. Managing Soil Salinity [Internet]. 2001. Available from: http://oaktrust.library.tamu.edu/bitstream/handle/1969.1/86985/pdf_1397.pdf [Accessed: 02 June 2020]

[126] Spark DL. Environmental Soil Chemistry. 2nd ed. UK: Academic Press; 2002. p. 352

[127] Roy S, Akhtaruzzaman M, Nath B. Spatio-seasonal variations of salinity and associated chemical properties in the middle section of Karnaphuli river water, Chittagong, Bangladesh using laboratory analysis and GIS technique. International Journal of Environmental Science and Development. 2020;**11**(8):372-382. DOI: 10.18178/ijesd.2020.11.8.1278

[128] Nasim W, Ahmad A, Amin A, Tariq M, Awais M, Saqib M, et al. Radiation efficiency and nitrogen fertilizer impacts on sunflower crop in contrasting environments of Punjab, Pakistan. Environmental Science and Pollution Research. 2018;**25**:1822-1836. DOI: 10.1007/s11356-017-0592-z

[129] Roy S, Kashem MA, Osman KT. The uptake of phosphorous and potassium of rice as affected by different water and organic manure management. Journal of Plant Sciences. 2018;**6**(2):31-40. DOI: 10.11648/j.jps.20180602.11

[130] Saud S, Fahad S, Yajun C, Ihsan MZ, Hammad HM, Nasim W, et al. Effects of nitrogen supply on water stress and recovery mechanisms in Kentucky bluegrass plants. Frontiers in Plant Science. 2017;**8**:983. DOI: 10.3389/fpls.2017.00983

[131] Leogrande R, Vitti C. Use of organic amendments to reclaim saline and sodic soils: A review. Arid Land Research and Management. 2019;**33**(1):1-21. DOI: 10.1080/15324982.2018.1498038

[132] Roy S, Kashem MA. Effects of organic manures in changes of some soil properties at different incubation periods. Open Journal of Soil Science. 2014;**4**:81-86. DOI: 10.4236/ojss.2014.43011

[133] Akram R, Turan V, Wahid A, Ijaz M, Shahid MA, Kaleem S, et al. Paddy land pollutants and their role in climate change. In: Hashmi MZ, Varma A, editors. Environmental Pollution of Paddy Soils. Springer International Publishing AG; 2018. pp. 113-124. DOI: 10.1007/978-3-319-93671-0_7

[134] Khan A, Tan DKY, Munsif F, Afridi MZ, Shah F, Wei F, et al. Nitrogen nutrition in cotton and control strategies for greenhouse gas emissions: A review. Environmental Science and Pollution Research. 2017;**24**:23471-23487. DOI: 10.1007/s11356-017-0131-y

[135] Bhatt MK, Labanya R, Joshi HC. Influence of long-term chemical fertilizers and organic manures on soil fertility - A review. Universal Journal of Agricultural Research. 2019;7(5):177-188. DOI: 10.13189/ujar.2019.070502

[136] Fahad S, Hussain S, Saud S, Tanveer M, Bajwa AA, Hassan S, et al. A biochar application protects rice pollen from high-temperature stress. Plant Physiology and Biochemistry. 2015;**96**:281-287. DOI: 10.1016/j.plaphy.2015.08.009

[137] Fahad S, Hussain S, Saud S, Hassan S, Tanveer M, Ihsan MZ, et al. A combined application of biochar and phosphorus alleviates heat-induced adversities on physiological, agronomical and quality attributes of rice. Plant Physiology and Biochemistry. 2016;**103**:191-198. DOI: 10.1016/j.plaphy.2016.03.001

[138] Liu X, Herbert SJ, Hashemi AM, Zhang X, Ding G. Effects of agricultural management on soil organic matter and carbon transformation - A review. Plant, Soil and Environment. 2006;**52**(12):531-543. DOI: 10.17221/3544-PSE

[139] Ding Z, Kheir AMS, Ali MGM, Ali OAM, Abdelaal AIN, Lin X, et al. The integrated effect of salinity, organic amendments, phosphorus fertilizers, and deficit irrigation on soil properties, phosphorus fractionation and wheat productivity. Scientific Reports. 2020;**10**:2736. DOI: 10.1038/s41598-020-59650-8

[140] Gunarathne V, Senadeera A, Gunarathne U, Biswas JK, Almaroai YA, Vithanage M. Potential of biochar and organic amendments for reclamation of coastal acidic-salt affected soil. Biochar. 2020;**2**:107-120. DOI: 10.1007/s42773-020-00036-4

[141] Ouni Y, Lakhdar A, Scelza R, Scotti R, Abdelly C, Barhoumi Z, et al. Effects of two composts and two grasses on microbial biomass and biological activity in a salt-affected soil. Ecological Engineering. 2013;**60**:363-369. DOI: 10.1016/j.ecoleng.2013.09.002

[142] Prapagar K, Indraratne SP, Premanandharajah P. Effect of soil amendments on reclamation of saline-sodic soil. Tropical Agricultural Research. 2012;**23**(2):168-176. DOI: 10.4038/tar.v23i2.4648

[143] Wu Y, Li Y, Zhang Y, Bi Y, Sun Z. Response of saline soil properties and cotton growth to different organic amendments. Pedosphere. 2018;**28**(3):521-529. DOI: 10.1016/S1002-0160(17)60464-8

[144] Yang L, Bian XG, Yang RP, Zhou CL, Tang BP. Assessment of

organic amendments for improving coastal saline soil. Land Degradation and Development. 2018;**29**(9):3204-3011. DOI: 10.1002/ldr.3027

[145] Yu Y, Chen Y, Guo C, Weidong J, Lu Z. Effects of organic amendments on the salt-affected soil. In: International Conference on Remote Sensing, Environment and Transportation Engineering; 24-26 June 2011; Nanjing, China. 2011. DOI: 10.1109/RSETE.2011.5964977

[146] Yu Y, Liu J, Liu C, Zong S, Lu Z. Effect of organic materials on the chemical properties of saline soil in the Yellow River Delta of China. Frontiers in Earth Science. 2015;**9**:259-267. DOI: 10.1007/s11707-014-0463-6

[147] Mahmood F, Khan I, Ashraf U, Shahzad T, Hussain S, Shahid M, et al. Effects of organic and inorganic manures on maize and their residual impact on soil physico-chemical properties. Journal of Soil Science and Plant Nutrition. 2017;**17**(1):22-32. DOI: 10.4067/S0718-95162017005000002

[148] Obi ME, Ebo POC. The effect of organic and inorganic amendments on soil physical properties and maize production in severely degraded sandy soil in southern Nigeria. Bioresource Technology. 1995;**51**(2-3):117-123. DOI: 10.1016/0960-8524(94)00103-8

[149] Tester CF. Organic amendment effects on physical and chemical properties of a sandy soil. Soil Science Society of America Journal. 1990;**54**(3):827-831. DOI: 10.2136/sssaj1990.03615995005400030035x

[150] Barzegar AR, Yousefi A, Daryashenas A. The effect of addition of different amounts and types of organic materials on soil physical properties and yield of wheat. Plant and Soil. 2002;**247**:295-301. DOI: 10.1023/A:1021561628045

[151] Gümüş İ, Şeker C. Effects of spent mushroom compost application on the physicochemical properties of a degraded soil. Solid Earth. 2017;**8**:1153-1160. DOI: 10.5194/se-8-1153-2017

[152] Kamal AM, Alam MA, Uddin MZ, Hossain MM, Islam MN. Impact of organic fertilizer on physical and chemical properties of soil as well as yield and quality of mango. Journal of the Bangladesh Society for Agricultural Science and Technology;**9**(1&2):167-170

[153] Xiaohou S, Min T, Ping J, Weiling C. Effect of EM Bokashi application on control of secondary soil salinization. Water Science and Engineering. 2008;**1**(4):99-106. DOI: 10.3882/j.issn.1674-2370.2008.04.011

[154] Im JN. Organic materials and improvement of soil physical characteristics. FAO/SIDA workshop: Organic materials and soil productivity. FAO Soils Bulletin. 1982;**45**:106-117

[155] Akhtaruzzaman M, Roy S, Mahmud MS, Shormin T. Soil properties under different vegetation types in Chittagong University campus, Bangladesh. Journal of Forest and Environmental Science. 2020;**36**(2):133-142. DOI: 10.7747/JFES.2020.36.2.133

[156] Tobiašová E. The effect of organic matter on the structure of soils of different land uses. Soil & Tillage Research. 2011;**114**:183-192. DOI: 10.1016/j.still.2011.05.003

[157] Šimanský V, Zaujec A. Suitable parameters for soil organic matter changes evaluation in agro-ecosystems. Folia Oecologica. 2009;**36**(1):50-57

[158] Adane A, Gebrekidan H, Kibret K. Effects of treatment application rates (fym and gypsum) on selected chemical properties of saline sodic soils under water limited condition in eastern lowlands Ethiopia. Forest Research and

Engineering. 2019;**3**(3):106-113. DOI: 10.15406/freij.2019.03.00086

[159] Bai Y, Yan Y, Zuo W, Gu C, Xue W, Mei L, et al. Coastal mudflat saline soil amendment by dairy manure and green manuring. International Journal of Agronomy. 2017;**2017**:1-9. DOI: 10.1155/2017/4635964

[160] Khatun M, Shuvo MAR, Salam MTB, Rahman SMH. Effect of organic amendments on soil salinity and the growth of maize (*Zea mays* L.). Plant Science Today. 2019;**6**(2):106-111. DOI: 10.14719/pst.2019.6.2.491

[161] Li J, Liang LN, Li X, Yang HF. Effects of compost amendment on soil chemical and biological properties in greenhouse soil. In: Proceedings of the 1st International Symposium on Organic Matter Management and Compost Use in Horticulture; 4-7 April 2011; Adelaide, South Australia, Australia. 2014. pp. 195-202. DOI: 10.17660/ActaHortic.2014.1018.19

[162] Bakayoko S, Soro D, Nindjin C, Dao D, Tschannen A, Girardin O, et al. Effects of cattle and poultry manures on organic matter content and adsorption complex of a sandy soil under cassava cultivation (*Manihot esculenta*, Crantz). African Journal of Environmental Science and Technology. 2009;**3**(8):190-197. DOI: 10.5897/AJEST09.072

[163] Shehzadi S, Shah Z, Mohammad W. Impact of organic amendments on soil carbon sequestration, water use efficiency and yield of irrigated wheat. BASE. 2017;**21**(1):36-49. DOI: 10.25518/1780-4507.13435

[164] Zhang ZJ, Wang XZ, Wang H, Huang E, Sheng JL, Zhou LQ, et al. Housefly larvae (*Musca domestica*) vermicompost on soil biochemical features for a chrysanthemum (*Chrysanthemum morifolium*) farm. Communications in Soil Science and Plant Analysis. 2020;**51**(10):1315-1330. DOI: 10.1080/00103624.2020.1763389

[165] Adnan M, Shah Z, Fahad S, Arif M, Alam M, Khan IA, et al. Phosphate-solubilizing bacteria nullify the antagonistic effect of soil calcification on bioavailability of phosphorus in alkaline soils. Scientific Reports. 2017;**7**:16131. DOI: 10.1038/s41598-017-16537-5

[166] Norton J, Strom C. Successful Restoration of Severely Disturbed Wyoming Lands: Reclamation on Salt/Sodium-Affected Soils, Bulletin-1231. University of Wyoming Extension and Wyoming Reclamation and Restoration Center (WRRC), Laramie; 2012

[167] Mkhabela M, Warman PR. The influence of municipal solid waste compost on yield, soil phosphorus availability and uptake by two vegetable crops, grown in a Pugwash sandy loam soil in Nova Scotia. Agriculture, Ecosystems and Environment. 2005;**106**:57-67. DOI: 10.1016/j.agee.2004.07.014

[168] Qadir M, Oster JD. Crop and irrigation management strategies for saline-sodic soils and waters aimed at environmentally sustainable agriculture. The Science of the Total Environment. 2004;**323**:1-19. DOI: 10.1016/j.scitotenv.2003.10.012

[169] Walker DJ, Bernal PM. The effects of olive mill waste compost and poultry manure on the availability and plant uptake of nutrients in a highly saline soil. Bioresource Technology. 2008;**99**(2):396-403. DOI: 10.1016/j.biortech.2006.12.006

[170] Hue NV. Correcting soil acidity of a highly weathered ultisol with chicken manure and sewage sludge. Communications in Soil Science and Plant Analysis. 1992;**23**:241-264. DOI: 10.1080/00103629209368586

[171] Tan KH, Dowling PS. Effect of organic matter on CEC due to permanent and variable charges in selected temperate region soils. Geoderma. 1984;**32**(2):89-101. DOI: 10.1016/0016-7061(84)90065-X

[172] Alharby HF, Fahad S. Melatonin application enhances biochar efficiency for drought tolerance in maize varieties: Modifications in physio-biochemical machinery. Agronomy Journal. 2020;**112**(4):2826-2847. DOI: 10.1002/agj2.20263

[173] Wichern F, Islam MR, Hemkemeyer M, Watson C, Joergensen RG. Organic amendments alleviate salinity effects on soil microorganisms and mineralisation processes in aerobic and anaerobic paddy rice soils. Frontiers in Sustainable Food System. 2020;**4**:30. DOI: 10.3389/fsufs.2020.00030

[174] Bouajila K, Sanaa M. Effects of organic amendments on soil physico-chemical and biological properties. Journal of Materials and Environmental Science. 2011;**2**(S1):485-490

[175] Larney FJ, Angers DA. The role of organic amendments in soil reclamation: A review. Canadian Journal of Soil Science. 2012;**92**:19-38. DOI: 10.4141/CJSS2010-064

[176] Fahad S, Nie L, Chen Y, Wu C, Xiong D, Saud S, et al. Crop plant hormones and environmental stress. In: Lichtfouse E, editor. Sustainable Agriculture Reviews. Vol. 15. Cham: Springer; 2015. pp. 371-400. DOI: 10.1007/978-3-319-09132-7_10

[177] Fahad S, Hussain S, Matloob A, Khan FA, Khaliq A, Saud S, et al. Phytohormones and plant responses to salinity stress: A review. Plant Growth Regulation. 2015;**75**:391-404. DOI: 10.1007/s10725-014-0013-y

[178] Kamran M, Cui W, Ahmad I, Meng X, Zhang X, Su W, et al. Effect of paclobutrazol, a potential growth regulator on stalk mechanical strength, lignin accumulation and its relation with lodging resistance of maize. Plant Growth Regulation. 2018;**85**:171-172. DOI: 10.1007/s10725-018-0367-7

[179] Egamberdieva D. Alleviation of salt stress by plant growth regulators and IAA producing bacteria in wheat. Acta Physiologiae Plantarum. 2009;**31**:861-864. DOI: 10.1007/s11738-009-02 97-0

[180] Fahad S, Hussain S, Saud S, Hassan S, Ihsan Z, Shah AN, et al. Exogenously applied plant growth regulators enhance the morpho-physiological growth and yield of rice under high temperature. Frontiers in Plant Science. 2016;**7**:1250. DOI: 10.3389/fpls.2016.01250

[181] Fahad S, Hussain S, Saud S, Hassan S, Chauhan BS, Khan F, et al. Responses of rapid viscoanalyzer profile and other rice grain qualities to exogenously applied plant growth regulators under high day and high night temperatures. PLoS One. 2016;**11**(7):e0159590. DOI: 10.1371/journal.pone.0159590

[182] Mahdy AM. Comparative effects of different soil amendments on amelioration of saline-sodic soils. Soil & Water Research. 2011;**6**(4):205-216. DOI: 10.17221/11/2011-SWR

[183] Lashari MS, Ye Y, Ji H, Li L, Kibue GW, Lu H, et al. Biochar-manure compost in conjunction with pyroligneous solution alleviated salt stress and improved leaf bioactivity of maize in a saline soil from Central China: A 2-year field experiment. Journal of the Science of Food and Agriculture. 2015;**95**(6):1321-1327. DOI: 10.1002/jsfa.6825

[184] Cha-um S, Pokasombat Y, Kirdmanee C. Remediation of salt-affected soil by gypsum and farmyard manure - Importance for the production

of Jasmine rice. Australian Journal of Crop Science. 2011;**5**(4):458-465

[185] Sun J, He F, Shao H, Zhang Z, Xu G. Effects of biochar application on *Suaeda salsa* growth and saline soil properties. Environment and Earth Science. 2016;**75**:630. DOI: 10.1007/s12665-016-5440-9

[186] El-Magd MMA, Zaki MF, Abou-Hussein SD. Effect of organic manure and different levels of saline irrigation water on growth, green yield and chemical content of sweet fennel. Australian Journal of Basic and Applied Sciences. 2008;**2**(1):90-98

[187] Esteban W, Pacheco P, Tapia L, Bastías E. Remediation of salt and boron-affected soil by addition of organic matter: An investigation into improving tomato plant productivity. IDESIA (Chile). 2016;**34**(3):25-32. DOI: 10.4067/S0718-34292016000300004

[188] Fahad S, Hussain S, Saud S, Khan F, Hassan S, Amanullah, et al. Exogenously applied plant growth regulators affect heat-stressed rice pollens. Journal of Agronomy and Crop Science. 2016;**202**:139-150. DOI: 10.1111/jac.12148

[189] Fahad S, Adnan M, Hassan S, Saud S, Hussain S, Wu C, et al. Rice responses and tolerance to high temperature. In: Hasanuzzaman M, Fujita M, Nahar K, Biswas JK, editors. Advances in Rice Research for Abiotic Stress Tolerance. Woodhead Publishing; 2019. pp. 201-224. DOI: 1016/B978-0-12-814332-2.00010-1

[190] Saud S, Li X, Chen Y, Zhang L, Fahad S, Hussain S, et al. Silicon application increases drought tolerance of Kentucky bluegrass by improving plant water relations and morph physiological functions. The Scientific World Journal. 2014;**2014**:1-10. DOI: 10.1155/2014/ 368694

[191] Saud S, Yajun C, Fahad S, Hussain S, Na L, Xin L, et al. Silicate application increases the photosynthesis and its associated metabolic activities in Kentucky bluegrass under drought stress and post-drought recovery. Environmental Science and Pollution Research. 2016;**23**:17647-17655. DOI: 10.1007/s11356-016-6957-x

[192] Zhao D, Wang Z, Yang F, Zhu W, An F, Ma H, et al. Amendments to saline-sodic soils showed long-term effects on improving growth and yield of rice (*Oryza sativa* L.). PeerJ. 2020;**8**:e8726. DOI: 10.7717/peerj.8726

[193] Abdel-Fattah MK, Fouda S, Schmidhalter U. Effects of gypsum particle size on reclaiming saline-sodic soils in Egypt. Communications in Soil Science and Plant Analysis. 2015;**46**:1112-1122. DOI: 10.1080/00103624.2015.1018528

[194] Khattak SG, Haq IU, Malik A, Khattak MJ. Naveedullah. Effect of various levels of gypsum application on the reclamation of salt affected soil grown under rice followed by wheat crop. Sarhad Journal of Agriculture. 2007;**23**(3):675-680

[195] Khosla BK, Gupta RK, Abrol IP. Salt leaching and the effect of gypsum application in a saline-sodic soil. Agricultural Water Management. 1979;**2**(3):193-202. DOI: 10.1016/0378-3774(79)90002-7

[196] Al-Busaidi A, Yamamoto T, Inoue M, Eneji AE, Mori Y, Irshad M. Effects of zeolite on soil nutrients and growth of barley following irrigation with saline water. Journal of Plant Nutrition. 2008;**31**(7):1159-1173. DOI: 10.1080/01904160802134434

[197] Abdi G, Khosh-Khui M, Eshghi S. Effects of natural zeolite on growth and flowering of strawberry (*Fragariax ananassa* Duch.). International Journal of Agricultural Research.

2006;**1**:384-389. DOI: 10.3923/ijar.2006.384.389

[198] Milosevic T, Milosevic N. The effect of zeolite, organic and inorganic fertilizers on soil chemical properties, growth and biomass yield of apple trees. Plant, Soil and Environment. 2009;**55**(12):528-535. DOI: 10.17221/107/2009-PSE

[199] Badora A. The influence of zeolites on quality indicators of soil-plant connection and food safety. In: Belviso C, editor. Zeolites - Useful Minerals. London, UK: IntechOpen; 2016. pp. 203-215. DOI: 10.5772/64590

[200] Jha B, Singh DN. Fly ash Zeolites: Innovations, Applications, and Directions. 1st ed. Singapore: Springer Science+Business Media; 2016. p. 211. DOI: 10.1007/978-981-10-1404-8_2

[201] Song JQ, Fujiyama H. Differences in response of rice and tomato subjected to sodium salinization to the addition of calcium. Soil Science & Plant Nutrition. 1996;**42**(3):503-510. DOI: 10.1080/00380768.1996.10416319

Chapter 20

TIP Aquaporins in Plants: Role in Abiotic Stress Tolerance

Marzena Małgorzata Kurowska

Abstract

Tonoplast Intrinsic Proteins (TIP) are one of five subfamilies of aquaporins in higher plants. Plants typically contain a large number of TIP genes, ranging from 6 to 35 compared to humans. The molecular weight of the TIP subfamily members ranges from 25 to 28 kDa. Despite their sequence diversity, all TIP monomers have the same structure, which consists of six transmembrane helices and five inter-helical loops that form an hourglass shape with a central pore. Four monomers form tetramers, which are functional units in the membrane. TIPs form channels in the tonoplast that basically function as regulators of the intracellular water flow, which implies that they have a role in regulating cell turgor. TIPs are responsible for precisely regulating the movement of not only water, but also some small neutral molecules such as glycerol, urea, ammonia, hydrogen peroxide and formamide. The expression of TIPs may be affected by different environmental stresses, including drought, salinity and cold. TIPs expression is also altered by phytohormones and the appropriate *cis*-regulatory motifs are identified in the promotor region of the genes encoding TIPs in different plant species. It was shown that manipulating TIP-encoding genes expression in plants could have the potential to improve abiotic stress tolerance.

Keywords: aquaporins, tonoplast intrinsic proteins, abiotic stress, drought, salinity, cold, phytohormones, gene expression analysis, *cis*-regulatory motifs, plants

1. Introduction

Aquaporins (aqua-pore, AQP), which are members of the major intrinsic proteins (MIPs) superfamily, facilitate the bidirectional flux of water and non-aqua substrates across the cell membranes [1–4]. These transmembrane proteins form pores in the lipid bilayers of archaea, bacteria, fungi, plants, non-mammalian metazoans and mammals including humans. In plants, the first AQP Noduline-26 (GmNOD26) was identified in soybean (*Glycina max* L.) [5], whereas their ability to conduct water in plants was demonstrated for the first time for an Arabidopsis tonoplast intrinsic protein homologue (AtTIP1; 1, gamma-TIP) *via* its heterologous expression in *Xenopus* oocytes [6]. In higher plants, MIPs are divided into five evolutionarily distinct subfamilies in terms of the similarities of their amino acid sequence and intercellular localisation: plasma membrane intrinsic proteins (PIPs), tonoplast intrinsic proteins (TIPs), nodulin-26 like intrinsic proteins (NIPs), small basic intrinsic proteins (SIPs) and X intrinsic proteins / uncharacterised-intrinsic proteins (XIPs) [7]. In turn, the TIP subfamily can be further divided into different subgroups. In Arabidopsis (*Arabidopsis thaliana*), the TIP subfamily has been divided into five subgroups from TIP1 to TIP5. Each subgroup is then again divided into different isoforms. In Arabidopsis, the TIP subfamily consists of 11 isoforms. Some of these have synonymous name such as TIP1;1 (gamma-TIP, γ-TIP, RB7), TIP1;2 (gamma-TIP2, γ-TIP2), TIP2;1 (delta-TIP, δ-TIP), TIP3;1 (alpha-TIP, α-TIP) and TIP3;2 (beta-TIP, β-TIP) [8].

Water, whose transport is facilitated by aquaporins and, among them TIPs, is their most important substrate. TIPs are responsible for precisely regulating the movement of not only water, but also some small neutral molecules that have great physiological significance such as glycerol, urea, ammonia, hydrogen peroxide (H_2O_2) and formamide [7]. The transport of these substrates was predicted using a bioinformatic analysis and some of them have also been proven experimentally by using their *in vitro* expression in yeast or in a *Xenopus* oocyte system in which the expression of gene of interest led to, for example, an increase water permeability (**Table 1**).

Species	**TIP name**	**Substrate**					**References**
		Water (H_2O)	**Hydrogen peroxide (H_2O_2)**	**Urea**	**Ammonia (NH_3)**	**Glycerol (GLY)**	
Arabidopsis (*Arabidopsis thaliana*)	AtTIP1;1	+	+	+	+		[6, 9–13]
	AtTIP1;2	+	+	+	+		
	AtTIP2;1	+		+			
	AtTIP2;3			+			
	AtTIP4;1						
	AtTIP5;1						
Barley (*Hordeum vulgare* L.)	HvTIP1;1	+					[14]
	HvTIP1;2	+					
	HvTIP2;3	+					
Bread wheat (*Triticum aestivum* L.)	TaTIP2;1				+		[15–17]
	TaTIP2;2				+		
Maize (*Zea mays* L.)	ZmTIP1;1	+	+	+	+	-	[18–20]
	ZmTIP1;2	+	+	+	+	-	
	ZmTIP2;3	+		+			
	ZmTIP4;4						
Rice (*Oryza sativa* L.)	OsTIP1;2	+				+	[21]
	OsTIP2;2	+				+	
	OsTIP3;2	+				+	
	OsTIP4;1	+					
	OsTIP5;1						
Radish (*Raphanus sativus*)	RsTIP1;1	+					[22]
	RsTIP2;1	+					
Sunflower (*Helianthus annuus* L.)	TIP7	+					[23]
	TIP20	+					
Tobacco (*Nicotiana tabacum* L.)	NtTIPa	+		+		+	[24]

+ transport of the substrate has been confirmed, – transport of the substrate has been excluded.

Table 1.
The transport of different substrates by TIPs, which was demonstrated via their in vitro expression in yeast or oocytes.

The central vacuole occupies most of the intracellular space of most mature plant cells [25]. The plant vacuole plays important roles not only in space filling, but also in osmotic adjustment, storage and digestion [26, 27]. TIPs form channels in the vacuolar membrane (tonoplast) that basically function as regulators of the intracellular water flow, which implies that they have a role in regulating cell turgor. It has been proposed that the presence of a specific TIP isoform may define the function of the vacuole and that specific TIP proteins could be used as markers for the different types of vacuoles that occur in different tissue types or in response to developmental and environmental stimuli [28, 29]. Vacuoles whose tonoplast contains TIP2;1 (delta-TIP, δ-TIP) are used by plant cells to store pigments and vegetative storage proteins (VSPs) [28]. Moreover, in vegetative tissues, the lytic or degradative vacuoles (LVs) are marked by the presence of TIP1;2 (gamma-TIP, γ-TIP) [26, 28, 30]. TIP1;2 may be expressed primarily at the time when the large central vacuoles are being formed during cell enlargement [30], while the protein storage vacuoles (PSVs) in mature seeds are marked by the presence of TIP3;1 (alpha-TIP, α-TIP) [26, 31, 32].

TIP proteins are most abundant in the tonoplast, but some isoforms have also been detected in the chloroplast, e.g. on the luminal site of the thylakoid, the chloroplast membranes or the envelope fraction using mass-spectrometry-based proteomics. Such a location was detected for TIP1;1, TIP1;2 and TIP2;1 in Arabidopsis [33–35].

2. Structure and diversity of TIPs

Although it has been suggested that AQPs and, among them, TIPs are multi-functional proteins, their structure is unique across all kingdoms of life. Despite their sequence diversity, all TIPs have a similar structure that consists of six trans-membrane helices (TM1 to TM6) with five connecting inter-helical loops (LA to LE) and two half helices that contain one highly conserved Asn-Pro-Ala (NPA) motif each of which forms its functional pore (**Figure 1**). The hydrophobic NPA motif is located at the first intracellular (loop B) and the third extracellular loop (loop E) and the overlap in the middle of the lipid bilayer ('hourglass' model) and forms two hemipores that generate a narrow part of the channel [7, 36]. A second filter region is the aromatic/Arginine (ar/R) selectivity filter, which includes four amino acids (aa), which are located at the non-cytosolic end of the pore [37]. These two constriction regions are very important for the transport selectivity of the channel, which have been predicted based on structural knowledge combined with

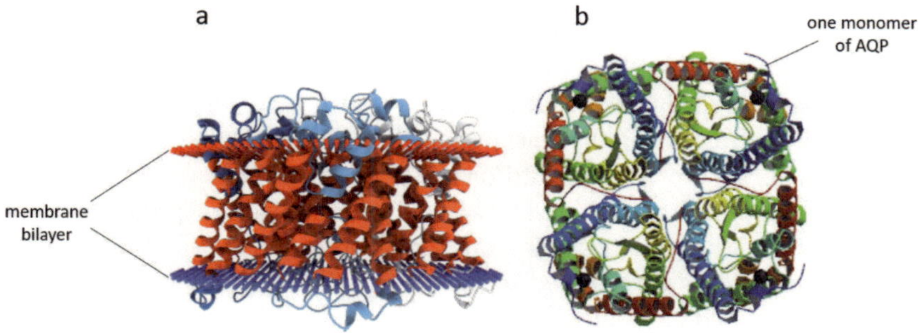

Figure 1.
The structure of aquaporin as represented by the crystal structure of spinach SoPIP2;1 (PDB 1Z98). The view from different sides of the holoprotein, which consists of four monomers. (a) the side view, (b) the tetrameric assembly from the cytoplasmic side. Each monomer functions as a single channel pore.

Species	Genome size/ ploidy x/no. of chromosomes*	AQP	PIP	TIP	NIP	SIP	XIP	References
Arabidopsis (*Arabidopsis thaliana*)	~135 Mbp 2n = 2x = 10	38	13	**11**	11	3	0	[8]
Barley (*Hordeum vulgare* L.)	~5.3 Gbp; 2n = 2x = 14	39	18	**11**	8	2	0	[51]
Bread wheat (*Triticum aestivum* L.)	~17 Gb 2n = 6x = 42, AABBDD, hexaploid	113 (65-A, 42-B, 36-D)	51	**29**	29	4	0	[52]
Cucumber (*Cucumis sativus* L.)	~350 Mbp 2n = 2x = 14	39	19	**8**	9	2	1	[53]
Foxtail millet (*Setaria italica* L.)	~490 Mb 2n = 2x = 18	39	12	**11**	13	3	0	[54]
Maize (*Zea mays* L.)	2.4 Gb 2n = 2x = 20	41	12	**18**	8	3	0	[55]
Potato (*Solanum tuberosum* L.)	~840 Mb 2n = 4x = 48	41	15	**11**	10	3	8	[56]
Rice Japonica (*Oryza sativa* L.)	500 Mb 2n = 2x = 24	33	11	**10**	10	2	0	[57]
Sorghum (*Sorghum bicolor*)	~730 Mb 2n = 2x = 20	37	13	**11**	11	2	0	[58]
Tomato (*Solanum lycopersicum* L.)	~828 Mb 2n = 2x = 24	47	14	11	12	4	6	[59]

**Based on the Ensemble Plants Database.*
The number of Tonoplast Intrinsic Proteins (TIP) has been marked in bold.

Table 2.
The number of AQP isoforms in the selected plant species.

simulation studies [38, 39]. Both filters at the pore "mouth" constitute barriers to the passage of inorganic cations (such as Na^+ and K^+) and protons [39]. The NPA motif not only plays a role in regulating membrane transport but also in protein localisation [40]. The substrate selectivity of AQPs is controlled by the amino acid residues of the NPA and ar/R filters as well as another conserved region, which is known as the Froger's residues, which consist of five conserved residues [41]. This region enables the aquaglyceroporins that transport glycerol from the AQPs that facilitate the transport of water to be distinguished [39].

All MIP sequences take on a typical hour-glass MIP helical fold with six transmembrane helices and two half-helices [42]. This fold has been maintained during evolution via its conservation around 40 positions within the transmembrane region [42]. The stability of the functionally important half-helix is modulated by a stabilising intra-helical salt-bridge interaction and/or two helix destabilising residues, glycine and proline, which was demonstrated by an analysis of the loop E region of 1468 MIP sequences and their structural models from six different groups of organisms [43]. In the cell membrane, MIPs are grouped as homotetramers that are located in the lipid bilayer (**Figure 1**). Each monomer functions independently as a single channel pore [44]. In plants, the only the crystal structure of the MIPs for two species is currently available in the Protein Data Bank (PDB: http://www.pdb.org), the spinach (*Spinacia oleracea*) aquaporin SoPIP2;1 was in its closed confirmation at a 2.1 Å resolution and its open confirmation at a 3.9 Å resolution [45] and the Arabidopsis aquaporins

AtPIP2;4 at a 3.7 Å resolution [46] all of which were determined using X-ray diffraction, which showed their structural identity. However, the functions and regulation of the majority of the MIP channels have not yet been fully characterised [42].

A large number of *MIP* genes and, among them *TIP* genes, have been identified in the genomes of different plant species. In plants, the MIP family contains around 2008 members while the TIP subfamily contains 585 members from different plant species, which have been deposited in the MIPDB [47, 48]. The length of the TIP sequences ranges from 55 to 780 aa. Plants typically contain a large number of *AQPs* genes and among them the *TIP* genes, when are compared with for example humans, ranges from 6 to 35 in moso bamboo (*Phyllostachys edulis*) and oilseed rape (*Brassica napus* L.), respectively [49, 50] (**Table 2**). The molecular weights of the TIP subfamily members ranges from 25 to 28 kDa [47].

3. Study of the tissue- and developmental-dependent expression of TIPs

Aquaporins are postulated to be the main water transporter protein in plants. However their presence and expression may vary among plant species, tissues and time during development. In plants, AQPs are present in almost all organs, including the roots, leaves, stems, flowers, fruits, seeds, dry seeds [60], pollen [13, 61–63], anther and specific cells such as the guard cells [64, 65]. Moreover, some *TIPs* are expressed in specific tissues, e.g. *AtTIP5;1* was localised in the pollen mitochondria, which are probably involved in the remobilisation of N via the transport of mitochondrial urea to the cytoplasm [61] as well as in a time-specific manner, e.g. the *RsTip1;1* (*gamma-VM23*) gene expression was high not only in the hypocotyls but also in the developing tap roots and young growing leaves and the pattern of its expression was connected with cell elongation [22] (**Table 3**).

Species	TIP name	Organ/tissue/process	References
Arabidopsis (*Arabidopsis thaliana*)	*AtTIP1;3, AtTIP5;1*	Pollen mitochondria	[61, 66, 67]
	AtTIP3;2	Senescence	
	AtTIP3;1	Seed-specific	
	AtTIP1;1	Entire vegetative body	
	AtTIP2;1	Developing vascular tissue, primary shoot	
Barley (*Hordeum vulgare* L.)	*HvTIP1;1, HvTIP2;3, HvTIP1;2, HvTIP3;1*	Leaf	[14]
	HvTIP1;1, HvTIP2;3, HvTIP1;2	Root	
Bread wheat (*Triticum aestivum* L.)	*TaTIP1;12*	Root-specific	[52]
	TaTIP2;22	Root-specific	
Eucalyptus (*Eucalyptus grandis* L.)	*EgTIP1;1, EgTIP2;1*	Flower bud	[68]
	EgTIP1;1, EgTIP1;4, EgTIP2;1, EgTIP2;2	Flower	
	EgTIP1;4	Root	
	EgTIP1;1, EgTIP1;2, EgTIP4;1	Stem	
	EgTIP1;3	Cambium	
	EgTIP1;2, EgTIP3;2	Leaf	
	EgTIP3;1, EgTIP3;2	Fruit	

Species	TIP name	Organ/tissue/process	References
Maize (*Zea mays* L.)	*ZmTIP2;3*	Root-specific	[20]
Radish (*Raphanus sativus*)	*RsTip1;1 (gamma-VM23)*	Hypocotyl of seedlings, developing tap roots and young growing leaves	[22]
	RsTIP2;1 (delta-VM23)	Not expressed in young roots or leaf mesophyll	
Rice (*Oryza sativa* L.)	*OsTIP3;2*	Mature seed, spikelet, callus	[21, 69]
	OsTIP1;1	Shoots, roots	
Sunflower (*Helianthus annuus* L.)	*TIP7 (delta-TIP)*	Guard cells	[23]
	TIP27 (delta-TIP)	Guard cells	
Tobacco (*Nicotiana tabacum* L.)	*NtTIP1;1 (TobRB7)*	Root-specific	[70]

Table 3.
The expression of the selected TIP genes in different plant tissues.

4. Study of the expression of *TIPs* under abiotic stress or phytohormone treatment, and their role in maintaining cell homeostasis under stress conditions

The gene expression is the process in which gene information is converted into a protein [7]. Analysing the gene expression profiles is a prerequisite for determining the physiological function of a gene [71]. The function of AQPs can generally be modified in two ways: by regulating the gene expression, which leads to changes in protein abundance or by changing its permeability – gating. However, it should be kept in mind that changes in the mRNA expression level often reflect the protein abundance, but that this is not necessarily strictly connected in any cell or tissue [37]. In term, AQP trafficking and activity could be affected by the post-translational modifications (PTMs), like deamination, phosphorylation, methylation, ubiquitination and acetylation [37]. Water homeostasis is fundamental for cell survival [72]. AQP facilitate the transport of water and play roles in response to abiotic stress conditions during which the cell homeostasis is disturbed. Different studies have suggested that a low abundance of the AQP proteins reduces the water permeability of membranes and that a high abundance of them increases it. Although it can be assumed that all AQPs are able to transport water, an experimental study was performed for only a few members of the TIP subfamily in different plant species (**Table 1**). The expression of TIPs may be affected by different environmental conditions, including drought, salinity and cold and this has been investigated in many plant species. Comparative transcriptome studies using RNA sequencing (RNA-seq), which is a next generation sequencing technology, could lead to the discovery of a differential expression of multiple aquaporin homologues in different tissues and/or in different conditions of growth, e.g. under abiotic stress. The transcriptome defines as a set of coding and non-coding RNA molecules that are present in a single cell or a population of cells [73]. Nowadays, RNA-seq technology is used more and more frequently [74, 75]. On the other hand, the reverse-transcription quantitative PCR (RT-qPCR) technique still serves as the method of choice to perform the expression analysis of specific genes in many researches.

Abiotic stresses, which have a significant impact on the plants, are involved in changes at the transcriptome, cellular and physiological levels. When the stress is prolonged, plant growth and productivity are severely diminished [76, 77]. The response of plants is very complicated and many mechanisms are initiated simultaneously to restore cellular homeostasis and promote survival [78]. Aquaporins are thought to be responsive to the stress signalling pathways and are also thought to be involved in the stress-coping mechanisms such as altering the tissue hydraulic conductivity [79, 80]. Drought and salinity are two major factors that significantly limit plant growth and productivity [81, 82]. The huge yield losses caused by drought stress ranging 45–92 (%) have been reported in major field crops like maize (*Zea mays* L.), wheat (*Triticum aestivum* L.), rice (*Oryza sativa* L.), chickpea (*Cicer arietinum* L.), soybean (*Glycine max* L.), sunflower (*Helianthus annuus* L.) [83]. Cold is also often associated with climate change [84]. Drought, salinity and freezing stresses cause a changing water status in plant cells and all of these result in cellular dehydration [85]. In order to improve the tolerance of crop plants to unfavourable environmental conditions, a detailed understanding of their stress responses is required.

4.1 Drought, salinity and cold stress

Aquaporins are responsible for precisely regulating the movement of water and therefore may play a crucial role in the drought-stress response as well as in drought-stress tolerance [78]. A transcriptome analysis of the leaf and root tissues in plant species revealed significant changes in the expression of the *TIP* genes in response to drought stress. Under drought stress, when water is limited, the down-regulation of the *TIP* genes can lead to a decrease in the water permeability of the tonoplast in order to avoid water loss and to minimise the water flow through the cell membranes in order to prevent the further loss of leaf turgor. This expression pattern has been shown in many studies. In *Nicotiana glauca*, the *NgMIP2* and *NgMIP3* genes, which are homologous to *TIP*, were down-regulated in leaves under drought stress when the leaves were wilting, usually three to four days after water was withheld [86] and in Arabidopsis, the levels of *AtTIP1;1, AtTIP1;2, AtTIP2;1* and *AtTIP2;2* were more than four-fold down-regulated after 12 days of drought [87]. In contrast to these results, the expression level of *AtTIP1;1* and *AtTIP2;1* under drought treatment was up-regulated, but the time that the stress was applied was only 24 hours [88]. Moreover, in barley (*Hordeum vulgare* L.), the expression of five *HvTIP* aquaporins in leaves: *HvTIP1;1, HvTIP1;2, HvTIP2;1, HvTIP2;2* and *HvTIP2;3* was down-regulated after ten days of severe drought treatment [89]. In turn, in *Festuca* species, the transcript level of the *TIP1;1* aquaporin decreased in leaves after 11 days of a water deficit [90]. The possible functional role of coffee (*Coffea arabica* L.) TIPs in in regulating the water balance has been explored by measuring the *TIP* gene expression in leaf and root tissues that had been subjected to drought [91]. In root tissue, the expression of three *CsTIP* genes (*CaTIP1;1, CsTIP2;1, CsTIP4;1*) decreased as the severity of the drought stress increased. In the leaf, a similar expression profile of their expression was obtained only for of *CaTIP1;1* [91]. By contrast, the level of *TIP* expression increased in response to drought in some plant species. In soybean (*Glycine max*), transcriptome analysis data revealed that under dehydration stress, the expression of *GmTIP2;1, GmTIP1;7* and *GmTIP1;8* was up-regulated. The authors suggested that these three genes are probably the major AQPs that are involved in the stress response [75].

Interestingly, the expression profile of *TIP* genes can be compared for cultivars that have contrasting responses to drought stress under control and stress conditions. In two cultivars of strawberry (*Fragaria* x *ananassa*) that have contrasting drought stress phenotypes, the gene expression of *FaTIPs* in the roots was analysed [79]. During severe drought stress, the *FaTIPs* were only up-regulated in the drought-tolerant cultivar. This was connected with a gradual stomatal closure in this cultivar, which was further associated with conservative water use [79]. In turn, in soybean (*Glycine max*), the difference in gene expression was investigated using RNA-seq and it was found that the expression level of seven *TIP* genes (*GmTIP2;1, 2;2, 2;3, 2;4, 2;5, 3;2* and *5;1*) in the leaves was lower in the slow-wilting soybean lines, but that it increased in the fast-wilting soybean lines after drought stress [74]. Slow wilting is an important physiological parameter that is used in stress tolerance studies of plants. A cultivar-specific *TIP* expression under drought has also been observed in the common bean (*Phaseolus vulgaris* L.) [92]. In this study, the expression of *PvTIP1;1* and *PvTIP4;1* decreased under drought compared to the control conditions in two cultivars, which has different phenotypes in terms of their tolerance to drought and yield at harvest, which were determined by time that was required for the plants to wilt after withholding water. Interestingly, the gene expression of *PvTIP4;1* during drought was cultivar specific with greater down-regulation of these gene in the drought-tolerant cultivar [92]. The differences in the relative water content (RWC) and water potential of leaves between these cultivars indicate that there is greater prevention of water loss in the tolerant cultivar during drought, which may be associated with the rapid and adequate down-regulation of the *AQPs* [92]. However, it should be kept in mind that the adaptive mechanism for reducing the effects of water deficit stress in which the *TIP* or other *AQP* expression is changed should be further analysed by a functional characterisation of these candidate genes or proteins [52].

The regulation of water transport has been postulated as being the most important and the best recognised function of all of the AQPs including TIPs under drought stress. A second important function could be their involvement in the movement of hydrogen peroxide (H_2O_2), which is one of the reactive oxygen species (ROS), which include the unstable, highly reactive molecules that are produced during abiotic stress, e.g. drought [83, 93]. This activity makes the AQPs important players in both the redox signalling network and in H_2O_2 detoxification [94]. TIPs can facilitate the movement of H_2O_2 across the membranes that have already been shown by their *in vitro* expression in the yeast or oocyte (**Table 1**). In barley under drought stress, the expression of the *HvTIP3;1* gene increase 5000-fold compared to their expression under optimal water conditions [89].

Soil salination and a high salt accumulation affects the growth, development and metabolism and yield of plants [95]. A wide range of physiological and biochemical alterations in plants are induced by salinity, which causes a lower water potential in the soil solution, ionic disequilibrium and also a higher accumulation of reactive oxygen species (ROS) [96]. Salinity stress leads to changes in the expression profiles of the *TIP* genes, which was shown in many plant species. In Arabidopsis (*Arabidopsis thaliana*), the plants were treated with 100 mm NaCl, which led to osmotic stress. Macroarray experiments with gene-specific tags were performed to investigate the expression of all of the *AQPs* including the *TIP* subfamily in roots that had been under salt treatment for 24 hours [97]. The *AtTIP1;1, AtTIP1;2, AtTIP2;2* and *AtTIP2;3* genes were highly expressed in the roots and their expression was maximally reduced at 6 h of treatment and remained low until 24 h of treatment, except for *AtTIP1;1*. In turn, in *Festuca arundinacea,* the plants were exposed to salt stress by irrigating them with 250 mM NaCl solution over 21 days, which resulted in an increase in the abundance of the *FaTIP1;1* transcript in two contrasting salt tolerance genotypes compared to the control conditions. This increase was noticed earlier on day six of the NaCl treatment in the high-salt tolerant (HST) genotypes and later on day 11 in the low-salt tolerant (LST) genotype [90].

Similar to water and salt stresses, cold stress is an important abiotic stress factor that significantly limits plant growth and development [98]. In *Festuca pratensis*, cold acclimation was associated with a lower transcript level of the *FpTIP1;1* gene in the leaves in both the high and low frost-tolerant genotypes. The authors concluded that the down-regulation of these isoforms during the exposure to cold may be one of the cold-acclimation components that prevents frost-induced cellular dehydration [90]. A similar pattern of expression was observed in roots in rice (*Oryza sativa* L.) under cold treatment. A 4°C treatment caused a decrease in the transcript level of the *OsTIP1;1* and *OsTIP2;2* genes. The mRNA levels of these genes started to decrease after 48 hours and continued to decrease after 72 hours of a chilling treatment [57]. One more example of a change in the expression of the *TIP* member genes comes from the cold treatment of cotton (*Gossypium hirsutum*). The *GhTIP1;1* transcripts mainly accumulated in the roots and hypocotyls under normal conditions, but were dramatically up-regulated in the cotyledons and down-regulated in the roots within a few hours after the cotton seedlings had been cold treated [99].

4.2 Phytohormone treatment

Phytohormones play vital roles in the ability of plants to acclimatise to varying environments, by mediating growth, development, and nutrient allocation [100]. Interestingly, it has been shown that the *TIPs* expression is also altered by phytohormones, including abscisic acid (ABA), cytokinin (CK), gibberellins (GA), and jasmonates (JA). Abscisic acid (ABA) is a plant growth regulator that plays an essential role in the abiotic stress response mainly during water stress [101]. It regulates the movement of the stomatal guard cells [102]. Jasmonates act as stress hormones that play an important role in the plant response to biotic and abiotic stresses [103]. In general, JA inhibits plant growth and modulate many growth and developmental processes of plants including root, shoot and leaf growth, trichome and tuber formation, fruit ripening and leaf senescence [104]. The gibberellins stimulate the growth of most organs via cell elongation and also respond to abiotic stresses [105]. Cytokinins play a role in regulating plant growth, development and their acclimation to environmental stresses [106, 107].

Cytokinin has been reported to influence the expression of the aquaporin genes. The expression of *PgTIP1* from ginseng (*Panax ginseng*) was positively regulated by applying cytokinin in suspension-cultured hormone-autotrophic ginseng cells [108]. Methyl jasmonate treatment caused a decrease in the nitrogen content in barley leaves, which was associated with an increased expression of the four tonoplast aquaporin genes (*HvTIP1;2, HvTIP2;2, HvTIP4;1* and *HvTIP4;2*), which are predicted to transport the nitrogen compounds from the vacuole to the cytosol. The up-regulation of the nitrogen-transporting *HvTIPs* may lead to a vacuolar unloading of ammonia and urea, which could both be remobilised when the nitrogen content in barley leaves decreases [109]. In barley, the changes in the expression of all of the investigated *HvTIPs* in response to MeJA treatment were associated with the presence of the *cis*-regulatory elements in their promotor regions, which are recognised by the jasmonate-related transcription factors [109]. Other types of elements that possibly participate in the response to phytohormones in the *HvTIP* promotor regions such as abscisic acid (ABA), gibberellins (GA) and auxin were also identified [89]. In barley, the effect of ABA and GA on the expression of the *HvTIP1;2, HvTIP2;1, HvTIP2;2, HvTIP2;3* and *HvTIP4;1* genes after 24 hours of treatment was studied in both the shoots and roots using quantitative real-time RT-PCR [110]. Interestingly, the exogenous application of ABA increased the expression of *HvTIP2;1* and *HvTIP4;1,* while the GA-treatment led to an increase of *HvTIP1;2* and *HvTIP4;1* in the roots compared to the control conditions. However, in the shoots, the ABA treatment decreased the expression of *HvTIP1;2*, while the GA treatment led to the up-regulation of *HvTIP4;1* [110].

5. Genetic manipulation of the TIP-encoding genes

The tonoplast intrinsic proteins genes have been reported to be involved in the increasing the abiotic stress tolerance in several plant species using genetic modification technology (**Table 4**). It was shown that manipulations of the expression of

TIP name/species	Method/expression in species	Phenotype	References
AQUA1 (TIP) *Populus alba* L.	OX *Populus alba*	Increased plant growth rate and water use efficiency under excess Zn conditions	[111]
SlTIP2;2 *Solanum lycopersicum* L.	OX *Solanum lycopersicum*	Increased drought tolerance due to the ability of plant to regulate its transpiration under drought stress conditions, improved CO_2 uptake and a balanced nutrient supply	[112]
SlTIP2;2 *Solanum lycopersicum* L.	OX *Arabidopsis thaliana*	Enhances the tolerance to salt stress	[113]
PgTIP1 *Panax ginseng*	OX *Glycine max*	Higher tolerance to salinity	[114]
PgTIP1 *Panax ginseng*	OX *Arabidopsis thaliana*	Acceleration of plant development, with faster growth, precocious flowering and a higher accumulation of biomass, increased seed size and seed mass, greatly increased growth rate	[85]
PgTIP1 *Panax ginseng*	OX *Arabidopsis thaliana*	Accumulated more Na + under salt stress, and exhibited superior performance under drought stress	[108]
CsTIP2;1 *Citrus spp.*	OX *Nicotiana tabacum*	Increased plant growth and tolerance to drought and salinity	[115]
JcTIP1;3 *Jatropha curcas*	OX *Arabidopsis thaliana*	Increased tolerance to drought and salinity, improved germination under high salt and mannitol stress	[116]
AtTIP5;1 *Arabidopsis thaliana*	OX *Arabidopsis thaliana*	Increased tolerance to boron toxicity	[117]
AtTIP1;1 *Arabidopsis thaliana*	Down-regulation /RNAi technology	Early senescence and plant death	[118]
AtTIP1;1 -mutant AtTIP1;2 and AtTIP1;2 -double mutant *Arabidopsis thaliana*	KO Transposon insertion	Single mutant -no significant effect on the metabolism or elemental composition of the plants Double mutant -minor increase in the anthocyanin content, and a decrease in catalase activity, no changes in the water status, mutant alive and thriving	[119]
AtTIP2;2 *Arabidopsis thaliana*	KO T-DNA insertion	Less sensitive to abiotic stresses (mannitol, NaCl and PEG)	[120]

OX-overexpression, KO- knockout, RNAi-RNA interference, PEG- polyethylene glycol.

Table 4.
Genetic manipulation of the TIP genes in different plant species and its effect on the phenotype.

the *TIP*-encoding genes in plants could have the potential to improve their abiotic stress tolerance (**Table 4**). In tomato (*Solanum lycopersicum* L.), overexpression of *SlTIP2;2* has been shown to substantially improve the water homeostasis under drought stress. Conversely, this overexpression also promoted more prodigal water use [112]. Another gene, *TIP1;1*, was artificially manipulated in a few studies *via* genetic engineering (**Table 4**). It is not only believed to be play an important role during plant growth and development, but also to be beneficial for growth under stress when it is overexpressed. The overexpression of the *PgTIP1;1* gene from *Panax ginseng* in Arabidopsis led to a general acceleration of plant development, which included a faster growth, precocious flowering and a higher accumulation of biomass compared to the wild type (wt), or the accumulation of more Na^+ under salt stress and also caused a superior performance under drought stress compared to wt [85, 108]. The authors concluded that PgTIP1;1 plays an important role in the growth and development of plant cells and suggested that regulating it in the water movement across the tonoplast has a great impact on plant vigour under favourable growth conditions and also in the responses of plants to drought, salt and cold stresses [85]. Therefore, *TIP1;1* might be a potential target in biotechnology and agriculture. It was interesting to determine whether silencing this gene would have the opposite effect on growth in control and stress condition. First, it was shown that the down-regulation of *AtTIP1;1* using RNA interference (RNAi) resulted in early senescence and plant death [118]. However, later research using transposon insertion did not confirm these results [119]. A single mutant (tip1;1–1) that had been generated using transposon insertion showed any significant effect on the metabolism or elemental composition of the plants. Additionally, a double mutant in the *AtTIP1;1* and *AtTIP1;2* genes (tip1;1–1 tip1;2–1) had a small increase in its anthocyanin content and a decrease in its catalase activity, but showed no changes in its water status. Moreover, the mutant was alive and thriving [119].

6. Conclusions

It is believed that aquaporins and, among them, the tonoplast intrinsic proteins may be important players in the plant water relations at the cell, tissue, organ and whole plant levels. TIP proteins could have a vast impact on the acquisition of knowledge about plant tolerance against abiotic stress and could serve as a target sequence for genetic modifications. A drawback in this approach is that in the genome of each plant species, many genes encode the same subgroup of isoforms. It is possible that some of these have a redundant function. Taken together, TIPs probably play complex and diversified roles in the response of plants to abiotic stresses depending on TIP isoforms and the type and degree of stress treatment.

Acknowledgements

This work was supported by the National Science Centre, Poland, SONATA project 2015/19/D/NZ1/03441.

Conflict of interest

The author declares no conflict of interest.

Author details

Marzena Małgorzata Kurowska
University of Silesia in Katowice, Katowice, Poland

*Address all correspondence to: marzena.kurowska@us.edu.pl

References

[1] Maurel C, Chrispeels MJ. Aquaporins. A molecular entry into plant water relations. Plant Physiol. 2001;125(1):135-138. DOI: 10.1104/pp.125.1.135 PMID: 11154316

[2] Li C, Wang W. Molecular Biology of Aquaporins. Adv Exp Med Biol. 2017;969:1-34. DOI: 10.1007/978-94-024-1057-0

[3] Sutka M, Amodeo G, Ozu M. Plant and animal aquaporins crosstalk: what can be revealed from distinct perspectives. Biophys Rev. 2017; 9(5):545-562. DOI: 10.1007/s12551-017-0313-3

[4] Bezerra-Neto JP, de Araújo FC, Ferreira-Neto JRC, da Silva MD, Pandolfi V, Aburjaile FF, et al. Plant Aquaporins: Diversity, Evolution and Biotechnological Applications. Curr Protein Pept Sci. 2019; 20 (4):368-395. DOI: 10.2174/13892037206661811020 95910

[5] Fortin MG, Morrison NA, Verma DP. Nodulin-26, a peribacteroid membrane nodulin is expressed independently of the development of the peribacteroid compartment. Nucleic Acids Res. 1987;15(2):813-824. DOI: 10.1093/nar/15.2.813

[6] Maurel C, Reizer J, Schroeder JI, Chrispeels MJ. The vacuolar membrane protein gamma-TIP creates water specific channels in Xenopus oocytes, EMBO J. 1993;12: 2241-2247

[7] Kapilan R, Vaziri M, Zwiazek JJ. Regulation of aquaporins in plants under stress. Biol Res. 2018; 16; 51 (1):4. DOI: 10.1186/s40659-018-0152-0

[8] Quigley F, Rosenberg JM, Shachar-Hill Y, Bohnert HJ. From genome to function: the Arabidopsis aquaporins. Genome Biol. 2002;3(1). DOI: 10.1186/gb-2001-3-1

[9] Liu LH, Ludewig U, Gassert B, Frommer WB, von Wirén N. Urea transport by nitrogen-regulated tonoplast intrinsic proteins in Arabidopsis. Plant Physiol. 2003;133(3):1220-1228. DOI: 10.1104/pp.103.027409

[10] Klebl F, Wolf M, Sauer N. A defect in the yeast plasma membrane urea transporter Dur3p is complemented by CpNIP1, a Nod26-like protein from zucchini (*Cucurbita pepo* L.), and by Arabidopsis thaliana delta-TIP or gamma-TIP. FEBS Lett. 2003;547(1-3):69-74. DOI: 10.1016/s0014-5793(03)00671-9

[11] Loqué D, Ludewig U, Yuan L, von Wirén N. Tonoplast intrinsic proteins AtTIP2;1 and AtTIP2;3 facilitate NH3 transport into the vacuole. Plant Physiol. 2005;137(2):671-680. DOI: 10.1104/pp.104.051268

[12] Bienert GP, Møller AL, Kristiansen KA, et al. Specific aquaporins facilitate the diffusion of hydrogen peroxide across membranes. J Biol Chem. 2007;282(2):1183-1192. DOI: 10.1074/jbc.M603761200

[13] Soto G, Alleva K, Mazzella MA, Amodeo G, Muschietti JP. AtTIP1;3 and AtTIP5;1, the only highly expressed Arabidopsis pollen-specific aquaporins, transport water and urea. FEBS Lett. 2008;582(29):4077-4082. DOI: 10.1016/j.febslet.2008.11.002

[14] Besse M, Knipfer T, Miller A, Verdeil JL, Jahn TP, Fricke W. Developmental pattern of aquaporin expression in barley (Hordeum vulgare L.) leaves. J Exp Bot. 2011; 62(12):4127-42. DOI: 10. 1093/jxb/err175

[15] Jahn TP, Møller AL, Zeuthen T, et al. Aquaporin homologues in plants and mammals transport ammonia.

FEBS Lett. 2004;574(1-3):31-36. DOI: 10.1016/j.febslet.2004.08.004

[16] Holm LM, Jahn TP, Møller AL, et al. NH3 and NH4+ permeability in aquaporin-expressing Xenopus oocytes. Pflugers Arch. 2005;450(6):415-428. DOI: 10.1007/s00424-005-1399-1

[17] Bertl A, Kaldenhoff R. Function of a separate NH3-pore in Aquaporin TIP2;2 from wheat. FEBS Lett. 2007;581(28):5413-5417. DOI: 10.1016/j.febslet.2007.10.034

[18] Bárzana G, Aroca R, Bienert GP, Chaumont F, Ruiz-Lozano JM. New insights into the regulation of aquaporins by the arbuscular mycorrhizal symbiosis in maize plants under drought stress and possible implications for plant performance. Mol Plant Microbe Interact. 2014;27(4):349-363. DOI: 10.1094/MPMI-09-13-0268-R

[19] Gu R, Chen X, Zhou Y, Yuan L. Isolation and characterization of three maize aquaporin genes, ZmNIP2;1, ZmNIP2;4 and ZmTIP4;4 involved in urea transport. BMB Rep. 2012;45(2):96-101. DOI: 10.5483/BMBRep.2012.45.2.96

[20] Lopez F, Bousser A, Sissoëff I, Hoarau J, Mahé A. Characterization in maize of ZmTIP2-3, a root-specific tonoplast intrinsic protein exhibiting aquaporin activity. J Exp Bot. 2004;55(396):539-541. DOI: 10.1093/jxb/052

[21] Li GW, Peng YH, Yu X, et al. Transport functions and expression analysis of vacuolar membrane aquaporins in response to various stresses in rice. J Plant Physiol. 2008;165(18):1879-1888. DOI: 10.1016/j.jplph.2008.05.002

[22] Higuchi T, Suga S, Tsuchiya T, et al. Molecular cloning, water channel activity and tissue specific expression of two isoforms of radish vacuolar aquaporin. Plant Cell Physiol. 1998;39(9):905-913. DOI: 10.1093/oxfordjournals.pcp.a029453

[23] Sarda X, Tousch D, Ferrare K, et al. Two TIP-like genes encoding aquaporins are expressed in sunflower guard cells. Plant J. 1997;12(5):1103-1111. DOI: 10.1046/j.1365-313x.1997.12051103.x

[24] Gerbeau P, Güçlü J, Ripoche P, Maurel C. Aquaporin Nt-TIPa can account for the high permeability of tobacco cell vacuolar membrane to small neutral solutes. Plant J. 1999;18(6):577-587. DOI: 10.1046/j.1365-313x.1999.00481.x

[25] Marty F. Plant vacuoles. Plant Cell. 1999;11(4):587-600. DOI: 10.1105/tpc.11.4.587

[26] Paris N, Stanley CM, Jones RL, Rogers JC. Plant cells contain two functionally distinct vacuolar compartments. Cell. 1996;85(4):563-572. DOI: 10.1016/s0092-8674(00)81256-8

[27] Wink M. The plant vacuole: a multifunctional compartment. J. Exp. Bot. (supplement), 44, 231-246 (1993).

[28] Jauh GY, Fischer AM, Grimes HD, Ryan CA Jr, Rogers JC. delta-Tonoplast intrinsic protein defines unique plant vacuole functions. Proc Natl Acad Sci U S A. 1998;95(22):12995-12999. DOI: 10.1073/pnas.95.22.12995

[29] Jauh GY, Phillips TE, Rogers JC. Tonoplast intrinsic protein isoforms as markers for vacuolar functions. Plant Cell. 1999;11(10):1867-1882. DOI: 10.1105/tpc.11.10.1867

[30] Ludevid D, Höfte H, Himelblau E, Chrispeels MJ. The Expression Pattern of the Tonoplast Intrinsic Protein gamma-TIP in Arabidopsis thaliana Is Correlated with Cell Enlargement. Plant Physiol. 1992;100(4):1633-1639. DOI: 10.1104/pp.100.4.1633

[31] Johnson KD, Herman EM, Chrispeels MJ. An abundant, highly conserved tonoplast protein in seeds. Plant Physiol. 1989;91(3):1006-1013. DOI: 10.1104/pp.91.3.1006

[32] Hoh B, Hinz G, Jeong BK, Robinson DG. Protein storage vacuoles form de novo during pea cotyledon development. J Cell Sci. 1995;108 (Pt 1):299-310.

[33] Zybailov, B.L.; Rutschow, H.; Friso, G.; Rudella, A.; Emanuelsson, O.; Sun, Q.; Van Wijk, K.J. Sorting Signals, N-Terminal Modifications and Abundance of the Chloroplast Proteome. PLoS ONE 2008; 3, e1994. DOI: 10.1371/journal.pone.0001994.

[34] Ferro, M.; Brugière, S.; Salvi, D.; Seigneurin-Berny, D.; Court, M.; Moyet, L.; Ramus, C.; Miras, S.; Mellal, M.; Le Gall, S.; et al. AT_CHLORO, a comprehensive chloroplast proteome database with subplastidial localization and curated information on envelope proteins. Mol. Cell. Proteom. 2010; 9, 1063-84. DOI: 10.1074/mcp.M900325-MCP200.

[35] Simm, S.; Papasotiriou, D.G.; Ibrahim, M.; Leisegang, M.S.; Müller, B.; Schorge, T.; Karas, M.; Mirus, O.; Sommer, M.S.; Schleiff, E. Defining the core proteome of the chloroplast envelope membranes Front. Plant Sci. 2013; 4, 11. DOI: 10.3389/fpls.2013.00011

[36] Jung JS, Preston GM, Smith BL, Guggino WB, Agre P. Molecular structure of the water channel through aquaporin CHIP. The hourglass model. J Biol Chem. 1994;269(20):14648-14654.

[37] Fox AR, Maistriaux LC, Chaumont F. Toward understanding of the high number of plant aquaporin isoforms and multiple regulation mechanisms. Plant Sci. 2017;264: 179-187. DOI: 10.1016/j.plantsci.2017.07.021

[38] Sui H, Han BG, Lee JK, Walian P, Jap BK. Structural basis of water-specific transport through the AQP1 water channel. Nature. 2001;414(6866):872-878. DOI: 10.1038/414872a

[39] Wu B, Steinbronn C, Alsterfjord M, Zeuthen T, Beitz E. Concerted action of two cation filters in the aquaporin water channel. EMBO J. 2009;28(15):2188-2194. DOI: 10.1038/emboj.2009.182

[40] Guan XG, Su WH, Yi F, et al. NPA motifs play a key role in plasma membrane targeting of aquaporin-4. IUBMB Life. 2010;62(3):222-226. DOI: 10.1002/iub.311

[41] Froger A, Tallur B, Thomas D, Delamarche C. Prediction of functional residues in water channels and related proteins. Protein Sci. 1998;7(6):1458-1468. DOI: 10.1002/pro.5560070623

[42] Verma RK, Gupta AB, Sankararamakrishnan *R. major* intrinsic protein superfamily: channels with unique structural features and diverse selectivity filters. Methods Enzymol. 2015a;557:485-520. DOI: 10.1016/bs.mie.2014.12.006

[43] Verma RK, Prabh ND, Sankararamakrishnan R. Intra-helical salt-bridge and helix destabilizing residues within the same helical turn: Role of functionally important loop E half-helix in channel regulation of major intrinsic proteins. Biochim Biophys Acta. 2015b;1848(6):1436-1449. DOI: 10.1016/j.bbamem.2015.03.013

[44] Gomes D, Agasse A, Thiébaud P, DelrotS,GerósH,ChaumontF.Aquaporins are multifunctional water and solute transporters highly divergent in living organisms. Biochim Biophys Acta. 2009;1788(6):1213-1228. DOI: 10.1016/j.bbamem.2009.03.009

[45] Törnroth-Horsefield S, Wang Y, Hedfalk K, et al. Structural mechanism

of plant aquaporin gating. Nature. 2006;439(7077):688-694. DOI: 10.1038/nature04316

[46] Wang H, Schoebel S, Schmitz F, Dong H, Hedfalk K. Characterization of aquaporin-driven hydrogen peroxide transport. Biochim Biophys Acta Biomembr. 2020;1862(2):183065. DOI: 10.1016/j.bbamem.2019.183065

[47] Data base for Major Intrinsic Proteins, http://mipdb.genouest.org/e107_plugins/mipdb_project/project.php, Accessed: 2020-07-23

[48] El Karkouri K, Gueuné H, Delamarche C. MIPDB: a relational database dedicated to MIP family proteins. Biol Cell. 2005;97(7):535-543. doi:10.1042/BC20040123

[49] Sun H, Li L, Lou Y, Zhao H, Gao Z. Genome-wide identification and characterization of aquaporin gene family in moso bamboo (*Phyllostachys edulis*). Mol Biol Rep. 2016;43(5):437-450. DOI: 10.1007/s11033-016-3973-3

[50] Yuan D, Li W, Hua Y, King GJ, Xu F, Shi L. Genome-Wide Identification and Characterization of the Aquaporin Gene Family and Transcriptional Responses to Boron Deficiency in *Brassica napus*. Front Plant Sci. 2017;8:1336. DOI: 10.3389/fpls.2017.01336

[51] Hove RM, Ziemann M, Bhave M. Identification and Expression Analysis of the Barley (Hordeum vulgare L.) Aquaporin Gene Family. PLoS One. 2015;10(6):e0128025. DOI: 10.1371/journal.pone.0128025

[52] Madrid-Espinoza J, Brunel-Saldias N, Guerra FP, Gutiérrez A, Del Pozo A. Genome-Wide Identification and Transcriptional Regulation of Aquaporin Genes in Bread Wheat (Triticum aestivum L.) under Water Stress. Genes (Basel). 2018;9(10):497. DOI: 0.3390/genes9100497

[53] Zhu YX, Yang L, Liu N, et al. Genome-wide identification, structure characterization, and expression pattern profiling of aquaporin gene family in cucumber. BMC Plant Biol. 2019;19(1):345. DOI: 10.1186/s12870-019-1953-1

[54] Singh RK, Shweta S, Muthamilarasan M, Rani R, Prasad M. Study on aquaporins of Setaria italica suggests the involvement of SiPIP3;1 and SiSIP1;1 in abiotic stress response. Funct Integr Genomics. 2019;19(4):587-596. DOI: 10.1007/s10142-018-00653-0

[55] Bari A, Farooq M, Hussain A, Tahir ul Qama Mr, Waseem Abbas M, Mustafa G, Karim A, Ahmed I, Hussain T. Genome-Wide Bioinformatics Analysis of Aquaporin Gene Family in Maize (Zea mays L.). Journal of Phylogenetics & Evolutionary Biology 2018;6:2. DOI: 0.4172/2329-9002.1000197

[56] Venkatesh J, Yu JW, Park SW. Genome-wide analysis and expression profiling of the Solanum tuberosum aquaporins. Plant Physiol Biochem. 2013;73:392-404. DOI: 10.1016/j.plaphy.2013.10.025

[57] Sakurai J, Ishikawa F, Yamaguchi T, Uemura M, Maeshima M. Identification of 33 rice aquaporin genes and analysis of their expression and function. Plant Cell Physiol. 2005;46(9):1568-1577. DOI: 10.1093/pcp/pci172

[58] Azad AK, Ahmed J, Alum MA, et al. Genome-Wide Characterization of Major Intrinsic Proteins in Four Grass Plants and Their Non-Aqua Transport Selectivity Profiles with Comparative Perspective. PLoS One. 2016;11(6):e0157735. DOI: 10.1371/journal.pone.0157735

[59] Reuscher S, Akiyama M, Mori C, Aoki K, Shibata D, Shiratake K. Genome-wide identification and expression analysis of aquaporins in

tomato. PLoS One. 2013;8(11):e79052. DOI: 10.1371/journal.pone.0079052

[60] Vander Willigen C, Postaire O, Tournaire-Roux C, Boursiac Y, Maurel C. Expression and inhibition of aquaporins in germinating Arabidopsis seeds. Plant Cell Physiol. 2006;47(9):1241-1250. DOI: 10.1093/pcp/pcj094

[61] Soto G, Fox R, Ayub N, Alleva K, Guaimas F, Erijman EJ, Mazzella A, Amodeo G, Muschietti J. TIP5;1 is an aquaporin specifically targeted to pollen mitochondria and is probably involved in nitrogen remobilization in Arabidopsis thaliana. Plant J. 2010; 64, 1038-1047 DOI: 10.1111/j.1365-313X.2010.04395

[62] Pérez Di Giorgio JA, Barberini ML, Amodeo G, Muschietti JP. Pollen aquaporins: What are they there for? Plant Signal Behav. 2016;11(9):e1217375. DOI: 10.1080/15592324.2016.1217375

[63] Sato R, Maeshima M. The ER-localized aquaporin SIP2;1 is involved in pollen germination and pollen tube elongation in Arabidopsis thaliana. Plant Mol Biol. 2019;100(3):335-349. DOI: 10.1007/s11103-019-00865-3

[64] Rodrigues O, Reshetnyak G, Grondin A, et al. Aquaporins facilitate hydrogen peroxide entry into guard cells to mediate ABA- and pathogen-triggered stomatal closure. Proc Natl Acad Sci U S A. 2017;114(34):9200-9205. DOI: 10.1073/pnas.1704754114

[65] Ding L, Chaumont F. Are Aquaporins Expressed in Stomatal Complexes Promising Targets to Enhance Stomatal Dynamics?. Front Plant Sci. 2020;11:458. DOI: 10.3389/fpls.2020.00458

[66] Höfte H, Hubbard L, Reizer J, Ludevid D, Herman EM, Chrispeels MJ. Vegetative and Seed-Specific Forms of Tonoplast Intrinsic Protein in the Vacuolar Membrane of Arabidopsis thaliana. Plant Physiol. 1992;99(2):561- DOI: 10.1104/pp.99.2.561

[67] Daniels MJ, Chaumont F, Mirkov TE, Chrispeels MJ. Characterization of a new vacuolar membrane aquaporin sensitive to mercury at a unique site. Plant Cell. 1996;8(4):587-599. DOI: 10.1105/tpc.8.4.587

[68] Rodrigues MI, Takeda AA, Bravo JP, Maia IG. The Eucalyptus Tonoplast Intrinsic Protein (TIP) Gene Subfamily: Genomic Organization, Structural Features, and Expression Profiles. Front Plant Sci. 2016;7:1810. DOI: 10.3389/fpls.2016.01810570. doi:10.1104/pp.99.2.561

[69] Liu Q, Umeda M, Uchimiya H. Isolation and expression analysis of two rice genes encoding the major intrinsic protein. Plant Mol Biol. 1994;26(6):2003-2007. DOI: 10.1007/BF00019511

[70] Yamamoto YT, Taylor CG, Acedo GN, Cheng CL, Conkling MA. Characterization of cis-acting sequences regulating root-specific gene expression in tobacco. Plant Cell. 1991;3(4):371-382. DOI: 10.1105/tpc.3.4.371

[71] Singh RK, Deshmukh R, Muthamilarasan M, Rani R, Prasad M. Versatile roles of aquaporin in physiological processes and stress tolerance in plants. Plant Physiol Biochem. 2020;149:178-189. DOI: 10.1016/j.plaphy.2020.02.009

[72] Nesverova V, Törnroth-Horsefield S. Phosphorylation-Dependent Regulation of Mammalian Aquaporins. Cells. 2019;8(2):82. DOI: 10.3390/cells8020082

[73] Yang, I.; Kim, S. Analysis of whole transcriptome sequencing data: Workflow and software. Genom.

Inform.2015, 13, 119-125. DOI: 10.5808/GI.2015.13.4.119

[74] Song L, Nguyen N, Deshmukh RK, et al. Soybean TIP Gene Family Analysis and Characterization of GmTIP1;5 and GmTIP2;5 Water Transport Activity. Front Plant Sci. 2016;7:1564. DOI: 10.3389/fpls.2016.01564

[75] Zhang DY, Kumar M, Xu L, et al. Genome-wide identification of Major Intrinsic Proteins in *Glycine soja* and characterization of GmTIP2;1 function under salt and water stress Sci Rep. 2017;7(1):4106. DOI: 10.1038/s41598-017-04253-z

[76] Atkinson NJ, Urwin PE. The interaction of plant biotic and abiotic stresses: from genes to the field. J. Exp. Bot. 2014;63, 3523-43. DOI: 10.1093/jxb/ers100

[77] Osakabe Y, Osakabe K, Shinozaki K, Tran LS. Response of plants to water stress. Front. Plant Sci. 2014; DOI: 10.3389/fpls.2014.00086

[78] Golldack D, Li C, Mohan H, Probst N. Tolerance to drought and salt stress in plants, unraveling the signaling networks. Front. Plant Sci. 2014; DOI: 10.3389/fpls.2014.00151.

[79] Merlaen B, De Keyser E, Ding L, Leroux O, Chaumont F, Van Labeke MC. Physiological responses and aquaporin expression upon drought and osmotic stress in a conservative vs prodigal Fragaria x ananassa cultivar. Plant Physiol Biochem. 2019;145:95-106. DOI: 10.1016/j.plaphy.2019.10.030

[80] Qiao Y, Ren J, Yin L, Liu Y, Deng X, Liu P, Wang S. Exogenous melatonin alleviates PEG-induced short-term water deficiency in maize by increasing hydraulic conductance. BMC Plant Biol. 2020;14,20(1):218. DOI: 10.1186/s12870-020-02432-1

[81] Cominelli E, Conti L, Tonelli C, Galbiati M. Challenges and perspectives to improve crop drought and salinity tolerance. N Biotechnol. 2013;30(4):355-361. DOI: 10.1016/j.nbt.2012.11.001

[82] Singh B, Kukreja S, Goutam U. Milestones achieved in response to drought stress through reverse genetic approaches. F1000Res. 2018;7:1311. DOI: 10.12688/f1000research.15606.1

[83] Fahad S, Bajwa AA, Nazir U, Anjum SA, Farooq A, Zohaib A, Sadia S, NasimW, Adkins S, Saud S, Ihsan MZ, Alharby H,Wu C,Wang D, Huang J. Crop production under drought and heat stress: Plant responses and Management Options. Front Plant Sci. 2017; 8:1147. DOI: 10.3389/fpls.2017.01147

[84] Dolferus R. To grow or not to grow: a stressful decision for plants. Plant Sci. 2014;229:247-261. DOI: 10.1016/j.plantsci.2014.10.002

[85] Peng Y, Lin W, Cai W, Arora R. Overexpression of a Panax ginseng tonoplast aquaporin alters salt tolerance, drought tolerance and cold acclimation ability in transgenic Arabidopsis plants. Planta. 2007;226(3):729-740. DOI: 10.1007/s00425-007-0520-4

[86] Smart LB, Moskal WA, Cameron KD, Bennett AB. MIP genes are down-regulated under drought stress in *Nicotiana glauca*. Plant Cell Physiol. 2001; 42(7):686-93. DOI: 10.1093/pcp/pce085

[87] Alexandersson E, Fraysse L, Sjo¨vall-Larsen S, Gustavsson S, Fellert M, Karlsson M, et al. Whole gene family expression and drought stress regulation of aquaporins. Plant Mol Biol. 2005; 59(3):469-84. DOI: 10.1007/s11103-005-0352-1

[88] Feng ZJ, Xu SC, Liu N, Zhang GW, Hu QZ, Xu ZS, et al. Identification of the AQP members involved in abiotic stress responses from Arabidopsis. Gene.

2018; 10;646:64-73. DOI: 10.1016/j.gene.2017.12.048

[89] Kurowska MM, Wiecha K, Gajek K, Szarejko I. Drought stress and re-watering affect the abundance of TIP aquaporin transcripts in barley. PLoS One. 2019;14(12):e0226423. DOI: 10.1371/journal.pone.0226423

[90] Pawłowicz I, Rapacz M, Perlikowski D, Gondek K, Kosmala A. Abiotic stresses influence the transcript abundance of PIP and TIP aquaporins in Festuca species. J Appl Genet. 2017; 58(4):421-435. DOI: 10.1007/s13353-017-0403-8

[91] Miniussi M, Del Terra L, Savi T, Pallavicini A, Nardini A. Aquaporins in *Coffea arabica* L.: Identification, expression, and impacts on plant water relations and hydraulics. Plant Physiol Biochem. 2015;95:92-102. DOI: 10.1016/j.plaphy.2015.07.024

[92] Zupin M, Sedlar A, Kidrič M, Meglič V. Drought-induced expression of aquaporin genes in leaves of two common bean cultivars differing in tolerance to drought stress. J Plant Res. 2017;130(4):735-745. DOI: 10.1007/s10265-017-0920-x

[93] Choudhury, F.K.; Rivero, R.M.; Blumwald, E.; Mittler, R. Reactive oxygen species, abiotic stress and stress combination. Plant J. 2017, 90, 856-867. DOI: 10.1111/tpj.13299

[94] Bienert GP, Chaumont F. Aquaporin-facilitated transmembrane diffusion of hydrogen peroxide. Biochim Biophys Acta. 2014; 1840(5):1596-604. DOI: 10.1016/j.bbagen.2013.09.017 PMID:24060746

[95] Nabati J., Kafi M., Nezami A., Moghaddam P.R., Ali M., Mehrjerdi M.Z. Effect of salinity on biomass production and activities of some key enzymatic antioxidants in Kochia (*Kochia scoparia*) Pak. J. Bot. 2011;43:539-548.

[96] Kamran M, Parveen A, Ahmar S, et al. An Overview of Hazardous Impacts of Soil Salinity in Crops, Tolerance Mechanisms, and Amelioration through Selenium Supplementation. Int J Mol Sci. 2019;21(1):148. DOI: 10.3390/ijms21010148

[97] Boursiac Y, Chen S, Luu DT, Sorieul M, van den Dries N, Maurel C. Early effects of salinity on water transport in Arabidopsis roots. Molecular and cellular features of aquaporin expression. Plant Physiol. 2005;139(2):790-805. DOI: 10.1104/pp.105.065029

[98] Afzal Z, Howton TC, Sun Y, Mukhtar MS. The Roles of Aquaporins in Plant Stress Responses. J Dev Biol. 2016;4(1):9. DOI: 10.3390/jdb4010009

[99] Li DD, Tai FJ, Zhang ZT, et al. A cotton gene encodes a tonoplast aquaporin that is involved in cell tolerance to cold stress. Gene. 2009;438(1-2):26-32. DOI: 10.1016/j.gene.2009.02.023

[100] Fahad S, Hussain S, Matloob A, Khan FA, Khaliq A, Saud S, Hassan S, Shan D, Khan F, Ullah N, Faiq M, Khan MR, Tareen AK, Khan A, Ullah A, Ullah N, Huang J. Phytohormones and plant responses to salinity stress: A review. Plant Growth Regul. 2014; 75(2):391-404. DOI: 10.1007/s10725-014-0013-y

[101] Yoshida T, Christmann A, Yamaguchi-Shinozaki K, Grill E, Fernie AR. Revisiting the Basal Role of ABA - Roles Outside of Stress. Trends Plant Sci. 2019;24(7):625-635. DOI: 10.1016/j.tplants.2019.04.008

[102] Cai S, Chen G, Wang Y, et al. Evolutionary Conservation of ABA Signaling for Stomatal Closure. Plant Physiol. 2017;174(2):732-747. DOI: 10.1104/pp.16.01848

[103] Jang G, Yoon Y, Choi YD. Crosstalk with Jasmonic Acid Integrates Multiple Responses in Plant Development. Int J Mol Sci. 2020;21(1):305. DOI: 10.3390/ijms21010305

[104] Yang J, Duan G, Li C, et al. The Crosstalks Between Jasmonic Acid and Other Plant Hormone Signaling Highlight the Involvement of Jasmonic Acid as a Core Component in Plant Response to Biotic and Abiotic Stresses. Front Plant Sci. 2019;10:1349. DOI: 10.3389/fpls.2019.01349

[105] Colebrook EH, Thomas SG, Phillips AL, Hedden P. The role of gibberellin signalling in plant responses to abiotic stress. J Exp Biol. 2014;217: 67-75. DOI: 10.1242/jeb.089938.

[106] Lubovská Z, Dobrá J, Storchová H, Wilhelmová N, Vanková R. Cytokinin oxidase/dehydrogenase overexpression modifies antioxidant defense against heat, drought and their combination in Nicotiana tabacum plants. J Plant Physiol. 2014;171(17):1625-1633. DOI: 10.1016/j.jplph.2014.06.021

[107] Ullah A, Manghwar H, Shaban M, et al. Phytohormones enhanced drought tolerance in plants: a coping strategy. Environ Sci Pollut Res Int. 2018;25(33):33103-33118. DOI: 10.1007/s11356-018-3364-5

[108] Lin W, Peng Y, Li G, et al. Isolation and functional characterization of PgTIP1, a hormone-autotrophic cells-specific tonoplast aquaporin in ginseng. J Exp Bot. 2007;58(5):947-956. DOI: 10.1093/jxb/erl255

[109] Kurowska MM, Daszkowska-Golec A, Gajecka M, Kościelniak P, Bierza W, Szarejko I. Methyl Jasmonate Affects Photosynthesis Efficiency, Expression of HvTIP Genes and Nitrogen Homeostasis in Barley. Int J Mol Sci. 2020;21(12):4335. DOI: 10.3390/ijms21124335

[110] Ligaba A, Katsuhara M, Shibasaka M, Djira G. Abiotic stresses modulate expression of major intrinsic proteins in barley (Hordeum vulgare). C R Biol. 2011; 334(2):127-139. DOI: 10.1016/j.crvi.2010.11.005 PMID: 21333943

[111] Ariani A, Francini A, Andreucci A, Sebastiani L. Over-expression of AQUA1 in Populus alba Villafranca clone increases relative growth rate and water use efficiency, under Zn excess condition. Plant Cell Rep. 2016;35(2):289-301. DOI: 10.1007/s00299-015-1883-9

[112] Sade N, Vinocur BJ, Diber A, et al. Improving plant stress tolerance and yield production: is the tonoplast aquaporin SlTIP2;2 a key to isohydric to anisohydric conversion?. New Phytol. 2009;181(3):651-661. DOI: 10.1111/j.1469-8137.2008.02689.x

[113] Xin S, Yu G, Sun L, Qiang X, Xu N, Cheng X. Expression of tomato SlTIP2;2 enhances the tolerance to salt stress in the transgenic Arabidopsis and interacts with target proteins. J Plant Res. 2014;127(6):695-708. DOI: 10.1007/s10265-014-0658-7

[114] An, J.; Hu, Z.; Che, B.; Chen, H.; Yu, B.; Cai, W. Heterologous expression of Panax ginseng PgTIP1 confers enhanced salt tolerance of soybean cotyledon hairy roots, composite, and whole plants. Front. Plant Sci., 2017, 8, 1232. DOI: 10.3389/fpls.2017.01232

[115] Martins, C.P.S.; Neves, D.M.; Cidade, L.C.; Mendes, A.F.S.; Silva,D.C.; Almeida, A.A. F.; Coelho-Filho, M.A.; Gesteira, A.S.; Soares-Filho, W.S.; Costa, M.G.C. Expression of the citrus CsTIP2;1 gene improves tobacco plant growth, antioxidant capacity and physiological adaptation under stress conditions. Planta, 2017, 245(5), 951-963. DOI: 10.1007/s00425-017-2653-4

[116] Khan, K.; Agarwal, P.; Shanware, A.; Sane, V.A. Heterologous expression of two jatropha aquaporins imparts drought and salt tolerance and improves seed viability in transgenic Arabidopsis thaliana. PLoS One, 2015, 10(6), e0128866. DOI: 10.1371/journal.pone.0128866

[117] Pang, Y.; Li, L.; Ren, F.; Lu, P.; Wei, P.; Cai, J.; Xin, L.; Zhang, J.; Chen, J.; Wang, X. Overexpression of the tonoplast aquaporin At-TIP5;1 conferred tolerance to boron toxicity in Arabidopsis. J. Genet. Genomics., 2010, 37(6), 389-397. DOI: 10.1016/S1673-8527(09)60057-6

[118] Ma S, Quist TM, Ulanov A, Joly R, Bohnert HJ. Loss of TIP1;1 aquaporin in Arabidopsis leads to cell and plant death. Plant J. 2004;40(6):845-859. DOI: 10.1111/j.1365-313X.2004.02265.x

[119] Schüssler MD, Alexandersson E, Bienert GP, et al. The effects of the loss of TIP1;1 and TIP1;2 aquaporins in Arabidopsis thaliana. Plant J. 2008;56(5):756-767. DOI: 10.1111/j.1365-313X.2008.03632.x

[120] Feng, Z.J.; Xu, S.C.; Liu, N.; Zhang, G.W.; Hu, Q.Z.; Xu, Z.S.; Gong, Y.M. Identification of the AQP members involved in abiotic stress responses from Arabidopsis. Gene, 2018, 646, 64-73 DOI: 10.1016/j.gene.2017.12.048

Chapter 21

Drought Responses on Physiological Attributes of *Zea mays* in Relation to Nitrogen and Source-Sink Relationships

Suphia Rafique

Abstract

Maize is the staple food crop and essential for world food security. Maize plants' water requirement is high for proper growth and development at vegetative stage and grain formation at reproductive stage. Drought is the major abiotic stress that affects almost all the growth stages of maize crop and it has a strong impact on all the physiological process of maize plants. Similarly, N metabolism is of central importance during drought stress. Nitrogen (N) is one of the macronutrients; it is a major requirement for crop growth and grain yield of maize. Therefore, nitrogen and water separately or in combination are the two most critical factors in maize production. Drought modifies source-sink relations and weakens the source and sink strength, which disturbs plant's growth, plant's adaptation to stress, and consequently affects crop productivity.

Keywords: maize, drought, nitrogen, source-sink relationship

1. Introduction

Plants are nonmotile, and they protect themselves from various biotic and abiotic stresses through structural and metabolic changes like altering leaf orientation, transpirational cooling, or adjustment of membrane lipid compositions [1]. Maize is a staple food crop that grows under optimal environmental conditions and throughout their life span is often exposed to various abiotic stresses [2] like either combined stresses like drought and heat, drought and salinity, drought and low-N stresses, or singly drought. In combined stress, plants show a wide range of responses that brings changes in growth and morphology, and the plants' ability to withstand these stresses greatly varies from species to species [3]. However, drought along with high frequencies, longer durations, and large ranges occur almost yearly across the globe, even in wet and humid regions [4, 5]. Water deficit is the most detrimental environmental stress that adversely affects the maize productivity. The effect of drought is manifested at cellular, physiological biochemical, metabolic, proteins, and gene levels. Water is an essential component for plants; it maintains cell turgidity and keeps the structure intact (by keeping the pressure inside cells high) so that they are fully expanded (i.e., turgid). In wilted plants, when turgor pressure approaches to zero, the cell membranes collapses, damage and becomes leaky, the key enzymes proteins denatured and their structure altered. To ensure high survival rates and production under drought conditions, maize plants rely on several strategies, including drought avoidance, escape, and tolerance [6–8]. Nitrogen plays a major role in plant nutrition and also combats with several abiotic stresses like salinity, drought, heat stress, etc. [9–13], but this mineral element is most often deficient in arable soils. Though the requirement of nitrogen is greatest among all mineral elements, its deficiency limits the plant's growth and development [14–16]. There is a tight regulation of carbon and nitrogen metabolism in photosynthesis and N uptake [14, 17–19]. In a review, Aziz Khan et al. [20] outline the nitrogen (N) responses in crop production and to ameliorate abiotic stresses for better crop production. Further they emphasize that nitrate and ammonium nitrogen are essential nutrients for successful crop production [21]. Moreover, shortage and N excess affect assimilate partitioning between vegetative and reproductive organs [22]. Mineral nutrition influences the physiological response of plants to water stress [23]. While Waraich et al. [24] discussed the role of macro- and micronutrients to decrease the adverse effect of drought in crop plants. Therefore, water and nitrogen affects the crop growth, development, and production either separately or in combination. Humbert et al. [25] observed the physio-molecular changes in response to water and nitrogen and finally concluded that the responses of plants to the combination of these two stresses may cause additional effect that was different from the individual effects, and hence, cannot be inferred from the end results obtained from different stresses applied individually. Drought affects maize grain yield to some degree at almost all growth stages, but the crop is the most susceptible during flowering [26, 27]. At the reproductive phase, N availability affects assimilate partitioning between vegetative and reproductive organs and N metabolism in young ear shoots [28]. Therefore, the timing and intensity of stress determine the extent to limits yield either due to source or sink limitations.

2. Maize response under drought stress

2.1 Effect of drought on morpho-physiological parameters

The water deficit conditions Boyer [29] showed the decrease in leaf area attributed to decrease rate leaf initiation and expansion or increase leaf senescence or shedding. Though the number of leaves per plant did not reduce in early vegetative maize growth in severe nonlethal water deficit conditions, Abrecht and Carberry [30] also demonstrated the gradual reduction in plant height in maize with increasing water deficit. Further supported by Moss and Downey [31] showed the significant reduction in plant height with increase in water stress. While, dry matter yield accumulation depends on the leaf area and leaf dry weight because leaf area is the major assimilatory surface for most crop any factor that affects the leaf area also affects dry matter [32]. In similar way, the short term responses of corn to a pre-anthesis water deficit, delayed leaf tip emergence and reduction in leaf area. While long-term consequences of water deficits have reduced final sizes of the leaves and internodes [33]. Drought stress significantly repressed relative leaf water content, leaf size, and photosynthesis-related parameters in maize seedling [34]. Drought affects various morpho-physiological processes including plant biomass, root length, shoot length, photosynthesis, water use efficiency (WUE) and leaf water content [35, 36]. At early growth stage 50 maize genotypes for drought tolerance was assessed. Principal component analysis revealed important morpho-physiological traits (FSL, FRL, DSW, DRW, RWC, and TDM) plays key role in drought tolerance. Later, Ali et al. [37] identified various morpho-physiological traits that have higher heritability, genetic advancement, and correlation and which contribute for improved maize grain yield. A set of 24 genotypes bred at different centers in India as well as in CIMMYT showing variability for drought tolerance were selected for molecular and morpho-physiological characterization. Phenotyping of inbred by morpho-physiological traits revealed that there was a positive relationship among root length, chlorophyll content, relative water content while anthesis-silking interval (ASI) have negative relationship with all these traits [38]. Water stress occurring during vegetative and tasseling stages reduced plant height, as well as leaf area development [39].

2.2 Photosynthesis under drought stress

In a study Anjum et al. [40] showed that drought stress in maize led to considerable decline in net photosynthesis (33.22%), transpiration rate (37.84%), stomatal conductance (25.54%), water use efficiency (50.87%), intrinsic water use efficiency (11.58%) and intercellular CO_2 (5.86%) as compared to well water (WW) control. The main cause of reduced photosynthesis under drought stress is due to decreases in both assimilation (A) and internal CO_2 concentration, that finally inhibits total photosynthetic metabolism [41]. Besides there are other several non-stomatal effects during drought which are responsible for stomatal closure. These include photophosphorylation, ribulose-1,5-bisphosphate (RuBP) regeneration [42, 43], rubisco activity, and ATP synthesis [44, 45]. The response of growth and some physiological characteristics were compared in two maize *(Zea mays* L.) cultivars, one drought-resistant (PNR473) and the other drought-sensitive (SR52). Drought-resistant (PNR473) had a higher growth rate and deeper rooting than the drought-sensitive cultivar under water stress treatment. But the drought-resistant cultivar had a higher transpiration rate and lower diffusive resistance during the onset of water stress, and higher relative water content and levels of abscisic acid and proline throughout the period of water stress [46]. While, Jama and Ottaman [47] reported delay in the irrigation during early growth stages of a corn plant (including anthesis) decreased plant dry weight. The responses of maize (*Zea mays* L.) from the third leaf stage to maturity for different soil water levels (well watered, moderately stressed, and severely stressed) indicated that drought stress relied on drought intensity and duration, with more severe drought stress creating more serious effects on maize. It is well known that leaf water status always interacts with stomatal conductance through plant hormone abscisic acid, increased production of ABA would help the plants to survive under drought stress [48]. According to Bray [49], dry soil induced and increased ABA concentrations in the roots to maintain root growth, increase root hydraulic conductivity, through these process shoot water uptake increases thus drought induced root-to-leaf signaling. Also direct correlation between the xylem ABA content and stomatal conductance has been demonstrated [50]. The role of other phytohormones like Brassinosteroid is for plant growth, as they are involved in main plant antioxidant processes encompassing the regulation and enhancement of plant tolerance against different environmental stresses [51].

2.3 Metabolic changes under drought stress

To maintain cell turgor and to lower the osmotic potential plants accumulate different types of organic and inorganic solutes in the cytosol [52]. Primary metabolites, such as sugars and amino acids, have also been reported to accumulate during water stress [53]. Two maize (*Zea mays* L.) cultivars, 704 and 301, accumulate soluble sugars, starch, and proline content in shoots and roots in response to drought stress. This suggests role of sugars in osmotic adjustment whereas, proline minimized the damage caused by dehydration [54]. The high salt concentration of soil causes a water deficit in crops [55], exogenous application of SA mitigated the adverse effects of salinity on maize plants by osmoregulation which is possibly mediated by increased production of sugar as well as proline [9]. Besides, the ROS (including oxygen ions, free radicals, and peroxidases) forms major by-product of normal metabolism and in abiotic stress conditions. However, during environmental stress such as drought, ROS levels increase dramatically resulting in oxidative damage to proteins, DNA, and lipids [56]. Among six maize genotypes (A638, B73, Grace-E5, Lo964, Lo1016, and Va35) studied, the drought-sensitive maize genotypes accumulated significantly more reactive oxygen species (ROS) and reactive nitrogen species (RNS). In addition, they showed rapid increases in enzyme activities involved in ROS and RNS metabolism compared to tolerant genotypes. However, antioxidant enzyme activities were higher of tolerant genotypes than in the sensitive genotypes [57]. In a study, the protective role of biochar (date and wheat) in combination with Me and P against water stress in two maize varieties were examined. The maize variety Mallika performed better and was more drought-tolerant with its increased osmolyte production and an efficient antioxidant defense system that eliminated ROS. Hence, ROS elimination further alleviated the damage to the photosynthesis system induced by water stress. On other hand, Azam variety was more sensitivity to drought [58]. Plants tolerate the water deficit conditions by adjusting their physiological and biochemical approaches. Maize leaves subjected to water deficit rigorously down-regulated NR activity and photosynthesis, also a correlation between maximal extractable foliar nitrate reductase (NR) activity and the rate of CO_2 assimilation was observed [59]. The assimilatory sulfate reduction in roots and leaves of the staple crop maize demonstrated the organ specific impact of drought upon sulfate metabolism. Under drought stress, the allocation of sulfate was significantly shifted to the roots allowed for significant increase of thiols derived from sulfate assimilation in roots. This enabled roots to produce biomass, while leaf growth was stopped. Therefore, sulfur metabolism related alterations at the transcriptional, metabolic and enzyme activity level are consistent with a promotion of root growth to search for water at the expense of leaf growth. The results provide evidence for the importance of antagonistic regulation of sulfur metabolism in leaves and roots to enable successful drought stress response at the whole plant level [60]. During the silking and blister stages moderate stress significantly change the relative water content (RWC) and also change the relative conductivity (RC) ($P < 0.05$) of the leaves; however, severe stress significantly decrease ($P < 0.01$) the leaf RWC and increase ($P < 0.01$) membrane permeability (leaf relative conductivity). Furthermore, under severe drought stress antioxidant enzyme activities declined ($P < 0.01$) in later stages, namely for superoxide dismutase (SOD) the tasseling and blister stages, for peroxidase (POD) the milk stage, and for catalase (CAT) during the tasseling, blister, and milk stages. Meanwhile, membrane lipid peroxidation (measured as malondialdehyde content) significantly increased ($P < 0.01$) in all stages [61].

2.4 Drought-tolerance mechanism

Phytohormones like, ABA, IAA, CKs, GA, SA, BRs, JA, ethylene, and triazole are directly or indirectly involved in plant responses to a wide range of stresses. They play critical roles in regulating plant growth and development, and stress tolerance to promote survival and acclimatize to varying environments [10]. Abscisic acid accumulates in response to drought stress and displays its well characterized effect on stomatal closure and promotes root growth in maize seedlings at low water potential [62]. Another possible role of endogenous ABA is to modulate glycinebetaine (GB) metabolism in maize particularly at the seedling stage in drought stress. The leaves of two maize cultivars, Zhengdan 958 (ZD958; drought-tolerant), and Jundan 20 (JD20; drought-sensitive) after exposure to integrated root-zone drought stress (IR-DS) shows increased betaine aldehyde dehydrogenase (BADH) activity and choline content the key enzyme and initial substrate, in GB biosynthesis. The peak of ABA content reached earlier than that of GB in the leaves of drought-stressed maize plants. Therefore, endogenous ABA seemed to be involved in modulating GB accumulation by enhancing BADH activity, thereby improving leaf RWC and enhancing shoot DM in drought-stressed maize plants, especially in the drought-sensitive cultivar (JD20) [63]. Similarly, Si application (400 mg L^{-1}) had positive effect on photosynthesis, water use efficiency, stomatal conductance, cholorophyll contents, Rubisco activity, and Rubisco activation state at 20 days of drought. Possibly Si may have direct or indirect role in maintenance of more active Rubisco enzyme and Rubisco activase and more stable proteins for carbon assimilation under stress conditions [64]. Maize was subjected to drought at the start of tasseling (6 days) followed by foliar spray of BR (0.1 mg l). Exogenous application of BR remarkably improved the gas exchange attributes, plant height, leaf area, cobs per plant, seedling dry weight both under drought and well-watered conditions [40]. The formation of cortical aerenchyma (RCA) in maize roots is associated with drought tolerance. RCA reduces root respiration by converting living cortical tissue to air volume. It has been hypothesized that RCA or large cortical cell size (CCS) increases drought tolerance by reducing root metabolic costs, permitting greater root growth and water acquisition from drying soil [65, 66]. Maize growth and yield responses were related to ROS production, osmolyte accumulation, and activation of anti-oxidative defense system under drought conditions. The regulation of these physio-biochemical responses of plants can be used as markers for drought stress tolerance [67]. Drought-tolerance mechanism is difficult in maize this may be due to complex genetic makeup. Metabolic traits provides better substitute to unravel the genetic mechanism of drought tolerance. The physiological status of plants can be monitored by metabolites that interconnect the visible phenotype with core genome, so that gene or gene loci could be identified and selected which are less affected by environmental factors [68–71]. However, the negative effects of abiotic stresses can be minimized by adopting the genetic approaches or by inducing resistance through transgenic approaches [3]. The importance of QTLs-based approach is associated to improve the maize crop performance under various abiotic stresses and in achieving increase in maize yield. Besides, QTLs with large effect and linked with stress-tolerance traits provides several modes to investigate the various components affecting source sink relationships of maize plants under abiotic stress [72]. Chen et al. [73] compares the role of drought resistance and drought recovery in drought adaptation in maize seedlings. After recovery most of the physiological parameters (like leaf water content, water potential, osmotic potential, gas exchange parameters, chlorophyll content, Fv/Fm and nitrogen content, and increased H_2O_2 accumulation and lipid peroxidation) rapidly return to normal level. Correlation analysis shows that physiological bases of drought adaptability closely related to drought recovery (r = 0.332) and not to drought resistance and (r = 0.714**) both are definitely different. Under drought stress higher chlorophyll content (r = 0.874***) and Fv/Fm (r = 0.626*) contributes to drought recovery. In addition, leaf water potential, chlorophyll content, and Fv/Fm could be used as efficient reference indicators in the selection of drought-adaptive genotypes.

3. Nutrient status in drought

The role of N metabolism regulation in maize drought tolerance was comprehensively studied [74]. Drought have strong impact on all the physiological process of plants and N metabolism is of central importance during drought stress because to survive drought and to maintain growth plants adaptive strategies includes such as improved nutrient uptake and transport, photosynthesis regulation, and to produced solutes and proteins that contains N compounds such as amino acids, amides and betaines. While Kant et al. [75] reported that relationship between drought and N nutrition consists of a complex network of regulatory interactions that affects almost all physiological processes in plants. Nitrogen (N) is one of the macronutrients, its major requirement for crop growth and grain yield of maize, whereas maize water requirement is highest in the reproductive stage [76]. Therefore, nitrogen and water separately or in combination are two are most critical factors in maize production. According to Saud et al. [77] the use of nutritional soil with a proper nitrogen rate remained effective in ameliorating the adverse influence of drought stress. However, a contrast study showed low field capacity irrigation and low nitrogen fertilizer rates application improves water use efficiency and nitrogen recovery efficiency (NRE) simultaneously [78]. Water and nutrient availability influence root system architecture and development [79, 80]. During vegetative stages of corn, under moderate drought stress rooting depth typically increases, allowing more efficient uptake of water and nutrients from deeper within the soil profile [81]. The increase in plant height related with increasing N fertilizer application under drought stress [82, 83]. Andrade [84] reported that N deprivation reduced leaf area index, leaf area duration, radiation interception, and radiation use efficiency. Improved tolerance to various abiotic stresses alternative breeding strategies are adopted, one of them is to select and inbreed maize plants under high population density to improved tolerance for low-N and drought [85]. Although under dense plant population maize prone to lodging due to increase in plant height and reduced culum diameters makes the stem weaker [86]. The mechanism underlying tolerance to high population density and low-N diverse tropical germplasm were grown under optimal, high plant population density and low-N. The association between ASI and grain yield, delayed senescence (expressed as chlorophyll concentration or number of green leaves above the ear) and ear/tassel weight ratio was observed in low-N. Also, grain yield negatively correlated with abortion rate. While under optimal and high population density, a positive association was reported between ovule number and abortion rate, suggesting a source limitation for C products [87]. In the same way selection of hybrid progenies for mid-season drought tolerance was due to improvements in morpho-physiological secondary traits such as reduced anthesis-silking interval, increased ears per plant, delayed senescence and relatively high leaf chlorophyll during late grain filling [88]. Drought at one of the sensitive growth stages, caused up to 40% grain yield losses. However, prolonged water stress during tasseling and ear formation stages leads to yield loss of 66–93% greater magnitude [39].

3.1 Effect of nitrogen at vegetative stage under drought stress

Variation in N supply affects both growth and development of maize plants [89]. Uhart and Andrade [90] reported that N deprivation reduced leaf area index, leaf area duration, radiation interception, and radiation use efficiency. However, nitrogen fertilizer changes canopy size which has an effect on the radiation use efficiency (RUE) of the crop thus affect plant growth and productivity [91]. According to Bänziger et al. [92], about 50% of all N in the leaf is directly involved in photosynthesis as either enzymes or chlorophyll. Thus, if the N supply is insufficient, photosynthesis is decreased by reducing the leaf area and photosynthesis rate as well as accelerating leaf senescence. Longer duration of green leaf area is one of the most important ways to improve maize yields. Bertin and Gallais [93] found that leaf senescence was highly correlated with a nitrogen nutrition index, mainly at low levels of nitrogen input. Using a comparative proteomic analysis of *Zea mays*, Prinsi et al. [94] found that the nitrogen status of plants may affect the post-translational modification of phosphoenolpyruvate carboxylase (PEPCase) that plays a role in phosphorylation in leaves. Several studies have reported that drought stress decreases N uptake and assimilation in plants [24, 95, 96]. Maize growth and many physiological processes associated with it are enhanced by N supply [97]. Nitrogen efficiency is a complex trait, to identify for N-efficiency 16 tropical maize cultivars, studied for leaf senescence under N deficiency (short-term nutrient solution experiment). Leaf chlorophyll contents and photosynthesis rates were used as measures for leaf senescence. The results show that the photosynthetic capacity of senescing leaves correlated with the N efficiency of the cultivars, rather than leaf chlorophyll content [98]. The effect of nitrogen on maize leaves at three level of drought stress (normal, mild, and severe stress) was examined, The NO content and nitrate reductase (NR) activity of maize leaves were significantly reduced under drought stress, while moderate nitrogen supply promoted the accumulation of NO and an increase in the Nitrate Reductase activity. Also, abscisic acid content increased and was positively correlated with the nitrogen concentration under drought stress. Together, these results indicate that moderate nitrogen supply increases plant resistance to drought stress, while high or low nitrogen concentrations increase the sensitivity of maize to drought stress [99]. Maize (*Zea mays* L.) crop was subjected to different periods of deficit irrigation and N rates in the field. The results indicated that optimum use of both water supply and N application will maximize the maize production. Deficit irrigation during early vegetative growth modestly reduced LAI, plant height, CGR, N uptake and total biomass production as compared to reproductive stage [100] also N uptake decreased with greater water and N deficits. Two maize hybrids differencing resistance to drought were fertilized with two different forms of nitrogen fertilizers, $Ca(NO_3)_2$ and $(NH_4)_2$ SO_4, and after fourth leaf stage they were exposed to drought stress. The results shows that two maize hybrids different in adaptability to two nitrogen treatment. NH_4 treated plants maintain the high turgor by improved osmotic adjustment under drought stress. Whereas, chlorophyll a, b values significantly higher in NH_4 treated plants compared to NH_3 treated plants in which chlorophyll content decreased throughout the drought stress [23]. Similar work Zhang et al. [101] indicated increased NO_3 nutrition played a favored anti-oxidative metabolic role, as compared with NH_4 nutrition, in the plants thereby increasing tolerance to drought stress. Further, two maize genotypes were accessed for effect of nitrogen rates and water stress. Under water stress, Shaadan 9 accumulates higher dry matter, grain yield, anti-oxidative enzyme activity, and lower MDA content than Shaadan 911. However, the addition of nitrogen increased dry matter and grain yield as well as activities of SOD, POD, and CAT to different levels and significantly decreased MDA content under water stress, higher for Shaandan 911 than for Shaandan 9. Thus, drought-sensitive variety showed its full potential after adding nitrogen in water stress condition [102]. The dry matter accumulation and nitrogen uptake were compared in sorghum and maize. Drought reduced nitrogen availability in soil for both the crops, although sensitivity of maize crop to nitrogen is more. Directly, maize crop shows less accumulation of dry matter in water deficit condition and indirectly because of nitrogen nutrition [103]. Twenty maize inbred lines were phenotype in response to two levels of water and nitrogen supply (control and stress) and combined nitrogen and water deficit. Image analyses study provide the opportunity for new traits, identified several color-related traits and kinetic chlorophyll fluorescence (PSII). For biomass production ability in maize, kinetic chlorophyll fluorescence (PSII) is relevant traits particularly under severe stress conditions. While architectural traits, like greater leaf area which provide good discrimination of resistant cultivars to abiotic stresses under climate change scenarios [104].

3.2 Effect of nitrogen at reproductive stage under water deficit condition

Maize crop tolerant to N-deficit conditions at early vegetative stage than later reproductive phase [105]. There is a synergistic relationship between water availability and N use efficiency in maize [106]. When N supply is limited, grain yield is more associated with N deficiency than drought stress, but with adequate N supply, drought stress is the main yield-limiting factor [97]. Similarly Halvorson et al. [107] described maize yield as a function of available water and nitrogen. Maize plants selected for tolerance to mid-season drought also provide tolerance toward nitrogen stress. This may be due to an increase in both the number of kernels per plant and kernel weight. Hence consistently increases grain yield across N-level [108]. The pre-anthesis drought significantly reduced the number of kernel rows, the number of kernels per row, as well as the 1000-kernel weight. The 80 kg N ha^{-1} was sufficient to achieve maximum grain yield under pre-anthesis drought. It is hypothesized that the adverse effects of pre-anthesis drought on grain yield can be mitigated if varieties are selected for roots which rapidly penetrate the soil and exploit the water resources in deep soil layers [109]. The effects of water stress imposed at low-sensitive growth stages (vegetative, reproductive, and both vegetative and reproductive) and level of nitrogen (N) supply (100 and 200 kg ha^{-1}) on the physiological and agronomic characteristics of the two hybrids of maize (*Zea mays* L.) were studied. The results showed that proline content increased and the relative water content, leaf greenness, 100-kernel weight, and grain yield decreased under conditions of water deficits. The limited irrigation imposed on maize during reproductive stage resulted in more yield reduction than that during vegetative stage. The 100-kernel weight was the most sensitive yield component to determine the yield variation in maize plant when the WD treatments were imposed in low-sensitive growth stages. The increase of N supply improved yield and IWUE when maize plant endured once irrigation shortage at vegetative stage. But, the performance of high N fertilizer reduced and eliminated when water deficit imposed once at reproductive stage and twice at vegetative and reproductive stages, respectively [110]. Further, the hybrids B73 × LH38, FS854, B73 × Mol7, and US13 were subjected to drought stress from the seventh leaf stage to evaluate differences in carbon and nitrogen accumulation and partitioning under drought. The results indicate that the greater drought tolerance of B73 × LH38 and FS854 to stress imposed during vegetative and early reproductive development resulted from their more active N uptake and assimilation and sugar production during the later portion of grain fill and from their more efficient partitioning of assimilate to the developing kernels [111]. Drought severity decreased grain yield sharply, but grain yield increased with nitrogen fertilizer. The time taken for pollination, grain weight, and total number of grains/ear were also affected. Increased proline content significantly under drought stress conditions shows activation of osmotic adjustment mechanism [112]. In a field study (carried out from 1995 to 1997) the effect of irrigation and water stress imposed at different development stages on vegetative growth, grain yield and other yield components of corn (*Zea mays* L.) was determined. The results of this 3-year study show that water stress occurring during vegetative and tasseling stages reduced plant height, and also leaf area development. Short-duration water deficits during the rapid vegetative growth period caused 28–32% loss of final dry matter weight. Whereas, single irrigation omission during one of the sensitive growth stages, caused up to 40% grain yield losses. Although prolonged water stress during tasseling and ear formation stages the predictable loss could be much greater (66–93%) [39]. Though, stay-green and kernel numbers are affected by nitrogen uptake and use efficiency, and by nitrogen remobilization [113]. Whereas, insufficient N supply combined with water deficit reduces per-plant kernel number and mass, total aboveground dry matter (DM) yield, and harvest index (HI) [114, 115].

4. Source-sink relationship under drought

Higher plants are heterotrophic, leaves are the major organs that are photosynthetically active sources with tissues that synthesize sugars and translocate mainly sucrose to other parts of the plants, which are photosynthetically less active or inactive sinks such as roots, fruits kernel, and tubers. This physiological dynamics is not static and changes with each different phonological stages. The plant life cycle begin from embryo (sink) that mobilize nutrient from storage organ seed (source). During vegetative stage the source organ are photosynthesizing leaves (fixed CO_2 + H_2O = sugars) that export carbohydrates to various developing sink organs, like, emerging leaves, roots for their utilization in growth and development. During reproductive stage, sugars and other nutrients are mobilized from mature leaves to developing seeds, kernel, fruits, and tubers. This sink transition and various competing sinks changes with respect to sink strength and all compete for carbohydrates allocation. This carbohydrate production (source) and carbohydrate partitioning (sink utilization of carbon) is also influence with various other factors such as nutrients, hormones, environmental factors. This coordinate regulation of source activity and sink strength determine the carbon © allocation throughout the plant and has been crucial for defining yield of the crop [116]. The balance between source-sink is disturbed due to insufficient sink strength or slow carbohydrates export leading to accumulation of carbohydrates in source organ causes feedback inhibition which in turn, down regulate the photosynthesis in leaves [117]. Drought, modify source-sink relations that disturb plant growth, also adaptation to stress and consequently affect crop productivity. Source-sink altered under water deficit condition because the growth primacies of plant changes. Since the source organ unable to supply assimilates to various competitive vegetative organ growth, to maintain reproductive structures and adaptation to stress. This decrease in photo assimilates reduces sink number and size [118]. They also emphasize the role of several metabolic and hormonal factors influencing not only the source strength, but especially the sink activity and their interrelations and their potential to improve yield stability under drought and salinity stresses. Further, Fahad et al. [10] highlighted the regulatory circuits of different phytohormones and cross talks among ABA, indole ABA, CKs, GA, SA, BRs, JA, ETHY, and TR at physiological and molecular levels on exposure to salinity. The decline in source and sink strengths during water deficit leads to important reductions in crop yield [3]. In the present climate change model [119], drought periods are becoming more frequent and severe [120, 121], and future crop varieties have to be more resilient to this stress. Maize is the third most important crop after wheat and rice, mainly grown for food, feed, and fodder. Plant breeding and agronomic practices have major role in genetic contribution to improved grain yield. Two main important physiological processes involved are sustained leaf photosynthesis during grain filling, which contributes to increases in dry matter accumulation, and second, an increase in kernel number due to higher partitioning to the kernels during the sensitive period of kernel number determination. Water use in corn is greatest during the late vegetative through early grain filling stages [76]. There are several factors that contribute to high plant performance under drought stress, like better portioning of biomass to the developing ear that results in faster spikelet growth which may results reduction in the number spikelet formed on the ear thus, facilitate overall seed set by reducing water and carbon stress per spikelet [122].

4.1 Source limitation and strength under drought stress

4.1.1 Source limitation

Growth conditions are favorable during pre-flowering and flowering, and a certain maize crop therefore establishes a large leaf area and many kernels and ears. Drought occurs after flowering causing the leaves to senesce early. The supply of assimilates will limit grain yield in this crop, and the plant will have many small kernels, thus limiting the source.

4.1.2 Regulation of source activity

Leaves are the main source organs; in maize, leaves ranged from 8 to 20 and these are present alternatively on nodes, and drought stress reduced the leaf size and number of leaves. Alternatively, reduced leaf areas under drought stress consider as an adaptive strategy. This may reduce the plant water requirement by reducing the leaf area and probability of plant survival is increased under limited water availability [123] but decrease in chlorophyll contents, chloroplast contents and photosynthetic activity reduced the grain yield [124, 125]. The important characteristics of superior germplasm include, more biomass allocation to leaf relative to stem weight, more leaf area, longer active leaf area duration [126] under different growing conditions. However features like morpho-physiological attributes such as leaf area, chlorophyll content, the rate and duration of photosynthesis, time of flowering, dry matter partitioning during silking, and leaf stay green plays key role in dry matter accumulation, harvest index and grain yield of corn [127]. Selection for stress tolerance has improved the grain yield in maize and it depends on higher leaf area per plant and higher harvest index (HI). Moreover, "stay green trait" or leaf greenness or reduction in the rate of leaf senescence during grain filling was one of the distinctions between older and newer hybrids [128]. Lafitte and Edmeades [129] selected maize cultivars in low and high N and secondary traits (Such as increased plant height, leaf area, chlorophyll concentrations, and delayed senescence) to improved grain yields. However, selection based in low-N to increase grain yield disturbs the balance of source and sink. Various secondary traits like water depletion pattern, leaf rolling, canopy temperature, reveals root potential and water extraction capacity. Whereas chlorophyll concentration quantifies the stay green trait of leaf. Flowering traits are associated with specific developmental stage, while, photosynthetic rate indicates the plant growth rate throughout the plant life cycle. Thus, these secondary traits are a related to specific mechanism and they contribute to enhance grain yield under drought stress [130]. Monneveux et al. [131] indicate that the traits of source organs contribute marginally to drought tolerance; variation of leaf or root traits seems to be less important than variation in tassel parameters for increasing drought tolerance. Further progress in drought tolerance in maize, the solution might reside in the manipulation of sink organs. It is therefore suggested that selection for even greater number of ears, bigger grains and smaller tassels may help to increase grain yield under water limited environments in the near future. Selection based on crop ideotype by Bolaños and Edmeades [132] was successful in producing drought-tolerant cultivars that were able to partition the assimilates to ear at flowering. However, Gambin et al. [133] indicated maize reproductive efficiency in kernel set is not constant across different plant growth rates (PGR) around flowering, therefore PGR per kernel used during this period as an indicator of source availability per kernel. The curvilinear response relating kernel number per plant and PGR around flowering, increased PGRs resulted in higher PGR per kernel around this period ($r2 = 0.86$; $p < 0.001$). Grain filling duration was partially explained ($r2 = 0.27$; $p < 0.01$) by the ratio between PGR per kernel during the effective grain filling period and kernel growth rate. Together, these results support the importance of source availability per kernel during early grain filling on the determination of maize potential sink capacity and final KW.

4.2 Sink limitation and strength under drought stress

4.2.1 Sink limitation

Growth conditions are favorable during pre-flowering and a certain maize crop therefore establishes a large leaf area. There is stress during flowering time and therefore the crop can establish only few ears and kernels. After flowering, the growing conditions may be favorable again, but the demand for assimilates by the kernels and their capacity to absorb the available assimilate will limit grain yield, thus limiting the sink.

4.2.2 Regulation of sink activity

Flowering is the most crucial stage for silk growth, pollination, and kernel setting in maize [134] and is the utmost susceptible stage to drought stress. An asynchrony between silk emergence and pollen shedding under drought stress before and during flowering increases the anthesis-silking interval (ASI) and primarily causes yield loss [135–138]. High-temperature stress also hampered the pollen characteristics. However, exogenous application of various PGRs significantly reduced the damaging effects of high temperature, compared with control. Further, pollen characteristics can be exploited as screening tools for varietal development but selection must focus for those germplasm sources which can tolerate temperature above 38°C [1, 11, 12]. Under drought stress, a positive correlation was observed between flowering and kernel number per plant ($r > 0.8$), barrenness ($r > 0.7$), and ASI (r = _0.4 to _0.7) in tropical maize [136]. In maize, increased abiotic stress tolerance increased yield and yield stability in some tropical and temperate maize germplasm [139, 140]. Canadian maize hybrids show increased tolerance toward various stress factors like high population density, weed interference, low night temperature, low soil moisture, and low-N stress due to the features such as increased leaf longevity, increased nutrient and water uptake, and greater assimilates supply during grain filling [140]. Similarly, tropical maize shows decreased kernel and ear abortion, selected for mid-season drought [141, 142]. Therefore, constitutive stress-tolerance mechanism may be operative in maize germplasm, which may be related to the creation of sink size [92]. The ovule, ear and kernels are the reproductive sinks, for their proper development assimilates supply above threshold level are necessary. In unstressed condition, ear abortion may occurs due to insufficient assimilates supply that led to barren plants. Thus, ovule fails to extrude silk due to slow growth, whereas kernel aborted following pollination [141, 143, 144]. The corn was subjected to low water potential (Cw) for 5 days at the time of pollination [145]. Zinselmeier et al. [145] observed that embryos formed but abortion occurred and kernel number decreased markedly. Besides, during this abortion, all of the intermediates in starch synthesis were depleted and the starch contained in the ovary almost disappeared. The period for ovule, kernel, and ear abortion falls 1 week before until 2 weeks after silking; during this period, stress factors like drought, shade, high density, and low-N stress accelerate these processes [27, 90, 146, 147]. Physiological reasons for abortion of reproductive structures under various stresses have shown that concurrent photosynthesis is required to maintain the flux of carbohydrates to the young ear around flowering but under drought it not possible to remobilize carbohydrate reserve for the support of ear development [148–150]. Drought also reduces invertase activity in ovaries and this will reduce the flux of hexose sugars; starch reserves of ovaries are depleted and it leads to the ovaries abortion [151]. In tropics, maize yield could be increased by improving the balance between the source supply and sink demand [152]. The main focus of maize breeding program is to increase the grain yield under high N; this practice resulted in an increase in both the source and sink in size and efficiency [153]. However, in N-limited conditions, not only the importance of C but also for N associates with source and sink [154] the N required during the exponential phase of grain growth to sustain the kernel number per plant and grain yield. The role of glutamine synthetase (GS) and asparagine synthetase (AS) (their main function in plant nitrogen remobilization) was determined in two maize varieties (ZD958 and NH101) in relation to post-silking drought stress (PD) nitrogen partitioning. The results indicate the PD stress increased nitrogen remobilization, and in ear leaves, the expression of *ZmGln1-3* was enhanced for both varieties. While under PD treatment, three AS genes (*ZmAS1, ZmAS2*, and 10 *ZmAS3*) were differentially regulated, of which the expression of *ZmAS3* was stimulated at the late stage of leaf senescence. In developing kernels, there were no significant differences in expression patterns of GS and AS genes between the well water (WW) and PD plants. Therefore, at the whole plant level, PD stress showed more influence on leaf nitrogen status, and the upregulation of GS and AS genes may contribute to the higher leaf nitrogen remobilization when exposed to PD treatments [155]. Genotypes selected for high grain yield under normal growing conditions also had a high level of dry-matter accumulation and partitioned more assimilates to the grain under water deficit conditions. The improved genotypes produced more numbers of ears/plant and kernels/ear [156]. Leaf growth and ASI are the main determinants of source and sink strengths of maize.

Author details

Suphia Rafique
Division of Plant Physiology, IARI, Pusa Campus, New Delhi, India

*Address all correspondence to: suphia123@gmail.com

References

[1] Fahad S, Muhammad ZI, Abdul K, Ihsanullah D, Saud S, Saleh A, et al. Consequences of high temperature under changing climate optima for rice pollen characteristics - Concepts and perspectives. Archives of Agronomy and Soil Science. 2018. DOI: 10.1080/03650340.2018.1443213

[2] Mittler R. Abiotic stress, the field environment and stress combination. Trends in Plant Science. 2006;**11**:15-19

[3] Fahad S, Bajwa AA, Nazir U, Anjum SA, Farooq A, Zohaib A, et al. Crop production under drought and heat stress: Plant responses and management options. Frontiers in Plant Science. 2017;**8**:1147. DOI: 10.3389/fpls.2017.01147

[4] Mishra AK, Singh VP. A review of drought concepts. Journal of Hydrology. 2010;**391**(1-2):202-216

[5] Chen H, Sun J. Changes in drought characteristics over China using the standardized precipitation evapotranspiration index. Journal of Climate. 2015;**28**(13):281-299

[6] Maiti RK, Satya P. Research advances in major cereal crops for adaptation to abiotic stresses. GM Crops & Food. 2014;**5**(4):259-279

[7] Banziger M, Edmeades GO, Beck D, Bellon M. Breeding for Drought and Nitrogen Stress Tolerance in Maize: From Theory to Practice. Mexico, D.F.: CIMMYT; 2000

[8] Song J, Zhou JC, Zhou WW, Xu HL, Wang FX, Xu YG, et al. Effects of salinity and nitrate on production and germination of dimorphic seeds both applied through the mother plant and exogenously during germination in *Suaeda salsa*. Plant Species Biology. 2016;**31**:19-28

[9] Fahad S, Bano A. Effect of salicylic acid on physiological and biochemical characterization of maize grown in saline area. Pakistan Journal of Botany. 2012;**44**:1433-1438

[10] Fahad S, Hussain S, Matloob A, Khan FA, Khaliq A, Saud S, et al. Phytohormones and plant responses to salinity stress: A review. Plant Growth Regulation. 2014b;**75**(2):391-404. DOI: 10.1007/s10725-014-0013-y

[11] Fahad S, Hussain S, Saud S, Tanveer M, Bajwa AA, Hassan S, et al. A biochar application protects rice pollen from high-temperature stress. Plant Physiology and Biochemistry. 2015a;**96**:281-287

[12] Fahad S, Nie L, Chen Y, Wu C, Xiong D, Saud S, et al. Crop plant hormones and environmental stress. Sustainable Agriculture Reviews. 2015b;**15**:371-400

[13] Fahad S, Hussain S, Saud S, Hassan S, Ihsan Z, Shah AN, et al. Exogenously applied plant growth regulators enhance the morphophysiological growth and yield of rice under high temperature. Frontiers in Plant Science. 2016c;**7**:1250

[14] Coruzzi G, Bush DR. Nitrogen and carbon nutrient and metabolite signalling in plants. Plant Physiology. 2001;**125**:61-64

[15] Coruzzi GM. Primary N-assimilation into amino acids in Arabidopsis. In: Somerville CR, Meyerowitz EM, editors. The Arabidopsis Book. Rockville, MD, USA: American Society of Plant Biologists; 2003. DOI: 10.1199/tab.0010. Available from: http://www.aspb.org/publications/Arabidopsis

[16] Miller AJ, Fan X, Orsel M, Smith SJ, Wells DM. Nitrate transport and signalling. Journal of Experimental Botany. 2007;**58**:2297-2306

[17] Thum KE, Shasha DE, Lejay LV, Coruzzi GM. Light and carbon-signaling pathways. Modeling circuits of interactions. Plant Physiology. 2003;**132**:440-452

[18] Urbanczyk-Wochniak E, Fernie AR. Metabolic profiling reveals altered nitrogen nutrient regimes have diverse effects on the metabolism of hydroponically-grown tomato (*Solanum lycopersicum*) plants. Journal of Experimental Botany. 2005;**56**:309-321

[19] Gutiérrez RA et al. Systems approach identifies an organic nitrogen-responsive gene network that is regulated by the master clock control gene CCA1. Proceedings of the National Academy of Sciences of the United States of America. 2008;**105**:4939-4944

[20] Aziz K, Daniel KYT, Muhammad ZA, Honghai L, Shahbaz AT, Mir A, et al. Nitrogen fertility and abiotic stresses management in cotton crop: A review. Environmental Science and Pollution Research. 2017b;**24**:14551-14566. DOI: 10.1007/s11356-017-8920-x

[21] Aziz K, Daniel KYT, Fazal M, Muhammad ZA, Farooq S, Fan W, et al. Nitrogen nutrition in cotton and control strategies for greenhouse gas emissions: A review. Environmental Science and Pollution Research. 2017a;**24**:23471-23487. DOI: 10.1007/s11356-017-0131-y

[22] Donald CM, Hamblin J. The biological yield and harvest index of cereal as agronomics and plant breeding criteria. Advances in Agronomy. 1976;**28**:361-405

[23] Mihailovic N, Jelic G, Filipovic R, Djurdjevic M, Dzeletovic Z. Effect of nitrogen form on maize response to drought stress. Plant and Soil. 1992;**144**:191-197

[24] Waraich EA, Ahmad R, Ashraf MY. Role of mineral nutrition in alleviating drought stress in plants. Australian Journal of Crop Science. 2011;**5**(6):764-777

[25] Humbert S, Subedi S, Cohn J, Zeng B, Bi YM, Chen X, et al. Genome-wide expression profiling of maize in response to individual and combined water and nitrogen stresses. BMC Genomics. 2013;**14**:3. DOI: 10.1186/1471-2164-14-3

[26] Claassen MM, Shaw RH. Water deficit effects on corn. I. Vegetative components. Agronomy Journal. 1970;**62**:652-655

[27] Grant RF, Jackson BS, Kiniry JR, Arkin GF. Water deficit timing effects on yield components in maize. Agronomy Journal. 1989;**81**:61-65

[28] Czyzewicz JR, Below FE. Genotypic variation for nitrogen uptake by maize kernels grown *in vitro*. Crop Science. 1994;**33**:491-497

[29] Boyer JS. Leaf enlargement and metabolic rates in corn, soybean and sunflower at various leaf water potentials. Plant Physiology. 1970;**46**:233-235

[30] Abrecht DG, Carberry PS. The influence of water deficit prior to tassel initiation on maize growth, development and yield. Field Crops Research. 1993;**3**(l):55-69

[31] Moss GI, Downey LA. Influence of drought stress on female gametophyte development in corn *(Zea mays* L.) and subsequent grain yield. Crop Science. 1971;**11**:368-372

[32] Watson DJ. The physiological basis of variation in yield. Advances in Agronomy. 1952;**4**:101-145

[33] NeSmith DS, Ritchie JT. Short-and long-term responses of corn to a pre-anthesis soil water deficit. Agronomy Journal. 1992a;**84**:107-113

[34] Xiangbo Z, Lei L, Lai J, Zhao H, Song W. Effects of drought stress and water recovery on physiological responses and gene expression in maize seedlings. BMC Plant Biology. 2018;**18**(68):2-16

[35] Egilla JN, Davies FT, Boutton TW. Drought stress influences leaf water content, photosynthesis, and water use efficiency of *Hibiscus rosa-sinensis* at three potassium concentrations. Photosynthetica. 2005;**43**(1):135-140

[36] Abdul Jaleel C, Manivannan P, Kishorekumar A, Sankar B, Gopi R, Somasundaram R, et al. Alterations in osmoregulation, antioxidant enzymes and indole alkaloid levels in *Catharanthus roseus* exposed to water deficit, colloids and surfaces. Biointerfaces. 2008;**59**:150-157

[37] Ali Q, Ali A, Waseem M, Muzaffar A, Ahmed S, Ali S, et al. Correlation analysis for morpho-physiological traits of maize (*Zea mays* L.). Life Science Journal. 2014, 2014;**11**(12s)

[38] Nepolean T, Singh I, Hussain F, Pandey N, Gupta HS. Molecular characterization and assessment of genetic variability of inbred line showing genetic variability for drought tolerance in maize. Journal of Plant Biochemistry and Biotechnology. 2013;**22**:71-79

[39] Çakir R. Effect of water stress at different development stages on vegetative and reproductive growth of corn. Field Crops Research. 2004;**89**: 1-16. DOI: 10.1016/j.fcr.2004.01.005

[40] Anjum SA, Wang LC, Farooq M, Hussain M, Xue LL, Zou CM. Brassinolide application improves the drought tolerance in maize through modulation of enzymatic antioxidants and Leaf gas exchange. Journal of Agronomy and Crop Science. 2011a;**197**:177-185

[41] Cornic G. Drought stress inhibits photosynthesis by decreasing stomatal aperture: Not by affecting ATP synthesis. Trends in Plant Science. 2000;**5**:187-188

[42] Meyer S, Genty B. Heterogenous inhibition of photosynthesis over the leaf surface of *Rosa rubinosa* L. during water stress and abscisic acid treatment: Induction of a metabolic component by limitation of CO_2 diffusion. Planta. 1999;**210**:126-131

[43] Lawlor DW. Carbon and nitrogen assimilation in relation to yield: Mechanism is the key to understanding production system. Journal of Experimental Botany. 2002;**53**:773-787

[44] Medrano H, Parry MA, Socias X, Lawlor DW. Long term water stress inactivates Rubisco in subterranean clover. The Annals of Applied Biology. 1997;**131**:491-501

[45] Tezara W, Mitchell V, Driscoll S, Lawlor D. Water stress inhibits plant photosynthesis by decreasing coupling factor and ATP. Nature. 1999;**401**:914-917. DOI: 10.1038/44842

[46] O'Regan BP, Cress t WA, van Staden J. Root growth, water relations, abscisic acid and proline levels of drought-resistant and drought-sensitive maize cultivars in response to water stress. South African Journal of Botany. 1993;**59**(1):98-104

[47] Jama AO, Ottman MJ. Timing of the first irrigation in corn and water stress conditioning. Agronomy Journal. 1993;**85**:1159-1164

[48] Seki M, Umezawa T, Urano K, Shinozaki K. Regulatory metabolic networks in drought stress responses. Current Opinion in Plant Biology. 2007;**10**:296-302. DOI: 10.1016/j.pbi.2007.04.014

[49] Bray EA. Classification of genes differentially expressed during

water-deficit stress in *Arabidopsis thaliana*: An analysis using microarray and differential expression data. Annals of Botany (London). 2002;**9**:803-811

[50] Socias X, Correia MJ, Medrano H. The role of abscisic acid and water relations in drought responses of subterranean clover. Journal of Experimental Botany. 1987;**48**(311):1281-1288

[51] Hussain MA, Fahad S, Rahat S, Muhammad FJ, Muhammad M, Qasid A, et al. Multifunctional role of brassinosteroid and its analogues in plants. Plant Growth Regulation. 2020. DOI: 10.1007/s10725-020-00647-8

[52] Rhodes D, Samaras Y. Genetic control of osmoregulation in plants. In: Strange K, editor. Cellular and Molecular Physiology of Cell Volume Regulation. Boca Raton: CRC Press; 1994. pp. 347-361

[53] Bartels D, Sunkar R. Drought and salt tolerance in plants. Critical Reviews in Plant Sciences. 2005;**24**:23-58

[54] Mohammadkhani N, Heidari R. Effects of drought stress on soluble proteins in two maize varieties. Turkish Journal of Biology. 2008;**32**:23-30

[55] Munns R. Genes and salt tolerance: Bringing them together. The New Phytologist. 2005;**167**:645-663

[56] Hesham FA, Fahad S. Melatonin application enhances biochar efficiency for drought tolerance in maize varieties: Modifications in physio-biochemical machinery. Agronomy Journal. 2020:1-22

[57] Apel K, Hirt H. Reactive oxygen species: Metabolism, oxidative stress, and signal transduction. Annual Review of Plant Biology. 2004;**55**:373-399

[58] Yang L, Fountain JC, Wang H, Ni X, Ji P, Lee RD, et al. Stress sensitivity is associated with differential accumulation of reactive oxygen and nitrogen species in maize genotypes with contrasting levels of drought tolerance. International Journal of Molecular Sciences. 2015;**16**:24791-24819. DOI: 10.3390/ijms161024791

[59] Foyer Christine H, Valadier M-H, Migge A, Becke TW. Drought-induced effects on nitrate reductase activity and mRNA and on the coordination of nitrogen and carbon metabolism in maize leaves. Plant Physiology. 1998;**117**:283-292

[60] Nisar A, Malagoli M, Wirtz M, Hell R. Drought stress in maize causes differential acclimation responses of glutathione and sulfur metabolism in leaves and roots. BMC Plant Biology. 2016;**16**:247

[61] Bai LP, Sui FG, Ge TD, Sun ZH, Lu YY, Zhou GS. Effect of soil drought stress on leaf water status, membrane permeability and enzymatic antioxidant system of maize. Pedosphere. 2006;**16**:326-332

[62] Saab IN, Sharp RE, Pritchard J, Voetberg GS. Increased endogenous abscisic acid maintains primary root growth and inhibits shoot growth of maize seedlings at low water potentials. Plant Physiology. 1990;**93**:1329-1336

[63] Zhang L, Gao M, Hu J, Zhang X, Wang K. Ashraf M (2012) modulation role of abscisic acid (ABA) on growth, water relations and glycinebetaine metabolism in two maize (*Zea mays* L.) cultivars under drought stress. International Journal of Molecular Sciences. 2012;**13**:3189-3202

[64] Saud S, Chen Y, Fahad S, Hussain S, Na L, Xin L, et al. Silicate application increases the photosynthesis and its associated metabolic activities in Kentucky bluegrass under drought stress and post-drought recovery. Environmental Science and Pollution

Research. 2016;**23**(17):17647-17655. DOI: 10.1007/s11356-016-6957-x

[65] Zhu J, Brown KM, Lynch JP. Root cortical aerenchyma improves the drought tolerance of maize (*Zea mays* L.) *Plant*. Cell and Environment. 2010;**33**:740-749

[66] Chimungu JG, Brown KM, Lynch JP. Large root cortical cell size improves drought tolerance in maize. Plant Physiology. 2014;**166**:2014

[67] Anjum SA, Ashraf U, Tanveer M, Khan I, Hussain S, Shahzad B, et al. Drought induced changes in growth, osmolyte accumulation and antioxidant metabolism of three maize hybrids. Frontiers in Plant Science. 2017;**8**:69

[68] Chan EK, Rowe HC, Hansen BG, Kliebenstein DJ. The complex genetic architecture of the metabolome. PLoS Genetics. 2010;**6**:e1001198

[69] Keurentjes JJ, Fu J, De Vos CR, Lommen A, Hall RD, Bino RJ, et al. The genetics of plant metabolism. Nature Genetics. 2006;**38**:842-849

[70] Meyer RC, Steinfath M, Lisec J, Becher M, Witucka-Wall H, Torjek O, et al. The metabolic signature related to high plant growth rate in *Arabidopsis thaliana*. Proceedings of the National Academy of Sciences of the United States of America. 2007;**104**:4759-4764

[71] Riedelsheimer C, Lisec J, Czedik-Eysenberg A, Sulpice R, Flis A, Grieder C, et al. Genome-wide association mapping of leaf metabolic profiles for dissecting complex traits in maize. Proceedings of the National Academy of Sciences of the United States of America. 2012;**109**:8872-8877

[72] Sharma S, Thakur S, Kaur N, Pathania A. QTL mapping for abiotic stress tolerance in maize: A brief review. International Journal of Chemical Studies. 2018;7(1):1206-1209

[73] Chen D, Wang S, Cao B, Cao D, Leng G, Li H, et al. Genotypic variation in growth and physiological response to drought stress and re-watering reveals the critical role of recovery in drought adaptation in maize seedlings. Frontiers in Plant Science. 2016;**6**:1241. DOI: 10.3389/fpls.2015.01241

[74] Wang H, Yang Z, Yu Y, Chen S, He Z, Wang Y, et al. Drought enhances nitrogen uptake and assimilation in maize roots. Agronomy Journal. 2017;**109**(1):39-46

[75] Kant S, Bi YM, Rothstein SJ. Understanding plant response to nitrogen limitation for the improvement of crop nitrogen use efficiency. Journal of Experimental Botany. 2011;**6**:1-11. DOI: 10.1093/jxb/erq297

[76] Kranz WL, Irmak S, van Donk SJ, Yonts CD, Martin DL. Irrigation Management for Corn. Neb Guide (G1367-A). University of Nebraska Extension; 2008. Available from: http://ianrpubs.unl.edu/live/g1850/build/g1850.pdf (Accessed: 18 January 2018)

[77] Saud S, Fahad S, Yajun C, Ihsan MZ, Hammad HM, Nasim W, et al. Effects of nitrogen supply on water stress and recovery mechanisms in Kentucky bluegrass plants. Frontiers in Plant Science. 2017;**8**:983. DOI: 10.3389/fpls.2017.00983

[78] Hafiz MH, Wajid F, Farhat A, Fahad S, Shafqat S, Wajid N, et al. Maize plant nitrogen uptake dynamics at limited irrigation water and nitrogen. Environmental Science and Pollution Research. 2016;**24**(3):2549-2557. DOI: 10.1007/s11356-016-8031-0

[79] Dwyer LM, Stewart DW. Effect of leaf age and position on net photosynthetic rates in maize (*Zea mays* L.). Agricultural and Forest Meteorology. 1986;**37**:29-46

[80] Lynch JP. Steep, cheap and deep: An ideotype to optimize water and

N acquisition by maize root systems. Annals of Botany. 2013;**112**:347-357

[81] Robertson WK, Hammond LC, Johnson JT, Boote KJ. Effects of plant-water stress on root distribution of corn, soybeans, and peanuts in sandy soil. Agronomy Journal. 1980;**72**:548-550

[82] Balko LG, Russell WA. Response of maize inbred lines to N fertilizer. Agronomy Journal. 1980;**72**:723-728

[83] Lucas EO. The effect of density and N fertilizer on growth and yield of maize (*Zea mays*, L.) in Nigeria. Journal of Agriculture Science (Cambridge). 1986;**107**:573-578

[84] Andrade FH. Analysis of growth and yield of maize, sunflower and soybean grown at Balcarce, Argentina. Field Crops Research. 1995a;**41**:1-12

[85] Vasal SK, Cordova HS, Beck DL, Edmeades GO. Choices among breeding procedures and strategies for developing stress tolerant maize germplasm. In: Edmeades GO, Bänziger M, Mickelson HR, Peña-Valdivia CB, editors. Developing Drought and Low N-Tolerant Maize. Proceedings of a Symposium; 25-29 March 1996; CIMMYT, El Batan, Mexico. Mexico D.F.: CIMMYT; 1997

[86] Kamarn M, Wenwen C, Irshad A, Xiangping M, Xudong Z, Wennan S, et al. Effect of paclobutrazol, a potential growth regulator on stalk mechanical strength, lignin accumulation and its relation with lodging resistance of maize. Plant Growth Regulation. 2017;**84**:317-332. DOI: 10.1007/s10725-017-0342-8

[87] Monneveux P, Zaidi PH, Sanchez C. Population density and low nitrogen affects yield-associated traits in tropical maize. Crop Science. 2005;**45**:535-545

[88] Zaidi PH, Srinivasan G, Cordova HS, Sanchez C. Gains from improvement for mid-season drought tolerance in tropical maize (*Zea mays* L.). Field Crops Research. 2004;**89**:135-152

[89] McCullough DE, Girardin P, Mihajlovic M, Aguilera A, Tollenaar. Influence of N supply on development and dry matter accumulation of an old and new maize hybrid. Canadian Journal of Plant Science. 1994;**74**:471-477

[90] Uhart SA, Andrade FH. Nitrogen deficiency in maize: I. Effects on crop growth, development, dry matter partitioning. Crop Science. 1995a;**35**:1376-1383

[91] Wajid N, Ashfaq A, Asad A, Muhammad T, Muhammad A, Muhammad S, et al. Radiation efficiency and nitrogen fertilizer impacts on sunflower crop in contrasting environments of Punjab, Pakistan. Environmental Science and Pollution Research. 2017;**25**:1822-1836

[92] Benziger M, Edmeades GO, Lefitte HR. Physiological mechanism contributing to the increased n stress tolerance of tropical maize selected for drought tolerance. Field Crops Research. 2002;**75**:223-233

[93] Bertin P, Gallais A. Genetic variation for nitrogen use efficiency in a set of recombinant maize inbred lines. I. Agro-physiological results. Maydica. 2000;**45**:53-66

[94] Prinsi B, Negri AS, Pesaresi P, Cocucci M, Espen L. Evaluation of protein pat-tern changes in roots and leaves of *Zea mays* plants in response to nitrate availability by two-dimensional gel electrophoresis analysis. BMC Plant Biology. 2009;**9**:113

[95] Cramer MD, Hawkins H-J, Verboom GA. The importance of nutritional regulation of plant water

flux. Oecologia. 2009;**161**:15-24. DOI: 10.1007/ s00442-009-1364-3

[96] Sardans J, Penuelas J. The role of plants in the effects of global change on nutrient availability and stoichiometry in the plant-soil system. Plant Physiology. 2012;**160**:1741-1761. DOI: 10.1104/pp.112.208785

[97] Eck HV. Irrigated corn yield response to nitrogen and water. Agronomy Journal. 1984;**76**:421-428

[98] Erley GS a'm, Begum N, Worku M, Bänziger M, Horst WJ. Leaf senescence induced by nitrogen deficiency as indicator of genotypic differences in nitrogen efficiency in tropical maize. Journal of Plant Nutrition and Soil Science. 2007;**170**:106-114

[99] Song H, Li YB, Zhou L, Xu ZZ, Zhou GS. Maize leaf functional responses to drought episode and rewatering. Agricultural and Forest Meteorology. 2018;**249**:57-70

[100] Pandey RK, Maranville JW, Admou A. Deficit irrigation and nitrogen effects on maize in a Sahelian environment. I. Grain yield and yield components. Agricultural Water Management. 2000a;**46**:1-13

[101] Zhang LX, Gao M, Li SQ, Li SX, Liang ZS. Water status and photosynthesis in two maize (*Zea mays* L.) cultivars as affected by supplied nitrogen form and drought stress. Pakistan Journal of Botany. 2011;**43**:1995-2001

[102] Zhang LX, Li SX, Zhang H, Liang ZS. Nitrogen rates and water stress effects on production, lipid peroxidation and antioxidative enzyme activities in two maize (*Zea mays* L.) genotypes. Journal of Agronomy and Crop Science. 2007a;**193**:387-397

[103] Lemaire G, Charrier X. Nitrogen uptake capacities of maize and sorghum crops in different nitrogen and water supply conditions. Agronomie. 1996;**16**(4):231-246

[104] Dodig D, Božinović S, Nikolić A, ZorićM, VančetovićJ, Ignjatović-MićićD, et al. Image-derived traits related to mid-season growth performance of maize under nitrogen and water stress. Frontiers in Plant Science. 2019;**10**:814. DOI: 10.3389/fpls.2019.00814

[105] Islam MR, Rahman SME, Rahman MM, Oh DH, Ra CS. The effects of biogas slurry on the production and quality of maize fodder. Turkish Journal of Agriculture and Forestry. 2010;**34**:91-99

[106] Kim KI, Clay DE, Carlson CG, Clay SA, Trooien T. Do synergistic relationships between nitrogen and water influence the ability of corn to use nitrogen derived from fertilizer and soil? Agronomy Journal. 2008;**100**:551-556

[107] Halvorson AD, Mosier AR, Reule CA, Bausch WC. Nitrogen and tillage effects on irrigated continuous corn yields. Agronomy Journal. 2006;**98**:63-71

[108] Banziger M, Edmeades GO, Lafitte HR. Physiological mechanisms contributing to the increased N stress tolerance of tropical maize selected for drought tolerance. Field Crops Research. 2002;**75**:223-233

[109] Moser SB, Feil B, Jampatong S, Stamp P. Effects of pre-anthesis drought, nitrogen fertilizer rate, and variety on grain yield, yield components, and harvest index of tropical maize. Agricultural Water Management. 2006;**81**:41-58. DOI: 10. 1016/j.agwat.2005.04.005

[110] Mansouri-Far C, Sanavy SA, Saberali SF. Maize yield response to deficit irrigation during low-sensitive growth stages and nitrogen rate

under semi-arid climatic conditions. Agricultural Water Management. 2010;**97**:12-22. DOI: 10.1016/j.agwat.2009.08.003

[111] Frederick JR, Below FE, Hesketh JD. Carbohydrate, nitrogen and dry matter accumulation and partitioning of maize hybrids under drought stress. Annals of Botany. 1990;**66**(4):407-415

[112] Tarighaleslami M, Zarghami R, Mashhadi M, Boojar A, Oveysi M. Effects of drought stress and different nitrogen levels on morphological traits of proline in leaf and protein of corn seed (*Zea mays* L.). American-Eurasian Journal of Agricultural & Environmental Sciences. 2012;**12**(1):49-56

[113] Gallais A, Hirel B. An approach to the genetics of nitrogen use efficiency in maize. Journal of Experimental Botany. 2004;**55**:295-306

[114] Boomsma CR, Santini JB, Tollenaar M, Vyn TJ. Maize morphophysiological responses to intense crowding and low nitrogen availability: An analysis and review. Agronomy Journal. 2009;**101**:1426-1452

[115] Chapuis R, Delluc C, Debeuf R, Tardieu F, Welcker C. Resiliences to water deficit in a phenotyping platform and in the field: How related are they in maize? European Journal of Agronomy. 2012;**42**:59-67

[116] Ruan YL. Sucrose metabolism: Gateway to diverse carbon use and sugar signaling. Annual Review of Plant Biology. 2014;**65**:33-67

[117] Rossi M, Bermudez L, Carrari F. Crop yield: Challenges from a metabolic perspective. Current Opinion in Plant Biology. 2015;**25**:79-89. DOI: 10.1016/j.pbi.2015. 05.004

[118] Albacete AA, Martínez-Andújar C, Pérez-Alfocea F. Hormonal and metabolic regulation of source-sink relations under salinity and drought: From plant survival to crop yield stability. Biotechnology Advances. 2013. DOI: 10.1016/j.biotechadv.2013.10.005

[119] IPCC. Climate change. Impacts, adaptation and vulnerability. In: Parry ML, Canziani OF, Palutikof JP, van der Linden PJ, Hanson CE, editors. Contribution of Working Group II to the Fourth Assessment Report of the Inter-Governmental Panel on Climate Change. Cambridge: Cambridge University Press; 2007. p. 976

[120] Harrison MT, Tardieu F, Dong Z, Messina CD, Hammer GL. Characterizing drought stress and trait influence on maize yield under current and future conditions. Global Change Biology. 2014;**20**(3):867-878

[121] Ray DK, Mueller ND, West PC, Foley JA. Yield trends are insufficient to double global crop production by 2050. PLoS One. 2013;**8**:e66428. DOI: 10.1371/journal.pone.0066428

[122] Bruce W, Edmeades G, Barker T. Molecular and physiological approaches to maize improvement for drought tolerance. Journal of Experimental Botany. 2002;**53**:13-25

[123] Belaygue C, Wery J, Cowan A, Tardieu F. Contribution of leaf expansion, rate of leaf appearance, and stolon branching to growth of plant leaf area under water deficit in white clover. Crop Science. 1996;**36**(5):1240-1246

[124] Flagella Z, Rotunno T, Tarantito E, Caterina RD, Caro AD. Changes in seed yield and oil fatty acid composition of high oleic sunflower (*Helianthus annuus* L.) hybrids in relation to the sowing date and the water regime. European Journal of Agronomy. 2002;**17**:221-230

[125] Göksoya AT, Demirb AO, Turana ZM, Dağüstü N. Responses of sunflower (*Helianthus annuus* L.) to full and limited irrigation at different growth stages. Field Crops Research. 2004;**87**:167-178

[126] Zaidi PH, Srinivasan G, Sanchez C. Relationship between line per se and cross performance under low nitrogen fertility in tropical maize (*Zea mays* L.). Maydica. 2003;**48**:221-231

[127] Tollenaar M, Lee EA. Dissection of physiological processes underlying grain yield in maize by examining genetic improvement and heterosis. Maydica. 2006;**51**:399-408

[128] Duvick DN, Smith JCS, Cooper M. Long-term selection in a commercial hybrid maize breeding program. Plant Breeding Reviews. 2004a;**24**:109-151

[129] Lafitte HR, Edmeades GO. Association between traits in tropical maize inbred lines and their hybrids under high and low soil nitrogen. Maydica. 1994;**40**:259-267

[130] Barker T, Campos H, Cooper M, Dolan D, Edmeades GO, Habben J, et al. Improving drought tolerance in maize. Plant Breeding Reviews. 2004;**25**, in press

[131] Monneveuxi P, Sanchez C, Tiessen A. Future progress in drought tolerance in maize needs new secondary traits and cross combinations. Journal of Agricultural Science. 2008;**146**:287-300

[132] Bolaños J, Edmeades GO. Eight cycles of selection for drought tolerance in lowland tropical maize. II. Responses in reproductive behavior. Field Crops Research. 1993b;**31**:253-268

[133] Gambín BL, Borrás L, Otegui ME. Source–sink relations and kernel weight differences in maize temperate hybrids. Field Crops Research. 2006;**95**:316-326

[134] Shaw RH. Climatic requirement. In: Sprague GF, editor. Maize and Maize Improvement. Agronomy Monographs 18. Madison, WI: ASA, CSSA, and SSSA; 1977. pp. 315-341

[135] Hall AJ, Vilella F, Trapani N, Chimenti C. The effects of water stress and genotype on the dynamics of pollen-shedding and silking in maize. Field Crops Research. 1982;**5**:349-363

[136] Bolaños J, Edmeades GO. The importance of the anthesis-silking interval in breeding for drought tolerance in tropical maize. Field Crops Research. 1996;**48**:65-80. DOI: 10.1016/0378-4290(96)00036-6

[137] Lu GH, Ren DL, Wang XQ, Wu JK, Zhao MS. Evaluation on drought tolerance of maize hybrids in China. Journal of Maize Sciences. 2010;**3**:20-24

[138] Ribaut JM, Hoisington DA, Deutsch JA, Jiang C, Gonzalez de Leon D. Identification of quantitative trait loci under drought conditions in tropical maize. 1. Flowering parameters and the anthesis-silking interval. Theoretical and Applied Genetics. 1996;**92**:905-914

[139] Byrne PF, Bolaños J, Edmeades GO, Eaton DL. Gains from selection under drought versus multilocation testing in related tropical maize populations. Crop Science. 1995;**35**:63-69

[140] Tollenaar M, Wu J. Yield improvement in temperate maize is attributable to greater water stress tolerance. Crop Science. 1999;**39**:1597-1604

[141] Edmeades GO, Bolaños J, Hernandez M, Bello S. Causes for silk delay in a lowland tropical maize population. Crop Science. 1993;**33**:1029-1035

[142] Chapman SC, Edmeades GO. Selection improves drought tolerance

in tropical maize populations. II. Direct and correlated responses among secondary traits. Crop Science. 1999;**39**:1315-1324

[143] Edmeades GO, Daynard TB. The relationship between final yield and photosynthesis at flowering in individual maize plants. Canadian Journal of Plant Science. 1979;**59**:585-601

[144] Westgate ME, Bassetti P. Heat and drought stress in corn: What really happens to the corn plant at pollination? In: Wilkinson D, editor. Proceedings of the 45th Annual Corn and Sorghum Research Conference; 5-6 December 1990; Chicago. Washington D.C.: ASTA; 1991. pp. 12-28

[145] Zinselmeier C, Jeong BR, Boyer JS. Starch and the control of kernel number in maize at low water potentials. Plant Physiology. 1999;**121**:25-36. DOI: 10.1104/pp.121.1.25

[146] Jacobs BC, Pearson CJ. Potential yield of maize determined by rates of growth and development of ears. Field Crops Research. 1991;**27**:281-298

[147] Reed AJ, Singletary GW, Schussler JR, Williamson DR, Christy AL. Shading effect on dry matter nitrogen partitioning, kernel number and yield of maize. Crop Science. 1988;**28**:819-825

[148] Schussler JR, Westgate ME. Assimilate flux determines kernel set at low water potential in maize. Crop Science. 1995;**35**:1074-1080

[149] Zinselmeier C, Lauer MJ, Westgate ME, Boyer JS. Reversing drought-induced losses in grain yield. Sucrose maintains embryo growth in maize. Crop Science. 1995a;**35**:1390-1400

[150] Zinselmeier C, Westgate ME, Schussler JR, Jones RJ. Low water potential disrupts carbohydrate metabolism in maize. Plant Physiology. 1995b;**107**:385-391

[151] Zinselmeier C, Habben JE, Westgate ME, Boyer JS. Carbohydrate metabolism in setting and aborting maize ovaries. Formation in maize hybrids representing three decades of grain 1-13. In: Westgate M, Boote K, editors. Physiology and Modeling. Kernel Set in Maize. CSSA Special Publications 29. Madison, WI: CSSA; 2000

[152] Fischer KS, Palmer AFE. Tropical maize. In: Goldsworthy PR, Fisher NM, editors. The Physiology of Tropical Field Crops. Imprint: New York John Wiley & Sons Ltd., United states US Publication. 1984. pp. 213-248

[153] Tollenaar M, Dwyer LM, Stewart DW. Ear and kernel. formation in maize hybrids representing three decades of grain yield improvement in Ontario. Crop Science. 1992;**32**:432-438

[154] Singletary GW, Below FE. Growth and composition of maize kernels cultured in vitro with varying supplies of carbon and nitrogen. Plant Physiology. 1989;**89**:341-346

[155] Li Y, Wang M, Zhang F, Xu Y, Chen X, Qin X, et al. Effect of post-silking drought on nitrogen partitioning and gene expression patterns of glutamine synthetase and asparagine synthetase in two maize (*Zea mays* L.) varieties. Plant Physiology and Biochemistry. 2016;**102**:62-69. DOI: 10.1016/j.plaphy.2016.02.002

[156] Kamara AY, Kling JG, Menkir A, Ibikunle O. Agronomic performance of maize (*Zea mays* L.) breeding lines derived from a low nitrogen maize population. The Journal of Agricultural Science. 2003;**141**:221-230

Chapter 22

Ecofriendly Marigold Dye as Natural Colourant for Fabric

Sujata F. Harlapur, Suneeta Harlapur and Shantabasavareddi F. Harlapur

Abstract

This chapter highlights on the applications of marigold plant extracts as an antibacterial and antimicrobial best dyer for textiles. *Tagetes erecta* usually known as Marigold is a vital wellspring of carotenoids and lutein, developed as a nursery plant. Marigold blossoms are yellow to orange red in colour. Now a days, lutein is transforming into an unquestionably common powerful fixing, used as a part of the medicines, food industry and textile coatings. This has increased more noticeable vitality of marigold and its exceptional concealing properties. Regardless of the way that marigold blooms; its extract has been used as a measure of veterinary supports. The examination was directed to contemplate the usage of a concentrate of marigold as a trademark shading, which is antibacterial and antimicrobial. The marigold extract ability was focused on colouring of the cotton fabrics. Investigations of the dye ability, wash fastness, light fastness, antibacterial tests and antimicrobial tests can be endeavoured. Studies have exhibited that surface concealing was not impacted by washing and drying in the shadow/sunlight. These surprises reveal that the concentrate of marigold extract can be used for cotton fabrics.

Keywords: dyeing, marigold, colour fastness, fabric, antibacterial, antimicrobial

1. Introduction

Marigold is a common name of Tagetes species, a genus of herbs and the member of the Asteraceae family. It is a native of Mexico and other warmer parts of America and naturalised elsewhere within tropics and sub tropics. In India it is cultivated as garden plant. The flowers which are yellow to orange in colour with corymbose clusters are much used for garlands and functions and also used to decorate households during celebration of festivals.

These plants do not require any additional care. They grow in all conditions of the environment. Marigold plants bare sun, heat, drought and grow in any well-drained soil. Marigolds are easy to grow even from transplants also. Marigold requires a mild climate for smooth growth and flowering. Mild climate is required during the growing period is 14.5° - 28.6°C and it improves the production of flowers at higher temperature (26.2° - 36.4°C). Depending on the climatic conditions, the marigold grows three times in a year – rainy, winter, and summer seasons. The different coloured flowers are shown in **Figure 1**.

Figure 1.
The types of marigold flowers available in India.

The earliest use of marigolds was by the Aztec people who attributed magical, religious and medicinal properties to marigolds. The first recorded use of marigolds is in the De La Crus-Badiano Aztec Herbal of 1552. The Herbal records the use of marigolds for treatment of hiccups.

The orange, yellow and reddish yellow (i.e. red) colour flowers contain several pigments which appear to vary with source and soil nature. The active ingredients found in marigolds are plant sterols called calendulin, calendic acid, polysaccharides, polyunsaturated fatty acid, carotenoids, flavonoids, triterpenes saponins, like triterpenoid, Tocopherols, oleanolic acid glycosides.

The aqueous extract of flowers showed activity against gram positive bacteria. An infusion of plant is employed against rheumatism, cold and bronchitis. Root extract is employed as laxatives. Marigold leaves are utilised in kidney troubles and muscular pains. The infusion of florets is prescribed as a diuretic and carminative. The florets are utilised in the treatment of eye diseases and ulcers. The oil obtained from the fresh flowers incorporates a paralysing effect on the spinal marrow. It is used as antiseptic also. The flower heads are said to possess stimulant and anti-helminthic properties. Their juice contains iodine and is employed on cuts and wounds.

India is producing a big amount of marigold flowers. It is mainly used for floral offerings to Goddess and also for decorative purpose. After its use, the flowers were thrown away as garbage in ponds and rivers, which creates a lot of pollution. Our main aim of this chapter is to bring an awareness about the employability of marigold in many applications.

2. Uses of Marigold

The flowers of marigolds are very beautiful with a lot of benefits make a garden as more attractive. Marigold and its extracts are highly useful in therapeutic treatments like headaches, swelling, toothache, wounds and numerous skin problems. In addition, they have been used in cooking. Marigold was one of the earliest cultivated flower. The ancient Greeks used the petals of marigold for decorations and other purposes like make-up, colouring food (stews, soups, pudding, etc), dying fabrics and as part of medicines. The significant uses of marigold flowers are deliberated as follows:

2.1 As a garden plant

2.1.1 Nematode control

The roots and stems of marigolds emit a chemical that will suppress the population of root-knot nematodes, tiny soil borne worms that prey on the roots of decorative plants and vegetables. It appears that French marigolds, particularly the 'Tangerine' variety, are the best against the destructive pests.

2.1.2 Beneficial to insects

Marigolds itself attracts many insects that are harmful to other plants of the garden, so that the other plants get protected from the pests and do not require any pesticides.

2.1.3 Beautifies the garden

Marigolds are capable of growing in hot shades with many colours like orange, yellow, red, mahogany or with combinations. Flowers are of a single, double or multi-layered and sizes vary from 6 inches to 3 feet. So makes the garden more gorgeous and increases the beauty of the garden.

2.1.4 Marigold companion benefits on planting

When marigold is planted nearby cruciferous and tomato plants, these plants get protected from worms. It is because of the scent released from marigold, which confuses the pests. So it avoids usage of chemical pesticides. Marigold is additionally a decent companion when planted near bush beans, squash, cucumbers and brinjal.

2.2 As a medicinal plant

The very important uses of marigold with respect to antioxidant properties as medicine are as follows.

2.2.1 Homemade skin treatment

Ointments made up of marigold are used to heal sunburns, acne and ulcerations in addition to healing of wounds, dry skin and blisters.

2.2.2 Digestive-soothing tea

Tea powder made with marigold flowers lowers the symptoms caused by inflammatory bowel diseases. Marigold tea is useful in solving gastric and ulcer problems as well as in reducing stomach cramps. The florets are consumed in tea form to assist ease digestion internally and improve liver health.

2.2.3 Immunity booster

Marigold extracts are accustomed to manage symptoms of coughs, sore throats or fever. So it can also be used as immunity booster.

2.3 Treatment for fungal infections

2.3.1 Lowers inflammation and oxidative damage

Research shows that active chemicals of marigold make it a natural cytotoxic, hepato-protective, and spasmogenic herb, that has been used in both animal and human trials. Extracts obtained from the flower contains C-reactive protein and cytokine levels which protect cells from being damaged like ageing and cell deterioration.

Marigold controls oxidative damage from free radicals which will affect delicate tissues of the skin, GI tract and genitals caused by viruses or bacteria [1]. In addition, research also shows that marigold helps to reduce the growth of bacteria's in wounds and might even be ready to reduce symptoms related to chemotherapy and cancer treatments [2].

2.3.2 Reduces eye inflammation and conjunctivitis

Marigold extracts are used for treating conjunctivitis and other chronic ocular inflammatory conditions of animals. As these extracts have shown antibacterial, antiviral, antifungal and immuno-stimulating properties and which can be used on delicate tissues of the eyes against oxidative damage [3].

2.3.3 As a natural antiseptic

The extract of Marigold is employed to treat bacterial ear infections and reduce pain. It has been identified that marigold drops can lower inner-ear swelling and inflammation very easily in a short period and even without the use of antibiotics. Marigold extract may be used as antifungal for treating fungal infections of ear [4].

2.3.4 Heals skin wounds, burns and rashes

From the traditional days the marigold is employed for treating irritated skin to cut back itchiness, redness, sensitivity, dryness and swelling. It found that marigold has the power to push the expansion of healthy new tissues, increasing the rate of blood flow to the affected areas and boost collagen production, which firms and strengthen the skin, hydrate dry skin and speed up the method of skin repair following surgery or damage.

Marigold's ability to support the treatment of wounds is believed to be the secretion due to stimulation of somatic cell production, which contains glycoproteins and nucleoproteins. It is also associated with increased cell turnover and improved collagen metabolism stimulation.

In folklore medicine, marigold products were applied to varied ulceration (both internal and external) and on wounds to stop infections, and even today marigold is employed during post-surgery to push faster incision healing.

Marigold extract is combined with natural lubricating products and applied on dry and rash-prone skin that will help to enhance skin hydration and firmness. The extract of marigold is employed topically to reduce rashes on the sensitive skin, discoloration, unsightly varicose veins, dermatitis, eczema, bruises and scalp dandruff. It used to treat swollen bug bites, burns, and infected cuts and decreases other signs of poor blood flow and inflammation affecting the skin.

2.3.5 Eases cramps and spasms

Marigold's analgesic property is useful for relieving muscle spasms, stomach cramps and PMS/menstrual cramps. It reduces cramping by improving blood circulation to the painful area and lowering inflammatory responses [5].

2.3.6 Mosquito repellent

Due to their pungent odour, antioxidant content and volatile oils, marigolds are often accustomed naturally repel mosquitoes, pests and other insects. This is often one reason as marigold commonly planted in vegetable gardens. The flowers of Marigold are employed as extract form in candles, room/ bug sprays, and lots of skin lotions so as to forestall mosquito bites. Use of marigold products on our skin helps to repel bites. The flowers' aroma works underground to stay away nematodes (microscopic worms) [6].

2.4 Marigold as perfume

Many Tagetes sp. of marigold yield strong aromatic oil and referred as Tagetes oil. The tagetes oil is principally used for the compounding of high grade perfumes. The oil from French marigold L. is produced chiefly in France and finds a decent market within the perfumery trade.

In India French marigold L. appears to be a promising source of the oil on account of its fairly high yield with true floral character and favourable tagetone content. The flowers are regularly distilled in India for the assembly of attars by absorbing their steam distillate over sandalwood oil or liquid paraffin. It also acts as fly repellent and is credited with larvicidal properties.

2.5 Natural dye

India has a rich, diversified plant kingdom, which is a treasure house of varied natural products. Natural dye is one such product. The natural colourants mainly flavonoids and carotenoids present in the Marigold were extracted by different techniques and used as natural colourants on different fabrics. The use of natural colourants exists from ancient period.

3. Dye of marigold

Dyes are one of the foremost important yields of plants. In most of the countries, natural dyeing is practiced only on handicrafts and artificial dyes are being employed altogether for commercial dyeing processes. The problems related to synthetic dyes arise during their application, manufacture, and process of synthesis, intermediates formed and other raw materials involved.

Now a days with the global concern on the use of eco-friendly and biodegradable materials, around the world the solicitation of natural dyes in the textile industry is enhanced. The interest in the utilisation of eco-friendly and biodegradable materials create the employment of natural dyes. In contrast, natural dyes are environmental friendly, exhibit better biodegradability and customarily have the ensuing compatibility with the environment than synthetic dyes [7]. Natural Dyeing may be carried out by using eco-friendly bath methods like alkaline bath, acidic bath, or neutral bath. The method is economically viable because the raw materials are available at low cost, which in turn reduces the cost of production. There are several articles available on different methods of mordanting on different fabrics, for dyeing with different natural dyes.

Dyeing of cotton and silk with marigold brings a growing interest within the revival of natural dyes in textile colouration [8]. The main idea of extracting dyes from marigold plant sources is to avoid environmental pollution and its characteristics like antibacterial, antifungal, etc. The application of various mordants along with marigold gives innumerable shades like brown, yellow, orange and so on to the fabric.

Dyes derived from marigold sources have emerged as a significant alternative to synthetic dyes. The dyes produced were dyed on cotton fabric and tested for colour fastness to washing properties. Mordants were used to fix the colour on the fabrics. The dyes obtained from the marigold plant are replaceable to synthetic dyes.

Light fastness of the numerous natural dyes, particularly which are extracted from flower parts is found to be poor to medium [9]. In earlier days India was a significant exporter of herbal dyes to the world. Now days the demand for natural dyes has augmented, because of the ban on the assembly of a variety of synthetic dyes and intermediates within the developed countries, thanks to the pollution problem [10].

3.1 Process of dyeing

The procedure involved in dyeing of cotton with marigold extract comprises the following steps

- Pre-Treatment
- Extraction of colourant
- Mordanting (fixing dye with fibre)
- Dyeing

3.1.1 Pre-treatment

Before carrying out the dying process. Fabric undergoes scouring. Scouring is a process used to remove the dust and other colourants present on fabric. During scouring, the cotton fabrics were washed in a solution containing x g/L soda ash and a couple of g/L non-ionic detergent solution at 50° C for 25 min, keeping the fabric to required liquor ratio. The scoured material was thoroughly washed with H_2O and dried at moderate temperature. The scoured material must be soaked in clean water for 40 min before dyeing or mordanting.

3.1.2 Extraction of colourant

Flowers from the plant source were dried in trays, in thin layers, in an exceedingly current of warm air immediately after picking. Then crushed and dissolved in water and allowed to boil over a water bath for two hours for the quick extraction. All the colour was extracted from flowers was filtered for immediate use.

3.1.3 Mordanting

It is a method where mordant is used to fix the dye on fabric. Mordant binds dye with fabrics by forming a complex, which then confers to the fabric. Most of the natural dyes have no substantively on textile fabrics without the utilisation of a mordant. All most all the natural dyes need a mordanting chemical to make an affinity between the fabric and dye [11].

During mordanting, the weighed cotton samples were treated with a different metal salt. The different percentage of mordant was dissolved in water to create a required liquor ratio. The wetted sample is immersed into the mordant solution and so dropped at heating. The temperature of the dye bath was raised to 80°C over a period of half an hour and left at that temperature for 30 minutes. The mordanted material should be rinsed with water and dried in shadow. Mordanted cotton must be used immediately for dyeing because some mordants are very sensitive to light.

Generally mordants are metallic salts form a metal complexes with the fibres and dyes [12]. After completion of mordanting, the metal salts fixed to the fabrics, attract the dye pigment molecules. For example if aluminium sulphate mordant is used, it induced into the fabric and helps to fix the natural dyes through either coordination bonds or hydrogen bonds and by other inter molecular forces [13].

3.1.4 Dyeing

The cotton samples must be dyed with dye extract, keeping required M:L ratio. The marigold dye extract has prepared by adding × gm dye powder in y ml of water. Dyeing was carried out by standard dyeing method. After dyeing, the dyed material was washed with cold water and dried at low temperature. Then it has been dipped in brine for dye fixing.

4. Purpose of natural dye

Mainly the colours generated from natural resources like marigold are vibrant, biodegradable, and non-allergic. These features indicate that they are far compatible with the environment and to be used around humans. It is easy to extract the natural colour from plants, fruits or flowers. Many natural dyes even have antimicrobial properties, making them safer for teenagers specifically.

Natural dyes will not contain harmful chemicals nor carcinogenic components, common to artificial or synthetic dyes [14]. Usage of natural dyes will help to preserve the environment and lowering human dependence on harmful products. When toxic runoff and residuals from the textile manufacturing and dyeing process often find yourself in our delicate oceans, we must always do and make sure that we are using the nontoxic alternative, natural dyes.

Furthermore, the products employed in producing natural dyes, particularly plants, produce no waste, unlike the products employed in the synthetic dyeing process. Because plants bypass the complete production process which takes to form synthetic dyes. This can be one more reason why natural dyes are infinitely better for the environment. By using natural dyes instead of synthetic dyes, we are ready to be closely connected to nature and recognise the importance it plays altogether of our lives. The main advantage of naturally dyed fabric is that they are capable of absorbing higher UV rays. By wearing clothes dyed naturally, you are able to protect your skin more from the UV rays. The natural dyes can be used on fabrics for different purposes.

4.1 Intention of Marigold dye

4.1.1 To provide different colour effect on fabric

Marigold flower is grown altogether parts of India throughout the year and maybe an honest source of material for natural dyes. Three different shades of

Marigold flowers are available in India; those are light yellow, golden yellow and orange mix yellow. These were taken as raw materials for natural dyeing of cotton fabrics. Depending upon the thickness of the colour, the amount of flower is soaked in a specific amount of water to extract the required colour. The 30 g of extract powder is mixed with 100 ml of water for about 30 minutes for extraction. However, in the case of orange-yellow and light yellow, the extraction time was 50 minutes.

Cotton material is used to dye with the extracts obtained from three different shades of marigold after mordanting with eight different mordants in each case. All the specified cases of attractive shades may be produced by such dyeing. The colour fastness property to washing of most of the dyed samples was within the range of 2–3 percent in general. Colour fastness to light of the dyed samples varied with the change of mordant and therefore the substrate.

4.1.2 To enhance antibacterial/antimicrobial properties of fabric

Cotton fabric is employed in many applications in the textile industry due to its unique properties. However, the antibacterial properties of marigold extract will help within its applications in hygienic textiles and employed in medical fields. Flowers of marigolds are generally used to worship in temples and decorate vehicles [15]. The dyed SPF fabrics were then evaluated for colour values, fastness properties, anti-bacterial activities in addition to the durability of the identical. The studies give us clarity regarding the benefits of using such mordants in case of achieving antibacterial/antimicrobial functionality with addition of eco-friendliness [16]. Marigold dyeing by using most typically found alum mordant was also distributed for comparison of the aim.

5. Materials and experimental processes

5.1 Materials

Plant material: Marigold (*Tagetes erecta*).
The scientific classifications are as follows.
Kingdom – Plantae
Order – Asterales
Family – Asteraceae
Genus – Tagetes
Species – T erecta

5.1.1 Plant description

A stout, branching herb as shown in **Figure 2** is 60 cm tall, native of Mexico, extensively cultivated as a border - annual in gardens all over India. Roots usually taproot system, Stems usually erect; Leaves strong scented, pinnately dissected: segments 1–5 cm long, oblong or lanceolate, serrate; flower - heads solitary, yellow to orange, 5-9 cm across, florets bisexual, pistillate, functionally staminate, rays many, long-clawed, sometimes two-lipped or quilled, involucre campanulate; fruit achenes, 6–7 mm long, usually dry with relatively thick, tough pericarps, sometimes beaked (rostrate) and/or winged (late), often dispersed with aid from pappi; Seeds 1 per fruit, exalbuminous; embryos straight. The flower samples of four cultivars of *T. erecta* (marigold) i.e., marigold orange (MGO), marigold yellow (MGY), marigold reddish yellow or red (MGR) were collected (**Figure 3**).

Figure 2.
Marigold plant.

Figure 3.
Different colours of marigold.

5.2 Experimental process

5.2.1 Extraction of marigold powder

The petals were separated, sun-dried, powdered, and stored at -20°C for further studies (**Figure 4**).

5.2.2 Extraction of lutein ester

The fine powder of the petals of marigold orange, marigold yellow, marigold reddish yellow of about 30 gm was used to extract with various solvents. The extraction was carried out under dark conditions by covering the apparatus with black cloth together with the condenser. The extract was filtered and concentrated to dryness in an exceedingly rotary evaporator under reduced pressure and at 40° C, this stored in a container at -20° C (**Figure 5**).

5.2.3 Identification and extraction of fatty acids

The concentrated extract of different cultivars of marigold flower petals containing lutein esters was saponified in presence of 40% methanolic potassium hydroxide at 60°C for 1 h under dark condition. The unsaponifiable matter containing free lutein was extracted by three times using hexane. The water layer was acidified to

Figure 4.
Extraction process of marigold.

liberate the free fatty acids and extracted three times with petroleum ether. The ether fractions were collected and the solvent was removed under vacuum. The fatty acids were converted to methyl ester according to the standard method [13] and fatty acid composition was analysed by using a gas liquid chromatography.

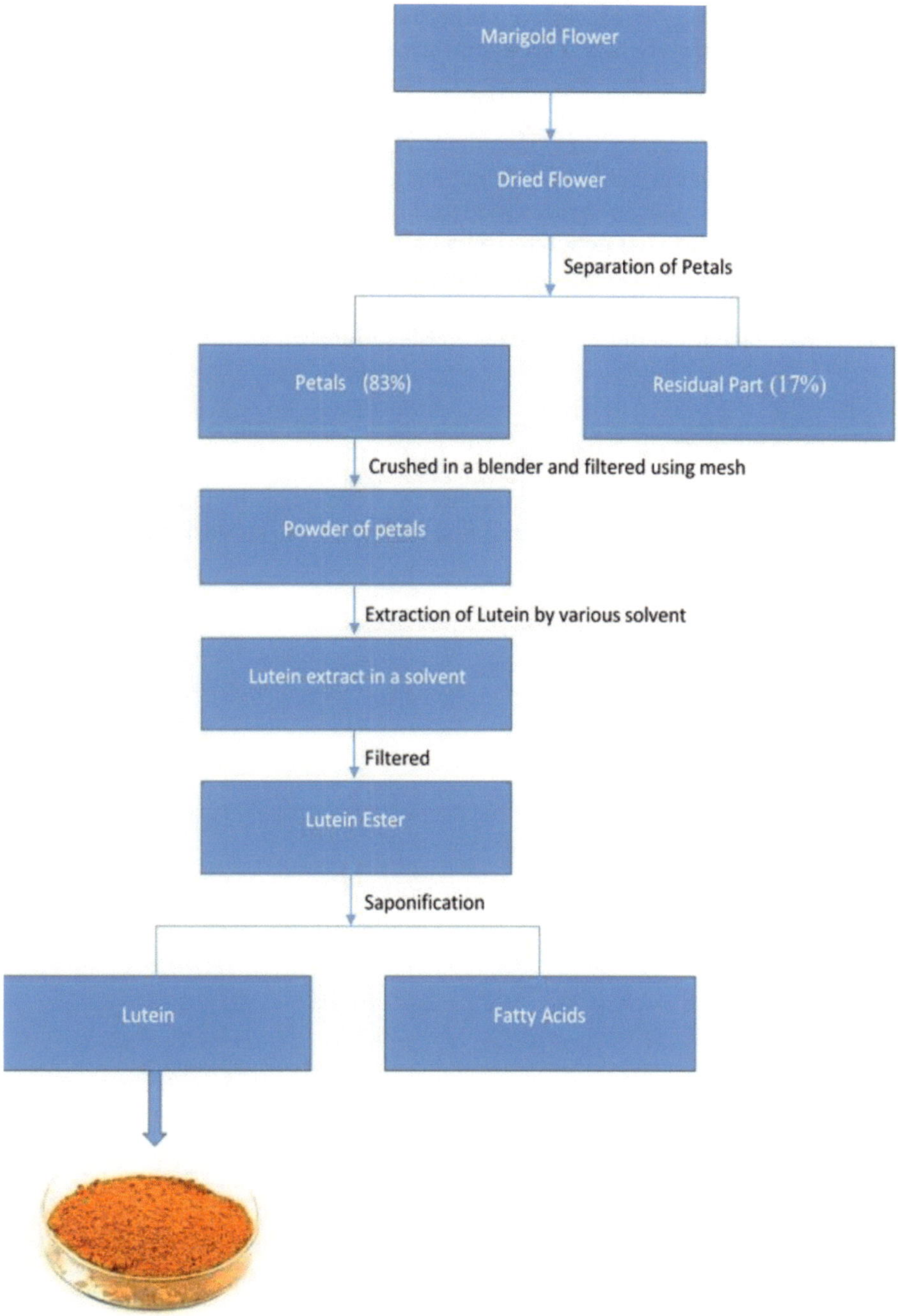

Figure 5.
Preparation and extraction of lutein.

6. Conclusion

The fabric dying process using marigold (*Tagetes erecta*) helps us to come to a general conclusion. Indian Marigold grows in all type of soil with different atmospheric condition and the required care to grow the marigold plant is very minimal. Marigold with rich quality of making garden beautiful have wide range of additional benefits. It has been used in the medicines, foods and in textile industry from ancient days. The marigold extracts are effective in protecting unsaturated fatty acids in the emulsion system. Marigold extract is used as a colouring agent.

Now days it is used in the fabrication of fabrics for colour dyeing and to provide antibacterial, antifungal and antimicrobial effects on fabric. Marigold dye arises as natural dye and free from harmful chemicals. Natural dye extraction is very simple and user friendly. Flower extract contains rich lutein esters. These extracts are used as antioxidant component in various food and pharmaceutical formulations.

Overall investigation has generated a lot of information on Indian marigold and its extract, which are highly beneficial in textile industry. This information can create commercially importance to produce viable products from Indian marigold for food, pharmaceutical industries and Textile industry.

Acknowledgements

Authors are thankful to Vemana Institute of Technology, Bangalore and Management of BVVs S R Vastrad Rural Polytechnic, Guledagudda for their support.

Author details

Sujata F. Harlapur[1*], Suneeta Harlapur[1] and Shantabasavareddi F. Harlapur[2]

1 Vemana Institute of Technology, Bangalore, Karnataka, India

2 BVV's S. R. Vastrad Rural Polytechnic, Guledagudda, Karnataka, India

*Address all correspondence to: sujataharlapur@vemanait.edu.in

References

[1] Preeti K C, Girija K, Ramadasan K, Anti-inflammatory activity of flower extract of *Calendula officinalis* Linn. and its possible mechanism of action. Indian J Exp Biol. 2009 47(2) 113-20.

[2] Xiao Q, Darwin R, Haiping H, Samuel L. Mackenzie, Patrick C, Identification and Analysis of a Gene from *Calendula officinalis* Encoding a Fatty Acid Conjugase, Plant physiology, 2001 125(2) 847-55.

[3] Disha A, Anita Rani, Anupam S, A review on phytochemistry and ethno-pharmacological aspects of genus Calendula. Pharmacogn Rev. 2013, 7(14) 179-187 doi: 10.4103/0973-7847.120520.

[4] Nahida T, Mariya H, Plants used to treat skin diseases. Pharmacogn Rev. 2014, 8(15) 52-60 doi: 10.4103/0973-7847.125531.

[5] Samra B, Khalid HJ, Qaiser J, Anwar HG, Studies on spasmogenic and spasmolytic activities of *Calendula officinalis* flowers. Phytother Res. 2006, 20(10) 906-10 doi: 10.1002/ptr.1980.

[6] https://www.almanac.com/plant/marigolds.

[7] Ahlström L, Eskilsson CS, Björklund E, Determination of banned azo dyes in consumer goods. Trends Anal. Chem. 2005 24 (1) 49-56.

[8] Mehanta ZA, Tiwari IA, Natural dye-yielding plants and indigenous knowledge on dye preparation in Arunachal Pradesh Northeast India. Curr. Sci., 2003 88 (4) 1474-1480.

[9] Samanta AK, Agarwal P, Application of natural dyes on textiles. Indian J Fibre Textile Res, 2009 34 384-399.

[10] Gaur RD, Traditional dye yielding plants of Uttarakhand, India. Nat. Prod. Rad, 2008 7(2) 154-165.

[11] Fatemeh SG, Ali SN, Sayed MM, Daryoush A, Javad M, The effect of mordant salts on antibacterial activity of wool fabric dyed with pomegranate and walnut shell extracts, Coloration Technology, 2012.

[12] Samanta AK, Konar A, Chakroborty S, Datta S, J Inst Engg (I), Text Engg, 2010 91 7.

[13] Senthikumar S, Umashankar P, Sujatha B, Indian Textile J, 2002 112(6) 15.

[14] Patil1 DB, Patil KN, Gaikwad PV, Patil PJ, Shewale UL, Bhamburdekar SB, Extraction of natural dye from rose flower for dyeing cotton fabrics. Int. J. Innovative Research in Multidisciplinary field, 2016 2(8) 133-135.

[15] M. D. Teli MD, Javed S, Maruti K, Ecofriendly dyeing and antibacterial finishing of soyabean protein fabric using waste flowers from temples. Textiles and Light Industrial Science and Technology (TLIST). 2013 2(2), 78.

[16] Sathyanarayanan, MP, Chaudhari MP, Bhat NV, Development of durable antibacterial agent from ban-ajwain seed (*Thymus serpyllum*) for cotton fabric, IJFTR, 2011, 36(3) 234-241.